- Metal
- Semimetal
- Nonmetal

10 VIII	11 IB	12 IIB	13 IIIA	14 IVA	15 VA	16 VIA	17 VIIA	18 VIIIA
								helium 2 **He** 4.00
			boron 5 **B** 10.81	carbon 6 **C** 12.01	nitrogen 7 **N** 14.01	oxygen 8 **O** 16.00	fluorine 9 **F** 19.00	neon 10 **Ne** 20.18
			aluminum 13 **Al** 26.98	silicon 14 **Si** 28.09	phosphorus 15 **P** 30.97	sulfur 16 **S** 32.07	chlorine 17 **Cl** 35.45	argon 18 **Ar** 39.95
nickel 28 **Ni** 58.69	copper 29 **Cu** 63.55	zinc 30 **Zn** 65.39	gallium 31 **Ga** 69.72	germanium 32 **Ge** 72.61	arsenic 33 **As** 74.92	selenium 34 **Se** 78.96	bromine 35 **Br** 79.90	krypton 36 **Kr** 83.80
palladium 46 **Pd** 106.42	silver 47 **Ag** 107.87	cadmium 48 **Cd** 112.41	indium 49 **In** 114.82	tin 50 **Sn** 118.71	antimony 51 **Sb** 121.75	tellurium 52 **Te** 127.60	iodine 53 **I** 126.90	xenon 54 **Xe** 131.29
platinum 78 **Pt** 195.08	gold 79 **Au** 196.97	mercury 80 **Hg** 200.59	thallium 81 **Tl** 204.38	lead 82 **Pb** 207.2	bismuth 83 **Bi** 208.98	polonium 84 **Po** (209)	astatine 85 **At** (210)	radon 86 **Rn** (222)

gadolinium 64 **Gd** 157.25	terbium 65 **Tb** 158.93	dysprosium 66 **Dy** 162.50	holmium 67 **Ho** 164.93	erbium 68 **Er** 167.26	thulium 69 **Tm** 168.93	ytterbium 70 **Yb** 173.04	lutetium 71 **Lu** 174.97
curium 96 **Cm** (247)	berkelium 97 **Bk** (247)	californium 98 **Cf** (251)	einsteinium 99 **Es** (252)	fermium 100 **Fm** (257)	mendelevium 101 **Md** (258)	nobelium 102 **No** (259)	lawrencium 103 **Lr** (260)

Charles H. Corwin
American River College

Alternate Edition

Chemistry
Concepts and Connections

Prentice Hall
Englewood Cliffs, New Jersey 07632

The Library of Congress has cataloged the Standard Edition of this work as follows:

Library of Congress Cataloging-in-Publication Data

Corwin, Charles H.
Chemistry : concepts and connections / Charles H. Corwin.
p. cm.
Includes index.
ISBN 0-13-481946-2
1. Chemistry. I. Title.
QD33.C815 1994
540--dc20 93-28087
CIP

Editor-in-Chief: Tim Bozik
Development Editor: Harriet Serenkin
Production Editor: Tom Aloisi
Marketing Manager: Kelly Albert
Copy Editor: Bill Thomas
Page Layout: Andy Zutis and Meryl Poweski
Designer: Judith A. Matz-Coniglio
Design Director: Florence Dara Silverman
Prepress Buyer: Paula Massenaro
Manufacturing Buyer Lori Bulwin
Supplements Editor: Mary Hornby
Photo Editor: Lori Morris-Nantz
Photo Research: Page Poore-Kidder

A Division of Simon & Schuster
Englewood Cliffs, New Jersey 07632

Printed in the United States of America
10 9 8 7 6 5 4 3 2 1

ISBN 0-13-482225-0

Prentice-Hall International (UK) Limited, *London*
Prentice-Hall of Australia Pty. Limited, *Sydney*
Prentice-Hall Canada Inc., *Toronto*
Prentice-Hall Hispanoamericana, S.A., *Mexico*
Prentice-Hall of India Private Limited, *New Delhi*
Prentice-Hall of Japan, Inc., *Tokyo*
Simon & Schuster Asia Pte. Ltd., *Singapore*
Editora Prentice-Hall do Brasil, Ltda., *Rio de Janeiro*

Brief Contents

Contents

15 Acids and Bases *458*

CHEMISTRY CONNECTION *Historical*

CHEMISTRY CONNECTION *Environmental*

16 Chemical Equilibrium *496*

CHEMISTRY CONNECTION *Environmental*

17 Oxidation and Reduction *535*

CHEMISTRY CONNECTION *Applied*

18 Advanced Chemical Calculations *569*

Appendices *610*

Photo Credit List *650*

Index *652*

Preface

Over the last 20 years, I have taught introductory chemistry to thousands of students. After much reflection, I have concluded that perhaps the most important variable in the learning process is positive reinforcement. I have found that it is in the best interest of students and teachers alike if instruction takes place in an environment that builds confidence and gives encouragement. It is my experience that students are more willing to put forth their best effort if given an expectation of success. In an atmosphere of optimism, students are more willing to accept the burden of learning, for which they must ultimately assume responsibility.

To this end, the tone of this text is friendly, affirmative, and inclusive. The text is especially sensitive to the needs of students who require review in some basic academic skills. The level of difficulty is gradual and progressive, and there are no math/chemistry skills taken for granted. To help students focus on a topic, a brief set of objectives introduces each chapter section. For emphasis, there are numerous example exercises, and there is frequent repetition of important points.

A primary concern of this text has been to generate and maintain student motivation. This concern has been addressed in a number of ways. Full-color photographs and line art are generously distributed throughout the book. In addition, the text emphasizes real-world applications of chemistry that help students relate the subject to their own lives.

Features

There are numerous key features that include the following:

Problem Solving. All problems involving calculations are solved systematically in three steps using the unit-analysis method. The unit analysis method of problem solving is first introduced in Chapter 1 and is then reinforced in Chapter 2. In the chapters that follow, unit analysis is applied to mole problems, stoichiometry, and solution calculations.

The use of algebra in solving problems is strictly optional. In selected instances, problem solving using an algebraic method is provided as an alternate solution. In Chapter 12, for example, gas law calculations are solved algebraically as well as by a modified unit analysis approach.

Recognizing that students require practice in order to learn to solve problems, there are 300 example exercises in the text, each paired with a self-test exercise. In addition, there are over 1600 end-of-chapter exercises arranged in a matched-pair format. Answers are provided for all of the odd-numbered exercises in Appendix J. Answers are found to the even-numbered exercises in the *Instructor's Resource Manual*.

Chemistry Connections. One of the ways that students are kept motivated and involved in the subject is through the inclusion of relevant vignettes on chemistry. These vignettes are entitled *Chemistry Connections* and range in subject matter from consumer to applied chemistry, and from historical biographies to environmental concerns.

Consumer

Applied

Historical

Environmental

Updates. Although introductory chemistry is more static than dynamic, there have been recent developments. For example, new elements have been synthesized, new group designations for the periodic table have been proposed, and there have been changes regarding systematic nomenclature. Special features called *Updates* discuss the latest developments.

Notes. One of the most difficult tasks of a textbook is to simplify explanations without overgeneralizing. To provide discussions free of interruption, a special note appears after many of the topics. This note qualifies any simplifications that may create a misimpression.

Conceptually Accurate Line Art. To help students conceptualize the atomic and subatomic levels, the text uses a magnification technique. For example, the individual molecules in an ice crystal are illustrated in Figure 3.3 and the atomic nucleus is portrayed in Figure 4.8.

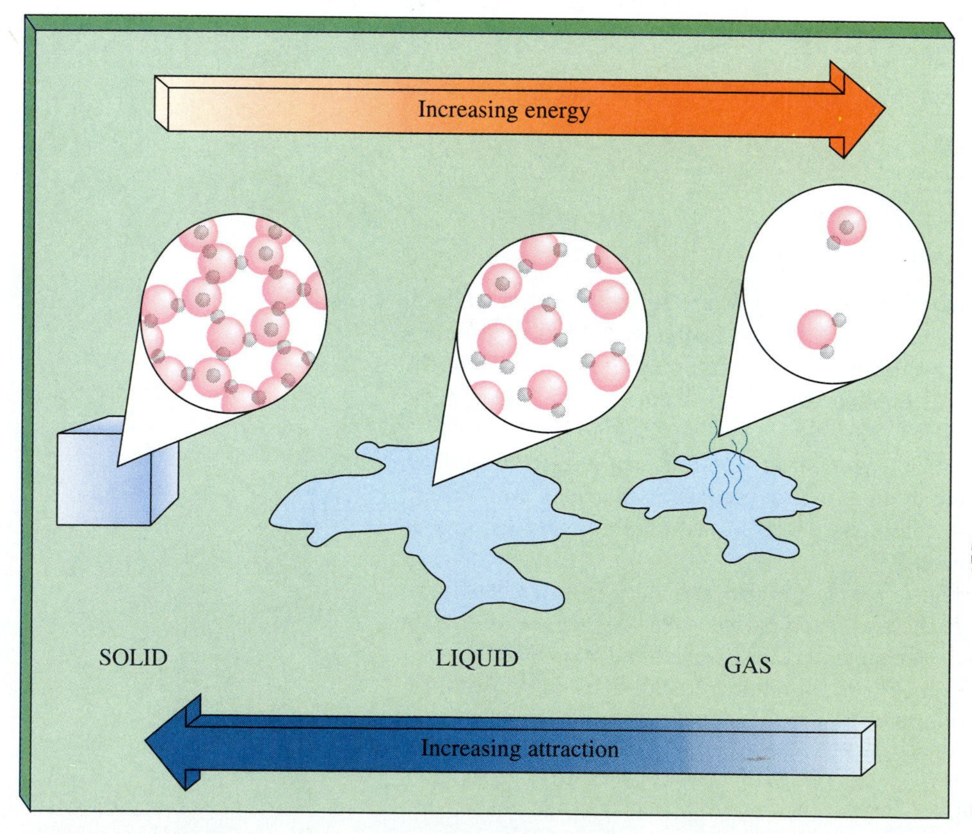

Figure 3.3

Nucleus (~10^{-13} cm diameter)

Electron orbits (~10^{-8} cm diameter, ~0.1 to 0.5 nm)

Figure 4.8

Concept Maps. To give students an overview of complex relationships, concept maps are used. A concept map is a diagram that illustrates how different aspects of a topic are related. For example, Figure 7.4 shows how the mole concept is related to Avogadro's number, molar mass, and molar volume.

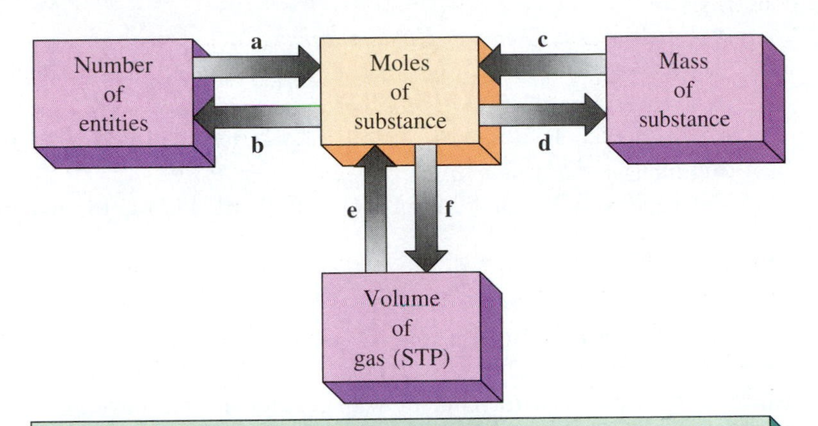

Figure 7.4

(a) Use N as a unit factor: multiply by 1 mol/6.02×10^{23}
(b) Use N as a unit factor: multiply by 6.02×10^{23}/1 mol
(c) Use MM as a unit factor: multiply by 1 mol/g
(d) Use MM as a unit factor: multiply by g/1 mol
(e) Use molar volume as a unit factor: multiply by 1 mol/22.4 L
(f) Use molar volume as a unit factor: multiply by 22.4 L/1 mol

Glossaries. All key terms appear in bold type in the text where they are first introduced, and a running glossary also appears in the accompanying margin. For reference, there is comprehensive glossary at the end of the text. There is also a matching key-term exercise at the end of each chapter.

Summaries. A chapter summary reviews important topics and provides a capsule view of each section.

Full-color Production. The primary use of color is to highlight important points and make the presentation more inviting. A secondary use of color is to convey relationships in a more subtle way. For example, elements, energy levels, and orbitals are consistently color-keyed in the art throughout the text. Another example of the functional use of color is the blue vertical page border that appears on pages of the end-of-chapter exercises. The blue border enables students to locate these sections easily.

Organization

This text maintains a traditional organization of topics for introductory chemistry. Recently, a national survey of chemistry teachers indicated that most professors prefer that a discussion of atomic orbitals and chemical bonding be delayed until later chapters. That preference is reflected herein. The Bohr atom concept is presented in Chapter 4, while the discussion of atomic orbitals is postponed until Chapter 10, and chemical bonding to Chapter 11. However, the flexible design of the text allows the experienced teacher to cover these topics earlier in the course at their discretion.

This book possesses an unusual degree of flexibility resulting from its early module development. Originally, each topic was assigned to a level in a hierarchy based on prerequisite topics. The initial premise was that the text should be sufficiently flexible to accommodate the objectives of different courses and preferences of various professors. A few of the possible chapter configurations are as follows.

- Emphasis on *Chemical Calculations:*
  Chapters 1, 2, 3, 4, 5, 6, 7, 12, 8, 9, 11, 13, 14, 15, 16, and 18.
- Emphasis on *Chemical Reactions:*
  Chapters 3, 1, 2, 4, 5, 6, 7, 8, 9, 11, 12, 13, 14, 15, 16, and 17.
- Emphasis on *Atomic and Molecular Structure:*
  Chapters 1, 2, 3, 4, 5, 10, 11, 6, 7, 8, 9, 12, 13, 14, 15, and 19.

Supplements

An alternate version of *Chemistry: Concepts and Connections* is available in a softcover edition. This edition is identical to the first 18 chapters of the hardcover edition and is intended for courses with restricted topic coverage. The following ancillaries are available for both versions.

- Instructor's Resource Manual
- Test Item File (with over 2000 class-tested questions)
- 3.5″ IBM Test Manager DOS
- 5.25″ IBM Test Manager DOS
- Mac Test Manager
- Laboratory Experiments
- I/M to Laboratory Experiments
- Student Study Guide
- Student Solutions Manual
- 120 Full-color Transparencies
- "How to Study Chemistry"
- *New York Times* Contemporary View Program

To assure that the teaching package is fully integrated, the author has personally written the *Instructor's Resource Manual, Test Item File*, and *Laboratory Manual*. The *Student Study Guide* and *Student Solutions Manual* were written in collaboration with Donald Lucas. Donald is an experienced chemistry tutor who has been selected for the AACJC Beacon Learning Project that is assessing peer-assisted instruction for at-risk students in math and science.

Acknowledgments

It is my pleasure to recognize several talented people who, directly or indirectly, were responsible for this project. First, it is my privilege to have outstanding colleagues, and I would be quite remiss not to mention each of the following: Ronald Backus, Nini Cardoza, James Cress, Sheila Epler, Melissa Green, Ronald Greider, Jim Morgan, John Newey, Luther Nolen, Karen Pesis, Nancy Reitz, Rina Roy, Michael Scott, Ronald Smedberg, Danny White, and Linda Zarzana. I am privileged to have Stephen Ruis for a friend and colleague as he has often served as an intellectual "sounding board." Steve is a dedicated teacher and frequently reviews textbooks in the subject area. Over a summer's vacation, Steve and I met daily and argued many issues including the hierarchy of concepts and the subtle implication of key terms. Our give-and-take sessions ultimately led to a better integration of topics and a more precise use of language.

I would like to thank Harriet Serenkin, developmental editor, who gave me much to ponder in her numerous queries, offered moral support, and drafted the environmental *Chemistry Connections*. I am equally appreciative for a refreshing stay in New York City hosted by Harriet and her husband, Patrick. I am also indebted to Bill Thomas, copy editor, for his advice and for his incredible ability to spot even the most minor inconsistency.

During peer review, I received many valuable insights. Beyond pointing out my oversights, these reviewers helped define that vague line between the simplifications that students require and the explanations that accuracy demands. Considerate and thoughtful comments were received from each of the following.

Caroline Ayers
East Carolina University

Paula Ballard
Jefferson St. College

Juliette Bryson
Chabot College

Joseph Cantrell
Miami U., Oxford

Robert Farina
Western Kentucky University

J. A. Grove
South Dakota St. U.

Arthur Hayes
Rancho Santiago College

Jack Healey
Chabot College

Dan Huchital
Seton Hall University

Jeffrey Hurlbut
Metropolitan State College

Leslie Kinsland
U. of Southern Louisiana

Ernest Kho, Jr.
University of Hawaii

R. M. Kren
U. of Michigan-Flint

David Lippmann
Southwest Texas St. U.

Leslie Lovett
Fairmont St. College

John McLean, Jr.
University of Detroit

Ronald Marcotte
Texas A & I U.

Gordon Parker
University of Toledo

James Petrich
San Antonio College

Dale Pigott
Victoria College

Dexter Plumlee
Northern Virginia CC

Erwin Richter
U. of Northern Iowa

Alan & Sharon Sherman
Middlesex CC

Kathleen Trahanovsky
Iowa State University

John Weyh
Western Washington U.

Finally, I want to acknowledge some individuals at Prentice Hall for going beyond the call of duty and giving a high-quality effort. Tim Bozik, Editor-in-Chief, coordinated many functions and played an integral role in the fresh and attractive design of the book. Tom Aloisi, Production Editor, met the deadlines of an accelerated schedule while insisting that the text be technically accurate and internally consistent. Mary Hornby, Supplements Editor, offered valuable advice while dependably coordinating the substantial ancillary program. Kelly Albert, Marketing Manager, took the time to understand the project from conception to completion before formulating an enthusiastic marketing plan. Ray Mullaney, Editor-In-Chief of College Development, afforded a seasoned and cohesive presence that was a resource for us all. To each of these people, singly and collectively, I extend my sincere appreciation.

Charles H. Corwin

Alternate Edition

Chemistry

Concepts and Connections

CHAPTER

1

A Mathematical Foundation

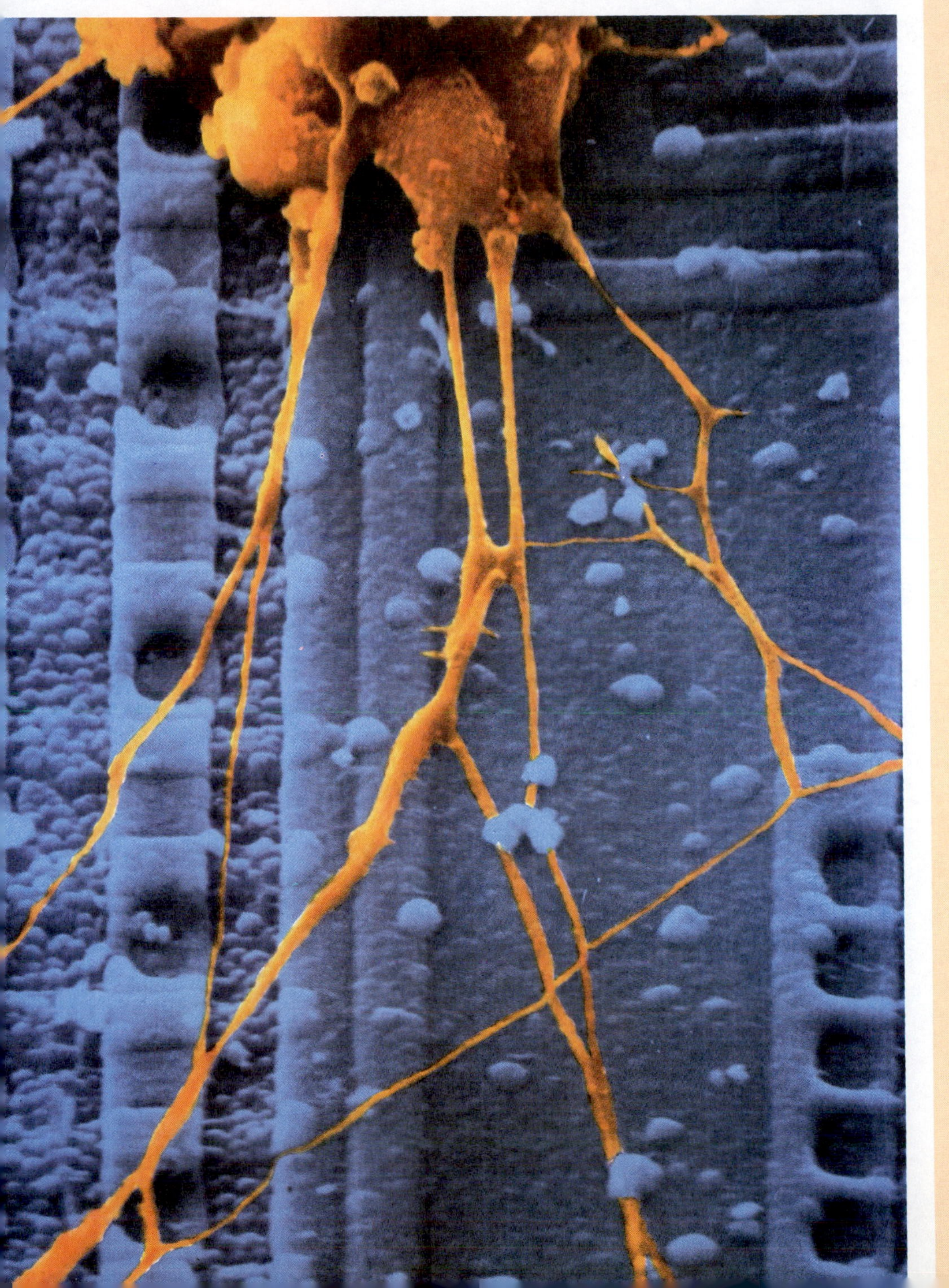

A human nerve cell is shown on the surface of a silicon computer chip and magnified over a hundred times. The nerve cell is smaller than the tiny computer chip and much more complex in the way it processes information.

To understand and use chemistry, you need to understand and be comfortable with mathematics. Calculators will help you perform simple calculations and computers will help you with long and tedious ones, but you still have to understand the basic principles of math. This goal is especially important as you advance in chemistry, because the math will take on a more important role.

This chapter presents basic mathematical principles, thereby laying a foundation for the chemical concepts and calculations in the following chapters. Specifically, this chapter shows you how to apply the basic operations of addition, subtraction, multiplication, and division to the numbers obtained from measurements.

Every measurement has units, such as centimeters, attached to a numerical value such as 25. For example, a measurement of length may be 25 centimeters. In general, this text will avoid English units such as inch, pound, and quart, although they will be used on occasion. Instead, we will use the metric units: centimeter, gram, and milliliter. The metric system and its units are discussed in detail in Chapter 2. For now, realize that a **centimeter** *is a unit of* length. *A* **gram** *is a unit of* mass, *and a* **milliliter** *is a unit of* volume. *Units are usually abbreviated using symbols. The symbol for centimeter is* **cm**, *for gram it is* **g**, *and for milliliter it is* **mL**. *It is interesting to note that a nickel is about 2 cm in diameter and its mass is about 5 g. Twenty drops from an eyedropper is approximately 1 mL.*

centimeter (cm) A common unit of length in the metric system of measurement that is equal to one-hundredth of a meter.

gram (g) The basic unit of mass in the metric system.

milliliter (mL) A common unit of volume in the metric system of measurement that is equal to one-thousandth of a liter.

1.1 Uncertainty in Measurement

To identify the following measuring instruments: metric ruler, balance, graduated cylinder, pipet, and buret.

To understand that no measurement is exact because all instruments have uncertainty.

A chemist observes, measures, and records the properties of matter. A measurement is obtained by using an **instrument** such as a balance. How exact the recorded measurements are depends on the instrument. A vast array of electronic instruments is now available using state-of-the-art technology. Some of these instruments provide extremely sensitive scientific **measurements**. For example, electronic balances are available that measure mass to one-millionth of a gram. A laser system can provide length measurements more precise than one-billionth of a meter. Chemists inject liquid volumes using hypodermic syringes that measure in millionths of a liter. Nevertheless, it is not possible—and never will be possible—to make an exact measurement. This is because no instrument can be calibrated exactly. Instruments

instrument A device for recording a measurement, such as length, mass, volume, time, or temperature.

measurement A numerical value with attached units that expresses a physical quantity such as length, mass, or volume.

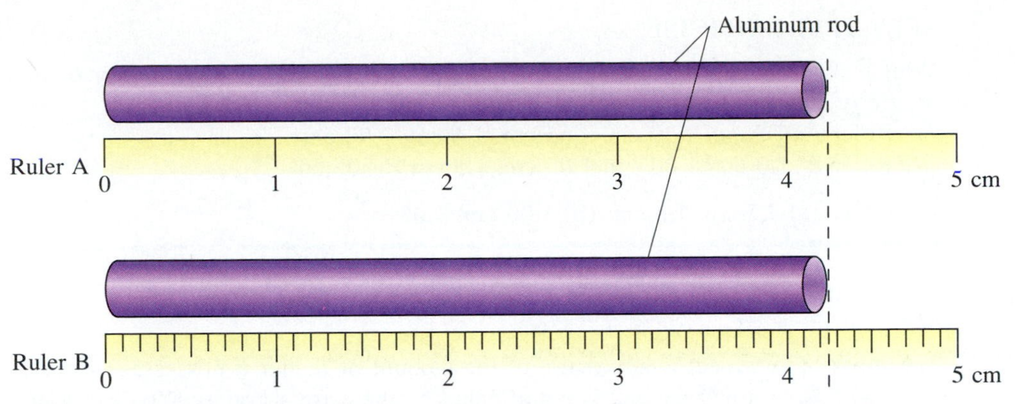

Figure 1.1 Metric Rulers for Measuring Length On Ruler A, each division is one centimeter, 1 cm. On Ruler B, a division is 1 cm, and each subdivision is one-tenth centimeter, 0.1 cm (not to scale).

may give very accurate measurements, but they never give exact measurements; in other words, all instruments have **uncertainty**.

uncertainty The degree of exactness in an instrumental measurement.

Length

To understand uncertainty, suppose you measure the length of an aluminum rod. You have two metric rulers available that differ by the number of divisions, as shown in Figure 1.1. Both rulers are satisfactory for the task. Ruler A, however, has more uncertainty than ruler B.

Ruler A has five divisions, each representing 1 cm. With it you can estimate the length of the aluminum rod with an uncertainty of ±0.1 cm. The aluminum rod measures about 4.2 cm. Therefore, if you record a length of 4.1 cm (4.2 − 0.1) or 4.3 cm (4.2 + 0.1), your value is acceptable.

Ruler B also has five divisions, each representing 1 cm. But within each of those divisions it has ten subdivisions, thus providing a more exact measurement. The exactness of the measurement is limited by the uncertainty of the ruler. Using ruler B, you see that the aluminum rod falls between 4.2 cm and 4.3 cm. It is unrealistic to try to estimate to a tenth of a subdivision. But you can estimate to a half of a subdivision. This provides an uncertainty of ±0.05 cm. The length of the aluminum rod using ruler B is therefore 4.25 cm (4.2 + 0.05). Since the uncertainty is ±0.05 cm, it is correct to record the length as 4.20 cm or 4.30 cm. Comparing the length and uncertainty obtained from metric rulers A and B gives the following:

Ruler A: 4.2 ± 0.1 cm **Ruler B:** 4.25 ± 0.05 cm

Thus, ruler B has less uncertainty and gives more exact measurements.

EXAMPLE EXERCISE 1.1

Which of the following measurements are consistent with the uncertainty of the metric rulers shown in Figure 1.1?

(a) Ruler A: 2 cm, 2.0 cm, 2.05 cm, 2.5 cm, 2.50 cm
(b) Ruler B: 3.0 cm, 3.3 cm, 3.33 cm, 3.35 cm, 3.50 cm

Solution: Ruler A has an uncertainty of ±0.1 cm and ruler B has an uncertainty of ±0.05 cm. Thus,

(a) Ruler A can give the measurements 2.0 cm and 2.5 cm.
(b) Ruler B can give the measurements 3.35 cm and 3.50 cm.

SELF-TEST EXERCISE

Which of the following measurements are consistent with the uncertainty of the metric rulers shown in Figure 1.1?

(a) Ruler A: 1.5 cm, 1.50 cm, 1.55 cm, 1.6 cm, 2.00 cm

(b) Ruler B: 1.0 cm, 1.00 cm, 1.05 cm, 1.5 cm, 1.505 cm

Answers: **(a)** 1.5 cm, 1.6 cm; **(b)** 1.00 cm, 1.05 cm

Mass

mass The quantity of matter in an object measured by a balance.

The **mass** of an object is a measure of the amount of matter it possesses. Mass is measured using a balance and is not affected by the earth's gravity. You can think of a balance as a two-pan teeter-totter. An object is placed on one pan and weight is added on the opposite pan until the balance is level. The measurement of mass also has uncertainty. This uncertainty varies with the balance. A typical mechanical balance in a chemistry laboratory may have an accuracy of one-hundredth of a gram. Thus, its mass measurements would have an uncertainty of ±0.01 g. Some laboratories have more expensive electronic balances with digital displays. These balances may have uncertainties ranging from ±0.1 to ±0.0001 g. The smaller the uncertainty, the more accurate the measurement is. Figure 1.2 shows three common types of balances.

weight The gravitational force of attraction between an object and the planet on which it is measured.

The term weight is often used when you really mean mass. Strictly speaking, mass and weight are not interchangeable. **Weight** is defined as mass times the acceleration due to gravity. Since the gravitational attraction is greater on earth than

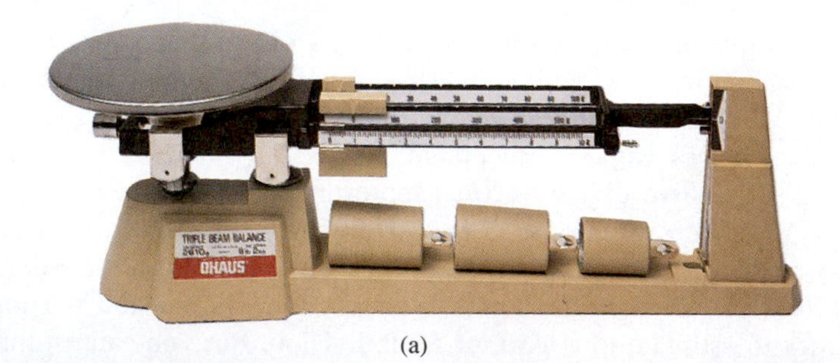

(a)

(b)

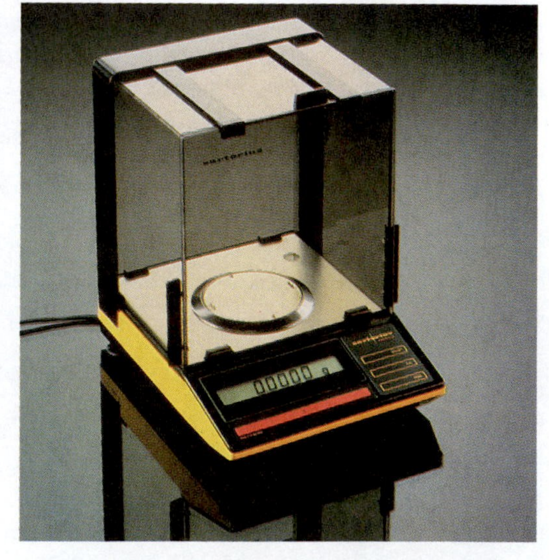

(c)

Figure 1.2 Balances for Measuring Mass (a) a platform balance with 0.1 g uncertainty; (b) a beam-type balance with 0.01 g uncertainty; and (c) an electronic balance with LED digital display.

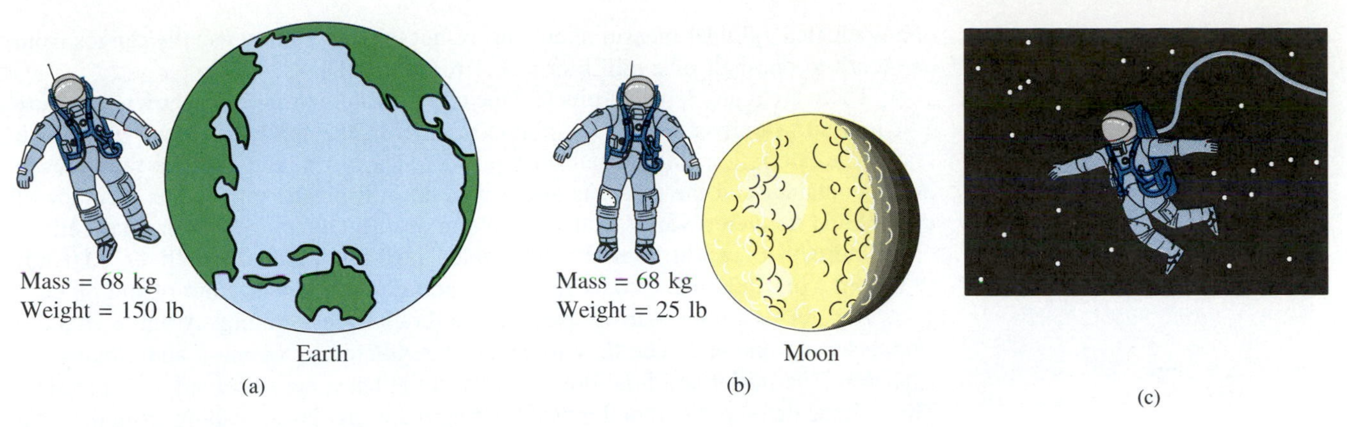

Figure 1.3 Mass versus Weight A mass measurement is independent of gravity while weight is not. (a) On earth, the astronaut has a mass of 68 kilograms and a weight of 150 pounds. (b) On the moon, the mass remains 68 kilograms, but the weight is only 25 pounds. (c) In space, the mass is still 68 kilograms even though the astronaut is weightless.

on the moon, objects weigh more on earth. Similarly, the same object weighs even more on the huge planet Jupiter than on earth. On the other hand, the mass of an object obtained using a balance is the same regardless of where the measurement is taken. That's because gravity operates equally on both pans of a balance, thereby canceling its effect. The mass of an object is therefore constant, whether it is measured on earth, the moon, or any other place. Figure 1.3 illustrates the distinction between mass and weight.

Volume

The amount of space occupied by a solid, gas, or liquid is its volume. Many pieces of laboratory equipment are available for measuring the volume of a liquid. Three of the most common are a graduated cylinder, a pipet, and a buret. Figure 1.4 shows common laboratory equipment for measuring volume.

A graduated cylinder is routinely used to measure a volume of liquid. The most common sizes of graduated cylinders are 10, 50, and 100 mL. The accuracy

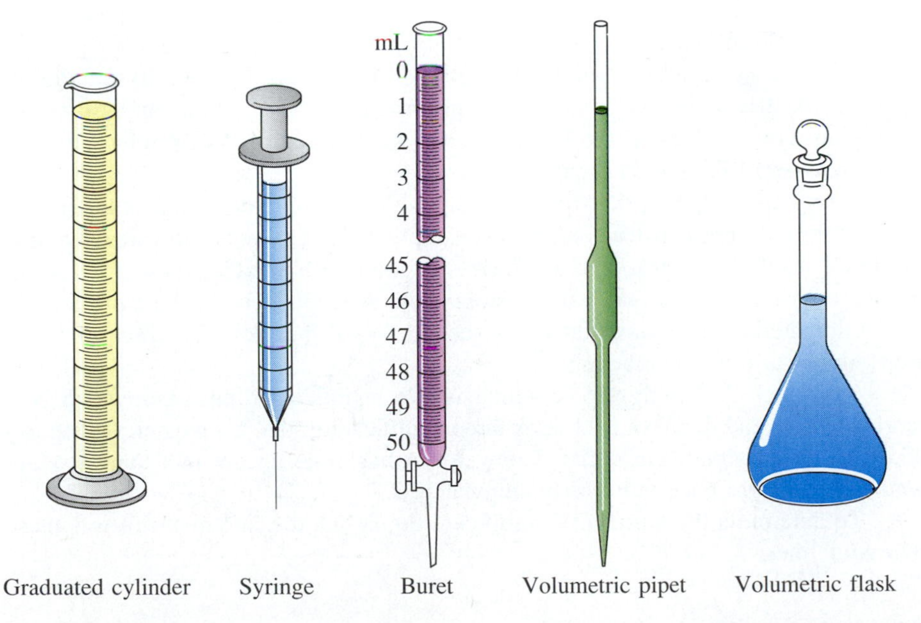

Figure 1.4 Instruments for Measuring Volume A graduated cylinder, syringe, and buret are calibrated to measure variable quantities of liquid, whereas a volumetric pipet and flask measure fixed quantities; for example, 10.0 mL or 250.0 mL.

of a graduated cylinder measurement varies, but the uncertainty usually ranges from one-tenth to one-half of a milliliter (±0.1 to ±0.5 mL).

There are many types of pipets. One type, a volumetric pipet, shown in Figure 1.4, is used to deliver a precise volume of liquid. The liquid is drawn up until it reaches a calibration line etched on the pipet. The tip of the pipet is then placed into a container and the liquid is allowed to drain from the pipet. The accuracy of the volume delivered varies, but the uncertainty usually ranges from one-tenth to one-hundredth of a milliliter. For example, a 10-mL pipet can deliver 10.0 mL (±0.1 mL) or 10.00 mL (±0.01 mL), depending on the uncertainty of the pipet.

A buret is a long, narrow piece of calibrated glass tubing. A valve called a stopcock is at one end. The flow of liquid is regulated by opening and closing the stopcock. The initial and final liquid levels in the buret are observed and recorded. The volume delivered is found by subtracting the initial buret reading from the final buret reading. Burets have uncertainties ranging from one-tenth to one-hundredth of a milliliter. For example, the liquid level in a 50-mL buret can read 22.5 mL (±0.1 mL) or 22.55 mL (±0.01 mL), depending on the uncertainty of the buret.

1.2 Significant Digits

To identify the number of significant digits in a given measurement.

significant digits The digits in a measurement that are known with certainty plus one digit that is estimated; also called significant figures.

In a recorded measurement such as that of length, mass, or volume, each numeral is a **significant digit** (also referred to as a significant figure). A digit is significant if its removal changes the uncertainty of the measurement. For example, suppose an object is weighed on three different balances. The masses recorded are 1.5, 1.50, and 1.500 g. The 1.5-g mass has an uncertainty of one-tenth of a gram, 1.50 g has an uncertainty of one-hundredth of a gram, and 1.500 g has an uncertainty of one-thousandth of a gram. The mass measurements have two, three, and four significant digits, respectively. Removing the last digit from any of them changes the uncertainty of the mass measurement.

Suppose an aluminum rod has a length of 14.2 cm. All three digits are significant, but the 2 is uncertain. If no further information is given, you can assume the digit 2 has a range of ±0.1 cm. Therefore, a rod that measures 14.2 cm has an exact length between 14.1 and 14.3 cm.

In any measurement the significant digits that are recorded express the uncertainty of the measuring instrument. As an example, let's observe the chemical reaction in Figure 1.5 that changes from a colorless solution to blue after about 35 seconds (s). To time the reaction we'll use a stopwatch. A stopwatch can be calibrated in tenths, hundredths, or thousandths of a second. We'll therefore use three different stopwatches to time the reaction.

Stopwatch A displays 35 s, which is two significant digits. Stopwatch B is more accurate and displays 35.1 s, or three significant digits. Stopwatch C displays 35.08 s, or four significant digits. Stopwatch A has more uncertainty than B; stopwatch B has more uncertainty than stopwatch C.

To determine the number of significant digits in a measurement, follow these two guidelines:

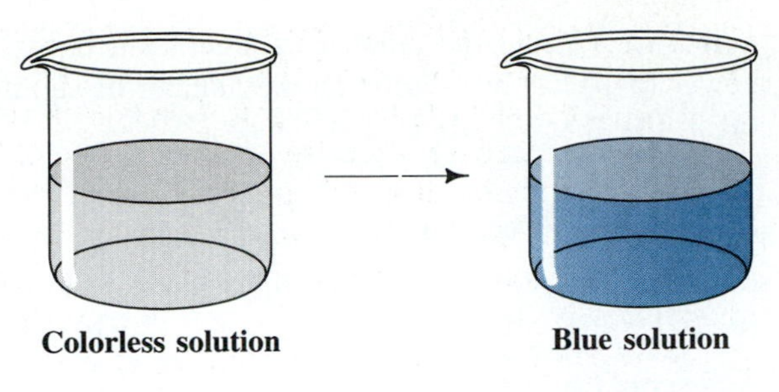

Stopwatch A:	0 s	35 s
Stopwatch B:	0.0 s	35.1 s
Stopwatch C:	0.00 s	35.08 s

Figure 1.5 Significant Digits and a Timed Reaction The data from the timed reaction demonstrate the uncertainty in three different stopwatches. Stopwatches A, B, and C have an uncertainty of ±1 s, ±0.1 s, and ±0.01 s, respectively.

1. Count the number of digits in a measurement from left to right. Start with the first nonzero digit and stop with the first uncertain digit.
2. Do not count zeros to the far right in a number having no decimal point (for example, 100 or 250). You will assume that they are simply decimal place holders and are not significant digits.

EXAMPLE EXERCISE 1.2

Determine the number of significant digits in the following measurements.

(a) 2.05 cm **(b)** 25.0 cm
(c) 0.025 g **(d)** 0.02050 g
(e) 250.0 mL **(f)** 2500 mL

Solution: To find the number of significant digits, count from left to right starting with the first nonzero digit; thus,
(a) 3; **(b)** 3; **(c)** 2; **(d)** 4; **(e)** 4; **(f)** 2 (assume the zeros are place holders)

SELF-TEST EXERCISE

State the number of significant digits in the following measurements.

(a) 2.50 cm **(b)** 0.0025 g
(c) 25.00 mL **(d)** 250 s

Answers: **(a)** 3; **(b)** 2; **(c)** 4; **(d)** 2

Note: In those special cases where one or more of the place-holder zeros is significant, you can write the value as a decimal number followed by a power of 10. For example, to indicate that 100 cm has only one significant digit, write 1×10^2 cm. To indicate that 100 cm has two significant digits, write 1.0×10^2 cm. To indicate that it has three significant digits, write 1.00×10^2 cm. Although a power of 10 is an elementary mathematical concept, it is reviewed for you in Section 1.6.

EXAMPLE EXERCISE 1.3

Determine the number of significant digits in the following measurements.

(a) 2.50×10^3 cm **(b)** 2.500×10^2 g
(c) 2.5×10^{-3} mL **(d)** 2.50×10^{-1} s

Solution: To find the number of significant digits, count from left to right. The power of 10 has no bearing on the number of significant digits; thus, **(a)** 3; **(b)** 4; **(c)** 2; **(d)** 3

SELF-TEST EXERCISE

State the number of significant digits in the following measurements.

(a) 5.0×10^3 cm

(b) 5.020×10^{-15} g

Answers: **(a)** 2; **(b)** 4

1.3 Rounding Off Nonsignificant Digits

OBJECTIVES

To round off a given measurement to a stated number of significant digits.

nonsignificant digits The digits in a measurement that exceed the uncertainty of the instrument.

rounding off The process of eliminating digits that are not significant.

All digits from a measuring instrument are significant. Frequently, however, you will use the data from measuring instruments to perform calculations. Calculations give significant digits as well as nonsignificant digits. **Nonsignificant digits** exceed the data from the measurement, but often appear in your calculator display. These nonsignificant digits should not be reported in your answer. Retain only the significant digits in your answer and eliminate the extra nonsignificant digits. The nonsignificant digits are eliminated by a procedure of **rounding off** to give only significant digits. Round off according to the following two rules:

1. If the first nonsignificant digit is less than 5, drop all nonsignificant digits. (As necessary, replace dropped digits with zeros to maintain the decimal place.)
2. If the first nonsignificant digit is 5 or more, increase the last significant digit by 1 and drop all nonsignificant digits.

On occasion, rounding off a calculated value can create a problem. For example, if we round off 151 to two significant digits using rule 2, we obtain 15. But 15 is only a fraction of the original value. Thus, we must insert a place-holder zero to obtain the correct rounded value; that is, 151 rounds to 150. Similarly, rounding off 1563 to two significant digits gives 1600.

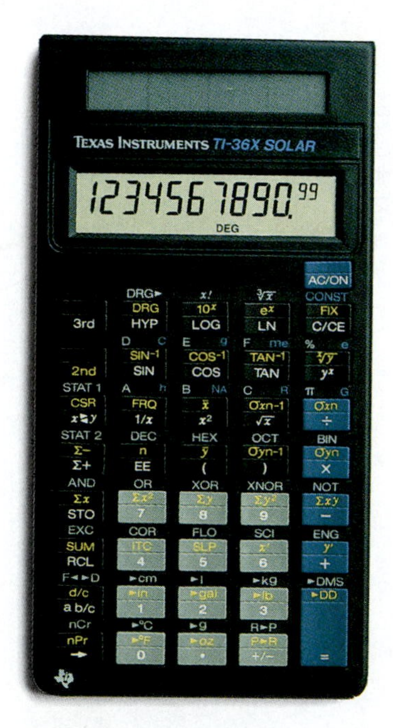

Calculators often give nonsignificant digits that must be rounded off.

EXAMPLE EXERCISE 1.4

Round off the following numbers to three significant digits. State which rule you used.

(a) 22.249

(b) 22.250

(c) 0.34548

(d) 0.034975

(e) 0.072038

(f) 72,267

Solution: To locate the first nonsignificant digit, count three significant digits from left to right. If the first nonsignificant digit is less than 5, drop all nonsignificant digits. If the first nonsignificant digit is 5 or greater, add 1 to the last significant digit.
(a) 22.2 (rule 1); **(b)** 22.3 (rule 2); **(c)** 0.345 (rule 1); **(d)** 0.0350 (rule 2); **(e)** 0.0720

(rule 1); (**f**) 72,300 (rule 2). Note that two zeros must be added to 723 to obtain the correct decimal place.

SELF-TEST EXERCISE

Round off the following numbers to three significant digits.

(a) 12.584 **(b)** 1922
(c) 0.06015 **(d)** 14,642

Answers: **(a)** 12.6; **(b)** 1920; **(c)** 0.0602; **(d)** 14,600

1.4 Adding and Subtracting Measurements

To add and subtract measurements and round off the answer to the proper number of significant digits.

When adding or subtracting measurements, the answer is limited by the value with the most uncertainty. Consider the measurements 5 cm, 5.0 cm, and 5.00 cm. The value 5 cm has the greatest uncertainty since it measures to the nearest centimeter. This means if you add the three lengths together the sum must also be rounded off to the nearest centimeter. Therefore, the sum of 5 + 5.0 + 5.00 cm is equal to 15 cm because the answer is rounded off to the nearest centimeter.

EXAMPLE EXERCISE 1.5

Perform the following addition and subtraction operations. Express the answer to the proper decimal place by rounding off nonsignificant digits.

(a) 106.61 g + 0.25 g + 0.195 g **(b)** 35.45 mL − 30.5 mL

Solution: In addition or subtraction, the significant digits in the answer are limited by the data with the most uncertainty.

(a) The masses 106.61 g and 0.25 g are measured to one-hundredth of a gram and 0.195 g to one-thousandth of a gram. Therefore, the sum of the three measurements is limited to hundredths of a gram. The answer is rounded off to two decimals.

$$\begin{array}{r} 106.61\ \text{g} \\ 0.25\ \text{g} \\ +\ \ 0.195\ \text{g} \\ \hline 107.055\ \text{g} \end{array}$$

The answer rounds off to 107.06 g.

The answer is read as "*one hundred seven point zero six grams.*"

(b) The volume 35.45 mL is measured to one-hundredth of a milliliter and 30.5 mL to one-tenth of a milliliter. Therefore, the difference of the two measurements is limited to tenths of a milliliter. The answer is rounded off to one decimal.

$$\begin{array}{r} 35.45\ \text{mL} \\ -\,30.5\ \ \text{mL} \\ \hline 4.95\ \text{mL} \end{array}$$

The answer rounds off to 5.0 mL.

The answer is read "*five point zero milliliters.*"

SELF-TEST EXERCISE

Peform the addition and subtraction and round off your answers.

(a) 8.6 g + 50.05 g

(b) 34.1 mL − 0.55 mL

Answers: **(a)** 58.7 g; **(b)** 33.6 mL

When adding or subtracting measurements, the answer is always *limited by* the measurement with the most *uncertainty*. In other words, the measurement that is least certain in a set of data always limits the answer. For example, if the least certain mass measurement in a set of data is to a tenth (for example, 2.5 g), the sum of the data must be rounded to one-tenth of a gram.

1.5 Multiplying and Dividing Measurements

To multiply and divide measurements and round off the answer to the proper number of significant digits.

The process of handling significant digits when multiplying and dividing measurements is treated differently than when adding and subtracting them. When adding and subtracting significant digits, the answer is limited by the measurement with the most uncertainty. When multiplying and dividing measurements, the answer is determined by the number of significant digits in the measurements. In a set of data, the measurement having the least number of significant digits limits the answer. Suppose a measurement having three significant digits is multiplied by a measurement having five significant digits. The calculated value is limited to the lesser number of significant digits, that is, three.

When multiplying measurements together, the units are also multiplied. For example, 2 ft times 3 ft is equal to 6 ft^2. The answer is read as "*six feet squared, or six square feet.*" When dividing measurements, the units are also divided. For

A laptop computer and most personal computer systems can perform tedious chemical calculations with the appropriate software.

example, 60 miles divided by 1 hour is equal to 60 miles/hour. The answer is read as "*sixty miles per hour.*"

EXAMPLE EXERCISE 1.6

Perform the following multiplication and division operations. Round off the answer to the proper number of significant digits.

(a) 50.5 cm × 12 cm **(b)** 103.37 g/20.5 mL

Solution: In multiplication and division the answer is limited by the data having the least number of significant digits.

(a) The length 50.5 cm represents three significant digits; 12 cm represents two. Thus, the product is limited to two significant digits.

$$(50.5 \text{ cm})(12 \text{ cm}) = 606 \text{ cm}^2 \qquad \text{The answer rounds off to } 610 \text{ cm}^2.$$

Note that 606 rounds off to 610 and not 61. Also note that units multiplied together are indicated by the superscript 2. The answer is read as "*six hundred ten square centimeters.*"

(b) Since the numerator (103.37 g) has five significant digits and the denominator (20.5 mL) has three, the answer is limited to three significant digits.

$$\frac{103.37 \text{ g}}{20.5 \text{ mL}} = 5.0424 \text{ g/mL} \qquad \text{The answer rounds off to } 5.04 \text{ g/mL}.$$

Notice that the calculated result is the ratio of two units. The anwer is read "*five point zero four grams per milliliter.*"

SELF-TEST EXERCISE

Perform the multiplication and division and round off your answers.

(a) (359 cm)(0.20 cm) **(b)** 73.950 g/25.5 mL

Answers: **(a)** 72 cm^2; **(b)** 2.90 g/mL

When multiplying or dividing measurements, the answer is always *limited by* the measurement having *the least number of significant digits*. For example, if the least number of significant digits in a multiplication problem is two (for example, 2.5 cm), then the product must be rounded to two digits.

1.6 Exponential Numbers

To become generally familiar with the concept of exponents and specifically with powers of 10.

To express a value both as a power of 10 and as an ordinary number.

When a value is multiplied by itself, the process is indicated by a number written as superscript. The superscript indicates the number of times the process is repeated. For example, if the number X is multiplied two times, the product is expressed as

X^2. Thus, $(X_1)(X_2) = X^2$. If the number X is multiplied n times, the product is expressed as X^n. Thus, $(X_1)(X_2)(X_3) \cdots X_n = X^n$.

exponent A number written as a superscript that indicates a value is multiplied times itself; for example, $10^4 = 10 \times 10 \times 10 \times 10$, or $cm^3 = cm \times cm \times cm$.

The superscript, such as 2 or n, is called an **exponent**. The value X^n is read *X to the nth power*. The value X^2 is read *X to the second power* or *X squared*. The value X^3 is read *X to the third power* or *X cubed*.

You can also multiply the value $1/X$ by itself. In that case, $1/X$ multiplied two times give $1/X^2$. Thus, $(1/X)(1/X) = 1/X^2$. The value $1/X$ multiplied n times gives the product $1/X^n$. Thus, $(1/X_1)(1/X_2)(1/X_3) \cdots 1/X_n = 1/X^n$.

EXAMPLE EXERCISE 1.7

Multiply the following and express the product using an exponent.

(a) $3 \times 3 \times 3 \times 3 =$ **(b)** $\frac{1}{4} \times \frac{1}{4} \times \frac{1}{4} =$

Solution: The exponent in the product corresponds to the number of times the value is multiplied times itself. Thus, **(a)** is 3^4, which is 81; and **(b)** is $(1/4)^3$, which equals 1/64.

SELF-TEST EXERCISE

Multiply the following and express the product using an exponent.

(a) 3 cm × 3 cm **(b)** 4 cm × 4 cm × 4 cm

Answers: **(a)** 9 cm^2; **(b)** 64 cm^3

The value $1/X^n$ is read as one over X to the nth power. The value $1/X^2$ is read as one over X squared. In chemistry, it is often necessary to multiply and divide exponential numbers. To divide by an exponential number, we simply invert the numbers and multiply.

To invert an exponential number, it is only necessary to change the sign of the exponent. Thus, the value X^2 can be expressed by inverting the fraction and changing the sign of the exponent 2:

$$X^2 = \frac{X^2}{1} = \frac{1}{X^{-2}}$$

In general, we can invert the fraction $1/X^n$ by changing the sign of the exponent n:

$$\frac{1}{X^n} = \frac{X^{-n}}{1} = X^{-n}$$

EXAMPLE EXERCISE 1.8

Invert each of the following exponential numbers.

(a) 10^4 **(b)** 10^{-10} **(c)** $\frac{1}{10^{-5}}$ **(d)** $\frac{1}{10^3}$

Solution: After inverting each of the fractions, we simply change the sign of the exponent; thus, **(a)** $1/10^{-4}$; **(b)** $1/10^{10}$; **(c)** 10^5; **(d)** 10^{-3}

SELF-TEST EXERCISE

Invert each of the following exponential numbers.

(a) $1/10^{22}$ **(b)** $1/10^{-12}$

Answers: **(a)** 10^{-22}; **(b)** 10^{12}

Powers of 10

A power of 10 is a number that results when 10 is raised to an exponential power. You know that an exponent raises any number to a higher power, but we are most interested in the base number 10. A **power of 10** has the general form

power of 10 A positive or negative exponent of 10.

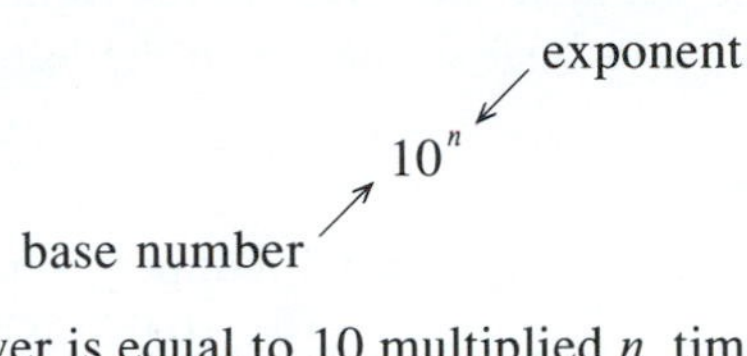

Ten to the *n*th power is equal to 10 multiplied *n* times. Written as an ordinary number, it is a 1 followed by *n* zeros. For example, 10 to the third power (that is, 10^3) is equal to 10 times 10 times 10. As an ordinary number, 10 to the third power is 1000. Notice that the number of zeros following the 1 in 1000 corresponds to the exponent 3. When the power is positive, the value is greater than 1. For example,

$$1 \times 10^3 = 1.000. = 1000$$

When the power is negative, the exponent indicates the number of times 1 is divided by 10. For example, 10 to the minus fourth power (10^{-4}) is equal to 1 divided by 10 four times. In ordinary numbers, 10 to the minus fourth power is 0.0001. The number 0.0001 is less than 1, which corresponds to a negative power.

$$1 \times 10^{-4} = .0001. = 0.0001$$

Table 1.1 lists selected powers of 10 for exponential numbers and the equivalent ordinary number.

Table 1.1 Powers of Ten

Exponential Number	Ordinary Number
$10^6 = 10 \times 10 \times 10 \times 10 \times 10 \times 10$	1,000,000
$10^3 = 10 \times 10 \times 10$	1000
$10^2 = 10 \times 10$	100
$10^1 = 10$	10
$10^0 = 1$	1
$10^{-1} = \frac{1}{10^1} = \frac{1}{10}$	0.1
$10^{-2} = \frac{1}{10^2} = \frac{1}{10} \times \frac{1}{10} = \frac{1}{100}$	0.01
$10^{-3} = \frac{1}{10^3} = \frac{1}{10} \times \frac{1}{10} \times \frac{1}{10} = \frac{1}{1000}$	0.001
$10^{-6} = \frac{1}{10^6} = \frac{1}{10} \times \frac{1}{10} \times \frac{1}{10} \times \frac{1}{10} \times \frac{1}{10} \times \frac{1}{10} \times \frac{1}{1{,}000{,}000}$	0.000 001

EXAMPLE EXERCISE 1.9

Convert the following ordinary numbers to exponential numbers, and vice versa.
(a) 10,000 **(b)** 0.00001 **(c)** 1×10^6 **(d)** 1×10^{-6}

Solution: An ordinary number is converted to the exponent that corresponds to the number of places the decimal point is moved.
(a) In 10,000 the decimal is moved four places to the left; thus, 1×10^4.
(b) In 0.00001 the decimal is moved five places to the right; thus, 1×10^{-5}.
A power of 10 is converted to an ordinary number by the number of places the decimal point is moved.
(c) For 1×10^6, move the decimal six places to the right of 1; thus, 1,000,000.
(d) For 1×10^{-6}, move the decimal six places to the left of 1; thus, 0.000 001.

SELF-TEST EXERCISE

Convert the ordinary numbers to exponential numbers, and vice versa.
(a) 0.000 01 **(b)** 10,000,000
(c) 1×10^9 **(d)** 1×10^{-7}

Answers: **(a)** 1×10^{-5}; **(b)** 1×10^7; **(c)** 1,000,000,000; **(d)** 0.000 000 1

You will have more self-confidence in performing calculations if you understand exponents and how to use them. You can, however, easily carry out math operations with exponents using an inexpensive calculator.

1.7 Scientific Notation

OBJECTIVES

To express any number in scientific notation.

scientific notation A method for expressing very large or small numbers by placing the decimal after the first significant digit and setting the magnitude using a power of 10.

Science frequently deals with numbers that are very large or very small. These numbers quickly become incomprehensible when they contain several zeros. To overcome this problem, a standard notation has been invented that places the decimal after the first significant digit and sets the magnitude of the number using a power of 10; this method is called **scientific notation**. In scientific notation, very large and very small numbers are condensed into a format using significant digits and a power of 10:

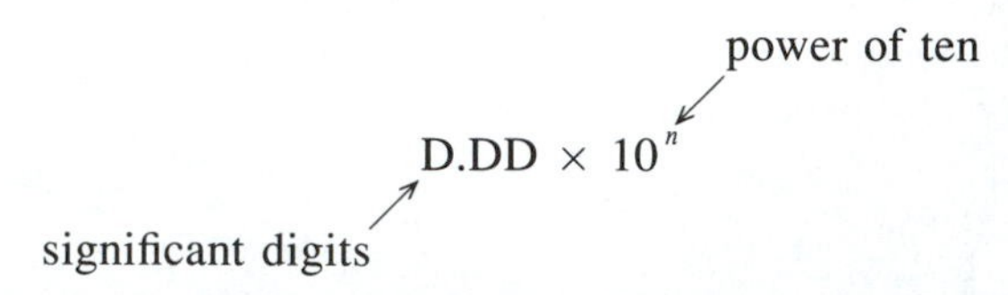

To use scientific notation, write down all the digits in the number. Then move the decimal to follow the first nonzero digit. Indicate the number of places the decimal is moved using power of 10 notation. For example, the large number 555,000 is written in scientific notation by first moving the decimal to the left five places to give 5.55 and, then, adding the appropriate power of 10, which is 10^5. Thus, 555,000 is expressed in scientific notation as 5.55×10^5.

A distance from earth of 1×10^{19} meters.

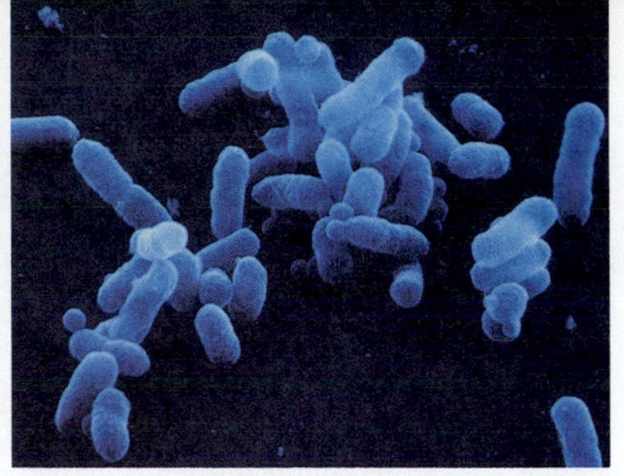

The sizes of cells are about 1×10^{-6} meters.

You can also express numbers smaller than 1 using scientific notation. For example, 0.000888 is written in scientific notation by first moving the decimal to the right four places to give 8.88 and, then, adding the appropriate power of 10, which is 10^{-4}. Thus, 0.000888 is expressed in scientific notation as 8.88×10^{-4}.

Regardless of the size of the number, in scientific notation the decimal point is always placed after the first nonzero digit. The magnitude of the number is indicated by a power of 10. To express a number in scientific notation, follow these two guidelines:

1. Move the decimal point after the first nonzero digit in the number, followed by the remaining significant digits.
2. Indicate how many places the decimal is moved using a power of 10. If the number is greater than 1, the power of 10 is positive. If the number is less than 1, the power of 10 is negative.

EXAMPLE EXERCISE 1.10

Express each of the following values in scientific notation.

(a) There are 26,900,000,000,000,000,000,000 helium atoms in 1 liter of helium gas at standard conditions.

(b) The mass of the helium atom is 0.000 000 000 000 000 000 000 006 65 g.

Solution: To write a value in scientific notation, we will follow the preceding guidelines.

(a) Move the decimal after the digit 2 and retain the remaining significant digits (2.69). Next, count the number of times the decimal is shifted. Since the decimal is shifted to the left 22 places, the exponent is +22. Thus, the value in scientific notation is 2.69×10^{22} helium atoms.

(b) Move the decimal after the digit 6 followed by the other significant digits (6.65). Next, count the number of times the decimal is shifted. Since the decimal is shifted to the right 24 places, the exponent is −24. Thus, the value in scientific notation is 6.65×10^{-24} g.

SELF-TEST EXERCISE

Express each of the following values as ordinary numbers.

(a) The mass of one mercury atom is 3.33×10^{-22} g.

(b) The number of atoms in 1 mL of liquid mercury is 4.08×10^{22}.

Answers: **(a)** 0.000 000 000 000 000 000 000 333 g;
(b) 40,800,000,000,000,000,000,000 atoms

1.8 Calculators and Significant Digits

OBJECTIVES

To use a calculator to multiply and divide ordinary and exponential numbers.

To express a value obtained from a calculator to the proper number of significant digits.

Initially, calculators were expensive and limited in what they could do. Gradually, their prices dropped and many more mathematical functions appeared on the calculator keypad. Today, the prices and functions of calculators vary widely, but many inexpensive calculators can perform all the operations you will encounter. In selecting a calculator for chemistry, choose one that has an exponent key, a log key, and a storage memory key.

Calculator Notation

On most calculators it is not possible to display exponents as superscripts. Typically, the exponent appears without the base number 10. For example, 5.55×10^5 entered into a calculator gives a display that looks like 5.55 05. The power of 10 (that is, 5) is represented in the calculator display as 05.

A special convention is available to write an exponential number from a calculator display. It is called calculator notation. In **calculator notation**, the exponent is indicated by an **E**. The exponent follows the E and is assumed to be a power of 10.

calculator notation A method of displaying exponential numbers without using superscripts; for example, 7.75 E−08.

If you enter the exponential number 5.55×10^5 into a calculator, the display reads 5.55 05. This number can be written in calculator notation as 5.55 E05. If you enter 3.33×10^{-12} into a calculator, the display reads 3.33 − 12. This number can be written in calculator notation as 3.33 E − 12. The following operations using exponential numbers illustrate calculator notation.

EXAMPLE EXERCISE 1.11

Perform the following math operations using a calculator with an exponent key. Express the answers in calculator notation and scientific notation.

(a) $(4.52 \times 10^3)(6.16 \times 10^2)$ **(b)** $(2.23 \times 10^5)(1.6 \times 10^{-8})$
(c) $(7.24 \times 10^4)/(1.36 \times 10^{-1})$ **(d)** $(3.2 \times 10^{-7})/(3.58 \times 10^3)$

Solution: In calculator notation the answers are written as
(a) 2.78 E06 **(b)** 3.6 E − 03 **(c)** 5.32 E05 **(d)** 8.9 E − 11
In scientific notation the answers are expressed as
(a) 2.78×10^6 **(b)** 3.6×10^{-3} **(c)** 5.32×10^5 **(d)** 8.9×10^{-11}

SELF-TEST EXERCISE

Perform the following operations using a calculator. Express the answers in calculator notation and scientific notation.
(a) $(8.99 \times 10^4)(1.26 \times 10^{-3})$ **(b)** $(6.5 \times 10^{-14})/(9.95 \times 10^9)$

Answers: **(a)** 1.13 E02, 1.13×10^2; **(b)** 6.5 E − 24, 6.5×10^{-24}

Calculators often display eight digits or more after performing multiplication or division. Therefore, each time we use a calculator we must assume the responsibility for reporting the correct number of significant digits. We must always determine the number of significant digits that are justified and then round off the number accordingly.

EXAMPLE EXERCISE 1.12

Multiply 3.14 times 5.54. Express the answer with the correct number of significant digits.

Solution: After multiplying, your digital display may read 17.3956. Since three significant digits are justified, the answer rounds to 17.4

SELF-TEST EXERCISE

Divide 200 by 3 using a calculator and express the answer with the correct number of significant digits.

Answer: 66.666667 rounds to 70

One other difficulty with significant digits is that the calculator usually drops all zeros to the right of the decimal. Even if the zeros are significant, they do not appear in the calculator display. A display that reads 5 suggests only one significant digit. If two significant digits are justified, the display must be interpreted as 5.0. If three significant digits are appropriate, the display must be interpreted as 5.00.

EXAMPLE EXERCISE 1.13

Divide 25.00 grams by 10.00 milliliters using a calculator. Express the answer using the correct number of significant digits.

Solution: After performing the calculation, the digital display may read 2.5. The quotient, however, should contain four significant digits. The correct answer is therefore written as 2.500 g/mL.

SELF-TEST EXERCISE

Multiply 42.5 cm times 7.5 cm and express the answer to the correct number of significant digits.

Answer: 318.75 rounds to 320 cm^2 or 3.2×10^2 cm^2.

Note You will frequently carry out a series of mathematical steps using a calculator. The question arises whether to round off after each calculation or only after the final answer. When you perform a series of multiplication and division steps, it is better to round off the final answer. Not only is it more accurate, but it is more convenient.

1.9 Unit Conversion Factors

To write a unit equation based on an equivalent relationship.

To write the two unit conversion factors related to a unit equation.

In the next section we will tackle our single most important task—problem solving. We will learn a simple but powerful method called unit analysis. To use the unit analysis method of problem solving, it is necessary to understand equivalent relationships. Equivalent relationships generate unit conversion factors, which enable you to solve the problem.

What is an equivalent relationship? An equivalent relationship is a relationship between two quantities that are equal. For example, 1 dollar equals 10 dimes and 1 dime equals 10 pennies.

$$1 \text{ dollar} = 10 \text{ dimes}$$

$$1 \text{ dime} = 10 \text{ pennies}$$

These equivalent relationships are called unit equations. A **unit equation** is a simple statement of two equivalent values.

What is a unit conversion factor? A unit conversion factor, or **unit factor**, is a ratio of two equivalent quantities; for example, 1 dollar/10 dimes. Since the numerator and denominator are equivalent or equal, the **reciprocal** of the ratio (10 dimes/1 dollar) is also a unit conversion factor. Therefore, for the unit equation 1 dollar = 100 pennies, we can write reciprocal unit factors that are equivalent:

unit equation A simple statement of two equivalent values written as an equation; for example, 1 m = 100 cm.

unit factor A ratio of two quantities that are equivalent and used to convert from one unit to another; for example, 1 m/100 cm.

reciprocal The relationship of a fraction and its inverse; for example, 2/3 and 3/2, or 1 m/100 cm and 100 cm/1 m.

$$\frac{1 \text{ dollar}}{100 \text{ pennies}} \quad \text{and} \quad \frac{100 \text{ pennies}}{1 \text{ dollar}}$$

A unit factor lets us convert from one unit to another unit. Suppose we want to convert five dollars into pennies. To perform the unit conversion, we choose the unit factor whose units in the denominator are the same as the units of the given value. Thus, we use the unit factor 100 pennies/1 dollar:

$$5 \text{ dollars} \times \frac{100 \text{ pennies}}{1 \text{ dollar}} = 500 \text{ pennies}$$

The dollar units cancel and we have the desired units, pennies.

There are 20 nickels in one dollar.

EXAMPLE EXERCISE 1.14

Write the unit equation and two unit conversion factors associated with each of the following relationships.

(a) dollars and nickels **(b)** minutes and seconds

Solution: First, write the unit equation and then the two factors.

(a) The unit equation is 1 dollar = 20 nickels. The two associated unit factors are

$$\frac{1 \text{ dollar}}{20 \text{ nickels}} \quad \text{and} \quad \frac{20 \text{ nickels}}{1 \text{ dollar}}$$

(b) The unit equation is 1 minute = 60 seconds. The two unit factors are

$$\frac{1 \text{ minute}}{60 \text{ seconds}} \quad \text{and} \quad \frac{60 \text{ seconds}}{1 \text{ minute}}$$

SELF-TEST EXERCISE

Write the two unit factors that correspond to each of the following unit equations.

(a) 1 lb = 16 oz **(b)** 1 gal = 4 qt

Answers: **(a)** 1 lb/16 oz and 16 oz/1 lb; **(b)** 1 gal/4 qt and 4 qt/1 gal

You should avoid using a decimal relationship in a unit equation. For example, there are 100 pennies in a dollar and we can write 1 penny = 0.01 dollar. The use of a decimal fraction is not recommended since it is easy to make an error. In general, you should write whole-number relationships, which are easier to grasp. In this example, we can write the unit equation 100 pennies = 1 dollar.

1.10 Problem Solving by Unit Analysis

OBJECTIVES

To state and apply the three steps in the unit analysis method of problem solving.

Chemistry has the reputation of dealing with tough mathematical problems. If chemists can create problems, why can't they create a simple method for solving these problems? They did. It is called the **unit analysis** (or dimensional analysis) **method** of problem solving. Unit analysis will not solve all problems, but it is very effective for most types of introductory chemistry problems.

unit analysis method A systematic procedure for solving problems that proceeds to a desired value from a related given value by the conversion of units.

Problem solving using unit analysis is as simple as one, two, three. Step 1: Read the problem and determine the units in the answer. Step 2: Analyze the problem and determine which given value is related to the answer. Step 3: Generate one or more unit factors that convert the given value to the units in the answer.

To use the unit analysis method of problem solving, follow these three steps.

Step 1: Write the units asked for in the answer.

Step 2: Write the given value related to the answer.

Step 3: Apply one or more unit factors to convert the given value to the answer.

The format for solving a problem using unit analysis is as follows:

$$\underset{(2)}{\text{related given value}} \times \underset{(3)}{\frac{\text{unit}}{\text{factor(s)}}} = \underset{(1)}{\text{units in answer}}$$

EXAMPLE EXERCISE 1.15

A roll of quarters has a value of 10 dollars. Find the number of quarters in a roll.

Solution: Step 1: Read the problem and write the units asked for in the answer (quarters). Step 2: Analyze the given information and select the related value (10 dollars). Step 3: Apply one or more unit factors correctly. The format is

$$\underset{(2)}{10 \text{ dollars}} \times \underset{(3)}{\frac{\text{unit}}{\text{factor(s)}}} = \underset{(1)}{\text{quarters}}$$

The unit equation that relates steps 1 and 2 is 1 dollar = 4 quarters. The two corresponding unit factors are

$$\underset{(A)}{\frac{1 \text{ dollar}}{4 \text{ quarters}}} \quad \text{and} \quad \underset{(B)}{\frac{4 \text{ quarters}}{1 \text{ dollar}}}$$

Select the unit factor that cancels the units in the given value, 10 dollars. It is unit factor (B), which has dollars in the denominator and allows for cancellation of units. Thus,

$$10 \cancel{\text{dollars}} \times \frac{4 \text{ quarters}}{1 \cancel{\text{dollar}}} = 40 \text{ quarters}$$

SELF-TEST EXERCISE

What is the dollar value of 560 nickels?

Answer: $28.00

EXAMPLE EXERCISE 1.16

Silver is a precious metal used in jewelry, coins, and silver bars. The mass of silver is usually measured in units of troy ounces when it is sold in large quantities. If silver sold for $6.50 per ounce in 1991, how many troy ounces of silver could be purchased for $29,575.00?

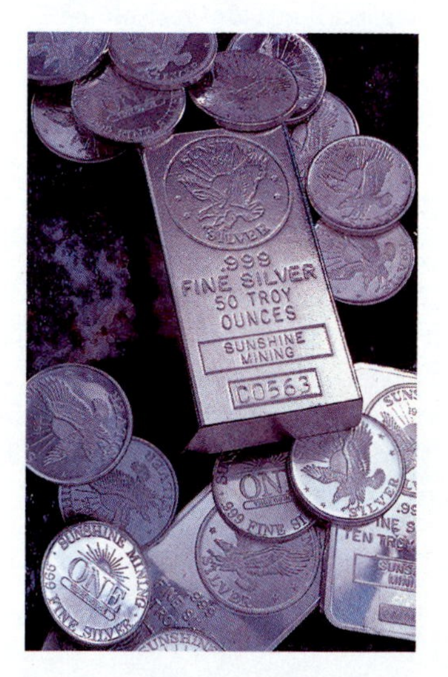

Silver coins and bars of silver.

Solution: Step 1: Read the problem and write the units asked for in the answer (troy ounces). Step 2: Analyze the given information and select the relevant value $29,575.00. Step 3: Apply one or more unit factors correctly. The format is

$$\underset{(2)}{\$29{,}575.00} \times \underset{(3)}{\frac{\text{unit}}{\text{factor(s)}}} = \underset{(1)}{\text{troy ounces}}$$

The unit equation that relates steps 1 and 2 is $6.50 = 1 troy ounce. From this equation two possible unit factors follow:

$$\underset{(A)}{\frac{\$6.50}{1 \text{ troy ounce}}} \quad \text{and} \quad \underset{(B)}{\frac{1 \text{ troy ounce}}{\$6.50}}$$

Which is the proper unit factor? The proper factor is the one that cancels the units in the given value. Thus, select unit factor (B), since the units we wish to cancel (dollars) appear in the denominator. This gives

$$\$29{,}575.00 \times \frac{1 \text{ troy ounce}}{\$6.50} = 4550 \text{ troy ounces}$$

SELF-TEST EXERCISE

What is the value of 5.00 troy ounces of platinum? (Assume platinum is selling for \$475.00 per ounce.)

Answer: \$2375.00

EXAMPLE EXERCISE 1.17

A gold nugget and bars of gold.

Gold is used to support the world's currencies. Currently, the United States has 263,000,000,000 troy ounces of gold bullion on reserve. Calculate the mass of gold in troy pounds. (Given: 12 troy ounces = 1 troy pound.)

Solution: Step 1: The units asked for are troy pounds. Step 2: The given value related to the answer is 263,000,000,000 troy ounces. Expressing this large number in scientific notation, we have

$$\underset{(2)}{2.63 \times 10^{11} \text{ troy ounces}} \times \underset{(3)}{\frac{\text{unit}}{\text{factor(s)}}} = \underset{(1)}{\text{troy pounds}}$$

The unit equation is 1 troy pound = 12 troy ounces. The two unit factors are

$$\underset{(A)}{\frac{1 \text{ troy pound}}{12 \text{ troy ounces}}} \quad \text{and} \quad \underset{(B)}{\frac{12 \text{ troy ounces}}{1 \text{ troy pound}}}$$

Step 3: To apply the proper unit factor, choose (A), because the units we wish to cancel (troy ounces) appear in the denominator of (A).

$$2.63 \times 10^{11} \cancel{\text{troy ounces}} \times \frac{1 \text{ troy pound}}{12 \cancel{\text{troy ounces}}} = 2.19 \times 10^{10} \text{ troy pounds}$$

SELF-TEST EXERCISE

Find the number of fluid ounces of liquid mercury in 4.00 quarts. (Given: 32 fl oz = 1 qt.)

Answer: 128 fl oz

Note

By definition, 1 foot is exactly equivalent to 12 inches. There is no uncertainty in the relationship. Unit factors derived from exact relationships have an infinite number of significant digits. Therefore, they have no effect on the number of significant digits in the answer. Examples of other exact equivalents include 1 mile = 5280 feet, 1 pound = 16 ounces, and 1 gallon = 4 quarts. From Example Exercise 1.17, 1 troy pound is exactly equal to 12 troy ounces. The answer was rounded to three significant digits to be consistent with the given value: 2.63×10^{11} troy ounces.

1.11 Percentage

To apply the concept of percentage.

percent (%) The ratio (times 100) of a single quantity compared to all quantities in a group; parts per hundred parts.

Percentage is a way of expressing the amount of a given quantity compared to all the quantities in a total sample. The ratio of parts per hundred parts is called a **percent** (symbol %). Thus, ten percent, written 10%, means 10 out of 100. A person in a 28% income tax bracket must pay 28 dollars in tax for every 100 dollars income. In other words, 28 cents of each taxable dollar (100 cents) goes to the government.

To calculate a percent, you divide one quantity by the total of all quantities in the sample. You then multiply that ratio by 100:

$$\frac{\text{one quantity}}{\text{all quantities}} \times 100 = \%$$

As an example, consider a college student who receives $200.00 in financial aid to purchase books and supplies for classes. If the student spends $50.00 for chemistry materials, what percent of the financial aid is spent for chemistry? Calculate the percent as follows:

$$\frac{\$50.00}{\$200.00} \times 100 = 25\%$$

One-fourth, or 25%, of the student's financial aid was spent for chemistry.

EXAMPLE EXERCISE 1.18

An ore sample taken from a Nevada silver mine weighed 5.330 g. After chemical analysis, the sample was found to contain 0.150 g of silver. Calculate the percent of silver in the ore sample.

Solution: In this example, the mass of silver is compared to the total mass of the sample. Thus,

$$\frac{0.150 \text{ g}}{5.330 \text{ g}} \times 100 = 2.81\%$$

The ore sample contained 2.81% silver.

SELF-TEST EXERCISE

A 14-karat gold ring contains 7.45 g of gold and 5.32 g of alloy. Calculate the percent of gold in the ring.

Answer: 58.3%

Percentages have equivalent decimal fractions. For example, 12% and 25% correspond to 0.12 and 0.25, respectively. The decimals 0.12 and 0.25 correspond to the fractions 12/100 and 25/100.

CHEMISTRY CONNECTION □ THE CONSUMER

The Coinage Metals

► **Can you name the metal that makes up most of the mass of a penny minted before 1982? after 1982?**

The earth has a solid metal core composed of only two metals, iron and nickel. On the other hand, there are dozens of metals that occur in the earth's crust. The metals that are most abundant in the earth's crust include aluminum, iron, calcium, sodium, potassium, magnesium, titanium, and manganese. However, these metals are found combined with other elements. Only a few metals occur naturally in the uncombined free state. Copper, silver, gold, and platinum are metals often found free in nature.

Silver and gold coins that were used by the Romans.

During the prehistoric Stone Age, copper was first used to make tools and weapons. The Bronze Age (~3000 B.C.) followed the Stone Age after it was discovered that copper and tin could be melted together to give an alloy of bronze. A bronze alloy has the advantage of being much stronger and more flexible than pure copper metal. Since ancient times, copper, silver, and gold have been used to make jewelry and precious objects because they have an attractive appearance and can be easily worked into various shapes.

Thousands of years ago in India and Egypt, silver and gold tokens were used for trade. Later, the Romans used silver and gold to make coins for monetary exchange. Today, copper, silver, and gold are often referred to as the **coinage metals**. The term coinage metals, however, is somewhat misleading. In 1934, the United States took gold coins out of circulation. In 1971, the United States stopped using silver in coins except for special collector issues. Currently, 5¢, 10¢, 25¢, and 50¢ coins are minted from an alloy that is 75% copper and 25% nickel. If you examine a 25¢ coin closely, the reddish-brown color of copper is clearly visible along the serrated edge.

The first 1¢ coin was minted in 1793 from pure copper and was the size of a half dollar. This 1¢ coin continued to be minted until 1856 when it was replaced by a coin made from a metal alloy of 88% copper and 12% nickel. The addition of nickel gave a white appearance to the coin. Five years later, the U.S. Mint substituted a new alloy of copper, tin, and zinc. In 1962, tin was eliminated from the penny and the coin was cast from an alloy of 95% copper and 5% zinc. In 1982, the price of copper rose dramatically and the U.S. Mint altered the composition once again. In lieu of copper pennies, the new 1¢ coins are mostly zinc. The current pennies are minted from a zinc disk plated with an ultra-thin coating of copper.

► **Pre-1982 pennies are mostly copper. Post-1982 pennies are almost entirely zinc.**

Consider the following problem example. A computer printout indicates that 45% of the students in a chemistry class are female. Find the number of females in the class if the total enrollment is 120 students. Since the given percentage is 45%, 45/100 are female students. To find the number of females, multiply the total number of students by the fraction representing the females. This gives

$$120\ \cancel{\text{students}} \times \frac{45\ \text{females}}{100\ \cancel{\text{students}}} = 54\ \text{females}$$

In the chemistry class of 120 students, the number of females is 54.

An Apollo astronaut on the moon.

EXAMPLE EXERCISE 1.19

The Apollo 11 spacecraft traveled 394,000 kilometers from earth to put the first person on the moon. How many kilometers had Apollo 11 gone when it had travelled 15.5% of the distance?

Solution: By definition, 15.5% means 15.5 parts per 100. In this case, 15.5% is 15.5 kilometers per 100 kilometers. Multiplying gives

$$394{,}000 \cancel{\text{kilometers}} \times \frac{15.5 \text{ kilometers}}{100 \cancel{\text{kilometers}}} = 61{,}100 \text{ kilometers}$$

SELF-TEST EXERCISE

Sterling silver is an alloy of 92.5% silver and 7.5% copper. What is the mass of a sterling silver chain that contains 27.75 g of silver?

Answer: 30.0 g

Summary

Sections 1.1–1.3 This chapter presented the math skills necessary for the calculations we will do in later chapters. We learned that no **measurement** is exact because all instruments have **uncertainty**. The number of **significant digits** in a measurement must be consistent with the uncertainty of the instrument. **Nonsignificant digits** are obtained from calculators and must be **rounded off**.

Sections 1.4–1.5 When adding or subtracting measurements, the answer is limited by the least accurate measurement, which can be tenths, hundredths, or thousandths of a unit, and so on. When multiplying or dividing measurements, the answer's precision is limited by the least number of significant digits in the data. Every measurement calculation must include the correct significant digits as well as the proper units.

Sections 1.6–1.7 In chemistry we often deal with numbers that are very large or very small. To avoid using many zeros as place holders, we use **scientific notation** to express the number. Scientific notation always uses the same format; that is, D.DD $\times$ 10^n. The significant digits (D.DD) are multiplied by a **power of ten** (10^n) to place the decimal point. If n is positive, the number is larger than 1. If n is negative, it is smaller than 1.

Section 1.8 With a little practice we can learn to add, subtract, multiply, and divide exponential numbers. Alternatively, we can use calculators to perform these operations. It is convenient to record an exponential number in the calculator display using **calculator notation**. If the display shows an exponential number such as 2.77 − 03, we can indicate the value in calculator notation as 2.77 E−03 or in scientific notation as 2.77×10^{-3}. When using a calculator, we must round off the numbers in the display to the proper number of significant digits.

Sections 1.9–1.10 Your single most important task has been introduced in this chapter. This is the task of problem solving. We will consistently use the **unit analysis method** of problem solving, which requires three simple steps. Step 1 is to write down the units of the quantity asked for in the answer. Step 2 is to select the given value that is related to the answer. Step 3 is to write a **unit equation** and the two corresponding **unit factors**. Select and apply the unit factor that allows the unit of the given value to be canceled. After rounding off your answer and attaching the proper units, check your answer for reasonableness. That is, your answer should seem reasonable based on the data given in the problem.

Section 1.11 Percentage is an important concept that we will later apply to the composition of substances and solutions. The ratio of a given quantity to the total—all multiplied by 100—gives a **percent**.

Key Terms

Select the key term that corresponds to the following definitions.

____ **1.** a numerical value with units that expresses a physical quantity such as length, mass, or volume

____ **2.** a common metric unit of length

____ **3.** a common metric unit of mass

____ **4.** a common metric unit of volume

____ **5.** a device for recording a measurement such as length, mass, or volume

____ **6.** the degree of precision or exactness of an instrumental measurement

____ **7.** the quantity of matter in an object; measured by a balance

____ **8.** the gravitational force of attraction between an object and the planet

____ **9.** the digits in a measurement that are known with certainty plus one digit that is estimated

____ **10.** the process of eliminating digits that are not significant

____ **11.** the digits in a measurement that exceed the uncertainty

____ **12.** a number written as a superscript that indicates a value is multiplied times itself; for example, $10^4 = 10 \times 10 \times 10 \times 10$, or $cm^3 = cm \times cm \times cm$

____ **13.** a positive or negative exponent of 10

____ **14.** a method of displaying exponential numbers without using superscripts; for example, 7.75 E−08

____ **15.** a method for expressing very large or small numbers by placing the decimal after the first significant digit and setting the magnitude using a power of 10

____ **16.** a simple statement of two equivalent values; for example, 1 foot = 12 inches

____ **17.** a ratio of two quantities that are equivalent and can be applied to convert from one unit to another; for example, 1 foot/12 inches.

____ **18.** the relationship of a fraction and its inverse; for example, 1 foot/12 inches and 12 inches/1 foot

____ **19.** a systematic procedure for solving problems that proceeds to a desired value from a related given value by the conversion of units

____ **20.** the ratio (times 100) of a single quantity compared to all quantities in a group; parts per hundred parts

(a) calculator notation *(Sec. 1.8)*
(b) centimeter (cm) *(Sec. 1.1)*
(c) exponent *(Sec. 1.6)*
(d) gram (g) *(Sec. 1.1)*
(e) instrument *(Sec. 1.1)*
(f) mass *(Sec. 1.1)*
(g) measurement *(Sec. 1.1)*
(h) milliliter (mL) *(Sec. 1.1)*
(i) nonsignificant digits *(Sec. 1.3)*
(j) percent (%) *(Sec. 1.11)*
(k) power of 10 *(Sec. 1.6)*
(l) reciprocal *(Sec. 1.9)*
(m) rounding off *(Sec. 1.3)*
(n) scientific notation *(Sec. 1.7)*
(o) significant digits *(Sec. 1.2)*
(p) uncertainty *(Sec. 1.1)*
(q) unit analysis method *(Sec. 1.10)*
(r) unit equation *(Sec. 1.9)*
(s) unit factor *(Sec. 1.9)*
(t) weight *(Sec. 1.1)*

Exercises

Uncertainty in Measurement (Sec. 1.1)

1. What state-of-the-art instrument is capable of making an exact measurement?
2. Using the latest electronic technology, what physical quantity can be measured with no uncertainty?
3. What quantity (length, mass, volume, time) is measured by the following instruments?
   (a) metric ruler (b) buret
   (c) balance (d) pipet
   (e) stopwatch (f) graduated cylinder
4. What quantity (length, mass, volume, time) is measured by the following units?
   (a) gram (g) (b) milliliters (mL)
   (c) centimeters (cm) (d) seconds (s)
5. Uncertainty is given for each of the following measurements. State the maximum and minimum values that are acceptable.
   (a) 6.5 ± 0.1 cm (b) 0.51 ± 0.01 g
   (c) 10.0 ± 0.1 mL (d) 35.5 ± 0.1 s
6. Uncertainty is given for each of the following measurements. State the maximum and minimum values that are acceptable.
   (a) 6.35 ± 0.05 cm (b) 1.556 ± 0.001 g
   (c) 30.05 ± 0.05 mL (d) 60.01 ± 0.01 s

Significant Digits (Sec. 1.2)

7. State the number of significant digits in the following measurements.
   (a) 0.05 cm (b) 12.0 cm
   (c) 0.707 g (d) 1.110 g
   (e) 24.60 mL (f) 100 mL
   (g) 3.71×10^3 s (h) 1.000×10^{-1} s
8. State the number of significant digits in the following measurements.
   (a) 2.50 cm (b) 5.05 cm
   (c) 2.00 g (d) 1000 g
   (e) 250 mL (f) 10.0 mL
   (g) 2.0×10^4 s (h) 1×10^{-3} s

Rounding Off Nonsignificant Digits (Sec. 1.3)

9. Round off the following numbers to three significant digits
   (a) 31.505 (b) 213,600
   (c) 5.155 (d) 77.504
   (e) 0.01842 (f) 0.000 000 484 500
   (g) 2.571×10^5 (h) 5.6954×10^{-2}
10. Round off the following numbers to three significant digits.
   (a) 61.15 (b) 362.01
   (c) 2155 (d) 0.3665
   (e) 12.59 (f) 35.55
   (g) 1.598×10^9 (h) 2.6514×10^{-4}

Adding and Subtracting Measurements (Sec. 1.4)

11. Perform the following addition operations and express the answer using the proper units and significant digits.
   (a) 31.15 cm + 0.5 cm
   (b) 50.2 cm + 5.25 cm
   (c) 0.4 g, 0.44 g, +0.444 g
   (d) 15.5 g, 7.50 g, + 0.050 g
   (e) 0.55 mL, 36.15 mL, +17.3 mL
   (f) 4 mL, 16.3 mL, + 0.95 mL
12. Perform the following subtraction operations and express the answer using the proper units and significant digits.
   (a) 45.35 cm − 41.1 cm
   (b) 24.9 cm − 2.55 cm
   (c) 242.167 g − 175 g
   (d) 27.55 g − 14.545 g
   (e) 22.10 mL − 10.5 mL
   (f) 10.0 mL − 0.15 mL

Multiplying and Dividing Measurements (Sec. 1.5)

13. Perform the following multiplication operations and express the answer using the proper units and significant digits.
   (a) 3.65 cm × 2.10 cm
   (b) 8.75 cm × 1.15 cm
   (c) 16.5 cm × 1.7 cm
   (d) 21.1 cm × 20 cm
   (e) 5.1 cm × 1.25 cm × 0.5 cm
   (f) 5.15 cm × 2.55 cm × 1.1 cm
   (g) 12.0 cm^2 × 1.00 cm
   (h) 22.1 cm^2 × 0.75 cm
14. Perform the following division operations and express the answer using the proper units and significant digits.
   (a) 66.3 g/7.5 mL (b) 12.5 g/4.1 mL
   (c) 42.620 g/10.0 mL (d) 91.235 g/10.00 mL
   (e) 26.0 cm^2/10.1 cm (f) 9.95 cm^2/0.15 cm
   (g) 131.78 cm^3/19.25 cm (h) 131.78 cm^3/19.26 cm

Exponential Numbers (Sec. 1.6)

15. Multiply the following values and express the product using an exponent.
   (a) 3 × 3
   (b) 2 × 2 × 2 × 2 × 2 × 2

(c) $1/2 \times 1/2 \times 1/2$
(d) $1/3 \times 1/3 \times 1/3 \times 1/3 \times 1/3$
(e) $10 \times 10 \times 10$
(f) $10 \times 10 \times 10 \times 10 \times 10 \times 10$
(g) $1/10 \times 1/10 \times 1/10$
(h) $1/10 \times 1/10 \times 1/10 \times 1/10$

16. Write the inverse for each of the following exponential numbers.
(a) 10^3 (b) 10^{23}
(c) 10^{-8} (d) 10^{-15}
(e) $1/10^3$ (f) $1/10^{12}$
(g) $1/10^{-7}$ (h) $1/10^{-22}$

17. Express the following ordinary numbers as powers of 10.
(a) 1 (b) 0.1
(c) 1,000,000 (d) 0.0001
(e) 10,000,000,000 (f) 0.000 001
(g) 100,000,000,000,000,000 (h) 0.000 000 000 000 001

18. Express the following powers of 10 as ordinary numbers.
(a) 1×10^1 (b) 1×10^{-1}
(c) 1×10^5 (d) 1×10^{-4}
(e) 1×10^8 (f) 1×10^{-9}
(g) 1×10^0 (h) 1×10^{-22}

Scientific Notation (Sec. 1.7)

19. Express the following ordinary numbers in scientific notation.
(a) 80,916,000 (b) 0.000 000 015
(c) 335,600,000,000,000 (d) 0.000 000 000 000 927

20. Express the following ordinary numbers in scientific notation.
(a) 1,010,000,000,000,000
(b) 0.000 000 000 000 456
(c) 94,500,000,000,000,000
(d) 0.000 000 000 000 000 019 50

21. There are 26,900,000,000,000,000,000,000 neon atoms in 1 liter of neon gas at standard conditions. Express this number of atoms in scientific notation.

22. Given that the mass of a neon atom is

0.000 000 000 000 000 000 000 0335 g

express the mass in scientific notation.

Calculators and Significant Digits (Sec. 1.8)

23. Perform the indicated mathematical operation. Express the answer in calculator notation and scientific notation using the correct number of significant digits.
(a) $(4.65 \times 10^5)(9.5 \times 10^2)$
(b) $(3.62 \times 10^4)/(1.60 \times 10^{-3})$
(c) $(2.33 \times 10^5)(11.21 \times 10^3)$
(d) $(1.11 \times 10^8)/(2.05 \times 10^{-11})$
(e) $(5.97 \times 10^{-3})(81.4 \times 10^{-6})$
(f) $(54.9 \times 10^6)/(7.87 \times 10^4)$
(g) $(1.16 \times 10^4)(3.14 \times 10^{-4})$
(h) $(11.17 \times 10^{-4})/(25.3 \times 10^{-5})$
(i) $(8.68 \times 10^{-1})(6.13 \times 10^{-2})$
(j) $(0.176 \times 10^{11})/(4.58 \times 10^{-19})$

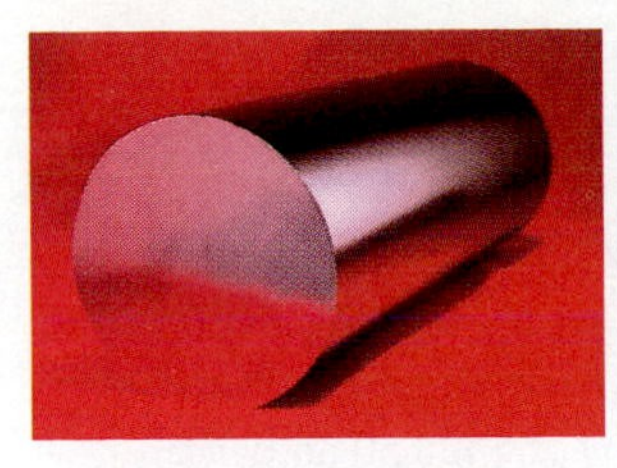
An ingot of pure silicion can be sliced into thin wafers.

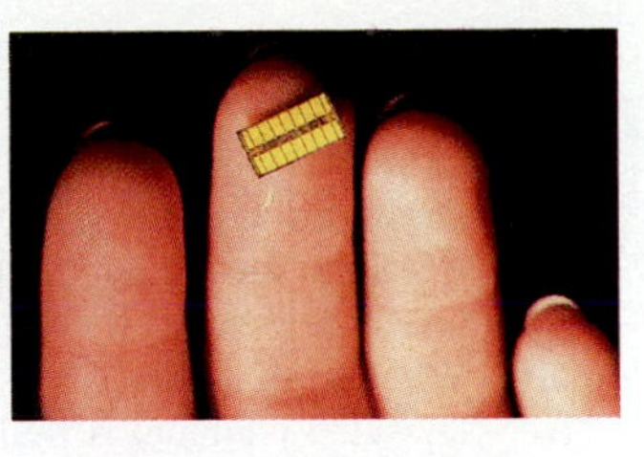
A silicon wafer is cut into small chips for integrated circuits in calculators.

24. Perform the indicated mathematical operation. Express the answer in calculator notation and scientific notation with the correct number of significant digits.

(a) $\dfrac{(4.97 \times 10^8)(3.24 \times 10^4)}{(56.71 \times 10^{-2})(1.31 \times 10^3)}$

(b) $\dfrac{(96.77 \times 10^{-7})(0.00181 \times 10^{-8})}{(3.49 \times 10^{10})(7.62 \times 10^{-12})}$

(c) $\dfrac{(1.78 \times 10^2)}{(3.69 \times 10^{-3})} \times \dfrac{(6.02 \times 10^{23})}{(1.6 \times 10^{20})}$

(d) $\dfrac{(16.4 \times 10^{-3})}{(1.186 \times 10^{-7})} \times \dfrac{(4.00 \times 10^6)}{(3.78 \times 10^{-4})}$

(e) $\dfrac{(3.111 \times 10^{-6})}{(2.08 \times 10^{-2})} \times \dfrac{(4.55 \times 10^5)}{(8.88 \times 10^9)}$

Unit Conversion Factors (Sec. 1.9)

25. Write a unit equation and the two associated unit factors for each of the following.
(a) dollars and dimes
(b) nickels and pennies
(c) dollars and quarters
(d) pennies and dimes
(e) nickels and dimes
(f) half-dollars and nickels
(g) pennies and half-dollars
(h) pennies and quarters
(i) quarters and nickels
(j) dimes and half-dollars

26. Write a unit equation and the two associated unit factors for each of the following.
(a) days and hours (b) inches and feet
(c) years and days (d) yards and inches
(e) years and centuries (f) feet and yards
(g) days and weeks (h) yards and miles
(i) months and years (j) miles and feet

Problem Solving by Unit Analysis (Sec. 1.10)

27. Apply the unit analysis method of problem solving to each of the following.
(a) How many nickels are in a \$2 roll?
(b) A coin collector has 2150 dimes. What is the dollar value?

(c) What is the number of pencils in 0.750 gross? (A gross of pencils is 12 dozen; that is, 144.)
(d) How many reams of paper can be packaged if 12,500 sheets are available? (A ream of paper is 500 sheets.)
(e) What is the dollar value of 75.0 ounces of silver? (Assume silver is trading at $6.50 an ounce.)
(f) What is the dollar value of 100.0 ounces of platinum? (Assume platinum is trading at $425.50 an ounce.)

28. Apply the unit analysis method of problem solving to each of the following.
(a) How many 40-hour work weeks are necessary to complete a job that is estimated to require 225 hours?
(b) How many yards long is a marathon race? A marathon is 26 miles, 385 yards, and one mile is 1760 yards.
(c) How many karats is a diamond weighing 0.305 gram? (One karat is 0.200 gram.)
(d) How many liters of gasoline are required to fill a 20.0-gallon tank? (One gallon is 3.78 liters.)
(e) How far does light travel in an hour? (The velocity of light is 186,000 miles per second.)
(f) How far does light travel in one year; that is, how far is a light-year? (The velocity of light is 3.00×10^{10} centimeters per second.)

Percentage (Sec. 1.11)

29. Perform the necessary percentage calculations to solve the following problems.
(a) In a freshman class of 5846 students, 101 select chemistry as a major. Calculate the percent of chemistry majors.
(b) Blood bank records indicate blood type and Rh factor for 55,368 patients as follows: 21,594 O+, 18,825 A+, 4706 B+, 1938 AB+, 3876 O−, 3322 A−, 831 B−, 276 AB−. Calculate the percent of each group.
(c) A 5.750-gram sample of bauxite ore was found by analysis to contain 34.1% aluminum. Find the mass of aluminum.
(d) Air is 20.9% oxygen by volume. Find the volume of air that contains 225 mL of oxygen.
(e) Mixing 250 mL of ethyl alcohol with 375 mL of water produces a solution that burns with a cool flame. Compute the percent of alcohol.
(f) Gasohol is a blend of ethanol in gasoline. If 12.5 gallons of gasohol contain 1.50 gallons of ethanol, what is the percent of ethanol?

30. Perform the necessary percentage calculations to solve the following problems.
(a) An introductory chemistry class began with 135 students. Of the original enrollment, 11.1% received a final grade of A. How many students earned an A?
(b) Water is composed of 11.2% hydrogen and 88.8% oxygen. What mass of water contains 15.0 g of oxygen?
(c) Ordinary table salt, sodium chloride, is 39.3% sodium. Calculate the mass of sodium in 0.375 g of salt.
(d) Iodized table salt contains 0.02% potassium iodide. Calculate the mass of salt that contains 0.100 g of potassium iodide.
(e) The earth's crust has a mass of 2.37×10^{25} g and contains 49.2% oxygen, 25.7% silicon, and 7.50% aluminum. Calculate the mass of each element.
(f) Uranus has five moons with a combined mass of 3.87×10^{21} kilograms. What is the mass of the largest moon, Titania, if it is 54.3% of the total?

31. Before 1982 the U.S. Mint cast penny coins from an alloy of copper and zinc. If a 1980 penny weighs 3.051 g and contains 2.898 g copper, what are the percentages of copper and zinc in the coin?

32. In 1982 the U.S. Mint stopped making copper pennies, because of the price of copper, and began phasing in pennies made of zinc plated with a thin layer of copper. If a 1990 penny weighs 2.554 g and contains 2.490 g zinc, what are the percentages of copper and zinc in the coin?

General Exercises

33. Draw a line on a separate sheet of paper and mark the length equal to line L below. Measure the line using ruler A and ruler B in Figure 1.1. Record the length of the line consistent with the uncertainty of each ruler.
L: ____________________

34. Draw a line on a separate sheet of paper and mark the length equal to line L below. Measure the line using ruler A and ruler B in Figure 1.1. Record the length of the line consistent with the uncertainty of each ruler.
L: ____

35. If a 10-mL pipet has an uncertainty of one-tenth milliliter, indicate its volume using ordinary numbers.

36. If a 1000-mL flask has an uncertainty of one milliliter, express the volume of the flask using scientific notation.

37. Round off the mass of a sodium atom, 22.989768 atomic mass units, to three significant digits.

38. Round off the velocity of light, 2.997925×10^{10} cm/s, to three significant digits.

39. Find the total mass of two brass cylinders that weigh 126.457 g and 131.6 g.

40. A 255-centimeter strip of magnesium has two 25.0-centimeter strips cut from it. Calculate the length of the strip of magnesium that remains.

41. Convert the following exponential numbers into scientific notation.
(a) 352×10^{4} (b) 416×10^{3}
(c) 0.170×10^{2} (d) 0.00125×10^{2}

42. Convert the following exponential numbers into scientific notation.
(a) 732×10^{-3} (b) 0.00350×10^{-1}
(c) 16.6×10^{-6} (d) 0.191×10^{-5}

43. The approximate number of carbon dioxide molecules expelled per human breath is 3.20×10^{20}. Express the value in numerical form.

44. A googol is 1×10^{100}. Design a way to enter a googol into your calculator and express the display in calculator notation. (*Hint:* The highest power that most calculators can ordinarily display is E99.)

45. The mass of an electron is 9.10953×10^{-28} g and a proton is 1.67265×10^{-24} g. Find the total mass of an electron and a proton.

46. The mass of a neutron is 1.67495×10^{-24} g and a proton is 1.67265×10^{-24} g. Find the mass difference between a neutron and a proton.

47. A metric ton is defined as 1000 kilograms or 2.20×10^3 pounds. An English ton is 2000 pounds. What is the difference in mass between a metric ton and an English ton expressed in pounds?

48. The distance from the earth to the moon is 2.39×10^5 miles; from the moon to Mars it is 4.84×10^7 miles. What is the total distance a space probe travels from the earth to the moon to Mars?

49. A crystal of salt has a cubic structure and each side measures 4.32×10^{-1} cm. What is the volume of the crystal? The formula for the volume of a cube is l^3, where l is the length of a side.

50. The radius of the sun is 4.32×10^5 miles. Assuming the sun is a sphere, calculate the volume of the sun. The formula for the volume of a sphere is $4\pi r^3/3$, where π is a constant equal to 3.14 and r is the radius.

51. The oldest rock found on earth was discovered in South Africa. The rock is estimated to be 4.0 billion years old using uranium-238 age dating. Express the age of the rock in hours using scientific notation. (*Hint:* More than one conversion step is necessary.)

52. The Apollo 15 brought back a lunar sample dubbed the Genesis Rock. The rock is estimated to be 4.1 billion years old using uranium-238 age dating. Express the age of the rock in minutes using scientific notation. (*Hint:* More than one conversion step is necessary.)

53. A troy ounce equals 31.1 grams; one troy pound equals 12 troy ounces. What is the mass in grams of a pound of gold? (*Note:* The mass of gold is measured using the troy system.)

54. An avoirdupois ounce equals 28.4 grams; one avoirdupois pound has 16 ounces. What is the mass in grams of a pound of feathers? (*Note:* The mass of feathers is measured in the avoirdupois system.)

55. Refer to Exercises 53 and 54 to answer the timeless question, *"Which weighs more, a pound of gold or a pound of feathers?"*

CHAPTER

2

Measurement

An Olympic ski jumper strives for maximum distance measured in meters. In fact, all Olympic events are recorded using the metric system of measurement, which is discussed in this chapter.

Since the beginning of civilization, there has been a need to convey measurement. When ancient traders exchanged goods, they had to agree on standards for measuring length, weight, and volume. Unfortunately, units of measurement varied from country to country and sometimes within a country. In the history of measurement, a unit of length has often been defined in terms of the human body because hands and feet are convenient and fairly uniform in size. The Bible, in fact, mentions units of digit, span, and cubit. A digit is the thickness of the index finger, and a span is the width of four fingers. In the book of Genesis, a cubit is defined as the distance from the elbow to the end of the middle finger.

The Greeks adopted the foot measurement from the Babylonians and divided it into 12 thumbnail breadths. The Romans, in turn, adopted the measurement and called it an unciae, *which means twelfths. The Anglo-Saxons changed* unciae *to inch and it was written into English law. Eventually, the inch passed on to the United States, where it is used to this day.*

A new system of measurement originated in 1215 with the signing of the Magna Carta. The Magna Carta was not only a political document; it was also an economic document, attempting to bring uniformity to world trade. According to the Magna Carta, "Throughout the kingdom there shall be standard measures of wine, ale, and corn." *To avoid variations in measurement, reference standards were selected. The inch was defined as the length of three barleycorns laid end to end. The foot was defined as 36 barleycorns. A bushel of barley was set equal to a mass of 50 pounds and a volume of 8 gallons. Thus, the* **English system** of measurement *was established, with reference standards for length, mass, and volume.*

English system A nondecimal system of measurement without any basic unit for length, mass, or volume.

2.1 The Metric System

OBJECTIVES

To understand the logic and simplicity of the metric system.

To know the three basic units and symbols of the metric system.

To write symbols for multiples and fractions of basic units.

By the late 1700s, there was dissatisfaction over the lack of uniformity in world measurement. Although the English system was prevalent, it was used primarily within the English empire. In 1790, the French government appointed a committee of scientists to investigate the possibility of a uniform measuring system. This committee spent nearly 10 years before presenting the metric system—a unified system of measurement. The delay was due to the French Revolution and even scientists were not immune from the revolution. A case in point was Antoine Lavoisier, a member of the metric committee and a brilliant chemist, who was guillotined because he was accused of collecting taxes. A short time before his death, he uttered his famous quote praising the metric system: *"Never has anything more grand and more simple, more coherent in all of its parts, issued from the hand of men."*

metric system A decimal system of measurement using prefixes and a basic unit to express physical quantities such as length, mass, and volume.

meter (m) The basic unit of length in the metric system of measurement.

liter (L) The basic unit of volume in the metric system equal to the volume of a cube 10 cm on a side.

The **metric system** is simple and coherent for two reasons. One, it uses a single basic unit for each quantity measured, as shown in Table 2.1. The basic unit of length is the **meter**. The basic unit of mass is the **gram**. The basic unit of volume is the **liter**. A second is considered the basic unit of time. Also, the metric system is a decimal system which uses prefixes to enlarge or reduce each basic unit; for example, a kilometer is 1000 meters and a centimeter is 0.01 meter.

The metric committee defined the meter, kilogram and liter as follows: The meter was equal to one ten-millionth the distance from the North Pole to the equator along a meridian passing through Dunkirk, France, and Barcelona, Spain. The kilogram was equal to the mass of a cube of water one-tenth meter on a side (at its temperature of maximum density, 4°C). Note that, although the gram is the basic unit in the metric system, the kilogram is the reference standard for mass. A gram is one-thousandth of the kilogram reference standard. The liter was defined as the volume occupied by 1 kg of water at 4°C.

The committee then constructed reference standards as shown in Figure 2.1. They cast a platinum bar as the reference standard for 1 meter. They cast a solid platinum cylinder as the reference standard for 1 kilogram. And they cast a platinum container as the reference for 1 liter. These platinum alloy castings became the original reference standards for world measurement. In 1875, all the major nations of the world, including the United States, signed the Treaty of the Meter. Duplicates of the original castings were made and distributed to all member nations.

Table 2.1 The Metric System

Quantity	Basic Unit	Symbol
length	meter*	m
mass	gram	g
volume	liter*	L

* The U.S. Metric Association recommends the spellings meter and liter. However, all other English-speaking nations use the spellings metre and litre.

Metric Prefixes

To express a multiple or a fraction of a basic unit, the metric system uses prefixes. Since all prefixes are related by a power of 10, the metric system is a decimal system. In other words, the prefix increases or decreases the basic unit by a power of 10. The prefix kilo, for example, increases a basic unit a thousandfold. Thus, a kilometer is 1000 meters. The prefix milli reduces a basic unit by a factor of 1000. A millimeter is therefore one one-thousandth of a meter.

Metric measurements are abbreviated using symbols for the prefix and basic unit. Table 2.2 lists some prefixes used with basic units in the metric system. Some metric unit abbreviations are kilometer, km; decimeter, dm; centimeter, cm; millimeter, mm; micrometer, μm; and nanometer, nm.

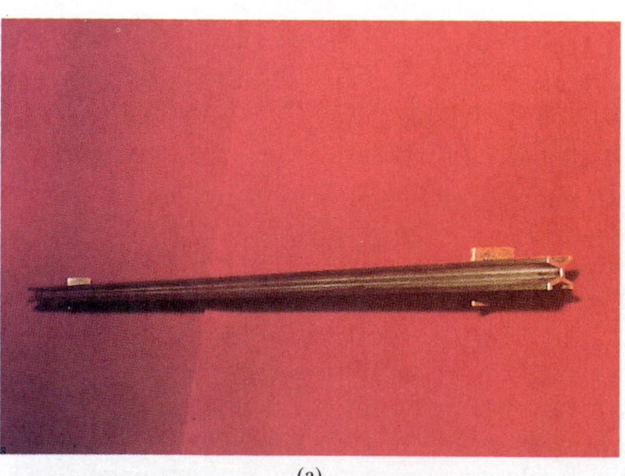

(a)

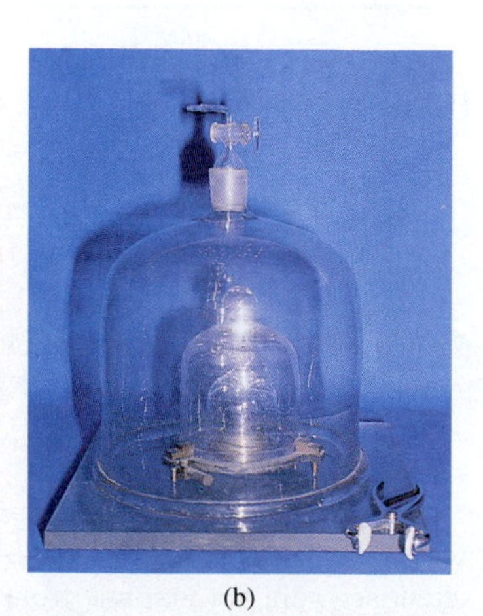

(b)

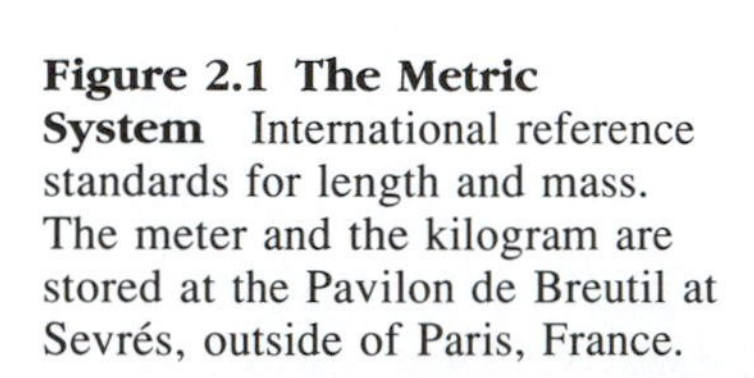

Figure 2.1 The Metric System International reference standards for length and mass. The meter and the kilogram are stored at the Pavilon de Breutil at Sevrés, outside of Paris, France.

Table 2.2 Metric Prefixes

Prefix	Symbol	Multiple/Fraction
kilo	k	$1000 = 10^3$
	basic unit (m, g, L)	
deci	d	$0.1 = 10^{-1}$
centi	c	$0.01 = 10^{-2}$
milli	m	$0.001 = 10^{-3}$
micro	μ*	$0.000\,001 = 10^{-6}$
nano	n	$0.000\,000\,001 = 10^{-9}$

* The Greek letter mu (symbol μ) is the abbreviation for micro.

The English system is neither a decimal system nor does it use basic units. In the English system, length can be expressed in inches, feet, yards, or miles, which are not related by factors of 10. Converting units in the English system is therefore cumbersome. Converting units in the metric system is, on the other hand, direct and simple.

EXAMPLE EXERCISE 2.1

Give the symbol for the following metric units. State the quantity (length, mass, or volume) measured by each.

(a) decimeter **(b)** nanometer **(c)** centigram
(d) kilogram **(e)** milliliter **(f)** microliter

Solution: The symbol for each unit is composed of the abbreviation for the prefix and for the basic unit. A decimeter is composed of the prefix deci (d) and the unit meter (m) to give dm. A decimeter is a unit of length.

(a) dm, length **(b)** nm, length **(c)** cg, mass
(d) kg, mass **(e)** mL, volume **(f)** μL, volume

SELF-TEST EXERCISE

Write the name for the metric symbols μs and kL.

Answer: microsecond and kiloliter

2.2 Metric Unit Conversion Factors

To write equivalent relationships between a basic unit and units having the following metric prefixes: kilo, deci, centi, milli, micro, and nano.

To write two unit conversion factors derived from an equivalent metric relationship.

Recall from Section 1.11 that a **unit equation** involves two quantities that are equal. As we said, converting units in the metric system is a simple process. That's because the metric system is a decimal system. As an example, let's find the relationship between kilometers and meters. Since the metric prefix kilo means 1000 basic units, 1 kilometer is 1000 meters. We can write this relationship as the unit equation 1 kilometer = 1000 meters. Using metric symbols, we have 1 km = 1000 m.

The prefix centi means 0.01 of a basic unit. A centimeter is one-hundredth of a meter. We could write the relationship as 1 cm = 0.01 m. Working with decimal fractions is, however, more difficult than working with whole multiples of basic units. You can avoid decimal fractions simply by restating the relationship. Since 1 m is divided into 100 cm, you can write the unit equation 1 m = 100 cm.

EXAMPLE EXERCISE 2.2

Complete the unit equation for each of the following metric relationships.

(a) 1 kg = ? g **(b)** 1 m = ? dm

Solution: Refer to the metric prefixes in Table 2.2 as necessary.

(a) The prefix kilo (k) is 1000 basic units; thus, 1 kg = 1000 g.

(b) The prefix deci (d) is 0.1 of a basic unit; therefore, there are 10 deci units in one basic unit. Thus, 1 m = 10 dm.

SELF-TEST EXERCISE

Complete the unit equation for each of the following metric relationships.

(a) 1 g = ? dg **(b)** 1 m = ? nm

(c) 1 L = ? mL **(d)** 1 s = ? μs

Answers:

(a) 1 g = 10 dg; **(b)** 1 m = 1,000,000,000 nm (1×10^9 nm)

(c) 1 L = 1000 mL; **(d)** 1 s = 1,000,000 μs (1×10^6 μs)

Converting from one basic unit to another is also a simple process. To do so, you use a unit conversion factor. A **unit factor** (or unit conversion factor) is a ratio of two equivalent quantities. That is, the quantity in the numerator is equal to the quantity in the denominator. Since the numerator and the denominator are equivalent, you can invert the ratio to produce a second unit factor. As an example, consider the unit equation 1 m = 100 cm. Its two corresponding unit factors are

$$\frac{1 \text{ m}}{100 \text{ cm}} \quad \text{and} \quad \frac{100 \text{ cm}}{1 \text{ m}}$$

EXAMPLE EXERCISE 2.3

Write the unit equation and two metric conversion factors for each of the following related metric units.

(a) kilometers and meters **(b)** milliliters and liters

Solution: Unit equations and unit factors are produced as follows.

(a) First, write an equivalent relationship. The prefix kilo multiplies the basic unit by 1000. Thus, 1 km is 1000 m or 1 km = 1000 m. Then write two unit factors that are associated with that unit equation:

$$\frac{1 \text{ km}}{1000 \text{ m}} \quad \text{and} \quad \frac{1000 \text{ m}}{1 \text{ km}}$$

(b) The prefix milli is one-thousandth of the basic unit. It follows that 1 L contains 1000 milliliters; 1 L = 1000 mL. The two corresponding unit factors are

$$\frac{1\text{ L}}{1000\text{ mL}} \quad \text{and} \quad \frac{1000\text{ mL}}{1\text{ L}}$$

SELF-TEST EXERCISE

Write the pair of unit factors associated with the following equivalences.

(a) 1 g = 10 dg
(b) 1 m = 100 cm
(c) 1 kL = 1000 L
(d) 1 s = 1,000,000 μs

Answers:

(a) 1 g/10 dg and 10 dg/1 g; **(b)** 1 m/100 cm and 100 cm/1 m
(c) 1 kL/1000 L and 1000 L/1 kL; **(d)** 1 s/1,000,000 μs and 1,000,000 μs/1 s

2.3 Metric Problems by Unit Analysis

To review the three steps in the unit analysis method of problem solving.

Sometimes it is necessary to change a measurement value from one unit to another. For example, you may wish to express a length measurement in millimeters, but your ruler is calibrated in centimeters. In those cases, you will have to convert one measurement value to a different one. Section 1.10 introduced the **unit analysis method** of problem solving. In this chapter, you will apply that same method to more complex examples. First, though, let's review the three steps in the unit analysis method of problem solving.

unit analysis method A systematic method of problem solving that converts a given value to an unknown value by applying one or more unit factors.

Step 1: Read the problem carefully, and write down the units of the unknown quantity. If the units of the answer are not specified, choose any units that are convenient. If, for example, a problem asks for the volume of a liquid without specifying the units, you may choose liters, deciliters, milliliters, or any other volume units to express your answers.

Step 2: Write down the given value that is related to the units of the unknown quantity. Some problems may include values that are not related. In these problems, you must sort through the given information to determine what is relevant and what is not.

Step 3: Apply one or more unit factors to convert the units of the given value to the units of the unknown quantity. In this chapter, the problems may require more conversions than the problems in Chapter 1. You may have to use two or more unit factors to convert a given value to the units of the answer.

Once again, write the three steps of the problem using the following format:

$$\underset{(2)}{\text{related given value}} \times \underset{(3)}{\frac{\text{unit}}{\text{factor(s)}}} = \underset{(1)}{\text{units in answer}}$$

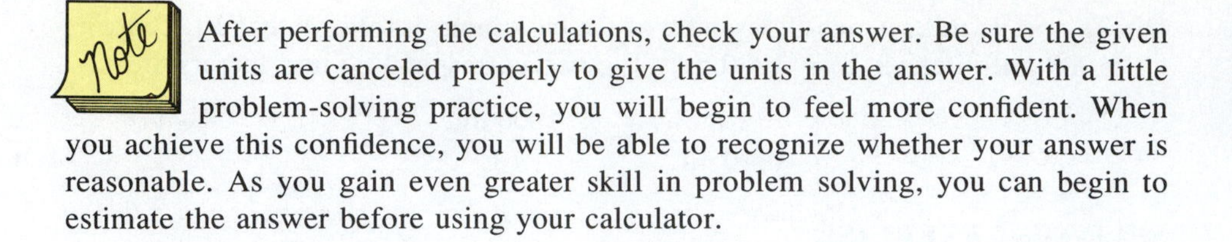

After performing the calculations, check your answer. Be sure the given units are canceled properly to give the units in the answer. With a little problem-solving practice, you will begin to feel more confident. When you achieve this confidence, you will be able to recognize whether your answer is reasonable. As you gain even greater skill in problem solving, you can begin to estimate the answer before using your calculator.

2.4 Metric–Metric Unit Conversions

OBJECTIVES

To convert a given metric measurement to another metric unit having a different prefix.

In this section you will sharpen your problem-solving skills. You will apply the three-step **unit analysis method** to metric problems. The following examples may require two or more unit factors to accomplish the metric conversion.

EXAMPLE EXERCISE 2.4

Convert 255 milligrams into units of grams.

Solution: Step 1: Write the unit of the unknown: g. Step 2: Write down the relevant given value: 255 mg. The problem solving format is

$$255 \text{ mg} \times \frac{\text{unit}}{\text{factor(s)}} = \text{g}$$

Step 3: Since 1 g = 1000 mg, the two possible unit factors are

$$\underset{(A)}{\frac{1 \text{ g}}{1000 \text{ mg}}} \quad \text{and} \quad \underset{(B)}{\frac{1000 \text{ mg}}{1 \text{ g}}}$$

Choose unit factor A because mg is in the denominator and it cancels the units of the given value. Use the problem-solving format

$$255 \cancel{\text{mg}} \times \frac{1 \text{ g}}{1000 \cancel{\text{mg}}} = 0.255 \text{ g}$$

Regarding significant digits, the equation 1 g = 1000 mg is derived from a definition. Definitions are exact and infinitely significant. Hence, the unit factor 1 g/1000 mg has no effect on the number of significant digits in the answer.

SELF-TEST EXERCISE

An automobile air-bag inflates in 30,500 microseconds. How many seconds are required for inflation?

Answer: 0.0305 s

EXAMPLE EXERCISE 2.5

A 10-μL syringe is used to inject 2.5 μL of ether into a gas analyzer. Express the volume of the sample in milliliters.

Solution: Step 1: The unknown unit is mL. Step 2: The relevant given value is 2.5 μL. You can ignore the syringe size of 10 μL. The format for solving the problem is

$$2.5\ \mu\text{L} \times \frac{\text{unit}}{\text{factor(s)}} = \text{mL}$$

Step 3: The unit factor(s) is not obvious because you do not know the equivalent relationship betweeen μL and mL. The prefixes micro and milli, however, are each related to the basic unit. Thus,

$$\mu\text{L} \longrightarrow \text{L} \longrightarrow \text{mL}$$

The prefix micro is one-millionth. Therefore, the first equivalent relationship is 1 L = 1,000,000 μL. The corresponding unit factors are

$$\frac{1\ \text{L}}{1{,}000{,}000\ \mu\text{L}} \quad \text{and} \quad \frac{1{,}000{,}000\ \mu\text{L}}{1\ \text{L}}$$

The prefix milli is one-thousandth. The unit equation is 1 L = 1000 mL. The two unit factors are

$$\frac{1\ \text{L}}{1000\ \text{mL}} \quad \text{and} \quad \frac{1000\ \text{mL}}{1\ \text{L}}$$

To solve this problem, you need two unit factors. Here's an outline for the format for solving this problem.

$$2.5\ \mu\text{L} \times \frac{\text{unit}}{\text{factor 1}} \times \frac{\text{unit}}{\text{factor 2}} = \text{mL}$$

Select 1 L/1,000,000 μL for unit factor 1 in order to cancel μL. Thus,

$$2.5\ \cancel{\mu\text{L}} \times \frac{1\ \text{L}}{1{,}000{,}000\ \cancel{\mu\text{L}}} \times \frac{\text{unit}}{\text{factor 2}} = \text{mL}$$

For unit factor 2, we will choose 1000 mL/1L:

$$2.5\ \cancel{\mu\text{L}} \times \frac{1\ \cancel{\text{L}}}{1{,}000{,}000\ \cancel{\mu\text{L}}} \times \frac{1000\ \text{mL}}{1\ \cancel{\text{L}}} = 0.0025\ \text{mL}$$

SELF-TEST EXERCISE

Perform the following metric conversions.

(a) 625 nm to mm **(b)** 0.776 kg to μg

Answers:
(a) 6.25×10^{-4} mm; **(b)** 7.76×10^{8} μg

note After you become proficient in problem solving, you can convert directly from μL to mL in one step. For now, use two unit factors to avoid errors. As an analogy, suppose you are traveling from London to Paris and must convert British pounds into French francs. To avoid confusion, you could first convert

British pounds into American dollars and then American dollars to French francs. The dollar is a basic monetary unit, just as the liter is a basic volume unit. With practice, you may be able to convert from μL to mL (or pounds to francs) directly.

2.5 Metric–English Unit Conversions

To state the metric equivalent of an inch, pound, and quart.

To perform metric–English unit conversions.

In 1975, President Gerald Ford signed an official metrification act. Thus, the United States became the last major power in the world to adopt the metric system formally. The metric system is now being taught in schools along with the English system. In truth, the United States is making only minimal progress in implementing the metric system. Economically, it will cost billions of dollars to retool machinery and convert industry to the metric system.

Despite its superiority, the metric system has been resisted from the beginning. When the French asked the Americans and the British for their cooperation in creating a worldwide system of standard measurement, both countries declined. The main objection was that the reference standard for the meter was based on a meridian passing through France and Spain.

Scientific measurements are exclusively metric. For everyday use, however, Americans are more familiar with inches, pounds, quarts, and many other English units. Therefore, to gain a practical appreciation for metric dimensions, we will compare the two systems. Table 2.3 shows units from both systems.

Table 2.3 Metric–English Equivalents

Quantity	English Unit	Metric Equivalent*
length	1 inch (in.)	1 in. = 2.54 cm
mass	1 pound (lb)	1 lb = 454 g
volume	1 quart (qt)	1 qt = 946 mL

* Since the metric and English systems of measurement are derived from different reference standards, these equivalents are not exact. However, in 1959 the U.S. National Bureau of Standards redefined the yard as exactly equal to 0.9144 m and the pound as exactly equal to 1/2.20462234 kg.

The following examples illustrate the application of the unit analysis method of problem solving to metric–English unit conversions.

EXAMPLE EXERCISE 2.6

What is the milliliter volume of a 12.0 fluid-ounce soft drink? (Given: 32 fluid ounces = 1 quart.)

Solution: The unit asked for is mL, and the related given value is 12.0 fl oz.

$$12.0 \text{ fl oz} \times \frac{\text{unit}}{\text{factor(s)}} = \text{mL}$$

We do not know the relationship between fl oz and mL, but both are related to quarts. Thus,

$$\text{fl oz} \longrightarrow \text{qt} \longrightarrow \text{mL}$$

First convert fl oz to qt using the given relationship, 32 fl oz = 1 qt. Second, convert qt to mL using the metric–English equivalency: 1 qt = 946 mL.

$$12.0 \text{ fl oz} \times \frac{\text{unit}}{\text{factor 1}} \times \frac{\text{unit}}{\text{factor 2}} = \text{mL}$$

For unit factor 1, write 1 qt/32 fl oz in order to cancel fluid ounces.

$$12.0 \cancel{\text{fl oz}} \times \frac{1 \text{ qt}}{32 \cancel{\text{fl oz}}} \times \frac{\text{unit}}{\text{factor 2}} = \text{mL}$$

For unit factor 2, select the ratio of 946 mL/1 qt in order to cancel quarts.

$$12.0 \cancel{\text{fl oz}} \times \frac{1 \cancel{\text{qt}}}{32 \cancel{\text{fl oz}}} \times \frac{946 \text{ mL}}{1 \cancel{\text{qt}}} = 355 \text{ mL}$$

In this problem the unit factor 1 qt/32 fl oz is exact, so the metric–English factor 946 mL/1 qt limits the answer to three significant digits. If you read the label on many soft drink cans, you will see the following: 12 fl oz (355 mL).

SELF-TEST EXERCISE

What is the volume in quarts of a 10.0-dL hospital saline solution?

Answer: 1.06 qt

EXAMPLE EXERCISE 2.7

The metric system is used internationally, so it is the system used at the Olympic Games. If an athlete runs 100 meters, what is the distance in yards?

Solution: The unknown unit is yd and the relevant known value is 100 m. This problem appears to be more difficult. Let's analyze the conversion of units as follows:

$$\text{m} \longrightarrow \text{cm} \longrightarrow \text{in.} \longrightarrow \text{yd}$$

Let's outline the general solution to the problem:

$$100 \text{ m} \times \frac{\text{unit}}{\text{factor 1}} \times \frac{\text{unit}}{\text{factor 2}} \times \frac{\text{unit}}{\text{factor 3}} = \text{yd}$$

The three equivalent relationships are as follows: 1 m = 100 cm, 1 in. = 2.54 cm, 1 yard = 36 in. We can systematically write each unit factor so as to cancel previous units.

$$100 \cancel{\text{m}} \times \frac{100 \cancel{\text{cm}}}{1 \cancel{\text{m}}} \times \frac{1 \cancel{\text{in.}}}{2.54 \cancel{\text{cm}}} \times \frac{1 \text{ yd}}{36 \cancel{\text{in.}}} = 109 \text{ yd}$$

A distance of 100 meters is equal to 109 yards. The world record for the 100-meter dash is a few tenths of a second greater than the record for the 100-yard event because it is 9 yards longer.

SELF-TEST EXERCISE

What is the diameter in millimeters of a tennis ball that measures 2.5 in?

Answer: 64 mm

CHEMISTRY CONNECTION □ THE CONSUMER

The Olympics

▶ Can you name the Olympic race that is nearly equal in length to one-quarter mile, that is, 440 yards?

The Olympic games originated in Ancient Greece as a series of running contests. According to tradition, the Greek games began in 776 B.C. and were held once every four years. Many countries participated, and the ancient games reached their height of popularity during the 5th and 4th centuries B.C. With the passage of time, the games declined in popularity and were eventually discontinued.

The modern Olympic games were initiated in Athens, Greece, in 1896. The Olympics that followed have been held at various cities around the world at intervals of every four years. The only exceptions were the World War years of 1916, 1940, and 1944. The Olympic summer games have been held three times in the United States. The Olympics took place in St. Louis in 1906, and in Los Angeles in 1932 and 1984.

In the past, United States track and field competitions have been routinely conducted using the English system of measurement. American athletes have raced distances of 100 yards, 440 yards, and 1 mile. Given that most other nations of the world employ the metric system, can you guess what system of measurement is used in the Olympics? Since the Olympics is an international competition, the international system of measurement is used, that is, the metric system. Olympic runners compete at distances of 100 meters, 400 meters, and 10 kilometers. It is an interesting to note that, in the Olympic shot-put event, athletes toss an official metal ball having a mass of 16 pounds (7.257 kg).

The modern Olympic games have steadily increased in the number of events, as well as the number of nations and participants. Women were first invited to compete in 1912, and the Olympic winter games were introduced at Chamonix, France, in 1924. Subsequently, the winter games have been held three times in the United States, twice at Lake Placid, and once at Squaw Valley. In cross-country skiing competitions, the men's races are 10, 30, and 50 kilometers while the equivalent women's races are 5, 15, and 30 kilometers. However, every Olympic competition—men's or women's—winter or summer—all employ the metric system of measurement.

Olympic weightlifters lift weights measured in kilograms.

▶ The Olympic 400 meter race is nearly identical to 440 yards, and the world-record times for the two races are within tenths of a second.

EXAMPLE EXERCISE 2.8

Many states have posted highway signs in both English and metric units. If the posted speed limit is 55 miles per hour, what is the metric speed limit in kilometers per hour? (Given: 1 mile = 1.61 km.)

Solution: This problem illustrates a variation from previous examples. The asked-for quantity, kilometers per hour, and the given value, miles per hour, are ratio units.

$$\frac{55 \text{ miles}}{1 \text{ hour}} \times \frac{\text{unit}}{\text{factor}} = \frac{\text{km}}{\text{hour}}$$

The equivalency 1 mile = 1.61 km is given. Applying the unit factor that cancels units gives

$$\frac{55 \text{ \cancel{miles}}}{1 \text{ hour}} \times \frac{1.61 \text{ km}}{1 \text{ \cancel{mile}}} = \frac{89 \text{ km}}{\text{hour}}$$

When metric speed limits are posted, the value is usually rounded to 90 kilometers per hour and posted as 90 km/h.

SELF-TEST EXERCISE

What is the speed in centimeters per second of a snail crawling 0.65 ft/min?

Answer: 0.33 cm/s

The key to successful problem solving is the careful and consistent application of the unit analysis method. There are three steps. One, determine the units in the answer. Two, analyze the problem for a relevant given quantity. Three, supply one or more unit factors chosen in such a way so as to cancel units.

2.6 Volume by Calculation

OBJECTIVES

To perform calculations that relate volume to length, width, and thickness of a rectangular solid.

To convert a given volume measurement to cubic centimeters, milliliters, or cubic inches.

The volume of a rectangular solid can be calculated. If the length (l), width (w), and thickness (t) are multiplied, the product equals the volume. The formula is

$$l \times w \times t = \text{volume}$$

When you perform volume calculations, you must express each dimension in the same units. Suppose, for example, the length and width are in centimeters, but the thickness is in millimeters. Before finding the volume, you will have to express all three dimensions in either centimeters or millimeters (cm or mm). The volume

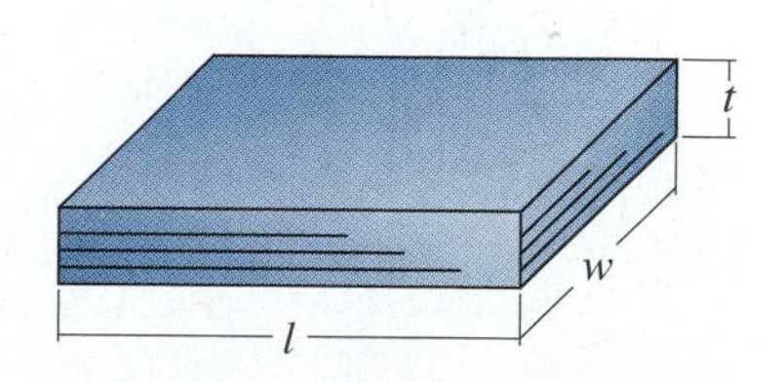

The volume of a rectangular solid is equal to length times width times thickness (or height).

cubic centimeter (cm^3) A unit of volume occupied by a cube one centimeter on a side; a volume exactly equal to one milliliter.

of a rectangular solid is typically expressed in **cubic centimeters (cm^3)** or cubic millimeters (mm^3). For example, if a rectangular solid measures 3 cm by 2 cm by 1 cm, the volume is 6 cm^3.

EXAMPLE EXERCISE 2.9

If a rectangular platinum bar is 5.55 cm long, 3.75 cm wide, and 2.25 cm thick and has a mass of 1.00 kg, what is its volume?

Solution: Since the known values are given in cm, calculate the volume in cubic centimeters (cm^3). The volume is the product of length times width times thickness.

$$(5.55\ \text{cm})(3.75\ \text{cm})(2.25\ \text{cm}) = 46.8\ \text{cm}^3$$

Notice that the mass of the platinum bar, 1.00 kg, is not relevant.

SELF-TEST EXERCISE

Calculate the volume of a rectangular solid having the dimensions

$$5.15\ \text{cm} \times 2.55\ \text{cm} \times 1.10\ \text{cm}$$

Answer: 14.4 cm^3

Let's consider a variation of the above problem. If you are given the volume and two dimensions of a rectangular solid, it is possible to calculate the third dimension. For example, suppose a rectangular solid has a volume of 24 cm^3 and measures 4 cm in length and 3 cm in width. We calculate the thickness of the solid by dividing 24 cm^3 by 4 cm and 3 cm; thus, the thickness of the solid is 2 cm.

EXAMPLE EXERCISE 2.10

A thin sheet of gold foil measures 10.0 cm by 25.0 cm. Its volume is 3.12 cm^3. Calculate the thickness of the foil in mm.

Solution: The volume is equal to the product of the three dimensions: volume = $l \times w \times t$. Therefore, to obtain the thickness of the foil, divide the volume by length and width.

$$\frac{3.12\ \text{cm}^3}{(10.0\ \text{cm})(25.0\ \text{cm})} = 0.0125\ \text{cm}$$

Note that the units reduce to centimeters. The cancellation of units is more clearly seen if we write out cm^3 as follows:

$$\frac{(\cancel{\text{cm}})(\cancel{\text{cm}})(\text{cm})}{(\cancel{\text{cm}})(\cancel{\text{cm}})} = \text{cm}$$

But the problem asked that the thickness be calculated in mm. The metric conversion of centimeter to millimeter is as follows:

$$0.0125\ \cancel{\text{cm}} \times \frac{1\ \cancel{\text{m}}}{100\ \cancel{\text{cm}}} \times \frac{1000\ \text{mm}}{1\ \cancel{\text{m}}} = 0.125\ \text{mm}$$

The gold foil is 0.125 mm thick. This thickness is slightly less than that of your fingernail.

SELF-TEST EXERCISE

The volume of a rectangular bronze solid is 64 mm^3. If the length is 8.0 mm and the width is 4.0 mm, what is the thickness of the rectangular solid?

Answer: 2.0 mm

In the metric system, the basic unit of volume is the liter. The **liter** is currently defined as the volume occupied by a cube exactly 10 cm on a side.

A cube is just a rectangular solid with sides of equal length. As such, we can calculate the volume of a 10-cm cube. We multiply length times width times height.

$$(10\ \text{cm})(10\ \text{cm})(10\ \text{cm}) = 1000\ \text{cm}^3$$

A volume of 1 L is calculated to be 1000 cm^3. Recall that, by definition, 1 L is 1000 mL. Thus,

$$1000\ \text{mL} = 1000\ \text{cm}^3$$

Therefore,

$$1\ \text{mL} = 1\ \text{cm}^3$$

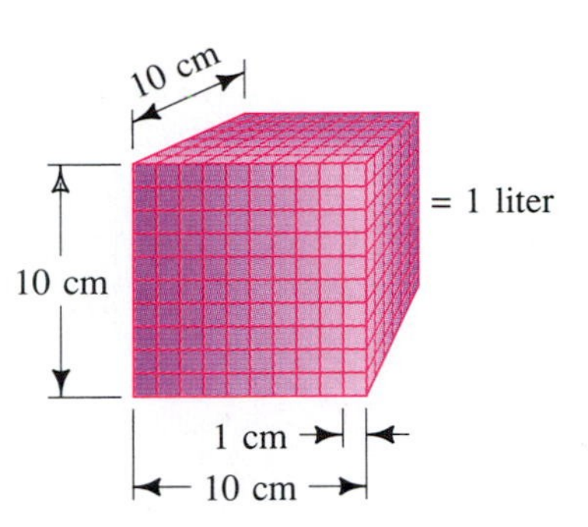

A liter of volume equals a cube 10 cm on a side.

A milliliter is exactly equal to a cubic centimeter (cm^3). A cubic centimeter is often abbreviated cc, such as in a 10-cc medical injection. Since the symbol cc is not a metric abbreviation, its use in chemistry is discouraged.

EXAMPLE EXERCISE 2.11

A German sports car has an engine whose size is 2990 cm^3. Find the engine volume in (a) milliliters, (b) liters, and (c) cubic inches.

Solution:

(a) Since 1 mL = 1 cm^3, you can easily make the conversion.

$$2990\ \cancel{\text{cm}^3} \times \frac{1\ \text{mL}}{1\ \cancel{\text{cm}^3}} = 2990\ \text{mL}$$

(b) The conversion from mL to L also requires only one unit factor.

$$2990\ \cancel{\text{mL}} \times \frac{1\ \text{L}}{1000\ \cancel{\text{mL}}} = 2.990\ \text{L}$$

(c) Conversion of cubic units from cm^3 to $in.^3$ is more challenging. We know that 1 in. = 2.54 cm. Therefore, begin the calculation as follows:

$$2990\ \text{cm}^3 \times \frac{1\ \text{in.}}{2.54\ \text{cm}} = \text{in.}^3$$

Notice that the units of cm^3 do not cancel, and $in.^3$ is not produced. To obtain cubic units, cube the unit factor:

$$2990\ \cancel{\text{cm}^3} \times \frac{1\ \text{in.}}{2.54\ \cancel{\text{cm}}} \times \frac{1\ \text{in.}}{2.54\ \cancel{\text{cm}}} \times \frac{1\ \text{in.}}{2.54\ \cancel{\text{cm}}} = \text{in.}^3$$

Simplifying, gives

$$2990\ \cancel{\text{cm}^3} \times \frac{1\ \text{in.}^3}{16.4\ \cancel{\text{cm}^3}} = 182\ \text{in.}^3$$

SELF-TEST EXERCISE

If an American economy car has a 2.00-L engine, what is the volume in cubic inches?

Answer: 122 in.^{3}

2.7 Volume by Displacement

To understand the laboratory technique of volume by displacement.

Volume is determined in three principal ways.

1. The volume of any liquid can be measured using calibrated glassware. Graduated cylinders, pipets, and burets are routinely used to measure the volumes of liquids.
2. The volume of a solid whose shape is regular can be determined by calculation. Rectangular objects and cylinders have regular shapes, so their volumes can be calculated.
3. The volume of an irregular solid is found indirectly by the amount of liquid it displaces. This technique is called **volume by displacement**.

volume by displacement A method for determining the volume of an object by measuring the increase in liquid level when the object is immersed in water.

Suppose we wish to determine the volume of a piece of jade. Since it is an irregular solid object, we cannot determine its volume directly. We have to use the technique of volume by displacement. We first fill the graduated cylinder in Figure 2.2 halfway with water and record the water level. Next, we carefully slip the piece of jade into the graduated cylinder and record the new water level. The difference between the initial and final water levels represents the volume of the piece of jade.

Volume by displacement is an important laboratory technique for measuring the volumes of gases. Although the volume of liquids can be measured using calibrated glassware, gases cannot. When a gas is produced from a chemical reaction, its volume can be measured by the amount of water it displaces. For example, heating baking soda releases carbon dioxide gas, which displaces water from a closed container. If we ignore that carbon dioxide is somewhat soluble in water, the volume

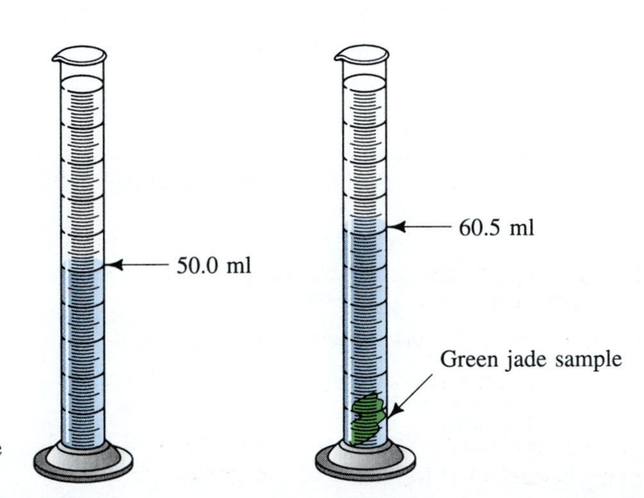

Figure 2.2 Volume of a Solid by Displacement The difference between the initial and final water levels is equal to the volume of the solid piece of jade.

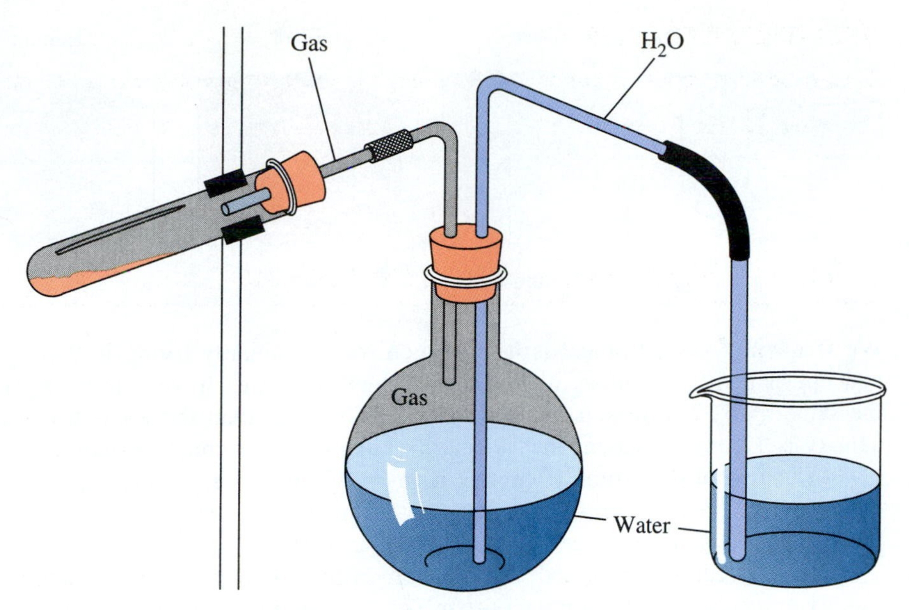

Figure 2.3 Volume of a Gas by Displacement A round-bottom flask is full of water before heating the compound in the test tube. As the compound decomposes with heat, a gas is released which displaces water from the flask into a beaker. The volume of water in the beaker corresponds to the volume of gas liberated.

of water displaced is equal to the volume of gas released. This technique, illustrated in Figure 2.3, is termed collecting a gas over water.

EXAMPLE EXERCISE 2.12

A 10.0-g amethyst gemstone dropped into a graduated cylinder raises the water level from 21.5 to 25.0 mL. What is its volume by displacement?

Solution: The volume by displacement is simply equal to the difference between the two water levels. Thus,

$$25.0 \text{ mL} - 21.5 \text{ mL} = 3.5 \text{ mL}$$

SELF-TEST EXERCISE

The carbon dioxide gas from a fire extinguisher was collected over water and displaced water into a beaker. If the water level in the beaker increased from 125 to 230 mL, what is the cubic centimeter volume of carbon dioxide?

Answer: 105 cm^3

2.8 Density

To understand the concept of density and state the value for the density of water: 1.00 g/mL.

To write the unit conversion factors related to a given density.

To perform calculations that relate density, mass, and volume.

The compactness of an object is expressed by the term **density** (symbol ***d***). The more compact, that is, the more concentrated the mass is, the greater the density.

density (*d*) The amount of mass in a unit volume of matter.

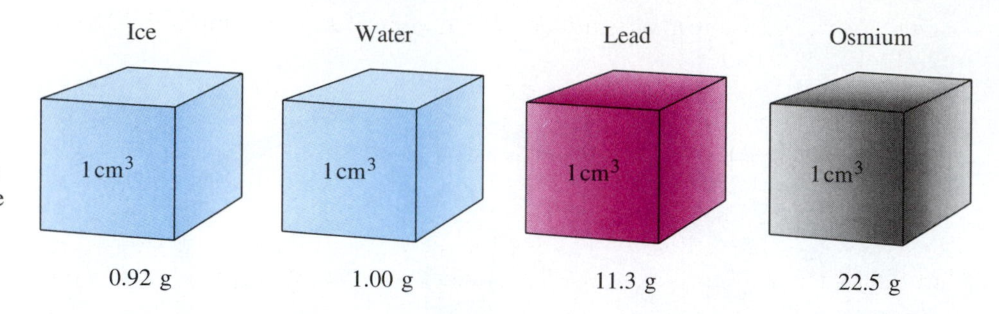

Figure 2.4 Illustration of Density Each cube represents a volume of 1.00 cm^3. Note that the mass of each cube increases as the density becomes greater. Osmium has the highest density of any element, 22.5 g/cm^3.

We frequently say an object is heavy when we are actually trying to convey the concept of density. A block of lead and a block of ice, each weighing 1 kg, have the same mass. The reason the lead may seem heavier than the ice is because its density is 12 times greater. That is, lead is 12 times more compact than ice.

The formal definition of density is mass of substance per unit volume. The term *per* indicates a ratio, such as in miles per hour or grams per liter. Thus, mass of a substance divided by its volume equals density.

Density can be expressed in many units. For solids and liquids, the units are usually grams per cubic centimeter (g/cm^3) or grams per milliliter (g/mL). For gases, which are far less dense, the density is expressed in grams per liter (g/L). Figure 2.4 illustrates the mass in a volume of 1 cm^3 of several substances.

Recall that water was used for the original metric standards for mass and volume. The mass of 1 kg of water was set equal to a volume of 1 L. Therefore, the density of water is 1 kg/L. An equivalent expression for the density of water is

Table 2.4 Densities of Selected Substances

Substance	Density (*d*)
Solids	
ice	0.917 g/cm^3 or g/mL
sucrose (table sugar)	1.58
magnesium	1.74
sodium chloride (table salt)	2.16
aluminum	2.70
iron	7.87
lead	11.3
gold	18.9
Liquids	
ethyl ether	0.714 g/cm^3 or g/mL
ethyl alcohol	0.789
water	1.000
carbon tetrachloride	1.594
mercury	13.6
Gases*	
hydrogen	0.090 g/L
helium	0.179
ammonia	0.760
air	1.293
oxygen	1.43
argon	1.783
carbon dioxide	1.964

* The densities of gases vary greatly with temperature and pressure. The given density values are at 0°C and normal atmospheric pressure.

1 g/mL. More precisely, the density of water is defined as 1.00 g/mL at 3.98°C, its temperature of maximum density.

Note that density varies slightly with temperature, but generally that variation is not significant for our calculations. Gases are an exception, since the density of a gas is altered significantly by temperature, as well as by pressure changes. Table 2.4 lists the densities of several common substances representing the three physical states.

A substance's behavior in water serves as a convenient way to compare its density to other substances. Substances that sink in water have a density greater than 1.00 g/mL. Substances that float on water have a density less than 1.00 g/mL.

EXAMPLE EXERCISE 2.13

A tall cylinder contains three liquids, water, ethyl ether, and mercury, separated into layers. Cubes of ice, aluminum, and gold are each dropped into the cylinder. Referring to Table 2.4, identify liquids L_1 and L_2. Identify the solid cubes S_1, S_2, and S_3.

Solution: Water is the reference because we know its density, 1.00 g/mL. Liquid L_1 must be ethyl ether (0.714 g/mL) since its density is less than that of water. Liquid L_2 must be mercury (13.6 g/mL) since its density is greater than that of water.

The densities of ice, aluminum, and gold are 0.917, 2.70, and 18.9 g/cm^3. Thus, S_1 is ice since it sinks in ether and floats on water. S_2 is aluminum, which sinks in water and floats on mercury. S_3 is gold, which sinks to the bottom because it is more dense than mercury.

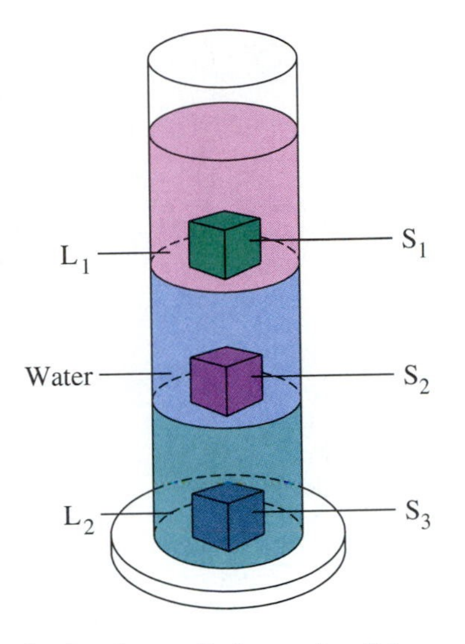

In the glass cylinder, each solid sinks in the liquids that are less dense.

Using the concept of density, let's try some calculations. Once again we will use the unit analysis method of problem solving. The secret to solving density problems is to realize that the units of measurement can be used as a unit factor. For example, since the density of mercury is 13.6 g/mL, we can write the unit equation 13.6 g = 1 mL. Two unit factors follow, 13.6 g/mL and its reciprocal.

$$\frac{13.6 \text{ g}}{1 \text{ mL}} \quad \text{and} \quad \frac{1 \text{ mL}}{13.6 \text{ g}}$$

EXAMPLE EXERCISE 2.14

A piece of jade has a mass of 26.382 g. Its volume by displacement is found to be 7.3 mL. Calculate the density of the jade sample.

Solution: Density is a ratio and is defined as mass per unit volume.

$$\frac{\text{mass}}{\text{volume}} = \text{density}$$

The mass is 26.382 g and the volume is 7.3 mL. Substituting and solving

$$\frac{26.382 \text{ g}}{7.3 \text{ mL}} = 3.6 \text{ g/mL}$$

Jade is a valuable mineral. It can be distinguished from other minerals that appear similar on the basis of its density.

SELF-TEST EXERCISE

Osmium is the most dense element. What is its density if 225 g of osmium occupy 10.0 cm^3?

Answer: 22.5 g/cm^3

EXAMPLE EXERCISE 2.15

An automobile battery contains 1250 mL of sulfuric acid. If the density of sulfuric acid is 1.84 g/mL, how many grams of acid are in the battery?

Solution: The problem asks for the answer in units of grams. The given value is 1250 mL. Therefore,

$$1250 \text{ mL} \times \frac{\text{unit}}{\text{factor}} = \text{g}$$

In this problem, the density, 1.84 g/mL, is interpreted as 1.84 g = 1 mL. This unit equality gives two unit factors:

$$\frac{1.84 \text{ g}}{1 \text{ mL}} \quad \text{and} \quad \frac{1 \text{ mL}}{1.84 \text{ g}}$$

We will select the first factor because it cancels units of milliliters. Thus,

$$1250 \cancel{\text{mL}} \times \frac{1.84 \text{ g}}{1 \cancel{\text{mL}}} = 2300 \text{ g}$$

Since three significant digits are justified, the answer may be written as 2.30×10^3 g.

SELF-TEST EXERCISE

An experiment calls for 150 g of ethyl ether ($d = 0.714$ g/mL). How many milliliters of ether are required?

Answer: 210 mL

EXAMPLE EXERCISE 2.16

The most abundant gases in our atmosphere are nitrogen, oxygen, and argon. What is the volume in liters of 1.00 kg of air? (Given: air density is 1.29 g/L at standard conditions.)

Solution: The unit asked for is L and the given value is 1.00 kg. Therefore,

$$1.00 \text{ kg} \times \frac{\text{unit}}{\text{factor(s)}} = \text{L}$$

Once again, density is used as a unit factor:

$$\frac{1.29 \text{ g}}{1 \text{ L}} \quad \text{and} \quad \frac{1 \text{ L}}{1.29 \text{ g}}$$

Since kg must be converted into g, we first use the relationship 1 kg = 1000 g. Systematically applying the two unit factors to cancel units, we have

$$1.00 \cancel{\text{kg}} \times \frac{1000 \cancel{\text{g}}}{1 \cancel{\text{kg}}} \times \frac{1 \text{ L}}{1.29 \cancel{\text{g}}} = 775 \text{ L}$$

The density of a gas varies with temperature and pressure. Therefore, the volume is best stated to be 775 L at standard conditions.

SELF-TEST EXERCISE

If the density of hydrogen gas at standard conditions is 0.0902 g/L, what is the mass of 50.0 mL?

Answer: 0.00451 g

EXAMPLE EXERCISE 2.17

What is the mass of a cube of copper that is 2.54 cm on a side? The density of copper is 8.96 g/cm^3.

Solution: We will choose g as the unknown unit of mass. The given value is a cube that is 2.54 cm on a side. We can calculate the volume by multiplying side times side times side. Thus,

$$(2.54 \text{ cm})(2.54 \text{ cm})(2.54 \text{ cm}) = 16.4 \text{ cm}^3$$

We are given that the density of copper is 8.96 g = 1 cm^3. We can apply the density unit factor and cancel units. This gives

$$16.4 \cancel{\text{cm}^3} \times \frac{8.96 \text{ g}}{1 \cancel{\text{cm}^3}} = 147 \text{ g}$$

Since 2.54 cm = 1 in., the volume of the copper cube is 1 in.^{3}. We found that the mass is 147 g, approximately one-third of a pound. Thus, a 1-in. cube of copper weighs one-third of a pound.

SELF-TEST EXERCISE

What is the density of a cube of silver that is 5.00 cm on a side and has a mass of 1312.5 g?

Answer: 10.5 g/cm^3

Specific Gravity

Sometimes the term specific gravity is used to convey the idea of density. **Specific gravity** (symbol **sp gr**) is defined as the ratio of the density of a substance to the density of water at 4°C. In other words,

specific gravity (sp gr) The ratio of the density of a substance compared to the density of water at 4°C; a unitless expression.

$$\text{sp gr} = \frac{d \text{ substance}}{d \text{ water}}$$

Notice that specific gravity is a ratio and therefore a quantity without units. That is, the units in the numerator and denominator cancel. Since the density of water is 1.00 g/mL, the specific gravity and the density are numerically equal. The density of a substance, however, must be expressed in units, usually g/mL.

Consider the following example. The acid in a fully charged automobile battery has a specific gravity of 1.25. We can convert the specific gravity of the battery acid to density as follows:

$$\text{sp gr}_{\text{acid}} = \frac{d_{\text{acid}}}{d_{\text{water}}}$$

Rearranging,

$$\text{sp gr}_{\text{acid}} \times d_{\text{water}} = d_{\text{acid}}$$

Since the sp gr of the acid is 1.25 and the density of water is 1.00 g/mL, we can write

$$1.25 \times \frac{1.00 \text{ g}}{\text{mL}} = 1.25 \text{ g/mL}$$

EXAMPLE EXERCISE 2.18

Diagnostic medical testing can include the determination of the specific gravity of body fluids. For the following, express the specific gravity as a density.

(a) urine (normal): sp gr = 1.02 **(b)** blood (normal): sp gr = 1.06

Solution: To convert the fluid from specific gravity to density, we simply multiply by 1.00 g/mL which is the density of water at 4°C.

(a) $1.02 \times \frac{1.00 \text{ g}}{\text{mL}} = 1.02 \text{ g/mL}$ **(b)** $1.06 \times \frac{1.00 \text{ g}}{\text{mL}} = 1.06 \text{ g/mL}$

A specific gravity value is often followed by a superscript; for example, a blood sample may be written as sp gr = 1.06^{20}. The superscript 20 refers to the Celsius temperature at which the specific gravity value is obtained.

SELF-TEST EXERCISE

If the density of a seawater sample is 1.04 g/mL, what is the specific gravity?

Answer: 1.04

2.9 Temperature

To state the values for the freezing point and boiling point of water on the Fahrenheit, Celsius, and Kelvin scales.

To express any temperature measurement using Fahrenheit degrees, Celsius degrees, or Kelvin units.

temperature A measure of the average energy in a system.

The hotness or coolness of the atmosphere is determined by how fast individual air molecules move. As molecules of air move faster, the temperature of the air increases; that is, it gets hotter. As the temperature increases, molecules possess more energy. **Temperature** is a measure of the average energy of motion of the molecules in a system. The system may be air molecules in the atmosphere, gas molecules collected over water, or even hydrogen molecules in a balloon.

The instrument for measuring temperature is a thermometer. Common thermometers are made from a narrow glass tube filled with liquid mercury. As the temperature increases, the mercury expands and its height in the thermometer increases. Early thermometers were filled with alcohol, but they were replaced because the alcohol boiled at too low a temperature.

The mercury thermometer was invented in 1724 by Daniel Gabriel Fahrenheit (1686–1736), a German physicist. In attempting to produce as cold a temperature as possible, Fahrenheit prepared an ice bath to which he added salt to lower the temperature further. He then assigned a value of zero to that temperature and marked his Fahrenheit scale accordingly.

Fahrenheit obtained a second reference point by recording his body temperature. He assigned that temperature a value of 96 units. The distance between the two reference points was divided into 96 equal units, and each division was termed a **Fahrenheit degree (°F)**. When calibrating other thermometers, it was difficult to reproduce these reference points accurately. For this reason the freezing point and boiling point of water were later selected as the standard reference points. After further refinements, the freezing point of water was assigned a value of 32°F and the boiling point of water 212°F on the Fahrenheit scale.

Fahrenheit degree (°F) The basic unit of temperature in the English system.

In 1742 Anders Celsius (1701–1744), a Swedish astronomer, invented a scale similar in principle to the Fahrenheit scale. On the Celsius scale, the freezing point of water was assigned a value of zero degrees and the boiling point of water (at normal atmospheric pressure) a value of 100 degrees. The scale was then divided into 100 equal divisions. Each division represented one **Celsius degree (°C)**. Since the two reference points are 100 divisions apart, the Celsius scale is sometimes referred to as the centigrade scale.

Celsius degree (°C) The basic unit of temperature in the metric system.

In science we often use two scales for measuring temperature: the Celsius scale and the Kelvin scale. The Kelvin scale was proposed by the English physicist William Thomson, who was later knighted and given the title Lord Kelvin. Although scientific thermometers are calibrated in degrees Celsius, many relationships in chemistry call for the use of the Kelvin scale. The unit of temperature is a **Kelvin unit** (symbol **K**, not **°K**).

Kelvin unit (K) The basic unit of temperature in the SI system.

The Kelvin scale assigns a value of zero kelvins (0 K) to the lowest possible temperature. That temperature represents a theoretical lowest limit for a system that has no energy of motion. In theory, atoms would form a motionless solid arranged in a perfect crystal lattice. This temperature is called **absolute zero** and corresponds to −273.15°C. Note that there is no highest temperature, although the interior of the sun is about 10,000,000 K.

absolute zero The theoretical temperature at which the kinetic energy of a gas is zero.

One division on the Kelvin scale is equivalent to one degree on the Celsius scale. Since 0 K corresponds to −273°C, the freezing point of water is 273 K and the boiling point 373 K. Figure 2.5 illustrates the relationship of the three temperature scales. Notice that 180 Fahrenheit units are equivalent to 100 Celsius units. Therefore, to convert from °F to °C, subtract 32 (the difference between the freezing point of water on the 2 scales) and then multiply by 100°C/180°F. The formula is

$$(°\text{F} - 32°\text{F})\,\frac{100°\text{C}}{180°\text{F}} = °\text{C}$$

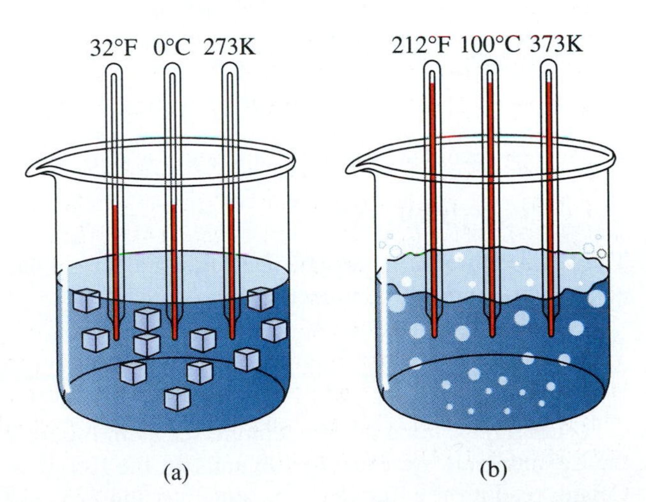

Figure 2.5 Fahrenheit, Celsius, and Kelvin Temperature Scales A Fahrenheit, Celsius, and Kelvin thermometer is placed in (a) ice water and (b) boiling water. Note the freezing point and boiling point on each scale. The difference in the number of divisions is 180 units on the Fahrenheit scale, 100 units on the Celsius scale, and 100 units on the Kelvin scale.

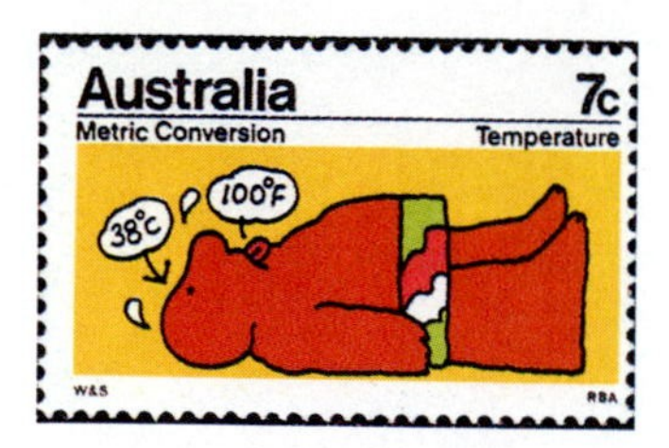

Australian postage stamp.

EXAMPLE EXERCISE 2.19

Normal human body temperature in degrees Fahrenheit is 98.6°F. What is a normal temperature in degrees Celsius?

Solution: To calculate °C, examine Figure 2.5 and compare the Celsius and Fahrenheit temperature scales. The conversion relationship is as follows:

$$(98.6°\text{F} - 32°\text{F}) \frac{100°\text{C}}{180°\text{F}} = °\text{C}$$

Simplifying and canceling units gives

$$(66.6°\cancel{\text{F}}) \frac{100°\text{C}}{180°\cancel{\text{F}}} = 37.0°\text{C}$$

Medical centers that record temperatures in degrees Celsius consider 37.0°C to be normal body temperature.

SELF-TEST EXERCISE

Liquid helium boils at −452°F. What is the boiling point temperature on the Celsius scale?

Answer: −269°C

To convert from °C to °F, simply reverse the procedure. Multiply the Celsius temperature by the ratio 180°F/100°C; then add 32°F. The formula is

$$°\text{C} \frac{180°\text{F}}{100°\text{C}} + 32°\text{F} = °\text{F}$$

EXAMPLE EXERCISE 2.20

The lowest recorded climatic temperature is −89.6°C at the Russian Antarctic Station. What is the equivalent Fahrenheit temperature?

Solution: To calculate °F, examine Figure 2.5 and compare the Celsius and Fahrenheit temperature scales. The conversion relationship is as follows:

$$(-89.6°\cancel{\text{C}}) \frac{180°\text{F}}{100°\cancel{\text{C}}} + 32°\text{F} = °\text{F}$$

Performing the math and canceling units,

$$-161°\text{F} + 32°\text{F} = -129°\text{F}$$

The lowest recorded climatic temperature is thus −129°F.

SELF-TEST EXERCISE

The temperature in the Mojave Desert in California has reached 51°C. What is the equivalent Fahrenheit temperature?

Answer: 124°F

Let's reexamine the temperature scales in Figure 2.5. Notice that 100 units on the Celsius scale are equal to 100 units on the Kelvin scale. Therefore to express a Celsius reading as a Kelvin temperature just add 273 units to the Celsius temperature.

Conversely, to express a Kelvin temperature as a Celsius temperature, subtract 273 units from the Kelvin temperature. It is helpful to remember that negative Kelvin temperatures are impossible. By definition, the coldest possible temperature is assigned a value of 0 K.

EXAMPLE EXERCISE 2.21

Actors and actresses perform under floodlights operating at a temperature of 3215 K. Find the corresponding **(a)** Celsius temperature and **(b)** Fahrenheit temperature.

Solution: To find the Celsius temperature, subtract 273 units from the Kelvin temperature.

(a) Given a temperature of 3215 K, the Celsius temperature is

$$3215 - 273 = 2942^{\circ}\mathrm{C}$$

(b) To find the Fahrenheit temperature, refer back to Figure 2.5. Follow this procedure:

$$(2942^{\circ}\cancel{\mathrm{C}})\,\frac{180^{\circ}\mathrm{F}}{100^{\circ}\cancel{\mathrm{C}}} + 32^{\circ}\mathrm{F} = {}^{\circ}\mathrm{F}$$

Simplifying and rounding off to four significant digits, the Fahrenheit temperature is

$$5296^{\circ}\mathrm{F} + 32^{\circ}\mathrm{F} = 5328^{\circ}\mathrm{F}$$

SELF-TEST EXERCISE

Liquid nitrogen boils at -196°C. What is the boiling point temperature on the **(a)** Fahrenheit scale and **(b)** Kelvin scale?

Answers: **(a)** -321°F; **(b)** 77 K

Liquid-nitrogen freezes the moisture in air into a white gas.

The best comprehension of temperature scales comes from an understanding of their relationship to one another (Figure 2.5). Simply memorizing an equation defeats the purpose of this exercise. In practice, temperatures are routinely converted using a wall chart. In addition, many inexpensive calculators convert temperatures with a touch of a key.

2.10 Heat and Specific Heat

OBJECTIVES

To describe the difference between heat and temperature.

To understand the concept of specific heat and state the value for the specific heat of water: 1.00 cal/g·°C.

To calculate an unknown quantity for a heat change, given any three of the following: heat gain or loss, mass of substance, specific heat of substance, or temperature change of substance.

heat A measure of the total energy in a system. Heat is the flow of energy from a hotter system to a cooler system.

Heat and temperature both measure the energy in a solid, liquid, or gaseous system. The distinction is that **heat** measures the *total energy* of all *particles* in a system,

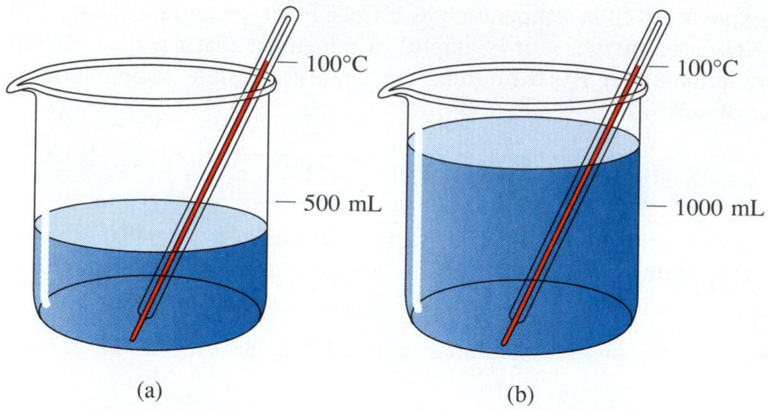

Figure 2.6 **Heat versus Temperature** In (a), 500 mL of water is heated to 100°C and in (b), 1000 mL is heated to 100°C. Although the temperatures are the same, the second beaker has twice the amount of heat.

calorie (cal) The amount of heat required to raise 1 g of water 1°C.

Calorie (Cal) A nutritional unit of heat energy equal to one kilocalorie.

whereas **temperature** measures the *average energy* of a single particle. To grasp the difference, consider the example in Figure 2.6. A beaker containing 500 mL of water is heated to 100°C. A second beaker containing 1000 mL of water is also heated to 100°C. The temperature, or average energy, is identical in each beaker. The heat, or total energy, is greater in the second beaker. That is, 1000 mL of water contains twice as much heat as does 500 mL of water.

Heat energy can be measured in units of calories or kilocalories. A **calorie (cal)** is defined as the amount of heat necessary to raise 1 gram of water 1 degree on the Celsius scale. A kilocalorie (kcal) is the amount of heat necessary to raise 1000 grams of water 1 degree on the Celsius scale. Although the heat necessary to raise the temperature one degree varies slightly with temperature, we will consider these variations to be negligible.

We are all familiar with nutritional Calories. A nutritional **Calorie (Cal)** is spelled with a capital letter to distinguish it from the metric calorie. One nutritional Calorie is identical to one kilocalorie or 1000 metric calories. Thus,

$$1 \text{ Cal} = 1 \text{ kcal} = 1000 \text{ cal}$$

EXAMPLE EXERCISE 2.22

One gram of natural gas undergoes combustion to produce 13.2 kcal of heat energy. Express the energy from the reaction in calories.

Solution: The unknown unit is cal, and the given value is 13.2 kcal. Since 1 kcal = 1000 cal, we convert as follows:

$$13.2 \cancel{\text{kcal}} \times \frac{1000 \text{ cal}}{1 \cancel{\text{kcal}}} = 13{,}200 \text{ cal}$$

In chemical reactions we can measure heat changes in kilocalories. We can also express energy changes in joules, a unit defined in Section 2.11.

SELF-TEST EXERCISE

A hot piece of steel is dropped into water and loses 1250 cal of heat energy. Express the heat loss in kilocalories.

Answer: 1.25 kcal

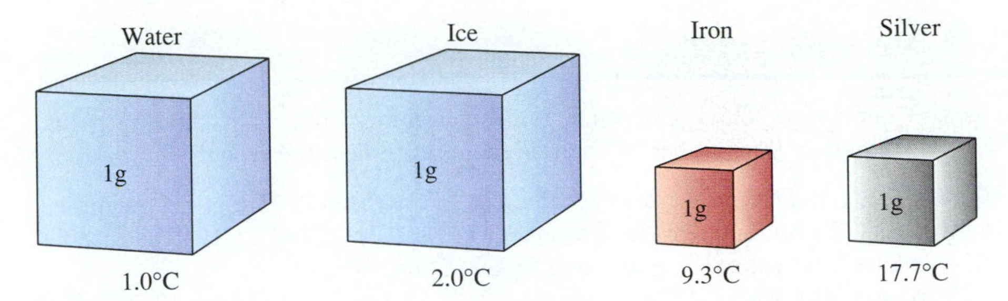

Figure 2.7 Specific Heat Each cube represents 1 g of the substance receiving 1 cal of heat. The temperature, which varies with the substance, increases the number of degrees shown.

Specific Heat

So far we have only considered the changes in heat for a total system. It is interesting to note, however, that each substance in a system has a unique resistance to heat changes. Water has a relatively high resistance to heat changes. That is, it has the ability to absorb or lose large amounts of heat without radically changing temperature. This property of water helps regulate the climate of our planet. It also explains why water is ideal for storing heat in homes with solar energy collectors.

Interestingly, ice has less resistance to heat changes than does water. Metals show very little resistance and are therefore good conductors of heat. Figure 2.7 shows the temperature change of four substances undergoing a heat change of one calorie.

Recall that 1 cal of heat is gained when the temperature of 1 g of water rises 1°C. The amount of heat necessary to raise 1 g of any substance 1°C is termed the **specific heat**. Table 2.5 lists the specific heat of several substances.

specific heat The amount of heat required to raise the temperature of 1 g of any substance 1°C.

Specific heat, like density, expresses a ratio. It is the ratio of heat compared to the mass of the substance times the change in temperature. In metric units, the ratio is calories per gram degree Celsius. The units can be expressed cal/g·°C. The dot (·) indicates multiplication and that g·°C is a product. Since the specific heat of water is 1.00 cal/g·°C, the two unit factors that naturally follow are

$$\frac{1.00 \text{ cal}}{1 \text{ g}\cdot°\text{C}} \quad \text{and} \quad \frac{1 \text{ g}\cdot°\text{C}}{1.00 \text{ cal}}$$

Table 2.5 Specific Heats of Selected Substances

Substance	Specific Heat (cal/g·°C)
water	1.00
ethyl alcohol	0.511
ice	0.492
Freon-12	0.232
aluminum	0.215
sodium chloride	0.210
iron	0.108
copper	0.0920
silver	0.0564
mercury	0.0331
lead	0.0308
gold	0.0305

Let's calculate the specific heat for silver given that a 15.5-g sample of the metal requires 66.9 cal of heat to be heated from 23.5° to 100.0°C. The specific heat (cal/g·°C) is equal to the heat required divided by the mass of the metal and the temperature change. Thus,

$$\frac{66.9 \text{ cal}}{15.5 \text{ g}(100.0 - 23.5)°\text{C}} = \text{cal/g}\cdot°\text{C}$$

Simplifying,

$$\frac{66.9 \text{ cal}}{15.5 \text{ g }(76.5)°\text{C}} = 0.0564 \text{ cal/g}\cdot°\text{C}$$

The following example exercises further illustrate calculations involving the concept of specific heat.

EXAMPLE EXERCISE 2.23

A beaker containing 225.1 g of water is heated from 21.0°C to its boiling point, 100.0°C. Calculate the number of calories required to heat the water.

Solution: Step 1: The answer asks for the unit of cal. Step 2: The given quantities are the mass (225.1 g) and temperature change (100.0°C − 21.0°C). Use the specific heat of water, 1.00 cal/g·°C, as the unit factor. Thus,

$$225.1\ \cancel{g} \times \frac{1.00\ \text{cal}}{1\ \cancel{g}\cdot°\cancel{C}} \times (100.0 - 21.0)°\cancel{C} = 17{,}800\ \text{cal}$$

The temperature change (79.0°C) limits the answer to three significant digits. Notice the problem stated that the water was heated to its boiling point, 100.0°C. To boil the water and change it to gaseous steam requires additional heat.

SELF-TEST EXERCISE

A flask contains 650.5 g of ethyl alcohol at 78.0°C. If the ethyl alcohol is cooled to 1.5°C, how many calories of energy are released? The specific heat of ethyl alcohol is 0.511 cal/g·°C.

Answer: 25,400 cal

EXAMPLE EXERCISE 2.24

Rock can be used to store solar heat energy. What is the specific heat of a rock sample, given the following experiment? A 25.0-g sample of a rock is heated to 250°C and dropped into 100.0 g of water at 19°C. After the rock cooled, the water temperature reached 31°C.

Solution: The unknown quantity is the specific heat of the rock. To find the heat loss by the rock, we must calculate the heat gain by the water. We are given the mass (100.0 g) and the temperature change (31° − 19°C). We know the specific heat (1.00 cal/g·°C) of water. Thus,

$$100.0\ \cancel{g} \times \frac{1.00\ \text{cal}}{1\ \cancel{g}\cdot°\cancel{C}} \times (31 - 19)°\cancel{C} = 1200\ \text{cal}$$

Since the water gains 1200 calories, the rock loses 1200 calories; that is, heat gain = heat loss. The rock cooled from 250° to 31°C. Therefore,

$$\frac{1200\ \text{cal}}{25.0\ \text{g}\ (250 - 31)°\text{C}} = 0.22\ \text{cal/g}\cdot°\text{C}$$

The specific heat of the rock is 0.22 cal/g·°C. The principle of rocks storing heat energy is demonstrated in sauna baths, where rocks collect heat that, in turn, vaporize water into steam.

SELF-TEST EXERCISE

A 1025-g steel horseshoe is heated to 425°C and dropped into 1550 g of water at 20°C. If the temperature of the water increases to 47°C, what is the specific heat of the steel horseshoe?

Answer: 0.11 cal/g·°C

2.11 The International System of Measurement (SI)

OBJECTIVES

To become familiar with the base units and prefixes of the International System of Units (SI).

To state the seven base units and symbols of SI.

To perform conversions between SI units and other measuring units.

The roots of the metric system can be traced back to the late 1700s. In 1875, representatives of several nations, including the United States, signed a treaty creating the International Bureau of Weights and Measures. This organization was given the authority to establish and maintain worldwide standards of measurement. In 1960, the Bureau approved a resolution to establish the International System of Units.

The **International System of Units** (abbreviated **SI** from the French Système Internationale d'Unités) is an extension of the metric system. SI, however, is much more comprehensive and sophisticated. Rather than three quantities and basic units, SI has seven quantities and base units. The base unit of length is the meter; the base unit of mass is the kilogram (not gram). The five other quantities are time, temperature, electric current, light intensity, and amount of substance. Their respective base units are the second, Kelvin, ampere, candela, and mole. As you will later learn, the mole is a crucial chemical unit for keeping track of atoms and molecules. In fact, the mole is the central unit for all the chemical calculations found in Chapter 7. The SI quantities and their base units are given in Table 2.6.

International System of Units (SI) A sophisticated scientific system of measurement having seven base units.

Except for mass, SI has redefined all metric reference standards. SI redefined the meter in 1960 and again in 1983. A meter is currently defined as the distance light travels in a vacuum in 1/299 792 458 second. A second is the official unit of time and is defined in terms of an atomic clock. Other SI base units have similarly complex definitions.

In SI, each unit is defined in terms of a natural phenomenon so that measurements can be reproduced anywhere in the world without the necessity of physical reference standards. Mass, which still relies on the International Prototype Kilogram in France, is the sole exception.

SI prefixes are more extensive than the prefixes for the metric system. Table 2.7 shows all the prefixes currently acceptable as multiples and fractions of the SI base units.

Table 2.6 The International System of Units (SI)

Quantity	Base Unit	Symbol
length	meter	m
mass	kilogram	kg
time	second	s
temperature	Kelvin	K
electric current	ampere	A
luminous intensity	candela	cd
amount of substance	mole	mol

Table 2.7 SI Prefixes and Symbols*

Prefix	Symbol	Multiple/ Fraction
exa	E	10^{18}
peta	P	10^{15}
tera	T	10^{12}
giga	G	10^{9}
mega	M	10^{6}
kilo	k	10^{3}
	basic unit	
milli	m	10^{-3}
micro	μ	10^{-6}
nano	n	10^{-9}
pico	p	10^{-12}
femto	f	10^{-15}
atto	a	10^{-18}

* SI recommends that digits be in groups of three about the decimal point and commas not be used between groups, for example, 101 325 and 0.010 015. The decimal point can be a dot or a comma (British versus French convention); for example, 100.0 and 100,0 are both acceptable. A period should not follow the symbol of a unit; for example, g is correct.

In addition to the seven base units, SI uses dozens of supplementary units. These additional units enable scientists to express any physical measurement. The units have special names and are derived from the base units. For example, the unit of energy, defined as a kilogram mass moving with a velocity of one meter per second, is the joule (J). The unit of pressure, defined as the force of one newton per square meter, is the pascal (Pa). Table 2.8 lists a few additional examples of derived SI units.

The United States Bureau of Standards of Weight and Measures is encouraging the use of SI units. Technical communications between scientists are increasingly reflecting SI convention. Since many instruments are calibrated in metric units, it is necessary to convert data into SI units. The following examples illustrate SI conversions.

EXAMPLE EXERCISE 2.25

A joule is a unit of mechanical energy, and a calorie is a unit of heat energy. How many joules are equivalent to 94.2 kcal of heat? (Given: 4.184 J = 1 cal.)

Solution: To do the conversion, first convert 94.2 kcal to cal. Then convert cal to joules. Since 1 kcal = 1000 cal, we have

$$94.2\ \cancel{\text{kcal}} \times \frac{1000\ \cancel{\text{cal}}}{1\ \cancel{\text{kcal}}} \times \frac{4.184\ \text{J}}{1\ \cancel{\text{cal}}} = 394{,}000\ \text{J}$$

The calculation reveals that 394 000 J (394 kJ) of mechanical energy are equivalent to 94.2 kcal of heat energy.

SELF-TEST EXERCISE

The density of gasoline is 0.781 g/mL. Express the density in SI units.

Answer: 781 kg/m^3

Although much thought and effort have gone into the International System, SI does have drawbacks. Foremost, SI is complicated, making it impractical for routine measurements. Second, most scientific instruments are calibrated in metric units, not SI. Third, SI disallows familiar, conventional units. For example, SI discourages liter as a unit of volume. In addition, the common metric prefixes deci and centi are to be eventually phased out.

Mastering the use of SI is a formidable task. Even professional scientists have to apply conscientious effort to use SI correctly. For our purposes, we will depend on the metric system and refer to SI units in those instances where SI currently enjoys widespread usage, for example, the use of joule as a unit of energy.

Table 2.8 SI Supplementary Units

Quantity	Derived Units	Symbol	SI Units
velocity	meter per second	v	m/s
density	kilogram per cubic meter	d	kg/m^3
force	newton	N	$kg{\cdot}m/s^2$
pressure	pascal	Pa	$kg/m{\cdot}s^2$
energy	joule	J	$kg{\cdot}m^2/s^2$

Summary

Section 2.1 The **metric system** is a simple, unified system of measurement. It uses three basic units: **meter** (m), **gram** (g), and **liter** (L). Physical standards for the quantities of length, mass, and volume were cast in plantinum alloy and serve as reference standards for the meter, kilogram, and liter. The metric system is a decimal system. Prefixes are applied to each of the basic units in order to create multiples and fractions of the measuring unit. The most common prefixes are kilo (k), deci (d), centi (c), milli (m), micro (μ), and nano (n).

Sections 2.2–2.4 All measurement conversion problems are solved by systematically applying the **unit analysis method**. Step 1: Find the unknown quantity and write down the units. Step 2: Select a given value that is related to the unknown quantity. Step 3: Convert the units of the given value to the units desired. Apply a unit factor in each conversion step so that units are canceled in sequence. As a final check, ask yourself if the answer is reasonable.

Section 2.5 With a few minor exceptions, every nation in the world uses the metric system. The metric system is superior because it is a decimal system, uses basic units, and is used for scientific measurement and international trade. It is also valuable to be familiar with the **English system** because it is common in the United States. You should be able to state the relationships 2.54 cm = 1 in., 454 g = 1 lb, and 946 mL = 1 qt. These metric–English equivalents can be used to convert from one system to the other.

Sections 2.6–2.7 The volume by calculation of a rectangular solid is equal to its length times width times thickness. The calculated volume is reported in cubic units, such as cm^3 or m^3. For irregularly shaped objects, as well as gases, we determine volume by a method known as **volume by displacement**. The amount of water displaced is equal to the volume of the object.

Section 2.8 **Density** expresses the ratio of mass to volume. The density of water is 1.00 g/mL and serves as a reference for other liquids and solids. If the density of a material is greater than 1 g/mL, it sinks in water. If the density is less than 1 g/mL, it floats. Since density is a ratio, it can be used as a unit factor. The density of mercury is 13.6 g/mL. The two associated unit factors are 13.6 g/1 mL and 1 mL/13.6 g.

Section 2.9 **Temperature** measures the average energy of molecular motion in a system. The **Fahrenheit, Celsius**, and **Kelvin** scales have related reference points for the freezing point and boiling point of water. The freezing point of water is 32°F, 0°C, and 273 K; the boiling point is 212°F, 100°C, and 373 K. Temperature conversion is accomplished by understanding this reference point relationship. Rather than memorizing equations that relate the different temperature scales, it is better to understand their relationship to one another. Refer to Figure 2.5.

Section 2.10 **Heat** is a measure of the total energy of a system; thus, it is the sum of all the energy of particles in motion. In the metric system, heat changes are measured in **calories (cal)** or kilocalories (kcal). **Specific heat** is the amount of heat necessary to raise 1 g of substance 1°C. For water, the value is 1.00 cal/g·°C. Substances with lower specific heats, such as metals, are better conductors of heat.

Section 2.11 The **International System of measurement** (SI) is a sophisticated extension of the metric system. It is capable of expressing any measurement value using seven base units and dozens of derived units. We will primarily use metric units. SI units will be used selectively and only where appropriate.

Key Terms

Select the key term that corresponds to the following definitions.

____ **1.** a nondecimal system of measurement without any basic unit for length, mass, or volume

____ **2.** a decimal system of measurement using prefixes and a basic unit for length, mass, and volume

____ **3.** a sophisticated scientific system of measurement having seven base units

____ **4.** the basic unit of length in the metric system

____ **5.** the basic unit of mass in the metric system

____ **6.** the basic unit of volume in the metric system equal to the volume of a cube 10 cm on a side

____ **7.** a simple statement relating two equivalent values; for example, 1 m = 100 cm or 1 qt = 946 mL

____ **8.** a ratio of two quantities that are equivalent; for example, 1 m/100 cm or 1 qt/946 mL

____ **9.** a method of problem solving that converts a given value to an unknown value by using unit conversion factors

____ **10.** a unit of volume occupied by a cube 1 centimeter on a side or a volume exactly equal to 1 milliliter

____ **11.** a method for determining the volume of an object by measuring the increase in liquid level when an object is immersed in water

____ **12.** the amount of mass in a unit volume

____ **13.** the ratio of the density of a substance compared to the density of water at 4°C; a unitless expression

____ **14.** a measure of the total energy in a system

____ **15.** a measure of the average energy in a system

____ **16.** the basic unit of temperature in the English system

____ **17.** the basic unit of temperature in the metric system

____ **18.** the base unit of temperature in the SI system

____ **19.** the amount of heat required to raise 1 g of any substance 1°C

____ **20.** the amount of heat required to raise 1 g of water 1°C

____ **21.** a nutritional unit of heat energy equal to 1 kilocalorie

(a) Calorie (Cal) *(Sec. 2.10)*
(b) calorie (cal) *(Sec. 2.10)*
(c) Celsius degree (°C) *(Sec. 2.9)*
(d) cubic centimeter (cm^3) *(Sec. 2.6)*
(e) density (*d*) *(Sec. 2.8)*
(f) English system *(Sec. 2.1)*
(g) Fahrenheit degree (°F) *(Sec. 2.9)*
(h) gram (*g*) *(Sec. 2.1)*
(i) heat *(Sec. 2.10)*
(j) International System (SI) *(Sec. 2.11)*
(k) Kelvin unit (K) *(Sec. 2.9)*
(l) liter (L) *(Sec. 2.1)*
(m) meter (m) *(Sec. 2.1)*
(n) metric system *(Sec. 2.1)*
(o) specific gravity (sp gr) *(Sec. 2.8)*
(p) specific heat *(Sec. 2.10)*
(q) temperature *(Sec. 2.9)*
(r) unit analysis method *(Sec. 2.3)*
(s) unit equation *(Sec. 2.2)*
(t) unit factor *(Sec. 2.2)*
(u) volume by displacement *(Sec. 2.7)*

Exercises

The Metric System (Sec. 2.1)

1. Which of the following statements concerning the metric system are true?
(a) It uses a basic unit for length, mass, and volume.
(b) It is a decimal system that uses prefixes related by a power of 10.
(c) It is used exclusively throughout the scientific community.
(d) It is an official system of measurement in the United States.

2. Which of the following statements concerning the English system are true?
(a) It uses a basic unit for length, mass, and volume.
(b) It is a decimal system that uses prefixes related by a power of 10.
(c) It is used exclusively throughout the scientific community.
(d) It is an official system of measurement in the United States.

3. State the basic metric unit and symbol for the following quantities.
(a) length
(b) time
(c) mass
(d) temperature
(e) volume
(f) heat

4. State the name of the metric prefix and its symbol corresponding to the following multiples and decimal fractions.
   (a) 10^{-2} (b) 0.001 (c) 10^{-3}
   (d) 1000 (e) 10^3 (f) 0.01
   (g) 10^{-1} (h) 0.1 (i) 10^{-9}
   (j) 0.000 001
5. Write the symbol for the following metric units.
   (a) microgram (b) micrometer
   (c) kilometer (d) millisecond
   (e) deciliter (f) kilocalorie
   (g) centiliter (h) cubic centimeter
   (i) nanogram (j) square meter
6. Write the name of the unit indicated by the following symbols.
   (a) m (b) μs (c) kg
   (d) mL (e) dm (f) cal
   (g) cL (h) m^2 (i) ng
   (j) mm^3

Metric Unit Conversion Factors (Sec. 2.2)

7. Give the unit equation that relates each of the following pairs. Write the equation so as to avoid decimal fractions.
   (a) mm and m (b) dL and L
   (c) g and μg (d) dm and m
   (e) kL and L (f) ns and s
8. Write the two unit factors corresponding to each unit equation in Exercise 7.
9. Give the unit equation that relates each of the following pairs. Write the equation so as to avoid decimal fractions.
   (a) m and nm (b) μm and m
   (c) cg and g (d) mL and cm^3
   (e) L and dL (f) cal and kcal
10. Write the two unit factors corresponding to each unit equation in Exercise 9.

Metric Problems by Unit Analysis (Sec. 2.3)

11. State the three steps used in the unit analysis method of problem solving.
12. In unit conversion, what is the guiding principle when choosing a unit factor from a pair of reciprocals?

Metric-Metric Unit Conversions (Sec. 2.4)

13. Perform the following metric system conversions.
   (a) 1.55 km to m (b) 10.6 dg to g
   (c) 0.486 g to cg (d) 125 mL to L
   (e) 1.885 L to dL (f) 100 ns to s
   (g) 0.388 m to mm (h) 94.6 kcal to cal
   (i) 0.000125 s to ms (j) 3.15×10^5 mg to g
14. Perform the following metric system conversions using two unit factors.
   (a) 120 mm to cm (b) 14.5 dL to cL
   (c) 255 mg to dg (d) 1.56×10^{-3} km to dm
   (e) 1.81 mL to μL (f) 8.15×10^4 ms to ns
   (g) 0.55 dm to mm (h) 3.5 kL to mL
   (i) 0.326 kg to cg (j) 0.000222 cm to nm

Metric-English Unit Conversions (Sec. 2.5)

15. State the following metric–English equivalents.
   (a) ? cm = 1 in. (b) ? g = 1 lb
   (c) ? mL = 1 qt (d) ? cm^3 = 1 qt
16. Calculate the following metric–English relationships.
   (a) ? cm = 12 in. (b) ? g = 2.20 lb
   (c) ? mL = 0.500 qt (d) ? cm^3 = 4.00 qt
17. Perform the following metric–English conversions.
   (a) 66 in. to cm (b) 86 cm to in.
   (c) 1.01 lb to g (d) 36 g to lb
   (e) 0.500 qt to mL (f) 750 mL to qt
18. Perform the following metric–English conversions using two or more unit factors.
   (a) 175 lb to kg (b) 1250 mL to gallons
   (c) 0.500 qt to L (d) 0.375 kg to lb
   (e) 72 in. to m (f) 800.0 m to yards
   (g) 152 cm to ft (h) 1.5×10^4 dg to lb
   (i) 2.20 lb to kg (j) 1/4 mile to km
19. The Environmental Protection Agency (EPA) highway mileage estimate for an economy car is 52 miles per gallon. What is the gasoline mileage estimate in km/L?
20. Assuming gasoline sells for $0.959 per gallon, calculate the price in cents per liter.
21. A .357 Magnum bullet has a muzzle velocity of 1200 ft/sec. Express the velocity in meters per second.
22. In the Olympic Games there is a 1500-meter event, but there is no 1-mile event. Calculate the number of meters in 1 mile. Which race would be longer, 1500 meters or 1 mile?

Volume by Calculation (Sec. 2.6)

23. A rectangular piece of solid mahogany measures 5.08 cm by 10.2 cm by 3.05 m. Calculate the volume in cubic centimeters.
24. A quartz rock was cut into a rectangular solid and made into a paperweight. If the paperweight has a mass of 165 g and measures 5.00 cm by 5.00 cm by 25.0 mm, what is its volume in cubic millimeters?
25. A rectangular piece of solid brass measures 4.95 cm by 2.45 cm and has a volume of 15.3 cubic centimeters. What is the thickness of the brass solid?
26. A sheet of aluminum foil measures 30.5 cm by 75.0 cm and has a mass of 9.94 g. Find the thickness of the foil in centimeters if the volume is 3.68 cm^3.
27. Complete the following volume equivalents.
   (a) ? mL = 1 L (b) ? cm^3 = 1 L
   (c) ? mL = 1 cm^3 (d) ? cm^3 = 1 $in.^3$
28. A hospital patient was given a 500.0-cc injection of saline solution. Calculate the volume in mL and L.
29. A large British automobile has an engine displacement of 415 $in.^3$. Calculate the engine volume in liters.
30. A small Japanese automobile has an engine size of 1.20 L. Express the engine volume in cubic inches.

Volume by Displacement (Sec. 2.7)

31. The water level in a 100-mL graduated cylinder reads 44.5 mL. After a sample of iron pyrite is dropped into the cylinder, the new water level reads 55.0 mL. What is the volume of the iron pyrite?

32. Zinc metal reacts with hydrochloric acid to produce hydrogen gas. The gas is collected over water and displaces 212 mL of water. Calculate the volume of the hydrogen gas produced by the reaction.

Density (Sec. 2.8)

33. Indicate whether the following substances will sink or float when dropped into water.
 (a) kerosene (d = 0.82 g/mL)
 (b) paraffin wax (d = 0.90 g/cm^3)
 (c) olive oil (d = 0.918 g/mL)
 (d) birch wood (d = 0.77 g/cm^3)
 (e) chloroform (d = 1.49 g/mL)
 (f) limestone (d = 2.79 g/cm^3)

34. A series of balloons is filled with the gases listed below. State whether the balloon will rise in the air or fall to the ground. (Assume that the mass of the balloon is negligible and the density of air is 1.29 g/L.)
 (a) helium (d = 0.178 g/L)
 (b) carbon monoxide (d = 1.25 g/L)
 (c) nitrous oxide (d = 1.96 g/L)
 (d) ammonia (d = 0.759 g/L)

35. Calculate the mass in grams for each of the following.
 (a) 250 mL of gasoline (d = 0.69 g/mL)
 (b) 36.5 mL of methanol (d = 0.791 g/mL)
 (c) 0.75 cm^3 of rock salt (d = 2.18 g/cm^3)
 (d) 455 cm^3 of borax (d = 1.715 g/cm^3)
 (e) 275 L of sea water (d = 1.025 g/cm^3)

36. Calculate the volume in milliliters for each of the following.
 (a) 10.0 g of nickel (d = 8.90 g/cm^3)
 (b) 0.500 g of bromine (d = 3.12 g/mL)
 (c) 33.5 g of carbon tetrachloride (d = 1.59 g/mL)
 (d) 0.899 kg of acetone (d = 0.792 g/mL)
 (e) 1.00 kg of cork (d = 0.25 g/cm^3)

37. Calculate the density in g/mL for each of the following.
 (a) 25.0 mL of ethyl alcohol having a mass of 19.7 g
 (b) 10.0 g of ether having a volume of 14.0 mL
 (c) 11.6-g marble whose volume is found by displacement to be 4.1 mL
 (d) 131.5-g bronze rectangular solid measuring 3.55 cm × 2.50 cm × 1.75 cm
 (e) 0.175-g piece of gold foil that measures 5.00 cm × 4.50 cm × 4.03 μm

38. Calculate the density in g/mL for each of the following.
 (a) jet fuel (sp gr = 0.775)
 (b) gasohol (sp gr = 0.801)
 (c) mercury (sp gr = 13.6)
 (d) salt water (sp gr = 1.05)
 (e) water (sp gr = 1.00)

Temperature (Sec. 2.9)

39. State the freezing point of water on the following temperature scales.
 (a) Fahrenheit **(b)** Kelvin **(c)** Celsius

40. State the boiling point of water on the following temperature scales.
 (a) Fahrenheit **(b)** Kelvin **(c)** Celsius

41. Express the following temperatures in degrees Celsius.
 (a) 100°F **(b)** −215°F **(c)** 2165°F **(d)** −40°F

42. Express the following temperatures in degrees Fahrenheit.
 (a) 19°C **(b)** −175°C **(c)** 1115°C **(d)** −273°C

43. Convert the following temperatures to Kelvin units.
 (a) 42°C **(b)** −20°C **(c)** 495°C **(d)** −185°C

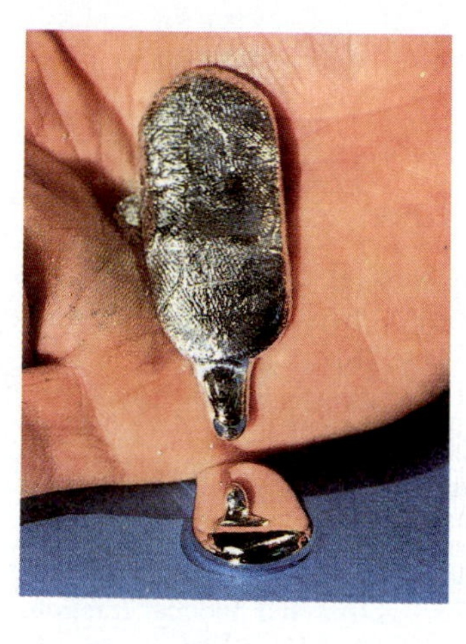

Gallium is a metal that melts at 30°C, in the palm of your hand.

44. Convert the following temperatures to degrees Celsius.
 (a) 273 K **(b)** 100 K **(c)** 298 K **(d)** 3 K

Heat and Specific Heat (Sec. 2.10)

45. Distinguish between temperature and heat in terms of the energy of a system.

46. State the value for the specific heat of water.

47. A beaker containing 250.0 g of water is heated from 23° to 100°C. How many calories of heat are required?

48. A shallow pond having 2.45 × 10^4 L of water cools from 28° to 17°C. What is the kilocalorie heat loss?

49. Calculate the heat required to raise the temperature of 25.0 g of iron (0.108 cal/g·°C) from 25.0° to 50.0°C.

50. Calculate the heat released as 35.5 g of copper metal (0.0920 cal/g·°C) cools from 50.0° to 45.0°C.

51. Compute the specific heat of gold if 25.1 cal are required to heat 30.0 g of gold from 27.7° to 54.9°C.

52. Find the specific heat of platinum if 35.7 cal are lost as 75.0 g of the metal cool from 43.9° to 28.9°C.

53. Calculate the mass of titanium (0.125 cal/g·°C) that requires 75.6 cal to raise the temperature of the metal from 20.7° to 31.4°C.

54. Calculate the mass of lead (0.0308 cal/g·°C) that releases 52.5 cal as the temperature of the metal cools from 35.7° to 25.1°C.

55. What is the temperature change when 10.5 g of silver (0.0566 cal/g·°C) absorb 35.2 cal of heat?

56. What is the drop in temperature when 8.92 g of sodium (0.293 cal/g·°C) lose 750 cal of heat energy?

The International System of Measurement (SI) (Sec. 2.11)

57. Perform the following SI unit conversions.
(a) 21 500 000 000 g to Eg
(b) 0.000 000 000 02 TL to L
(c) 9.83×10^{-17} m to am
(d) 1.04×10^{14} fs to s.

58. Perform the following SI unit conversions.
(a) 143 000 Mg to Gg
(b) 10.0 nL to μL
(c) 0.000 000 035 Pm to km
(d) 75 ms to ps

59. The density of mercury is 13.6 g/mL. Express the density of mercury in SI units (kg/m^3).

60. The specific heat of mercury at 20°C is 0.0332 cal/g·°C. Express the specific heat in SI units (J/kg·K).

General Exercises

61. According to the Bible, Noah's ark measured 300 by 50 by 30 cubits. Estimate the dimensions in feet. A cubit was defined as the distance from the elbow to the tip of the middle finger, about 18 inches.

62. Excluding end zones, a football field measures 300 feet by 160 feet. Calculate the playing area in square meters.

63. State how many significant digits are represented in the following unit factors.
(a) 1 g/10 dg (b) 1 lb/454 g
(c) 1 L/1000 mL (d) 1 qt/946 mL

64. The mass of the earth is 1.31×10^{25} pounds. Calculate its mass in kilograms.

65. The velocity of an oxygen molecule at room temperature is 975 miles per hour. What is the velocity in meters per second?

66. Calculate the number of 5-grain aspirin tablets that can be manufactured from a 2.50-kg batch of powdered aspirin. (1 grain = 64.8 mg)

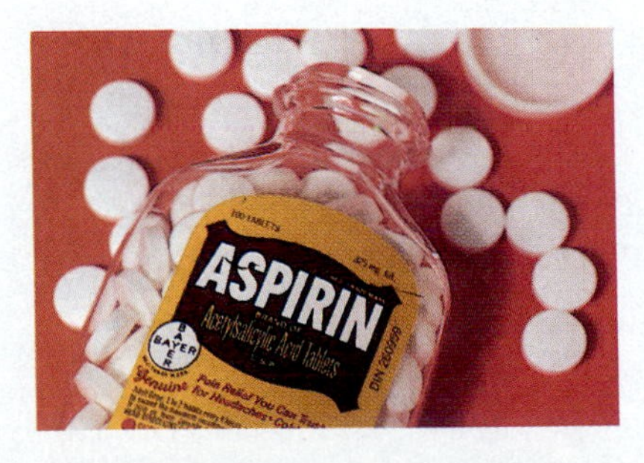

A bottle of 5-grain aspirin tablets.

67. An AM ratio station broadcasts at a frequency of 1380 kilocycles per second. If the radiowaves have a velocity of 3.00×10^8 m/s, what is the AM broadcast wavelength in cm/cycle?

68. An FM radio station sends out radiowaves measuring 305 cm from crest to crest. If the radiowaves travel 3.00×10^8 m/s, what is the FM frequency in megacycles per second?

69. A computer floppy disk holds 800 kilobytes of memory. If a hard disk stores 40 megabytes, how many floppy disks of information can be loaded onto the hard disk?

70. A substance containing sodium produces a yellow flame when heated with a Bunsen burner. If the yellow light has a wavelength of 589 nanometers, what is the wavelength in Ångstrom units? (Given: 1 Å = 10^{-8} cm.)

71. Express the density of water in the English units of pounds per cubic foot.

72. A 15.357-g opal gemstone is dropped into a graduated cylinder containing 50.0 mL of water. If the water level increased to 57.5 mL, what is the density of the opal?

73. Calculate the mass of a 25.0 mL sample of methanol whose specific gravity is 0.791.

74. One milliliter of water contains about 20 drops. One gram of water contains 3.34×10^{22} molecules. How many molecules of water are in one drop?

75. A scientist invented a new temperature scale (°N) and assigned the freezing point and boiling point of water at −500°N and 500°N, respectively. What is (a) normal body temperature (37°C) in °N and (b) absolute zero in °N?

76. One gram of propane undergoes combustion with oxygen to produce 11.95 kilocalories of energy. Express the heat of combustion in joules.

77. An unknown metal is heated to 250°C and dropped into 500.0 g of water at 22°C. If the water temperature rises to 71°C, what is the heat loss of the metal in (a) calories and (b) joules?

78. A 10.0-g sample of copper (0.0920 cal/g·°C) absorbs 27.8 cal of heat. If the initial temperature is 22.7°C, what is the final temperature?

79. A 15.5-g sample of titanium (0.125 cal/g·°C) loses 32.9 cal of heat. What is the final temperature if the metal was initially at 28.9°C?

80. A 10.0-g sample of metal at 20.5°C was dropped into 75.0 g of water at 80.0°C. The maximum temperature reached by the metal was 78.3°C. Refer to Table 2.5 to identify the metal.

81. A parsec is an astronomical unit, which is the distance light travels in 3.26 years. Express this distance in meters given the velocity of light, 3.00×10^8 m/s.

82. The International Prototype Kilogram is a platinum–iridium cylinder with a radius (r) equal to 1.95 cm. Calculate the height (h) of the standard cylinder given its density, 21.50 g/cm^3. The volume of a cylinder equals $\pi r^2 h$.

CHAPTER

3

Matter and Energy

The eruption of Mt. Etna in Italy provides evidence that temperature and states of matter are related. As the red-hot lava from the volcano cools, it changes from a molten liquid to a solid black mass.

In the seventeenth century, the British scientist Robert Boyle (1627–1691) established the importance of laboratory experiments for the study of science. Boyle, realizing the value of carefully planned experiments, described his work thoroughly so that others might repeat and confirm his discoveries. In 1661, he published The Sceptical Chemist, *which essentially said that theories were no better than the laboratory methods they were based upon. Boyle rejected the Greek view of matter that said air, earth, fire, and water were basic elements. Instead, he proposed that an element was a substance that could be identified by experiment.*

Following the lead of Boyle, scientists began to systematically study matter and its changes. Although scientists continued to propose theories about the behavior of matter, their theories were based upon experimental evidence. Today, the study of matter and the changes it undergoes is the definition of chemistry.

In this chapter we will study physical changes and chemical changes of matter. If we alter the shape, mass, or volume of matter, it is considered a physical change. If we alter the composition of matter, it is a chemical change. Every chemical change is accompanied by a change in energy; and every chemical change involves a physical change of matter. In 1905, Albert Einstein (1879–1955) proposed that matter and energy are interrelated. That theory is evident in his now famous equation $E = mc^2$, *which states that energy (E) equals mass (m) times the velocity of light squared* (c^2). *The first dramatic example of mass being converted into energy took place on July 16, 1945. On that date a mushroom cloud rose above the white sands of the New Mexico desert from the first atomic bomb explosion (see Figure 3.1).*

Figure 3.1 Atomic Bomb Explosion In an atomic bomb, unstable matter splits apart and releases energy. A small amount of mass is converted to an enormous amount of energy. This illustrates that mass and energy are related; that is, $E = mc^2$.

3.1 Physical States of Matter

OBJECTIVES

To describe the three physical states of matter in terms of the motion of particles.

physical state Refers to the condition of a substance existing as a solid, liquid, or gas.

Matter is anything that has mass and occupies volume. It exists in one of three **physical states**: solid, liquid, or gas. In the solid state, matter has a fixed shape and definite volume. A solid cannot be compressed because its particles are tightly packed. In the liquid state, particles are free to move past one another. The particles are closely packed and can be compressed only slightly. The shape of a liquid may change, but its volume is fixed. In the gaseous state, the particles are widely spaced and distributed uniformly throughout the container. If the volume expands, the gas particles move away from each other. If the volume decreases, the gas compresses, and particles move closer to one another. Table 3.1 summarizes the properties of the physical states of matter.

Even though we often describe a substance as being a solid, liquid, or gas, the same substance can appear in more than one physical state. By changing the temperature, we can change the physical state of a substance. Water, for example, can be changed from solid ice, to liquid water, to gaseous steam. Other substances behave similarly at different temperatures.

sublimation A direct change of state from a solid to a gas without forming a liquid.

Scientists use specific terms to describe a change of physical state. As temperature increases, a solid *melts* to produce a liquid, and a liquid *vaporizes* to produce a gas. A direct change from a solid to a gas is called **sublimation**. Dry Ice and mothballs, for example, sublime as they change from a solid to a gas.

Table 3.1 Physical States of Matter

Property	Solid	Liquid	Gas
shape	definite	indefinite	indefinite
volume	fixed	fixed	variable
compressibility	negligible	negligible	high

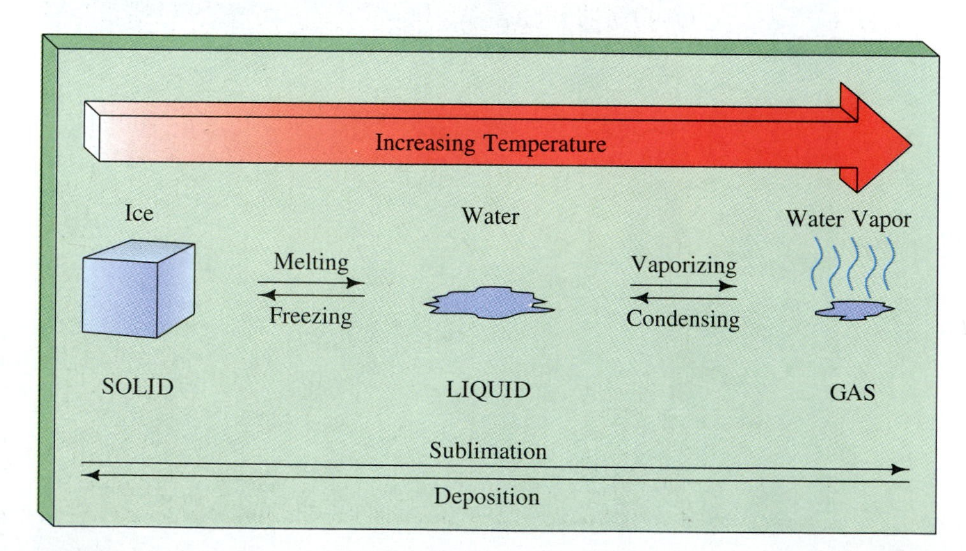

Figure 3.2 Change in Physical State As temperature increases, a solid substance changes to a liquid and then to a gas. As temperature decreases, a gas condenses to a liquid and then freezes to a solid.

As temperature decreases, a gas *condenses* to a liquid, and a liquid *freezes* to a solid. A direct change from a gas to a solid is called *deposition.* A refrigerator demonstrates deposition by collecting ice. Opening the refrigerator door allows in moist air, which deposits frost without a trace of liquid water. Figure 3.2 shows the relationship of temperature and physical state.

Liquid bromine vaporizing to a gas.

EXAMPLE EXERCISE 3.1

State the term for the following changes of state.

(a) solid to a liquid
(b) liquid to a gas
(c) solid to a gas
(d) gas to a liquid
(e) liquid to a solid
(f) gas to a solid

Solution: Refer to Figure 3.2 for the changes of physical state.

(a) melting
(b) vaporizing
(c) sublimation
(d) condensing
(e) freezing
(f) deposition

SELF-TEST EXERCISE

Identify the physical state of matter described below.

(a) widely spaced arrangement of particles; indefinite shape and volume
(b) ordered arrangement of particles; definite shape and fixed volume
(c) disordered arrangement of particles; indefinite shape and fixed volume

Answers: **(a)** gas; **(b)** solid; **(c)** liquid

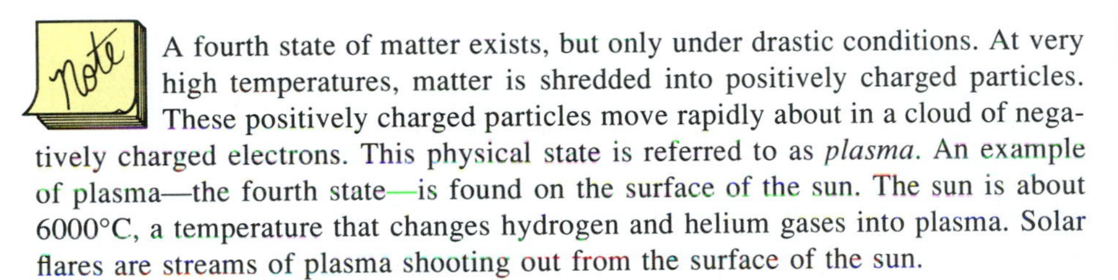

A fourth state of matter exists, but only under drastic conditions. At very high temperatures, matter is shredded into positively charged particles. These positively charged particles move rapidly about in a cloud of negatively charged electrons. This physical state is referred to as *plasma.* An example of plasma—the fourth state—is found on the surface of the sun. The sun is about 6000°C, a temperature that changes hydrogen and helium gases into plasma. Solar flares are streams of plasma shooting out from the surface of the sun.

Solid iodine undergoing sublimation.

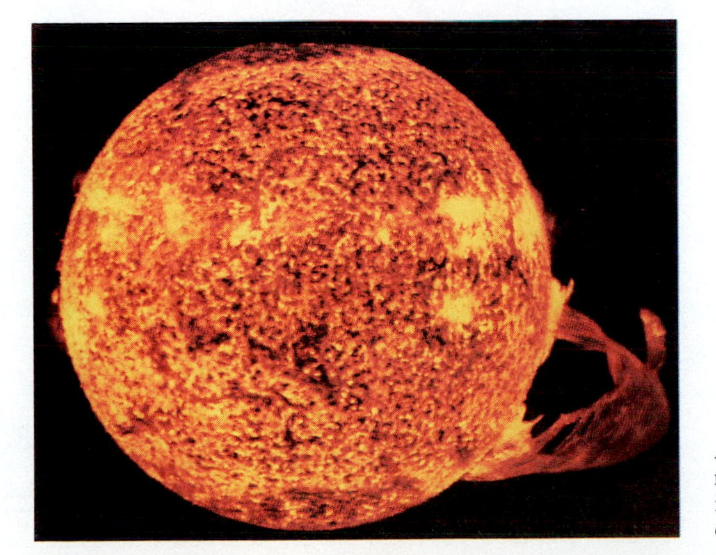

A spectacular solar flare (bottom right) at the surface of the sun illustrates plasma, the fourth state of matter.

3.2 Physical State and Energy

OBJECTIVES

To understand the relationship of physical state and the amount of attraction between individual particles.

To relate the temperature of a gas to the kinetic energy and velocity of its molecules.

We can view matter as a large collection of very small particles that are attracted to one another. The degree of attraction between particles is a criterion that determines the physical state. A high degree of attraction is observed for solids. The attractive force holds the particles in a fixed position and allows little or no movement. When a substance in the *solid state* is heated, the particles acquire a greater energy and begin to vibrate more intensely.

When a solid has sufficient heat energy, particles will vibrate so intensely that the force of attraction between particles will be overcome and particles will move past one another. This characterizes the *liquid state*. A liquid is therefore a substance in which particles have sufficient energy to move about randomly.

When a liquid has enough energy to completely overcome the attractive force, its particles are free to fly about within the limits of the container. At this point, the liquid has changed into the *gaseous state*. For now, we will refer to one of these particles as a molecule. Although not strictly true, for the present we will define a **molecule** as the smallest particle that represents a compound. Figure 3.3 illustrates the relationship of the energy of molecules versus the attraction between molecules.

molecule The smallest particle that represents a compound.

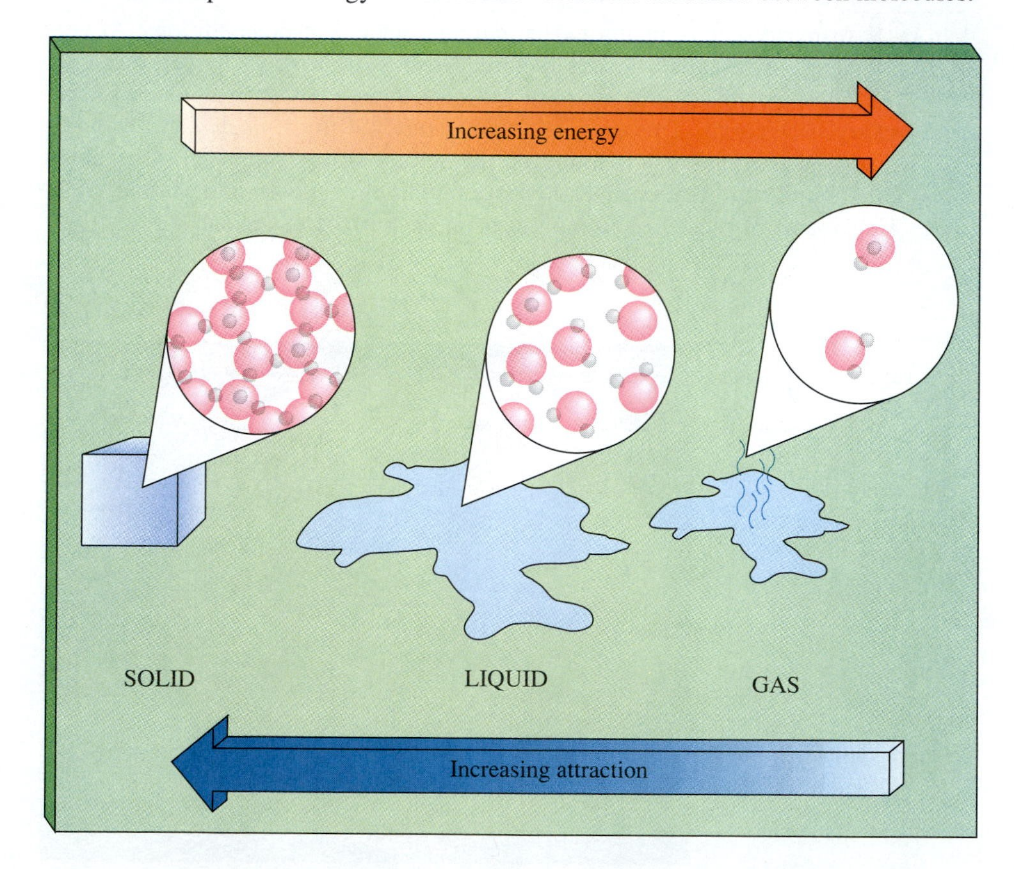

Figure 3.3 Energy vs Molecular Attraction The energy of individual molecules in a solid increases as heat is added. With additional heat energy, the attraction between molecules is overcome. Gradually, molecules acquire greater motion and eventually become a gas.

Table 3.2 Properties and Physical States

Property	Solid	Liquid	Gas
kinetic energy of molecules	very low	low	very high
attraction between molecules	very high	high	very low
confinement of molecules	very restricted	restricted	unrestricted

kinetic energy The energy associated with the motion of molecules.

The energy associated with the motion of molecules is called **kinetic energy**. Raising the temperature of a solid substance gives the molecules more kinetic energy. When the heat is sufficient, a change of state will occur from a solid to a liquid. Raising the temperature further provides the kinetic energy necessary for the molecules to break free and have complete freedom of movement. They are then in the gaseous state. Table 3.2 summarizes this relationship of physical state and molecule properties.

The kinetic energy of a gas molecule is directly related to temperature. As the temperature increases, the kinetic energy increases. As the temperature decreases, the kinetic energy decreases. At the same temperature, smaller gas molecules move faster than larger gas molecules. This relationship is shown by the fact that the kinetic energy (KE) of a molecule is equal to half its mass (m) times its velocity squared (v^2). This statement can be expressed as the following equation:

$$\text{KE} = \frac{1}{2}mv^2$$

At a given temperature, the kinetic energy is the same for all gas molecules. In the same container, for example, hydrogen and oxygen molecules have equal kinetic energies. We can deduce from the equation, however, that the gas with the lowest mass will have the highest velocity. Because hydrogen molecules are lighter, they will have a greater velocity than the heavier oxygen molecules.

EXAMPLE EXERCISE 3.2

In 1783 a French scientist ascended to 10,000 feet in a hydrogen-filled balloon. Assume the temperature decreased from 25° to −5°C during the ascent and state the change in each of the following.

(a) kinetic energy of the gas molecules **(b)** velocity of the gas molecules

Solution: The temperature, kinetic energy, and velocity of a gas are related as follows:

(a) Since temperature and kinetic energy are directly related, the *kinetic energy* of the hydrogen in the balloon *decreases* as the gas cools.

(b) A drop in temperature produces a decrease in kinetic energy. Since the mass of molecules remains constant, the *velocity* of hydrogen molecules *decreases*.

SELF-TEST EXERCISE

A sample of fluorine gas in a Teflon-lined cylinder is heated from 25° to 250°C. State the change in each of the following.

(a) kinetic energy of the gas **(b)** velocity of fluorine molecules

Answers: **(a)** increases; **(b)** increases

Not all molecules of a given gas have the same velocity. That is, some molecules move faster than others. Strictly speaking, it is the *average velocity* of a gas molecule that is proportional to its temperature. In a gaseous mixture, it is the average velocity of small molecules that is faster than the average velocity of large molecules.

3.3 Classification of Matter

To classify a sample of matter as an element, compound, or mixture.

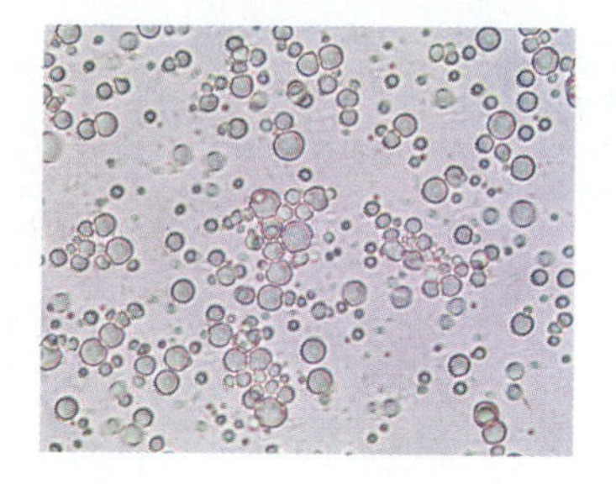

A microscopic view of milk reveals a heterogeneous mixture of fat droplets in a clear liquid.

Some matter has properties that are consistent throughout the sample. Other matter has properties that vary within the sample. One way to tell if the properties of matter are consistent is to melt a sample. Consider pure gold, which melts at 1064°C. It makes no difference whether we have a large nugget or a small flake; gold melts at 1064°C. Now consider gold ore, which often contains quartz that melts at 1610°C. When we melt gold ore, the quartz and pure gold melt at different temperatures. Therefore, gold and gold ore are two different types of matter. Matter such as gold whose properties are uniform throughout is said to be a **homogeneous substance**. Matter such as gold ore whose properties are not uniform is said to be *heterogeneous*.

homogeneous substance Matter having constant composition, definite and consistent properties.

element A pure substance that cannot be broken down any further by ordinary chemical reaction.

compound A pure substance that can be broken down into two or more simpler substances by chemical reaction.

heterogeneous mixture Matter having variable composition and indefinite properties; matter composed of two or more substances that can be physically separated.

homogeneous mixture Matter having variable composition, but definite and consistent properties; examples include gas mixtures, solutions, and alloys.

Pure substances, such as gold, appear homogeneous. That is, they are consistent throughout the sample, and their physical and chemical properties are constant. A pure substance can be either an element or a compound. An **element** is a substance that cannot be broken down any further by chemical reaction. Examples of elements are carbon, hydrogen, and oxygen. Elements combine to form compounds. A **compound** is a substance that can be broken down into elements by a chemical reaction. Examples of compounds formed from the elements carbon, hydrogen, and oxygen are sugar and alcohol. In summary, elements and compounds are pure homogeneous substances having constant properties.

When you combine two or more pure substances, you produce a *mixture*. There are two types of mixtures: heterogeneous and homogeneous. **Heterogeneous mixtures** contain two or more pure substances. A coarse mixture of iron and salt, for example, is a heterogeneous mixture. Iron is affected by a magnet, but salt is not. Salt dissolves in water, but iron does not. A coarse mixture of iron and salt has properties that are not constant throughout a given sample; therefore, it is heterogeneous.

Homogeneous mixtures also contain two or more pure substances. But unlike heterogeneous mixtures, the substances mix uniformly throughout the sample. Homogeneous mixtures include solutions such as seawater, metal alloys such as brass, and mixtures of gases such as air.

Homogeneous mixtures have properties that are uniform throughout a given sample, but are not necessarily constant from sample to sample. Seawater from the Pacific Ocean and Dead Sea differ. Seawater from the Dead Sea has a higher density and greater concentration of dissolved salt. Brass is a homogeneous mixture, but the amount of copper varies from 60% to 80%. Although a given brass sample is homogeneous, the properties can vary from sample to sample. Figure 3.4 classifies matter and shows the relationship of pure substances and mixtures.

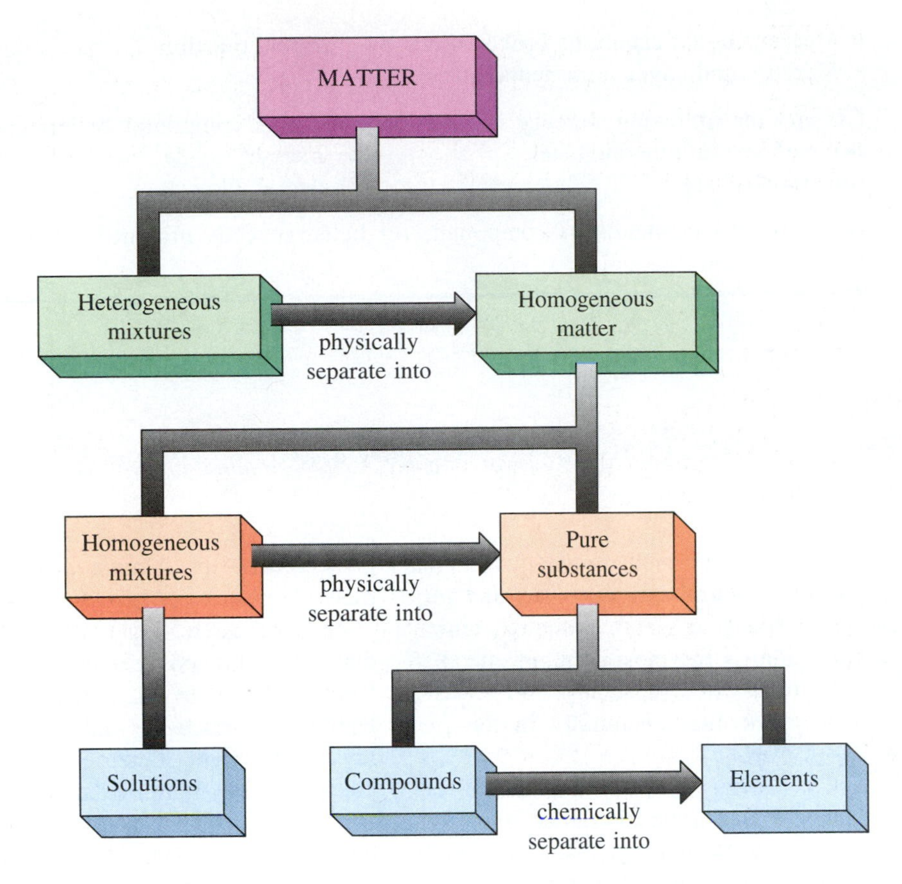

Figure 3.4 Classification of Matter Matter consists of heterogeneous mixtures and homogeneous matter. The properties of heterogeneous mixtures vary within a sample while homogeneous matter has constant properties. Homogeneous matter can be a solution, a compound, or an element.

Example Exercise 3.3 illustrates the distinction between different types of matter, that is, the differences among mixtures, compounds, and elements.

EXAMPLE EXERCISE 3.3

Consider the following properties of copper.

- Copper metal cannot be broken down by chemical reaction.
- Copper oxidizes, when heated in air, to give copper oxide.
- Copper, in the form of malachite ore, is found in South America.
- Copper and tin form bronze alloys.

Classify the following copper samples as an element, compound, or mixture.

(a) copper wire
(b) copper oxide
(c) copper ore
(d) bronze alloy

Solution: Refer to Figure 3.4 to classify each sample.

(a) Copper wire is a metallic *element*.
(b) Copper oxide is a *compound* of the two elements copper and oxygen.
(c) Copper ore is a *heterogeneous mixture* of several substances.
(d) Bronze alloy is a *homogeneous mixture* of copper and tin.

SELF-TEST EXERCISE

Consider the following properties of mercury.

- Mercury oxide decomposes to give mercury and oxygen gas.
- Mercury, in the form of cinnabar ore, is found in Spain and Italy.

Solid mercury oxide is an orange powder, but decomposes to give metallic liquid mercury.

- Mercury liquid cannot be broken down by chemical reaction.
- Mercury and silver form dental alloys.

Classify the following mercury samples as an element, compound, or mixture.
(a) mercury in a thermometer (b) mercury oxide
(c) cinnabar ore (d) dental alloy

Answers: (a) element; (b) compound; (c) heterogeneous mixture; (d) homogeneous mixture.

CHEMISTRY CONNECTION □ THE CONSUMER

A Student Success Story

▶ Can you name the element that in 1800 was so valuable that Napoleon Bonaparte spent a fortune to have dining utensils made from this metal?

In 1885, Charles Hall was a 22-year-old chemistry major at Oberlin College in Ohio. One day his chemistry teacher told the class that anyone who could discover an inexpensive way to produce aluminum metal from its ore would grow rich and benefit humanity. At that time aluminum was a rare and very expensive metal. Although aluminum is the most common metal in the earth's crust, it is too reactive to be found naturally in the free state.

Charles Martin Hall (1863–1914)

After graduation, Charles Hall set up a home laboratory in a woodshed behind his father's church. Using homemade batteries and carbon electrodes, he devised a method for producing aluminum metal by passing an electric current through a molten mixture of the two minerals, bauxite and cryolite. After only eight months of constant experimenting, Hall had invented a method for reducing the aluminum oxide in a mineral to elemental aluminum metal. In February, 1886, Charles Hall walked into his former teacher's office with a handful of aluminum globules.

As his chemistry teacher had predicted, within a short period of time, Hall became rich and famous for his discovery. In 1911, he received the prestigious Perkin medal. Upon his death, he left five million dollars to his alma mater, Oberlin College. The process for making aluminum metal gave rise to a huge industry, second only to steel as a construction metal. Hall's original globules of aluminum have been preserved by the Aluminum Company of America (ALCOA).

It is an interesting coincidence that the French chemist Paul Héroult, without knowledge of Hall's work, made the same discovery at the same time. Thus, the method for obtaining aluminum metal from its mineral form is referred to as the Hall-Héroult process. Following the discovery of the Hall-Héroult process, the price of aluminum metal dropped dramatically as shown in the following table.

The Hall-Héroult Process and the Value of Aluminum

Year	Price of Aluminum
1885	$100.00/lb
1890	5.00
1895	0.70
1915	0.18
1970	0.30
1980	0.80
1990	0.50

▶ Napoleon had utensils made from aluminum. At the time, aluminum was a rare and valuable metal.

3.4 Names and Symbols of the Elements

OBJECTIVES

To identify a few of the most abundant elements in the earth's crust, oceans, and atmosphere.

To state the names and symbols of several common elements.

There are 81 naturally occurring elements that are stable. There are a few other elements, such as uranium, that occur in nature but are radioactively unstable. Of all the elements, ten account for 99% of the earth's crust, water, and atmosphere. Three of those ten elements (oxygen, silicon, and aluminum) account for more than 80%. And of those three, oxygen is the most abundant element. Oxygen is combined with silicon in sand and rocks, combined with hydrogen in water, and is found free in the atmosphere. In all its various forms, oxygen is about half of the total mass of all these elements. Table 3.3 shows the abundance of 10 elements in the earth's crust.

Oxygen, silicon, and aluminum are the three most abundant elements in the earth's crust. If we include the core of the earth, the percentages of iron and nickel increase dramatically. In the human body, oxygen, carbon, hydrogen, nitrogen, calcium, and phosphorus account for over 99% of the mass. The remaining 1% consists of trace elements, most of which are essential to human life.

Each element has an individual name and symbol. The names of the elements are derived from various sources. Many are named for their properties. For example, hydrogen is derived from the Greek word for water, *hydro*, and means 'water former.' Argon, an inert gas, comes from the Greek word for 'inactive.' Carbon is derived

Table 3.3 Abundance of the Elements in the Earth's Crust

Element	Mass Percent	Element	Mass Percent
oxygen	49.5%	potassium	2.4%
silicon	25.7%	magnesium	1.9%
aluminum	7.5%	hydrogen	0.9%
iron	4.7%	titanium	0.6%
calcium	3.4%	all other elements	0.5%
sodium	2.6%		

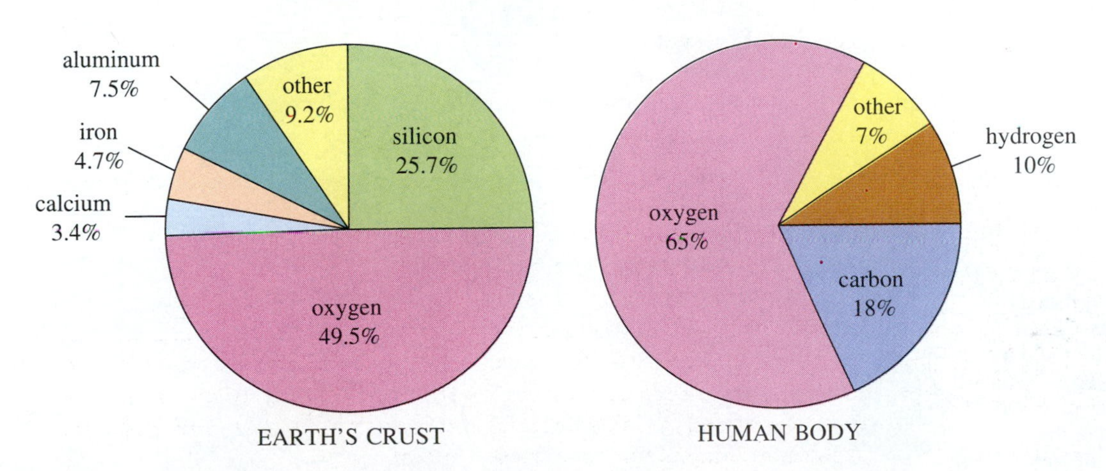

Distribution of elements in the earth's crust and human body.

from the Latin *carbo*, meaning coal; calcium is derived from the Latin *calcis*, which translates as lime, a mineral source for calcium. Some elements are named for their region of discovery. For example strontium comes from Strontian, Scotland, and scandium from Scandinavia. Other elements are named for a scientist, such as curium for Marie Curie and einsteinium for Albert Einstein.

chemical symbol An abbreviation for the name of a chemical element; for example, Cu is the symbol for copper.

atom The smallest particle that represents an element.

The names of the elements are abbreviated using **chemical symbols**. The first symbols were invented by alchemists in the Middle Ages so as to maintain the secrecy of their work. Alchemists dabbled in the occult and attempted to convert base metals, such as iron and lead, into gold. In 1803, the English chemist John Dalton (1766–1844) proposed that elements are composed of indivisible, spherical particles. Dalton called these individual particles **atoms**, from the Greek *atomos*, meaning indivisible. He suggested using circles with different markings as symbols for the elements. Dalton reasoned that a circular symbol seemed to best represent his proposed spherical atoms. Figure 3.5 shows selected symbols chosen by Dalton to represent elements.

In 1813, the Swedish chemist J. J. Berzelius (1779–1848) proposed our current system of symbols for the elements. He suggested that the symbol correspond to the first letter of the name of the element, for example, H for hydrogen, O for oxygen, and C for carbon. If two elements start with the same letter, another letter in the name is used, for example, Ca for calcium, Cd for cadmium, and Cl for chlorine.

In some instances the symbol of the element corresponds to the original Latin name of the element. For example, the symbol for lead is Pb, because the Latin name for lead is *plumbum*. Lead pipe was used to transport water in Roman times and that is the derivation of our word plumbing. Similarly, the symbol for gold is Au from the Latin *aurum*, meaning shining dawn. Other symbols derived from the Latin name of the element include silver, Ag; copper, Cu; iron, Fe; mercury, Hg; potassium, K; sodium, Na; antimony, Sb; and tin, Sn. The Latin names are argentum, cuprum, ferrum, hydrargyrum, kalium, natrium, stibium, and stannum, respectively.

When you write the symbols for the elements, it is important to follow convention. The first letter is always capitalized and the second is lowercase. Thus, the symbol for the element cobalt is Co. In contrast, note that the formula for the compound carbon monoxide is CO. Table 3.4 lists the names and symbols of 48 common elements.

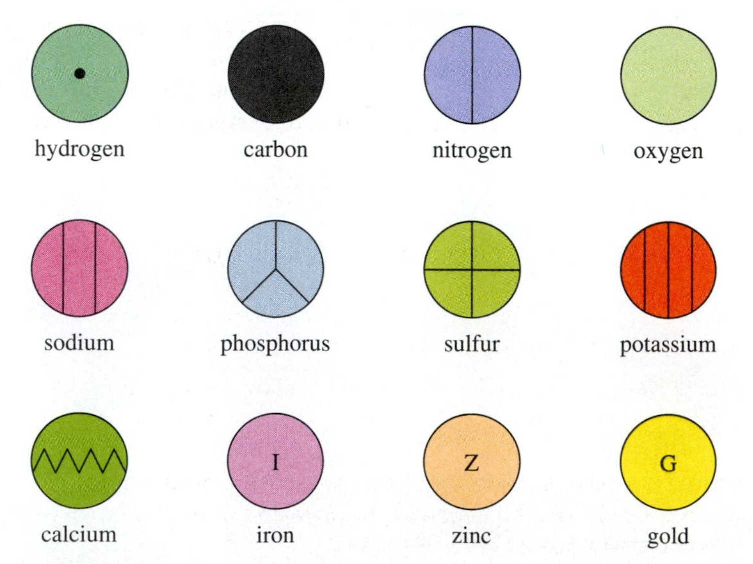

Figure 3.5 Dalton's Symbols for Selected Elements Each element is symbolized by a circle with an inscribed marking. Dalton's system was not practical, but it did convey the idea that an element was composed of tiny spherical particles called atoms.

Table 3.4 Names and Symbols of Common Elements

Element	Symbol	Element	Symbol	Element	Symbol	Element	Symbol
aluminum	Al	chlorine	Cl	lead	Pb	radium	Ra
antimony	Sb	chromium	Cr	lithium	Li	selenium	Se
argon	Ar	cobalt	Co	magnesium	Mg	silicon	Si
arsenic	As	copper	Cu	manganese	Mn	silver	Ag
barium	Ba	fluorine	F	mercury	Hg	sodium	Na
beryllium	Be	germanium	Ge	neon	Ne	strontium	Sr
bismuth	Bi	gold	Au	nickel	Ni	sulfur	S
boron	B	helium	He	nitrogen	N	tellurium	Te
bromine	Br	hydrogen	H	oxygen	O	tin	Sn
cadmium	Cd	iodine	I	phosphorus	P	titanium	Ti
calcium	Ca	iron	Fe	platinum	Pt	xenon	Xe
carbon	C	krypton	Kr	potassium	K	zinc	Zn

EXAMPLE EXERCISE 3.4

Supply the symbol for each of the following elements.

(a) lithium
(b) nitrogen
(c) chromium
(d) chlorine
(e) phosphorus
(f) arsenic

Solution: Refer to Table 3.4 as necessary for symbols you have not yet learned.
(a) Li; **(b)** N; **(c)** Cr; **(d)** Cl; **(e)** P; **(f)** As

SELF-TEST EXERCISE

Supply the name for each of the following elements.

(a) K
(b) F
(c) Kr
(d) Mn
(e) Ra
(f) Zn

Answers: **(a)** potassium; **(b)** fluorine; **(c)** krypton; **(d)** manganese; **(e)** radium; **(f)** zinc

3.5 Metals, Nonmetals, and Semimetals

To distinguish between the properties of metallic and nonmetallic elements.

To predict whether an element is a metal, a nonmetal, or a semimetal using the periodic table.

To predict the physical state of an element under normal conditions of temperature and atmospheric pressure using the periodic table.

Recall from the preceding sections that elements are the building blocks of matter. That is, all matter is composed of combinations of elements. Now you will see that different elements have different properties.

malleable The property of a metal that allows it to be beaten or rolled into a foil.

ductile The property of a metal that allows it to be drawn into a wire.

metal An element that is generally shiny in appearance, has a high density and high melting point, and is a good conductor of heat and electricity.

nonmetal An element that is generally dull in appearance, has a low density and low melting point, and is not a good conductor of heat and electricity.

semimetal An element that is generally metal-like in appearance and has properties that are between that of a metal and a nonmetal; also called a metalloid.

If an element can be hammered into a thin sheet, it is said to **malleable**. If it can be drawn into a fine wire, it is said to be **ductile**. Some elements are good conductors of heat and electricity. Elements having these properties are called **metals**. Although there are exceptions, metals typically have a high density, a high melting point, and a bright metallic luster. Aluminum, iron, and silver are familiar metals.

Some elements do not possess the above properties. They are not conductors of heat and electricity, and they crush to a powder if hammered. These elements have a dull appearance in the solid state. Several occur as gases under normal conditions. Elements having these properties are referred to as **nonmetals**. Examples of nonmetals are solid carbon and sulfur and gaseous hydrogen and oxygen. Table 3.5 lists general characteristics of metals and nonmetals.

Still other elements have properties in between the metals and nonmetals. They are referred to as **semimetals** or metalloids. Silicon and germanium are semimetals used in the semiconductor industry for making transistors and integrated circuits.

EXAMPLE EXERCISE 3.5

Which of the following properties is characteristic of a metal?

(a) good conductor of heat
(b) malleable
(c) high melting point
(d) reacts with other metals

Solution:

(a) Metals are good conductors of heat.
(b) Metals are malleable.
(c) Metals usually have high melting points.
(d) Metals do not react with other metals, but rather form homogeneous mixtures called alloys.

SELF-TEST EXERCISE

Which of the following properties is characteristic of a nonmetal?

(a) good conductor of electricity
(b) ductile
(c) high density
(d) reacts with nonmetals

Answer: **(d)** Nonmetals react with metals and nonmetals.

Table 3.5 General Characteristics of Elements*

Property	Metals	Nonmetals
physical state	solid	solid, gas
appearance	metallic luster	dull
pliability	malleable, ductile	brittle
conductivity	heat, electricity	nonconductor
density	usually high	usually low
melting point	usually high	usually low
chemical reactivity	react with nonmetals	react with metals and nonmetals

* These are general characteristics and there are exceptions. For example, the metals lithium and magnesium have low densities; gallium and cesium melt at room temperature.

Periodic Table of the Elements

Later we will devote an entire chapter to the systematic arrangement of the elements. For now, accept that each element is identified by an atomic number in a similar way that individuals are identified by a social security number. The number that identifies an element is different for each element. For example, the identifying number for hydrogen is 1, helium is 2, carbon is 6, silver is 47, and uranium is 92.

The elements are arranged numerically by atomic number and placed into a special table. This table is called the *periodic table of the elements*. In the periodic table of elements, the metals are on the left side and the nonmetals are on the right side. Hydrogen is an exception. Although it is a nonmetal, hydrogen has been placed in the center of the periodic tables that appear in this textbook. Semimetals separate the metals from the nonmetals. They lie along a stairstep in the middle of the periodic table. The semimetals include boron, silicon, germanium, arsenic, antimony, and tellurium. The radioactive elements polonium and astatine are also considered semimetals. Figure 3.6 shows the arrangement of metals, nonmetals, and semimetals in the periodic table.

Physical States of the Elements

The periodic table can help you master a large amount of information. With very little effort you can correctly predict the physical state of most elements. Excluding mercury, all the metals are in the solid state at normal conditions of 25°C and normal atmospheric pressure. All the semimetals are also in the solid state at normal temperature and pressure.

Nonmetals, on the other hand, show great diversity in physical state. At 25°C and normal pressure, 5 nonmetals are solids, bromine is a liquid, and 11 are gases.

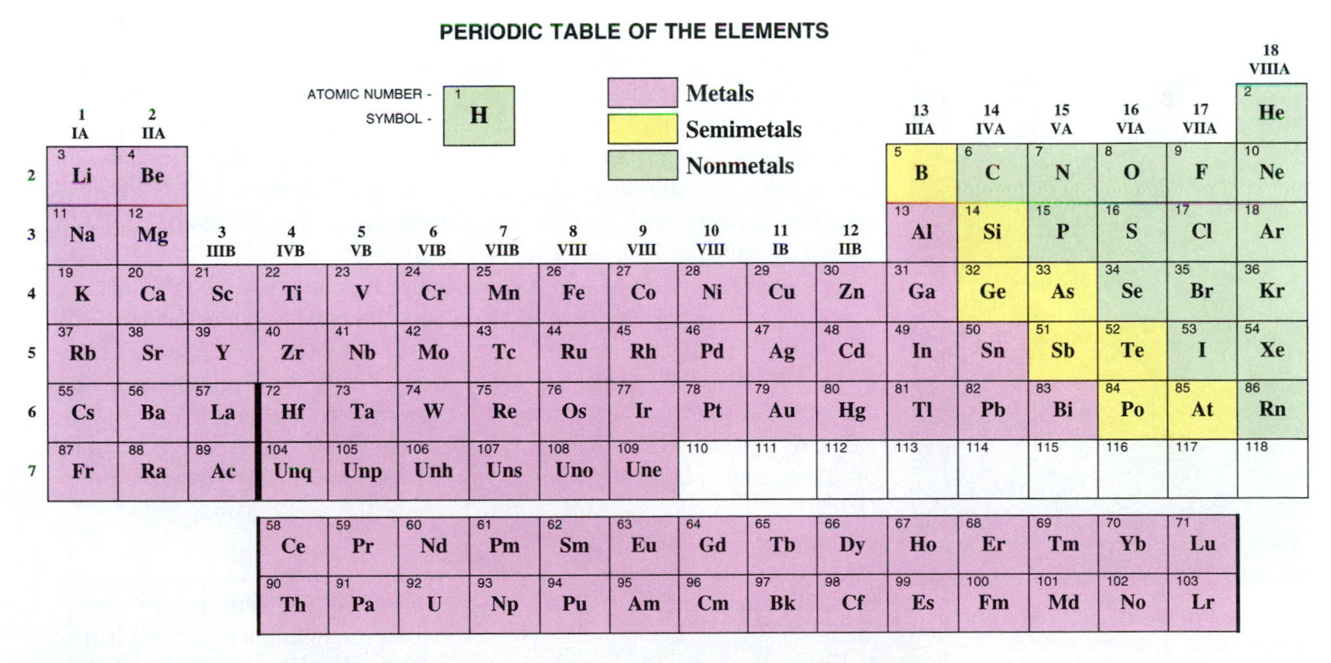

Figure 3.6 Metals, Nonmetals, and Semimetals Arrangement of metals, semimetals, and nonmetals in the periodic table.

The 11 gases are hydrogen, nitrogen, oxygen, fluorine, chlorine, helium, neon, argon, krypton, xenon, and radon. All these gases are colorless except fluorine and chlorine, which are greenish-yellow. Figure 3.7 illustrates the physical states of the elements.

UPDATE

Elements 104 and Beyond

▶ **Can you guess the element named for the English scientist who discovered that an atom has a nucleus?**

Currently, there are 109 elements that have either been discovered in nature or synthesized in the laboratory. Of the 109 elements, those with atomic number 93, or greater, have been artificially produced from the high-energy collision of simpler elements.

In 1964 a Russian team claimed to be the first to synthesize element 104. They named the element kurchatovium (Ku) in honor of Igor Kurchatov, a Soviet physicist. American scientists at the University of California, Berkeley, attempted to reproduce the Russian results, but were unsuccessful. In 1967, however, the American group synthesized element 104 by a different method. The Americans proposed the name rutherfordium (Rf) after the English physicist Ernest Rutherford. An argument ensued over the discovery and name of element 104 which still persists.

Elements 104, 105, and 106 were synthesized using this particle accelerator at the University of California, Berkeley.

IUPAC Names for Elements 104 and Beyond

Element	IUPAC Name*	Symbol
104	unnilquadium	Unq
105	unnilpentium	Unp
106	unnilhexium	Unh
107	unnilseptium	Uns
108	unniloctium	Uno
109	unnilennium	Une

* Prefix roots: nil = 0, un = 1, bi = 2, tri = 3, quad = 4, pent = 5, hex = 6, sept =7, oct = 8, enn = 9

A few years later, a similar controversy arose over the discovery of element 105. As a result, element 105 also has both Russian and American names and symbols. Recently, the International Union of Pure and Applied Chemistry (IUPAC) decided to resolve the controversy by suggesting systematic names for elements 104 and beyond. According to IUPAC, the name is to be constructed from Latin prefixes for the number of the element, plus an -ium suffix.

The name and symbol for element 104, for example, is derived from the Latin prefix un = 1, nil = 0, and quad = 4. The name of element 104 is formed as follows: un + nil + quad + ium, or unnilquadium. A symbol for the element is composed from the first letter of each prefix. For unnilquadium, the symbol is Unq.

Scientists have predicted that element 114 should be unusually stable and have properties similar to lead. If element 114 is discovered, what would be its IUPAC name and symbol? The IUPAC name is composed of the roots of the number 114. Thus, element 114 is named as follows: un + un + quad + ium = ununquadium. The three letter symbol for **ununquadium** is **Uuq**. The table above lists IUPAC names and symbols for elements 104 and beyond.

▶ **Element 104 was named rutherfordium in honor of Ernest Rutherford who discovered the atomic nucleus in 1911.**

PERIODIC TABLE OF THE ELEMENTS

ATOMIC NUMBER - 1
SYMBOL - H

Solids
Liquids
Gases

	1 IA	2 IIA	3 IIIB	4 IVB	5 VB	6 VIB	7 VIIB	8 VIII	9 VIII	10 VIII	11 IB	12 IIB	13 IIIA	14 IVA	15 VA	16 VIA	17 VIIA	18 VIIIA
																		2 He
2	3 Li	4 Be											5 B	6 C	7 N	8 O	9 F	10 Ne
3	11 Na	12 Mg											13 Al	14 Si	15 P	16 S	17 Cl	18 Ar
4	19 K	20 Ca	21 Sc	22 Ti	23 V	24 Cr	25 Mn	26 Fe	27 Co	28 Ni	29 Cu	30 Zn	31 Ga	32 Ge	33 As	34 Se	35 Br	36 Kr
5	37 Rb	38 Sr	39 Y	40 Zr	41 Nb	42 Mo	43 Tc	44 Ru	45 Rh	46 Pd	47 Ag	48 Cd	49 In	50 Sn	51 Sb	52 Te	53 I	54 Xe
6	55 Cs	56 Ba	57 La	72 Hf	73 Ta	74 W	75 Re	76 Os	77 Ir	78 Pt	79 Au	80 Hg	81 Tl	82 Pb	83 Bi	84 Po	85 At	86 Rn
7	87 Fr	88 Ra	89 Ac	104 Unq	105 Unp	106 Unh	107 Uns	108 Uno	109 Une	110	111	112	113	114	115	116	117	118

58 Ce	59 Pr	60 Nd	61 Pm	62 Sm	63 Eu	64 Gd	65 Tb	66 Dy	67 Ho	68 Er	69 Tm	70 Yb	71 Lu
90 Th	Pa	92 U	93 Np	94 Pu	95 Am	96 Cm	97 Bk	98 Cf	99 Es	100 Fm	101 Md	102 No	103 Lr

Figure 3.7 Physical States of the Elements At normal conditions, all of the metals are in the solid state except Hg. Most of the nonmetals are gases except C, P, S, Se, and I, which are solids. The only two elements in the liquid state are Hg and Br.

Consider the element magnesium. If you examine the periodic table of elements in Figure 3.6, you observe that Mg is a metal. The illustration of magnesium in Figure 3.7 shows that it is in the solid physical state.

EXAMPLE EXERCISE 3.6

Classify each of the following elements as a metal, nonmetal, or semimetal and indicate its physical state at 25°C and normal atmospheric pressure.

(a) barium
(b) boron
(c) bismuth
(d) bromine

Solution: Referring to Figures 3.6 and 3.7 we see that:

(a) Ba is on the left side of the periodic table. It is a *solid metal* at normal conditions.
(b) B is found on the stairstep in the periodic table; it is a *solid semimetal*.
(c) Bi is below the stairstep in the periodic table. It is a *solid metal* at normal conditions.
(d) Br is on the right side of the periodic table. It is a *liquid nonmetal* at normal conditions.

SELF-TEST EXERCISE

Classify each of the following elements as a metal, nonmetal, or semimetal and indicate the physical state at 25°C and normal atmospheric pressure.

(a) aluminum
(b) hydrogen
(c) helium
(d) arsenic

Answers:
(a) solid metal; **(b)** gaseous nonmetal; **(c)** gaseous nonmetal; **(d)** solid semimetal

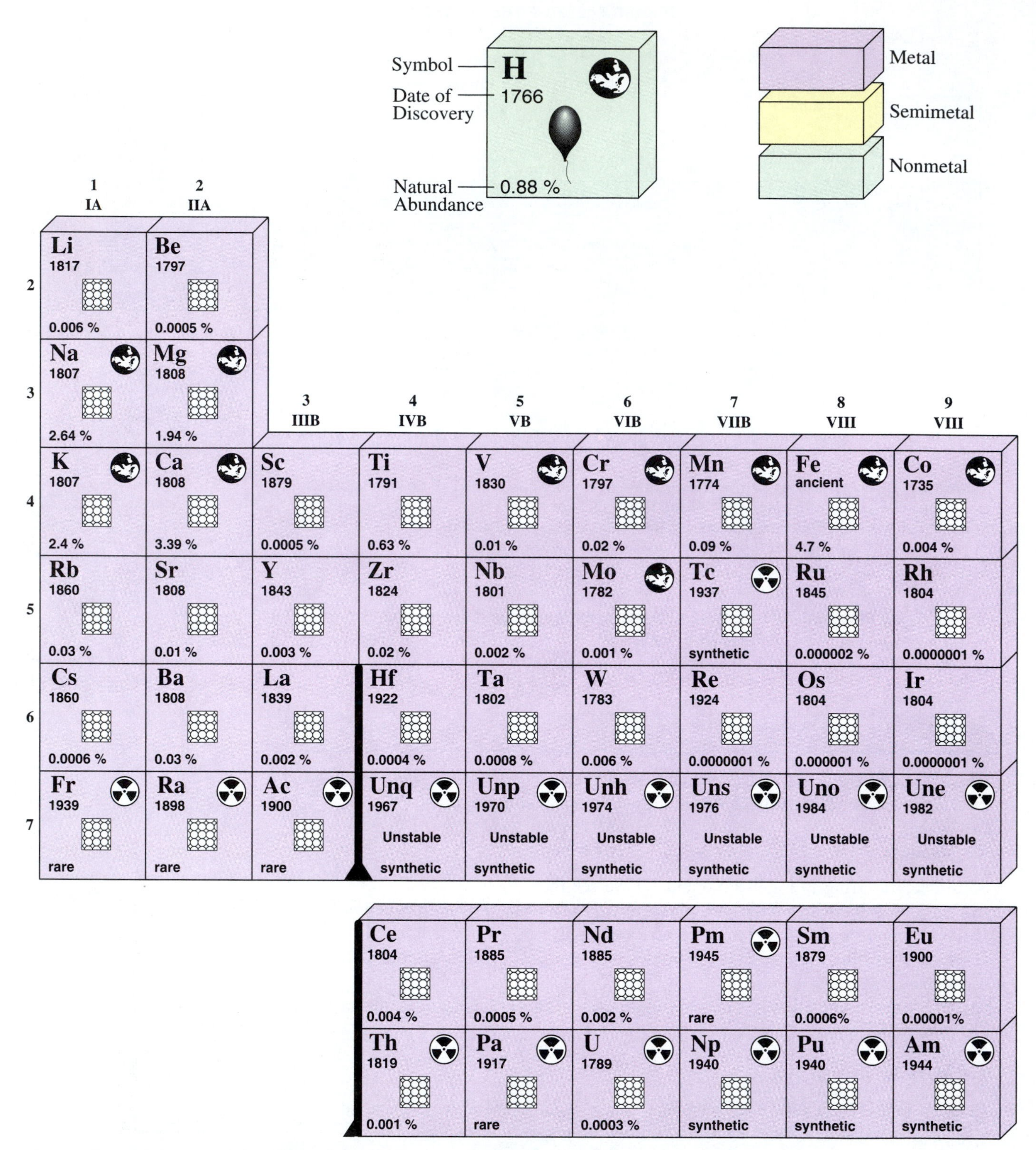

Figure 3.8 Pictorial Periodic Table of Elements The natural abundance is the percent by mass of the elements in the earth's crust, oceans, and atmosphere. The natural abundance of an element listed as rare is less than 1 mg per metric ton (1000 kg). An element listed as synthetic is made artificially and does not occur naturally. An element listed as unstable disintegrates in a fraction of a second.

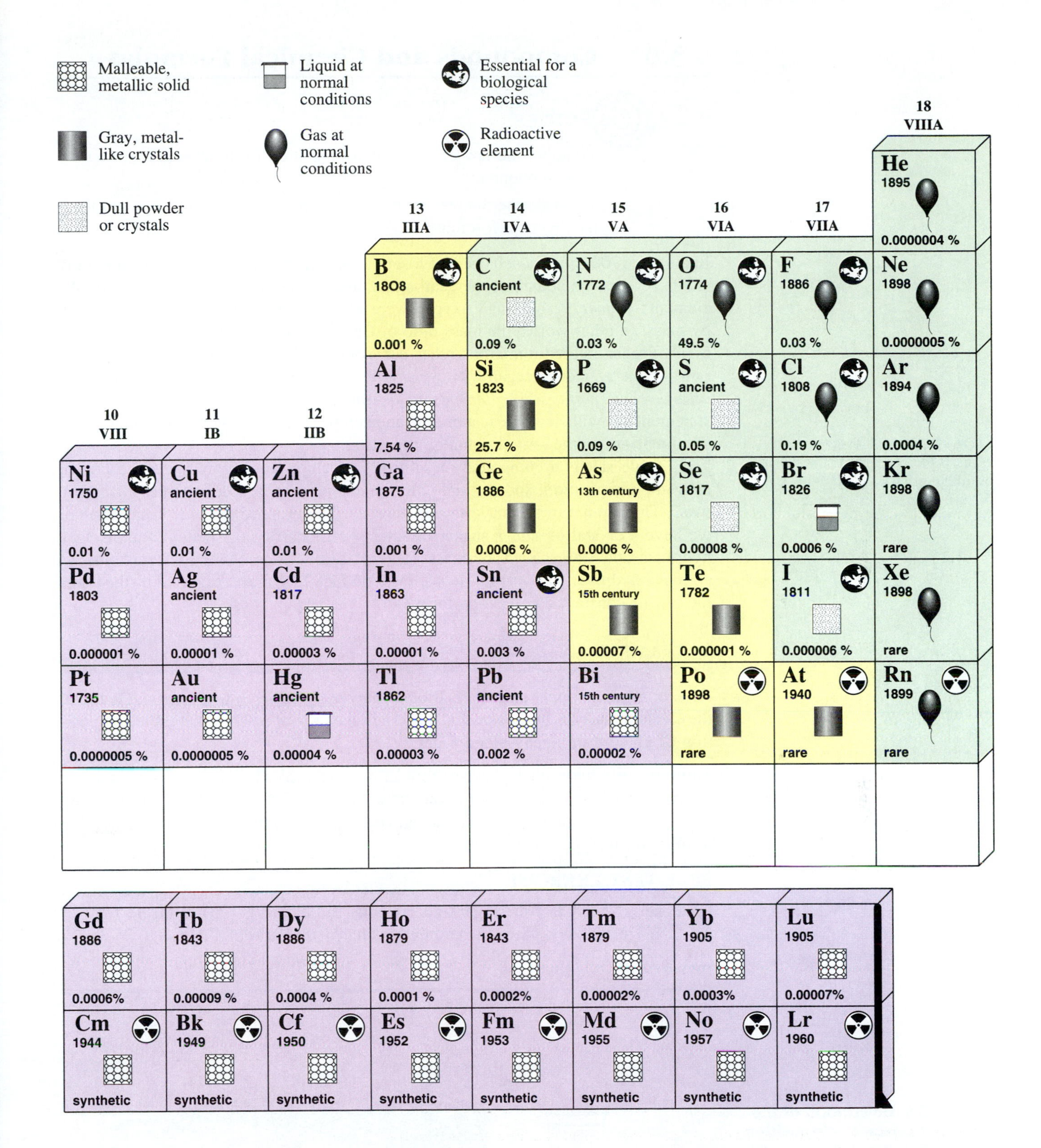

Malleable, metallic solid
Gray, metal-like crystals
Dull powder or crystals
Liquid at normal conditions
Gas at normal conditions
Essential for a biological species
Radioactive element
10 VIII
11 IB
12 IIB
13 IIIA
14 IVA
15 VA
16 VIA
17 VIIA
18 VIIIA
He 1895 0.0000004 %
B 1808 0.001 %
C ancient 0.09 %
N 1772 0.03 %
O 1774 49.5 %
F 1886 0.03 %
Ne 1898 0.0000005 %
Al 1825 7.54 %
Si 1823 25.7 %
P 1669 0.09 %
S ancient 0.05 %
Cl 1808 0.19 %
Ar 1894 0.0004 %
Ni 1750 0.01 %
Cu ancient 0.01 %
Zn ancient 0.01 %
Ga 1875 0.001 %
Ge 1886 0.0006 %
As 13th century 0.0006 %
Se 1817 0.00008 %
Br 1826 0.0006 %
Kr 1898 rare
Pd 1803 0.000001 %
Ag ancient 0.00001 %
Cd 1817 0.00003 %
In 1863 0.00001 %
Sn ancient 0.003 %
Sb 15th century 0.00007 %
Te 1782 0.000001 %
I 1811 0.000006 %
Xe 1898 rare
Pt 1735 0.0000005 %
Au ancient 0.0000005 %
Hg ancient 0.00004 %
Tl 1862 0.00003 %
Pb ancient 0.002 %
Bi 15th century 0.00002 %
Po 1898 rare
At 1940 rare
Rn 1899 rare
Gd 1886 0.0006%
Tb 1843 0.00009 %
Dy 1886 0.0004 %
Ho 1879 0.0001 %
Er 1843 0.0002%
Tm 1879 0.00002%
Yb 1905 0.0003%
Lu 1905 0.00007%
Cm 1944 synthetic
Bk 1949 synthetic
Cf 1950 synthetic
Es 1952 synthetic
Fm 1953 synthetic
Md 1955 synthetic
No 1957 synthetic
Lr 1960 synthetic

3.6 Compounds and Chemical Formulas

OBJECTIVES

To apply the law of constant composition to the chemical formula of a compound.

To relate the number of atoms of each element in a compound to its chemical formula.

In the late 1700s, the French chemist Joseph Louis Proust (1754–1826) painstakingly analyzed the compound copper carbonate. No matter how it was prepared, the elements copper, carbon, and oxygen were always present in the same proportion by mass. In other words, the mass ratio of Cu:C:O was constant, and that ratio was 5:1:4. Proust studied numerous other compounds and always obtained a constant ratio of the elements, regardless of sample size. In 1799, he stated, "*Compounds always contain the same elements in a constant proportion by mass.*" Proust's statement has withstood the test of time and is now referred to as the **law of constant composition**, or the law of definite proportion.

law of constant composition The principle which states that a compound always contains the same elements in the same proportion by mass.

Table salt is a compound of sodium and chlorine. According to the law of constant composition, sodium and chlorine are present in a definite proportion by mass. The amount of salt makes no difference. The proportion is the same whether we have a crystal of salt, a shaker of salt, or a block of salt. Water contains 11% hydrogen and 89% oxygen by mass. Regardless of the amount of water, it always contains hydrogen and oxygen in the ratio of 11:89 by mass. Figure 3.9 illustrates the law of constant composition for water.

EXAMPLE EXERCISE 3.7

After analyzing copper carbonate, Proust analyzed glucose in grape sugar. He found the elements carbon, hydrogen, and oxygen in the ratio of 6:1:8. What is the composition of glucose in human blood sugar?

Solution: We know that glucose from grape sugar contains the elements carbon, hydrogen, and oxygen (C:H:O) in the mass ratio of 6:1:8. From the law of constant composition, we can predict that glucose from blood sugar must also contain C:H:O in the ratio of 6:1:8 by mass.

SELF-TEST EXERCISE

Fructose is found in fruit sugar and contains the elements C:H:O in the ratio of 6:1:8. If fructose is obtained from honey, what is the ratio of C:H:O?

Answer: 6:1:8 by mass

Figure 3.9 Illustration of the Law of Constant Composition A drop of water, a glass of water, and a lake of water all contain hydrogen and oxygen in the same proportion by mass; that is, 11% H and 89% O.

Chemical Formulas

We have seen how chemists use symbols for the names of elements. In a similar manner, they use chemical formulas for the names of compounds. **Chemical formulas** symbolically express the number of atoms of each element in a compound. The number of atoms is indicated with a subscript to the right of the element's symbol, unless the subscript is 1; subscripts of 1 are omitted for simplicity. For example, water contains two atoms of hydrogen and one atom of oxygen. Its chemical formula is therefore H_2O. Ammonia has one atom of nitrogen and three atoms of hydrogen; its chemical formula is NH_3. The chemical formula of sulfuric acid, H_2SO_4, is interpreted in Figure 3.10.

chemical formula An abbreviation for the name of a chemical compound that indicates the number of atoms of each element; for example H_2O is the chemical formula for water.

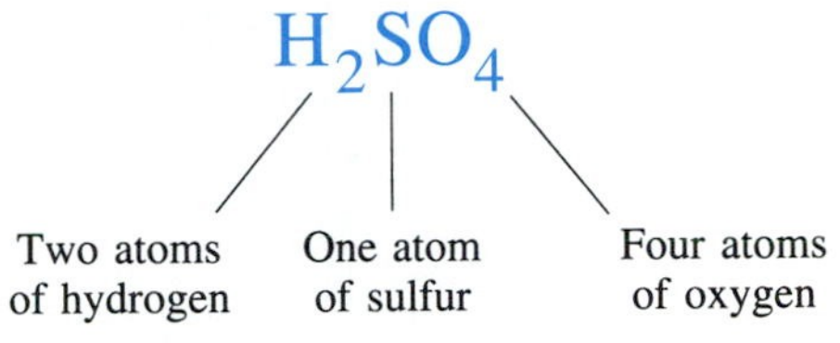

Figure 3.10 Interpretation of a Chemical Formula In the chemical formula for sulfuric acid, H_2SO_4, there are 2 atoms of hydrogen (H), 1 atom of sulfur (S), and 4 atoms of oxygen (O). Sulfuric acid is a strong corrosive liquid found in automobile batteries.

EXAMPLE EXERCISE 3.8

State the composition of atoms in the following formulas.

(a) chloroform, $CHCl_3$ **(b)** vitamin B_3, $C_6H_6N_2O$

Solution:

(a) Chloroform's formula contains 1 atom of carbon, 1 atom of hydrogen, and 3 atoms of chlorine; thus, $CHCl_3$ represents a total of 5 atoms.

(b) The formula of vitamin B_3 contains 6 carbon atoms, 6 hydrogen atoms, 2 nitrogen atoms, and 1 oxygen atom; thus, $C_6H_6N_2O$ represents a total of 15 atoms.

SELF-TEST EXERCISE

Write the chemical formula for the following compounds, given their composition of atoms.

(a) Freon-12; 1 carbon atom, 2 chlorine atoms, 2 fluorine atoms

(b) vitamin B_6: 8 carbon atoms, 11 hydrogen atoms, 1 nitrogen atom, 3 oxygen atoms

Answers: **(a)** CCl_2F_2; **(b)** $C_8H_{11}NO_3$

Some chemical formulas use parentheses to make them clear. Antifreeze is a solution of ethylene glycol; it has the chemical formula $C_2H_4(OH)_2$. The formula is composed of 2 atoms of carbon, 4 atoms of hydrogen, and 2 units of OH. The parentheses emphasize that OH is a fundamental part in the formula and is so written in the chemical formula. The total number of atoms are, therefore, 2 atoms of carbons, 6 atoms of hydrogen, and 2 atoms of oxygen. Figure 3.11 illustrates the chemical formula for the explosive trinitrotoluene (TNT), $C_7H_5(NO_2)_3$.

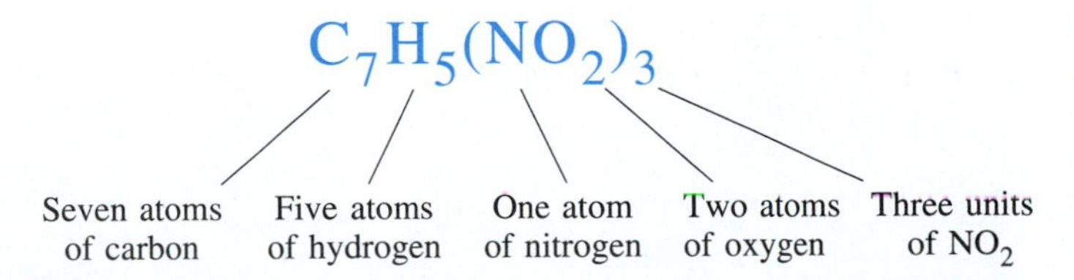

Figure 3.11 Interpretation of a Chemical Formula In TNT there are 7 carbon atoms, 5 hydrogen atoms, and 3 NO_2 units. The nitrogen and two oxygen atoms are a single unit, which appear three times in the chemical formula.

EXAMPLE EXERCISE 3.9

State the composition of atoms in the following chemical formulas.

(a) glycerine, $C_3H_5(OH)_3$ **(b)** nitroglycerine, $C_3H_5O_3(NO_2)_3$

Solution:

(a) The formula for glycerine contains 3 carbon atoms, 5 hydrogen atoms, and 3 OH units. Thus, $C_3H_5(OH)_3$ has a total of 14 atoms.

(b) The formula for nitroglycerine contains 3 carbon atoms, 5 hydrogen atoms, 3 oxygen atoms, and 3 NO_2 units. Thus, $C_3H_5O_3(NO_2)_3$ has a total of 20 atoms.

SELF-TEST EXERCISE

Write the chemical formula for the following compounds given their composition of atoms.

(a) resorcinol: 6 carbon atoms, 4 hydrogen atoms, and 2 (OH) units

(b) xylene: 6 carbon atoms, 4 hydrogen atoms, and 2 (CH_3) units

Answers: **(a)** $C_6H_4(OH)_2$; **(b)** $C_6H_4(CH_3)_2$

Note: In Chapter 6, you will learn to name compounds systematically. This task is made easier if you begin simply by pronouncing the chemical formula. Even chemists refer to compounds by their formulas. For example, water has the formula H_2O and is verbalized as "*H two O*." Sulfuric acid is H_2SO_4 and is verbalized as "*H two, S, O four*." Glucose, $C_6H_{12}O_6$, is verbalized as "*C six, H twelve, O six*."

3.7 Physical and Chemical Properties

OBJECTIVES

To classify a property of a substance as physical or chemical.

Previously, we classified matter as either a substance or a mixture. Recall that substances have consistent physical and chemical properties regardless of the sample size. In addition, a substance may occur as an element or a compound. Finally, recall that elements combine chemically to form compounds.

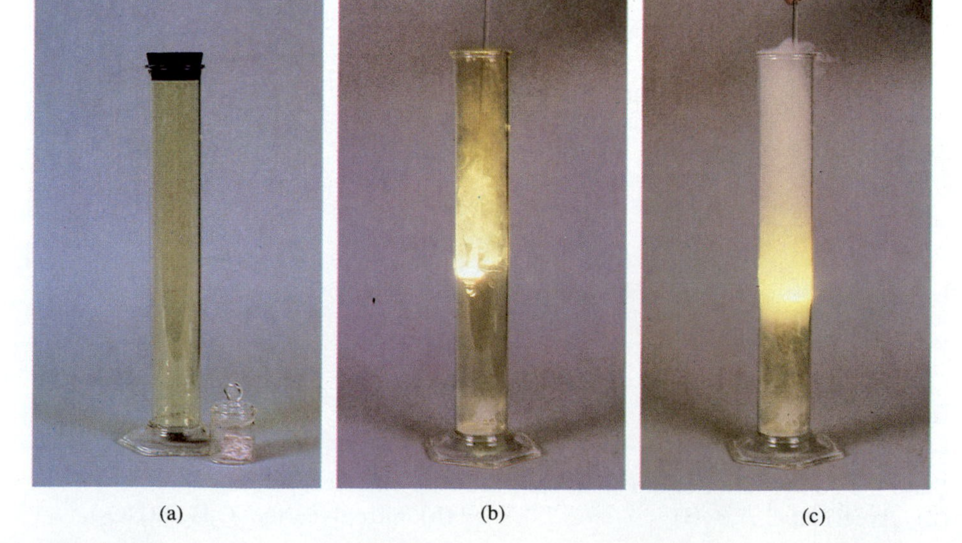

(a) (b) (c)

Figure 3.12 Sodium, Chlorine, and Sodium Chloride (a) Chlorine, a yellow poisonous gas, is shown in the tall cylinder next to a small piece of sodium metal. (b) Sodium and chlorine react to give sodium chloride and (c) a few minutes later, the tall cylinder contains smoke and table salt.

Table 3.6 Physical Properties of Sodium, Chlorine, and Sodium Chloride at Normal Temperature and Pressure

Property	Sodium	Chlorine	Sodium Chloride
appearance	silver metal	yellowish gas	white crystals
density	0.97 g/cm^3	2.90 g/L	2.165 g/cm^3
melting point	97.8°C	−101.0°C	801°C
boiling point	882.9°C	−34.6°C	1413°C
solubility in 100 g of water	reacts with water	0.51 g at 30°C	35.7 g at 0°C

The gold foil (left) and copper wire (right) illustrate respectively the malleability and ductility of metals.

The physical and chemical properties of a compound bear no resemblance to the properties of its elements. For example, sodium is a soft reactive metal, and chlorine is a toxic greenish-yellow gas. Yet these two elements combine to form white crystalline table salt (Figure 3.12). In the laboratory, it is possible to identify an element or compound. This is because no two substances have the same physical and chemical properties. Some elements and compounds have certain properties in common, but no two elements or compounds have all properties that are identical.

We have been talking about the physical and chemical properties of substances. Now let's formally define what those properties are. The **physical properties** of a substance refer to those characteristics we can observe without the substance changing composition. The list of physical properties is quite long. Those properties usually considered important are physical state (solid, liquid, gas), color, density, crystalline form, melting point, boiling point, electrical conductivity, heat conductivity, solubility in water, malleability, ductility, and hardness. Odor is usually classified as a physical property. Table 3.6 shows the physical properties of sodium, chlorine, and sodium chloride at room temperature and atmospheric pressure.

physical property A property that can be observed without changing the chemical formula of a substance.

The **chemical properties** of a substance describe its reactions with other substances. The chemical properties of sodium, for example, include its reaction with oxygen to form sodium oxide and its reaction with water to produce hydrogen gas. There are many reactions for the elements, more than we can possibly memorize. We can, however, use the periodic table of elements to predict chemical behavior. The periodic table is arranged according to families of elements that, in general, give similar reactions. In Chapter 5, we will study families of elements. For now, Table 3.7 illustrates how elements in the same families have similar chemical properties.

chemical property A property of a substance that cannot be observed without changing the chemical formula of the substance.

The distinction between a physical property and chemical property is clear. Physical properties are observed without altering the composition of the substance. Chemical properties, on the other hand, always involve a chemical reaction. When a chemical reaction occurs, the composition of a substance changes, and other substances are formed.

EXAMPLE EXERCISE 3.10

Classify the following properties of water as physical or chemical.

(a) is a colorless, odorless liquid at 20°C
(b) dissolves table sugar
(c) reacts with sodium metal
(d) occupies a 1-mL volume for each gram

Solution: If a reaction occurs, there is a change in the chemical formula and the property is chemical. Otherwise, the property is physical.
(a) Color and odor are physical properties.
(b) Solubility is a physical property.
(c) A chemical reaction is a chemical property.
(d) Density is a physical property.

SELF-TEST EXERCISE

Classify the following properties of water as physical or chemical.
(a) produces a gas with calcium metal
(b) is a colorless, odorless gas at 100°C
(c) is a colorless crystalline solid at 0°C
(d) is insoluble in ether
(e) does not conduct electricity
(f) combines with sulfur trioxide gas to give sulfuric acid

Answers: **(a)** chemical; **(b)** physical; **(c)** physical; **(d)** physical; **(e)** physical; **(f)** chemical

Table 3.7 Chemical Properties for Families of Elements at Normal Temperature and Pressure

Element Reacts with:	Oxygen	Water	Hydrochloric Acid
Family IA/1			
lithium	Li_2O	H_2 + LiOH	H_2 + LiCl
sodium	Na_2O	H_2 + NaOH	H_2 + NaCl
potassium	K_2O	H_2 + KOH	H_2 + KCl
Family IIA/2			
calcium	CaO	H_2 + $Ca(OH)_2$	H_2 + $CaCl_2$
strontium	SrO	H_2 + $Sr(OH)_2$	H_2 + $SrCl_2$
barium	BaO	H_2 + $Ba(OH)_2$	H_2 + $BaCl_2$
Family IVA/14			
carbon	CO_2	NR	NR
silicon	SiO_2	NR	NR
germanium	GeO_2	NR	NR
Family VIIIA/18			
helium	NR	NR	NR
neon	NR	NR	NR
argon	NR	NR	NR

* NR is an abbreviation for no reaction.

Note

There is an analogy between the properties of substances and the properties of people. People have physical properties, such as height, weight, and hair and eye color, that we can observe without changing the person. We can, however, only study the behavior of people when they interact with other people. In a similar fashion, we can observe physical properties of substances without changing the substance. In general, you will study the chemical behavior of a substance by observing the substance interacting with another substance.

3.8 Physical and Chemical Changes

OBJECTIVES

To classify a change in a substance as physical or chemical.

As we have been discussing, when a substance is altered, the change is classified as physical or chemical. Another way to distinguish the two changes is to say that **chemical changes** alter the formula of a substance; **physical changes** do not. Hammering gold into a thin foil, melting table salt, boiling wood alcohol, and dissolving table sugar in water are physical changes. The gold, salt, alcohol, and sugar all retain their same chemical formulas. In other words, the formulas of these substances (Au, NaCl, CH_3OH, and $C_{12}H_{22}O_{11}$) are identical before and after the physical change. Thus, changing shape, changing physical state (melting, boiling), and dissolving are all observations that indicate a physical change.

chemical change The process of undergoing a change in chemical formula or chemical composition.

physical change The process of undergoing a change without altering the chemical formula of a substance.

Chemical changes involve a chemical reaction in which a new substance is produced. The products of a chemical reaction have different formulas and properties than the starting substances. Gasoline burning, iron rusting, and an antacid tablet fizzing are examples of a chemical change. The gasoline, iron, and antacid tablet produce new substances from their reactions. Observations for a chemical change include the oxidation of matter (rapid burning or slow rusting); the release of gas bubbles (effervescing, that is, fizzing); the formation of an insoluble solid in solution (precipitating); the release of heat or light; and a change in color or odor. Figure 3.13 illustrates some chemical changes.

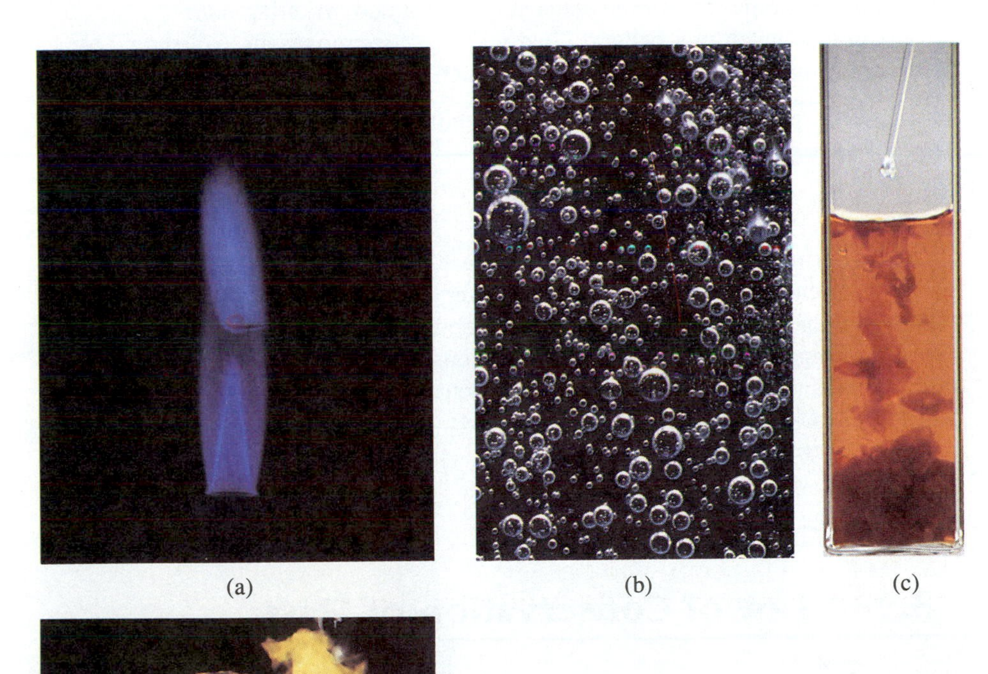

(a) (b) (c) (d)

Figure 3.13 Examples of Chemical Change Evidence of a chemical change includes (a) burning, (b) fizzing, (c) precipitating, and (d) exploding.

EXAMPLE EXERCISE 3.11

Classify each of the following observations as a physical change or a chemical change.

(a) burning sulfur gives a light blue flame
(b) heating water in a flask produces moisture on the glass
(c) adding together two colorless solutions gives a yellow solid
(d) dissolving white crystals in water gives a colorless solution
(e) pouring vinegar onto baking soda produces gas bubbles

Solution: The observations that indicate a physical change include changing shape, volume, or physical state. The observations that suggest a substance is undergoing a chemical change include burning, fizzing, precipitating, and changing color permanently.

(a) Sulfur is burned; thus, the change is chemical.
(b) Water is boiled; thus, the change is physical.
(c) Two solutions give a precipitate; thus, the change is chemical.
(d) Crystals are dissolved; thus, the change is physical.
(e) Baking soda fizzes; thus, the change is chemical.

SELF-TEST EXERCISE

Classify each of the following observations as a physical change or a chemical change.

(a) freezing water in a refrigerator to make cubes of ice
(b) adding silver nitrate solution to tap water gives a cloudy solution
(c) adding ammonia to chlorine water produces a new irritating odor
(d) touching a match to a balloon filled with hydrogen gives an explosion
(e) grinding aspirin tablets produces a powder

Answers: **(a)** physical; **(b)** chemical; **(c)** chemical; **(d)** chemical; **(e)** physical

Note: For reasons of simplicity, we have classified changes as either physical or chemical. Physical changes never involve a chemical change, that is, a change in chemical formula. Chemical changes, on the other hand, are always accompanied by a physical change. Consider the rusting of iron. Rusting not only involves a change in the formula of iron, but also a change in its color and density. Therefore, when we classify a change as chemical, it is understood that a physical change has also occurred.

3.9 Law of Conservation of Mass

To apply the conservation of mass law to a chemical reaction.

In 1789, the French chemist Antoine Lavoisier (1743–1794) announced the conservation of mass principle. Through careful laboratory experiments, Lavoisier found that the mass of substances before a chemical change was always equal to the mass of substances after the change. He concluded that matter was neither created nor destroyed during a chemical reaction. This principle has become known as the **law of conservation of mass**.

law of conservation of mass The principle which states that mass can neither be created nor destroyed.

As an example of conservation of mass, consider the reaction of hydrogen and oxygen. Hydrogen and oxygen always combine in the same 1:8 ratio by mass to form water. That is, 1 g of hydrogen combines with 8 g of oxygen; 2 g of hydrogen combines with 16 g of oxygen; 3 g of hydrogen reacts with 24 g of oxygen; and so on. We can predict the mass of water produced from each of the above-mentioned reactions. From the conservation of mass law, we know that the mass of hydrogen plus the mass of oxygen must equal the mass of water produced. That is,

$$1.0 \text{ g hydrogen} + 8.0 \text{ g oxygen} = 9.0 \text{ g water}$$

Conversely, we can predict the masses of hydrogen and oxygen produced from the decomposition of water. Passing an electric current through water produces hydrogen gas and oxygen gas. This process is called electrolysis. If the electrolysis of 45.0 g of water produces 5.0 g of hydrogen, how many grams of oxygen are evolved?

The conservation of mass law states that the mass of hydrogen and oxygen is equal to the mass of the water, 45.0 g. Therefore, the mass of oxygen must equal the mass of the water minus the mass of the hydrogen. That is,

$$45.0 \text{ g water} - 5.0 \text{ g hydrogen} = 40.0 \text{ g oxygen}$$

EXAMPLE EXERCISE 3.12

A flashbulb weighs 11.250 g and contains magnesium metal and oxygen gas. The metal and gas are ignited electrically to produce magnesium oxide, heat, and a flash of light. What is the mass of the flashbulb after the reaction?

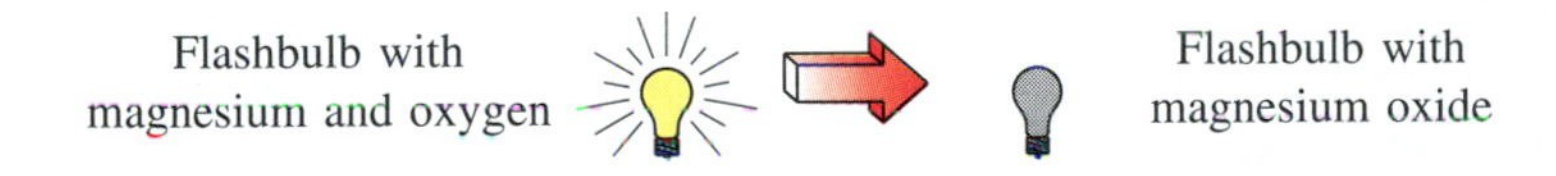

Solution: The conservation of mass law states that the mass of products is equal to the mass of reactants. Since the reaction takes place inside a sealed flashbulb, the mass is the same after the reaction, 11.250 g.

SELF-TEST EXERCISE

A coil of magnesium ribbon has a mass of 0.243 g. The ribbon is ignited in air to produce 0.403 g of magnesium oxide. Calculate the mass of oxygen in the air that reacts according to the conservation of mass law.

Answer: 0.160 g of oxygen

Note The law of conservation of mass is one of the most important principles in chemistry. Historically, it was used to develop the first table of atomic weights and to determine the formulas of compounds. As we will discuss in Chapter 9, it is the cornerstone for all quantitative relationships in chemistry.

3.10 Law of Conservation of Energy

To distinguish between potential energy and kinetic energy.

To identify forms of energy, including heat, light, chemical, electrical, mechanical, and nuclear energy.

To apply the law of conservation of energy to a chemical reaction.

law of conservation of energy The principle which states that energy can neither be created nor destroyed.

potential energy The stored energy that matter possesses owing to its position or chemical composition.

kinetic energy The energy associated with the mass and velocity of a particle. The kinetic energy of a particle is equal to one-half its mass times its velocity squared.

In the middle of the 1800s, a principle known as the conservation of energy gradually became recognized. The German physicist Hermann Helmholtz (1821–1894), the English physicist James Joule (1818–1889), and others contributed to its recognition. It was stated in different ways, but essentially it said that energy cannot be created or destroyed. Energy can, however, be converted from one form into another. The principle is known as the **law of conservation of energy**. Furthermore, this principle applies to a chemical change. The energy of all the substances before a reaction is the same as the total energy of the new substances after a reaction.

There are two types of energy, potential energy and kinetic energy. They are related to each other as follows. **Potential energy** (PE) is stored energy that matter has as a result of its position or chemical composition. **Kinetic energy** (KE), as we said earlier, is the energy associated with motion. A boulder perched at the top of a mountain has potential energy. If the boulder rolls down the mountain, it loses potential energy but gains kinetic energy. A sample of uranium has potential energy. As it disintegrates, it emits radioactive particles moving at high velocities. Thus, in the process of disintegration, potential energy is converted to kinetic energy. Figure 3.14 illustrates the change from potential energy to kinetic energy. Since energy cannot be created or destroyed, the total of potential and kinetic energy remains constant for every physical and chemical change.

To see how the conservation of energy law works, consider the properties of water. When water is boiled at 100°C, energy is required. Since energy can neither be created nor destroyed, it follows that the same amount of heat energy used for boiling is released when the steam condenses to water at 100°C. It requires 540 calories of heat to convert 1 g of water at 100°C to steam at 100°C. It therefore follows that 540 calories of heat energy are released when 1 g of steam condenses to liquid water at 100°C.

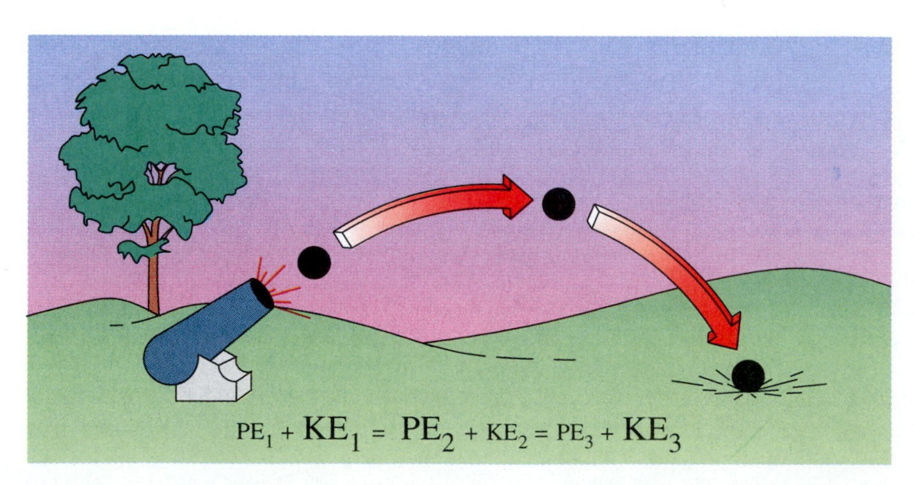

$PE_1 + KE_1 = PE_2 + KE_2 = PE_3 + KE_3$

Figure 3.14 Relationship of Potential and Kinetic Energy A cannonball shot into the air loses kinetic energy while gaining potential energy. At the top of its flight, PE is at a maximum and the KE is at a minimum. As it falls to earth, the potential energy decreases as the kinetic energy increases.

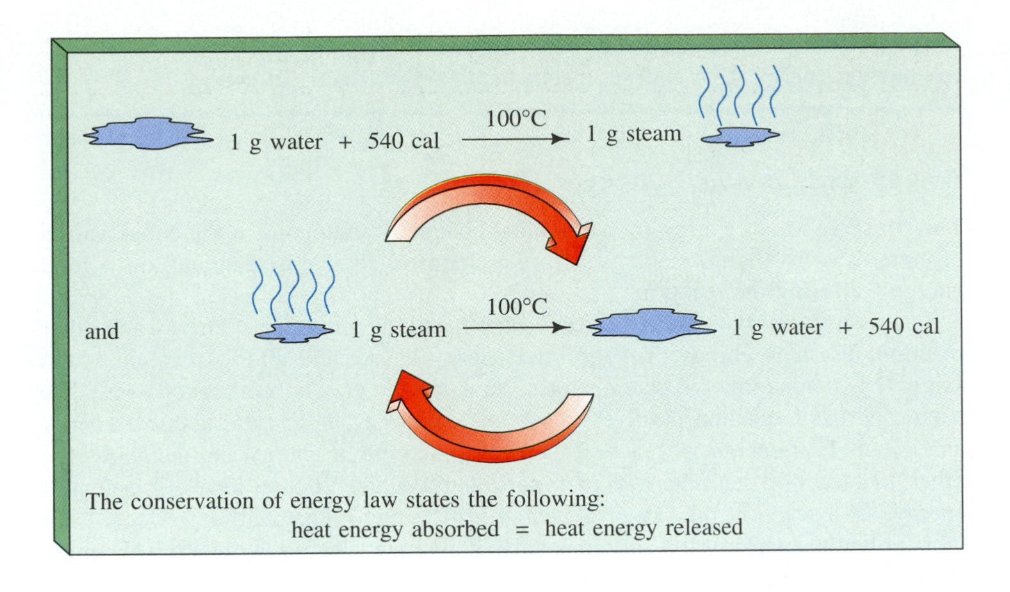

Forms of Energy

There are six forms of energy: heat, light, chemical, electrical, mechanical, and nuclear. Furthermore, each of these forms of energy can be converted into another form. For example, solar energy panels convert light energy from the sun to heat energy, which we can store as hot water. Solar energy can also be converted to electrical energy using photocells, but currently this process is only 15% efficient. Mechanical energy can also be converted to electrical energy, which is a much more efficient process. For example, the mechanical energy of water falling over a dam can be used to drive a turbine, which is turn drives an electrical generator. A hydroelectric plant operates at nearly 90% efficiency.

EXAMPLE EXERCISE 3.13

Classify two forms of energy involved in each of the following processes.
(a) radioactive uranium converts water into steam
(b) steam drives a turbine
(c) turbine revolves to drive an electrical generator
(d) electrical generator recharges a lead storage battery

Solution: Refer to the six forms of energy listed above.
(a) nuclear energy is changed into heat energy
(b) heat energy is changed into mechanical energy
(c) mechanical energy is changed into electrical energy
(d) electrical energy is changed into chemical energy

SELF-TEST EXERCISE

State at least two forms of energy involved in operating each of the following.
(a) flashlight
(b) solar calculator
(c) electric golf cart
(d) nuclear power plant

Answers:
(a) chemical energy, light energy
(b) light energy, electrical energy

(c) chemical energy, electrical energy, and mechanical energy
(d) nuclear energy, heat energy, mechanical energy, electrical energy

Energy and Chemical Changes

The conservation of energy for a chemical change is analogous to the conservation of mass. That is, the total energy before a chemical change is identical to the total energy following the reaction.

It is important to understand that some reactions release heat energy and other reactions use heat energy. Yet the total energy before and after a reaction is the same. If reactions involve heat changes, how can the energy remain constant? The reason is that a reaction involves two types of energy, *potential* chemical energy and *kinetic* heat energy. A reaction increases one type of energy and decreases the other. In a reaction that frees heat to its surroundings, the products have less potential energy. In a reaction that absorbs heat from the environment, the products have more potential energy. All reactions involve conversions between potential chemical energy and kinetic heat energy.

Applying this law to chemical changes, we can say that the heat energy released from a given reaction is equal to the energy absorbed for the reverse reaction. For example, 3200 calories of heat energy are released when 1 g of water is produced from the gases hydrogen and oxygen. For the reverse process, how much energy is required to break down 1 g of water into hydrogen and oxygen? The energy must be the same, 3200 calories. If we break down water by passing an electric current though the liquid, the electrical energy required is the equivalent of 3200 calories.

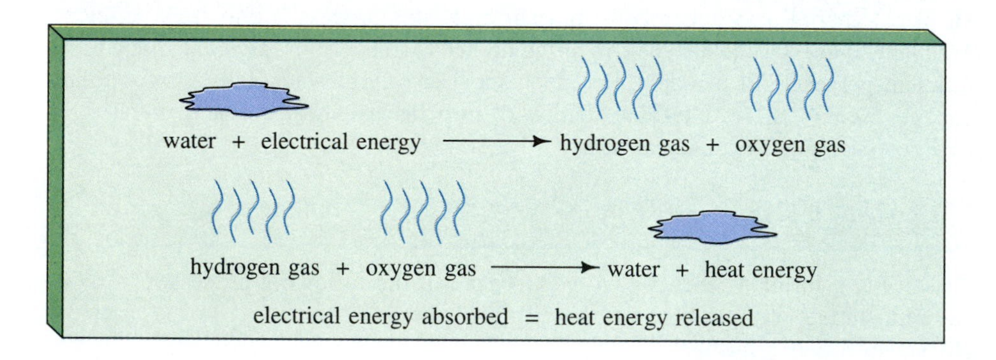
water + electrical energy → hydrogen gas + oxygen gas
hydrogen gas + oxygen gas → water + heat energy
electrical energy absorbed = heat energy released

3.11 Law of Conservation of Mass and Energy

OBJECTIVES

To apply the conservation of mass and energy law to a chemical change.

In 1905, a 26-year-old Swiss patent clerk had three scientific papers published simultaneously. One paper dealt with the photoelectric effect, the ability of light to knock electrons out of metals. The second paper was a mathematical explanation of Brownian motion, the random movement of particles in a fluid. The third paper laid out the special theory of relativity. Relativity theory constructed a new model of the universe. Although the third paper revolutionized physics, the 1921 Nobel prize in physics was awarded for the paper on the photoelectric effect. The year 1905 was an eventful one for the Swiss patent clerk, since he also received his doctoral degree. The young man would arguably become the most famous scientist of the twentieth century. His name was Albert Einstein (1879–1955).

In a portion of his paper on the special theory of relativity, Einstein suggested that matter and energy are interrelated. He went on further to work out the relationship in the form of his famous equation $E = mc^2$. In the equation, Einstein proposed that energy (E) and mass (m) are related by the velocity of light (c) squared. Since mass and energy are related, it follows that the conservation of mass and conservation of energy laws are related. A more accurate statement combines these two laws into the **law of conservation of mass and energy**. The combined law states that the total of mass and energy for all substances before a chemical change is exactly equal to the total of mass and energy following the chemical change.

law of conservation of mass and energy The principle which states that the total mass and energy of substances before a chemical change are equal to the total mass and energy after a chemical change.

If there is a loss in mass, the amount of energy must increase. Even very slight changes in mass bring about enormous changes in energy. A mass loss of 0.001 g, for example, produces enough heat energy to raise the temperature of a large swimming pool about 30°C. Since the mass loss in an ordinary chemical reaction is less than a microgram, the change in mass is undetectable. Therefore, from a laboratory point of view, the original conservation of mass law is still valid today.

As a radioactive substance disintegrates, it converts a small portion of its mass to energy. The fact that a small change in mass produces an enormous amount of energy is very significant. It's that fact that enables us to use nuclear energy to support the needs of civilization. Although the use of nuclear energy in our society is controversial because of the potential hazards, our future may well depend on this source of energy.

Summary

Section 3.1 Chemistry is the study of matter and energy. Matter can exist in any of three **physical states**. The *solid state* is characterized by a definite shape and fixed volume. The *liquid state* has a variable shape but a fixed volume. the *gaseous state* has neither a definite shape nor a fixed volume. Gases diffuse uniformly throughout their containers. Gases can be compressed or expanded, whereas changes in volume for liquids and solids are negligible.

Section 3.2 The physical state of matter is related to the energy and attraction of individual particles. At lower temperatures, a substance has a minimum amount of energy. This means the individual particles cannot overcome their natural attraction for one another. As the temperature increases, the particles gain energy and begin to break down the force of attraction. At some point, the particles have sufficient energy to break free from some of their neighboring particles. This condition is called *melting*. A further increase in temperature gives particles enough energy to overcome the force of attraction for all their neighboring particles. This is called *vaporizing* in which the liquid changes to a gas.

Section 3.3 We classified matter according to its properties. A **homogeneous substance** can be either an **element** or a **compound**. It has fixed properties. Physically combining two or more pure substances produces a **heterogeneous mixture**, which has a variable composition. **Homogeneous mixtures** such as alloys, solutions, and mixtures of gases have variable proportions, but their properties are uniform and constant throughout the sample.

Sections 3.4–3.5 Elements can be classified as **metals** or **nonmetals**. The characteristic properties of metals include metallic luster, malleable, ductile, conductor of heat and electricity, and usually high density and high melting point. Metals react only with nonmetals; nonmetals can react with other nonmetals. The periodic table of the elements shows a general trend in properties, from metallic to nonmetallic. Metals are to the left in the periodic table, and nonmetals are to the right. Elements having properties between metals and nonmetals are **semimetals** and are found toward the middle of the periodic table.

Section 3.6 A given compound, regardless of its origin, always contains the same elements in the same proportion by mass. This is a statement of the **law of constant composition**. The composition of a compound is given by its **chemical formula**. The formula is written using the **chemical symbol** for each element, along with a subscript number to indicate the number of atoms. For example, vitamin E is a fat-soluble vitamin that promotes fertility and has the formula $C_{29}H_{50}O_2$. The formula of vitamin E represents 29 carbon atoms, 50 hydrogen atoms, and 2 oxygen atoms. The total number of **atoms** in the formula of vitamin E is 81.

Section 3.7 Properties of substances are divided into two categories: physical and chemical. **Physical properties** include color, density, melting point, boiling point, conductivity, and solubility. **Chemical properties** refer only to the reactivity of a substance. The properties of compounds are not related to the properties of their constituent elements.

Section 3.8 Changes in compounds are classified as physical or chemical. **Physical changes** include dissolving and changing shape, amount, or physical state. **Chemical changes** involve a change in composition. Evidence for a chemical change includes a substance burning, a mixture bubbling, a solid forming from solution, a change in color or odor, or the release of heat or light energy.

Sections 3.9–3.11 Two basic laws describe chemical changes. One is that the total mass of substances before a chemical change is equal to the total mass of substances after a reaction. The other is that the total energy of the substances before a chemical change is equal to the total energy of the substances after the change. Moreover, the amount of energy released from a reaction is exactly equal to the amount of energy necessary to reverse the reaction. These two laws, the **law of conservation of mass** and the **law of conservation of energy**, are the working principles of laboratory chemistry. In nuclear chemistry, these two laws are considered in their combined form, the **law of conservation of mass and energy**.

Key Terms

Select the key term that corresponds to the following definitions.

____ **1.** refers to the condition of a substance existing as a solid, liquid, or gas

____ **2.** the direct change of state from a solid to a gas without forming a liquid

____ **3.** the energy associated with the mass and velocity of a particle

____ **4.** the smallest particle that represents a compound

____ **5.** matter having constant composition, definite and consistent properties

____ **6.** matter having variable composition and indefinite properties; matter composed of two or more substances that can be separated using physical methods

(a) atom *(Sec. 3.4)*
(b) chemical change *(Sec. 3.8)*
(c) chemical formula *(Sec. 3.6)*
(d) chemical property *(Sec. 3.7)*
(e) chemical symbol *(Sec. 3.4)*

____ 7. matter having variable composition, but definite and consistent properties; examples include natural gas, gasoline, and bronze

____ 8. a pure substance that can be broken down into two or more simpler substances by chemical reaction

____ 9. a pure substance that cannot be broken down any further by ordinary chemical reaction

____ 10. an element that is generally shiny in appearance, has a high density and high melting point, and is a good conductor of heat and electricity

____ 11. an element that is generally dull in appearance, has a low density and low melting point, and is not a good conductor of heat and electricity

____ 12. an element that is generally metal-like in appearance and has properties that are between that of a metal and a nonmetal

____ 13. the property of a metal that allows it to be beaten or rolled into a foil

____ 14. the property of a metal to be drawn into a wire

____ 15. the smallest particle that represents an element

____ 16. the principle that states that a compound always contains the same elements in the same proportion by mass

____ 17. a property that can be observed without changing the chemical formula of a substance

____ 18. a property that cannot be observed without changing the chemical formula of a substance

____ 19. the process of undergoing a change in chemical formula or composition

____ 20. the process of undergoing a change without altering the chemical formula of a substance; for example, a change in shape

____ 21. a symbol representing an element

____ 22. a symbolic representation of a compound indicating the number of atoms of each element

____ 23. the stored energy that matter possesses owing to its position or composition

____ 24. the principle which states that mass cannot be created or destroyed

____ 25. the principle which states that energy cannot be created or destroyed

____ 26. the principle which states that the total mass and energy of substances before a chemical change are equal to the total mass and energy after a chemical change

(f) compound *(Sec. 3.3)*
(g) ductile *(Sec. 3.5)*
(h) element *(Sec. 3.3)*
(i) heterogeneous mixture *(Sec. 3.3)*
(j) homogeneous mixture *(Sec. 3.3)*
(k) homogeneous substance *(Sec. 3.3)*
(l) kinetic energy *(Sec. 3.2)*
(m) law of conservation of energy *(Sec. 3.10)*
(n) law of conservation of mass *(Sec. 3.9)*
(o) law of conservation of mass and energy *(Sec. 3.11)*
(p) law of constant composition *(Sec. 3.6)*
(q) malleable *(Sec. 3.5)*
(r) metal *(Sec. 3.5)*
(s) molecule *(Sec. 3.2)*
(t) nonmetal *(Sec. 3.5)*
(u) physical change *(Sec. 3.8)*
(v) physical property *(Sec. 3.7)*
(w) physical state *(Sec. 3.1)*
(x) potential energy *(Sec. 3.10)*
(y) semimetal *(Sec. 3.5)*
(z) sublimation *(Sec. 3.1)*

Exercises

Physical States of Matter (Sec. 3.1)

1. Indicate the shape (definite or indefinite) and volume (fixed or variable) for each of the following.
   (a) solids **(b)** liquids
   (c) gases

2. Indicate the degree of compressibility for each of the following states of matter.
   (a) solids **(b)** liquids
   (c) gases

3. Describe the movement of individual particles in the following.
   (a) solids **(b)** liquids
   (c) gases

4. A substance is considered to be in the solid, liquid, or gaseous physical state under normal conditions. Define what is meant by normal conditions.

5. Supply the term that describes the following changes of physical state.
   (a) liquid to gas **(b)** solid to gas
   (c) liquid to solid **(d)** solid to liquid
   (e) gas to liquid **(f)** gas to solid

6. Indicate whether heat energy is absorbed or released for the following changes of physical state.

(a) solid to liquid
(b) liquid to solid
(c) gas to liquid
(d) liquid to gas
(e) solid to gas
(f) gas to solid

Physical State and Energy (Sec. 3.2)

7. What is the relationship between kinetic energy and temperature?
8. What is the relationship between kinetic energy and molecular motion?
9. A sample of ammonia gas in a glass sphere is heated from 20° to 50°C. State whether the following increase or decrease.
   (a) kinetic energy of the gas
   (b) velocity of ammonia molecules
10. A sample of carbon dioxide gas is cooled from 25° to −50°C. State whether the following increase or decrease.
    (a) kinetic energy of the gas
    (b) velocity of carbon dioxide molecules
11. At the same temperature, is the kinetic energy of lighter gas molecules less than, equal to, or greater than heavier gas molecules?
12. At the same temperature, is the average velocity of lighter gas molecules less than, equal to, or greater than heavier gas molecules?
13. If two gases have the same kinetic energy, is the average temperature of lighter gas molecules less than, equal to, or greater than heavier gas molecules?
14. If two gases have the same kinetic energy, is the average velocity of lighter gas molecules less than, equal to, or greater than heavier gas molecules?

Classification of Matter (Sec. 3.3)

15. Distinguish between and give examples for each of the following pairs of terms.
    (a) homogeneous substance and heterogeneous mixture
    (b) homogeneous mixture and heterogeneous mixture
    (c) element and compound
16. Classify each of the following as an element, compound, or mixture.
    (a) iodine
    (b) silver ore
    (c) nitrogen dioxide
    (d) ocean water
    (e) auto emissions
    (f) alloy wheels
    (g) iron
    (h) steel (Mn in Fe)
17. Classify each of the following as an element, compound, or mixture.
    (a) ice
    (b) distilled water
    (c) silicon
    (d) Lake Michigan water
    (e) 18K gold jewelry
    (f) aluminum
    (g) stainless steel (Cr in Fe)
    (h) uranium ore
18. Classify each of the following as an element, compound, or mixture.
    (a) earth's crust
    (b) earth's atmosphere
    (c) oxygen
    (d) ozone (O_3)
    (e) seawater
    (f) sulfur dioxide
    (g) soft drink
    (h) titanium

Names and Symbols of the Elements (Sec. 3.4)

19. Name the three most abundant elements in the earth's crust.
20. Which of the following elements is not one of the earth's ten most abundant: aluminum, calcium, hydrogen, iron, oxygen, magnesium, uranium, potassium, silicon, sodium, titanium?
21. Provide the chemical symbol corresponding to each of the following elements.
    (a) bromine
    (b) oxygen
    (c) antimony
    (d) lithium
    (e) argon
    (f) magnesium
    (g) fluorine
    (h) sodium
    (i) bismuth
    (j) nickel
    (k) helium
    (l) tellurium
    (m) hydrogen
    (n) tin
    (o) calcium
    (p) platinum
    (q) carbon
    (r) potassium
    (s) iodine
    (t) titanium
    (u) iron
    (v) xenon
    (w) krypton
    (x) zinc

The element copper in the form of tubing.

22. Provide the name corresponding to each of the following chemical symbols.
    (a) As (b) Ba (c) Be (d) Cd
    (e) Cl (f) Ge (g) Co (h) Cu
    (i) Al (j) B (k) Au (l) Cr
    (m) Mn (n) Hg (o) Ne (p) P
    (q) Ra (r) Sr (s) Si (t) Ag
    (u) Pb (v) N (w) S (x) Se
23. Refer to the periodic table and find the atomic numbers of the following elements.
    (a) bismuth
    (b) silver
    (c) cadmium
    (d) sodium
    (e) mercury
    (f) copper
    (g) potassium
    (h) tin

24. Write the name of the element corresponding to each of the following atomic numbers. Refer to the periodic table to find the symbol of the element.

- **(a)** at. no. 1
- **(b)** at. no. 15
- **(c)** at. no. 79
- **(d)** at. no. 53
- **(e)** at. no. 6
- **(f)** at. no. 38
- **(g)** at. no. 82
- **(h)** at. no. 20

Metals, Nonmetals, and Semimetals (Sec. 3.5)

25. List the six semimetals that are not radioactive.

26. Silicon is used in the semiconductor industry to make transistors and integrated-circuit chips. Use the periodic table to predict another element that is used in the semiconductor industry.

27. Refer to the periodic table and give the physical state for each of the following elements at 25°C and normal atmospheric pressure.

- **(a)** Li
- **(b)** N
- **(c)** P
- **(d)** Sn
- **(e)** Cd
- **(f)** Mn
- **(g)** Mg
- **(h)** K
- **(i)** Hg
- **(j)** Br

28. State whether each of the following properties is more typical of a metal or a nonmetal.

- **(a)** heat conductor
- **(b)** gaseous state
- **(c)** malleable
- **(d)** dull appearance
- **(e)** low density
- **(f)** electrical insulator
- **(g)** high melting point
- **(h)** ductile
- **(i)** reacts with metals
- **(j)** reacts with nonmetals

29. Refer to the periodic table of the elements and classify the following elements as metals, nonmetals, or semimetals.

- **(a)** aluminum
- **(b)** boron
- **(c)** phosphorus
- **(d)** manganese
- **(e)** beryllium
- **(f)** krypton
- **(g)** radium
- **(h)** fluorine
- **(i)** hydrogen
- **(j)** uranium

30. Refer to the periodic table of the elements and indicate the physical state of the following elements at 25°C and normal atmospheric pressure.

- **(a)** cobalt
- **(b)** germanium
- **(c)** argon
- **(d)** titanium
- **(e)** nitrogen
- **(f)** bromine
- **(g)** hydrogen
- **(h)** mercury
- **(i)** barium
- **(j)** antimony

Compounds and Chemical Formulas (Sec. 3.6)

31. Joseph Proust stated the law of constant composition after analyzing the compound copper carbonate. He prepared the compound experimentally and found the mass ratio of Cu:C:O to always be 5:1:4. Predict the mass ratio of elements in copper carbonate isolated from a mineral.

32. The ratio of C:H:O in natural vitamin C is 9:1:12. Predict the composition of synthetic ascorbic acid (vitamin C) produced in a laboratory.

33. State the number of atoms of each element in the following formulas of compounds.

- **(a)** aspirin, $C_9H_8O_4$
- **(b)** lactose, $C_{12}H_{22}O_{11}$
- **(c)** glycerin, $C_3H_5(OH)_3$
- **(d)** picric acid, $C_6H_2(NO_3)_3OH$
- **(e)** ether, $(C_2H_5)_2O$

34. Write the chemical formula for each of the following compounds, given their atomic compositions.

- **(a)** grain alcohol: 2 carbon, 6 hydrogen, 1 oxygen
- **(b)** Freon-12: 2 carbon, 2 chlorine, 4 fluorine
- **(c)** vitamin A: 20 carbon, 30 hydrogen, 1 oxygen
- **(d)** methionine: 5 carbon, 11 hydrogen, 2 oxygen, 1 nitrogen, 1 sulfur
- **(e)** aspartame: 14 carbon, 18 hydrogen, 5 oxygen, 2 nitrogen

35. Find the total number of atoms in each of the following chemical formulas.

- **(a)** calciferol (vitamin D), $C_{27}H_{44}O$
- **(b)** conine (poison in hemlock), $C_8H_{17}N$
- **(c)** quinine (an antimalarial), $C_{20}H_{24}N_2O_2$
- **(d)** ethylene glycol (in antifreeze), $C_2H_4(OH)_2$
- **(e)** adipic acid (for making nylon), $HO_2C(CH_2)_4CO_2H$

36. Find the total number of atoms in each of the following chemical formulas.

- **(a)** cholesterol (a steroid), $C_{27}H_{45}OH$
- **(b)** stearic acid (a fatty acid), $CH_3(CH_2)_{16}CO_2H$
- **(c)** tristearin (a triglyceride), $C_3H_5(C_{17}H_{35}O_2)_3$
- **(d)** lysine (an amino acid), $NH_2(CH_2)_4CH(CO_2H)NH_2$
- **(e)** glutamic acid (in MSG), $HO_2C(CH_2)_2CH(NH_2)COOH$

Physical and Chemical Properties (Sec. 3.7)

37. Classify the following properties of silver as physical or chemical.

- **(a)** good conductor of heat and electricity
- **(b)** no reaction with hydrochloric acid
- **(c)** forms a liquid at 962°C

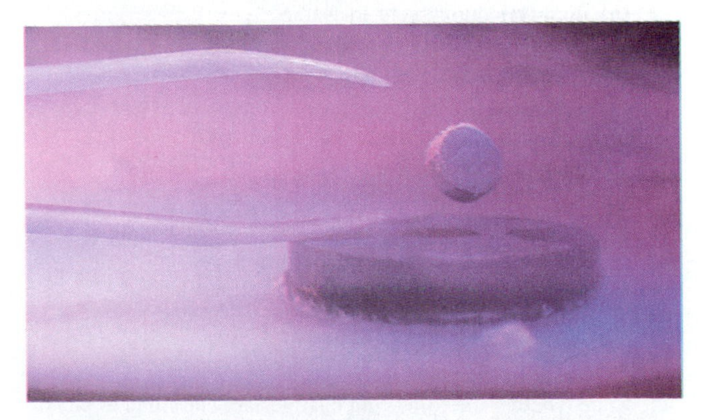

A small magnet dropped by tweezers appears to float above a ceramic disk. After the disk has been cooled to −196°C with liquid nitrogen, it becomes a superconductor. The ceramic disk produces a magnetic field which causes the small magnet to levitate.

(d) oxidizes slowly at 100°C
(e) can be drawn into fine wires

38. Classify the following properties of aluminum as physical or chemical.
(a) can be machined into thin foils
(b) generates a colorless, odorless gas with sulfuric acid
(c) vaporizes at 2467°C
(d) conducts heat energy
(e) produces copper metal from a blue copper solution

The recycling of aluminum is a physical property.

39. Classify the following properties as physical or chemical.
(a) sugar ferments to alcohol
(b) silicon seed crystals grow into silicon ingots
(c) iron oxidizes to give rust
(d) alcohol vaporizes
(e) lithium reacts with oxygen in air
(f) benzene dissolves in chloroform
(g) copper wire conducts an electric current
(h) iodine crystals undergo sublimation
(i) brandy produces an aromatic bouquet
(j) plutonium disintegrates by radioactive emission

40. Classify the following properties of ethyl alcohol as physical or chemical.
(a) dissolves completely in water
(b) produces a gas with sodium metal
(c) becomes a colorless gas at 78.5°C
(d) a mass of 0.789 g occupies a volume of 1 mL
(e) becomes a crystalline solid at −117°C
(f) flammable in air
(g) requires 0.511 calorie of heat to raise 1 g 1°C
(h) combines with acetic acid to give ethyl acetate
(i) bubbles when poured on sodium metal
(j) produces ether and water with sulfuric acid

Physical and Chemical Changes (Sec. 3.8)

41. State whether the following changes are classified as physical or chemical.
(a) changing from a gas to a liquid
(b) changing from a solid to a gas
(c) changing shape
(d) changing mass or volume
(e) changing by exploding
(f) changing by dissolving
(g) changing crystal shape

42. State whether the following observations are classified as physical or chemical.
(a) burning, combustion (rapid oxidation)
(b) rusting, tarnishing (slow oxidation)
(c) changing color, odor, or taste
(d) evolving gas bubbles from a mixture
(e) forming an insoluble substance upon mixing two solutions
(f) changing by melting or boiling
(g) releasing energy as heat or light

43. Classify the following changes as physical or chemical.
(a) converting gasoline to carbon dioxide and water vapor
(b) pouring baking soda on spilled car battery acid and producing gas
(c) grinding sugar crystals into powder
(d) observing a deep-blue solution upon combining two colorless liquids
(e) increasing the volume of nitrogen gas by increasing its temperature
(f) cutting magnesium ribbon into strips
(g) combining ethylene glycol and water to produce automotive antifreeze
(h) burning a log in a fire
(i) digesting carbohydrates for energy
(j) distilling brandy to obtain a higher percentage of alcohol

44. Classify the following changes as physical or chemical.
(a) tearing off sheets of aluminum foil
(b) heating sugar to a black carbon residue
(c) adding air to a tire to raise the pressure
(d) dividing powdered zinc into two equal masses
(e) igniting magnesium metal to produce a flare
(f) dissolving iodine crystals in ether
(g) combining the miscible liquids benzene and chloroform
(h) producing an orange solid from solutions of mercury nitrate and sodium iodide
(i) cooling mercury until it forms a silvery solid
(j) heating baking soda until it produces a suffocating gas

Law of Conservation of Mass (Sec. 3.9)

45. If 2.50 g of iron filings reacts completely with 1.44 g of yellow sulfur powder, what is the mass of the iron sulfide formed?

46. How much potassium nitrate was decomposed if 85 g of potassium nitrite and 16 g of oxygen were formed by the decomposition?

47. Orange mercury oxide powder decomposes upon heating to give droplets of mercury and oxygen gas. If heating 0.750 g of mercury oxide produces 0.695 g of liquid residue, what is the mass of oxygen evolved?

48. What is the mass of ammonia gas that must react with 36.451 g of hydrogen chloride gas to give 53.453 g of solid ammonium chloride?

Law of Conservation of Energy (Sec. 3.10)

49. Using a roller coaster car as an example, discuss the relationship between potential and kinetic energy.

50. Using a playground swing as an example, discuss the relationship between potential and kinetic energy.

51. Classify the following as an example of potential energy or kinetic energy.
(a) lead plates and sulfuric acid in a storage battery
(b) electricity produced by the battery in (a)
(c) magnesium metal and oxygen gas in a flashbulb
(d) heat and light from a magnesium/oxygen flashbulb
(e) a sample of radioactive radium
(f) radioactive emission from radium
(g) steam expanding to drive a piston
(h) steam contained at high pressure
(i) gasoline in a 20-gallon tank
(j) gasoline undergoing combustion

52. Steam at 100°C passing through a radiator releases 210,000 kilocalories of heat. How much heat energy was initially required to convert the water, at 100°C, to steam?

53. A 10.0-g sample of mercury abosrbs 110 calories as it is heated from 25°C to its boiling point at 356°C. It then requires an additional 697 calories to vaporize. How much heat is released as the mercury vapor cools to a liquid at 25°C?

54. Nitrogen and hydrogen combine to form ammonia gas and 649 calories of heat per gram of ammonia.
(a) How much heat energy is required to break down 1.000 g of ammonia into hydrogen and nitrogen?
(b) If 0.824 g of nitrogen gas is evolved, what is the mass of hydrogen gas produced?

55. List six different forms of energy.

56. Classify the forms of energy involved in the following energy conversions relating to an automobile.
(a) battery discharging electricity
(b) electricity turning a starter motor
(c) starter motor turning a flywheel
(d) spark igniting a gasoline mixture, producing gaseous carbon dioxide and water
(e) gaseous products expand and force down pistons
(f) pistons rotating the crankshaft
(g) rotating crankshaft turning the driveline
(h) crankshaft belt driving the generator
(i) generator charging the battery
(j) battery powering the headlights

Law of Conservation of Mass and Energy (Sec. 3.11)

57. Define each symbol in Einstein's equation $E = mc^2$.

58. According to the conservation of mass and energy law, can matter or energy be destroyed? Explain.

59. State the conservation of mass and energy law as it applies to a chemical reaction.

60. Gasoline reacts with oxygen to produce carbon dioxide, water, and heat energy. An experiment reveals that the mass of gasoline and oxygen is equal to the mass of the carbon dioxide and water. Discuss this in relation to the conservation of mass and energy law.

General Exercises

61. Describe the fourth state of matter. Where is plasma being constantly produced?

62. Assuming the temperature remains constant, should you increase or decrease the pressure to convert a liquid to a gas?

63. Write the equation that relates the kinetic energy, mass, and velocity of a particle. Define each symbol in the equation.

64. In a steel cylinder of oxygen gas, is the velocity of all oxygen molecules identical? Explain.

65. State whether each of the following is an example of a homogeneous mixture or heterogeneous mixture.
(a) a bronze alloy of copper and tin
(b) an ore sample of quartz and gold
(c) a gas mixture of methane, carbon dioxide, and ammonia
(d) a dry cleaning liquid containing chlorohydrocarbons
(e) an ice cube floating on water

66. State whether the following describes a physical change or chemical change.
(a) a change in physical state but not chemical formula
(b) a change in chemical formula but not physical state

67. Boron nitride, BN, is used in manufacturing silicon computer chips. However, it is unstable and decomposes to boron and nitrogen while releasing 3.5 kcal/g. How much energy is necessary to produce 1 gram of BN from the elements?

68. In a nuclear power plant, a radioactive substance decays and releases heat. The heat vaporizes water and the steam drives a turbine that creates electricity. State four forms of energy involved in the process.

69. Give the English name and symbol for each of the following elements.
(a) hydrargyrum (b) natrium
(c) cuprum (d) kalium
(e) stibium (f) plumbum
(g) ferrum (h) aurum
(i) stannum (j) argentum

70. Using the recent recommendations of IUPAC, provide a name and symbol for the following.
(a) element 104 (b) element 105
(c) element 106 (d) element 107
(e) element 108 (f) element 109
(g) element 117 (h) element 130
(i) element 146 (j) element 168

CHAPTER

4

Atomic Theory and Structure

This Tiffany landscape scene uses stained glass to portray the colors of iris and magnolia flowers. A piece of glass appears colored because atoms in the glass absorb one color of light and transmit another. If the glass absorbs yellow light, the color observed is blue.

The true nature of atoms is of interest to physicists as well as philosophers. Our present understanding of atoms is based on experiment and theory. Atomic theory is sufficiently complex, in fact, that there may be no ultimate answers. When lecturing on the subject, the Danish physicist Niels Bohr was fond of saying to his classes, "Every sentence that I utter should be regarded by you not as an assertion but as a question." *As we shall see, Niels Bohr made a significant contribution to our knowledge of atomic structure. In 1922, he was awarded the Nobel prize in physics for his efforts.*

The concept of atoms was born in Greece about 450 B.C. *The Greeks proposed scientific explanations of nature but did not test their theories. The only experiments they performed were thoughtful mental exercises.*

One of the great issues of the day was whether the universe was composed of continuous or discontinuous events. Zeno, a Greek philosopher, used a paradox to argue that motion is discontinuous (Figure 4.1). He pointed out that in order to travel any distance, first, half the distance must be traveled; then half of the remaining distance must be traveled; and so on. Zeno's paradox suggests that if motion is continuous you can never arrive at your destination. He therefore concluded that motion is discontinuous and occurs by a series of tiny jumps.

The continuity controversy was applied to matter as well as to motion. The Greek philosopher Democritus argued that matter is discontinuous and could not be infinitely divided. He said that at some point a fundamental, indivisible particle would emerge. A century later, Aristotle, the most influential scientist of his time, argued for continuity in the universe. Motion is continuous. Matter is continuous. Atoms do not exist! So powerful was Aristotle's influence that atoms became a closed issue and Aristotle's point of view prevailed for the next 21 centuries.

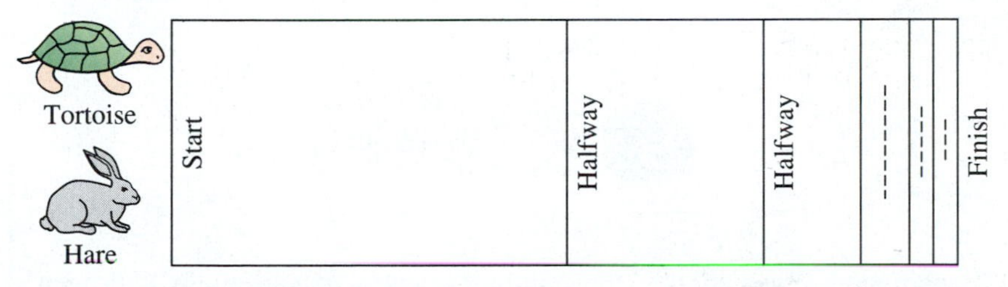

Figure 4.1 Zeno's Paradox Motion must occur by discontinuous jumps, otherwise a race can never by completed. Neither the tortoise nor the hare can complete the race if they continue to move half the distance to the finish.

4.1 Evidence for Atoms: The Dalton Model

To describe an atom based upon Dalton's five proposals.

In 1803, a modest Quaker schoolteacher named John Dalton (1766–1844) presented evidence for the particle, or discontinuous, nature of matter. Dalton was familiar with the theory of Democritus and used the word atom for a particle of matter. (Greek *atomos* means indivisible.) Unlike the Greeks, however, Dalton offered experimental evidence for the particle nature of matter. The color-blind Dalton was not an outstanding experimenter, but his evidence was based on the work of great scientists who preceded him, that is, Robert Boyle, Joseph Proust, and Antoine Lavoisier.*

John Dalton, a mild-mannered devout man, overturned twenty centuries of Greek thinking regarding matter.

Dalton's Atomic Theory

Dalton knew the relative atomic masses of carbon, hydrogen, and oxygen. He used that information to predict the combining ratios in methane, CH_4. According to the law of constant composition, elements always react in the same proportion by mass. Carbon and oxygen, for example, always react in the same ratio to produce carbon dioxide. Dalton proposed that an atom of carbon combined with two atoms of oxygen to produce carbon dioxide.

carbon dioxide, CO_2

Furthermore, if two atoms of hydrogen combine with each atom of oxygen in carbon dioxide, then two molecules of water result.

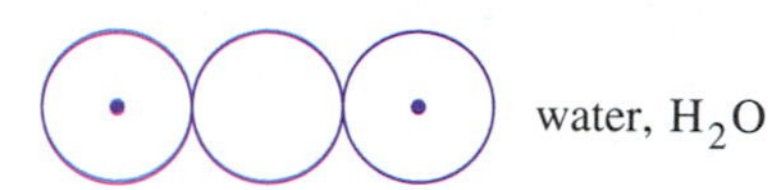

water, H_2O

Finally, he argued that, if two atoms of hydrogen in water replace each of the oxygen atoms in carbon dioxide, a molecule having one carbon atom and four hydrogen atoms would result. The resulting compound is called methane.

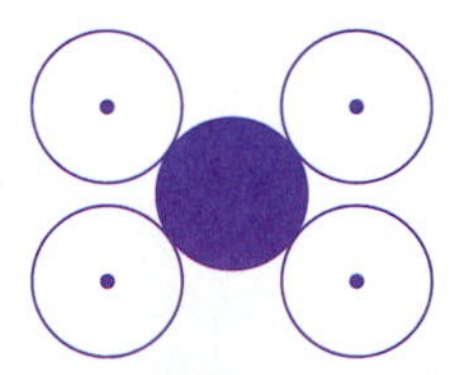

methane, CH_4

Experiments revealed that the combining ratio of carbon to hydrogen in methane agreed perfectly with the prediction. Thus, Dalton was the first to use experimental evidence to support the atomic theory.

*Robert Boyle (1627–1691), English physicist and chemist, is given credit for the scientific method and establishing the importance of experiments to support theories. Joseph Proust (1754–1826) established the law of constant composition as mentioned in Section 3.6. The French chemist Antoine Lavoisier (1743–1794) is considered the founder of modern experimental chemistry.

Dalton, like Robert Boyle, was interested in gases and kept a daily journal of atmospheric conditions in his native Manchester, England. In the seventeenth century, Boyle concluded from experiments that a gas was made up of tiny particles. Dalton expanded on Boyle's theory and proposed that all matter was composed of particles.

In addition to Boyle's work, Dalton relied on two other important scientific principles. One was the *law of conservation of mass*, demonstrated in 1789 by Antoine Lavoisier. By carefully weighing substances before and after chemical reaction. Lavoisier proved that matter was neither created nor destroyed. The other was the *law of constant composition*, established in 1799 by another French chemist, Joseph Louis Proust. Proust had shown experimentally that a compound always contains the same elements in the same proportion by mass.

Dalton first presented his evidence to the Literary and Philosophical Society of Manchester. He proposed that an element is composed of tiny, indivisible, indestructible particles called atoms. Furthermore, he proposed that compounds are simply combinations of two or more atoms of different elements. In 1808, Dalton published his atomic theory in the classic text *A New System of Chemical Philosophy*. In a surprisingly short period of time, atomic theory became generally accepted throughout the scientific community.

Dalton's atomic theory contained the following five proposals:

1. An element is composed of tiny, indivisible, indestructible particles called atoms.
2. All atoms of an element are identical and have the same properties.
3. Atoms of different elements combine to form compounds.
4. Compounds contain atoms in small, whole-number ratios.
5. Atoms may combine in more than one ratio to form different compounds.

4.2 Evidence for Subatomic Particles: The Thomson Model

OBJECTIVES

To state the relative charge and mass of the electron and proton.

To describe the Thomson model of the atom.

About fifty years after Dalton's proposal, there was disturbing evidence that the atom was *not* indivisible. In 1855, glass tubes were invented from which most of the gas had been evacuated. When electricity was applied to one end, the tubes appeared to glow. This phenomenon was referred to as fluorescence and the glowing ray was thought to be a type of light energy. Since the ray emanated from the negative electrode, which is called the cathode, the radiation was referred to as a **cathode ray** (Figure 4.2).

cathode ray A stream of negative particles produced in a cathode-ray tube.

In the late 1870s, the English physicist William Crookes (1832–1919) observed that cathode rays were attracted by a magnetic field. This observation suggested that cathode rays were actually particles, not radiation. When different gases were put in the tubes, the results did not change. These results led to the theory that cathode rays were composed of tiny, negatively charged subatomic particles. The particles were named **electrons**. *Elektron* is Greek for amber, a material that becomes charged when rubbed with fur. The symbol for the electron is e^-.

electron (e^-) A subatomic particle having a negligible mass and a relative charge of one minus.

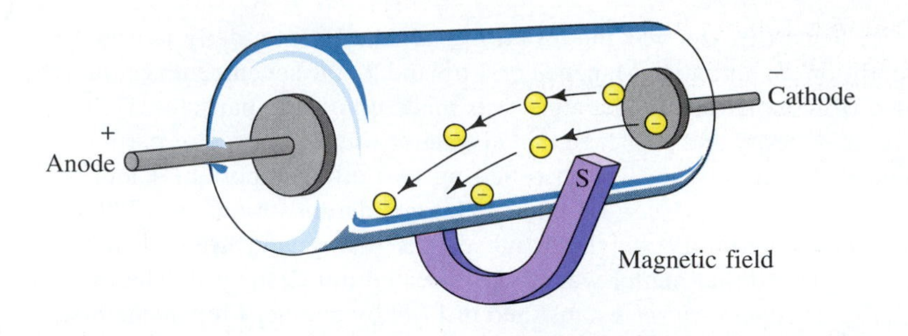

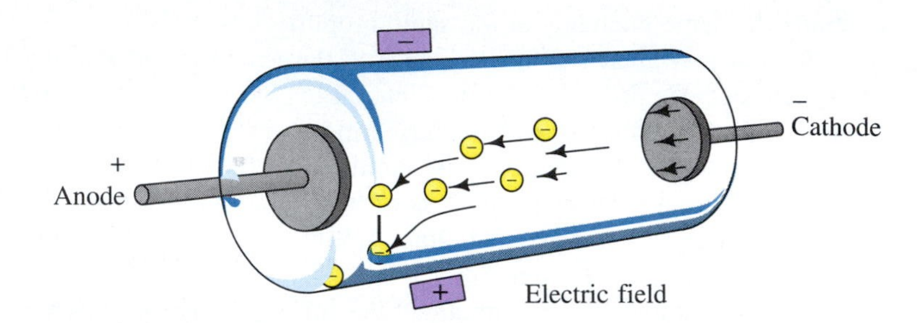

Figure 4.2 Cathode-Ray Tubes Note the influence on cathode-rays by a magnetic field and an electric field. Several English and German physicists share the credit for these experiments.

In 1886, the German physicist Eugen Goldstein (1850–1930) experimented with a cathode-ray tube having small holes, or channels, in the cathode (Figure 4.3). He discovered that positive as well as negative rays were produced at the cathode. The positive rays moved in the opposite direction from the negative rays through the holes in the cathode. Goldstein referred to these positive rays as channel rays. Through language translation, channel rays became known as canal rays (*kanal* is German for channel). Further experiments revealed that canal rays were composed of small positive particles. The smallest particle had a charge equal, but opposite in sign, to that of an electron. It was later called a **proton**. Proton comes from the Greek *proteios*, meaning of first importance. The symbol for the proton is p^+.

proton (p^+) A subatomic particle having an approximate mass of 1 amu and a relative charge of one plus.

In 1897, J. J. Thomson (1856–1940) demonstrated that cathode rays are deflected by an electric field as well as by a magnetic field. This seemed to be the final piece of evidence to support the notion that electrons were particles. Even though evidence had accumulated for 20 years, Thomson is given the credit for the discovery of the electron. Since the particle is so tiny, Thomson could not measure the actual charge or mass of the electron. It was possible, however, to determine the ratio of charge to mass for an electron. Thomson continued his experiments and obtained values for the charge-to-mass ratio for both the electron and proton.

The actual charge on an electron was determined a few years later by the American physicist Robert Millikan (1868–1953). This important experiment al-

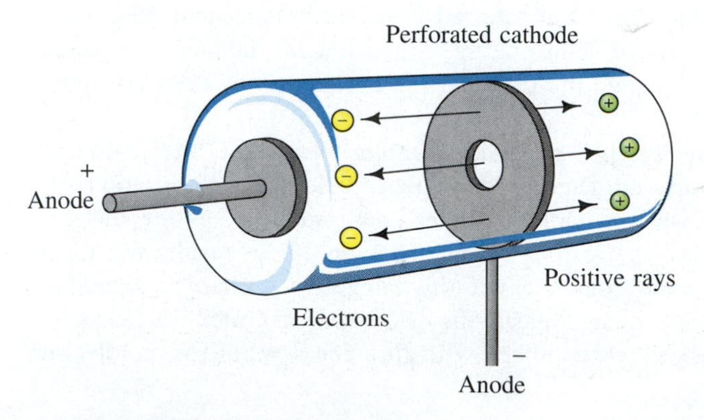

Figure 4.3 Canal Rays This cathode-ray tube shows the positive canal rays discovered by Goldstein and the cathode rays observed earlier.

Table 4.1 Relative Charge and Mass of the Electron and Proton

Subatomic Particle	Relative Charge	Relative Mass
electron, e^-	1−	1/1836
proton, p^+	1+	1

lowed Thomson to calculate the mass of the electron and proton. Using the charge-to-mass ratio for the electron and Millikan's value for the charge, Thomson calculated the mass of the electron to be 9.11×10^{-28} g. From the charge-to-mass ratio for the proton, he calculated the mass of the proton to be 1.67×10^{-24} g. The mass of the electron, as the calculations reveal, is only about 1/1836 of the proton. Table 4.1 lists the charge and mass of the electron and proton relative to each other.

In 1903, Thomson proposed a subatomic model of the atom. He visualized the atom as a sphere of positive charge in which negatively charged electrons were embedded. The Thomson proposal was popularly known as the *plum pudding model*. The model pictured electrons embedded in homogeneous spheres of positive charge analogous to raisins embedded in plum pudding (Figure 4.4). Although the plum pudding model was incorrect, it was consistent with the evidence at the time. In 1911, though, one of Thomson's former students would supervise a revolutionary experiment that changed the picture of the atom dramatically.

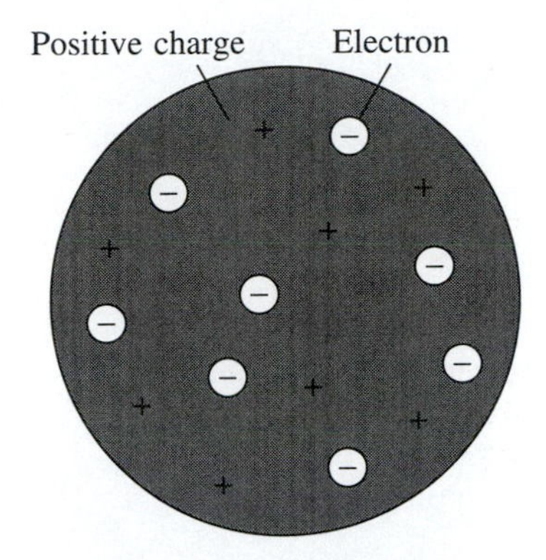

Figure 4.4 Plum Pudding Model of the Atom Atoms are pictured as homogeneous spheres of positive charge. The small negative particles in the spheres represent electrons.

Note J. J. Thomson (Figure 4.5) spent his academic career at Cambridge University and, at the age of 27, was director of the prestigious Cavendish Laboratory. In 1906, he won the Nobel prize in physics for his work on the electron; two years later he was knighted. Seven of Thomson's former students and assistants went on to receive Nobel prizes.

Figure 4.5 Sir Joseph John Thomson J. J. Thomson is pictured at the apparatus he built to measure the charge-to-mass ratio for the electron and proton.

4.3 Evidence for a Nuclear Atom: The Rutherford Model

To draw a diagram of the nuclear atom.

To state the diameters for an atom and its nucleus.

To state the relative charge and approximate mass of the electron, proton, and neutron.

Ernest Rutherford (1871–1937) was digging potatoes on his father's farm in New Zealand when he received the news. He had won a scholarship to Cambridge University. He had actually placed second in a contest, but the winner declined the scholarship to get married. The 24-year-old Rutherford postponed his own wedding plans and immediately set out for England.

At Cambridge, Rutherford studied subatomic particles and earned his doctorate under J. J. Thomson. Upon graduation, he went to McGill University in Canada and began work in the field of radioactivity. After a year he went back to New Zealand, married, and then returned to Manchester University in England. There he continued to study radioactivity and coined the terms alpha and beta for two types of radiation. He discovered a third type of radiation that was not affected by a magnetic field and gave it the name gamma. In 1908, Rutherford won the Nobel prize in chemistry for his work on radioactivity.

In 1906, Rutherford found that alpha particles were identical to helium atoms stripped of electrons. He experimented with alpha particles by firing them at thin gold foils (Figure 4.6). As expected, most of the particles passed through. On occasion, however, some alpha particles were deflected slightly by the foil. This observation seemed reasonable, since the plum pudding model pictured atoms as low-density spheres. But in 1911 the true picture of the atom was unveiled. The person in charge of the experiment was Hans Geiger, Rutherford's assistant, and the inventor of the Geiger counter. Here's a description of the experiment in Rutherford's own words:

> One day Geiger came to me and said, "Don't you think that young Marsden, whom I am training in radioactive methods, ought to begin a small research?" Now I had thought that too, so I said, "Why not let him see if any alpha particles can be scattered through a large angle?" I may tell you in confidence

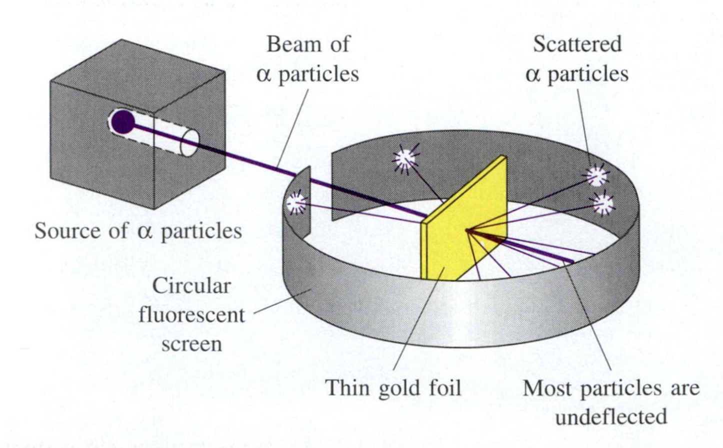

Figure 4.6 Rutherford's Alpha Scattering Experiment The diagram shows the deflection of alpha particles by a thin gold foil. Although the foil was only 2000 atoms thick, about 0.5 micrometer, a few alpha particles actually rebounded backwards.

that I did not believe that they would be, since we knew that the alpha particle was a very fast massive particle, with a great deal of energy. . . . Then I remember two or three days later Geiger coming to me in great excitement and saying, “We have been able to get some of the alpha particles coming backwards.” It was quite the most incredible event that has ever happened to me in my life. It was almost as incredible as if you fired at 15-inch shell at a piece of tissue paper and it came back and hit you.

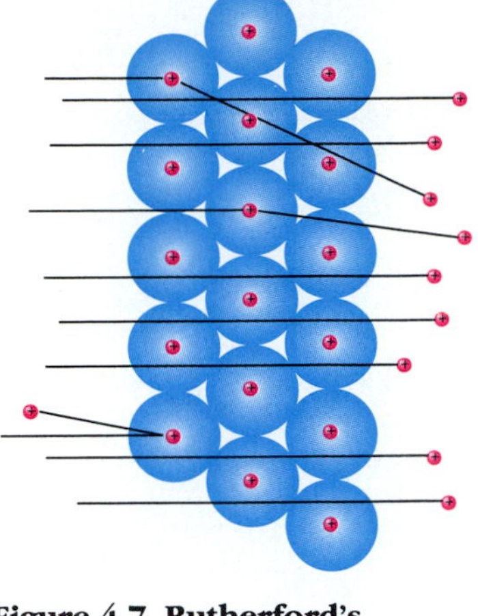

Figure 4.7 Rutherford's Nuclear Atom The atoms are represented by circles and the nuclei by small dots. Alpha particles are positively charged and deflected as they approach the large, positive, gold nucleus.

Rutherford interpreted the alpha scattering results as follows. He said that most of the alpha particles pass directly through the foil because an atom is largely empty space with electrons moving about. But in the center of the atom is the **atomic nucleus** containing protons. Rutherford reasoned that, compared to the atom, the nucleus is tiny and has a very high density. The alpha particles that bounced backward were recoiling after striking the dense nucleus. Figure 4.7 illustrates the scattering of alpha particles by the atomic nuclei in the gold foil.

In 1911, Rutherford proposed a planetary model of the atom. He theorized that negatively charged electrons travel in circular orbits about the positive nucleus. Since the nucleus appeared to be unusually heavy, Rutherford predicted that it contained an undiscovered neutral particle. Twenty years later, James Chadwick, a former student of Rutherford, discovered that neutral particle. In 1935, Chadwick was awarded the Nobel prize in physics for his discovery of the **neutron**. The symbol for the neutron is n^0.

From his experiments, Rutherford was able to estimate the size of the atom and its nucleus. He calculated the diameter of the atom was about 10^{-8} can and the nucleus about 10^{-13} cm. It is cumbersome to use negative exponents when referring to the size of an atom. Therefore, scientists express atomic sizes in a more convenient unit, such as the nanometer (one-billionth of a meter). The approximate diameters of atoms range from 0.1 to 0.5 nm.

atomic nucleus A region of very high density in the center of the atom.

neutron (n^0) A subatomic particle having a charge of zero.

Figure 4.8 illustrates the Rutherford planetary model of the atom. Note that it distorts the relative proportions of the nucleus and the atom. The actual ratio of the diameter of the nucleus to the diameter of the atom is a centimeter to a kilometer. By analogy, if the nucleus was the size of a small BB, •, the atom would be as large as the Houston Astrodome. This observation makes it clear why so few alpha particles were scattered by the atoms in the gold foil.

In summary, Table 4.2 presents data about the electron, proton, and neutron. The mass of atomic and subatomic particles is expressed in atomic mass units (amu). An amu is defined to be equal to 1/12 the mass of a carbon atom having six protons and six neutrons in its nucleus.

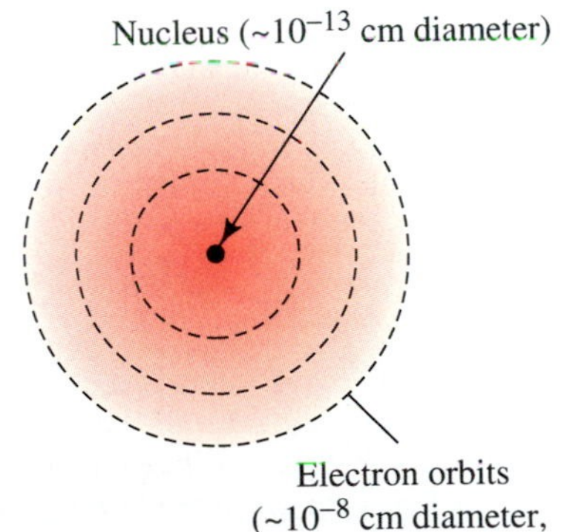

Figure 4.8 Planetary Model of the Atom Protons and neutrons are in the nucleus. Rutherford proposed that electrons orbit the nucleus just as the planets circle the sun.

Table 4.2 Fundamental Subatomic Particles

Particle	Symbol	Location	Charge	Mass (amu)
electron	e^-	outside nucleus	1−	0.00055 ~ 0
proton	p^+	inside nucleus	1+	1.00728 ~ 1
neutron	n^0	inside nucleus	0	1.00866 ~ 1

EXAMPLE EXERCISE 4.1

Give the relative charge and approximate mass in amu for each of the following subatomic particles.

(a) electron
(b) proton
(c) neutron

Solution: Based on experiments by Thomson, Rutherford, Chadwick, and others, we have the following data:

(a) The electron (e^-) has a relative charge of one negative ($1-$) and an approximate mass of 0 amu.

(b) The proton (p^+) has a relative charge of one positive ($1+$) and an approximate mass of 1 amu.

(c) The neutron (n^0) is a neutral particle and has a relative charge of 0 and an approximate mass of 1 amu.

SELF-TEST EXERCISE

Give the approximate diameter in centimeters for each of the following.

(a) atomic diameter **(b)** atomic nucleus

Answers: **(a)** 10^{-8} cm; **(b)** 10^{-13} cm

Currently, it is the nucleus of the atom that is of greatest interest to atomic scientists. As a result of atom-smashing experiments, we are aware of 100's of subatomic particles. Current research suggests that subatomic particles are composed of particles called quarks. Quarks, antiquarks, and strange quarks have all been identified.

Atomic scientists are still searching for the ultimate particle from which all matter is composed. But each time a new candidate for the ultimate particle is found, it separates into finer particles or flashes of energy. Although experiments during the past two centuries have helped us understand matter, perhaps Aristotle was correct. Matter may be continuous. Indivisible particles may not exist.

4.4 Atomic Notation

To draw a diagram of an atom, given its atomic notation.

To define and illustrate an isotope.

atomic number (Z) A number characteristic of an element, which indicates the number of protons found in the nucleus of one of its atoms.

mass number (A) A number that represents the total of the number of protons and neutrons in the nucleus of a given isotope.

atomic notation A symbolic method for expressing the composition of an atomic nucleus; the mass number and atomic number are indicated to the left of the chemical symbol for the element.

Every element has a characteristic number of protons in its atomic nuclei. This value is called the **atomic number**. The total number of protons and neutrons in the nucleus of an atom is called the **mass number**. A formal shorthand method for keeping track of protons and neutrons in an atom of a particular element is called **atomic notation**. By convention, the symbol of the element (E) is preceded by a superscript and subscript. The superscript, designated as the A value, represents the mass number. The subscript, designated as the Z value, represents the atomic number. Thus,

$$\begin{array}{r} \text{mass number} \longrightarrow \\ \text{atomic number} \longrightarrow \end{array} {}^{A}_{Z}\mathrm{E} \nwarrow \text{symbol of the element}$$

As an example, an atom of sodium may be written ${}^{23}_{11}\mathrm{Na}$. Here the atomic number (protons) is 11 and the mass number (protons plus neutrons) is 23. From this information, we can determine the number of neutrons by subtracting the atomic number from the mass number. The nucleus of this sodium atom contains 12 neutrons ($23 - 11 = 12$). Since atoms are neutral, the number of negative electrons must equal the number of positive protons. Therefore, 11 electrons surround the nucleus. A simple diagram of the sodium atom is as follows.

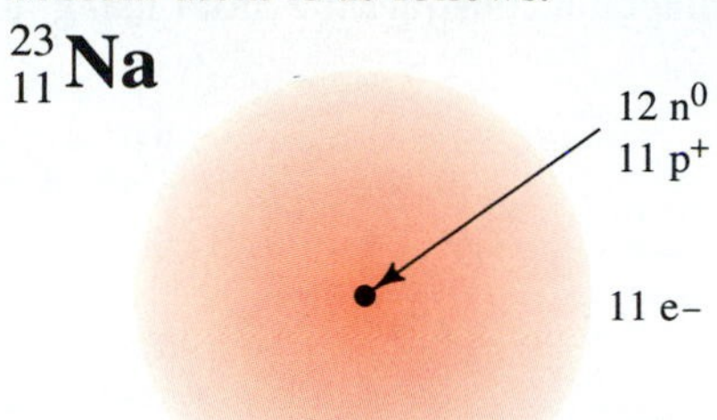

EXAMPLE EXERCISE 4.2

Given the atomic notation for the following atoms, draw a diagram showing the arrangement of protons, neutrons, and electrons.

(a) $^{1}_{1}H$ **(b)** $^{19}_{9}F$ **(c)** $^{197}_{79}Au$

Solution: We will draw a diagram of each atom by placing the protons and neutrons in the nucleus surrounded by electrons.

(a) Since the atomic number is the same as the mass number, this hydrogen nucleus has one proton and no neutrons. Since it has one proton, it also has one electron.

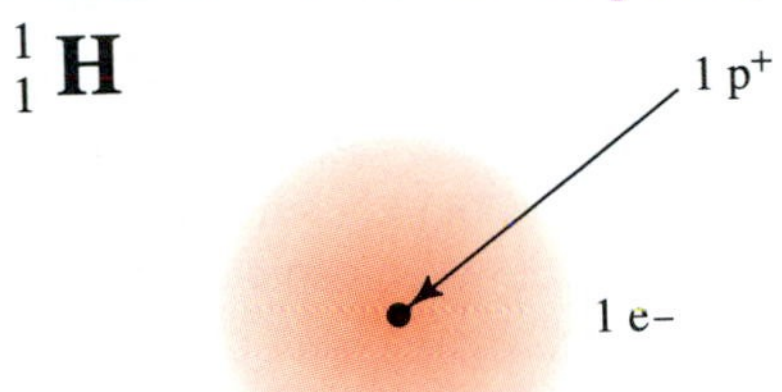

(b) This fluorine atom has nine protons and ten neutrons (19 − 9 = 10). Surrounding the nucleus are nine electrons.

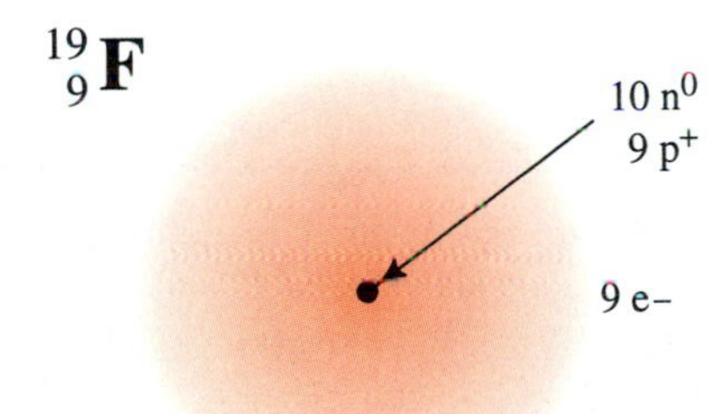

(c) This atom of gold has 79 protons in its nucleus. The number of neutrons is found by subtraction: 197 − 79 = 118. The number of electrons is the same as the number of protons: 79.

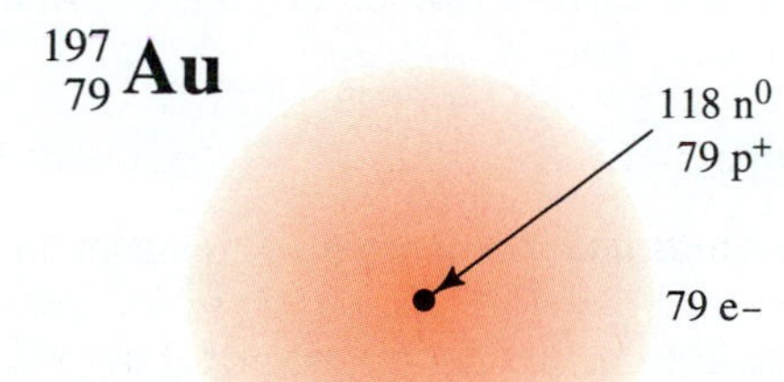

SELF-TEST EXERCISE

Given the following diagram, illustrate the atom using atomic notation.

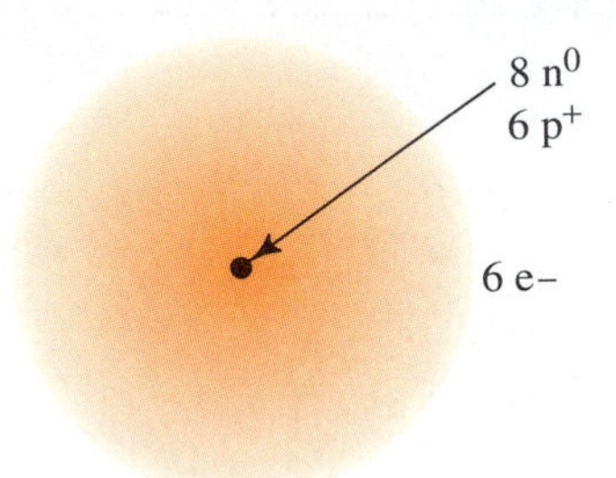

Answer: $^{14}_{6}C$

Often, the convention is used whereby the name of an element is followed by its mass number, such as uranium-235. This notation refers to an atom of uranium having a total of 235 protons and neutrons. Carbon-14, cobalt-60, and strontium-90 are common examples of this notation. Sometimes this convention uses the symbol of the element followed by the mass number, for example, U-235 and C-14.

Isotopes

Only about 20 elements occur in nature with a fixed number of neutrons. For most elements, the number of neutrons in the nucleus varies. Atoms of the same element that have a different number of neutrons in the nucleus are called **isotopes**.

isotopes Atoms having the same atomic number but a different mass number. Atoms of the same element that have a different number of neutrons in the nucleus.

Hydrogen occurs in nature as two stable isotopes. The stable isotopes are protium ($^{1}_{1}H$) and deuterium ($^{2}_{1}H$). A protium atom has only a proton in its nucleus, while deuterium has a proton and neutron. There is a third isotope of hydrogen, called tritium ($^{3}_{1}H$), that has a proton and two neutrons in its nucleus. Tritium is unstable and it radioactively decays into $^{3}_{2}He$. Table 4.3 lists all the stable isotopes for the first 12 elements.

EXAMPLE EXERCISE 4.3

State the number of protons and the number of neutrons in an atom of each of the following isotopes.

(a) $^{37}_{17}Cl$ **(b)** $^{120}_{50}Sn$

Solution: The subscript value refers to the atomic number (p^+) and the superscript value refers to the mass number (p^+ and n^0).

(a) The number of protons in every isotope of chlorine corresponds to its atomic number of 17. To find the number of neutrons in the given isotope, we subtract the atomic number from the mass number. Thus, there are 20 neutrons (37 − 17).

(b) The number of protons in every isotope of tin is 50. In the tin-120 isotope there are 70 neutrons (120 − 50).

SELF-TEST EXERCISE

State the number of protons and the number of neutrons in an atom of each of the following isotopes.

(a) mercury-202 **(b)** U-233

Answers: **(a)** 80 p^+ and 122 n^0; **(b)** 92 p^+ and 141 n^0

Table 4.3 Isotopic Abundance of the First 12 Elements

Atomic Number	Mass Number	Isotope	Isotopic Mass	Percent Abundance*	Atomic Number	Mass Number	Isotope	Isotopic Mass	Percent Abundance*
1	1	$^{1}_{1}H$	1.008	99.985	8	16	$^{16}_{8}O$	15.995	99.759
1	2	$^{2}_{1}H$	2.014	0.015	8	17	$^{17}_{8}O$	16.999	0.037
2	3	$^{3}_{2}He$	3.016	trace					
2	4	$^{4}_{2}He$	4.003	100	8	18	$^{18}_{8}O$	17.999	0.204
3	6	$^{6}_{3}Li$	6.015	7.42	9	19	$^{19}_{9}F$	18.998	100
3	7	$^{7}_{3}Li$	7.016	92.58	10	20	$^{20}_{10}Ne$	19.992	90.92
4	9	$^{9}_{4}Be$	9.012	100	10	21	$^{21}_{10}Ne$	20.994	0.26
5	10	$^{10}_{5}B$	10.013	19.6	10	22	$^{22}_{10}Ne$	21.991	8.82
5	11	$^{11}_{5}B$	11.009	80.4	11	23	$^{23}_{11}Na$	22.990	100
6	12	$^{12}_{6}C$	12.000	98.89	12	24	$^{24}_{12}Mg$	23.985	78.70
6	13	$^{13}_{6}C$	13.003	1.11	12	25	$^{25}_{12}Mg$	24.986	10.13
7	14	$^{14}_{7}N$	14.003	99.63	12	26	$^{26}_{12}Mg$	25.983	11.17
7	15	$^{15}_{7}N$	15.000	0.37					

*The percent abundance is the average amount of the given isotope as it occurs in various samples of the element. The abundance of isotopes is nearly constant from sample to sample throughout the world.

In general, the properties of isotopes of the same element are similar. An extra neutron in the nucleus of an atom has little effect on the properties of the element. Consider an atom of carbon-12, which has six protons and six neutrons in the nucleus. If we add one neutron, we would have an atom of carbon-13. Other than a slight mass change, the properties of C-12 and C-13 are nearly identical.

If, on the other hand, we add one proton to the nucleus, the properties change radically. The atoms would no longer be isotopes. In fact, the addition of one proton would change the element from carbon to nitrogen! Carbon occurs naturally as diamond or graphite, whereas nitrogen is a colorless, odorless gas in our atmosphere.

4.5 Atomic Mass Scale

OBJECTIVES

To gain an understanding of the atomic mass scale.

To become familiar with the mass spectrometer, the instrument for determining the mass of an isotope.

Atoms are much too small to weigh directly. Atoms are so small it is awkward to refer to their mass in grams. Consider that the mass of one carbon atom is 1.99×10^{-23} g. What we need is a more convenient unit for expressing the tiny amount of mass in an atom. The term **atomic mass unit (amu)** was created to meet this need.

atomic mass unit (amu) A unit of mass equal to exactly 1/12 the mass of a C-12 atom.

An atomic mass unit is a relative measure. That is, masses of atoms are compared to carbon-12 and expressed in atomic mass units. Carbon-12 was chosen as a reference and assigned a value of exactly 12 atomic mass units (12 amu). In other words, one atomic mass unit is equal to 1/12 the mass of a carbon-12 atom. The mass of all other atoms is measured relative to the mass of a carbon-12 reference.

Originally, chemists defined an atomic mass unit as 1/16 the mass of an oxygen-16 atom. Physicists, unfortunately, chose naturally occurring oxygen as a reference. Since oxygen occurs naturally as three isotopes, the atomic mass scale differed for chemists and physicists. In 1961, this discrepancy was resolved when all scientists agreed upon carbon-12 as a reference standard.

If atoms are far too light to weigh directly, how can we determine their masses relative to carbon-12? The answer to this question evolved over a period of time. In the late 1700s, experiments revealed the masses of the elements combining in a chemical reaction. John Dalton used this information to devise a relative atomic mass scale. He chose hydrogen as a reference and assigned it a value of 1. Other elements were assigned atomic masses based on the combination of their mass ratios relative to hydrogen. This idea of relative atomic masses continued into the twentieth century.

The present atomic mass scale is a relative scale; that is, it compares the mass of an atom to the mass of carbon-12. These relative atomic masses are determined using an instrument called a **mass spectrometer**. The mass spectrometer measures the deflection of atoms in a magnetic field. Heavier atoms are deflected less by the magnetic field; lighter atoms are influenced more.

mass spectrometer An instrument used to determine the atomic mass of an isotope relative to carbon-12.

The principle of the mass spectrometer can be understood by analogy. Imagine you are standing on a bridge and dropping rocks into a river. The heaviest rocks are only slightly affected by the force of the current and drop straight to the bottom. Medium-size rocks are somewhat affected by the force of the water and are swept a few meters downstream. The lightest rocks are most affected by the current and are carried several meters downstream before striking the bottom (Figure 4.9). The magnetic field in a mass spectrometer has an effect similar to the current in the river. Atoms passing through a magnetic field, like the rocks passing through the water, are thus separated according to their mass.

The mass spectrometer operates as follows. First, an element is injected into the instrument and vaporized to individual atoms. Next, an electron beam knocks electrons out of the passing atoms. This creates positively charge atoms called ions. The positive ions are accelerated between two charged plates and shot through a

Figure 4.9 Analogy for the Mass Spectrometer The force of the water current has a greater effect on lighter rocks than on heavier ones. Thus, rocks of different mass are separated as they fall through the water. The force of the current on the rocks is analogous to the effect of a magnetic field on isotopes in a mass spectrometer.

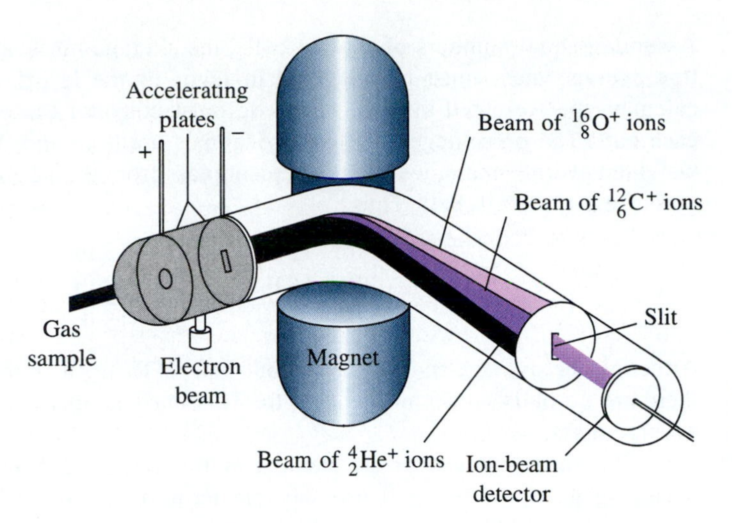

Figure 4.10 The Principle of a Mass Spectrometer Carbon-12 is being detected in the presence of the lighter isotope He-4 and the heavier isotope O-16. Decreasing the magnetic field allows He-4 to be detected. By increasing the magnetic field, O-16 is detected.

magnetic field to strike a detector. Heavier charged atoms are affected less by the magnetic field. Lighter ions are affected more. A simplified diagram of a mass spectrometer is shown in Figure 4.10.

The mass spectrometer is a very sensitive instrument. It can even distinguish between two isotopes of the same element. For instance, the mass spectrometer can separate carbon-12 and carbon-13. After detecting carbon-12, a slight increase of the magnetic field allows carbon-13 to be detected. Continuing to adjust the magnetic field allows other isotopes to be detected. The mass of an isotope is determined by comparing the strength of its magnetic field to that of carbon-12. The masses of all isotopes are calculated relative to carbon-12.

4.6 Atomic Mass

OBJECTIVES

To calculate the atomic mass for an element, given the mass and percent abundance of each naturally occurring isotope.

To state the atomic mass of an element by referring to the periodic table of the elements.

By the beginning of the twentieth century, the mass spectrometer was available to scientists. This meant the mass of an atom could be found with great accuracy. Even the masses of isotopes could be determined. To calculate the atomic mass of an element, it is necessary to consider the natural abundance of each isotope of that element. Averaging the masses of all isotopes gives the atomic mass of the element. There is a subtle trick, however, for averaging the isotopic masses. We must find an average mass that is weighted in favor of the most abundant isotope. Before proceeding, let's digress to an analogy.

A company manufactures both 12-pound and 16-pound metal balls for the shotput, a track and field event. Of the total number of balls manufactured, 25% are 12 lb and 75% are 16 lb. What is the average mass of a shotput ball? If we add 12 lb + 16 lb and divide by 2, we find

$$\frac{12 \text{ lb} + 16 \text{ lb}}{2} = 14 \text{ lb}$$

Assuming equal numbers of shotput balls, the average mass is 14 lb. However, the true average mass must be weighted in favor of the larger number of balls. To calculate the weighted average mass, we must consider the percent abundance of each ball. The percentages written as decimals are 0.25 and 0.75. To arrive at the weighted average mass, we add the percent mass from the 12-lb ball plus the percent mass from the 16-lb ball. Thus,

$$\begin{aligned} 12\text{ lb:} \quad 12\text{ lb} \times 0.25 &= 3\text{ lb} \\ 16\text{ lb:} \quad 16\text{ lb} \times 0.75 &= \underline{12\text{ lb}} \\ & 15\text{ lb} \end{aligned}$$

Although the average mass of a shotput ball is 15 lb, it is important to note that there are no balls with a mass of 15 lb. This mass simply represents the weighted average mass.

atomic mass The weighted average mass of all the naturally occurring isotopes of an element.

The **atomic mass** of an element is the weighted average of all naturally occurring isotopes. We will use separate terms to distinguish atomic mass from isotopic mass. The isotopic mass is the mass of a given isotope. The atomic mass represents the average mass of all naturally occurring isotopes of an element. The weighted atomic mass of an element is frequently referred to as *atomic weight*.

To calculate the weighted average mass, we will use a method similar to the one in the shotput example. To calculate the weighted atomic mass, we must know the mass of each isotope and its percent abundance.

EXAMPLE EXERCISE 4.4

Carbon occurs naturally as two isotopes. Calculate the atomic mass for carbon, given the mass and natural abundance of each isotope.

Isotope	Mass (amu)	Abundance (%)
C-12	12.00000	98.89
C-13	13.00335	1.11

Solution: The mass contribution from C-12 plus the mass contribution from C-13 equals the atomic mass of an average carbon atom. The natural abundances expressed as decimals are 0.9889 and 0.0111.

$$\begin{aligned} \text{C-12:} \quad 12.00000\text{ amu} \times 0.9889 &= 11.87\text{ amu} \\ \text{C-13:} \quad 13.00335\text{ amu} \times 0.0111 &= \underline{0.144\text{ amu}} \\ & 12.01\text{ amu} \end{aligned}$$

The weighted atomic mass of carbon is 12.01 amu. Is there a single carbon atom anywhere in nature with a mass of 12.01 amu? No. The atomic mass is a weighted average of all naturally occurring carbon atoms.

SELF-TEST EXERCISE

Copper occurs naturally as two isotopes. Calculate the atomic mass for copper given the isotopic data.

Isotope	Mass (amu)	Abundance (%)
Cu-63	62.930	69.09
Cu-65	64.928	30.91

Answer: 63.55 amu

EXAMPLE EXERCISE 4.5

Silicon, the second most abundant element in the earth's crust is used extensively in the electronics industry. Calculate the atomic mass of silicon, given the mass and abundance of its three natural isotopes.

Isotope	Mass (amu)	Abundance (%)
Si-28	27.977	92.21
Si-29	28.976	4.70
Si-30	29.974	3.09

Solution: We can find the atomic mass by adding the weighted mass contribution from each isotope.

$$\begin{aligned} \text{Si-28: } & 27.977 \text{ amu} \times 0.9221 = 25.80 \text{ amu} \\ \text{Si-29: } & 28.976 \text{ amu} \times 0.0470 = 1.36 \text{ amu} \\ \text{Si-30: } & 29.974 \text{ amu} \times 0.0309 = \underline{0.926 \text{ amu}} \\ & \qquad\qquad\qquad\qquad\quad 28.09 \text{ amu} \end{aligned}$$

The average mass of a single silicon atom is 28.09 amu. Note, however, that there are no silicon atoms with a mass of 28.09 amu.

SELF-TEST EXERCISE

Zinc occurs naturally as five isotopes. Calculate the atomic mass for zinc, given the isotope data.

Isotope	Mass (amu)	Abundance (%)
Zn-64	63.9291	48.89
Zn-66	65.9260	27.81
Zn-67	66.9271	4.11
Zn-68	67.9249	18.57
Zn-70	69.9253	0.62

Answer: 65.39 amu

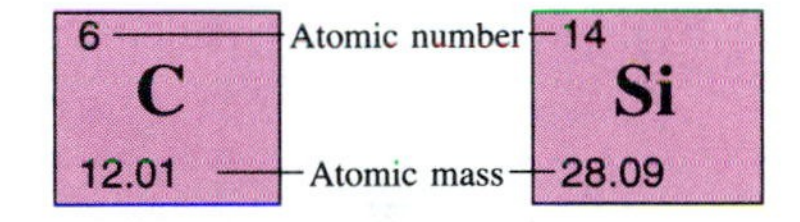

Figure 4.11 Carbon and Silicon from the Perodic Table of the Elements The atomic number designates the number of protons. The atomic mass is the weighted average mass of all naturally occurring isotopes of the element.

The Periodic Table

The atomic mass for each element has been calculated from the masses and abundances of their naturally occurring isotopes. These atomic mass values are usually listed in a table in chemistry textbooks. Even more conveniently, the atomic masses accompany the symbols of the elements in the periodic table. In this textbook the atomic number is printed above the symbol, and the atomic mass is printed below the symbol. Figure 4.11 illustrates carbon and silicon as they appear in the periodic table.

The atomic mass number of an element represents the average value for all its naturally occurring isotopes. From the periodic table, however, it is not possible to tell if an element has more than one isotope. Of the first 83 elements, 81 have one or more stable isotopes. Only technetium (element 43) and promethium (element 61) are radioactive and unstable.

The periodic table of elements in the front of this book lists an important isotope for radioactive elements as shown in Figure 4.12. The value in parentheses below the symbol of the element represents the mass number of the most stable or best known radioisotope. The mass number is not an atomic mass, but rather the

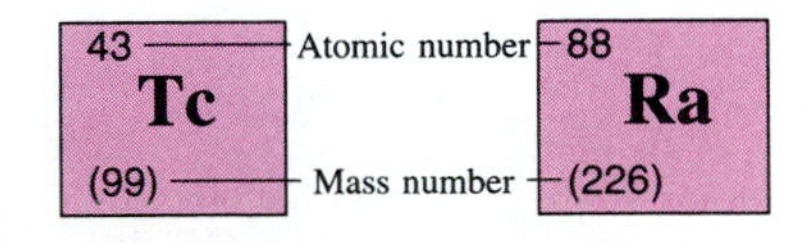

Figure 4.12 Radioactive Elements from the Periodic Table of Elements An important isotope of technetium, Tc, has a mass number of 99. The radioactive isotope shown for radium has a mass number of 226.

total number of protons and neutrons in the nucleus. From the periodic table, you can easily distinguish the elements that are stable from those that are unstable. Those elements that are radioactive and unstable list a whole-number value in parentheses below the symbol of the element.

EXAMPLE EXERCISE 4.6

Refer to the periodic table and obtain the following information.
(a) atomic number and atomic mass of iron
(b) atomic number and mass number of plutonium (Pu)

Solution: In the periodic table of the elements we observe:

26
Fe
55.85

(a) The atomic number of iron is 26; that is, Fe has 26 protons in the nucleus. The atomic mass of Fe is 55.85 amu.
(b) The atomic number of plutonium is 94; that is, Pu has 94 protons in the nucleus. The mass number shown for radioactive Pu is 244.

SELF-TEST EXERCISE

Refer to the periodic table and predict which of the following elements are radioactive and have no stable isotopes.
(a) promethium, Pm
(b) mercury, Hg
(c) cobalt, Co
(d) radon, Rn

Answers: **(a)** Pm and **(d)** Rn are radioactive.

4.7 Evidence for Electron Energy Levels: The Bohr Model

To describe the Bohr model of the atom.

To understand the relationship between electron energy levels in an atom and the lines in an emission spectrum.

In 1911, a brilliant 25-year-old Dane completed his doctorate in physics and left for England to begin his postdoctoral work under J.J. Thomson. After a few months, however, the Dane left Cambridge to join Ernest Rutherford at Manchester. This was the same year that Rutherford and his co-workers discovered the atomic nucleus. It would take only two years before the young Dane would raise our understanding of the atom to yet another level. This reflective young intellectual from Denmark was Niels Bohr (1885–1962).

Before continuing with Niels Bohr, we have to regress to 1900. In that year Max Planck (1858–1947), a German professor of physics, presented a revolutionary concept. Planck's concept was that since matter comes in lumps, energy might also come in lumps. We call a lump of matter an atom. Planck called a lump of light energy a **quantum** (*pl.* quanta). What made this idea so revolutionary was that, previously, light was thought to be a continuous wave of energy. By analogy, consider a wave of water.

quantum (*pl.* quanta) A single particle of light energy.

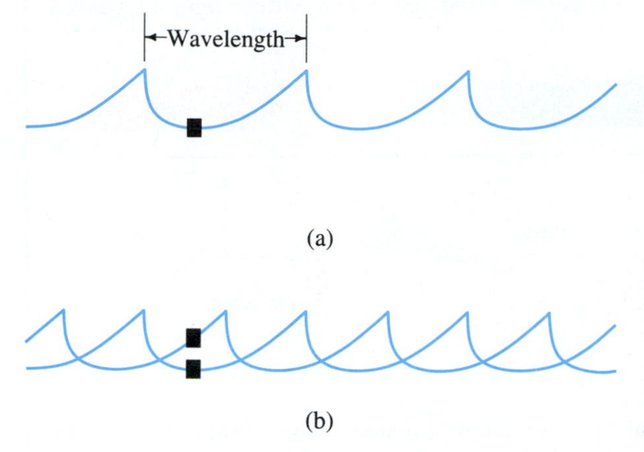

A water wave and light wave are similar. (a) The distance between peaks is the *wavelength*. (b) The number of times per second the cork bobs up and down is the *frequency*.

We can think of light as traveling in a wavelike fashion similar to an ocean wave. The **wavelength** of light is the distance the wave travels to complete one cycle. Short-wavelength light is more energetic than long-wavelength light. The **frequency** of light is the number of cycles completed in 1 second. High-frequency light is more energetic than low-frequency light. Figure 4.13 illustrates the wave nature of light.

wavelength The distance a light wave travels to complete one cycle.

frequency The number of times a light wave completes a cycle in 1 second.

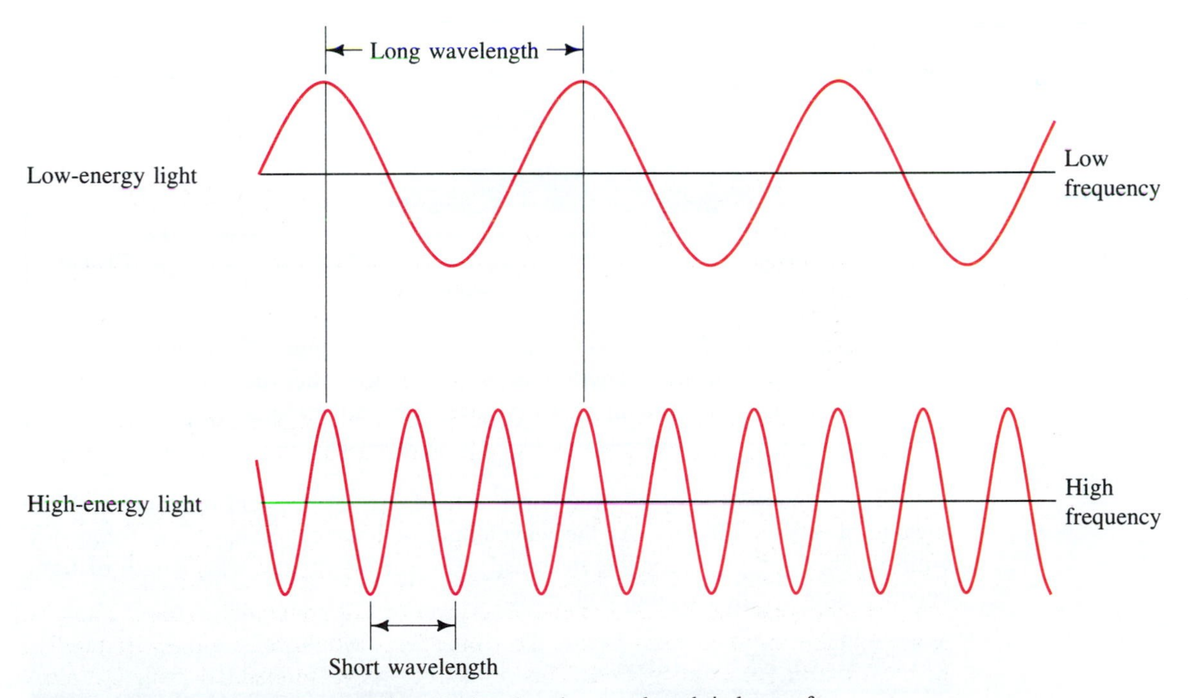

Figure 4.13 Wave Nature of Light Notice that the wavelength is longer for low energy light than it is for high energy light. As the wavelength gets shorter, the frequency increases and the energy is greater.

radiant energy spectrum Light energy extending from short-wavelength X-rays through long-wavelength microwaves; also termed light energy spectrum or electromagnetic spectrum.

visible spectrum Light energy that is observed as violet, blue, green, yellow, orange, or red; the radiant energy spectrum from approximately 400–700 nm.

continuous spectrum A broad, uninterrupted band of light energy.

Radiant Energy: A Continuous Spectrum

Light is a form of radiant energy that travels with wavelike motion. Its velocity is constant at 300,000,000 meters per second, regardless of its wavelength. What we observe as white light is actually several colors of different wavelength. The various wavelengths are separated when white light is passed through a prism.

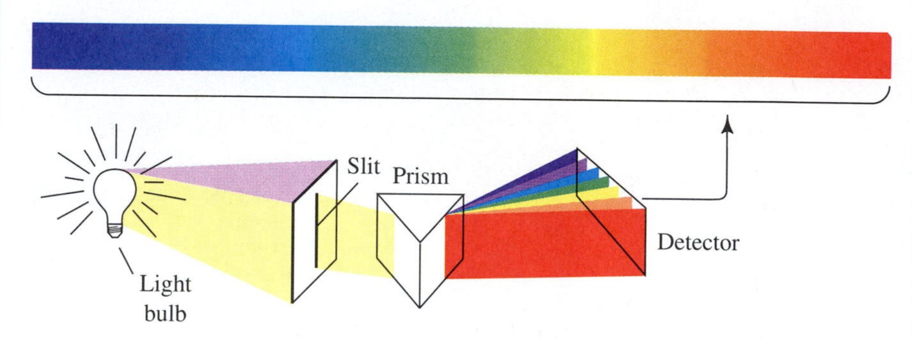

A rainbow is a natural phenomenon created when sunlight passes through raindrops. Individual drops of rain act as miniature prisms to separate the various bands of color in sunlight. Visible light is only one small portion of an entire spectrum. The **radiant energy spectrum** includes X-ray, ultraviolet, visible, infrared, and microwave radiation. Our eyes can detect light in the **visible spectrum** but not in the ultraviolet or infrared. The wavelength of ultraviolet radiation is too short to be seen; infrared radiation is too long to be seen. The wavelength of light is expressed in nanometers (nm). Recall that a nanometer is one-billionth of a meter. The range of visible light, violet to red, is about 400–700 nm.

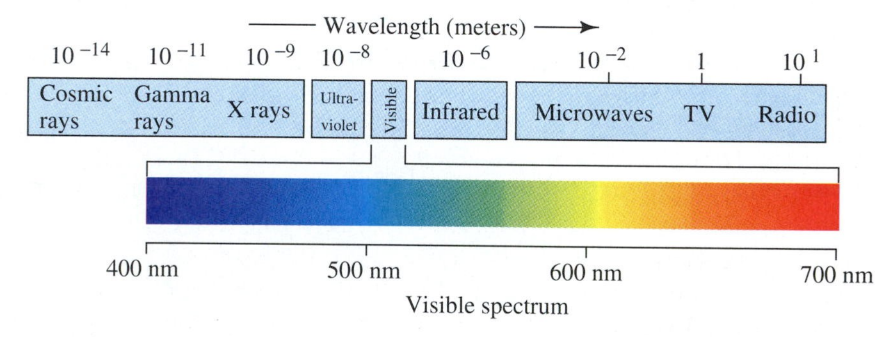

Light, a **continuous spectrum** of energy, ranges from high-energy X-rays through long-wavelength microwaves. Notice that the visible spectrum is only a selected window in a vast panorama of radiant energy.

The velocity of light in a vacuum is a fundamental constant in nature and has a value of 3.00×10^8 m/s. As the wavelength of light gets longer, the frequency with which it completes a cycle decreases. Conversely, as the wavelength of light becomes shorter, the frequency increases.

EXAMPLE EXERCISE 4.7

What is the relationship between each of the following?

(a) wavelength and frequency **(b)** wavelength and energy

Solution: Refer to Figure 4.13.

(a) As the wavelength *decreases,* the frequency *increases* in order for the light to maintain a constant velocity. If the wavelength increases, the frequency simultaneously decreases.

(b) The energy of light radiation *decreases* as the wavelength gets *longer.* Conversely, the energy increases as the wavelength becomes shorter. Thus, violet light is more energetic than red light. As the wavelength gets shorter, the frequency increases and the energy is greater.

SELF-TEST EXERCISE

Select the wavelength of light that is most energetic in each of the following.

(a) blue, yellow, red **(b)** 625 nm, 575 nm, 470 nm

Answers: **(a)** blue; **(b)** 470 nm

The Bohr Atom

Soon after Rutherford established the nuclear atom, he speculated that electrons revolved around the nucleus just as the planets revolve around the sun. This was called the planetary model of the atom. Bohr became a proponent of this model and speculated that electrons revolved about the nucleus in a circular orbit. He further suggested that the electron orbits were at a fixed distance from the nucleus. Since the orbit distance was fixed, the electron possessed a definite energy. In other words, the electron was said to occupy an orbit of discrete energy, which was referred to as an electron **energy level**. Electrons could be found only in specified energy levels and nowhere else. Figure 4.14 illustrates the model of the hydrogen atom as visualized by Bohr. This model is usually referred to as the Bohr model of the atom, or simply the **Bohr atom**.

energy level An orbit of specific energy that electrons occupy at a fixed distance from the nucleus; designated 1, 2, 3, 4. . . .

Bohr atom A model of the atom that pictures the electron circling the nucleus in an orbit of specific energy.

The Bohr model was a beautiful mental picture of electrons in atoms. But no one knew whether the model was right or wrong because there was no experimental evidence to support Bohr's theory. At this time, quite by coincidence, Bohr received a research paper on the emission of light from excited hydrogen gas. The paper had been written 25 years earlier by the Swiss mathematician and physicist Johann Jakob Balmer (1825–1898). After examining the paper, Bohr found that excited hydrogen gas emits separate discrete wavelengths of light, rather than a continuous spectrum. The three most prominent emission lines are red, blue-green, and violet.

Figure 4.15 shows the emission spectrum of hydrogen. An emission spectrum is produced when hydrogen gas is excited by an electrical voltage. To do so, first, hydrogen gas is sealed in a gas discharge tube. Second, an electrical voltage is used

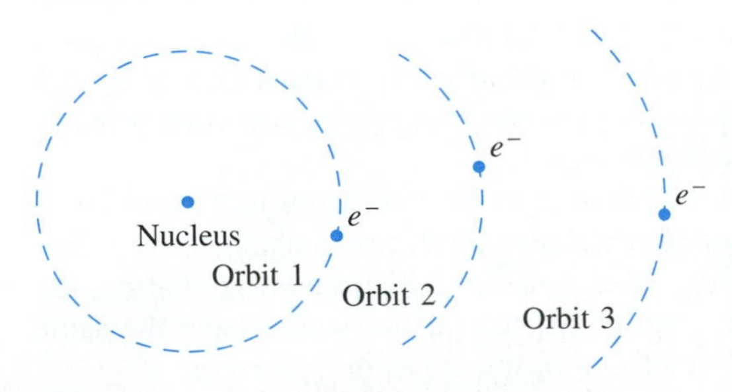

Figure 4.14 The Bohr Model of the Hydrogen Atom The electron is a specific distance from the nucleus and occupies an orbit of discrete energy. According to the Bohr model, the electron is found only in a given energy level.

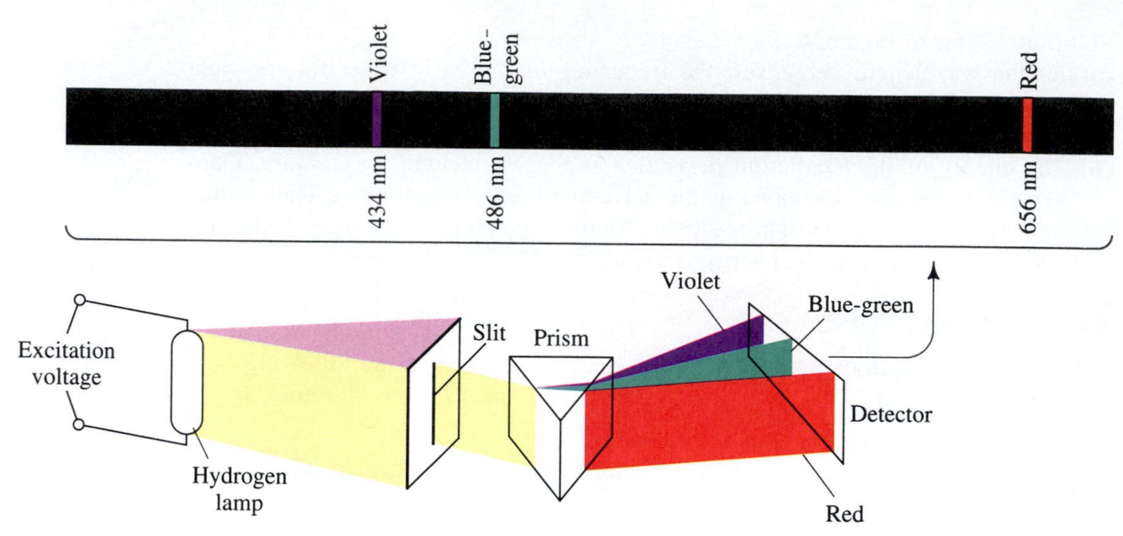

Figure 4.15 Hydrogen Emission Spectrum An emission line spectrum is produced when hydrogen gas is excited by an electrical voltage. After the emitted light is passed through a prism, three discrete vivid lines are observed.

emission line spectrum A collection of narrow slits of light energy that results from excited atoms of a given element releasing energy.

spectral lines The individual narrow slits of light in an emission line spectrum.

to energize the hydrogen gas. The discharge tube then releases light, which separates into a series of narrow lines when passed through a prism. This collection of narrow slits of light energy is referred to as an **emission line spectrum**. The individual slits of light are called **spectral lines**.

After reading the paper on the spectrum of hydrogen, Bohr realized he had the experimental evidence to support his model of the atom. The concept of electron energy levels was supported by the line spectrum of hydrogen. When an electrical voltage is applied to hydrogen, its atoms are excited in the gas discharge tube. An excited atom forces the electron to a higher energy orbit, for example, from 1 to 2 or from 1 to 3. Since this is an unstable state, the electron quickly drops from the higher level back to a lower level, for example, from 4 to 3 or from 3 to 2. In the process, the electron loses a discrete amount of energy. This discrete energy loss corresponds to a quantum of light energy. That is, the emitted quantum of light has the same amount of energy as that lost by the electron as it dropped from a higher to a lower energy level.

The experimental evidence fit Bohr's model of the atom perfectly. Recall, however, that it was Planck's idea that light was composed of particles, or quanta. Planck, therefore, opened the door for Bohr's insight. In the hydrogen spectrum, Figure 4.15, there are three bright lines that are red, blue-green, and violet. The red line has the longest wavelength of the three and the lowest energy. Further experiments revealed that the red line corresponds to an excited electron dropping from energy level 3 to 2. The blue-green line is more energetic than the red and corresponds to an excited electron dropping from energy level 4 to 2. The violet line is the most energetic of the three and is produced when an electron drops from energy level 5 to 2. Figure 4.16 illustrates the correlation between Bohr's electron energy levels and the observed lines in the hydrogen spectrum.

Since a single quantum is emitted each time an electron drops to a lower level, it follows that several quanta are emitted when several electrons change levels. For example, if ten electrons drop from the level 5 to level 2 in a hydrogen atom, ten quanta would be emitted. Moreover, each of the ten quanta would have the same energy and would be observed as a violet line in the emission spectrum.

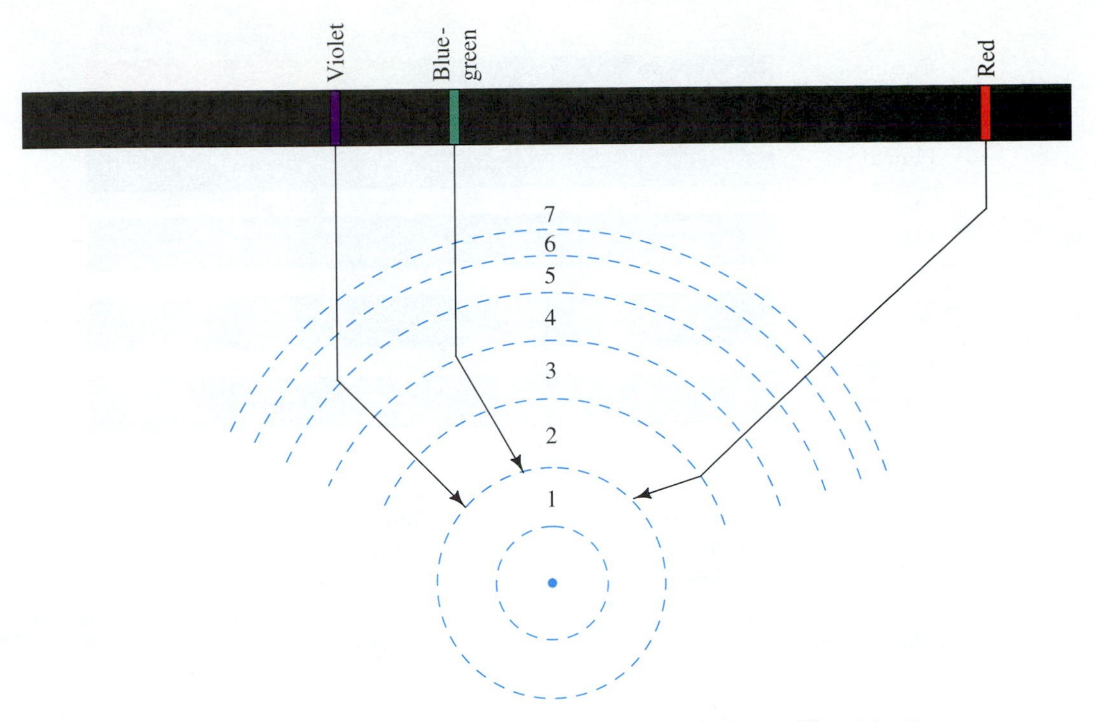

Figure 4.16 Relating Energy Levels in Hydrogen to Spectral Lines The violet line is emitted when an electron drops from the 5th to 2nd orbit. The blue-green line corresponds to the transition from the 4th to 2nd; the red line is produced when the electron drops from the 3rd to 2nd energy level.

EXAMPLE EXERCISE 4.8

Explain the relationship between an observed emission line and electron energy levels.

Solution: When an electron drops from a higher energy level to a lower energy level, light is emitted. For each electron that drops between levels, a single quantum of light energy is emitted. Moreover, the energy lost by the electron as it drops corresponds exactly to the energy of the quantum of light that is emitted. Several quanta of light having the same energy are observed as a spectral line.

SELF-TEST EXERCISE

Indicate the number of quanta emitted and the color of the spectral line for each of the following electron transitions occurring in hydrogen.

(a) One electron dropping from energy level 3 to 2
(b) 100 electrons dropping from energy level 3 to 2
(c) 100 electrons dropping from energy level 4 to 2
(d) 500 electrons dropping from energy level 5 to 2

Answers: **(a)** 1 quantum, red; **(b)** 100 quanta, red; **(c)** 100 quanta, blue-green; **(d)** 500 quanta, violet.

Further study of emission spectra revealed that each element produced a unique set of spectral lines. This observation indicated that the energy levels must be unique for atoms of each element. A line spectrum is therefore sometimes referred to as an ''atomic fingerprint.'' In 1868, the atomic fingerprint of a new element was observed

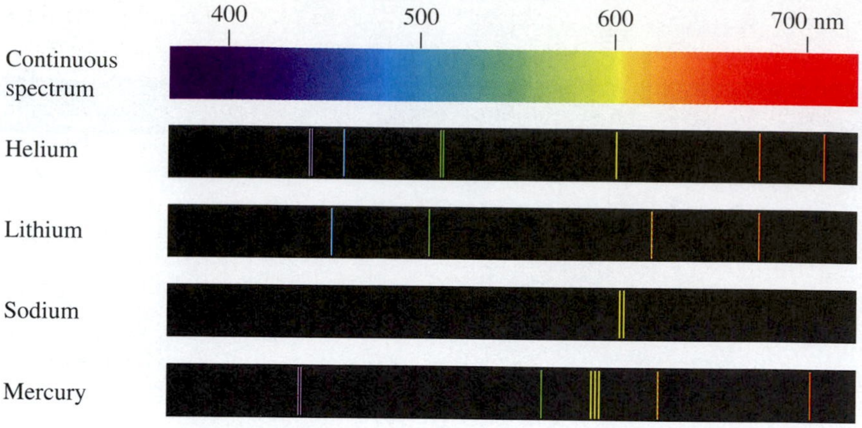

Figure 4.17 Continuous Spectrum versus Line Spectrum A continuous spectrum is produced from an ordinary light bulb. The emission line spectra are produced from the excited elements in the gaseous state.

in the spectrum from the sun. The element was named helium, after *helios*, the Greek word for sun. In 1895, an element was discovered in uranium ore with an identical atomic fingerprint. Thus, helium was discovered on earth, 27 years after it had first been observed in the sun's spectrum. Figure 4.17 compares the emission line spectra of four elements.

The Balmer Formula

Balmer formula A mathematical equation for calculating the emitted wavelength of light from an excited hydrogen atom when an electron drops to the second energy level.

Niels Bohr was born in 1885, the same year Balmer published his paper on the emission spectrum of hydrogen. In addition, his paper included a formula that accounted for the visible lines in the spectrum of hydrogen. After a simple mathematical analysis, Balmer showed that the wavelength (λ) of a hydrogen spectral line was related to a small whole number (n). The **Balmer formula** allows for the calculation of wavelengths of light as follows:

$$\frac{1}{\lambda} = \frac{1}{91\text{ nm}}\left(\frac{1}{2^2} - \frac{1}{n^2}\right)$$

When Balmer set $n = 3$, the calculated wavelength (λ) was equal to 650 nm. The calculation is as follows:

$$\frac{1}{\lambda} = \frac{1}{91\text{ nm}}\left(\frac{1}{2^2} - \frac{1}{3^2}\right)$$

$$= \frac{1}{91\text{ nm}}\left(\frac{1}{4} - \frac{1}{9}\right)$$

$$= \frac{1}{91\text{ nm}}(0.25 - 0.11)$$

$$= \frac{0.14}{91\text{ nm}}$$

Taking the reciprocal of each fraction, we obtain the calculated wavelength value.

$$\lambda = \frac{91\text{ nm}}{0.14} = 650\text{ nm}$$

The calculated wavelength, 650 nm, agrees with the observed wavelength for the red line in the emission spectrum of hydrogen. That is, the red line in the hydrogen spectrum is experimentally observed to have a wavelength of 650 nm.

Substituting $n = 4$ and $n = 5$ gave values for the blue-green and violet lines observed in the spectrum of hydrogen. When Bohr read Balmer's paper, he realized that the n value could represent an electron energy level. Furthermore, the 2 in the formula corresponded to the second energy level. Thus, the lines in the emission spectra of the elements provided Bohr with the necessary experimental evidence to support his concept of electron energy levels.

Although the emission spectrum of hydrogen has three vivid lines, there are other lines that are visible but faint. Let's perform a sample calculation to find the wavelength of one of these lines.

EXAMPLE EXERCISE 4.9

Calculate the wavelength of light corresponding to the energy released when the electron drops from $n = 6$ to the second level in a hydrogen atom.

Solution: We can use the Balmer formula to calculate the wavelength of this line in the spectrum. Substituting 6 for n gives

$$\frac{1}{\lambda} = \frac{1}{91 \text{ nm}}\left(\frac{1}{2^2} - \frac{1}{6^2}\right)$$

Simplifying, we have

$$\frac{1}{\lambda} = \frac{1}{91 \text{ nm}}\left(\frac{1}{4} - \frac{1}{36}\right)$$

$$= \frac{1}{91 \text{ nm}}(0.25 - 0.03)$$

$$= \frac{0.22}{91 \text{ nm}}$$

Taking the reciprocal of each fraction, we obtain

$$\lambda = \frac{91 \text{ nm}}{0.22} = 410 \text{ nm}$$

The calculation reveals that there is a line in the hydrogen spectrum at 410 nm. A line of this wavelength is in the violet portion of the spectrum.

SELF-TEST EXERCISE

Calculate the wavelength of light corresponding to the energy released when the electron drops from $n = 10$ to the second level in a hydrogen atom.

Answer: 380 nm

Several years after Balmer had introduced his formula, the Swedish physicist Johannes Rydberg (1854–1919) derived a more general equation for calculating the wavelengths of spectral lines. The Rydberg equation accounts for an electron dropping from a higher energy level (n_H) to any lower energy level (n_L). The Rydberg equation can be written as

$$\frac{1}{\lambda} = \frac{1}{91 \text{ nm}}\left(\frac{1}{n_L^2} - \frac{1}{n_H^2}\right)$$

The Balmer formula works well for the visible lines in the spectrum of hydrogen. This is because visible lines are produced when the electron drops to $n = 2$. However, the Rydberg equation also accounts for spectral lines in the ultraviolet and infrared portion of the spectrum. When the electron drops to $n = 1$, the lines are not visible to the eye. These lines are in the ultraviolet region of the spectrum. When the electron falls to $n = 3$, the energy released corresponds to lines in the infrared region of the spectrum.

4.8 Principal Energy Levels and Sublevels

To designate the energy sublevels within a given electron energy level.

To determine the maximum number of electrons that can occupy a given energy level or sublevel.

In 1913, Niels Bohr had proposed a model for the atom that included electrons around the nucleus in levels of fixed energy. He proposed that electrons occupied circular orbits and traveled about the nucleus. His proposal was supported experimentally by the lines in the emission spectrum of hydrogen. The emission spectra of other elements, however, had far too many lines to interpret. Although Bohr could not explain the fine detail in spectra other than hydrogen, he did suggest the idea that electrons occupied sublevels within main energy levels. The picture that gradually emerged had electrons distributed in one or more energy sublevels within a **principal energy level**. An electron **energy sublevel** is designated *s*, *p*, *d*, or *f*, which is a reference to the *sharp*, *principal*, *diffuse*, and *fine* lines in the emission spectra of the elements.

principal energy level A main energy level composed of sublevels.

energy sublevel An electron energy level resulting from the splitting of a principal energy level; designated *s*, *p*, *d*, *f*. . . .

The number of sublevels in an energy level corresponds to the number of the principal energy level. That is, the first principal energy level (1) has one sublevel, which is designated 1*s*. The second principal energy level (2) is split into two sublevels, designated 2*s* and 2*p*. The third principal energy level (3) is split into three sublevels, designated 3*s*, 3*p*, and 3*d*. The fourth principal energy level (4) is composed of 4*s*, 4*p*, 4*d*, and 4*f* sublevels (Figure 4.18).

The maximum number of electrons that can be placed into each energy sublevel is determined from mathematical equations. For now, accept that an *s* sublevel can have a maximum of two electrons. A *p* sublevel can have six electrons. A *d* sublevel can have ten electrons. And an *f* sublevel can have a maximum of fourteen electrons.

To find the maximum number of electrons in a principal energy level, we add

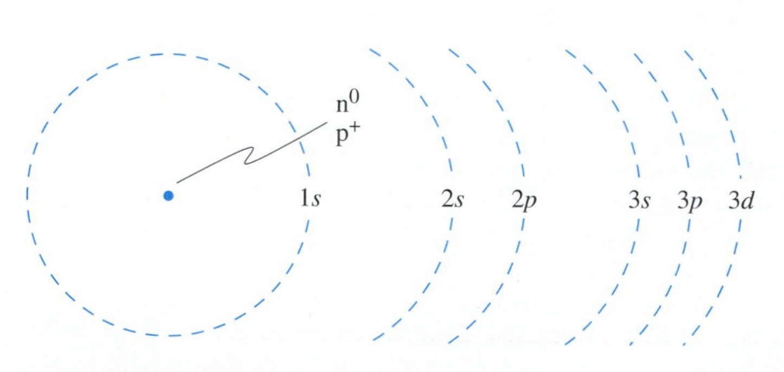

Figure 4.18 A Cross-Section of an Atom The first principal energy level has only one sublevel (1*s*). The second principal energy level has two sublevels (2*s* and 2*p*). The third principal energy level has three sublevels (3*s*, 3*p* and 3*d*). Although the diagram suggests electrons travel in circular orbits, this is a simplification and is not actually the case.

up the electrons in each sublevel. The first energy level has one s sublevel; it can only contain 2 e^-. The second major energy level has two sublevels, a $2s$ and $2p$. The $2s$ can hold 2 e^- and the $2p$ can hold 6 e^-. Thus, the second energy level can hold a maximum of 8 e^-.

The following example further illustrates the relationship of principal energy levels, sublevels, and the maximum number of electrons.

EXAMPLE EXERCISE 4.10

State the sublevels in the third principal energy level. Find the maximum number of electrons that can occupy the third principal energy level.

Solution: The number of sublevels within an energy level corresponds to the number of the principal energy level. The third principal energy level is split into three sublevels: $3s$, $3p$, and $3d$. The maximum number of electrons that can occupy a given sublevel is as follows:

$$\begin{aligned} s \text{ sublevel} &= 2\ e^- \\ p \text{ sublevel} &= 6\ e^- \\ d \text{ sublevel} &= 10\ e^- \end{aligned}$$

The maximum number of electrons in the third principal energy level is, therefore, found by adding the three sublevels together:

$$\begin{aligned} 3s + 3p + 3d &= \text{total electrons} \\ 2\ e^- + 6\ e^- + 10\ e^- &= 18\ e^- \end{aligned}$$

The third principal energy level can have a maximum of 18 electrons.

SELF-TEST EXERCISE

What sublevels are in the fourth principal energy level? What is the maximum number of electrons that can occupy the fourth principal energy level?

Answers: $4s$, $4p$, $4d$, $4f$; 32 e^- (2 e^- + 6 e^- + 10 e^- + 14 e^-)

Electrons are arranged about the nucleus in sublevels that represent a specific energy value. Electrons further from the nucleus are considered to be more energetic because they are less influenced by the dense positive nucleus. Table 4.4 summarizes

Table 4.4 Distribution of Electrons by Energy Level

Principal Energy Level	Energy Sublevel	Maximum Electrons in Sublevel	Maximum Electrons in Principal Level
1	$1s$	2 e^-	2 e^-
2	$2s$	2 e^-	
	$2p$	6 e^-	8 e^-
3	$3s$	2 e^-	
	$3p$	6 e^-	
	$3d$	10 e^-	18 e^-
4	$4s$	2 e^-	
	$4p$	6 e^-	
	$4d$	10 e^-	
	$4f$	14 e^-	32 e^-

the relationships of principal energy levels, sublevels, and the number of electrons.

4.9 Arrangement of Electrons by Energy Sublevel

To list the order of sublevels according to increasing energy.

To write the predicted electron configuration for selected elements.

Electrons are arranged about the nucleus in a regular manner. The first electrons fill the energy sublevel closest to the nucleus. Additional electrons fill energy sublevels farther and farther from the nucleus. In other words, each energy level is filled sublevel by sublevel. The *s* sublevel is filled before a *p* sublevel, a *p* sublevel is filled before a *d* sublevel, and a *d* sublevel is filled before an *f* sublevel in each energy level.

In general, the sublevels are higher in energy as the principal energy level increases. Therefore, we would expect the order of sublevel filling to be 1*s*, 2*s*, 2*p*, 3*s*, 3*p*, 3*d*, 4*s*, and so on. This is not quite accurate. There are exceptions. For instance, the 4*s* sublevel is lower in energy than the 3*d*, and the 5*s* is lower than the 4*d*. A partial list of sublevels in the order of increasing energy is $1s < 2s < 2p < 3s < 3p < 4s < 3d < 4p < 5s < 4d < 5p < 6s$.

In Chapter 5 you will learn to predict how energy sublevels are filled from the position of the element in the periodic table. In fact, the unusual shape of the periodic table is the result of arranging sublevels according to increasing energy. For now, you should memorize the order of sublevel filling or refer to Figure 4.19.

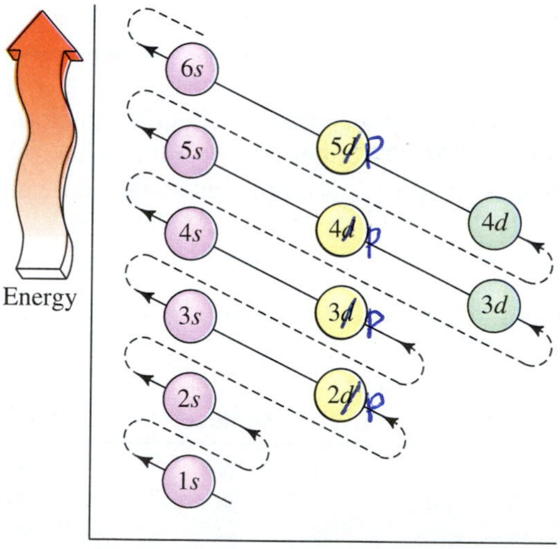

Figure 4.19 Filling Diagram for Energy Sublevels The order of sublevel filling is arranged according to increasing energy. Electrons first fill the 1*s* sublevel followed by the 2*s*, 2*p*, 3*s*, 3*p*, 4*s*, 3*d*, 4*p*, 5*s*, 4*d*, 5*p*, and 6*s*.

EXAMPLE EXERCISE 4.11

State the energy sublevel that accepts electrons immediately after each of the following sublevels are filled.

(a) 3*p* **(b)** 4*d*

Solution: If you have not memorized the order of sublevels, refer to the filling diagram in Figure 4.19.

(a) Although the third principal energy level has 3*s*, 3*p*, and 3*d* sublevels, the 3*d* does not immediately follow the 3*p*. Instead, the 4*s* sublevel follows the 3*p* and precedes the 3*d*. Thus,

3*s*, 3*p*, 4*s*

(b) Although the fourth energy level has 4*s*, 4*p*, 4*d*, and 4*f* sublevels, the 4*f* does not immediately follow the 4*d*. Instead, the 5*p* sublevel begins accepting electrons after the 4*d* is filled. Thus,

4*p*, 5*s*, 4*d*, 5*p*

SELF-TEST EXERCISE

State the energy sublevel that immediately follows each of the following sublevels according to increasing energy.

(a) 2*s* **(b)** 5*p*

Answers: **(a)** 2*p*; **(b)** 6*s*

Note: An alternate mnemonic device for keeping track of sublevels according to increasing energy is given in Figure 4.20. In this memory aid, the bottom tip of the triangle points toward the nucleus. Notice that *s* sublevels are all on the left and *p* sublevels on the right. Moreover, the 3*d* sublevel follows the 4*s* and precedes the 4*p*. Likewise, the 4*d* sublevel fills after the 5*s* and before the 5*p*.

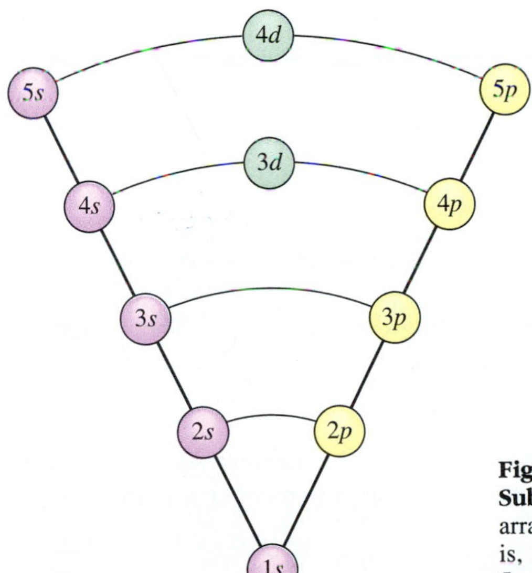

Figure 4.20 Filling Diagram for Energy Sublevels The order of sublevel filling is arranged according to increasing energy; that is, 1*s*, 2*s*, 2*p*, 3*s*, 3*p*, 4*s*, 3*d*, 4*p*, 5*s*, 4*d*, and 5*p*.

Electron Configuration

The **electron configuration** of an atom is a shorthand statement for describing the arrangement of electrons by sublevel. First, the sublevel is written, followed by a superscript that indicates the number of electrons. For example, if the 2*p* sublevel contains two electrons, the standard notation is $2p^2$. Thus,

electron configuration A shorthand description of the arrangement of electrons by sublevels according to increasing energy.

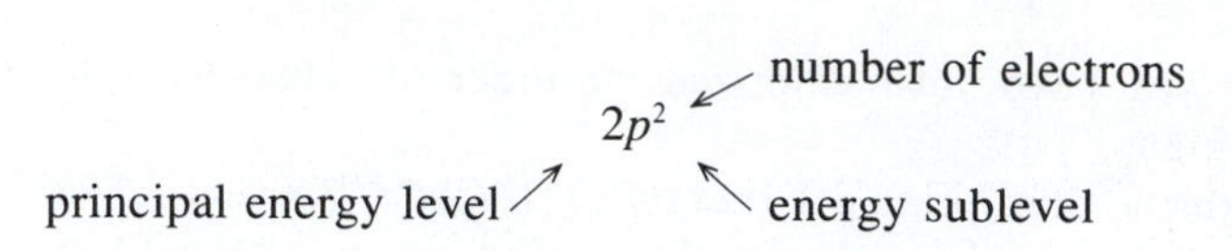

The procedure for writing the electron configuration for an atom of an element is straightforward. First, find the atomic number, which corresponds to the number of electrons in a neutral atom. Then write the sublevels according to increasing energy. Remember, each sublevel is filled with electrons in sequence until the total equals the atomic number of the element.

By way of example, we can write the electron configuration for cobalt, whose atomic number is 27. Given the atomic number, we know the cobalt nucleus has 27 protons and is surrounded by 27 electrons. The order in which the electron sublevels are arranged according to increasing energy is as follows:

$$1s\ 2s\ 2p\ 3s\ 3p\ 4s\ 3d \cdots$$

The energy sublevels in cobalt are filled beginning with the $1s$ sublevel and ending when there is a total of 27 electrons. The electron configuration for cobalt is as follows:

$$_{27}\text{Co: } 1s^2\ 2s^2\ 2p^6\ 3s^2\ 3p^6\ 4s^2\ 3d^7$$

The following example exercise illustrates writing the electron configuration for other elements.

EXAMPLE EXERCISE 4.12

Write the electron configuration for each of the following elements, given the atomic number.

(a) $_{10}\text{Ne}$ **(b)** $_{38}\text{Sr}$

Solution: The atomic number of an element designates the number of protons in the nucleus of an atom. This number is also the number of electrons in the atom.

(a) For an atom of neon, we see the atomic number is 10. Therefore, the number of electrons is also 10. Filling sublevels until 10 electrons are present gives

$$_{10}\text{Ne: } 1s^2\ 2s^2\ 2p^6$$

As always, an s sublevel may have two electrons and a p sublevel can have a maximum of six electrons.

(b) For an atom of strontium, we see the atomic number is 38. The number of electrons in an atom of strontium is 38. Thus,

$$_{38}\text{Sr: } 1s^2\ 2s^2\ 2p^6\ 3s^2\ 3p^6\ 4s^2\ 3d^{10}\ 4p^6\ 5s^2$$

To check your answer, find the total number of electrons by adding up superscripts. The total is 38, which agrees with the atomic number for strontium.

SELF-TEST EXERCISE

Write the electron configuration for each of the following elements. Use standard notation, grouping electrons together according to sublevels.

(a) fluorine **(b)** argon
(c) nickel **(d)** cadmium

Answers: **(a)** $1s^2\ 2s^2\ 2p^5$; **(b)** $1s^2\ 2s^2\ 2p^6\ 3s^2\ 3p^6$; **(c)** $1s^2\ 2s^2\ 2p^6\ 3s^2\ 3p^6\ 4s^2\ 3d^8$; **(d)** $1s^2\ 2s^2\ 2p^6\ 3s^2\ 3p^6\ 4s^2\ 3d^{10}\ 4p^6\ 5s^2\ 4d^{10}$

Chemistry Connection □ The Consumer

Neon Lights

The colors of the dragon are produced by different, excited gaseous elements.

► **Can you name the element in the sign that is emitting the reddish-orange light?**

At the turn of the 20th century, J. J. Thomson discovered the electron using a cathode-ray tube. Thomson constructed his cathode-ray tube out of thin glass and placed a metal electrode in each end. After evacuating air from the glass tube, he introduced a small amount of gas. When the metal cathode and anode electrodes were electrically excited, he noticed that the tube glowed. He was able to identify the glowing rays from the cathode as a stream of small negative particles. These particles were named electrons and Thomson is given credit for their discovery. The early experiments by Thomson were a forerunner to modern cathode-ray tubes (CRT) that are used today in television sets and computer monitor screens.

In 1913, Niels Bohr explained that exciting gases with electricity caused electrons to be temporarily promoted to higher energy states within the atom. The excited electrons, however, quickly lose energy and return to their original state. These electrons lose energy by emitting light. The light emitted by different gases varies because the energy levels within atoms vary for each element. For example, a gas discharge tube containing mercury vapor gives off a blue glow, while hydrogen gas gives off a reddish-purple glow. The following figures show the light emission from (a) mercury vapor and (b) nitrogen gas discharge tubes.

In 1898, the Scottish chemist William Ramsay discovered the inert gas neon. Unlike argon, which comprises about 1% of air, neon is much more rare. It is about a thousand times less concentrated in our atmosphere. When neon gas is placed into a narrow glass tube and electrically excited, it produces a reddish-orange light that is very arresting to the eye. The fact that gas discharge tubes produce an attractive array of colors led naturally to their use as advertising lights. Light from excited neon gas is very intense, and the term "neon light" has become a generic term for all advertising lights.

Obviously, not all advertising lights are the same color. That is, "neon lights" can be red, green, blue, and so on. In order to produce a particular color light, a gas discharge tube must be filled with a specific gas. To produce a purple light, argon gas can be used. To obtain a pink light, helium gas is used. Only if we wish to produce a reddish-orange light, is the advertising sign actually filled with neon gas.

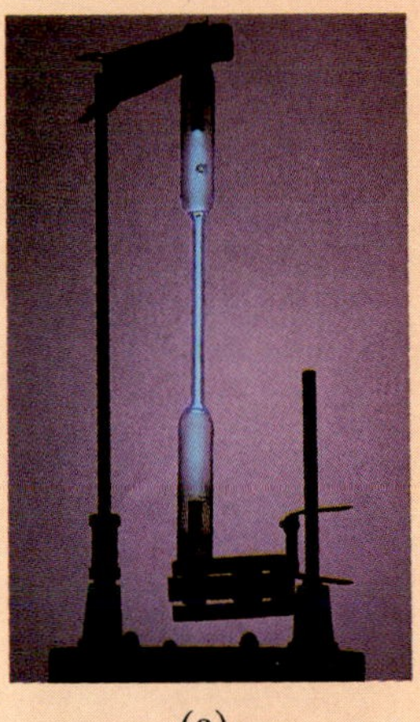

(a)

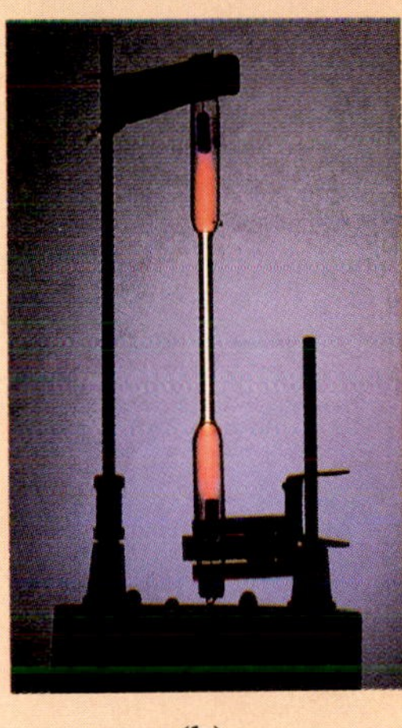

(b)

(a) Hg gas discharge tube; (b) N gas discharge tube

► **The element is neon gas.**

This concludes our simplified treatment of early atomic theory according to the Bohr model of the atom. In the 1920s, our understanding of the atom went through a revolutionary change. A new model of the atom emerged in which the position of an electron was uncertain. However, the electron was said to have a high probability of occupying a given region in space about the nucleus. These spatial regions of high electron probability were called orbitals, and the new model was referred to as the quantum mechanical model of the atom. In Chapter 10, modern atomic theory is presented which gives electrons and energy levels a much more rigorous and sophisticated treatment.

Summary

Section 4.1 At the beginning of the nineteenth century, John Dalton offered a strong argument for the existence of atoms. Unlike the Greek philosopher Democritus, Dalton supported his argument using the experimental laws of *constant composition* and *conservation of mass*.

Section 4.2 Toward the end of the century, evidence accumulated that the atom itself was divisible. When a voltage was applied to a sealed glass tube containing a gas, rays were observed to stream from the cathode to the anode. These **cathode rays** were affected by both magnetic and electric fields. It was concluded that cathode rays were composed of tiny, negatively charged particles. These particles were named **electrons**. When holes were made in the cathode, positive particles were observed moving in the opposite direction. The smallest positive particle was produced when the cathode-ray tube contained hydrogen gas. This particle was named a **proton**.

In 1897, J. J. Thomson performed an experiment that yielded the charge-to-mass ratio of the electron. Soon afterward, the actual charge and mass of the electron and proton were calculated. The charge on the electron is equal, but opposite, to that of the charge on the proton. The mass of the proton is about 2000 times heavier than the mass of the electron.

Section 4.3 In 1911, Ernest Rutherford performed a classic experiment in which alpha particles were fired at a thin sheet of gold foil. Much to his astonishment, some of the alpha particles actually bounced backward. He interpreted the results as evidence for a tiny dense **atomic nucleus** at the center of the atom. The nucleus contained positively charged protons surrounded by negatively charged electrons. Twenty years later, the nucleus was found to also contain neutral particles called **neutrons**.

Section 4.4 Atoms of the same element contain a constant number of protons, but the number of neutrons in the nucleus may vary. Atoms with the same atomic number but a different mass number are called **isotopes**. Thus, Dalton's original proposal that all atoms of an element are identical proved to be incorrect.

Sections 4.5–4.6 Although the mass of an atom is much too small to measure directly, we can determine its relative mass with a **mass spectrometer**. Carbon-12 is used as the reference isotope and is assigned a mass of exactly 12 **atomic mass units** (amu) on the atomic mass scale. The mass of all other atoms is related to the mass of carbon-12. Most elements exist naturally as two or more isotopes. To find a representative value for the atomic mass of an element, we have to average the masses of the isotopes. The weighted average mass of all isotopes is termed the **atomic mass** of the element. This weighted atomic mass is also referred to as the atomic weight.

Section 4.7 The Rutherford model of the atom pictured electrons circling about the nucleus. In 1913, Niels Bohr suggested that electrons travel in orbits at a specific distance from the nucleus. An electron traveling in such an orbit would have a specific energy. If an electron changed orbits, there would be a stepwise change in energy. These orbits were referred to as electron **energy levels**. Bohr found evidence for his theory in the **spectral lines** emitted from gas discharge tubes. The **emission line spectrum** of hydrogen contained three vivid lines: a violet line, a blue-green line, and a red line. Bohr argued as follows. Electrons are found in electron energy levels that have a specific energy. If an electron drops from a higher energy level to a lower level, a particle of light is released. The energy of the particle of light is equal to the difference in energy between the two energy levels. Since the particle of light has a specific energy, it has a discrete **wavelength**. Spectral lines of discrete wavelength are experimentally observed. Therefore, electrons must occupy discrete energy levels. Furthermore, the observed wavelengths of the spectral lines correspond to the calculated wavelengths using the **Balmer formula**.

Section 4.8 A closer examination of emission spectra revealed **sublevels** within **principal energy levels**. The number of sublevels equals the number of the principal energy level. For example, the fourth principal energy level has four sublevels (*s*, *p*, *d*, *f*). The number of electrons in each sublevel varies. An *s* sublevel can have a maximum of 2 electrons, a *p* sublevel can have 6, a *d* sublevel can have 10, and an *f* sublevel has a maximum of 14 electrons. Electrons begin filling the sublevels closest to the nucleus. Sublevels farther from the nucleus are filled in sequence. Sublevel filling is in order of increasing energy. The following sequence is the actual order of sublevels according to increasing energy: $1s < 2s < 2p < 3s < 3p < 4s < 3d < 4p < 5s < 4d < 5p < 6s$. Notice that the $4s$ sublevel fills before the $3d$. You should either memorize this sequence or be able to draw a filling diagram similar to the one shown in Figure 4.19.

Section 4.9 A detailed description of the distribution of electrons by sublevel is given by the **electron configuration**. A superscript following each sublevel indicates the number of electrons in the energy sublevel. For instance, $1s^2\ 2s^2\ 2p^6\ 3s^1$ is the electron configuration for sodium, element number 11. Note that the total of the superscripts is equal to the atomic number of the element.

Key Terms

Select the key term that corresponds to the following definitions.

____ **1.** a stream of negative particles produced in a cathode-ray tube

____ **2.** a subatomic particle having a negligible mass and a relative charge of one minus

____ **3.** a subatomic particle having an approximate mass of 1 amu and a relative charge of one plus

____ **4.** a neutral subatomic particle having an approximate mass of 1 amu

____ **5.** a region of very high density in the center of the atom

____ **6.** a characteristic number for an element that indicates the number of protons found in the nucleus of one of its atoms

____ **7.** a number that represents the total of the number of protons and neutrons in the nucleus of an atom

____ **8.** a symbolic method for expressing the composition of an atomic nucleus; the mass number and atomic number are indicated to the left of the chemical symbol for the element

____ **9.** atoms having the same atomic number but a different mass number

____ **10.** a unit of mass exactly equal to 1/12 the mass of a C-12 atom

____ **11.** an instrument used to determine the atomic mass of an isotope relative to carbon-12

____ **12.** the weighted average mass of all the naturally occurring isotopes of an element

____ **13.** a single particle of light energy emitted when an electron drops from a higher energy level to a lower energy level

____ **14.** a broad, uninterrupted band of light energy

____ **15.** the distance light must travel for a wave to complete one cycle

____ **16.** the number of times a light wave completes a cycle in 1 second

____ **17.** light energy extending from short-wavelength X-rays through long-wavelength microwaves

____ **18.** light energy that is observed as violet, blue, green, yellow, orange, or red; that portion of the radiant energy spectrum from approximately 400 to 700 nm

____ **19.** an orbit having a specific energy that electrons can occupy as they circle the nucleus; designated 1, 2, 3, 4 . . .

____ **20.** a model picture of the atom that describes the electron circling the nucleus in an orbit of specific energy

____ **21.** a collection of narrow slits of light energy that results from excited atoms of a given element releasing energy

____ **22.** the individual narrow slits of light in an emission line spectrum

____ **23.** a mathematical equation for calculating the emitted wavelength of light when an excited electron drops to the second level in a hydrogen atom

____ **24.** an electron energy level that results from the splitting of a principal energy level; designated *s*, *p*, *d*, *f* . . .

____ **25.** a main energy level containing one or more sublevels; designated 1, 2, 3, 4 . . .

____ **26.** a shorthand description of the arrangement of electrons by sublevels according to increasing energy

(a) atomic mass *(Sec. 4.6)*
(b) atomic mass unit (amu) *(Sec. 4.5)*
(c) atomic notation *(Sec. 4.4)*
(d) atomic nucleus *(Sec. 4.3)*
(e) atomic number (*Z*) *(Sec. 4.4)*
(f) Balmer formula *(Sec. 4.7)*
(g) Bohr atom *(Sec. 4.7)*
(h) cathode ray *(Sec. 4.2)*
(i) continuous spectrum *(Sec. 4.7)*
(j) electron (e^-) *(Sec. 4.2)*
(k) electron configuration *(Sec. 4.9)*
(l) emission line spectrum *(Sec. 4.7)*
(m) energy level *(Sec. 4.7)*
(n) energy sublevel *(Sec. 4.8)*
(o) frequency *(Sec. 4.7)*
(p) isotopes *(Sec. 4.4)*
(q) mass number (*A*) *(Sec. 4.4)*
(r) mass spectrometer *(Sec. 4.5)*
(s) neutron (n°) *(Sec. 4.3)*
(t) principal energy level *(Sec. 4.8)*
(u) proton (p^+) *(Sec. 4.2)*
(v) quantum *(Sec. 4.7)*
(w) radiant energy spectrum *(Sec. 4.7)*
(x) spectral lines *(Sec. 4.7)*
(y) visible spectrum *(Sec. 4.7)*
(z) wavelength *(Sec. 4.7)*

Exercises

Evidence for Atoms: The Dalton Model (Sec. 4.1)

1. State Dalton's five proposals regarding the atom.
2. State the two experimental laws Dalton used to support his atomic theory.
3. Which of Dalton's proposals were later shown to be invalid?
4. Are atoms indestructible? Explain.

Evidence for Subatomic Particles: The Thomson Model (Sec. 4.2)

5. State the simplest particle that was observed in cathode rays.
6. State the simplest particle that was observed in canal rays.
7. State the relative charge and mass of the electron.
8. State the relative charge and mass of the proton.
9. In the Thomson model of the atom, the raisins in plum pudding were analogous to what part of the atom?
10. In the Thomson model of the atom, the homogeneous consistency of plum pudding was analogous to what part of the atom?

Evidence for a Nuclear Atom: The Rutherford Model (Sec. 4.3)

11. What did Rutherford conclude about the atom when alpha particles recoiled backward after striking a thin gold foil?
12. Describe an atom according to the Rutherford model.
13. State the location of electrons, protons, and neutrons in the planetary model of the atom.
14. State the approximate diameter in centimeters for an atom and its nucleus.
15. State the relative charge for the electron, proton, and neutron.
16. State the approximate mass (amu) for the electron, proton, and neutron.

Atomic Notation (Sec. 4.4)

17. State the number of neutrons in an atom of each of the following isotopes.
    (a) ^{4}He **(b)** ^{32}S **(c)** ^{10}B
    (d) ^{44}Ca **(e)** ^{15}N **(f)** ^{52}Cr
    (g) ^{26}Mg **(h)** ^{58}Ni
18. State the number of neutrons in an atom of each of the following isotopes.
    (a) hydrogen-3 **(b)** Sr-90
    (c) carbon-14 **(d)** U-235
    (e) cobalt-60 **(f)** Pt-145
    (g) aluminum-27 **(h)** Pu-239
19. Draw a diagram of the arrangement of protons, neutrons, and electrons for an atom of each of the following isotopes.
    (a) ^{2}H **(b)** ^{3}He
    (c) ^{7}Li **(d)** ^{13}C
    (e) ^{16}O **(f)** ^{20}Ne
20. Draw a diagram of the arrangement of protons, neutrons, and electrons for an atom of each of the following isotopes.
    (a) ^{24}Mg **(b)** ^{31}P
    (c) ^{35}Cl **(d)** ^{40}Ar
    (e) ^{56}Fe **(f)** ^{131}I
21. Is it possible for atoms of two different elements to have the same atomic number? Explain.
22. Is it possible for atoms of two different elements to have the same mass number? Explain.
23. Given the atomic notation for a specific isotope, complete the following table.

Atomic Notation	Atomic Number	Mass Number	Number of Protons	Number of Neutrons	Number of Electrons
$^{4}_{2}He$					
$^{13}_{6}C$					
$^{21}_{10}Ne$					
$^{28}_{14}Si$					
$^{40}_{18}Ar$					
$^{50}_{22}Ti$					

24. Complete the following table by providing the missing information.

Atomic Notation	Atomic Number	Mass Number	Number of Protons	Number of Neutrons	Number of Electrons
Fe		56			
	30			37	
Se		78			
	38			50	
Sn		120			
	54			77	

Atomic Mass Scale (Sec. 4.5)

25. Give the reference isotope and its assigned mass for our current atomic mass scale.
26. Explain why atomic masses are expressed on a *relative* atomic mass scale.

27. A mass spectrometer is adjusted to detect the carbon-12 isotope. State whether the magnetic field must be increased or decreased to detect the following isotopes.
(a) $^{4}_{2}He$ (b) $^{20}_{10}Ne$ (c) $^{11}_{5}B$ (d) $^{14}_{6}C$

28. A mass spectrometer is adjusted to detect the potassium-40 isotope. State whether the magnetic field must be increased or decreased to detect the following isotopes.
(a) $^{27}_{13}Al$ (b) $^{45}_{21}Sc$ (c) $^{68}_{30}Zn$ (d) $^{90}_{40}Zr$

Atomic Mass (Sec. 4.6)

29. Distinguish between atomic mass and isotopic mass using lithium as an example. Refer to Table 4.3 for isotopic data.

30. Beryllium, aluminum, phosphorus, and arsenic have only one naturally occurring isotope. Refer to the atomic masses in the periodic table of elements and give the isotopic mass for each of the following.
(a) $^{9}_{4}Be$ (b) $^{27}_{13}Al$ (c) $^{31}_{15}P$ (d) $^{75}_{33}As$

31. A marble collection has large marbles with a mass of 5 g each and small marbles with a mass of 2 g each. Calculate the simple average mass and the weighted average mass if there are 100 large marbles and 200 small marbles in the collection.

32. A chemistry professor states that the course grade depends on quizzes (10%), homework (10%), tests (40%), lab experiments (20%), and a final exam (20%). If a student's scores were quizzes (73), homework (95), tests (75), experiments (92), and final exam (58), what is the student's weighted average score?

33. Calculate the atomic mass for silver given the data for its two isotopes.

Ag-107:	106.905 amu	51.82%
Ag-109:	108.905 amu	48.18%

34. Calculate the atomic mass for iron given the following data.

Fe-54:	53.940 amu	5.82%
Fe-56:	55.935 amu	91.66%
Fe-57:	56.935 amu	2.19%
Fe-58:	57.933 amu	0.33%

35. Calculate the atomic mass for titanium given the data for its five naturally occurring isotopes.

Ti-46:	45.953 amu	7.93%
Ti-47:	46.952 amu	7.28%
Ti-48:	47.950 amu	73.94%
Ti-49:	48.948 amu	5.51%
Ti-50:	49.945 amu	5.34%

36. Calculate the atomic mass for platinum, which has six naturally occurring isotopes.

Pt-190:	189.960 amu	0.0127%
Pt-192:	191.961 amu	0.78%
Pt-194:	193.963 amu	32.9%
Pt-195:	194.965 amu	33.8%
Pt-196:	195.965 amu	25.3%
Pt-198:	197.968 amu	7.21%

37. Calculate the atomic mass for magnesium given the isotopic mass and natural abundance found in Table 4.3.

38. Calculate the atomic mass for neon given the isotopic mass and natural abundance found in Table 4.3.

39. Chlorine occurs as two stable isotopes: Cl-35 and Cl-37. Predict which isotope is more abundant given the atomic mass of chlorine, 35.453 amu.

40. Bromine occurs as two stable isotopes in approximately equal abundance. Given one of the isotopes is Br-79, predict the other naturally occurring isotope.

41. Polonium (element 84) was discovered by Marie Curie and named after her native Poland. Referring only to the periodic table, state whether polonium is stable or radioactive.

42. Promethium (element 61) has been identified in the Andromeda galaxy by spectroscopic observation. Referring only to the periodic table, state whether promethium has any stable earthly isotopes.

Evidence for Electron Energy Levels: The Bohr Model (Sec. 4.7)

43. Which of the following wavelengths of light is most energetic: violet, green, or orange?

44. Which of the following wavelengths of light is longest: blue, yellow, or red?

45. Which of the following wavelengths of light is most energetic: 650 nm, 550 nm, or 450 nm?

46. Which of the following wavelengths of light is least energetic: 1150 nm, 520 nm, or 210 nm?

47. Draw a diagram of the Bohr planetary model of the atom.

48. What is the experimental evidence for electron energy levels in an atom?

49. When a voltage is applied to a gas discharge tube, electrons gain energy and jump up to an excited state. What form of energy is released when the electron returns to the original energy level?

50. The conservation of energy law is obeyed when electrons are energized by electrical discharge. What is the relationship between the electrical energy required to excite the electron and the energy released?

51. Three vivid lines are observed in the emission line spectrum of hydrogen. State the energy levels involved in producing the following.
(a) the red line (b) the blue-green line
(c) the violet line

52. If an electron drops from the fifth main level in a hydrogen atom, predict an energy level it could drop to and produce the following.
(a) infrared line (b) ultraviolet line

53. Which of the following energy-level changes for an electron is most energetic: 5 to 2, 4 to 2, or 3 to 2?

54. Which of the following energy-level changes for an electron is least energetic: 7 to 1, 3 to 1, or 2 to 1?

55. Using the Balmer formula, calculate the wavelength of light emitted when an electron drops from the fourth to the second level in a hydrogen atom.

56. What color is the emission line that is observed when the electron drops from the fourth to the second level in a hydrogen atom?

57. Using the Balmer formula, calculate the wavelength of light emitted when an electron drops from the fifth to the second level in a hydrogen atom.

58. What color is the emission line that is observed when the electron drops from the fifth to the second level in a hydrogen atom?

Principal Energy Levels and Sublevels (Sec. 4.8)

59. What experimental evidence was used to support the concept of electrons in energy levels?

60. What experimental evidence suggested that energy levels consisted of sublevels?

61. Designate all the sublevels within each of the following principal energy levels.
(a) first **(b)** second
(c) third **(d)** fourth

62. State the number of sublevels in each of the following principal energy levels.
(a) first **(b)** third
(c) fifth **(d)** sixth

63. What is the maximum number of electrons in each of the following sublevels?
(a) $2s$ **(b)** $3s$ **(c)** $2p$ **(d)** $4p$
(e) $3d$ **(f)** $5d$ **(g)** $4f$ **(h)** $5f$

64. What is the maximum number of electrons in each of the following?
(a) an *s* sublevel **(b)** a *p* sublevel
(c) a *d* sublevel **(d)** an *f* sublevel

65. What is the maximum number of electrons in each of the following principal energy levels?
(a) first **(b)** second
(c) third **(d)** fourth

66. What is the theoretical maximum number of electrons in each of the following principal energy levels?
(a) fifth **(b)** sixth

Arrangement of Electrons by Energy Sublevel (Sec. 4.9)

67. List the order of sublevels from $1s$ through $5p$ according to increasing energy. (*Hint:* Draw a filling diagram.)

68. Draw a filling diagram and predict the sublevel that follows $5p$.

69. Write the predicted electron configuration for the following elements.
(a) He **(b)** Be **(c)** C **(d)** Mg
(e) S **(f)** K **(g)** Co **(h)** Cd

70. Write the predicted electron configuration for the following elements.
(a) boron **(b)** fluorine
(c) manganese **(d)** nickel
(e) aluminum **(f)** argon
(g) arsenic **(h)** tin

71. Which element has the following predicted electron configuration?
(a) $1s^2\ 2s^1$
(b) $1s^2\ 2s^2\ 2p^6\ 3s^2\ 3p^2$
(c) $1s^2\ 2s^2\ 2p^6\ 3s^2\ 3p^6\ 4s^2\ 3d^2$
(d) $1s^2\ 2s^2\ 2p^6\ 3s^2\ 3p^6\ 4s^2\ 3d^{10}\ 4p^6\ 5s^2$

72. Which element has the following predicted electron configuration?
(a) $1s^2\ 2s^2\ 2p^5$
(b) $1s^2\ 2s^2\ 2p^6\ 3s^2\ 3p^6$
(c) $1s^2\ 2s^2\ 2p^6\ 3s^2\ 3p^6\ 4s^2\ 3d^{10}\ 4p^6\ 5s^2\ 4d^5$
(d) $1s^2\ 2s^2\ 2p^6\ 3s^2\ 3p^6\ 4s^2\ 3d^{10}\ 4p^6\ 5s^2\ 4d^{10}\ 5p^5$

General Exercises

73. The scattering of alpha particles by a thin gold foil has been described by the analogy *"like missiles shot through our solar system."* If a missile represents an alpha particle, what do the planets represent?

74. Atomic dimensions can be described by the analogy "*like a small BB in the middle of the Astrodome.*" If the Houston Astrodome represents the size of a typical atom, what does the BB represent?

75. J. J. Thomson found the electron charge-to-mass ratio to be 1.76×10^8 coulomb/g, and Robert Millikan measured the absolute charge as 1.60×10^{-19} coulomb. Calculate the mass of an electron using the experimental data.

76. J. J. Thomson found the proton charge-to-mass ratio to be 9.57×10^4 coulomb/g, and Robert Millikan measured the absolute charge as 1.60×10^{-19} coulomb. Calculate the mass of a proton using the experimental data.

77. Suppose an atom were enlarged to the size of a golf ball. If a golf ball was magnified the same number of times, it would be approximately the size of which of the following: a basketball, ~30 cm; the New Orleans Superdome, ~200 m; the earth, ~12,000 km; the solar system, ~6,000,000,000 km?

78. Use atomic notation to designate the three isotopes of hydrogen.
(a) protium **(b)** deuterium
(c) tritium

79. Gallium exists as two isotopes: Ga-69 and Ga-71. Given the abundance and isotopic mass of Ga-69 (60.1% and 68.92558 amu), what is the isotopic mass of Ga-71? The atomic mass of gallium is 69.723 amu.

80. Boron occurs naturally as B-10 and B-11. Given the mass of each isotope (10.013 amu and 11.009 amu), what is the percentage abundance of each isotope? The atomic mass of boron is 10.811 amu. (*Hint:* algebra required.)

81. We commonly refer to advertising lights that are red, blue, green, and other colors as neon lights. If neon lights are essentially gas discharge tubes, is the gas used in the advertising lights always neon?

82. The most dominant lines in the emission spectrum of neon occur between 600 and 650 nm. Refer to the continuous light spectrum in Section 4.7 and predict the observed color of a neon light.

83. Indicate the region of the spectrum (infrared, visible, or ultraviolet) for each of the following wavelengths of light.
   (a) 200 nm **(b)** 500 nm
   (c) 1200 nm **(d)** 1500 nm

84. Which of the following light frequencies has the higher energy: 5×10^{10} cycles/s or 5×10^{11} cycles/s?

85. When electrons drop from the second to the first quantum level in the hydrogen atom, the spectral line emitted is invisible. Calculate the wavelength of the emission line using the Rydberg equation.

$$\frac{1}{\lambda} = \frac{1}{91\ \text{nm}}\left(\frac{1}{n_L^2} - \frac{1}{n_H^2}\right)$$

86. Indicate the region of the spectrum (infrared, visible, or ultraviolet) for the emitted spectral line in Exercise 85.

87. When electrons drop from the fourth to the third quantum level in the hydrogen atom, the emitted quanta are not visible. Calculate the wavelength of the emission line using the Rydberg equation.

88. Indicate the region of the spectrum (infrared, visible, or ultraviolet) for the emitted spectral line in Exercise 87.

89. The third main energy level can hold a maximum of 18 electrons, yet the third row in the periodic table has only 8 elements. Explain.

90. The fourth main energy level can hold a maximum of 32 electrons, yet the fourth row in the periodic table has only 18 elements. Explain.

CHAPTER 5

The Periodic Table

In the New York City Marathon, runners are classified in groups with similar athletes according to their ability. In the periodic table, elements are classified in groups with similar elements according to their reactivity.

As early as 600 B.C., the Greeks began to speculate that the universe was composed of a single fundamental element. Thales, the founder of Greek science, mathematics, and philosophy, suggested that water was the fundamental element and that the earth was a flat disc floating on a universe of water. He explained that air and space were less dense forms of water, and earth was a more dense form. This was a radical departure from the accepted view that gods and demons were the forces that shaped nature.

A few years later, the Greek philosopher Anaximenes suggested that air was the universal element. This theory was soon followed by Heraclitus' proposal that fire was the element and Xenophanes' suggestion that earth was the basic element. About 450 B.C., the Greek philosopher Empedocles noted that, when wood burns, smoke is released (air), followed by a flame (fire). He also noted that a cool surface held over the fire collects moisture (water), and the only remains are ashes (earth). Empedocles interpreted his experiment as evidence for air, fire, water, and earth as four basic elements. He further speculated that other substances were examples of these same elements combined in varying proportions and arrangements (Figure 5.1).

The notion of air, earth, fire, and water as basic elements was later adopted by Aristotle. He modified the theory to include the concept of ether, which he believed to fill all space. The influence of Aristotle was so great that the elements he identified became an integral part of chemical theory for the next twenty centuries.

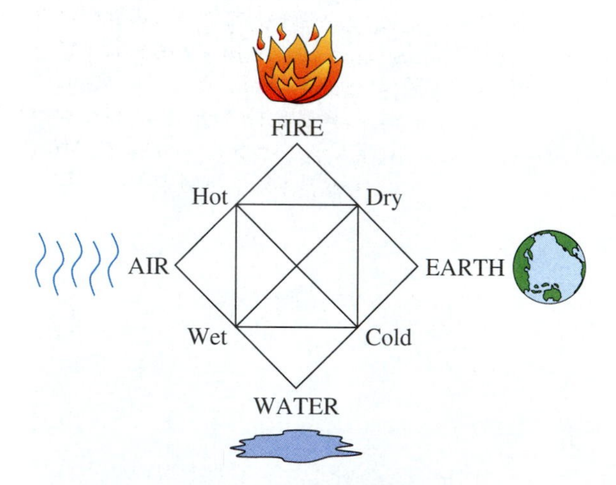

Figure 5.1 The Four Greek Elements A representation of the four elements—air, fire, earth, water—proposed by the Greeks. Notice the relationship of the four properties—hot, dry, cold, wet—associated with each element.

5.1 Systematic Arrangement of the Elements

OBJECTIVES

To identify the contribution of Döbereiner, Newlands, Meyer, and Mendeleev in establishing the periodic table of the elements.

Although several metals, including copper, iron, gold, silver, and zinc, had been known since prehistoric time, it was not until the 1700s that these metals were classified as elements. By 1800, thirty elements had been isolated and identified. By 1870, there were more than sixty.

In 1829, the German chemist J. W. Döbereiner (1780–1849) observed that several elements could be classified into groups of three, called *triads*. Each of the three elements in a triad showed remarkable similarity in their chemical reactions. In addition, they showed an orderly trend in physical properties such as density, melting point, and especially atomic mass. Figure 5.2 shows some of the triads Döbereiner suggested based on their similar properties.

In the Cl/Br/I triad, chlorine is the most chemically reactive, followed by bromine and then iodine. In this triad, the physical state progresses from gaseous chlorine to liquid bromine to solid crystals of iodine. The color of Cl/Br/I follows the spectrum from greenish-yellow to reddish-brown to violet. Most importantly, in the Cl/Br/I triad, the trend in atomic mass is 35/80/127. Notice that the atomic mass of bromine is about halfway between that of chlorine and iodine.

In 1865, the English chemist J. A. R. Newlands (1837–1898) presented another arrangement of elements shown in Figure 5.3. Newlands suggested that the 62 known elements be arranged into groups of seven according to increasing atomic mass. An eighth element would then repeat the properties of the first element in the previous group. Interestingly, his theory, called the *law of octaves* was received with ridicule and was not accepted for publication. Although Newlands' insight into the periodic relationships of the elements was essentially correct, it took 20 years for him to

Triad	Li	Ca	S	Cl	Cr
	Li	Ca	S	Cl	Cr
	Na	Sr	Se	Br	Mn
	K	Ba	Te	I	Fe

Figure 5.2 Döbereiner's Classification of Elements (1829) Each triad, for example, Li, Na, and K, contained three elements with similar chemical properties and an ordered trend of physical properties.

Octave	1	2	3	4	5	6	7	8
	H	F	Cl	Co, Ni	Br	Pd	I	Pt, Ir
	Li	Na	K	Cu	Rb	Ag	Cs	Tl
	Be	Mg	Ca	Zn	Sr	Cd	Ba, V	Pb
	B	Al	Cr	Y	Ce, La	U	Ta	Th
	C	Si	Ti	In	Zr	Sn	W	Hg
	N	P	Mn	As	Di, Mo	Sb	Nb	Bi
	O	S	Fe	Se	Ru, Rh	Te	Au	Os

Figure 5.3 Newlands' Classification of 62 Elements (1865) Each octave, for example, F through S, contained seven elements with the eighth element, Cl, having properties similar to F. Notice that Newlands placed two elements, such as Co and Ni, in a single position; this was one of the reasons his table was rejected.

			Ti = 50	Zr = 90	? = 180
			V = 51	Nb = 94	Ta = 182
			Cr = 52	Mo = 96	W = 186
			Mn = 55	Rh = 104,4	Pt = 197,4
			Fe = 56	Ru = 104,4	Ir = 198
			Ni = Co = 59	Pd = 106,6	Os = 199
H = 1			Cu = 63,4	Ag = 108	Hg = 200
	Be = 9,4	Mg = 24	Zn = 65,2	Cd = 112	
	B = 11	Al = 27,4	? = 68	Ur = 116	Au = 197?
	C = 12	Si = 28	? = 70	Sn = 118	
	N = 14	P = 31	As = 75	Sb = 122	Bi = 210?
	O = 16	S = 32	Se = 79,4	Te = 128?	
	F = 19	Cl = 35,5	Br = 80	J = 127	
Li = 7	Na = 23	K = 39	Rb = 85,4	Cs = 133	Tl = 204
		Ca = 40	Sr = 87,6	Ba = 137	Pb = 207
		? = 45	Ce = 92		
		?Er = 56	La = 94		
		?Yt = 60	Di = 95		
		?In = 75,6	Th = 118?		

Figure 5.4 Mendeleev's Periodic Table of Elements (1869) In Mendeleev's original periodic table, he arranged the elements in vertical columns according to increasing atomic mass. Mendeleev also predicted undiscovered elements, which he indicated by a ?. He named three of these elements ekaboron, ekaaluminum, and ekasilicon, which correspond to atomic masses equal to 45, 68, and 70, respectively.

receive professional recognition. In 1887, Newlands was awarded the Davy medal by the Royal Society of Great Britain.

In the 1860s, the German chemist Lothar Meyer (1830–1895) and the Russian chemist Dmitri Mendeleev (1834–1907) independently developed similar concepts concerning the relationship of the elements. After studying the physical and chemical properties of the elements, both chemists suggested that the properties repeat at regular intervals when the elements are arranged in order of *increasing atomic mass*. Mendeleev had the good fortune of publishing his periodic table (see Figure 5.4) a short time before Meyer published a similar table. Today, Mendeleev is regarded as the architect of the periodic table, whereas Meyer is all but forgotten.

Mendeleev's brilliance in creating the periodic table was the result of his patient and systematic study of several properties. After repeatedly ordering the elements according to various properties, he concluded that the elements should be arranged according to increasing atomic mass. Furthermore, he had the insight and courage to predict the existence and properties of three elements before their discovery. His most famous prediction was the element he called ekasilicon. In 1886, ekasilicon was discovered in Germany and given the name germanium. The observed properties of germanium and the properties of ekasilicon predicted by Mendeleev are compared in Table 5.1.

Table 5.1 Mendeleev's Prediction of Properties for Ekasilicon (Ek) versus Germanium (Ge)

Property	Predicted (1869)	Observed (1886)
color	gray	gray
atomic mass	72 amu	72.6 amu
density	5.5 g/mL	5.32 g/mL
melting point	very high	937°C
formula of oxide	EkO_2	GeO_2
density of oxide	4.7 g/mL	4.70 g/mL
formula of chloride	$EkCl_4$	$GeCl_4$
boiling point of chloride	100°C	86°C

Chemistry Connection ◻ Historical

Dmitri Mendeleev

▶ Can you state the element that is named in honor of Dmitri Mendeleev?

Dmitri Ivanovich Mendeleev was born in Siberia, the youngest of 14 to 17 children (records vary). Mendeleev's father was a high school principal, but blindness ended his career when he was quite young. Mendeleev's mother, a woman of remarkable energy and determination, subsequently started a glass factory to support the family. About the time Mendeleev graduated from high school, his father died and his mother's factory burned down. Left destitute, his mother used her influence to get him into college just a few months before her own death.

Mendeleev enrolled at the University of St. Petersburg where he graduated first in his class. Following graduate work in Europe, he returned to St. Petersburg as a professor of chemistry. Mendeleev's consuming interest was finding a common thread that linked the rapidly growing number of elements. In 1869, he published a periodic table that related properties of elements to their atomic masses. He even predicted the existence and properties of three undiscovered elements for which he left blank spaces in his table. In time, each of his predictions was verified. In 1874, *gallium* was discovered in France (Gallia); in 1876 *scandium* was found in Scandinavia; and in 1886 *germanium* was discovered in Germany.

Gradually, scientists recognized the value of the periodic table, and Mendeleev became the most famous chemist in the world. He was invited to major universities throughout Europe and gave lectures in the United States and Canada. He also divorced his wife and married a young art student. In the eyes of the Russian Orthodox Church, he was a bigamist because he did not wait the required seven years. When Czar Alexander II was questioned about Mendeleev's bigamy, he replied: "Yes, Mendeleev has two wives, but I have only one Mendeleev." Apparently, Mendeleev's popularity and stature placed him above the law.

In the late 19th century, Russia was engulfed in political turmoil. The government felt that academic freedom contributed to unrest and imposed repressive measures on the university. Mendeleev was an outspoken liberal and voiced his concern for the rights of students, and women in particular. This led to his forced resignation from the university in 1890. Fortunately, Mendeleev still had influential friends and he was named director of the Bureau of Weights and Measures.

In 1906, just a few months before his death, he missed winning the Nobel prize by a single vote. Historians speculate that Mendeleev was more deserving of the recognition, but his controversial personality caused the prize to be awarded to Henri Moissan, the French chemist who first isolated fluorine.

Dmitri Mendeleev (1834–1907)

▶ Element 101, synthesized in 1955, is named mendelevium, Md, in honor of Mendeleev.

Although Mendeleev's original periodic table arranged the elements in vertical columns, two years later he published another version showing the elements in horizontal rows. He arranged the elements by increasing atomic mass and began each new row with an element that repeated the properties of a previous element. Figure 5.5 illustrates a portion of Mendeleev's 1871 periodic table of elements.

Group	I	II	III	IV	V	VI	VII	VIII
Formula of Oxide	R_2O	RO	R_2O_3	RO_2	R_2O_5	RO_3	R_2O_7	RO_4
	H							
	Li	Be	B	C	N	O	F	
	Na	Mg	Al	Si	P	S	Cl	
	K	Ca	**eka-**	Ti	V	Cr	Mn	Fe, Co
	Cu	Zn	**eka-**	**eka-**	As	Se	Br	& Ni
	Rb	Sr	Yt	Zr	Nb	Mo	—	Ru, Rh
	Ag	Cd	In	Sn	Sb	Te	I	& Pd
	Cs	Ba	Di	Ce	—	—	—	
	—	—	—	—	—	—	—	
	—	—	Er	La	Ta	W	—	Os, Ir
	Au	Hg	Tl	Pb	Bi	—	—	& Pt
	—	—	—	Th	—	U	—	

Figure 5.5 Mendeleev's Periodic Table of Elements (1871) The formula for the oxide of an element (R) is indicated by the notation R_2O, RO, and so on. For example, the group I oxides include H_2O, Li_2O, and Na_2O; the group II oxides include BeO, MgO, and CaO. In groups III and IV, he indicated the existence of three undiscovered elements, which he named ekaboron, ekaaluminum, and ekasilicon.

The Noble Gases

The periodic table took on a significant addition with the discovery of the group of elements on the far right side of the table. Argon was found in 1894, quickly followed by helium, neon, krypton, and xenon. Originally, this group was referred to as the *inert gases* because they demonstrated no chemical reactivity.

In 1962, a compound containing xenon was synthesized at the University of British Columbia. Actually, a reaction of platinum and fluorine was accidentally contaminated with oxygen, and chemists reasoned correctly that the reaction would also occur if xenon were substituted for oxygen. After the first compound was made, $PtXeF_6$, several others quickly followed. Although compounds of xenon and krypton have been prepared, the elements argon, neon, and helium are yet to be combined. More recently, the term **noble gas** has been substituted for inert gas in order to communicate the unreactive nature of each gas in this group. For the same reason, copper, silver, and gold are referred to as noble metals because of their resistance to chemical reaction.

noble gases The relatively unreactive Group VIIIA/18 elements.

5.2 The Periodic Law

To state the original periodic law proposed by Mendeleev.

To state the modern periodic law based on Moseley's discovery.

periodic law The properties of the elements recur in a repeating pattern when arranged according to increasing atomic number.

In 1869, the **periodic law** stated by Mendeleev proposed that elements showed recurring properties according to *increasing atomic mass*. In 1913, the work of a 25-year-old postdoctoral student led to a revised statement of the periodic law. H. G. J. Moseley (1887–1915), working under Ernest Rutherford, bombarded atomic nuclei with high-energy radiation. By studying the X-rays that were subsequently emitted, Moseley discovered that the nuclear charge increased by 1 for each element in the periodic table. Thus,

H = 1 He = 2

Li = 3 Be = 4 B = 5 C = 6 N = 7 O = 8 F = 9 Ne = 10

Na = 11 Mg = 12

Soon thereafter, chemists concluded that arranging elements according to increasing nuclear charge, rather than atomic mass, more clearly explained the repeating properties of elements. That is, arranging elements according to atomic number better explained the trends in the periodic table. The order of the elements, therefore, should be arranged according to the number of protons in their nucleus. Thus, the periodic law was rewritten. Rather than the elements being arranged according to increasing atomic mass, they are arranged according to *increasing atomic number*. As it so happens, with only a few exceptions, the trends in atomic mass and atomic number are identical.

In the 1920s, after the periodic law had been rewritten and Niels Bohr had introduced the concept of electron energy levels, the periodic table took on a new shape. It then resembled the familiar arrangement used today, as shown in Figure 5.6. A vertical column in the periodic table is called a **group** of elements, and a horizontal row is called a **period** of elements. If you closely examine the fourth period (row 4) in the table, you will notice that the sequence of atomic masses does not increase for Co and Ni. That is, Ni has a *lower* atomic mass than Co. The reason that cobalt precedes nickel in the periodic table is because the atomic number of Co (27) is lower than the atomic number of Ni (28). It is also true, of course, that the properties of cobalt are more consistent with the elements in its group. Likewise, the properties of Ni are similar to the elements Pd and Pt found in the nickel group.

group A vertical column in the periodic table; a family of elements with similar properties.

period A horizontal row in the periodic table; a series of elements with properties that vary from metallic to nonmetallic.

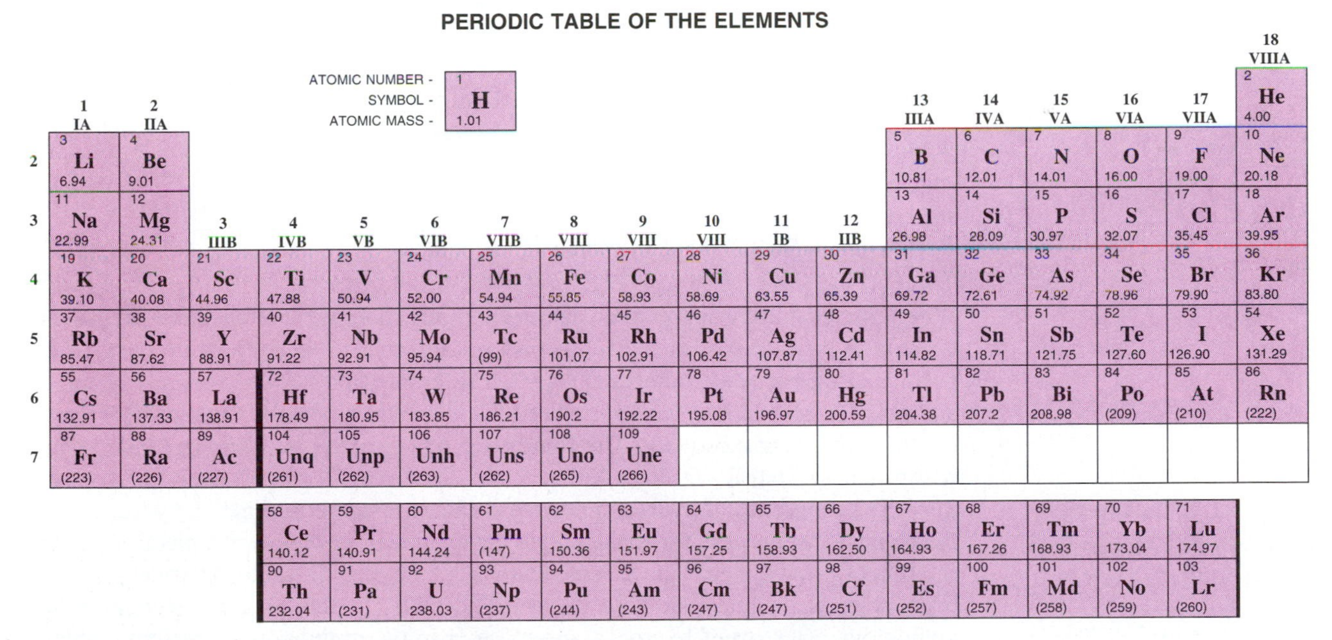

Figure 5.6 The Periodic Table of the Elements Atomic numbers increase stepwise throughout the periodic table. The atomic masses, with few exceptions such as Co and Ni, also increase. However, atomic charge, and not atomic mass, accounts for recurring properties.

EXAMPLE EXERCISE 5.1

Find the two elements in the fifth period of the periodic table that violate the original periodic law proposed by Mendeleev.

Solution: Mendeleev proposed that the elements be arranged according to increasing atomic mass. Beginning with Rb, each of the elements in the fifth period increases in atomic mass until iodine. Although the atomic numbers of Te (52) and I (53) increase, the atomic masses of Te (127.60) and I (126.90) do not. Experimentally, it is I, and not Te, whose properties are similar to F, Cl, and Br.

SELF-TEST EXERCISE

Find a pair of elements in the periodic table with atomic numbers less than 20 that are an exception to the original periodic law.

Answer: Ar and K

5.3 The Periodic Table of the Elements

To use the following terms in reference to the arrangement of elements in the periodic table:
(a) groups (families) and periods (series)
(b) representative elements and transition elements
(c) metals, semimetals, and nonmetals
(d) alkali metals, alkaline earth metals, halogens, and noble gases
(e) lanthanide series and actinide series
(f) rare earth elements and transuranium elements

To designate a group of elements in the periodic table using either the American convention (IA through VIIIA) or the IUPAC convention (1 through 18).

When we examine the periodic table in Figure 5.7, we notice there are seven horizontal rows of elements. These rows are referred to as **periods** or *series*. The periods of elements are numbered 1 to 7. The first period has only two elements, H and He. The second and third periods each have eight elements, Li through Ne, and Na through Ar, respectively. The fourth and fifth periods have eighteen elements, K through Kr, and Rb through Xe.

The properties of hydrogen make it difficult to position. That is, hydrogen has the physical properties of a nonmetal but the chemical properties of a metal. In many references you will find H placed on the far left of the periodic table. Alternately, some texts place H on boths sides of the periodic table. In this text, we recognize the ambiguous behavior of hydrogen and place H in the middle of the periodic table.

There are 18 vertical columns in the periodic table. Each column is referred to as a **group** or *family* of elements. In the past, American chemists have used a Roman numeral followed by the letter A or B to designate a group of elements; for example, IA, IIA, IIB, and so on. Refer to the periodic table in the front of the text

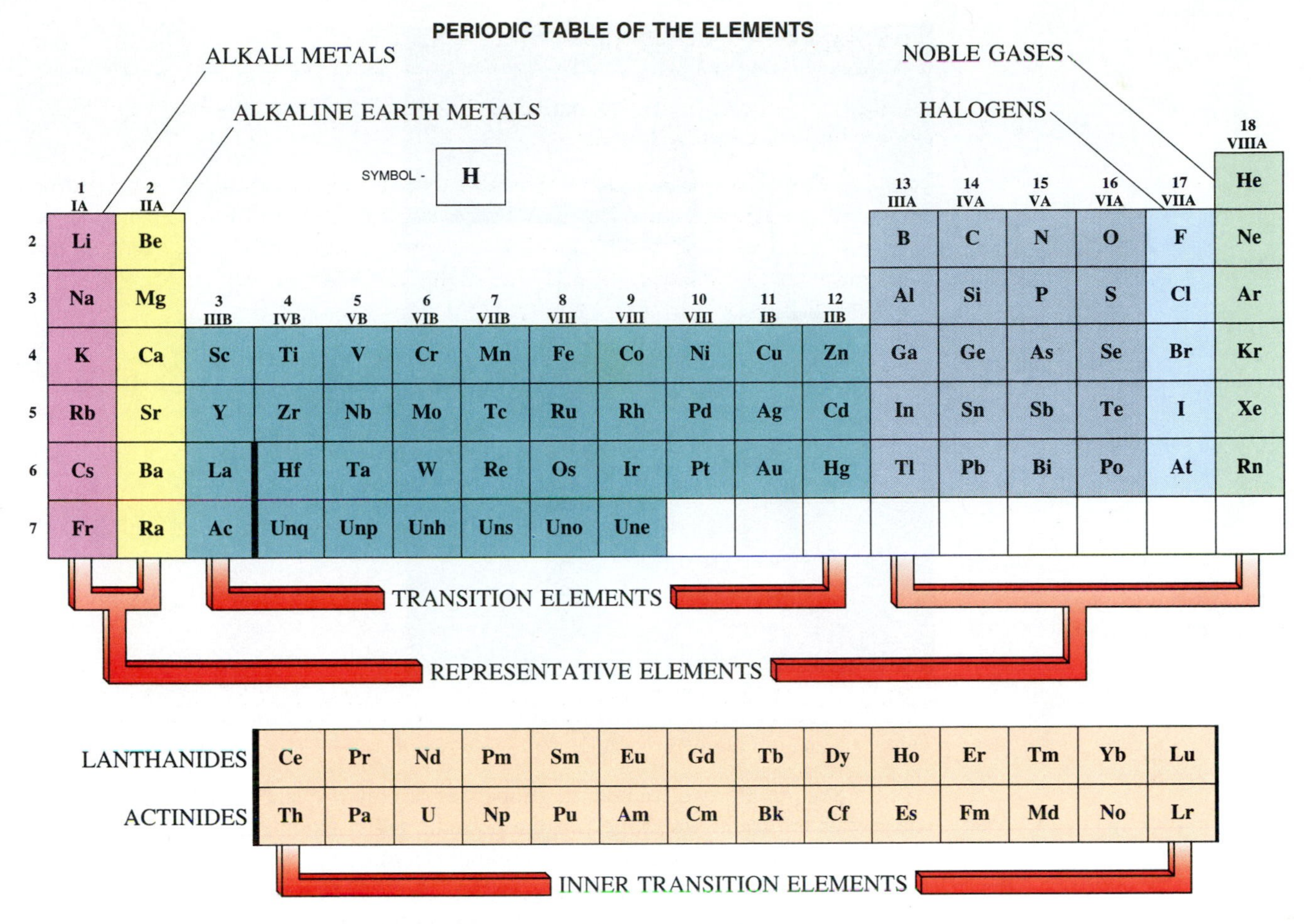

Figure 5.7 Names of Groups and Periods The common names of groups and periods are shown for selected families and series.

and you will notice that Group IA has the elements Li to Fr, Group IIA has the elements Be to Ra, and Group IIB has the elements Zn, Cd, Hg.

Since the 1920s, a controversy has existed over the numbering of groups in the periodic table. (See the *Update* on page 147.) Recently, the International Union of Pure and Applied Chemistry (IUPAC) resolved the dispute by proposing to number the groups using the numerals 1 to 18. For example, IUPAC recommends that Group IA be designated Group 1, Group IIB be designated Group 12, and Group VIIIA be designated Group 18. Currently, we are in the process of adopting the IUPAC convention. Hence, the periodic tables in this text will list both conventions, for example, IA/1, IIB/12, and VIIIA/18.

Groups of elements in the periodic table can also be referred to by their family name. The family name for Group IA/1 is the **alkali metals**. The elements in Group IIA/2 are called the **alkaline earth metals**. The family names of some groups of elements in the periodic table use the first element in the group. For example, Group IB/11 elements are the copper group (Cu, Ag, Au), and Group IIIA/13 elements are the boron group. The elements in Group VIIA/17 are called the **halogens**. The word halogen is derived from the Greek for "salt former," and compounds containing a halogen are usually saltlike. Group VIIIA/18 elements are all gases that at normal conditions are usually unreactive; they are called the noble gases.

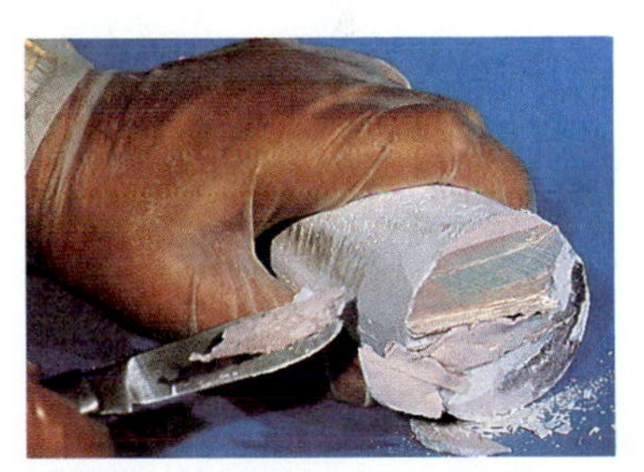

Lithium and potassium, similar to sodium shown above, are each soft, shiny metals.

alkali metals The Group IA/1 elements, excluding hydrogen.

alkaline earth metals The Group IIA/2 elements.

halogens The Group VIIA/17 elements.

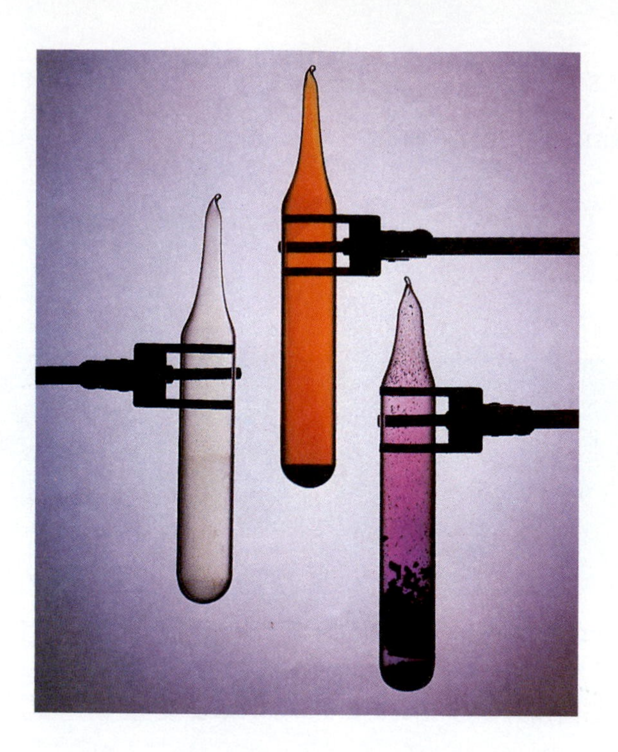

The Group VIIA/17 family of elements – Cl_2, Br_2, I_2 – the halogens.

In Section 3.5, we studied the properties of metallic and nonmetallic elements. The metals are found on the left of the periodic table. Nonmetals are found on the right of the periodic table but include the element hydrogen (H). Metals and nonmetals react with each other by transferring electrons. A metal reacts by losing one or more electrons to a nonmetal that is gaining the electrons. Two nonmetals react with each other by sharing electrons.

Between the metals and nonmetals lie eight elements having properties that are somewhat metallic and somewhat nonmetallic. These elements B, Si, Ge, As, Sb, Te, radioactive Po and At, are semimetals (or metalloids). Two of these elements, silicon and germanium, are used in the semiconductor industry to make transistors for electronic devices.

representative elements The Group A (1, 2, 13–18) elements in the periodic table; also termed main-group elements.

transition elements The group B (3–12) elements in the periodic table.

We can place the groups of elements in the periodic table into one of two categories, either representative elements or transition elements. **Representative elements** (also called main-group elements) are found in the A groups, to the left and to the right, of the periodic table. As a rule, the chemical behavior of representative elements is predictable. For example, magnesium always reacts with oxygen to produce MgO. **Transition elements** are found in the B groups in the middle of the periodic table. The chemical behavior of B group elements is not usually predictable and varies depending on reaction conditions. For example, in the presence of limited oxygen gas, iron reacts to form FeO; if excess oxygen is available, the product is Fe_2O_3.

inner transition elements The elements in the lanthanide and actinide series.

lanthanide series The elements with atomic numbers 58 to 71.

Beneath the body of the periodic table are two series of metals referred to as the **inner transition elements**. These two series of elements are placed below so as to avoid an unduly wide periodic table. Specifically, the first series, elements Ce to Lu, belong between La and Hf in the periodic table and are considered a part of Period 6. This series is called the **lanthanide series** because it follows element 57, lanthanum. Since the elements from Ce to Lu are so similar—they occur together in nature and are difficult to separate—they are sometimes considered to belong to the Group IIIB/3 family of elements. As the natural abundance of most of these

UPDATE

The Periodic Table of Elements

► **Can you state the two groups in the periodic table in which hydrogen is often placed?**

Recently, the designation of groups of elements in the periodic table has been a topic of interest. In the 1970s a lingering controversy surfaced that was addressed by the International Union of Pure and Applied Chemistry (IUPAC). Previously, in the United States, elements on the left side and right side of the periodic table had been designated as A groups. The elements in the middle of the periodic table were labeled as B groups. This so-called American convention is shown on the right (top).

In England and in the rest of Europe, elements on the left side of the periodic table had been previously designated A groups, and those on the right side had been labeled B groups. The so-called European convention is shown to the right.

After considerable discussion and compromise, IUPAC resolved the controversy between the American and the European conventions in 1985. IUPAC proposed that groups of elements be designated by the numerals 1 through 18. Currently, IUPAC recommends that groups of elements be numbered as shown in the table to the right.

One of the reasons for IUPAC's recommendation is that the last digit in the group number corresponds to the Roman numeral in both the American and European conventions. For example, in group 13, the number 3 corresponds to the Roman numeral III in both the American (IIIA) and European versions (IIIB).

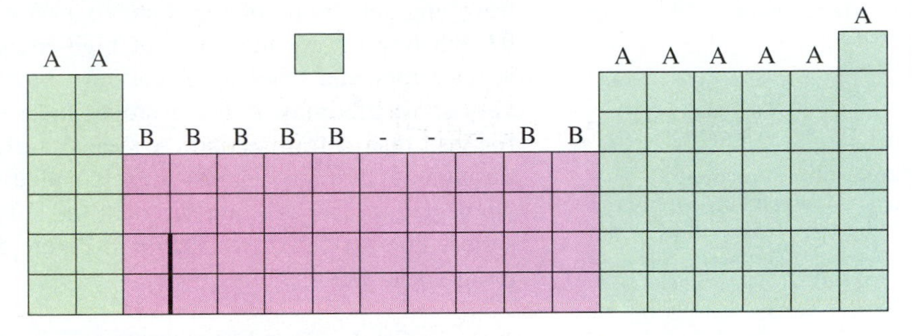

Periodic Table of Elements—American Convention: *Representative elements are in A groups, and transition elements are in B groups.*

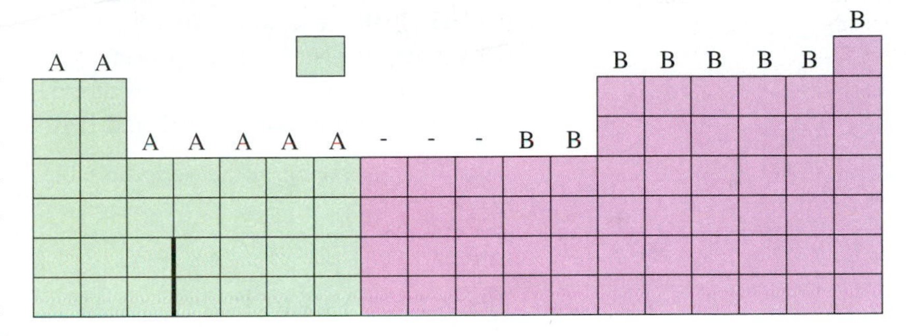

Periodic Table of Elements—European Convention: *These group designations resulted directly from the horizontal expansion of Mendeleev's periodic table of 1871.*

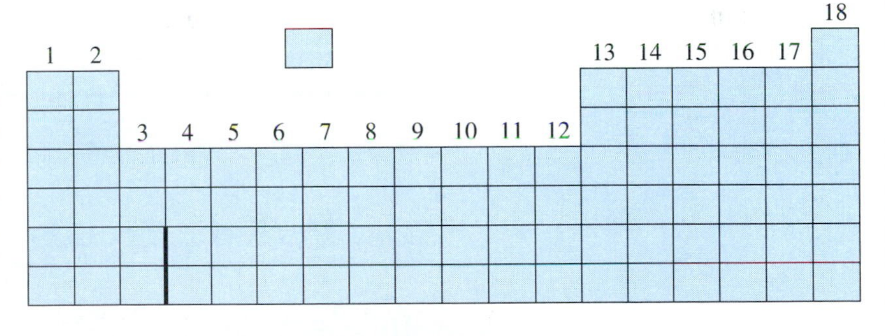

Periodic Table of Elements—IUPAC Convention: *Groups of elements are indicated by sequential numerals 1 through 18.*

► **Since hydrogen can either lose or gain 1 e$^-$, it is often placed in groups IA and VIIA; however, it is neither an alkali metal nor a halogen.**

rare earth elements The elements with atomic numbers 21, 39, 57, 58 through 71.

actinide series The elements with atomic numbers 90 to 103.

transuranium elements The elements beyond atomic number 92. All the elements following uranium are synthetic.

elements is less than 0.0001% in the earth's crust, they are collectively referred to as the **rare earth elements**.

The second series of inner transition elements, Th to Lr, is called the **actinide series** because it follows element 89, actinium. This series, Th to Lr, is considered a part of Period 7. All the elements in this series are radioactive. In fact, except for trace amounts, none of the elements past uranium is naturally occurring. Element 93 and beyond are the result of high-energy syntheses using cyclotrons, particle accelerators, and other atom smashers. Most of the isotopes of these elements have very short lifetimes. A few isotopes have been reported, even though they lasted for less than a millisecond. A special series of elements following uranium, Np through Lr through element 109, is called the **transuranium elements**. If a new element past 109 is synthesized in the future, it will belong to the transuranium series. The periodic table shown in Figure 5.7 summarizes the names of important groups and periods.

EXAMPLE EXERCISE 5.2

Select the symbol of the element that fits the following descriptions.
(a) the alkali metal in the fourth period
(b) the halogen in the third period
(c) the rare earth with the lowest atomic mass
(d) the metal in Group VIIB/7 and Period 4

Solution: Referring to the periodic table in Figure 5.7, we find
(a) K **(b)** Cl
(c) Sc **(d)** Mn

SELF-TEST EXERCISE

Select the symbol of the element that fits the following descriptions.
(a) the alkaline earth metal in the sixth period
(b) the noble gas in the third period
(c) the actinide with the highest atomic mass
(d) the semimetal in Group IIIA/13

Answers: **(a)** Ba; **(b)** Ar; **(c)** Lr; **(d)** B

5.4 Periodic Trends

To predict the trend in atomic size for groups and periods of elements in the periodic table.

To predict the trend in metallic character for groups and periods of elements in the periodic table.

We can visualize atoms as spheres and express their size in terms of the atomic radius. The atomic radius is the distance from the nucleus to the outermost electrons. As this distance is very small, we will express the atomic radius in units of nanometers (nm).

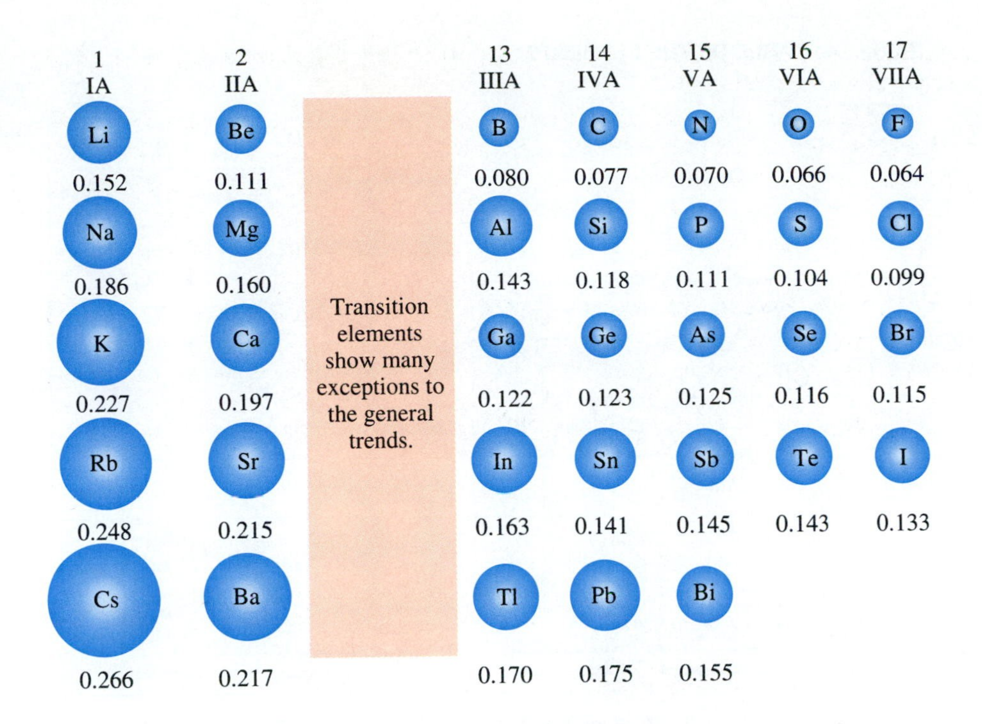

Figure 5.8 Atomic Radii of Selected Representative Elements The atomic radii decrease up a group and across a period. The values for atomic radii are given in nm; that is, 10^{-9} m.

In the periodic table, there are two general trends in atomic size. First, as we move up a group of elements from bottom to top, the radii of the atoms decrease. We can explain the decrease in radius by fewer energy levels of electrons surrounding the nucleus. With fewer energy levels, the distance from the nucleus to the outermost electrons is less. Therefore, the trend in *atomic radius decreases up a group*.

Second, as we move left to right within a period of elements, the radii of the atoms decrease. The explanation for this decrease is as follows. The atomic number increases from left to right and, therefore, the number of protons increases. If the nucleus has more protons, the nuclear charge of the elements increases. This has the effect of pulling the electrons closer to the nucleus and reducing the size of the atom. Although the number of electrons is also increasing across a period, the electrons are filling sublevels about the same distance from the nucleus. Thus, the trend in *atomic radius decreases across a period* from left to right.

Figure 5.8 illustrates the trends in atomic radii for a portion of the periodic table. The transition elements and inner transition elements have been omitted because they have numerous exceptions to the general trends. Moreover, the atomic radii of transition elements in a given sublevel are reasonably similar.

The periodic table also shows two general trends for the metallic character of the elements. First, recall that the metals are on the left side of the periodic table and the nonmetals are on the right side. Thus, as we move from left to right, the trend in *metallic character decreases across a period.*

Second, the chemical properties of an element are defined by its chemical behavior toward other elements. For example, metals react with nonmetals to give saltlike compounds. Metals react with acids to liberate hydrogen gas. To react, a metal must lose one or more of its outermost electrons. As we move up a group, the outermost electrons are closer to the nucleus. Although it is a simplification, if the electrons and nucleus are closer together, it is more difficult to remove an electron from an atom. That is, as the distance between the negatively charged electrons and the positively charged nucleus becomes less, the tendency for the metal to lose an

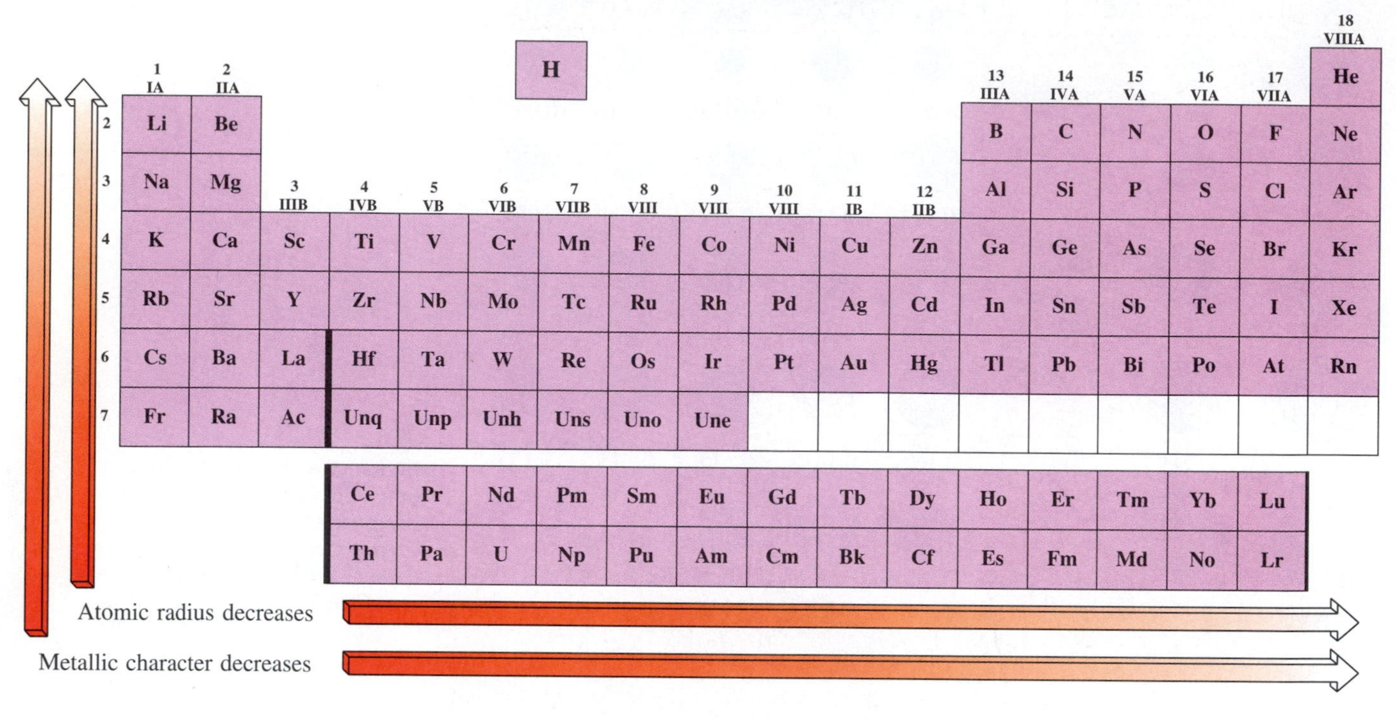

Figure 5.9 General Trends in Atomic Radii and Metallic Character The general trend for the atomic radius is to decrease up a group and across a period from left to right. The metallic character trend in the periodic table is similar.

outermost electron is less. Therefore, the trend in *metallic character decreases up a group*. Figure 5.9 illustrates the general trends in the periodic table for atomic radius and metallic character.

EXAMPLE EXERCISE 5.3

According to the general trends in the periodic table, predict the element in each pair that has the smaller atomic radius.

(a) Na or K **(b)** P or N **(c)** Ca or Ni **(d)** Si or S

Solution: The general trend in atomic radii decreases up a group and across a period from left to right. Referring to the periodic table in Figure 5.9,

(a) Na is above K in Group IA/1; the atomic radius of Na is smaller.
(b) N is above P in Group VA/15; the atomic radius of N is smaller.
(c) Ni is to the right of Ca in Period 4; the atomic radius of Ni is smaller.
(d) S is to the right of Si in Period 3; the atomic radius of S is smaller.

SELF-TEST EXERCISE

According to the general trends in the periodic table, predict the element in each pair that has the most metallic character.

(a) Sn or Pb **(b)** Ag or Sr **(c)** Al or B **(d)** Br or As

Answers: **(a)** Pb; **(b)** Sr; **(c)** Al; **(d)** As

5.5 Predicting Properties of Elements

To predict a reasonable value for a physical property, given the values for other elements in the same group.

To predict the general chemical properties of an element, given the chemical behavior of other elements in the same group.

Predicting Physical Properties

When Mendeleev was designing the periodic table, he placed elements with similar properties into the same group. He was then able to predict yet undiscovered elements from the gaps produced in his periodic table. he was successful in predicting the properties of ekasilicon, for example, after noticing trends in properties such as color, density, melting point, and atomic mass.

Today, chemists have most of the information regarding properties cataloged in reference books such as the *Handbook of Chemistry and Physics.* If we don't have a value for a particular property, however, we can use the same procedure as Mendeleev to estimate the numerical value for a physical property. For example, if we have the density of two elements in the same group, we can make a reasonable estimate of the density for another element in the same group. Table 5.2 lists selected physical properties of the alkali metals that appear in Group IA/1 in the periodic table.

Table 5.2 Physical Properties of the Alkali Metals

Element	Atomic Radius (nm)	Density at 20°C (g/mL)	Melting Point (°C)	Atomic Mass (amu)
Li	0.152	0.53	180.5	6.9
Na	0.186	0.97	97.8	23.0
K	0.227	0.86*	63.3	39.1
Rb	0.248	1.53	38.9	85.5
Cs	0.266	1.87	28.4	132.9

* Notice that the density of K is less than that of Na. Small irregularities in group trends are not unusual.

Notice that the radioactive element francium, Fr, is not included with the other alkali metals in Table 5.2. We can, however, make some reasonable predictions about its physical properties based on the trends shown by the other elements. Since Fr is below Cs in Group IA/1, we can predict that its atomic radius is greater than 0.266 nm, its density is greater than 1.87 g/mL, its melting point is less than 28.4°C, and its atomic mass is greater than 132.9 amu.

EXAMPLE EXERCISE 5.4

Predict the missing value (?) for each physical property shown below. The **(a)** atomic radius, **(b)** density, and **(c)** melting point are given for two of three alkaline earth metals in Group IIA/2.

Element	Atomic Radius (nm)	Density at 20°C (g/mL)	Melting Point (°C)
Ca	0.197	1.54	(?)
Sr	0.215	(?)	769
Ba	(?)	3.65	725

Solution: We can estimate a value by observing the group trend.

(a) To determine the atomic radius value for Ba, we first find the increase from Ca to Sr; that is, 0.215 nm − 0.197 nm = 0.018 nm. We then add this difference (0.018 nm) to the atomic radius of Sr and obtain 0.215 nm + 0.018 nm = 0.233 nm. Note that we assumed the atomic radius increased the same amount from Sr to Ba as it did from Ca to Ba. *(The literature value is 0.218* nm.)

(b) Notice that Sr lies between Ca and Ba in Group IIA/2. We can estimate that the density of Sr lies midway between Ca and Ba. To find the density of Sr, we calculate the average value for Ca and Ba; that is, (1.54 + 3.65)/2 = 2.60 g/mL. *(The literature value is 2.63 g/mL.)*

(c) From the general trend, we can predict that the melting point of Ca is greater than that of Sr. To determine the value, let's find the increase in melting point from Ba to Sr. It is 769°C − 725°C = 44°C. Now we add 44°C to the value of Sr: 769°C + 44°C = 813°C. We therefore predict the melting point of Ca as 813°C. *(The literature value is 839°C.)*

SELF-TEST EXERCISE

Predict the missing value (?) for each physical property shown below. The atomic radius, density, and melting point are given for two or three metals in group VIII/10.

Element	Atomic Radius (nm)	Density at 20°C (g/cm^3)	Melting Point (°C)
Ni	0.125	8.91	(?)
Pd	0.138	(?)	1554
Pt	(?)	21.5	1772

Answers: 0.151 nm; 15.2 g/cm^3; 1336°C

Although we have calculated numerical values for physical properties, we should realize that the values are only estimates. In some cases, the actual literature value may not agree very well with the calculated value.

Predicting Chemical Properties

In Chapter 8, we will systematically study chemical reactions. For now, it is possible to predict the products of chemical reactions by understanding the periodic table. For example, if we know that magnesium and oxygen react to give magnesium oxide (MgO), we can predict that the other Group IIA/2 elements react in a similar fashion. That is, calcium, strontium, and barium should react with oxygen to give similar oxides (CaO, SrO, and BaO), and they do. There are exceptions to the general rule, but the principle of using the periodic table is helpful to our understanding.

EXAMPLE EXERCISE 5.5

Metallic sodium reacts with chlorine gas to give sodium chloride, NaCl. Predict the products formed when **(a)** lithium and **(b)** potassium react with chlorine gas.

Solution: Since Li and K are in the same group as Na (Group IA/1), we can predict that the products are similar to NaCl. Thus,
(a) lithium metal should react with chlorine gas to give LiCl.
(b) potassium metal should react with chlorine gas to give KCl.

SELF-TEST EXERCISE

The chemical formulas for the oxides of potassium, calcium, gallium, and germanium are, respectively, K_2O, CaO, Ga_2O_3, and GeO_2. Refer to the periodic table and predict the chemical formula for each of the following compounds.

(a) rubidium oxide
(b) strontium oxide
(c) indium oxide
(d) tin oxide

Answers: **(a)** Rb_2O; **(b)** SrO; **(c)** In_2O_3; **(d)** SnO_2

5.6 The *s*, *p*, *d*, *f* Blocks of Elements

OBJECTIVES

To correlate the *s*, *p*, *d*, *f* sublevels with the positions of the elements in the periodic table.

To write the predicted electron configuration for selected elements using the periodic table.

In Section 4.8, we saw that the order of electron energy sublevels according to increasing energy is $1s < 2s < 2p < 3s < 3p < 4s < 3d < 4p < 5s < 4d < 5p$. In this section, we will see that the unusual shape for the border of the periodic table is the result of the ordering of energy sublevels. That is, the order of energy sublevels follows the systematic arrangement of elements by groups. Elements in Groups IA/1 and IIA/2 are filling *s* sublevels. Elements in groups IIIA/13 through VIIIA/18 are filling *p* sublevels. Transition elements in Groups IIIB/3 through IIB/12 are filling *d* sublevels. The inner transition elements (Ce to Lu and Th to Lr) are filling *f* sublevels. The elements filling the *s* sublevels are on the left in the periodic table and are collectively called the *s block* of elements. Similarly, the *p block* is composed of elements completing *p* sublevels; the *d block* contains the *d* sublevels; and the *f block* contains the lanthanide and actinide series. Figure 5.10 illustrates the *s*, *p*, *d*, and *f* block elements in the periodic table.

In Section 4.8, you had to memorize, for example, that the 3*d* energy sublevel came after the 4*s* energy sublevel. Now you can simply refer to the periodic table and see that there are four sublevels of *s* block elements before there is a *d* sublevel. Also, recall that the number of sublevels corresponds to the main energy level. In the third main level, there are three sublevels (3*s*, 3*p*, and 3*d*). In the fourth main level, there are four sublevels (4*s*, 4*p*, 4*d*, and 4*f*). We can see in Figure 5.10 that the lanthanide series and actinide series are *f* block elements. The lanthanides are filling the 4*f* energy sublevel and the actinides are filling the 5*f* sublevel.

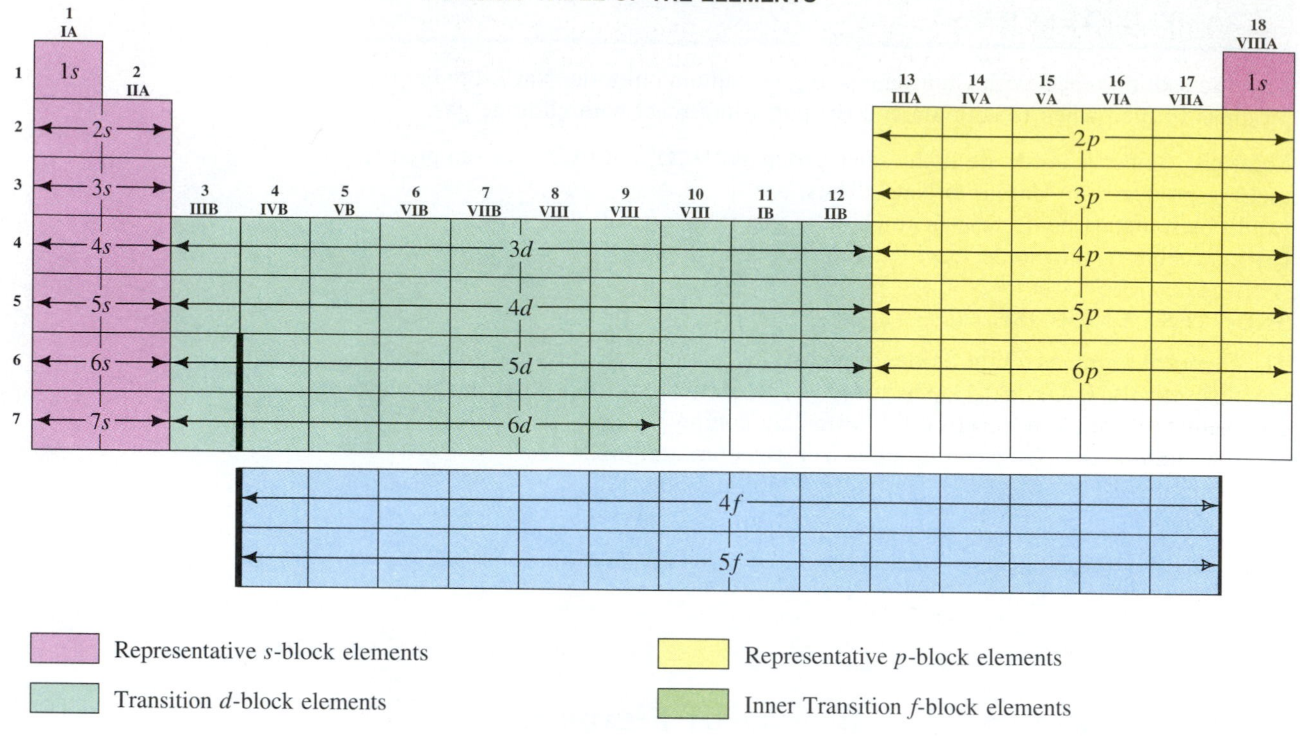

Figure 5.10 Energy Sublevels Relationship of energy sublevels and the *s*, *p*, *d*, and *f* blocks of elements in the periodic table.

EXAMPLE EXERCISE 5.6

State the highest-energy sublevel being filled by the following series of elements.

(a) H **(b)** S **(c)** Ni **(d)** U

Solution: Refer to a periodic table and determine the sublevel based on the period and the block of elements.

(a) Although hydrogen is placed in the middle of our table, it has only one electron, and H is thus filling a 1*s* energy sublevel.

(b) Sulfur is in third period and is a *p* block element; S is filling a 3*p* sublevel.

(c) Nickel is in the first series of *d* block elements; therefore, Ni is filling a 3*d* sublevel.

(d) Uranium is in the second series of *f* block elements; it is filling a 5*f* energy sublevel.

SELF-TEST EXERCISE

State the highest-energy sublevel being filled by the following series of elements.

(a) Cs to Ba **(b)** Y to Cd **(c)** In to Xe **(d)** Ce to Lu

Answers: **(a)** 6*s*; **(b)** 4*d*; **(c)** 5*p*; **(d)** 4*f*

In Section 4.9, you learned to write electron configurations for elements after memorizing the order of sublevel filling. Now that you understand the relationship of sublevels in the periodic table, you can write electron configurations using the periodic table. Although there are a few exceptions to the predicted electron configurations for some of the elements, these minor exceptions do not concern us at this point.

The electron configuration for Na (atomic number 11) is $1s^2\ 2s^2\ 2p^6\ 3s^1$. As a shorthand method, we can abbreviate the electron configuration by indicating the innermost electrons with the symbol of the corresponding noble gas. The first noble gas with an atomic number less than 11 is neon (atomic number 10). The symbol for neon is placed in brackets, [Ne], followed by the outermost electrons. That is, the electron configuration for Na can be written as [Ne] $3s^1$.

EXAMPLE EXERCISE 5.7

Refer to a periodic table and write the predicted electron configuration for each of the following elements.

(a) Li **(b)** P **(c)** Co **(d)** I

Solution: Now that you understand blocks of elements in the periodic table, you can predict the order of sublevels according to increasing energy.

(a) We can see that Li has one electron in the 2*s* sublevel; thus, the electron configuration is $1s^2\ 2s^1$, or simply [He] $2s^1$.

(b) Phosphorus is the third element in the 3*p* sublevel. The electron configuration for P is $1s^2\ 2s^2\ 2p^6\ 3s^2\ 3p^3$, or [Ne] $3s^2\ 3p^3$.

(c) Cobalt is the seventh element in the 3*d* sublevel. The electron configuration for Co is $1s^2\ 2s^2\ 2p^6\ 3s^2\ 3p^6\ 4s^2\ 3d^7$, or [Ar] $4s^2\ 3d^7$.

(d) Iodine is the fifth element in the 5*p* sublevel. Although its electron configuration is lengthy, we can follow the order of sublevels in blocks in the periodic table. The electron configuration for I is $1s^2\ 2s^2\ 2p^6\ 3s^2\ 3p^6\ 4s^2\ 3d^{10}\ 4p^6\ 5s^2\ 4d^{10}\ 5p^5$, or simply [Kr] $5s^2\ 4d^{10}\ 5p^5$.

SELF-TEST EXERCISE

Refer to a periodic table and write the predicted electron configuration for each of the following elements.

(a) Zn **(b)** La

Answers:

(a) $1s^2\ 2s^2\ 2p^6\ 3s^2\ 3p^6\ 4s^2\ 3d^{10}$, or simply [Ar] $4s^2\ 3d^{10}$.

(b) $1s^2\ 2s^2\ 2p^6\ 3s^2\ 3p^6\ 4s^2\ 3d^{10}\ 4p^6\ 5s^2\ 4d^{10}\ 5p^6\ 6s^2\ 5d^1$, or [Xe] $6s^2\ 5d^1$.

Several elements have actual electron configurations that vary from their predicted arrangements. Under some circumstances, electrons in a lower-energy sublevel may exist in an unfilled sublevel of slightly higher energy. Some of these exceptions are easy to explain; others are not. For our purposes, we will write electron configurations based on the general predictions of sublevel filling from the periodic table and not concern ourselves with the exceptions.

5.7 Predicting Valence Electrons

To predict the number of valence electrons for a representative element in the periodic table.

When an element undergoes a chemical reaction, only the outermost electrons are involved. These electrons are highest in energy and farthest from the nucleus. These

valence electrons The electrons that occupy the outermost *s* and *p* sublevels of an atom.

outermost electrons are called **valence electrons**. They form chemical bonds between atoms and are responsible for the chemical behavior of the element. For our purposes, we will ignore the transition elements and only consider the main-group elements. In that case, the number of valence electrons is equal to the total number of electrons in the outermost *s* and *p* sublevels.

To see how to predict the number of valence electrons, examine the group numbers in the periodic table. Using the Roman numeral convention, we can see that the group number is identical to the total number of valence electrons. That is, elements in Group IA have 1 valence electron, and elements in Group IIA have 2 valence electrons. Group IIIA has 2 electrons from an *s* sublevel plus 1 electron from a *p* sublevel for a total of 3 valence electrons. Group IVA has 4 valence electrons; Group VA has 5 valence electrons, and so on.

If we use the IUPAC designation for group numbers, the last digit (for example, the 3 in 13) indicates the number of valence electrons. Group 1 has 1 valence electron, and Group 2 has 2 valence electrons. Group 13 has 3 valence electrons; Group 14 has 4 valence electrons, and Group 18 has 8 valence electrons. Note that the *s* and *p* sublevels can have a maximum of 2 and 6 electrons, respectively. Thus, the maximum number of valence electrons is equal to 8 (2 + 6).

Lithium is in Group IA/1, so a lithium atom has only 1 valence electron. Oxygen is in Group VIA/16 and has 6 valence electrons. The following example further illustrates the relationship between group number and valence electrons.

EXAMPLE EXERCISE 5.8

Refer to the periodic table and predict the number of valence electrons for an atom of each of the following representative elements.

(a) Na **(b)** Al
(c) S **(d)** Xe

Solution: Find the element in the periodic table, observe the group number, and indicate the number of valence electrons.

(a) Since sodium is in Group IA/1, Na has 1 valence electron.
(b) Aluminum is in Group IIIA/13, so Al has 3 valence electrons.
(c) Sulfur is in Group VIA/16, so S has 6 valence electrons.
(d) Xenon is in Group VIIIA/18, so Xe has 8 valence electrons.

SELF-TEST EXERCISE

Refer to the periodic table and state the number of valence electrons for any element in each of the following groups.

(a) Group IIA **(b)** Group VA
(c) Group 14 **(d)** Group 17

Answers: **(a)** 2; **(b)** 5; **(c)** 4; **(d)** 7

Note: We can quickly predict the number of valence electrons for any representative element by referring to its group number in the periodic table. Notice, however, that the transition elements are filling *d* sublevels, and not an *s* or *p* sublevel. This complicates the discussion and, therefore, we cannot easily predict the number of valence electrons for a transition element.

5.8 Electron Dot Formulas of Atoms

OBJECTIVES

To write the electron dot formula for a representative element in the periodic table.

We said previously that only valence electrons are involved when an element undergoes a chemical reaction. That's because valence electrons are farthest from the nucleus and available to form the chemical bonds that hold atoms together. To keep track of valence electrons, chemists have devised a notation called the electron dot formula. Electron dot formulas are also referred to as Lewis structures in honor of G. N. Lewis (1875–1946), a famous American chemist, who first proposed the idea.

An **electron dot formula** shows an atom of an element surrounded by its valence electrons. The symbol of the element represents the **core** (or kernel) of the atom. That is, it represents the nucleus and the inner electrons. Dots are placed about the symbol to represent the valence electrons. Figure 5.11 illustrates the general symbol for an electron dot formula.

electron dot formula A representation for an atom and its valence electrons that shows the chemical symbol surrounded by a dot for each valence electron.

core The portion of the atom that includes the nucleus and inner electrons that are not available for bonding; also termed the kernel of the atom.

In actual practice, we will use the following convention for writing electron dot formulas.

1. Write the symbol of the element to represent the core of the atom.
2. For convenience, assume that the atom has four sides. Each side can have a maximum of two dots. Each dot represents one valence electron. The maximum number of valence electrons is eight (*s* sublevel + *p* sublevel = $2\ e^- + 6\ e^- = 8\ e^-$).
3. Determine the number of valence electrons from the group number of the element in the periodic table. Draw one dot about the symbol for each valence electron.

Although there is no absolute convention for placing dots, we will place the first dot to the right of the symbol, the second dot beneath the symbol, the third dot to the left of the symbol, and the fourth dot above the symbol. The fifth dot will pair with the first dot. The sixth, seventh, and eighth dots will pair with the second, third, and fourth dots, respectively, moving clockwise about the symbol.

As an example, consider the element phosphorus. Phosphorus is in Group VA/15 and therefore has 5 valence electrons. We can write the electron dot formula for phosphorus as follows. First, write the symbol for phosphorus; then add five dots, one at a time. Thus,

$$\text{P} \;>\; \text{P}\cdot \;>\; \underset{\cdot}{\text{P}}\cdot \;>\; \cdot\underset{\cdot}{\text{P}}\cdot \;>\; \cdot\overset{\cdot}{\underset{\cdot}{\text{P}}}\cdot \;>\; \cdot\overset{\cdot}{\underset{\cdot}{\text{P}}}:$$

$$\text{core} + 1\ e^- + 2\ e^- + 3\ e^- + 4\ e^- + 5\ e^-$$

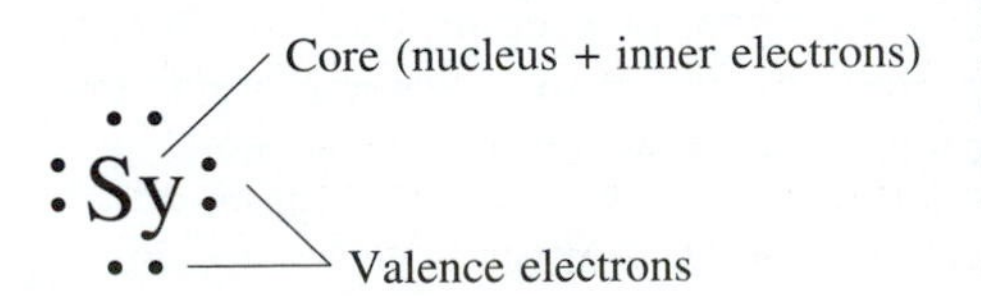

Figure 5.11 Representation of an Electron Dot Formula The electron dot formula for an atom of an element uses the symbol for the element to represent the core of the atom. Each valence electron is represented by a dot arranged about the symbol.

Be aware that there are different conventions for writing electron dot formulas. But, more important, keep in mind that the placement of dots is *not* intended to show the actual positions of the electrons about the core of the atom. Actually, electrons are in constant motion, moving about the nucleus. Electron dot formulas are only intended to help us keep track of the number of valence electrons.

EXAMPLE EXERCISE 5.9

Write the electron dot formula for each of the following elements.

(a) Na **(b)** Si

(c) S **(d)** Xe

Solution: Find the group number of the element in the periodic table and note the number of valence electrons. Write the symbol of the element to represent its core, and place the same number of dots about the symbol as there are valence electrons. In the examples, Na has 1 valence electron, Si has 4, S has 6, and Xe has 8. The electron dot formulas are as follows.

(a) Na· **(b)** ·Ṡi· (one dot above, one below)

(c) ·Ṡ: (two dots below) **(d)** :Ẍe: (two dots above, two below)

SELF-TEST EXERCISE

Write the electron dot formula for each of the following elements.

(a) nitrogen **(b)** iodine

Answers:

(a) ·Ṅ: (one dot below) **(b)** :İ: (two dots below)

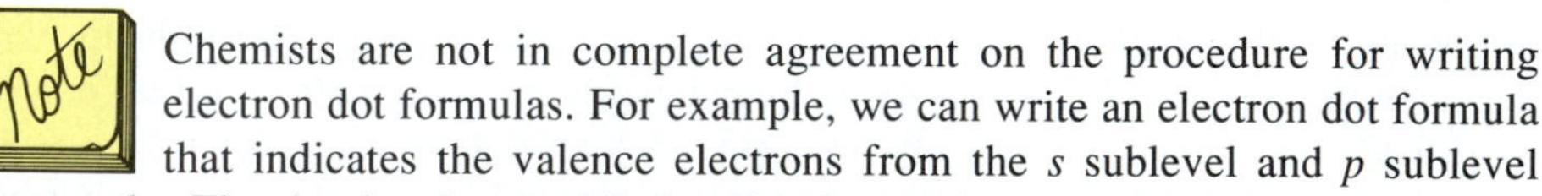

Chemists are not in complete agreement on the procedure for writing electron dot formulas. For example, we can write an electron dot formula that indicates the valence electrons from the *s* sublevel and *p* sublevel separately. That is, the element Mg has 2 valence electrons that fill the *s* sublevel. Therefore, we can write the 2 valence electrons together on one side of the symbol. Thus, the Lewis structure of Mg can be written

Mg: which may be preferable to Mg· (one dot below)

Similarly, Al has 3 valence electrons, 2 in an *s* sublevel and 1 in a *p*, which can be shown as

Al: (one dot below) or the simplified version ·Al· (one dot below)

Either of these methods is acceptable as long as you understand that valence electrons are not stationary particles on the four sides of the nucleus. Electrons are in constant motion, and the electron dot formula is only a device for keeping track of valence electrons. This will be important when we are considering the chemical bonds that are formed by the interaction of valence electrons between atoms involved in a chemical reaction.

5.9 Ionization Energy

OBJECTIVES

To state which group of elements has the highest and which has the lowest ionization energy in a given row in the periodic table.

To predict which element in a pair has the higher ionization energy based on the general trends in the periodic table.

Electrons can be removed from metals and nonmetals, but metals lose them more easily. In fact, metals undergo chemical reaction by losing one or more valence electrons. Since electrons are negatively charged, metals become positively charged after losing an electron. A charged metal atom is called an **ion**.

ion An atom that bears a charge as the result of gaining or losing valence electrons.

It always requires energy to remove an electron from an atom. By definition, the amount of energy necessary to remove an electron from a neutral atom in the gaseous state is called the **ionization energy**, or ionization potential. Thus, for sodium

ionization energy The amount of energy necessary to remove an electron from a neutral atom in the gaseous state.

$$Na \xrightarrow{\text{ionization energy}} Na^+ + e^-$$

Figure 5.12 shows the relative energy required to remove a single electron from an atom of the elements through atomic number 86.

Notice in Figure 5.12 that the elements having the highest ionization energy belong to the same group. This group of elements consists of the noble gases. One reason this group shows little tendency to undergo reaction is the difficulty with which an electron is removed from a noble gas element. We can also reason that the noble gases have high ionization energies because their valence sublevels are completely filled. They do not need to gain or lose electrons in order to become stable.

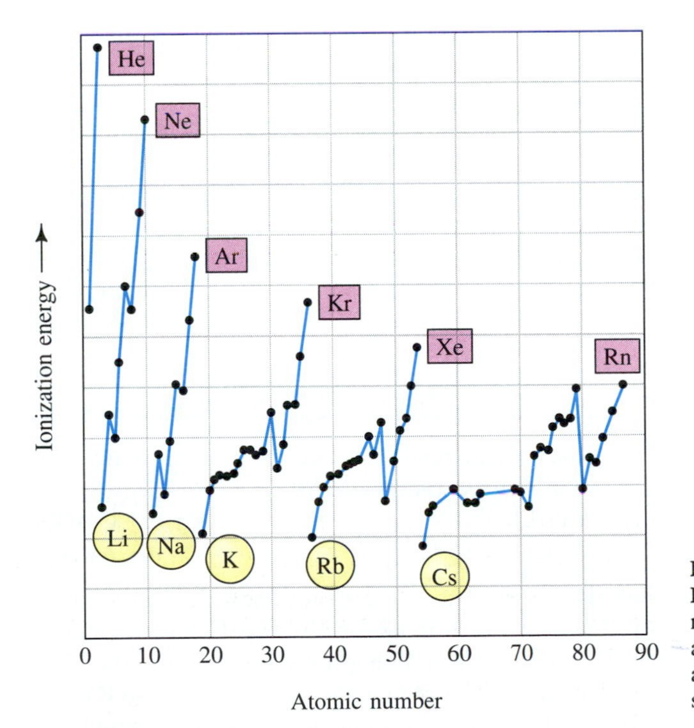

Figure 5.12 Ionization Energy The diagram shows the relative energy required to remove a single electron from a neutral atom of an element in the gaseous state.

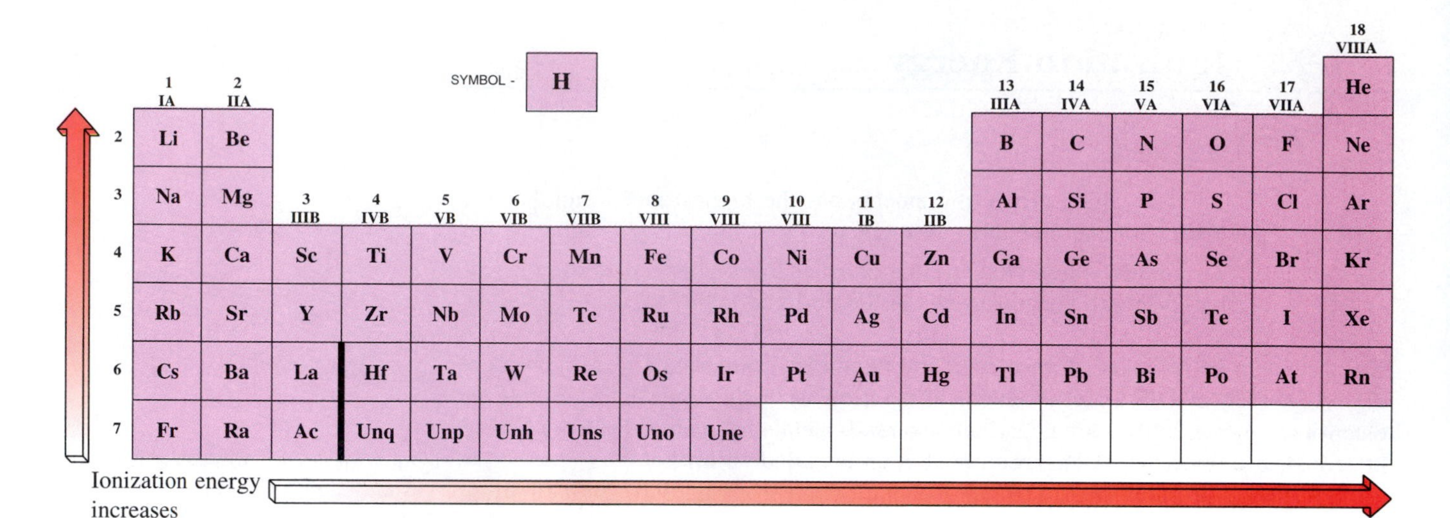

Figure 5.13 General Trends in Ionization Energy The general trend for the ionization energy of elements is to increase up a group and across a period from left to right.

In Figure 5.12, notice that the elements having the lowest ionization energy belong to the alkali metals. All the alkali metals have a similar electron configuration. That is, they all have one electron in an *s* sublevel. If we remove an electron from an alkali metal, the resulting ion has the same number of electrons as the preceding noble gas. For example, a lithium ion has 2 electrons, which is the same as a helium atom. A sodium ion has 10 electrons (same as Ne), a potassium ion has 18 electrons (same as Ar), a rubidium ion has 36 electrons (same as Kr), and a cesium ion has 54 electrons (same as Xe). When each alkali metal loses an electron, it assumes a noble gas configuration. Hence, removing a second electron is exceedingly difficult, which is even more difficult because the atom is now positively charged.

There are general trends in the periodic table regarding ionization energy, as shown by the periodic table in Figure 5.13. Within a group of elements, the energy required to remove an electron increases as the atomic radius decreases. As we proceed up a group, the valence electrons are closer to the nucleus and are more tightly held. Thus, the ionization energy *increases up a group of elements*. As we move from left to right in the periodic table, the nuclear charge becomes greater. The energy required to remove an electron, therefore, is greater. In general, the ionization energy *increases across a period of elements*.

EXAMPLE EXERCISE 5.10

Using the general trends in the periodic table, predict which of the following elements has the higher ionization energy.

(a) Na or Li
(b) C or Ge
(c) F or O
(d) Al or Na

Solution: Referring to the trends shown in Figure 5.13, we see that

(a) Li is above Na in Group IA/1, so Li has the higher ionization energy.
(b) C is above Ge in Group IVA/14, so C has the higher ionization energy.
(c) F is to the right of O in the second period, so F has the higher ionization energy.
(d) Al is to the right of Na in the third period, so Al has the higher ionization energy.

SELF-TEST EXERCISE

Using only the general trends in the periodic table, predict which of the following elements has the higher ionization energy.

(a) Ca or Ba **(b)** F or Ne

Answers: **(a)** Ca; **(b)** Ne

In this section, we have considered the energy required to remove an electron from a neutral atom, that is, ionization energy. The amount of energy involved when a neutral atom gains an electron is called *electron affinity*. Owing to the observation that the noble gases are neither reactive nor are they easily ionized, we conclude that they are unusually stable. In fact, nonmetals have a strong tendency to gain electrons in order to assume a noble gas electron configuration. Thus, nonmetals have a high electron affinity and metals show little tendency to gain an electron.

5.10 Predicting Ionic Charges

To predict an ionic charge for a representative element in the periodic table.

To predict metal and nonmetal ions that are isoelectronic with a given noble gas element.

In general, when metals and nonmetals react, metals lose electrons and nonmetals gain electrons. More specifically, metals lose electrons from their valence level and nonmetals add electrons to their valence level. Recall that an atom that bears a charge as the result of gaining or losing electrons is called an ion. Thus, metal atoms become positive ions, and nonmetal atoms become negative ions; in other words, they gain an **ionic charge**.

ionic charge Refers to the positive charge on an atom that has lost electrons or the negative charge on an atom that has gained electrons.

The positive ionic charge on a metal ion is related to its number of valence electrons. Metals in Group IA/1 give up 1 valence electron to produce a positive ionic charge of 1+. Elements in Group IIA/2 give up their 2 valence electrons to produce a positive ionic charge of 2+. Metals in Group IIIA/13 usually lose 3 electrons and have an ionic charge of 3+. Group IVA/14 metals can lose 4 electrons, producing a charge of 4+.

The negative charge on a nonmetal ion is also governed by its number of valence electrons. Nonmetals in Group VIIA/17 have 7 valence electrons and tend to add 1 electron to assume a noble gas configuration. After gaining 1 electron, the nonmetal has a negative ionic charge of 1−. Elements in Group VIA/16 gain 2 valence electrons, which produces an ionic charge of 2−. Nonmetals in Group VA/15 add 3 valence electrons, which gives an ionic charge of 3−.

Figure 5.14 shows several common elements and their ionic charge as predicted by the position of the element in the periodic table.

Lithium is in Group IA/1, so a lithium ion has an ionic charge of 1+. The lithium ion is written Li^+, with the one (in 1+) understood. Oxygen is in Group

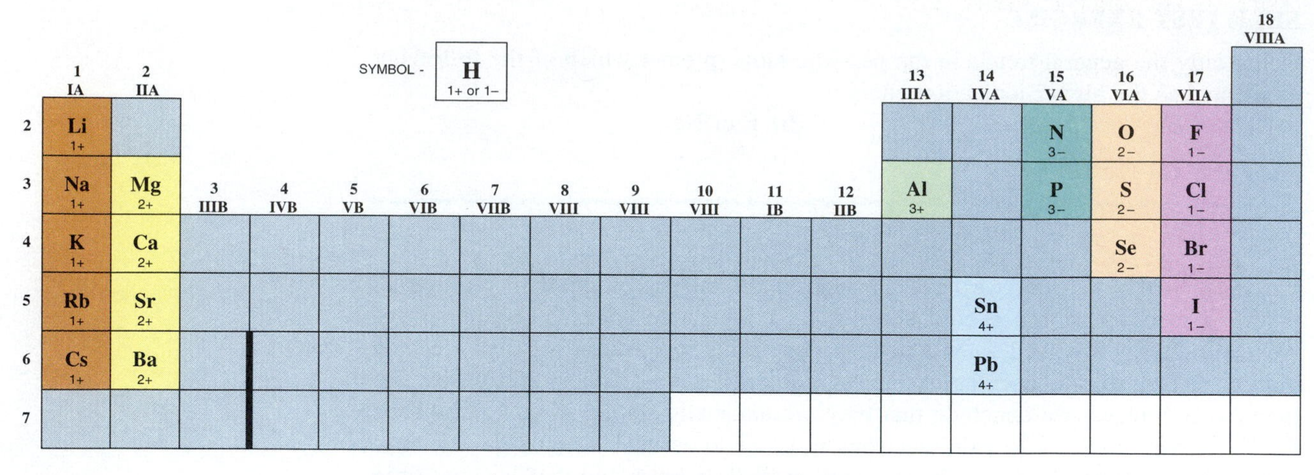

Figure 5.14 Periodic Table of Ionic Charges Notice that the metals have a positive ionic charge equal to their group number. Nonmetals have a negative ionic charge equal to 8 minus their number of valence electrons.

VIA/16 and has an ionic charge of 2 − and is shown as O^{2-}. The following examples further illustrate the relationship between group number and ionic charge.

EXAMPLE EXERCISE 5.11

Predict the ionic charge for the following ions based on the group number of the element in the periodic table.

(a) Mg ion **(b)** S ion
(c) Al ion **(d)** Br ion

Solution: Referring to the periodic table, we find that

(a) magnesium is in Group IIA/2, so the ionic charge is two plus, Mg^{2+}.
(b) sulfur is in Group VIA/16, so the ionic charge is two minus, S^{2-}.
(c) aluminum is in Group IIIA/13, so the ionic charge is three plus, Al^{3+}.
(d) bromine is in Group VIIA/17, so the ionic charge is one minus, Br^{-}.

SELF-TEST EXERCISE

Predict the ionic charge for the following ions based on the group number of the element in the periodic table.

(a) Ga ion **(b)** N ion

Answers: **(a)** 3 +, Ga^{3+}; **(b)** 3 −, N^{3-}

In Section 5.9, you learned that Group IA/1 elements form ions by losing 1 electron. We discussed that each Group IA/1 ion has an electron configuration identical to the previous noble gas element. Similarly, Group VIIA/17 elements gain 1 electron to assume an electron configuration that is identical to the next noble gas element.

For example, a sodium ion (Na^{+}) and a fluoride ion (F^{-}) each have 10 electrons, as does a neutral neon atom. Although their properties are not related, their electron configurations are identical. By definition, two or more ions having the same number of electrons are said to be **isoelectronic**. The sodium ion, fluoride ion, and a neon atom are said to be members of an isoelectronic series.

isoelectronic Refers to two or more ions having the same electron configuration; for example, O^{2-} and Mg^{2+} each have 10 electrons, and their electron configurations are identical to the noble gas neon.

EXAMPLE EXERCISE 5.12

Refer to the periodic table and predict which of the following ions are isoelectronic with the noble gas element argon.

(a) K^+ (b) Br^-
(c) Ca^{2+} (d) S^{2-}

Solution: Referring to the periodic table, we find that

(a) K^+ has 18 electrons (19 − 1); it is isoelectronic with argon (18 e^-).
(b) Br^- has 36 electrons (35 + 1); it is *not* isoelectronic with argon (18 e^-), but rather with krypton (36 e^-).
(c) Ca^{2+} has 18 electrons (20 − 2); it is isoelectronic with argon.
(d) S^{2-} has 18 electrons (16 + 2); it is isoelectronic with argon.

SELF-TEST EXERCISE

Refer to the periodic table and predict which of the following ions are in the isoelectronic series with the noble gas element xenon.

(a) Cs^+ (b) I^-
(c) La^{3+} (d) Se^{2-}

Answers: **(a)** Cs^+, **(b)** I^-, and **(c)** La^{3+} are isoelectronic with Xe; **(d)** Se^{2-} is isoelectronic with Kr.

5.11 Electron Configuration of Ions

To write the predicted electron configuration for ions of selected elements.

In Section 4.9, we first learned to write the electron configurations for elements. In Section 5.7, we learned how to predict the electron configurations of elements from the arrangement of blocks of elements in the periodic table. Now we will learn how to write the electron configuration of an ion. To do so, first find the element in the periodic table. Write out the electron configuration as we did in Section 5.6. Then, for positive ions, remove the number of electrons that corresponds to its positive ionic charge. For negative ions, add the number of electrons that corresponds to the negative ionic charge.

For example, the electron configuration for a sodium atom is $1s^2\ 2s^2\ 2p^6\ 3s^1$. Since the sodium ion is Na^+, the atom loses 1 electron. The electron configuration is $1s^2\ 2s^2\ 2p^6$. The electron configuration for a fluorine atom is $1s^2\ 2s^2\ 2p^5$. Since fluoride ion, F^-, has one more electron, the electron configuration is $1s^2\ 2s^2\ 2p^6$. We can show the loss and gain of electrons as

$$\underset{1s^2\ 2s^2\ 2p^6\ 3s^1}{\text{Na}} \xrightarrow{\text{loses } 1\ e^-} \underset{1s^2\ 2s^2\ 2p^6}{\text{Na}^+}$$

$$\underset{1s^2\ 2s^2\ 2p^5}{\text{F}} \xrightarrow{\text{gains } 1\ e^-} \underset{1s^2\ 2s^2\ 2p^6}{\text{F}^-}$$

There is one exception to the periodic table trend that we should point out. The transition elements first lose 2 electrons from the highest *s* sublevel before losing electrons from their outer *d* sublevel. For example, in the fourth period the 4*s* electrons are given up *before* the 3*d*. The electron configuration for a manganese atom is $1s^2\ 2s^2\ 2p^6\ 3s^2\ 3p^6\ 4s^2\ 3d^5$. The electron configuration for Mn^{2+} after losing 2 electrons is $1s^2\ 2s^2\ 2p^6\ 3s^2\ 3p^6\ 3d^5$. Note that manganese formed the Mn^{2+} ion by losing 2 electrons from the 4*s* sublevel, rather than the 3*d* sublevel.

Electron configurations can be simplified using a noble gas symbol to represent the core or kernel of the atom, that is, the nucleus and inner electrons. This method of showing the core of the atom is called **core notation**. In the preceding example using Mn, the electron configuration can be written using core notation as [Ar] $4s^2\ 3d^5$. For the Mn^{2+}, we can write the electron configuration in core notation as [Ar] $3d^5$.

core notation A method of writing electron configuration where all the inner electrons are represented by a noble gas symbol in brackets followed by valence electrons, for example, [Ne] $3s^2$.

EXAMPLE EXERCISE 5.13

Refer to the periodic table and write the predicted electron configuration for each of the following ions using core notation.

(a) Be^{2+} **(b)** Fe^{3+} **(c)** Cl^- **(d)** Se^{2-}

Solution: Referring to the periodic table and recalling the blocks of elements, we can write the electron configuration for the element:

(a) Be is $1s^2\ 2s^2$ or, in core notation, [He] $2s^2$. The electron configuration for Be^{2+} is $1s^2$, or simply [He].

(b) Fe is $1s^2\ 2s^2\ 2p^6\ 3s^2\ 3p^6\ 4s^2\ 3d^6$, or [Ar] $4s^2\ 3d^6$. The electron configuration for the ion Fe^{3+} is simply [Ar] $3d^5$.

(c) Cl is [Ne] $3s^2\ 3p^5$. For Cl^-, the electron configuration is written as [Ne] $3s^2\ 3p^6$, or [Ar].

(d) Se is [Ar] $4s^2\ 3d^{10}\ 4p^4$. For Se^{2-}, the electron configuration can be written as [Ar] $4s^2\ 3d^{10}\ 4p^6$, or simply [Kr].

SELF-TEST EXERCISE

Refer to the periodic table and write the predicted electron configuration for each of the following ions using core notation.

(a) Cd^{2+} **(b)** P^{3-}

Answers: **(a)** [Kr] $4d^{10}$; **(b)** [Ne] $3s^2\ 3p^6$, or [Ar]

We should emphasize that the electron configurations that we have predicted may or may not agree exactly with the actual arrangement of the electrons. Although our predictions are reasonable and generally true, there are a number of slight exceptions to the predicted arrangement of electrons.

Summary

Section 5.1 At the beginning of the nineteenth century, the German chemist J. W. Döbereiner began to organize the rapidly growing number of discovered elements into triads. Each group of three elements had similar physical and chemical properties.

In 1865, the English chemist J. A. R. Newlands organized more than 50 elements according to his law of octaves. In 1869, Dmitri Mendeleev proposed a periodic law that stated an arrangement for more than 60 elements according to increasing atomic mass. Mendeleev published his table of elements, which even included yet undiscovered elements and their predicted properties. He is generally given credit for being the architect of the modern periodic table.

Section 5.2 At the turn of the twentieth century, the nature of the atom was being revealed. In 1911, Ernest Rutherford showed that the mass of an atom was located in its positively charged nucleus. Two years later, Moseley showed that the nuclear charge on each element increases progressively. The **periodic law** had to be rewritten to state that the physical and chemical properties repeat periodically when the elements are arranged in order of increasing atomic number.

Sections 5.3–5.4 Today, the periodic table is a valuable resource for chemists because of the enormous amount of information it contains. The table is organized by vertical columns called **groups** and by horizontal rows called **periods** or series. Groups of elements are also referred to as families because they have similar chemical properties. The table is divided into metals on the left and nonmetals on the right. Semimetals have intermediate properties and separate the metals from the nonmetals. The trend of atomic radius decreases up a group and decreases from left to right. The trend of metallic character also decreases up a group and decreases from left to right.

Section 5.5 Trends in the periodic table enable us to predict physical and chemical properties. If we know the values of physical properties for atomic radius, density, and melting point for two elements in a group, we can make a reasonable prediction of these values of another element in the same group.

Section 5.6 In Section 4.8, we memorized the order in which electrons fill sublevels. In this chapter we learned that the periodic table is arranged by sublevels of increasing energy. We can state the energy sublevel being filled by the outermost electrons by referring to the periodic table. We can write electron configuration by following the sublevels in *s*, *p*, *d*, and *f* blocks of elements.

Sections 5.7–5.8 The periodic table is valuable for predicting the number of electrons an element has available for bonding, that is, its number of **valence electrons**. The number of valence electrons corresponds to the group number in the periodic table. For example, a Group IA/1 element has 1 valence electron and a Group VIIA/17 has 7 valence electrons. Dots are drawn about the symbol of the element to show the **core**, or kernel, of the atom and its valence electrons. This structure is referred to as the **electron dot formula** of the element.

Section 5.9 After an element has lost or gained electrons, it has a positive or negative charge. Positively or negatively charged atoms are called **ions**. The amount of energy required to remove an electron from an atom in the gaseous state is called the **ionization energy**. Since the noble gases have filled valence levels, the ionization energy is extremely high for that group of elements. Each alkali metal has one electron in addition to a noble gas structure. We can correctly predict that the ionization energy for the alkali metals is the lowest of any elements in their period. In the periodic table, the trend in ionization energy increases up a group and increases from left to right.

Section 5.10 We can predict the usual **ionic charge** on the representative elements after examining the group number of the element. As a rule, metals lose all their valence electrons and become positively charged. Nonmetals gain electrons in order to obtain a noble gas structure. For example, barium metal is in Group IIA/2, has 2 valence electrons and forms Ba^{2+}. The nonmetal sulfur is in Group VIA/16, has

6 valence electrons, and gains 2 more electrons to obtain an argon electron structure. The resulting ion is S^{2-}.

Section 5.11 To write the electron configuration for an ion, we first write the electron configuration for the neutral element. If the ion has a positive charge, we subtract electrons from the electron configuration of the element. For the barium ion, Ba^{2+}, we subtract 2 electrons. If the ion has a negative charge, we add electrons. For the bromide ion, Br^{-}, we add 1 electron to the electron configuration of the element.

Key Terms

Select the key term that corresponds to the following definitions.

____ **1.** the properties of the elements recur in a repeating pattern when arranged according to increasing atomic number

____ **2.** a vertical column in the periodic table; a family of elements with similar properties

____ **3.** a horizontal row in the periodic table; a series of elements with properties varying from metallic to nonmetallic

____ **4.** the Group IA/1 elements, excluding hydrogen

____ **5.** the Group IIA/2 elements

____ **6.** the Group VIIA/17 elements

____ **7.** the relatively unreactive Group VIIIA/18 elements

____ **8.** the elements with atomic numbers 58 to 71

____ **9.** the elements with atomic numbers 90 to 103

____ **10.** the Group A (1, 2, 13 to 18) elements in the periodic table; also termed main-group elements

____ **11.** the Group B (3 to 12) elements in the periodic table

____ **12.** the elements in the lanthanide and actinide series

____ **13.** the elements with atomic numbers 21, 39, 57, 58 through 71

____ **14.** the elements beyond atomic number 92

____ **15.** the portion of the atom that includes the nucleus and inner electrons, which are not available for bonding; also termed the kernel of the atom

____ **16.** electrons that occupy the outermost *s* and *p* sublevels of an atom, which are involved in chemical reactions

____ **17.** a symbolic diagram for an element and its valence electrons; the chemical symbol is surrounded by a dot for each valence electron; also termed Lewis structures

____ **18.** the amount of energy necessary to remove an electron from a neutral atom in the gaseous state

____ **19.** an atom that bears a charge as the result of gaining or losing valence electrons

____ **20.** the relationship of ions having the same electron configuration; for example, O^{2-} and Mg^{2+} each have 10 electrons and their electron configurations are identical to the noble gas neon

____ **21.** refers to the positive charge on an atom that has lost electrons or the negative charge on an atom that has gained electrons

____ **22.** a method of writing electron configuration where all the inner electrons are represented by a noble gas symbol in brackets, followed by valence electrons; for example, [Ne] $3s^2$

(a) actinide series *(Sec. 5.3)*
(b) alkali metals *(Sec. 5.3)*
(c) alkaline earth metals *(Sec. 5.3)*
(d) core *(Sec. 5.8)*
(e) core notation *(Sec. 5.11)*
(f) electron dot formula *(Sec. 5.8)*
(g) group *(Sec. 5.1)*
(h) halogens *(Sec. 5.3)*
(i) inner transition elements *(Sec. 5.3)*
(j) ion *(Sec. 5.9)*
(k) ionic charge *(Sec. 5.10)*
(l) ionization energy *(Sec. 5.9)*
(m) isoelectronic *(Sec. 5.10)*
(n) lanthanide series *(Sec. 5.3)*
(o) noble gases *(Sec. 5.1)*
(p) period *(Sec. 5.1)*
(q) periodic law *(Sec. 5.2)*
(r) rare earth elements *(Sec. 5.3)*
(s) representative elements *(Sec. 5.3)*
(t) transition elements *(Sec. 5.3)*
(u) transuranium elements *(Sec. 5.3)*
(v) valence electrons *(Sec. 5.7)*

Exercises

Systematic Arrangement of the Elements (Sec. 5.1)

1. According to Döbereiner's triad, which element has properties that lie between those of chlorine and iodine? between those of calcium and barium?

2. According to Newlands' law of octaves, which element would complete the octave beginning with hydrogen? beginning with fluorine?

3. According to Mendeleev's periodic table of 1871, which element would begin a new period following the potassium series? the rubidium series?

4. Why did Mendeleev not include the noble gases in his periodic table of 1871?

The Periodic Law (Sec. 5.2)

5. Newlands, Meyer, and Mendeleev all suggested that properties tend to repeat periodically when the elements are arranged according to what trend?

6. Consider only the third and fourth rows in the periodic table and find two pairs of elements that obey the modern periodic law but violate the original periodic law as stated by Mendeleev.

7. Following Moseley's discovery in 1913, the periodic law states that physical and chemical properties tend to recur periodically when the elements are arranged according to what trend?

8. By studying the X-ray emission from excited nuclei, Moseley discovered that the elements have a stepwise increase in what property?

The Periodic Table of the Elements (Sec. 5.3)

9. Vertical columns in the periodic table are referred to by what two terms?

10. Horizontal rows in the periodic table are referred to by what two terms?

11. What is the collective term for the main-group elements that appear in Group IA through VIIIA (that is, Groups 1, 2, and 13 through 18)?

12. What is the collective term for the elements that belong to Groups IIIB through IIB (that is, Groups 3 through 12)?

13. What is the collective term for the two series of elements that include Ce–Lu and Th–Lr?

14. What is the collective term for the elements on the left side of the periodic table?

15. What is the collective term for the elements on the right side of the periodic table?

16. What is the collective term for the elements having properties that lie between those of the metals and nonmetals?

17. Identify the group number corresponding to each of the following families of elements.
(a) alkali metals **(b)** alkaline earth metals
(c) halogens **(d)** noble gases

18. Identify the group number corresponding to each of the following families of elements.
(a) boron group **(b)** oxygen group
(c) nickel group **(d)** copper group

19. What is the collective term for the elements in the series that follows element 57?

20. What is the collective term for the elements in the series that follows element 89?

21. What is the collective term for all the Group IIIB/3 elements, that is, Sc, Y, La, and Ce Lu?

22. What is the collective term for all the synthetic elements past uranium (currently, elements 93 through 109)?

23. According to the 1985 recommendation of IUPAC, what is the group number for each of the following groups designated by the American convention?
(a) Group IA **(b)** Group IB
(c) Group IIIA **(d)** Group IIIB
(e) Group VA **(f)** Group VB
(g) Group VIIA **(h)** Group VIIB

24. According to the American convention using Roman numerals, what is the group number designation for each of the following?
(a) Group 2 **(b)** Group 4
(c) Group 6 **(d)** Group 8
(e) Group 9 **(f)** Group 10
(g) Group 12 **(h)** Group 14
(i) Group 16 **(j)** Group 18

25. Refer to the periodic table and select the symbol of the element that fits the following descriptions.
(a) the Group IVA/14 semimetal in the fourth period
(b) the third-period alkali metal
(c) the nonradioactive halogen that normally exists as a solid
(d) the lanthanide that is not naturally occurring
(e) the rare earth element whose atomic number is greatest
(f) the sixth-period representative element with properties similar to Be
(g) the fifth-period transition element with properties similar to Ti
(h) a lighter-than-air noble gas used in blimps and balloons

26. Refer to the periodic table and select the symbol of the element that fits the following descriptions.
(a) the Group VA/15 semimetal in the fifth period
(b) the third-period alkaline earth metal
(c) the halogen that exists as a reddish-brown liquid at normal conditions
(d) the actinide with properties similar to Ce
(e) the rare earth element whose atomic mass is lowest

(f) the fourth-period representative element with properties similar to O
(g) the sixth-period transition element with properties similar to Ni
(h) the radioactive noble gas

Periodic Trends (Sec. 5.4)

27. According to the general trend, the atomic radius (increases/decreases) proceeding down a group of elements in the periodic table.
28. According to the general trend, the atomic radius for a period of elements (increases/decreases) proceeding from left to right in the periodic table.
29. According to the general trend, metallic character (increases/decreases) proceeding down a group of elements in the periodic table.
30. According to the general trend, metallic character for a period of elements (increases/decreases) proceeding from left to right in the periodic table.
31. According to general trends in the periodic table, predict which element in each pair has the larger atomic radius.
    (a) Li or Na (b) N or P
    (c) Mg or Ca (d) Ar or Kr
    (e) Rb or Sr (f) As or Se
    (g) Pb or Bi (h) I or Xe
32. According to general trends in the periodic table, predict which element in each pair has greater metallic character.
    (a) B or Al (b) Na or K
    (c) Mg or Ba (d) H or Fe
    (e) K or Ca (f) Mg or Al
    (g) Fe or Cu (h) S or Ar

Predicting Properties of Elements (Sec. 5.5)

33. Predict the missing value (?) for each physical property shown below. The atomic radius, density, and melting point are given for two of three elements in Group IA/1.

Element	Atomic Radius (nm)	Density at 20°C (g/mL)	Melting Point (°C)
K	(?)	0.86	63.3
Rb	0.248	(?)	38.9
Cs	0.266	1.90	(?)

34. Predict the missing value (?) for each physical property shown below. The atomic radius, density, and boiling point are given for two of three elements in Group VIIA/17.

Element	Atomic Radius (nm)	Density at bp (g/mL)	Boiling Point (°C)
Cl	(?)	1.56	−34.6
Br	0.115	(?)	58.8
I	0.133	4.97	(?)

35. Predict the missing value (?) for each physical property shown below. The atomic radius, density, and melting point are given for two of three elements in Group VIB/6.

Element	Atomic Radius (nm)	Density at 20°C (g/mL)	Melting Point (°C)
Ar	0.180	1.66	(?)
Kr	(?)	3.48	−152
Xe	0.210	(?)	−107

36. Predict the missing value (?) for each physical property shown below. The atomic radius, density, and boiling point are given for two of three elements in Group VIIIA/18.

Element	Atomic Radius (nm)	Density at STP (g/L)	Boiling Point (°C)
Ar	0.180	1.66	(?)
Kr	(?)	3.48	−152
Xe	0.210	(?)	−107

37. The formulas for the oxides of sodium, magnesium, aluminum, and silicon are, respectively, Na_2O, MgO, Al_2O_3, and SiO_2. Using the periodic table, predict the chemical formulas of the following similar compounds.
    (a) lithium oxide (b) calcium oxide
    (c) gallium oxide (d) tin oxide
38. The formulas for the chlorides of potassium, calcium, boron, and germanium are, respectively, KCl, $CaCl_2$, BCl_3, and $GeCl_4$. Using the periodic table, predict the chemical formulas of the following similar compounds.
    (a) potassium fluoride (b) calcium fluoride
    (c) boron bromide (d) germanium iodide
39. The chemical formula for zinc oxide is ZnO. Predict the formulas for the following similar compounds.
    (a) cadmium oxide (b) zinc sulfide
    (c) mercury sulfide (d) cadmium selenide
40. The chemical formula for barium chloride is $BaCl_2$. Predict the formulas for the following similar compounds.
    (a) strontium chloride (b) strontium bromide
    (c) magnesium iodide (d) beryllium fluoride
41. Selenium reacts with oxygen to produce SeO_3. Predict the chemical formulas of the following similar compounds.
    (a) sulfur oxide (b) tellurium oxide
    (c) selenium sulfide (d) tellurium sulfide
42. Phosphorus reacts with oxygen to produce both P_2O_3 and P_2O_5. Predict two formulas for each of the following compounds.
    (a) nitrogen oxide (b) arsenic oxide
    (c) phosphorus sulfide (d) antimony sulfide

The s, p, d, f Blocks of Elements (Sec. 5.6)

43. Which energy sublevels (*s*, *p*, *d*, or *f*) are being filled by the elements in Groups IA/1 and IIA/2?
44. Which energy sublevels (*s*, *p*, *d*, or *f*) are being filled by the elements in Groups IIIA/13 through VIIIA/18?

45. Which energy sublevels (s, p, d, or f) are being filled by the elements in Groups IIIB/3 through IIB/12?

46. Which energy sublevels (s, p, d, or f) are being filled by inner transition elements?

47. Which specific energy sublevel is being filled by the lanthanide series?

48. Which specific energy sublevel is being filled by the actinide series?

49. Refer to the periodic table and state the highest-energy sublevel that contains electrons for each of the following elements.
(a) H (b) Na
(c) Sm (d) Br
(e) Sr (f) C
(g) Sn (h) Cs
(i) Tl (j) Unq

50. Refer to the periodic table and state the highest-energy sublevel that contains electrons for each of the following elements.
(a) He (b) K
(c) U (d) Pd
(e) Be (f) Co
(g) Si (h) Pt
(i) Ho (j) Unp

51. Refer to the periodic table and write the predicted electron configuration for each of the following elements.
(a) Li (b) F
(c) Mg (d) P
(e) Ca (f) Mn
(g) Ga (h) Rb
(i) Tc (j) Xe

52. Refer to the periodic table and write the predicted electron configuration for each of the following elements.
(a) B (b) Ti
(c) Na (d) O
(e) Ge (f) Ba
(g) Pd (h) Kr
(i) Al (j) Si

Predicting Valence Electrons (Sec. 5.7)

53. State the number of valence electrons in each of the following groups as predicted from the periodic table.
(a) Group IA/1 (b) Group IIIA/13
(c) Group VA/15 (d) Group VIIA/17

54. State the number of valence electrons in each of the following groups as predicted from the periodic table.
(a) Group IIA/2 (b) Group IVA/14
(c) Group VIA/16 (d) Group VIIIA/18

55. State the number of valence electrons for each of the following representative elements.
(a) H (b) B
(c) N (d) F
(e) Ca (f) Si
(g) O (h) Ar

56. State the number of valence electrons for each of the following representative elements.
(a) He (b) Pb
(c) Se (d) Ne
(e) Cs (f) Ga
(g) Sb (h) Br

Electron Dot Formulas of Atoms (Sec. 5.8)

57. Write the electron dot formula for each of the following representative elements.
(a) H (b) B
(c) N (d) F
(e) Ca (f) Si
(g) O (h) Ar

58. Write the electron dot formula for each of the following representative elements.
(a) He (b) Pb
(c) Se (d) Ne
(e) Cs (f) Ga
(g) Sb (h) Br

Ionization Energy (Sec. 5.9)

59. According to the general trend, the first ionization energy for a group of elements (increases/decreases) proceeding down a group in the periodic table.

60. According to the general trend, the first ionization energy for a period of elements (increases/decreases) proceeding from left to right in the periodic table.

61. Which group of elements in the periodic table has the highest ionization energy?

62. Which group of elements in the periodic table has the lowest ionization energy?

63. Refer to the periodic table and predict which of the following pairs of elements has the highest ionization energy.
(a) Mg or Ca (b) S or Se
(c) Sn or Pb (d) N or P
(e) Sn or Sb (f) Se or Br
(g) Ga or Ge (h) Si or P
(i) Br or Cl (j) As or Sb

64. Refer to the periodic table and predict which of the following pairs of elements has the lowest ionization energy.
(a) Rb or Cs (b) He or Ar
(c) B or Al (d) F or I
(e) Mg or Si (f) Pb or Bi
(g) Ca or Ge (h) P or Cl
(i) Sc or Zn (j) H or Li

Predicting Ionic Charges (Sec. 5.10)

65. State the predicted ionic charge for metallic ions in each of the following groups of elements.
(a) Group IA/1 (b) Group IIA/2
(c) Group IIIA/13 (d) Group IVA/14

66. State the predicted ionic charge for nonmetallic ions in each of the following groups of elements.
(a) Group IVA/14 (b) Group VA/15
(c) Group VIA/16 (d) Group VIIA/17

67. Write the ionic charge for the following ions as predicted from the group number in the periodic table.
(a) K ion (b) Be ion
(c) Al ion (d) Sn ion
(e) C ion (f) Se ion
(g) O ion (h) F ion

68. Write the ionic charge for the following ions as predicted from the group number in the periodic table.
(a) Ga ion (b) Cs ion
(c) Ra ion (d) Pb ion
(e) Br ion (f) S ion
(g) P ion (h) I ion

69. Refer to the periodic table and predict which of the following ions are isoelectronic with the noble gas element argon.
(a) K^+ (b) Ca^{2+}
(c) Al^{3+} (d) Cl^-
(e) S^{2-} (f) N^{3-}

70. Refer to the periodic table and predict which of the following ions are isoelectronic with the noble gas element krypton.
(a) Rb^+ (b) Ca^{2+}
(c) Y^{3+} (d) Br^-
(e) Se^{2-} (f) As^{3+}

Electron Configuration of Ions (Sec. 5.11)

71. Refer to the periodic table and write the predicted electron configuration for each of the following positive ions.
(a) Mg^{2+} (b) K^+
(c) Fe^{2+} (d) Cs^+

72. Refer to the periodic table and write the predicted electron configuration for each of the following positive ions.
(a) Sr^{2+} (b) Y^{3+}
(c) Fe^{3+} (d) Ti^{4+}

73. Refer to the periodic table and write the predicted electron configuration for each of the following negative ions.
(a) F^- (b) S^{2-}
(c) N^{3-} (d) I^-

74. Refer to the periodic table and write the predicted electron configuration for each of the following negative ions.
(a) Br^- (b) Te^{2-}
(c) As^{3-} (d) O^{2-}

General Exercises

75. Examine Figure 5.5 and determine the present-day names for the three elements that Mendeleev predicted in 1871 and named ekaboron, ekaaluminum, and ekasilicon.

76. Use Döbereiner's concept to identify the triad containing three elements at normal conditions that show a transition from a yellow gas to a reddish-brown liquid to a violet solid.

77. Use the American convention to designate the group number corresponding to each of the following listed by the European convention.
(a) Group IA (b) Group IB
(c) Group IIIA (d) Group IIIB
(e) Group VA (f) Group VB

78. Use the 1985 recommendation of IUPAC to designate the group number corresponding to each of the following listed by the European convention.
(a) Group IIA (b) Group IIB
(c) Group IVA (d) Group IVB
(e) Group VIA (f) Group VIIIB

79. Refer to Table 5.2 for the physical properties of Rb and Cs. Predict the numerical value for the atomic radius, density, and melting point for Fr.

80. Refer to the periodic table and write the predicted electron configuration for each of the following elements. To simplify, indicate core electrons by the symbol of the corresponding noble gas in brackets; for example, Ba may be written [Xe] $6s^2$.
(a) Sr (b) Ru
(c) Sb (d) Cs
(e) W (f) Bi
(g) Ra (h) Ac
(i) Pu (j) Unh

81. Explain why the ionization energy for the alkaline earth metals is higher than that of the alkali metals.

82. Propose an explanation for why the ionization energy of aluminum is less than that of magnesium, contrary to the general trend.

83. Explain why the ionization energy for hydrogen is much higher than that of the other Group IA/1 elements.

84. Some periodic tables place hydrogen in both Groups IA/1 and VIIA/17. Predict two ionic charges for hydrogen. Write the formulas of the two ions and explain the ionic charges.

CHAPTER

6

Naming Chemical Compounds

A name of an ice cream indicates the ingredient flavor, for example, vanilla or chocolate. Similarly, a name of a compound indicates the ingredient elements, for example, sodium chloride or zinc oxide.

During the Middle Ages, alchemists identified hundreds of mysterious substances. The method they used to name these substances and assign symbols to them was somewhat haphazard. But since there were only hundreds of substances, this method did not pose a great problem. Toward the end of the 1700s, however, chemists had identified more than 10,000 elements and compounds. And the number of substances was growing rapidly. Chemists then faced the staggering task of providing names and symbols for all of these substances. This problem was eventually solved by using a set of systematic rules.

The French chemist Antoine Lavoisier was responsible for developing the first systematic method of naming substances. He proposed that chemical names refer to the composition of the compound and be derived from Latin or Greek. In 1787, Lavoisier, with the aid of others, published Methods of Chemical Nomenclature. *The naming system he proposed was so clear and logical that it was universally accepted in a short time. In fact, it was so thoughtful that it became the basis for our current system of naming.*

6.1 IUPAC Systematic Nomenclature

OBJECTIVES

To classify a compound as a binary ionic compound, ternary ionic compound, binary molecular compound, binary acid, or ternary oxyacid.

To classify an ion as a monoatomic cation, monoatomic anion, polyatomic cation, or polyatomic anion.

By the early 1900s, the number of known compounds had increased dramatically. There was, however, more than one way to name these compounds. There was therefore no easy way for chemists to communicate about them. To help chemists communicate, several suggestions were offered to eliminate naming problems. In 1921, the International Union of Pure and Applied Chemistry (IUPAC) formed the Commission on the Nomenclature of Inorganic Chemistry. In 1938, the IUPAC Committee on the Reform of Inorganic Nomenclature met in Berlin. Two years later it released a comprehensive set of rules. Although the rules have been expanded and revised, the 1940 Rules still stand as the official international system for naming chemical formulas. These rules are referred to as **IUPAC nomenclature**. We will first learn to classify five types of compounds and then we will apply these IUPAC rules to each of these classes.

IUPAC nomenclature The system of rules set forth by the International Union of Pure and Applied Chemistry for naming chemical formulas.

Table 6.1 History of Inorganic Nomenclature

Year	Reference for Names of Substances
Middle Ages	Alchemical names and symbols
1787	Lavoisier publishes *Methods of Chemical Nomenclature*
1921	IUPAC forms Commission on Inorganic Nomenclature
1938	IUPAC convenes first Nomenclature Convention
1940	IUPAC issues *1940 Rules* on naming inorganic compounds
1957	IUPAC amends rules and issues report in English and French
1990	IUPAC publishes recommendations on *Nomenclature of Inorganic Chemistry*, referred to as the *Red Book**

* The *Red Book* is the definitive guide for naming chemical compounds and writing chemical formulas.

organic compound A compound containing the element carbon.

inorganic compound A compound not containing the element carbon.

binary ionic compound Refers to a compound with one metal and one nonmetal.

ternary ionic compound Refers to a compound containing three elements including at least one metal.

binary molecular compound Refers to a compound containing two nonmetals.

aqueous solution A substance dissolved in water.

binary acid A compound containing hydrogen and a nonmetal dissolved in water.

ternary oxyacid A compound containing hydrogen, a nonmetal, and oxygen dissolved in water.

Classification of Compounds

Compounds are classified as either organic or inorganic. With a few exceptions, **organic compounds** contain the element carbon and **inorganic compounds** do not contain carbon. According to the 1940 Rules, inorganic compounds can be placed into one of several categories. Five common classes are binary ionic, ternary ionic, binary molecular, binary acid, or ternary oxyacid. A **binary ionic compound** contains two elements, a metal and nonmetal. Examples of binary ionic compounds are NaCl, $CaBr_2$, and Al_2O_3. A **ternary ionic compound** contains three elements, with at least one metal and one nonmetal. Examples of ternary ionic compounds are $KMnO_4$, $BaSO_4$, and $Pb(NO_3)_2$.

A **binary molecular compound** contains two elements that are both nonmetals. Water, H_2O, is a common example, as are NH_3 and CCl_4. An **aqueous solution** is produced when a compound is dissolved in water. Aqueous solutions are indicated using the symbol (aq). A **binary acid** is an aqueous solution of a compound containing hydrogen and one other nonmetal. Formulas of acids begin with H, and examples of binary acids are HCl(aq) and H_2S(aq).

Ternary oxyacids are aqueous solutions of compounds containing hydrogen, oxygen, and one other element. Examples of ternary oxyacids are HNO_3(aq), and H_2SO_4(aq). Figure 6.1 illustrates the relationship of the five types of compounds we are classifying according to IUPAC rules.

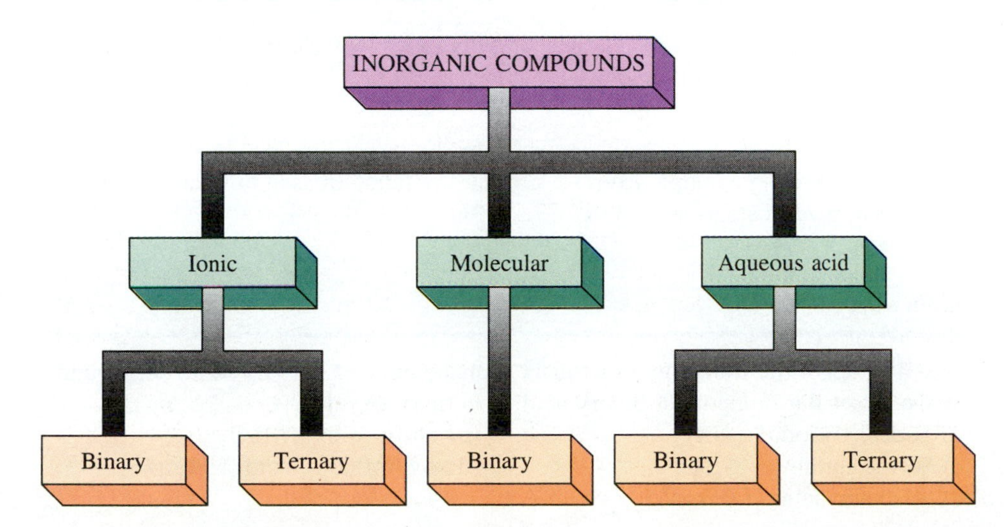

Figure 6.1 Classification of Compounds According to IUPAC inorganic nomenclature, compounds are divided into categories. Five of the classes are binary ionic, ternary ionic, binary molecular, binary acid, and ternary oxyacid.

UPDATE

Systems of Nomenclature

► **Can you give the systematic name for stomach acid?**

The forerunners of modern chemistry were the alchemists who were usually associated with attempts at converting base metals into gold. The alchemists did, however, produce a rich vocabulary describing chemical substances. Unfortunately, their names for substances did not offer a clue to their chemical compositions.

The Official IUPAC Red Book

Antoine Lavoisier, the founder of modern chemistry, was responsible for the first published naming system that indicated the chemical composition of substances. Lavoisier, along with several collaborators, published *Méthode de Nomenclature Chimique* in 1787. The great Swedish chemist J. J. Berzelius adopted Lavoisier's ideas and extended the system into the Germanic languages that, in turn, were translated into English.

In 1892, a conference was held in Geneva, Switzerland, that laid the foundation for an internationally accepted system of nomenclature for *organic compounds*. Subsequently, a group formed and was designated the International Union of Pure and Applied Chemistry (IUPAC).

In 1921, IUPAC addressed the issue of systematic nomenclature for *inorganic compounds*. Almost 20 years elapsed before IUPAC issued a systematic set of rules. The so-called 1940 IUPAC rules were consistent with the methods indicated in Lavoisier's text published 150 years earlier. The 1940 recommendations suggest more than one acceptable system of nomenclature. For example, $CuSO_4$ can be named copper(II) sulfate according to the preferred Stock system of nomenclature or cupric sulfate according to the Latin system. Although common names still persist, such as baking soda for $NaHCO_3$, IUPAC disapproves of the use of common names and discourages the use of names that do not indicate the chemical composition.

In truth, the chemistry community is not in complete agreement on subtle aspects of nomenclature. Currently, the most authoritative guide to systematic naming is the 1990 IUPAC recommendations contained in *Nomenclature of Inorganic Chemistry*. This publication is referred to as the *Red Book* (see Figure).

► **Stomach acid is aqueous HCl and its IUPAC systematic name is hydrochloric acid.**

The following example exercise provides practice in assigning compounds to one of five given categories.

EXAMPLE EXERCISE 6.1

Classify each of the following as a binary ionic compound, ternary ionic compound, binary molecular compound, binary acid, or ternary oxyacid.

(a) nickel(II) iodide, NiI_2
(b) sulfur dioxide, SO_2
(c) silver chromate, Ag_2CrO_4
(d) hydrofluoric acid, HF(aq)
(e) carbonic acid, H_2CO_3(aq)

Solution: We can classify each compound or solution as follows.

(a) NiI_2 contains two elements, a metal and nonmetal. Thus, NiI_2 is a *binary ionic compound*.

(b) SO_2 contains two elements, both nonmetals. Thus, SO_2 is a *binary molecular compound*.

(c) Ag_2CrO_4 contains three elements, two metals and a nonmetal. Thus, Ag_2CrO_4 is a *ternary ionic compound*.

(d) HF(aq) is a compound of hydrogen and a nonmetal dissolved in water. Thus, HF(aq) is a *binary acid*.

(e) H_2CO_3(aq) is a compound containing three elements, including hydrogen and oxygen, dissolved in water. Thus, H_2CO_3(aq) is a *ternary oxyacid*. Even though this common acid contains carbon, it is generally considered to be an inorganic acid.

The beakers contain an acid-base indicator and show that lime juice, vinegar, and soda water are all acidic.

SELF-TEST EXERCISE

Classify each of the following as a binary ionic compound, ternary ionic compound, binary molecular compound, binary acid, or ternary oxyacid.

(a) carbon disulfide, CS_2

(b) lithium dichromate, $Li_2Cr_2O_7$

(c) magnesium iodide, MgI_2

(d) nitric acid, HNO_3(aq)

(e) hydrochloric acid, HCl(aq)

Answers:

(a) binary molecular compound; **(b)** ternary ionic compound; **(c)** binary ionic compound; **(d)** ternary oxyacid; **(e)** binary acid

Classification of Ions

According to IUPAC nomenclature, ions are named systematically, depending on the category into which they are placed. Positive ions are collectively referred to as **cations**. Negative ions are referred to as **anions**. A single atom bearing a positive or negative charge is called a **monoatomic ion**. A particle containing two or more atoms having a positive or negative charge is called a **polyatomic ion**. Figure 6.2 illustrates the classification and relationship of these ions.

cation A positively charged ion.

anion A negatively charged ion.

monoatomic ion A single atom bearing a positive or negative charge as the result of gaining or losing valence electrons.

polyatomic ion A unit containing two or more atoms bearing a positive or negative charge.

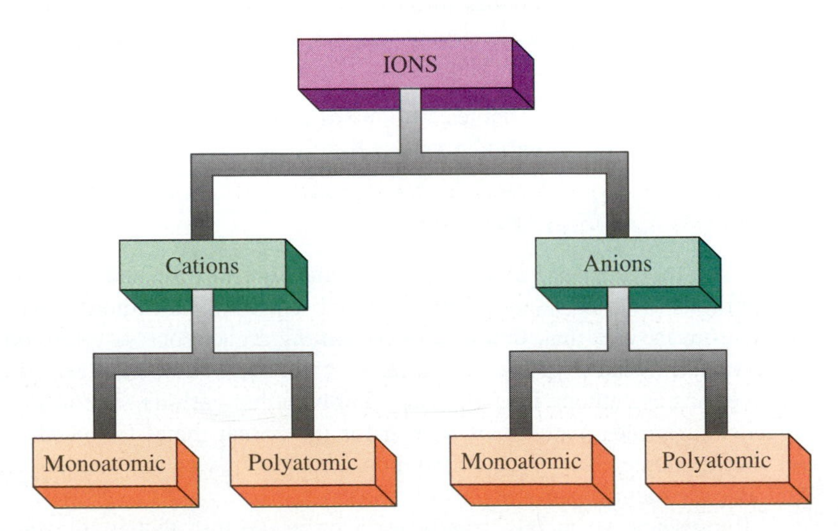

Figure 6.2 Classification of Ions According to the IUPAC nomenclature, ions are divided into four categories. They are monoatomic cations, monoatomic anions, polyatomic cations, and polyatomic anions.

The following example exercise illustrates the classification of different types of ions.

EXAMPLE EXERCISE 6.2

Classify each of the following ions as a monoatomic cation, monoatomic anion, polyatomic cation, or polyatomic anion.

(a) bismuth ion, Bi^{3+}
(b) chloride ion, Cl^-
(c) mercurous ion, Hg_2^{2+}
(d) sulfate ion, SO_4^{2-}

Solution: We can classify each ion as follows.

(a) Bi^{3+} is a single atom with a positive charge. Thus, Bi^{3+} is a *monoatomic cation.*
(b) Cl^- is a single atom with a negative charge. Thus, Cl^- is a *monoatomic anion.*
(c) Hg_2^{2+} contains two atoms and has a positive charge. Thus, Hg_2^{2+} is a *polyatomic cation.*
(d) SO_4^{2-} contains five atoms and has a negative charge. Thus, SO_4^{2-} is a *polyatomic anion.*

SELF-TEST EXERCISE

Classify each of the following ions as a monoatomic cation, monoatomic anion, polyatomic cation, or polyatomic anion.

(a) ammonium ion, NH_4^+
(b) sulfide ion, S^{2-}
(c) permanganate ion, MnO_4^-
(d) lithium ion, Li^+

Answers:
(a) polyatomic cation; **(b)** monoatomic anion; **(c)** polyatomic anion; **(d)** monoatomic cation

6.2 Monoatomic Ions

To write both the Stock system and Latin system names for common monoatomic cations.

To write systematic names and formulas for common monoatomic anions.

To predict charges for ions of representative metals and nonmetals using the periodic table.

Naming Monoatomic Cations

As we now know, metal atoms can lose valence electrons and become positively charged ions. Another name for positively charged ions is cathode ions. This name comes from the fact that, in a battery containing an aqueous solution, positive ions are attracted to the negative electrode, or cathode. Rather than referring to these positive ions as cathode ions, they are simply called **cations**. According to IUPAC nomenclature rules, cations are named for the parent metal followed by the word ion. For example, Na^+ is called sodium ion, Mg^{2+} is called magnesium ion, and Al^{3+} is called aluminum ion.

Representative metals usually form only one ion; that is, they lose valence

Table 6.2 Common Monoatomic Cations

Cation	Stock System	Latin System	Cation	Stock System	Latin System
Al^{3+}	aluminum ion		Pb^{4+}	lead(IV) ion	plumbic ion
Ba^{2+}	barium ion		Li^{+}	lithium ion	
Bi^{3+}	bismuth ion		Mg^{2+}	magnesium ion	
Cd^{2+}	cadmium ion		Mn^{2+}	manganese(II) ion	manganous ion
Ca^{2+}	calcium ion		Hg_2^{2+}	mercury(I) ion*	mercurous ion
Co^{2+}	cobalt(II) ion	cobaltous ion	Hg^{2+}	mercury(II) ion	mercuric ion
Co^{3+}	cobalt(III) ion	cobaltic ion	Ni^{2+}	nickel(II) ion	nickelous ion
Cu^{+}	copper(I) ion	cuprous ion	K^{+}	potassium ion	
Cu^{2+}	copper(II) ion	cupric ion	Ag^{+}	silver ion	
Cr^{3+}	chromium(III) ion		Na^{+}	sodium ion	
H^{+}	hydrogen ion		Sr^{2+}	strontium ion	
Fe^{2+}	iron(II) ion	ferrous ion	Sn^{2+}	tin(II) ion	stannous ion
Fe^{3+}	iron(III) ion	ferric ion	Sn^{4+}	tin(IV) ion	stannic ion
Pb^{2+}	lead(II) ion	plumbous ion	Zn^{2+}	zinc ion	

* Note that mercury(I) is a polyatomic ion.

electrons predictably. The transition metals, however, often form more than one ion. Iron, for example, can form Fe^{2+} and Fe^{3+}. Other examples of metal ions having two or more charges are copper (Cu^{+} and Cu^{2+}) and lead (Pb^{2+} and Pb^{4+}). To name a metal ion having more than one ionic charge, it is necessary to specify the charge. IUPAC recommends that the ion be named for the parent metal followed by its charge specified by Roman numerals in parentheses. Thus, Fe^{2+} is named iron(II) ion and Fe^{3+} is named iron(III) ion. Similarly, Cu^{+} is named copper(I) ion, and Cu^{2+} is named copper(II) ion; Pb^{2+} is named lead(II) ion, and Pb^{4+} is named lead(IV) ion. This method of naming ions is referred to as the **Stock system**. It is named after the German chemist Alfred Stock (1876–1946), who was influential in drafting the 1940 Rules.

Stock system A naming system that designates the variable charge on a metal cation using Roman numerals in parentheses.

IUPAC allows another, older method for naming metal cations having two charges. It is called the **Latin system** or suffix system. This system takes the Latin name of the metal and adds an *-ous* or -ic *suffix*. The lower of the two ionic charges receives the -ous suffix, and the higher charge receives the -ic suffix. For example, the Latin name for iron is ferrum. We take the ferr- stem and add either an -ous or -ic suffix. Thus, Fe^{2+} is named ferrous ion and Fe^{3+} is named ferric ion. The Latin name for copper is cuprum. We add the -ous or -ic suffix to the cupr- stem. Thus, we have cuprous ion (Cu^{+}) and cupric ion (Cu^{2+}). The Latin name for lead is plumbum. We add the -ous or -ic suffix to the plumb- stem. Thus, we have plumbous ion (Pb^{2+}) and plumbic ion (Pb^{4+}).

Latin system A naming system that designates the variable charge on a metal cation using an -ic or -ous suffix attached to the stem of the Latin name.

Mercury is an exception. When mercury forms Hg^{+}, the resulting ion becomes more stable by combining with another Hg^{+} ion. Thus, the mercury(I) ion is written Hg_2^{2+}. The Latin name for mercury, hydrargyrum, is difficult to pronounce with a suffix. IUPAC recommends adding the suffix to the English name. Thus, Hg_2^{2+} is named mercurous ion, and Hg^{2+} is named mercuric ion. Table 6.2 lists systematic names for several common cations.

Naming Monoatomic Anions

As we now know, nonmetal atoms can gain valence electrons and become negatively charged ions. Another name for negatively charged ions is anode ions. This name

Table 6.3 Common Monoatomic Anions

Anion	IUPAC Name
F^-	fluoride ion
Cl^-	chloride ion
Br^-	bromide ion
I^-	iodide ion
O^{2-}	oxide ion
S^{2-}	sulfide ion
N^{3-}	nitride ion
P^{3-}	phosphide ion

comes from the fact that, in a battery containing an aqueous solution, negative ions are attracted to the positive electrode, or anode. Rather than referring to these negative ions as anode ions, they are simply called **anions**. According to IUPAC rules, nonmetallic ions are named as follows: *nonmetal stem + -ide suffix*, followed by the word *ion*. Examples of nonmetal ions are Cl^-, chloride ion, S^{2-}, sulfide ion, and P^{3-}, phosphide ion. Table 6.3 lists systematic names for several common anions.

Mastering Formulas of Ions

We can use the periodic table to help learn the names and formulas of the ions in Tables 6.2 and 6.3. We can apply what we discussed in Section 5.10 about predicting ionic charge based on the group number of the element. Recall that Group IA/1 metals always form 1+ ions. Thus, we have Li^+, Na^+, K^+, Rb^+, and Cs^+. Also recall that the Group IIA/2 elements always form 2+ ions. Thus, we have Be^{2+}, Mg^{2+}, Ca^{2+}, Sr^{2+}, and Ba^{2+}.

Not all elements are predictable from the periodic table. Group IIIA/13 metals usually form 3+ ions, although In^+ and Tl^+ are exceptions. Aluminum, however, always loses 3 electrons to give Al^{3+}. Tin and lead in Group IVA/14 are not predictable. That is, tin can form either Sn^{2+} or Sn^{4+}. Similarly, lead forms Pb^{2+} or Pb^{4+}. Therefore, you will have to memorize the names and formulas of these ions.

The way in which nonmetals gain electrons to form negative ions is predictable. All the halogens in Group VIIA/17 gain 1 electron to become isoelectronic with the nearest noble gas. The halogens form 1− ions. Thus, the halides are written F^-, Cl^-, Br^-, and I^-. The elements in Group VIA/16 require 2 electrons to assume a noble gas configuration. These elements form 2− ions. Thus, we have O^{2-}, S^{2-}, Se^{2-}, and Te^{2-}. Nitrogen and phosphorus in Group VA/15 form the two ions N^{3-} and P^{3-}.

With few exceptions, all the transition elements have two *s* electrons as well as a variable number of *d* electrons. In general, a transition metal loses its two *s* electrons to form an ion with a 2+ charge, for example, Ni^{2+}. Many of the transition metals form additional ions having charges that are unpredictable. You will have to memorize the names and formulas of the unpredictable ions. Figure 6.3 shows the relationship of ionic charge to the position of the element in the periodic table.

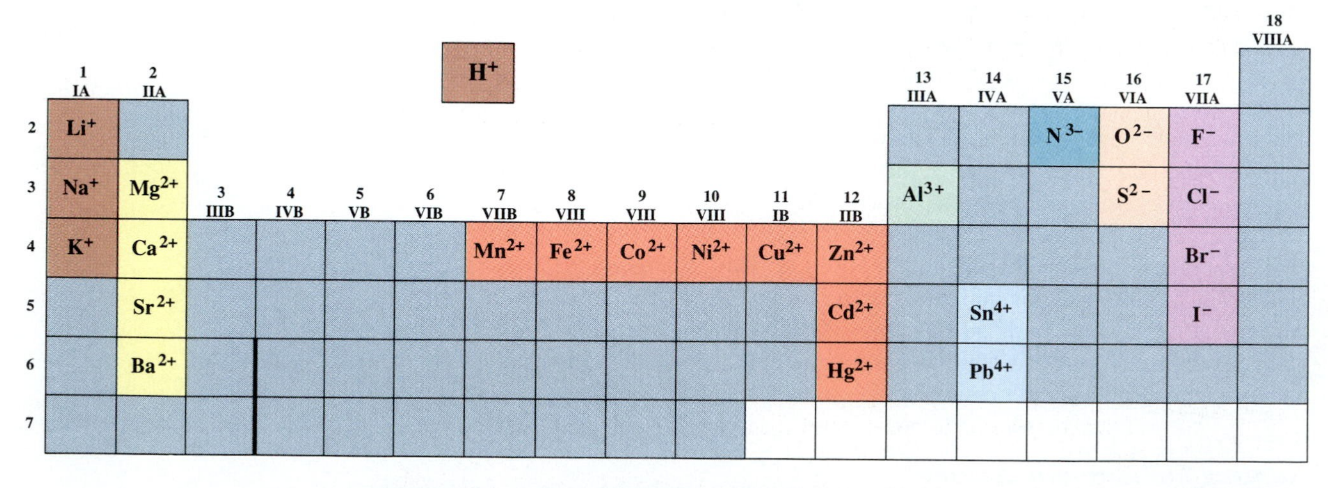

Figure 6.3 Periodic Table of Selected Ions Note the correlation of ionic charge and group number. Although the transition elements exhibit more than one ionic charge, nearly all these metals can have a common charge of 2 plus.

Although it is necessary to memorize the names and formulas of ions, the periodic table is a valuable resource for verifying ionic charges.

EXAMPLE EXERCISE 6.3

Provide the formula for each of the following monoatomic ions.

(a) sodium ion **(b)** barium ion **(c)** manganese(II) ion
(d) aluminum ion **(e)** bromide ion **(f)** sulfide ion

Solution: We can use the periodic table to help provide the formula.

(a) Sodium is found in Group IA/1, which has 1 valence electron. We can correctly predict the formula as Na^{+}.
(b) Barium is found in Group IIA/2, which has 2 valence electrons. We can correctly predict the formula, Ba^{2+}.
(c) Manganese is a transition metal and often has an ionic charge of 2+. In this case the name is followed by (II), so we are certain that the charge on the ion is 2+; thus, Mn^{2+}.
(d) Aluminum is found in Group IIIA/13, which has 3 valence electrons. We can correctly predict the formula, Al^{3+}.
(e) The bromide ion is found in Group VIIA/17, which has 7 valence electrons. We would predict the addition of 1 electron and the correct formula, Br^{-}.
(f) The sulfide ion is found in Group VIA/16, which has 6 valence electrons. We can easily predict the correct formula, S^{2-}.

SELF-TEST EXERCISE

Supply a systematic name for each of the following monoatomic ions.

(a) Sr^{2+} **(b)** Cd^{2+} **(c)** Co^{2+}
(d) Hg^{2+} **(e)** S^{2-} **(f)** N^{3-}

Answers:
(a) strontium ion; **(b)** cadmium ion; **(c)** cobalt(II) or cobaltous ion; **(d)** mercury(II) or mercuric ion; **(e)** sulfide ion; **(f)** nitride ion

Note: The ionic charge of an element is sometimes referred to as an *oxidation number*. Oxidation number is, however, a broad term that indicates more than just ionic charge. It describes the assigned positive or negative charge on an atom in a molecule or polyatomic ion. In Chapter 17, we will formally study oxidation numbers and learn to calculate values for elements in different substances.

6.3 Polyatomic Ions

To write systematic names and formulas for common polyatomic ions.

Most polyatomic anions have an *-ate* ending. Examples include the nitrate ion NO_3^- and the sulfate ion SO_4^{2-}. A few polyatomic anions have an *-ite* ending.

Table 6.4 Common Polyatomic Ions

Cation	IUPAC Name
NH_4^+	ammonium ion
Anion	**IUPAC Name**
$C_2H_3O_2^-$	acetate ion
ClO^-	hypochlorite ion
ClO_2^-	chlorite ion
ClO_3^-	chlorate ion
ClO_4^-	perchlorate ion
CN^-	cyanide ion*
OH^-	hydroxide ion*
NO_2^-	nitrite ion
NO_3^-	nitrate ion
HCO_3^-	hydrogen carbonate ion
HSO_3^-	hydrogen sulfite ion
HSO_4^-	hydrogen sulfate ion
MnO_4^-	permanganate ion
CrO_4^{2-}	chromate ion
$Cr_2O_7^{2-}$	dichromate ion
CO_3^{2-}	carbonate ion
SO_3^{2-}	sulfite ion
SO_4^{2-}	sulfate ion
PO_4^{3-}	phosphate ion

* Note the -ide suffix, which is an exception to the general rules.

Examples include the nitrite ion NO_2^- and the sulfite ion SO_3^{2-}. Notice that in each case the formula for the -ite ending has one less oxygen than that for the -ate ending.

This pattern of -ate and -ite supplies a key to help name some polyatomic ions. Given the formula for the chlorate ion ClO_3^-, we can predict the formula for the chlorite ion. Since the -ate ending has changed to an -ite ending, the formula has one less oxygen. Therefore, the formula for the chlorite ion is ClO_2^-. This general relationship between -ate and -ite endings allows us to simplify our task of memorizing formulas of ions.

There is only one common positive polyatomic ion. It is the ammonium ion, NH_4^+, and its formula should be memorized. There are two polyatomic ions that have an -ide ending. They are the cyanide ion CN^- and the hydroxide ion OH^-. They also should be memorized. Table 6.4 lists several common polyatomic ions whose formulas should be learned.

You may find it helpful to make flashcards to help you memorize ions. Write the name of the ion on one side of a card and the formula of the ion on the other side. The following example exercise illustrates how general principles can be used to assist you in memorizing the names and formulas of ions.

EXAMPLE EXERCISE 6.4

Provide a systematic name for each of the following polyatomic ions.

(a) CO_3^{2-} **(b)** CrO_4^{2-} **(c)** ClO_2^- **(d)** HSO_4^-

Solution: We can usually make reasonable predictions for the names of polyatomic ions. We can verify our prediction if we have memorized the names and formulas of the ions in Table 6.4.

(a) CO_3^{2-} contains the nonmetal carbon. We can predict that the name has an -ate ending; CO_3^{2-} is correctly named the *carbonate ion.*

(b) CrO_4^{2-} contains the metal chromium. We can predict that the name has an -ate ending; CrO_4^{2-} is correctly named the *chromate ion.*

(c) ClO_2^- contains the nonmetal chlorine. The ClO_2^- ion is related to ClO_3^-, which is named the chlorate ion. Since ClO_2^- has one less oxygen atom, we can predict that the suffix ending changes to -ite. Thus, ClO_2^- is named the *chlorite ion.*

(d) HSO_4^- is related to the sulfate ion, SO_4^{2-}. With the hydrogen present, the correct name becomes the hydrogen *sulfate ion.*

SELF-TEST EXERCISE

Provide the formula for each of the following polyatomic ions.

(a) acetate ion
(b) dichromate ion
(c) perchlorate ion
(d) hydrogen sulfite ion

Answers:
(a) $C_2H_3O_2^-$; **(b)** $Cr_2O_7^{2-}$; **(c)** ClO_4^-; **(d)** HSO_3^-

IUPAC has stated systematic rules for naming polyatomic ions. Unfortunately, they are too complex to help us learn this introductory list of ions. Therefore, it is necessary to memorize the ions in Table 6.4. Your task is easier, however, if you recall the following:

1. There is only one common polyatomic cation, NH_4^+.
2. Most polyatomic anions have an -ate ending.
3. The -ate ending changes to an -ite ending with one less oxygen, for example, nitrate, NO_3^-, and nitrite, NO_2^-.
4. The OH^- and CN^- ions are exceptions and have an -ide ending.

6.4 Writing Chemical Formulas

OBJECTIVES

To write neutral formula units for compounds containing monoatomic ions.

To write neutral formula units for compounds containing a polyatomic ion.

Ionic compounds are composed of positive and negative ions. The smallest representative particle of an ionic compound is called a **formula unit**. In a formula unit the total positive charge is equal to the total negative charge. That is, the positive charge from the metal ions must be the same as the negative charge from the nonmetal ions.

formula unit The smallest representative particle in an ionic compound.

In a formula unit containing Ca^{2+} and O^{2-}, the positive and negative charges are the same. Thus, the formula of the compound is CaO. In a formula unit containing Ca^{2+} and Cl^-, the charges are not the same. It is necessary to have two Cl^- ions for each Ca^{2+} ion. The formula of this compound is $CaCl_2$. Example Exercise 6.5 provides additional illustrations of writing formulas for ionic compounds.

EXAMPLE EXERCISE 6.5

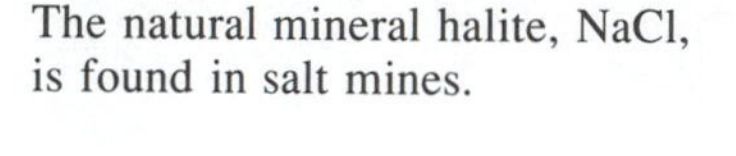

The natural mineral halite, NaCl, is found in salt mines.

Write the chemical formula for the following binary compounds, given their constituent ions.

(a) sodium chloride, Na^+ and Cl^-
(b) magnesium bromide, Mg^{2+} and Br^-
(c) bismuth oxide, Bi^{3+} and O^{2-}

Solution:

(a) Since the positive and negative ions each have a charge of 1, the ratio is 1:1, Na_1Cl_1. The subscript 1 is understood, and the formula unit of sodium chloride is written NaCl.

(b) The positive ion has a charge of 2+, and the negative ion has a charge of 1−. Therefore, two negative ions are required to produce a neutral formula unit. The formula of magnesium bromide is written $MgBr_2$.

(c) This example is more difficult. The positive ion has a charge of 3+, and the negative ion has a charge of 2−. The lowest common multiple is 6.

Therefore, two positive ions are required for every three negative ions to produce a neutral unit. A bismuth oxide formula unit is written Bi_2O_3.

SELF-TEST EXERCISE

Write the chemical formula for the following binary compounds, given their constituent ions.

(a) iron(II) sulfide, Fe^{2+} and S^{2-}
(b) mercury(I) fluoride, Hg_2^{2+} and F^-
(c) lead(IV) oxide, Pb^{4+} and O^{2-}

Answers: **(a)** FeS; **(b)** Hg_2F_2; **(c)** PbO_2

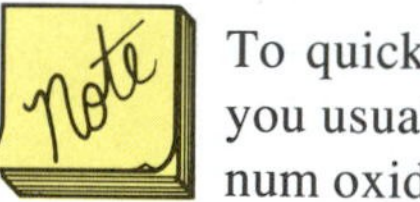

To quickly verify that you have written the chemical formula correctly, you usually can simply crossover the charge on each ion. Consider aluminum oxide, which contains Al^{3+} and O^{2-}. The 3+ charge on the aluminum ion becomes the subscript for the oxygen, and the 2− charge on the oxide ion gives the subscript for the aluminum ion.

$$Al^{(3+)} O^{(2-)} = Al_2O_3$$

Magnesium oxide is composed of Mg^{2+} and O^{2-}. The 2+ charge on the magnesium ion becomes the subscript for the oxygen, and the 2− charge on the oxide ion becomes the subscript for the magnesium ion; thus, we have Mg_2O_2. However, ionic compounds are written in their simplest ratios, and Mg_2O_2 simplifies to MgO.

Alternatively, some students create a formula unit of Al_2O_3 an ion at a time. That is, Al^{3+} and O^{2-} cannot produce a neutral formula unit because of the extra positive charge. Adding another O^{2-} makes an extra negative charge. An additional Al^{3+} gives a positive charge of 6 and a negative charge of 4. Finally, after an additional O^{2-}, the total positive charge is 6+, and the total negative charge is 6−. Thus, the formula unit, Al_2O_3, is electrically neutral.

Formula Units Containing Polyatomic Ions

Formula units are the simplest particles that represent an ionic compound. In a neutral formula unit the total positive charge is equal to the total negative charge. Previously, we learned to write neutral formula units for binary ionic compounds. Similarly, in a formula unit containing K^+ and SO_4^{2-}, two K^+ ions are required for each SO_4^{2-}. Thus, a neutral formula unit is written K_2SO_4.

Magnesium sulfate is found in Epsom salts and contains Mg^{2+} and SO_4^{2-}. Since the charge is the same on each ion, the ratio of positive ion to negative ion is 1:1. The neutral formula unit is written $MgSO_4$.

Ammonium sulfate is a nitrogen-supplying component in fertilizer and contains the NH_4^+ and SO_4^{2-} ions. Since the negative charge is greater, two NH_4^+ ions are necessary to give a neutral formula unit. To avoid misunderstanding, we will place parentheses around the NH_4^+ ion. The correct formula is written $(NH_4)_2SO_4$. Example Exercise 6.6 provides additional illustrations of writing formulas for ionic compounds.

EXAMPLE EXERCISE 6.6

Write the chemical formula for the following ternary compounds, given their constituent ions.

(a) calcium carbonate, Ca^{2+} and CO_3^{2-}
(b) calcium hydroxide, Ca^{2+} and OH^-
(c) calcium phosphate, Ca^{2+} and PO_4^{3-}

Solution:

(a) Since the positive and negative ions each have a charge of 2, the ratio is 1:1, $CaCO_3$. This compound occurs naturally as chalk, limestone, or marble.

(b) The positive ion has a charge of 2+, and the negative ion has a charge of 1−. Therefore, two negative ions are required to produce a neutral formula unit. Parentheses are required around the polyatomic ion, and the formula of calcium hydroxide is written $Ca(OH)_2$. Calcium hydroxide is also known as slaked lime and is sometimes used to mark the boundaries of an athletic field.

(c) The positive ion has a charge of 2+, and the negative ion has a charge of 3−. The lowest common multiple is 6. Three positive ions are required for every two negative ions to produce a neutral formula unit. A calcium phosphate unit is written $Ca_3(PO_4)_2$. Calcium phosphate is found in tooth enamel.

SELF-TEST EXERCISE

Write the chemical formula for the following ternary compounds, given their constituent ions.

(a) silver permanganate, Ag^+ and MnO_4^-
(b) iron(III) carbonate, Fe^{3+} and CO_3^{2-}
(c) ammonium sulfate, NH_4^+ and SO_4^{2-}

Answers: **(a)** $AgMnO_4$; **(b)** $Fe_2(CO_3)_3$; **(c)** $(NH_4)_2SO_4$

As before, we can usually verify that the formula is correct by simply crossing over the charge on each ion. Consider calcium phosphate, which contains Ca^{2+} and PO_4^{3-}. The 2+ charge on the calcium ion becomes the subscript for the phosphate ion. Conversely, the 3− charge on the phosphate ion becomes the subscript for the calcium ion.

Remember that parentheses are required around a polyatomic ion each time two or more polyatomic ions appear in the chemical formula. Moreover, the formula is reduced to the simplest ratio. The Ca^{2+} ion and O^{2-} CO_3^{2-} ion combine to give $Ca_2(CO_3)_2$ which simplifies to $CaCO_3$.

6.5 Binary Ionic Compounds

OBJECTIVES

To determine the ionic charge on a cation in a binary ionic compound.

To write systematic names and formulas for binary ionic compounds.

A metal and nonmetal react to form a binary ionic compound. The metal loses its valence electrons to become a positive cation, and the nonmetal gains electrons to become a negative anion. The oppositely charged cation and anion are attracted to each other in ratios that create compounds that are neutral. That is, even though ionic compounds contain charged ions, overall the compound has a charge of zero.

According to IUPAC rules, the positive cation is always written before the negative anion. Therefore, in naming binary ionic compounds, the metal cation is named first, followed by the nonmetal anion. The nonmetal anion bears an *-ide* suffix. For example, NaCl is named sodium chloride and $CaBr_2$ is named calcium bromide.

Some compounds contain metals having a variable ionic charge, such as Fe^{2+} and Fe^{3+} for iron. To specify the charge on the ion in the compound, IUPAC rules use either the Stock or Latin system of naming. To name the compound correctly, we first have to be able to determine the ionic charge of the cation. Example Exercise 6.7 illustrates the calculation of an ionic charge.

EXAMPLE EXERCISE 6.7

Determine the ionic charge for iron in iron oxide, Fe_2O_3.

Solution: The charge on an oxide ion is 2− and there are three oxide ions. The total negative charge must be equal to 6 negative.

$$O^{2-} + O^{2-} + O^{2-} = 6 \text{ negative}$$

Since all compounds are electrically neutral, the total positive charge must equal the total negative charge: 6 negative = 6 positive. Thus, the two iron ions have a charge of 6 positive.

$$Fe^{x+} + Fe^{x+} = 6 \text{ positive}$$
$$Fe^{x+} = 3 \text{ positive}$$

The iron ion is therefore Fe^{3+}, and Fe_2O_3 is named iron(III) oxide according to the Stock system. It is named ferric oxide according to the Latin system.

SELF-TEST EXERCISE

Determine the ionic charge for the transition metal cation in each of the following compounds.

(a) Cu_3P **(b)** CoN

Answers: **(a)** Cu^+; **(b)** Co^{3+}

Naming Binary Ionic Compounds

Binary ionic compounds are named by designating the cation followed by the anion. KI, for example, is composed of the potassium ion and the iodide ion. Thus, its name is potassium iodide. Mg_3N_2 is composed of magnesium ions and nitride ions; its name is magnesium nitride. PbO_2 is composed of a lead ion and two oxide ions. Lead, however, has two possible ions, Pb^{2+} and Pb^{4+}. Since we can calculate the ionic charge to be 4+, the name is lead(IV) oxide. We could also name PbO_2 according to the Latin system, that is, plumbic oxide.

EXAMPLE EXERCISE 6.8

Supply a systematic name for each of the following binary ionic compounds.

(a) ZnO **(b)** BaI_2
(c) $MnCl_2$ **(d)** SnF_2

Solution: We name each compound by designating the two ions.

(a) ZnO is named for the zinc ion and the oxide ion. Thus, ZnO is named *zinc oxide*.
(b) BaI_2 is named for the barium ion and the iodide ion. Thus, BaI_2 is named *barium iodide*.
(c) $MnCl_2$ is named for the manganese(II) ion and the chloride ion. Thus, $MnCl_2$ is named *manganese(II) chloride*.
(d) SnF_2 is named for the tin ion and the fluoride ion. However, tin has two possible ions, Sn^{2+} and Sn^{4+}. Since there are two F^- ions, the ionic charge on tin must be 2+; thus, the name is *tin(II) fluoride*. When we name the compound using the Latin system, stannous fluoride, we recognize the compound as a popular toothpaste ingredient.

SELF-TEST EXERCISE

Supply a systematic name for each of the following binary ionic compounds.

(a) Cd_3P_2 **(b)** Fe_2S_3

Answers: **(a)** cadmium phosphide; **(b)** iron(III) sulfide or ferric sulfide

EXAMPLE EXERCISE 6.9

Provide the formula for each of the following binary ionic compounds.

(a) lithium fluoride **(b)** strontium chloride
(c) lead(II) sulfide **(d)** mercuric nitride

Solution: To write the formula, we combine the cation and anion into a neutral formula unit. See Section 6.4 to review writing formula units.

(a) Lithium fluoride is composed of Li^+ and F^-. A neutral formula unit and the formula of the compound are written LiF.

The natural mineral galena is lead sulfide, PbS.

(b) Strontium chloride contains Sr^{2+} and Cl^-. A neutral formula unit and the formula of the compound are written $SrCl_2$.
(c) Lead(II) sulfide contains Pb^{2+} and S^{2-}. A neutral formula unit and the formula of the compound are written PbS.
(d) Mercuric nitride is composed of Hg^{2+} and N^{3-}. A neutral formula unit and the formula of the compound are written Hg_3N_2.

SELF-TEST EXERCISE

Provide the formula for each of the following binary ionic compounds.
(a) copper(II) iodide **(b)** ferric oxide

Answers: **(a)** CuI_2; **(b)** Fe_2O_3

6.6 Ternary Ionic Compounds

To determine the ionic charge on a cation in a ternary ionic compound.

To write systematic names and formulas for ternary ionic compounds.

Compounds containing a metal and two other elements are classified as ternary ionic compounds. Typically, ternary ionic compounds are combinations of monoatomic metal cations and polyatomic anions containing oxygen. As with all ionic compounds, the cation is written first in the formula. The names of ternary ionic compounds usually have an -ate or -ite ending. For example, $KClO_3$ is named potassium chlorate and $KClO_2$ is named potassium chlorite.

Some compounds contain metals having a variable ionic charge, such as Fe^{2+} and Fe^{3+}. To specify the charge on the ion in the compound, IUPAC rules allow the use of either the Stock system or Latin system of naming. To name the compound correctly, we first have to be able to determine the ionic charge of the cation. Example Exercise 6.10 illustrates the calculation of the ionic charge.

EXAMPLE EXERCISE 6.10

Determine the ionic charge for iron in iron phosphate, $Fe_3(PO_4)_2$.

Solution: The charge on a phosphate ion is 3− and there are two phosphate ions. Therefore, the total negative charge must be equal to 6 negative.

$$PO_4^{3-} + PO_4^{3-} = 6 \text{ negative}$$

Since all compounds are electrically neutral, the total positive charge must equal the total negative charge: 6 negative = 6 positive. Thus, the three iron ions have a charge of 6 positive.

$$Fe^{x+} + Fe^{x+} + Fe^{x+} = 6 \text{ positive}$$
$$Fe^{x+} = 2 \text{ positive}$$

The iron ion is Fe^{2+} and $Fe_3(PO_4)_2$ is named iron(II) phosphate according to the Stock system and ferrous phosphate according to the Latin system.

SELF-TEST EXERCISE

Determine the ionic charge for the metal cation in each of the following compounds.

(a) $Hg(OH)_2$ **(b)** $Co(ClO_3)_3$

Answers: **(a)** Hg^{2+}; **(b)** Co^{3+}

Naming Ternary Ionic Compounds

Ternary ionic compounds are named by designating the cation, followed by the anion. Li_2CO_3 is composed of the lithium ion and the carbonate ion. Thus, its name is lithium carbonate. $Ba(NO_3)_2$ is composed of the barium ion and nitrate ions. Its name is barium nitrate.

The compound $CuSO_4$ is composed of a copper ion and a sulfate ion. However, copper has two possible ions, Cu^+ and Cu^{2+}. Since the sulfate charge is 2−, the ionic charge on copper must be 2+. Because compounds are electrically neutral, we have Cu^{2+}. The name of $CuSO_4$ is copper(II) sulfate or cupric sulfate.

The Taj Mahal in India is made of marble, $CaCO_3$, a ternary ionic compound.

EXAMPLE EXERCISE 6.11

Supply a systematic name for each of the following ternary ionic compounds.

(a) $KMnO_4$ **(b)** $Sr(ClO_2)_2$

(c) $Hg(NO_3)_2$ **(d)** NaCN

Solution: We name each compound by designating the two ions.

(a) $KMnO_4$ is named for the potassium ion and the permanganate ion. Thus, $KMnO_4$ is named *potassium permanganate*.

(b) $Sr(ClO_2)_2$ is named for the strontium ion and the chlorite ion. Thus, $Sr(ClO_2)_2$ is named *strontium chlorite*.

(c) $Hg(NO_3)_2$ is named for the mercury ion and nitrate ion. However, mercury has two possible ions, Hg_2^{2+} and Hg^{2+}. Since there are two NO_3^- ions, the ionic charge on mercury must be 2+. Thus, $Hg(NO_3)_2$ is named *mercury(II) nitrate*, or mercuric nitrate.

(d) NaCN is named for the sodium ion and the cyanide ion. Thus, NaCN is named *sodium cyanide*. Notice this compound contains three elements but has an -ide ending. Cyanide compounds are ternary naming exceptions.

SELF-TEST EXERCISE

Supply a systematic name for each of the following ternary ionic compounds.

(a) $BaCrO_4$ **(b)** $Cu(NO_2)_2$

Answers: **(a)** barium chromate; **(b)** copper(II) nitrite or cupric nitrite

EXAMPLE EXERCISE 6.12

Provide the formula for each of the following ternary ionic compounds.

(a) sodium nitrite **(b)** nickel(II) acetate

(c) ferric sulfate **(d)** potassium hydroxide

Solution: To write the formula, we combine the cation and anion into a neutral formula unit.

(a) Sodium nitrite is composed of Na^+ and NO_2^-. A neutral formula unit and the formula of the compound are written $NaNO_2$.

Caustic soda pellets are sodium hydroxide, NaOH.

(b) Nickel(II) acetate contains Ni^{2+} and $C_2H_3O_2^-$. A neutral formula unit and the formula of the compound are written $Ni(C_2H_3O_2)_2$.

(c) Ferric sulfate is composed of Fe^{3+} and SO_4^{2-}. A neutral formula unit and the formula of the compound are written $Fe_2(SO_4)_3$.

(d) Potassium hydroxide contains K^+ and OH^-. A neutral formula unit and the formula of the compound are written KOH.

SELF-TEST EXERCISE

Provide the formula for each of the following ternary ionic compounds.

(a) mercury(I) acetate **(b)** tin(IV) permanganate

Answers: **(a)** $Hg_2(C_2H_3O_2)_2$; **(b)** $Sn(MnO_4)_4$

6.7 Binary Molecular Compounds

To write systematic names and formulas for binary molecular compounds.

Two nonmetals can combine to form a binary molecular compound. In general, the more nonmetallic element is written second in the chemical formula. There are, however, a number of exceptions. IUPAC specifically prescribes the following order for the elements in a compound: C, P, N, H, Se, S, I, Br, Cl, O, F. Notice that hydrogen is in the middle of the series. Thus, the binary compounds of hydrogen are written as follows: CH_4, PH_3, NH_3, H_2Se, H_2S, HI, HBr, HCl, H_2O, and HF.

Naming Binary Molecular Compounds

In naming binary molecular compounds, IUPAC specifies that the second element have an *-ide* suffix. It also specifies that the subscripts of the elements are to be indicated using Greek prefixes. The number of atoms of each element is indicated by the Greek prefix in Table 6.5.

According to the *1990 Red Book* recommendations, the prefix *mono-* is always omitted unless its presence is necessary to avoid confusion. The only common exceptions you are responsible for are CO and NO. The name of CO includes the

Table 6.5 Greek Prefixes for Atomic Ratios in Binary Molecular Compounds

No. of Atoms	Prefix	No. of Atoms	Prefix
1	mono	6	hexa
2	di	7	hepta
3	tri	8	octa
4	tetra	9	nona*
5	penta	10	deca

* Nona, a Latin prefix, is commonly used. IUPAC, however, prefers the Greek prefix ennea.

mono prefix in front of oxygen and is written carbon monoxide. Similarly, NO is named nitrogen monoxide. The name for laughing gas, N_2O, does not require the mono prefix and is systematically named dinitrogen oxide.

Let's consider the binary molecular compound P_4S_3. This compound is found on match tips and ignites in air when struck on a rough surface (Figure 6.4). Since the ratio is four phosphorus atoms to three sulfur atoms, the Greek prefixes are tetra and tri, respectively. Thus, the name of P_4S_3 is tetraphosphorus trisulfide. A different compound, P_4S_7, has the same elements but is named tetraphosphorus heptasulfide.

Compounds containing a semimetal and a nonmetal are named according to the same IUPAC rules for binary molecular compounds containing two nonmetals. For example, the systematic name for SiF_4 is silicon tetrafluoride and Sb_2S_5 is diantimony pentasulfide.

Figure 6.4 Chemistry of Matches The substance P_4S_3 is on match tips. When the match is struck on a rough surface, the P_4S_3 ignites, and the heat initiates a reaction that produces a flame.

EXAMPLE EXERCISE 6.13

Give the IUPAC systematic name for each of the following binary molecular compounds.

(a) IF_6 **(b)** Cl_2O_5
(c) P_4S_{10} **(d)** Br_3O_8

Solution: We always name binary molecular compounds by attaching an -ide suffix to the second nonmetal and indicating the atomic ratios with Greek prefixes.

(a) IF_6 is first named iodine fluoride. After supplying the Greek prefixes for the atomic ratios, we have *iodine hexafluoride*.
(b) Cl_2O_5 is first named chlorine oxide. After supplying the Greek prefixes for the atomic ratios, we have *dichlorine pentaoxide*.
(c) P_4S_{10} is first named phosphorus sulfide. After supplying the Greek prefixes for the atomic ratios, we have *tetraphosphorus decasulfide*.
(d) Br_3O_8 is first named bromine oxide. After supplying the Greek prefixes for the atomic ratios, we have *tribromine octaoxide*.

SELF-TEST EXERCISE

Give the IUPAC systematic name for each of the following binary molecular compounds.

(a) BCl_3 **(b)** N_2O_4

Answers: **(a)** boron trichloride; **(b)** dinitrogen tetraoxide

EXAMPLE EXERCISE 6.14

Provide the formula for each of the following binary molecular compounds.

(a) oxygen difluoride **(b)** bromine monochloride
(c) diphosphorus pentasulfide **(d)** tetraiodine nonaoxide

Solution: To write the formula, we give the symbol for each element, followed by a subscript indicating the number of atoms.

(a) Oxygen difluoride is composed of one oxygen atom and two fluorine atoms. The formula of the compound is written OF_2.
(b) Bromine monochloride is composed of one bromine atom and one chlorine atom. The formula of the compound is written BrCl.
(c) Diphosphorus pentasulfide is composed of two phosphorus atoms and five sulfur atoms. The formula of the compound is written P_2S_5.

(d) Tetraiodine nonaoxide is composed of four iodine atoms and nine oxygen atoms. The formula of the compound is written I_4O_9.

SELF-TEST EXERCISE

Provide the formula for each of the following binary molecular compounds.
(a) diphosphorus tetraiodide **(b)** sulfur hexafluoride

Answers: **(a)** P_2I_4; **(b)** SF_6

In past years, for ease of pronunciation, double vowels were avoided in naming compounds with Greek prefixes. For example, if the Greek prefix ended in an "a" or "o" and the nonmetal was oxygen, the first vowel was dropped. Thus, tetroxide, not tetraoxide, was preferred. According to the *1990 Red Book*, however, vowels are not to be dropped, with one exception. If mono is used as a prefix before oxygen, then monoxide, not monooxide, is recommended.

6.8 Binary Acids

OBJECTIVES

To write systematic names and formulas for binary acids.

A binary acid is an aqueous solution of a compound containing hydrogen and one other nonmetal. Formulas of acids begin with H, and examples of binary acids are HF(aq) and HCl(aq). A binary acid is produced by dissolving a binary molecular compound, such as HF, in water. The resulting aqueous solution, HF(aq), is a binary acid.

Binary acids are systematically named by using the prefix *hydro-* before the *nonmetal stem* and adding an *-ic acid* ending. As an example, consider muriatic acid, which is used to control the acidity of swimming pools. Muriatic acid is the binary molecular compound hydrogen chloride dissolved in water; that is, HCl(aq). The IUPAC name for muriatic acid is systematically formed in the following manner: hydrochlor(ine) -ic acid; the systematic name for HCl(aq) is hydrochloric acid.

EXAMPLE EXERCISE 6.15

Give the IUPAC systematic name for HF(aq), a binary acid.

Solution: We name a binary acid as hydro- nonmetal stem -ic acid. HF(aq) contains the nonmetal fluorine. We construct its name as follows: hydro + fluor + ic acid: HF(aq) is named *hydrofluoric acid*.

SELF-TEST EXERCISE

Give the IUPAC systematic name for HI(aq).

Answer: hydroiodic acid

EXAMPLE EXERCISE 6.16

Provide the formula for hydrosulfuric acid, a binary acid.

Solution: Binary acids are aqueous solutions of compounds containing hydrogen and another nonmetal. Hydrosulfuric acid contains the sulfide ion, S^{2-}. Since the hydrogen ion is H^+, we write the neutral compound as H_2S. Thus, the formula for the aqueous binary acid is $H_2S(aq)$.

SELF-TEST EXERCISE

Provide the formula for hydroselenic acid.

Answer: $H_2Se(aq)$

Be careful not to confuse the names of binary acids and binary molecular compounds. A binary acid is a compound of hydrogen and another nonmetal dissolved in water. For example, HCl(aq) is a binary acid and is named hydrochloric acid. On the other hand, gaseous HCl is a binary molecular compound and is named hydrogen chloride.

6.9 Ternary Oxyacids

OBJECTIVES

To write systematic names and formulas for ternary oxyacids.

A ternary oxyacid is a compound of hydrogen and a polyatomic ion dissolved in water. Most ternary oxyacids are named systematically by attaching an *-ic acid* ending to the *nonmetal stem* in the polyatomic ion. For example, the compound $HNO_3(aq)$ is a ternary oxyacid. The IUPAC name for $HNO_3(aq)$ is systematically formed as follows: *nonmetal stem* + *ic acid*. The systematic name for $HNO_3(aq)$ is therefore *nitr* + *ic acid*, or nitric acid.

Some ternary oxyacids are named systematically using an *-ous acid* ending on the *nonmetal stem* in the polyatomic ion. For example, the compound $HNO_2(aq)$ is a ternary oxyacid. The IUPAC name for $HNO_2(aq)$ is systematically formed as follows: *nonmetal stem* + *ous acid*. The systematic name for $HNO_2(aq)$ is therefore *nitr* + *ous acid*, or nitrous acid.

A ternary oxyacid with an -ic acid ending contains a polyatomic ion ending in an -ate suffix. Consider chloric acid, $HClO_3(aq)$, which contains the chlorate ion, ClO_3^-. In an analogous way, a ternary oxyacid with an -ous acid ending contains a polyatomic ion ending in an -ite suffix. Consider chlorous acid, $HClO_2(aq)$, which contains the chlorite ion, ClO_2^-. We can further illustrate this principle with four ternary oxyacids containing chlorine.

Ternary Oxyacid	Polyatomic Ion
perchlor**ic acid**, $HClO_4$	perchlor**ate ion**, ClO_4^-
chlor**ic acid**, $HClO_3$	chlor**ate ion**, ClO_3^-
chlor**ous acid**, $HClO_2$	chlor**ite ion**, ClO_2^-
hypochlor**ous acid**, $HClO$	hypochlor**ite ion**, ClO^-

The following example exercise further illustrates the naming of ternary oxyacids.

EXAMPLE EXERCISE 6.17

Give the IUPAC systematic name for $H_2SO_4(aq)$, the common acid found in an automobile battery.

Solution: Ternary oxyacids are named as -ic acids or -ous acids. Since $H_2SO_4(aq)$ contains the sulfate polyatomic ion, it is an -ic acid. We can construct the name as follows:

sulfur + ic acid: $H_2SO_4(aq)$ is named *sulfuric acid*

Notice that we use the entire word *sulfur* for the stem when forming the name of this acid. Sulfur is the exception as most acid names are formed from an abbreviated nonmetal stem. The names of ternary oxyacids can be mastered only with practice and experience.

SELF-TEST EXERCISE

Give the IUPAC systematic name for $H_2SO_3(aq)$, a ternary oxyacid.

Answer: sulfurous acid

EXAMPLE EXERCISE 6.18

Provide the formula for acetic acid, the common ternary oxyacid found in vinegar.

Solution: Ternary acids with an -ic acid ending contain polyatomic ions ending in -ate. Acetic acid contains the acetate ion, $C_2H_3O_2^-$. Since the hydrogen ion is H^+, we can write the neutral compound $HC_2H_3O_2$. The formula of acetic acid is written $HC_2H_3O_2(aq)$.

SELF-TEST EXERCISE

Provide the formula for carbonic acid, a ternary oxyacid.

Answer: $H_2CO_3(aq)$

6.10 Acid Salts

To write systematic names and formulas for acid salts.

acid A hydrogen-containing compound that releases hydrogen ions (H^+) when dissolved in water.

base A hydroxide-containing compound that releases hydroxide ions (OH^-) when dissolved in water.

A hydrogen-containing compound that releases hydrogen ions (H^+) when dissolved in water is termed an **acid**. A hydroxide-containing compound that releases hydroxide ions (OH^-) when dissolved in water is termed a **base**.

CHEMISTRY CONNECTION □ THE CONSUMER

Household Chemicals

► **Can you name two common household chemicals that, when mixed, produce a deadly gas?**

Table salt and table sugar are two familiar household chemicals. Vinegar, another household chemical, is a solution of acetic acid. Citrus fruit contains citric acid and is responsible for the sour taste of lemons and limes. Aspirin contains acetylsalicylic acid that, if taken in excess, can irritate the lining of the stomach.

Perhaps the most dangerous chemical in the home is caustic soda (lye), sold under various tradenames as a drain cleaner. If it contacts your skin, it gives a slippery feeling, and this is quickly followed by the loss of tissue. If taken internally by a child, caustic soda could be lethal. Household ammonia is potentially dangerous and should not be used in conjunction with bleach. Together, these two chemicals produce a poisonous gas.

Common household chemicals.

Common Household Chemicals

Name	Formula	Product/Use	Safety*†
Acids:			
acetic acid	$HC_2H_3O_2$	vinegar	
carbonic acid	H_2CO_3	carbonated drinks	
hydrochloric acid	HCl	swimming pools	avoid contact*
sulfuric acid	H_2SO_4	battery acid	avoid contact*
Bases:			
ammonia	NH_3	cleaning solutions	
magnesium hydroxide	$Mg(OH)_2$	milk of magnesia	
sodium bicarbonate	$NaHCO_3$	antacid, fire extinguisher	
sodium hydroxide	$NaOH$	drain and oven cleaner	avoid contact*
Miscellaneous:			
aluminum hydroxide	$Al(OH)_3$	antacid tablets	
carbon dioxide (solid)	CO_2	Dry Ice	avoid frostbite
Epsom salts	$MgSO_4{\cdot}7H_2O$	cathartic, laxative	
sodium hypochlorite	$NaClO$	bleach	avoid contact*
Organic:			
ethylene glycol	$C_2H_4(OH)_2$	antifreeze	avoid ingestion†
methanol	CH_3OH	solvent, antifreeze	avoid ingestion†
naphthalene	$C_{10}H_8$	mothballs	avoid ingestion†
trichloroethane	$C_2H_3Cl_3$	spot remover	avoid ingestion†

* In the event of contact, flush with water.
† Seek immediate medical attention.

Hydrochloric acid, used to acidify swimming pools, is sold in the supermarket as muriatic acid. Sulfuric acid, found in lead-storage batteries, is a dangerous chemical that should be handled with great caution. Sulfuric acid is strongly corrosive and can cause skin ulcers. Another potential danger of sulfuric acid is associated with "jump-starting" an automobile. When an electric current passes through battery acid, hydrogen gas is evolved. If the "jumper cables" spark, hydrogen can react explosively with oxygen in air.

► **Solutions of household ammonia and ordinary bleach react to give a lethal gas.**

Table 6.6 Selected Acids and Acid Anions

Acid	Name	Acid Anion	Name
$H_2CO_3(aq)$	carbonic acid	HCO_3^-	hydrogen carbonate
$H_3PO_4(aq)$	phosphoric acid	$H_2PO_4^-$	dihydrogen phosphate
		HPO_4^{2-}	monohydrogen phosphate
$H_2SO_4(aq)$	sulfuric acid	HSO_4^-	hydrogen sulfate
$H_2SO_3(aq)$	sulfurous acid	HSO_3^-	hydrogen sulfite

neutralization reaction A type of reaction in which an acid and a base react to produce a salt and water.

salt An ionic compound produced from the reaction of an acid and a base that does not contain the hydrogen ion or the hydroxide ion.

acid salt An ionic compound that results from the partial neutralization of an acid; the compound contains one or more hydrogen atoms bonded to the anion, for example, $NaHCO_3$ and NaH_2PO_4.

When an acid and a base react, a salt and water are produced. This is called a **neutralization reaction**. The salt that is produced is formed from the cation of the base and the anion of the acid. For example, the reaction of HCl(aq) and KOH(aq) produce KCl and water. In this example KCl is the salt. In general, a **salt** can be defined as an ionic compound containing neither the hydrogen ion nor hydroxide ion.

Acids that contain more than one hydrogen ion, for example, $H_2CO_3(aq)$, can be partially neutralized. That is, we can add just enough sodium hydroxide base to neutralize only one of the hydrogens and obtain $NaHCO_3$. An acidic compound that is only partially neutralized is called an **acid salt**.

The compound $NaHCO_3$ is called an acid salt because it can react with additional base. Notice the presence of hydrogen in the acid salt $NaHCO_3$. The compound Na_2CO_3 does not react with additional base. An ordinary salt, such as Na_2CO_3, results from the complete neutralization of an acid.

Table 6.6 lists some acids and the corresponding polyatomic acid anions found in acid salts.

An acid salt is named by first giving the positive metal ion, followed by the name of the acid anion. For example, $NaHCO_3$ is named sodium hydrogen carbonate. Another example is Na_2HPO_4, which is named sodium monohydrogen phosphate. Its common name is sodium phosphate dibasic.

EXAMPLE EXERCISE 6.19

Give the systematic name for each of the following acid salts.

(a) $LiHSO_3$ **(b)** KH_2PO_4

Solution: Acid salts are named for the metal and the acid anion.

(a) $LiHSO_3$ is composed of a lithium ion and the hydrogen sulfite ion. The name of the acid salt is *lithium hydrogen sulfite*.

(b) KH_2PO_4 is composed of a potassium ion and a dihydrogen phosphate ion. The name of the acid salt is *potassium dihydrogen phosphate*.

SELF-TEST EXERCISE

Provide the chemical formula for each of the following acid salts.

(a) aluminum hydrogen sulfate

(b) ammonium monohydrogen phosphate

Answers: **(a)** $Al(HSO_4)_3$; **(b)** $(NH_4)_2HPO_4$

The hydrogen in an acid salt is often referred to using the prefix *bi*. For example, ordinary household baking soda, $NaHCO_3$, has the common name sodium bicarbonate as well as the systematic name sodium hydrogen carbonate. Similarly, $KHSO_3$, is called potassium bisulfite as well as potassium hydrogen sulfite.

6.11 Predicting Chemical Formulas

To predict the chemical formula of a compound given the formula of a similar compound.

We have learned the value of the periodic table for mastering the formulas of ions. Now we are going to use the periodic table to predict the chemical formulas of compounds.

We can predict the formulas of compounds based on the formula of a similar compound. Let's begin with the alkali metal chlorides. Sodium is in Group IA/1. The formula of sodium chloride is NaCl. All the alkali metal chlorides, in fact, have a similar chemical formula. For the alkali metal family, the chloride compounds have the following formulas: LiCl, KCl, RbCl, and CsCl.

We can predict the formulas for the alkaline earth metal sulfates in a similar manner. Barium is in Group IIA/2, and the formula for barium sulfate is $BaSO_4$. Therefore, the formulas for the other alkaline earth metal sulfates are $BeSO_4$, $MgSO_4$, $CaSO_4$, and $SrSO_4$.

We can also predict the formulas of compounds having similar, but different, polyatomic ions. For example, calcium carbonate has the formula $CaCO_3$. If we substitute silicate for carbonate, we can predict the formula of the compound because carbon and silicon are in the same group. The formula of calcium silicate is $CaSiO_3$. We can predict the formula for sodium periodate given that the formula for sodium perchlorate is $NaClO_4$. Periodate contains iodine. Since iodine and chlorine are both elements in Group VIIA/17, the chemical formula for sodium periodate is $NaIO_4$.

EXAMPLE EXERCISE 6.20

Predict the chemical formula for the following binary compounds, given the formula of a similar compound.

(a) cesium bromide — Given: sodium bromide, NaBr
(b) beryllium chloride — Given: strontium chloride, $SrCl_2$
(c) gallium oxide — Given: aluminum oxide, Al_2O_3
(d) silver telluride — Given: silver oxide, Ag_2O

Solution: To predict the chemical formula, we will compare the elements that are different in the similar compounds.

(a) The two elements Cs and Na are both in Group IA/1. Therefore, the formula for cesium bromide is CsBr.

(b) The two elements Be and Sr are both in Group IIA/2. Therefore, the formula for beryllium chloride is $BeCl_2$.

(c) The two elements Ga and Al are both in Group IIIA/13. Therefore, the formula for gallium oxide is Ga_2O_3.

(d) Silver telluride contains the element tellurium. The two elements Te and O are both in Group VIA/16. Therefore, the formula for silver telluride is Ag_2Te.

SELF-TEST EXERCISE

Predict the chemical formula for the following ternary compounds, given the formula of a similar compound.

(a) francium nitrate Given: sodium nitrate, $NaNO_3$

(b) gallium sulfate Given: aluminum sulfate, $Al_2(SO_4)_3$

Answers: **(a)** $FrNO_3$; **(b)** $Ga_2(SO_4)_3$

We can also predict the names of ternary oxyacids having similar polyatomic ions. For example, the formula of chloric acid is $HClO_3(aq)$. If we substitute bromine for chlorine, we can predict the formula of the resulting acid because Cl and Br are in the same group. The name of $HBrO_3(aq)$ is bromic acid. Similarly, the name of $HIO_3(aq)$ is iodic acid. Table 6.7 lists the formulas and names of the ternary oxyacids that contain a halogen.

Table 6.7 Ternary Oxyacids Containing Halogens

Ion	Name	Acid	Name
ClO^-	hypochlorite	$HClO(aq)$	hypochlorous acid
ClO_2^-	chlorite	$HClO_2(aq)$	chlorous acid
ClO_3^-	chlorate	$HClO_3(aq)$	chloric acid
ClO_4^-	perchlorate	$HClO_4(aq)$	perchloric acid
BrO^-	hypobromite	$HBrO(aq)$	hypobromous acid
BrO_2^-	bromite	$HBrO_2(aq)$	bromous acid*
BrO_3^-	bromate	$HBrO_3(aq)$	bromic acid
BrO_4^-	perbromate	$HBrO_4(aq)$	perbromic acid
IO^-	hypoiodite	$HIO(aq)$	hypoiodous acid
IO_2^-	iodite	$HIO_2(aq)$	iodous acid*
IO_3^-	iodate	$HIO_3(aq)$	iodic acid
IO_4^-	periodate	$HIO_4(aq)$	periodic acid

* These acids are not stable.

The periodic table is a powerful tool for predicting the formula for an unknown compound that is similar to one that is known. However, there are often exceptions. For example, given that the formula of potassium nitrate is KNO_3, you would predict that the formula of potassium phosphate is KPO_3 because N and P are in the same group. This is not the case. Even though N and P belong to the same family, the correct formulas are KNO_3 and K_3PO_4, respectively.

Summary

Section 6.1 All inorganic compounds can be named according to systematic **IUPAC nomenclature**. Before applying systematic IUPAC rules, first classify the compound by placing it into one of five categories. After classifying the substance, apply the IUPAC rules. The five categories are **binary ionic**, **ternary ionic**, **binary molecular**, **binary acid**, and **ternary oxyacid**. An outline of this classification scheme appears in Figure 6.1.

Section 6.2 The names of most **monoatomic cations** are derived from the parent metal. Cations having two possible ionic charges require further identification. The ionic charge may be indicated using the Stock system or Latin system. The **Stock system** indicates the charge on the metal using Roman numerals in parentheses. The **Latin system** or suffix system attaches an *-ous* or *-ic* ending to the Latin name of cations having a variable ionic charge. Thus, Cu^{2+} is named copper(II) ion according to the Stock system and cupric ion using the Latin system.

The names of the **monoatomic anions** are derived from the parent nonmetal. According to IUPAC rules, the anion is named as follows: *nonmetal stem* + *-ide* suffix. Examples of nonmetal ions include fluoride ion, F^-, and oxide ion, O^{2-}.

Section 6.3 The names of most negative **polyatomic ions** have an *-ate* ending. Examples include the chlorate ion, ClO_3^-, and the acetate ion, $C_2H_3O_2^-$. A few negative polyatomic ions have an *-ite* ending. An example is the chlorite ion, ClO_2^-. Notice that the chlorite ion has one less oxygen atom than the chlorate ion. Polyatomic ions having one less oxygen than an ion with an -ate ending have an -ite ending. The CN^- and OH^- ions are exceptions. These ions have an *-ide* ending. Their names are cyanide ion and hydroxide ion, respectively.

Section 6.4 Ionic compounds are composed of positive and negative ions. The smallest representative particle of an ionic compound is called a **formula unit**. For every formula unit, the net ionic charge must be zero. That is, the total positive charge from the metal ions must be the same as the total negative charge from the nonmetal ions.

Sections 6.5–6.6 **Binary ionic compounds** have names ending with an *-ide* suffix; for example, $CaCl_2$ is named calcium chloride. Ternary compounds have names that end in *-ate* or *-ite*. For example, $Mg(ClO_3)_2$ is named magnesium chlorate, and $Mg(ClO_2)_2$ is named magnesium chlorite. Ternary compounds containing the cyanide ion or hydroxide ion are exceptions and have *-ide* endings.

Sections 6.7–6.8 **Binary molecular compounds** are named using an *-ide* suffix; for example, SO_2 is named sulfur dioxide. **Binary acids** are named *hydro* + *nonmetal stem* + *ic acid*. An aqueous solution of hydrogen fluoride, HF(aq), is named hydrofluoric acid.

Section 6.9 **Ternary oxyacids** are usually named *nonmetal stem* + *ic acid*. An aqueous solution of hydrogen nitrate, HNO_3(aq), is named nitric acid. An *-ic acid* is neutralized to give an *-ate salt*. A few ternary acids are named *nonmetal stem* + *ous acid*. An aqueous solution of hydrogen nitrite, HNO_2(aq), is named nitrous acid. An *-ous acid* is neutralized to give an *-ite salt*.

Section 6.10 The partial neutralization of an **acid** with a **base** produces an acid salt. **Acid salts** are characterized by polyatomic ions containing hydrogen. Examples include $NaHSO_4$ and KH_2PO_4. The acid salt is named by specifying the metal ion and polyatomic ion that compose the compound. These acid salt examples are named sodium hydrogen sulfate and potassium dihydrogen phosphate.

Section 6.11 The periodic table can be used to predict the formulas of compounds. For example, given that the formula of potassium nitrate is KNO_3, you can predict the formulas of other nitrate compounds containing an element in the same group as potassium. Since sodium and lithium are also alkali metals, we can correctly predict that sodium nitrate is $NaNO_3$ and lithium nitrate is $LiNO_3$.

Key Terms

Select the key term that corresponds to the following definitions.

____ **1.** international system of rules for naming chemical formulas
____ **2.** a compound containing the element carbon
____ **3.** a compound not containing the element carbon
____ **4.** a compound containing one metal and one nonmetal
____ **5.** a compound containing three elements, including at least one metal
____ **6.** a compound containing two nonmetals
____ **7.** a substance dissolved in water
____ **8.** a compound containing hydrogen and a nonmetal dissolved in water
____ **9.** a compound containing hydrogen, a nonmetal, and oxygen dissolved in water
____ **10.** a positively charged ion
____ **11.** a negatively charged ion
____ **12.** a single atom bearing a positive or negative charge
____ **13.** a unit containing two or more atoms bearing a positive or negative charge
____ **14.** a naming system that designates the variable charge on a metal cation using Roman numerals in parentheses
____ **15.** a naming system that designates the variable charge on a metal cation using an -ic or -ous suffix attached to the stem of the Latin name
____ **16.** the smallest representative particle of an ionic compound
____ **17.** a hydrogen-containing compound that releases hydrogen ions (H^+) when dissolved in water
____ **18.** a hydroxide-containing compound that releases hydroxide ions (OH^-) when dissolved in water
____ **19.** a chemical reaction of an acid and a base to produce a salt and water
____ **20.** an ionic compound produced from a neutralization reaction that does not contain a hydrogen ion or a hydroxide ion
____ **21.** an ionic compound that results from the partial neutralization of an acid; the compound contains one or more hydrogen atoms bonded to the anion, for example, $NaHCO_3$ and NaH_2PO_4

(a) acid *(Sec. 6.10)*
(b) acid salt *(Sec. 6.10)*
(c) anion *(Sec. 6.1)*
(d) aqueous solution *(Sec. 6.1)*
(e) base *(Sec. 6.10)*
(f) binary acid *(Sec. 6.1)*
(g) binary ionic *(Sec. 6.1)*
(h) binary molecular *(Sec. 6.1)*
(i) cation *(Sec. 6.1)*
(j) formula unit *(Sec. 6.4)*
(k) IUPAC nomenclature *(Sec. 6.1)*
(l) inorganic compound *(Sec. 6.1)*
(m) Latin system *(Sec. 6.2)*
(n) monoatomic ion *(Sec. 6.1)*
(o) neutralization reaction *(Sec. 6.10)*
(p) organic compound *(Sec. 6.1)*
(q) polyatomic ion *(Sec. 6.1)*
(r) salt *(Sec. 6.10)*
(s) Stock system *(Sec. 6.2)*
(t) ternary ionic *(Sec. 6.1)*
(u) ternary oxyacid *(Sec. 6.1)*

Exercises

IUPAC Systematic Nomenclature (Sec. 6.1)

1. Classify each of the following as a binary ionic compound, ternary ionic compound, binary molecular compound, binary acid, or ternary oxyacid.
(a) silver nitrate, $AgNO_3$
(b) lithium nitride, Li_3N
(c) ammonia, NH_3
(d) hydrobromic acid, $HBr(aq)$
(e) sulfurous acid, $H_2SO_3(aq)$

2. Classify each of the following as a binary ionic compound, ternary ionic compound, binary molecular compound, binary acid, or ternary oxyacid.
(a) carbon dioxide, CO_2
(b) hydrofluoric acid, $HF(aq)$
(c) phosphoric acid, $H_3PO_4(aq)$
(d) cobalt(III) oxide, Co_2O_3
(e) zinc phosphate, $Zn_3(PO_4)_2$

3. Classify each of the following ions as a monoatomic cation, monoatomic anion, polyatomic cation, or polyatomic anion.
(a) bromide ion, Br^- **(b)** ammonium ion, NH_4^+
(c) acetate ion, $C_2H_3O_2^-$ **(d)** mercuric ion, Hg^{2+}

4. Classify each of the following ions as a monoatomic cation, monoatomic anion, polyatomic cation, or polyatomic anion.

(a) bismuth Bi^{3+}
(b) chlorate ion, ClO_3^-
(c) hydronium ion, H_3O^+
(d) nitride ion, N^{3-}

Monoatomic Ions (Sec. 6.2)

5. Supply a systematic name for each of the following monoatomic cations.
(a) K^+
(b) Ba^{2+}
(c) Ag^+
(d) Cd^{2+}

6. Provide the formula for each of the following monoatomic cations.
(a) lithium ion
(b) aluminum ion
(c) zinc ion
(d) strontium ion

7. Supply the Stock system name for each of the following monoatomic cations.
(a) Hg^{2+}
(b) Cu^{2+}
(c) Fe^{2+}
(d) Co^{3+}

8. Provide the formula for each of the following monoatomic cations.
(a) lead(II) ion
(b) nickel(II) ion
(c) tin(IV) ion
(d) manganese(II) ion

9. Supply the Latin system name for each of the following monoatomic cations.
(a) Cu^+
(b) Fe^{3+}
(c) Sn^{2+}
(d) Pb^{4+}

10. Provide the formula for each of the following monoatomic cations.
(a) cupric ion
(b) mercuric ion
(c) ferrous ion
(d) plumbic ion

11. Supply a systematic name for each of the following monoatomic anions.
(a) F^-
(b) I^-
(c) O^{2-}
(d) P^3

12. Provide the formula for each of the following monoatomic anions.
(a) chloride ion
(b) bromide ion
(c) sulfide ion
(d) nitride ion

Polyatomic Ions (Sec. 6.3)

13. Supply a systematic name for each of the following polyatomic ions.
(a) ClO^-
(b) SO_3^{2-}
(c) $C_2H_3O_2^-$
(d) CO_3^{2-}

14. Supply a systematic name for each of the following polyatomic ions.
(a) CN^-
(b) NO_3^-
(c) CrO_4^{2-}
(d) HSO_4^-

15. Provide the formula for each of the following polyatomic ions.
(a) hydroxide ion
(b) nitrite ion
(c) dichromate ion
(d) hydrogen carbonate ion

16. Provide the formula for each of the following polyatomic ions.
(a) hydrogen sulfite ion
(b) chlorite ion
(c) sulfate ion
(d) perchlorate ion

Writing Chemical Formulas (Sec. 6.4)

17. Write the chemical formula for the following binary compounds, given their constituent ions.
(a) sodium chloride, Na^+ and Cl^-
(b) aluminum bromide, Al^{3+} and Br^-
(c) silver oxide, Ag^+ and O^{2-}
(d) bismuth oxide, Bi^{3+} and O^{2-}
(e) tin(IV) iodide, Sn^{4+} and I^-

18. Write the chemical formula for the following binary compounds, given their constituent ions.
(a) barium chloride, Ba^{2+} and Cl^-
(b) lithium nitride, Li^+ and N^{3-}
(c) cadmium sulfide, Cd^{2+} and S^{2-}
(d) magnesium phosphide, Mg^{2+} and P^{3-}
(e) lead(IV) sulfide, Pb^{4+} and S^{2-}

19. Write the chemical formula for the following ternary compounds, given their constituent ions.
(a) potassium nitrate, K^+ and NO_3^-
(b) ammonium dichromate, NH_4^+ and $Cr_2O_7^{2-}$
(c) aluminum sulfite, Al^{3+} and SO_3^{2-}
(d) bismuth hypochlorite, Bi^{3+} and ClO^-

20. Write the chemical formula for the following ternary compounds, given their constituent ions.
(a) iron(III) hydrogen carbonate, Fe^{3+} and HCO_3^-
(b) cuprous carbonate, Cu^+ and CO_3^{2-}
(c) mercury(II) cyanide, Hg^{2+} and CN^-
(d) stannic acetate, Sn^{4+} and $C_2H_3O_2^-$

21. Write the chemical formula for the following ternary compounds given their constituent ions.
(a) strontium nitrite, Sr^{2+} and NO_2^-
(b) zinc permanganate, Zn^{2+} and MnO_4^-
(c) calcium chromate, Ca^{2+} and CrO_4^{2-}
(d) chromium(III) perchlorate, Cr^{3+} and ClO_4^-

22. Write the chemical formula for the following ternary compounds, given their constituent ions.
(a) lead(IV) sulfate, Pb^{4+} and SO_4^{2-}
(b) stannous chlorite, Sn^{2+} and ClO_2^-
(c) cobalt(II) hydroxide, Co^{2+} and OH^-
(d) mercurous phosphate, Hg_2^{2+} and PO_4^{3-}

Binary Ionic Compounds (Sec. 6.5)

23. Supply a systematic name for each of the following binary ionic compounds.
(a) MgO
(b) $AgBr$
(c) $CdCl_2$
(d) Al_2S_3

24. Provide the formula for each of the following binary ionic compounds.
(a) lithium oxide
(b) strontium oxide
(c) zinc chloride
(d) nickel(II) fluoride

25. Supply a Stock system name for each of the following binary ionic compounds.
(a) CuO
(b) FeO
(c) HgO
(d) SnO

26. Provide the formula for each of the following binary ionic compounds.
(a) manganese(II) nitride (b) iron(II) bromide
(c) tin(IV) fluoride (d) cobalt(II) sulfide

27. Supply a Latin system name for each of the following binary ionic compounds.
(a) Cu_2O (b) Fe_2O_3
(c) Hg_2O (d) SnO_2

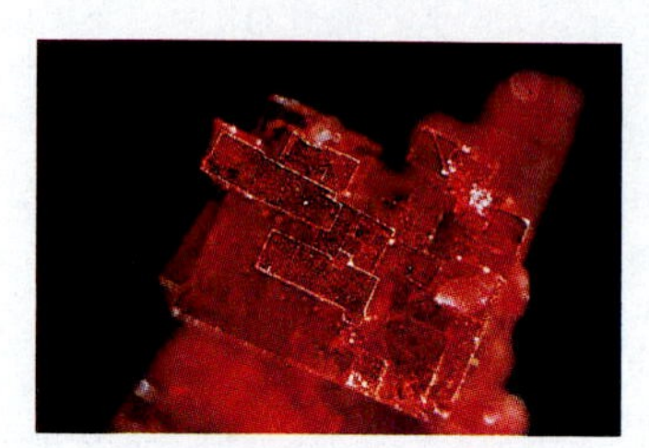

The natural mineral cuprite is Cu_2O.

28. Provide the formula for each of the following binary ionic compounds.
(a) plumbous oxide (b) ferric phosphide
(c) mercuric iodide (d) cuprous chloride

Ternary Ionic Compounds (Sec. 6.6)

29. Supply a systematic name for each of the following ternary ionic compounds.
(a) $KMnO_4$ (b) $Sr(ClO_4)_2$
(c) $BaCrO_4$ (d) $Cd(CN)_2$

30. Provide the formula for each of the following ternary ionic compounds.
(a) potassium nitrate (b) magnesium perchlorate
(c) silver sulfate (d) aluminum dichromate

31. Supply a Stock system name for each of the following ternary ionic compounds.
(a) $CuSO_4$ (b) $FeCrO_4$
(c) $Hg(NO_2)_2$ (d) $Pb(C_2H_3O_2)_2$

32. Provide the formula for each of the following ternary ionic compounds.
(a) manganese(II) acetate (b) copper(II) chlorite
(c) tin(II) phosphate (d) iron(III) hypochlorite

33. Supply a Latin system name for each of the following ternary ionic compounds.
(a) Cu_2SO_4 (b) $Fe_2(CrO_4)_3$
(c) $Hg_2(NO_2)_2$ (d) $Pb(C_2H_3O_2)_4$

34. Provide the formula for each of the following ternary ionic compounds.
(a) cuprous chlorite (b) plumbic sulfite
(c) mercuric chlorate (d) ferrous chromate

Binary Molecular Compounds (Sec. 6.7)

35. Give a systematic name for each of the following binary molecular compounds.
(a) S_2Cl_2 (b) P_2O_3
(c) N_2O (d) C_3O_2

36. Give a systematic name for each of the following binary molecular compounds.
(a) HCl (b) BrF_3
(c) I_2O_4 (d) Cl_2O_3

37. Provide the formula for each of the following binary molecular compounds.
(a) nitrogen dioxide (b) carbon tetrachloride
(c) iodine bromide (d) hydrogen sulfide

38. Provide the formula for each of the following binary molecular compounds.
(a) chlorine dioxide
(b) tribromine octaoxide
(c) tetraphosphorus heptasulfide
(d) sulfur hexafluoride

Binary Acids (Sec. 6.8)

39. Give the IUPAC systematic name for each of the following binary acids.
(a) $HF(aq)$ (b) $HBr(aq)$
(c) $H_2Se(aq)$

40. Provide the formula for each of the following binary acids.
(a) hydrochloric acid (b) hydroiodic acid
(c) hydrosulfuric acid

Ternary Oxyacids (Sec. 6.9)

41. Give the IUPAC systematic name for each of the following ternary oxyacids.
(a) $HClO_2(aq)$ (b) $HNO_3(aq)$
(c) $H_3PO_4(aq)$ (d) $H_2SO_3(aq)$

42. Give the IUPAC systematic name for each of the following ternary oxyacids.
(a) $HClO_4(aq)$ (b) $HIO(aq)$
(c) $HBrO_2(aq)$ (d) $H_2SeO_4(aq)$

43. Provide the formula for each of the following ternary oxyacids.
(a) acetic acid (b) hypochlorous acid
(c) phosphorous acid (d) chloric acid

44. Provide the formula for each of the following ternary oxyacids.
(a) sulfuric acid (b) nitrous acid
(c) iodous acid (d) bromic acid

Acid Salts (Sec. 6.10)

45. Give the systematic name for each of the following acid salts.
(a) $CaHPO_4$ (b) LiH_2PO_4
(c) $Zn(HCO_3)_2$ (d) $Ni(HSO_3)_2$

46. Provide the formula for each of the following acid salts.
(a) barium dihydrogen phosphate
(b) silver monohydrogen phosphate
(c) cuprous hydrogen carbonate
(d) iron(III) hydrogen sulfate

Predicting Chemical Formulas (Sec. 6.11)

47. Predict the chemical formula for the following binary compounds, given the formula of a similar compound.
(a) rubidium chloride   Given: sodium chloride, NaCl
(b) beryllium oxide   Given: calcium oxide, CaO

48. Predict the chemical formula for the following binary compounds, given the formula of a similar compound.
(a) gallium nitride   Given: aluminum nitride, AlN
(b) zirconium oxide   Given: titanium oxide, TiO_2

49. Predict the chemical formula for the following ternary compounds, given the formula of a similar compound.
(a) francium sulfate
Given: sodium sulfate, Na_2SO_4
(b) barium bromate
Given: barium chlorate, $Ba(ClO_3)_2$

50. Predict the chemical formula for the following ternary compounds, given the formula of a similar compound.
(a) iron(III) selenate
Given: iron(III) sulfate, $Fe_2(SO_4)_3$
(b) lanthanum nitrate
Given: scandium nitrate, $Sc(NO_3)_3$

51. Predict the chemical formula for the following ternary oxyacids, given the formula of a similar ternary oxyacid.
(a) hypobromous acid
Given: hypochlorous acid, HClO(aq)
(b) periodic acid
Given: perchloric acid, $HClO_4(aq)$

52. Predict the chemical formula for the following ternary oxyacids, given the formula of a similar ternary oxyacid.
(a) arsenous acid
Given: phosphorous acid, $H_3PO_3(aq)$
(b) selenic acid
Given: sulfuric acid, $H_2SO_4(aq)$

General Exercises

53. State the ionic charge for each of the following.
(a) iron metal atoms   **(b)** ferrous ions
(c) iron(III) ions   **(d)** iron compounds

54. State the ionic charge for each of the following.
(a) chlorine gas molecules   **(b)** chloride ions
(c) hypochlorite ions   **(d)** chlorine compounds

55. Polyatomic ions require an even number of valence electrons to be stable. Predict which of the following polyatomic anions could have a charge of 2−. (*Hint:* Add up the total number of valence electrons.)
(a) periodate ion, $IO_4^{?-}$ **(b)** thiosulfate ion, $S_2O_3^{?-}$
(c) silicate ion, $SiO_3^{?-}$ **(d)** thiocyanate ion, $CNS^{?-}$

56. Polyatomic ions require an even number of valence electrons to be stable. Predict which of the following polyatomic anions could have a charge of 1−. (*Hint:* Add up the total number of valence electrons.)
(a) hypobromite ion, $BrO^{?-}$
(b) peroxysulfate ion, $S_2O_8^{?-}$
(c) formate ion, $HCO_2^{?-}$
(d) cyanate ion, $CNO^{?-}$

57. Complete the table by combining cations and anions into chemical formulas. Give a systematic name for each of the resulting compounds. For example, Ag^+ and Br^- combine to give AgBr, which is named silver bromide.

Ions	Ag^+	Hg_2^{2+}	Al^{3+}	Sn^{4+}
Br^-	AgBr silver bromide			
F^-				
O^{2-}				
N^{3-}				

58. Complete the table by writing the formulas of the cations and anions. Combine the cations with each of the anions to give a correct chemical formula. For example, the lithium ion (Li^+) and chloride ion (Cl^-) combine to give LiCl.

Ions	lithium ion	copper(II) ion	iron(III) ion	lead(IV) ion
chloride ion	Li^+Cl^- LiCl			
iodide ion				
phosphide ion				
sulfide ion				

59. Complete the table by combining cations and anions into chemical formulas. Give a systematic name for each of the resulting compounds. For example, K^+ and CrO_4^{2-} combine to give K_2CrO_4, which is named potassium chromate.

Ions	K^+	Cd^{2+}	Cr^{3+}	Bi^{3+}
CrO_4^{2-}	K_2CrO_4 potassium chromate			
MnO_4^-				
ClO_4^-				
SO_3^{2-}				

60. Complete the table by writing the formulas of the cations and anions. Combine the cations with each of the anions to give a correct chemical formula. For example, the sodium ion (Na^+) and hydroxide ion (OH^-) combine to give NaOH.

Ions	sodium ion	iron(II) ion	calcium ion	cobalt(III) ion
hydroxide ion	Na^+OH^- NaOH			
sulfate ion				
acetate ion				
phosphate ion				

61. State the suffix ending for each of the following.
(a) Na_2S **(b)** $H_2S(aq)$
(c) Na_2SO_3 **(d)** $H_2SO_3(aq)$
(e) Na_2SO_4 **(f)** $H_2SO_4(aq)$

62. State the suffix ending for each of the following.
(a) KI **(b)** HI(aq)
(c) KIO **(d)** HIO(aq)
(e) KIO_2 **(f)** $HIO_2(aq)$
(g) KIO_3 **(h)** $HIO_3(aq)$
(i) KIO_4 **(j)** $HIO_4(aq)$

63. Write the chemical formula for each of the following household chemicals.
(a) hydrogen dioxide (hydrogen peroxide)
(b) sodium hypochlorite (bleach)
(c) sodium hydroxide (caustic soda)
(d) sodium bicarbonate (baking soda)

64. Write the chemical formula for each of the following household chemicals.
(a) acetic acid (vinegar solution)
(b) aqueous nitrogen trihydride (ammonia solution)
(c) aqueous magnesium hydroxide (milk of magnesia)
(d) aqueous sodium bisulfate ("bowl cleaner")

65. Give a systematic name for each of the following binary compounds containing a semimetal.
(a) BF_3 **(b)** $SiCl_4$
(c) As_2O_5 **(d)** Sb_2O_3

66. Provide the formula for each of the following binary compounds.
(a) boron tribromide **(b)** trisilicon tetranitride
(c) diarsenic trioxide **(d)** diantimony pentaoxide

67. "Canned heat" is a flammable gel containing calcium acetate dissolved in an alcoholic solution. Write the formula for calcium acetate.

68. Manganese(IV) oxide is mixed with chlorate compounds to assure their safe thermal decomposition. Write the formula of the manganese compound.

69. Oxalic acid, $H_2C_2O_4$, is used to clean jewelry. Its acid salt is found in spinach and produces an abrasive sensation on teeth. What is the formula for the sodium acid salt?

70. The transuranium element lawrencium is unstable and has a lifetime of only a few minutes. The chemical formula of lawrencium chloride was determined at the University of California by rapidly weighing small quantities. Predict the formula for lawrencium chloride given the formula of lutetium chloride, $LuCl_3$.

Chemical Formula Calculations

CHAPTER 7

Each flask contains a weighed amount of compound dissolved in solution. Chemical formula calculations are based on measured quantities of substance and are discussed in this chapter.

In Chapters 1 and 2, we developed problem-solving skills. In Chapters 3 through 6, we learned the nature of chemical substances. Now, in this chapter, we will combine what we learned in the first six chapters. We will apply the unit analysis method of calculation to the formulas of substances.

In general, when we see extremely large numbers, we are unable to appreciate the significance of their size. In Section 7.1, we are going to encounter a very large number: 602,000,000,000,000,000,000,000. It refers to a collection of particles and is called Avogadro's number. To appreciate the enormity of this number, consider the following analogy. If you had a collection of Avogadro's number of marbles, its volume would equal the size of the moon. This analogy is portrayed in Figure 7.1.

7.1 Avogadro's Number

OBJECTIVES

To state the value of Avogadro's number: 6.02×10^{23}.

To state the mass of Avogadro's number of atoms of any element.

Is it possible to keep track of atoms by counting them? The answer is yes, although not directly. Atoms are much too small to count individually, so we count them in groups. That is, we count atoms in the same way we count eggs by the dozen (12), pencils by the gross (144), and sheets of paper by the ream (500). Since atoms are so tiny, the groups contain a very large number of atoms.

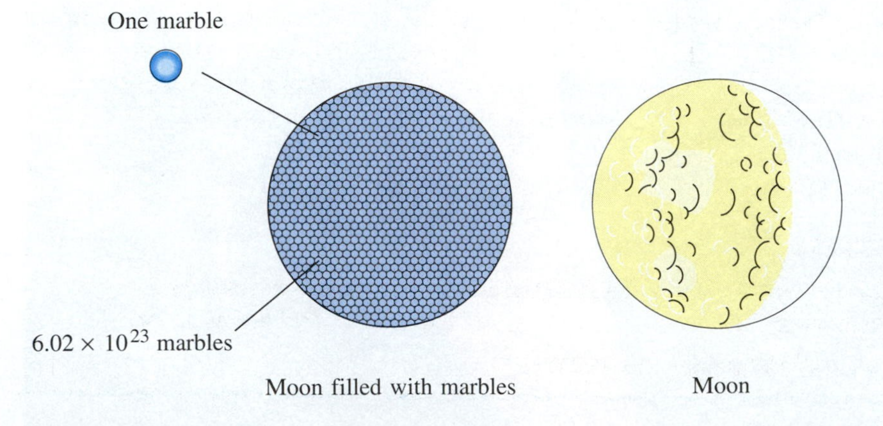

Figure 7.1 Analogy for Avogadro's Number If you had Avogadro's number of marbles, each having a diameter of about an inch, you would have a pile of marbles equal to the size of the entire moon.

There's a related problem to keeping track of atoms: their mass. In Section 4.6, we discussed the concept of atomic mass. Recall that, since the mass of an individual atom is much too light to measure directly, we determined its relative mass. An atom of carbon-12 was chosen as a reference and was given a mass of exactly 12 atomic mass units (12 amu). The masses of other atoms were then calculated *relative* to carbon-12. But atomic mass units do not register on a laboratory balance. Typically, balances measure hundredths (0.01 g) or thousandths (0.001 g) of a gram. There is a gigantic void between the practical world in the laboratory that measures to 0.01 g or 0.001 g and the theoretical world of atoms that have masses approaching 0.000 000 000 000 000 000 000 01 g!

How then can we keep track of atoms if they are too light to weigh on a balance? Let's state the problem differently. How can we relate the mass of an atom in atomic mass units to the mass of an atom in grams? The solution is that we must consider a very large number of atoms. We must consider enough atoms so that the mass in grams has the same value as the atomic mass in atomic mass units. For example, we must consider the number of atoms necessary to provide a mass of 12.01 g of carbon (atomic mass = 12.01 amu), 55.85 g of iron (atomic mass = 55.85 amu), 207.2 g of lead (atomic mass = 207.2 amu), and so on.

Now we have another question. How many atoms of an element equal its atomic mass in grams? Chemists have determined that this number of atoms is extremely large. Experiments have shown that 12.01 g of carbon contains 6.02×10^{23} carbon atoms. Similarly, 55.85 g of iron contain 6.02×10^{23} iron atoms, and 207.2 g of lead contains 6.02×10^{23} lead atoms. In fact, the atomic mass in grams of every element contains 6.02×10^{23} atoms of that element. This very large number is referred to as **Avogadro's number** (symbol *N*) in honor of the Italian physicist Amedeo Avogadro (1776–1856).

Avogadro's number (*N*) The number of atoms, molecules, or formula units that constitute 1 mole of a given substance; 6.02×10^{23} individual particles.

To determine the mass of Avogadro's number of atoms, we first find the element in the periodic table. Below the symbol of the element is the atomic mass. The mass of 6.02×10^{23} atoms of an element is equal to the atomic mass value expressed in grams. Example Exercise 7.1 illustrates the relationship between the mass of an individual atom and the mass of Avogadro's number of atoms.

EXAMPLE EXERCISE 7.1

Refer to the periodic table and find the atomic mass of the following elements. State the mass of Avogadro's number of atoms for each element.

(a) magnesium **(b)** silicon
(c) argon **(d)** tin

Solution: The atomic mass of each element is listed below the symbol of the element in the periodic table: Mg = 24.31 amu, Si = 28.09 amu, Ar = 39.95 amu, Sn = 118.71 amu. The mass of Avogadro's number of atoms is the atomic mass expressed in grams. Therefore, the atomic mass rounded to the nearest 0.1 g is

(a) Mg = 24.3 g **(b)** Si = 28.1 g
(c) Ar = 40.0 g **(d)** Sn = 118.7 g

SELF-TEST EXERCISE

Refer to the periodic table and state the mass for each of the following.

(a) one atom of Au **(b)** 6.02×10^{23} atoms of Au

Answers: **(a)** 197.0 amu; **(b)** 197.0 g

So how do chemists count atoms? First, they weigh samples of elements on a laboratory balance. Second, they reason that the atomic mass in grams of every element contains Avogadro's number of atoms. If the sample weighs less than the atomic mass, it contains less than 6.02×10^{23} atoms. If the sample weighs more than the atomic mass, it contains more than 6.02×10^{23} atoms. By weighing a sample, it is therefore possible to keep track of the number of atoms. In practice, Avogadro's number is sometimes referred to as the chemist's dozen.

7.2 The Mole Concept

To state the number of entities (atoms, molecules, formula units) in 1 mole of any substance.

To perform calculations that relate moles of a substance to the number of particles of substance.

mole (mol) The amount of substance containing the same number of entities as there are atoms in exactly 12 g of C-12.

In 1971, the mole was adopted into the International System of Units (SI). The **mole** (symbol **mol**) is the measure of the amount of chemical substance. SI officially defines 1 mole as the amount of substance that contains as many individual entities as there are atoms in exactly 12 grams of carbon-12.

The individual entities may be atoms, molecules, formula units, or any other particle. The number of atoms in 12 grams of carbon-12 is Avogadro's number. Therefore, a mole must contain Avogadro's number of atoms. For any substance

$$\textit{1 mole} = \text{Avogadro's number } (N) = 6.02 \times 10^{23} \text{ entities}$$

We can perform calculations that relate moles and number of entities. For instance, we can find how many molecules of oxygen are present in a 0.250-mol sample of the gas. By applying the unit analysis method of problem solving, we have

$$0.250 \cancel{\text{mol O}_2} \times \frac{6.02 \times 10^{23} \text{ molecules O}_2}{1 \cancel{\text{mol O}_2}} = \text{molecules O}_2$$

$$= 1.51 \times 10^{23} \text{ molecules O}_2$$

The following examples further illustrate the mole concept.

EXAMPLE EXERCISE 7.2

Calculate the number of entities in the following amounts of substance.

(a) 0.125 mol of sodium atoms, Na

(b) 4.25×10^{-3} mol of chlorine molecules, Cl_2

(c) 0.0763 mol of sodium chloride formula units, NaCl

Solution: Regardless of the entity, 1 mole contains 6.02×10^{23} entities. The unit equation is 1 mol = 6.02×10^{23} atoms, molecules, or formula units. As we learned previously, we will apply a unit factor to cancel the units of the given quantity.

UPDATE

Avogadro's Number

▶ Can you guess whether or not Amedeo Avogadro was the first to determine Avogadro's number?

In 1911, Ernest Rutherford determined a value for Avogadro's number based upon the principle of radioactivity. Rutherford was studying the emission of alpha particles from radioactive radium and was able to count the alpha particles using a Geiger counter. (The Geiger counter had been invented by his postdoctoral assistant, Hans Geiger.) After the alpha particles decayed into helium gas, he measured the volume of the helium. From the data, Rutherford calculated a value of 6.11×10^{23} atoms/mol for Avogadro's number.

Currently, the most accurate reported value for Avogadro's number is 6.0221367×10^{23}. This experimental value was found by using an X-ray method. X-rays were used to determine the spacing of atoms in an ultra-pure crystal of silicon (see figure). After striking the nuclei of adjacent silicon atoms, the X-rays were diffracted, that is, the path of the X-ray beam was changed. After measuring the angle of diffraction, the volume occupied by a single atom was calculated. The value for Avogadro's number was obtained by dividing the individual volume of a silicon atom into the total volume of a one-mole silicon crystal.

We can illustrate how to calculate a value for Avogadro's number as follows. In the figure below, notice that the volume occupied by each silicon atom is 0.020116 nm^3, and the volume of the 1-mole crystal is 12.114 cm^3. To find Avogadro's number, we simply divide the volume of one atom of silicon into the volume of one mole of silicon. However, we must be careful to use the same units for the volumes of the atom and the cyrstal. For example, we can convert the volume of the crystal from cm^3 to nm^3.

$$12.114\ cm^3 \times \left(\frac{1\ m}{100\ cm}\right)^3 \times \left(\frac{10^9\ nm}{1\ m}\right)^3 = nm^3$$

$$12.114\ \cancel{cm^3} \times \frac{1\ \cancel{m^3}}{10^6\ \cancel{cm^3}} \times \frac{10^{27}\ nm^3}{1\ \cancel{m^3}} = 1.2114 \times 10^{22}\ nm^3$$

If we divide the volume of the silicon atom into the volume of the one-mole crystal, we have the number of atoms in one mole of silicon.

$$\frac{1.2114 \times 10^{22}\ \cancel{nm^3}}{1\ mol\ Si} \times \frac{1\ atom\ Si}{0.020116\ \cancel{nm^3}} = \frac{6.0221 \times 10^{23}\ atoms\ Si}{1\ mol\ Si}$$

In this example, we obtained a value of 6.0221×10^{23} atoms in one mole. This calculation illustrates a highly accurate method for determining Avogadro's number.

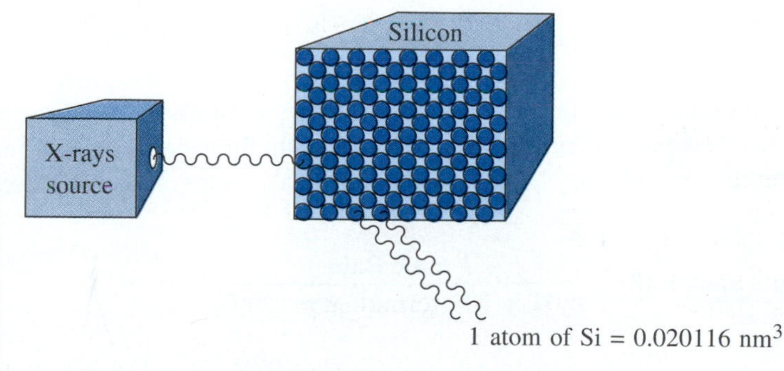

X-ray diffraction of a silicon crystal

▶ In 1865, the Austrian chemist Johann Loschmidt first determined there were ~1×10^{24} molecules in a mole of gas. Although Avogadro had died nine years earlier, Avogadro's name is synonymous with 6.02×10^{23}.

$$\textbf{(a)}\ 0.125\ \cancel{\text{mol Na}} \times \frac{6.02 \times 10^{23}\ \text{atoms Na}}{1\ \cancel{\text{mol Na}}} = \text{atoms Na} = 7.53 \times 10^{22}\ \text{atoms Na}$$

$$\textbf{(b)}\ 4.25 \times 10^{-3}\ \cancel{\text{mol Cl}_2} \times \frac{6.02 \times 10^{23}\ \text{molecules Cl}_2}{1\ \cancel{\text{mol Cl}_2}} = \text{molecules Cl}_2$$
$$= 2.56 \times 10^{21}\ \text{molecules Cl}_2$$

$$\textbf{(c)}\ 0.0763\ \cancel{\text{mol NaCl}} \times \frac{6.02 \times 10^{23}\ \text{formula units NaCl}}{1\ \cancel{\text{mol NaCl}}} = \text{formula units NaCl}$$
$$= 4.59 \times 10^{22}\ \text{formula units NaCl}$$

SELF-TEST EXERCISE

Calculate the number of entities in the following amounts of substance.
(a) 0.335 mol of calcium, Ca
(b) 0.112 mol of carbon dioxide, CO_2
(c) 0.00527 mol of calcium carbonate, $CaCO_3$

Answers: **(a)** 2.02×10^{23} atoms of Ca; **(b)** 6.74×10^{22} molecules of CO_2; **(c)** 3.17×10^{21} formula units of $CaCO_3$

Each of these are 1-mole samples comprised of 6.02×10^{23} entities. Clockwise from top: table sugar, water, mercury, sulfur, table salt, copper, and lead. Potassium dichromate is in the center.

Alternatively, we can find the amount of substance given the number of entities of substance. For instance, we can calculate the moles of barium fluoride corresponding to 2.50×10^{23} formula units of BaF_2. Applying the unit analysis method of problem solving, we have

$$2.50 \times 10^{23}\ \cancel{\text{formula units BaF}_2} \times \frac{1\ \text{mole BaF}_2}{6.02 \times 10^{23}\ \cancel{\text{formula units BaF}_2}}$$
$$= \text{mol BaF}_2 = 0.415\ \text{mol BaF}_2$$

Now let's consider some additional example exercises to further illustrate the mole concept.

EXAMPLE EXERCISE 7.3

Calculate the number of moles of substance in each of the following.
(a) 7.77×10^{22} atoms of potassium, K
(b) 5.34×10^{25} molecules of iodine, I_2
(c) 1.25×10^{21} formula units of potassium iodide, KI

Solution: One mole of any substance contains 6.02×10^{23} particles. The unit equation is 1 mol = 6.02×10^{23} atoms, molecules, or formula units. In the unit analysis method of problem solving, we systematically apply unit factors in order to have cancellation of units.

(a) $$7.77 \times 10^{22}\ \cancel{\text{atoms K}} \times \frac{1\ \text{mol K}}{6.02 \times 10^{23}\ \cancel{\text{atoms K}}} = \text{mol K} = 0.129\ \text{mol K}$$

(b) $$5.34 \times 10^{25}\ \cancel{\text{molecules } I_2} \times \frac{1\ \text{mol } I_2}{6.02 \times 10^{23}\ \cancel{\text{molecules } I_2}} = \text{mol } I_2 = 88.7\ \text{mol } I_2$$

(c) $$1.25 \times 10^{21}\ \cancel{\text{formula units KI}} \times \frac{1\ \text{mol KI}}{6.02 \times 10^{23}\ \cancel{\text{formula units KI}}}$$
$$= \text{mol KI} = 2.08 \times 10^{-3}\ \text{mol KI}$$

Even though the amount of KI is small, 0.00208 mol, the number of formula units is a very large number.

SELF-TEST EXERCISE

Calculate the number of moles of substance in each of the following.
(a) 1.31×10^{21} atoms of carbon, C
(b) 2.15×10^{22} molecules of carbon monoxide, CO
(c) 8.88×10^{24} formula units of lithium carbonate, Li_2CO_3

Answers: **(a)** 2.18×10^{-3} (0.00218) mol C; **(b)** 3.57×10^{-2} (0.0357) mol CO; **(c)** 1.48×10^{1} (14.8) mol Li_2CO_3

Avogadro's number is so huge it is difficult to imagine the immensity of 6.02×10^{23}. Chemists often provide analogies to communicate just how large the number is. To appreciate the magnitude of Avogadro's number, consider the following analogies.

Number:

If 6.02×10^{23} dollars were placed in a savings account paying 5% interest per year, the account would earn interest at the rate of about 1 million dollars every nanosecond!

(a)

Length:

If 6.02×10^{23} hydrogen atoms were laid side-by-side, the total length would be long enough to encircle the earth at the equator about 1 million times.

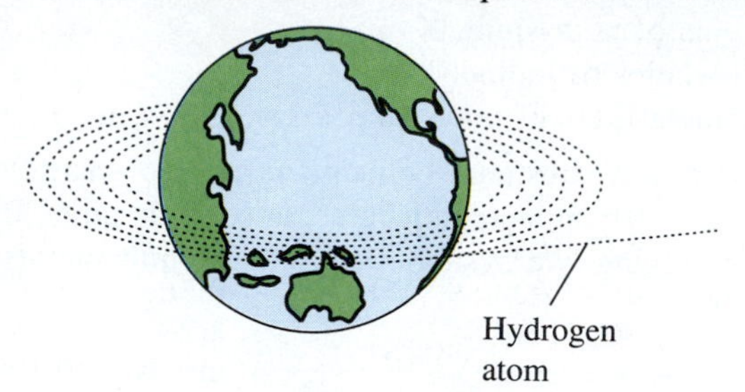

Mass:

The total mass of 6.02×10^{23} Olympic shotput balls would be about the mass of the earth.

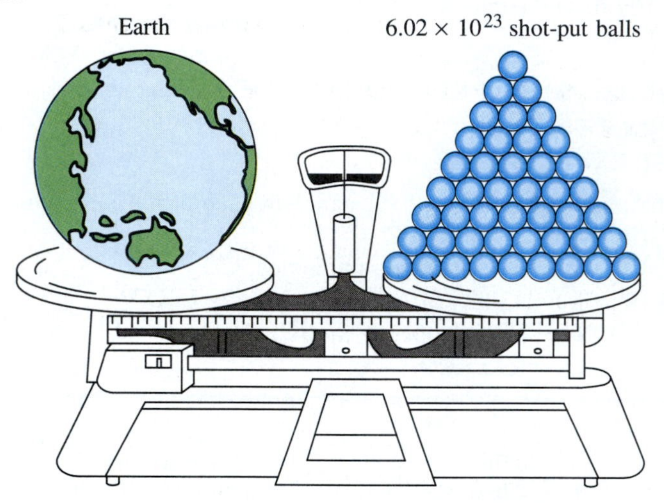

Volume:

The total volume occupied by 6.02×10^{23} softballs would be about the size of the earth.

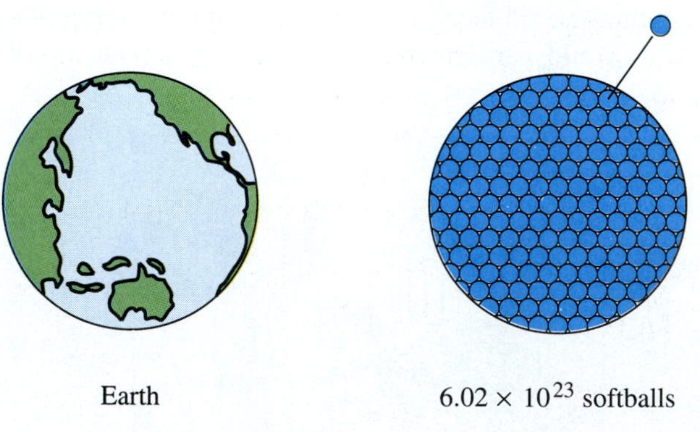

7.3 Molar Mass

To calculate the molar mass of a substance given its chemical formula.

As we said, a mole is the SI unit for an amount of a substance. It tells us the number of particles in a sample of substance. A mole also indicates the mass of a sample of substance. For example, 1 mole of carbon contains 6.02×10^{23} atoms. By definition, we know that 1 mole of carbon has a mass of 12.01 g. In fact, the atomic mass of any element, expressed in grams, corresponds to 1 mole of substance. This mass is referred to as the **molar mass (MM).**

molar mass (MM) The mass of 1 mole of pure substance expressed in grams; the mass of Avogadro's number of atoms, molecules, or formula units.

The molar mass of an element is equal to its atomic mass. By referring to the periodic table, we find that the atomic mass of iron is 55.8 amu. Thus, the molar mass of iron is 55.8 g/mol. Since naturally occurring oxygen is O_2, the molar mass of oxygen gas is equal to twice 16.0 g, or 32.0 g/mol.

We can calculate the molar mass of iron(III) oxide, Fe_2O_3, by adding together the atomic masses of 2 moles of iron atoms and 3 moles of oxygen atoms. Thus,

$$Fe_2O_3\text{:}\quad 2(55.8\text{ g Fe}) + 3(16.0\text{ g O}) = 159.6\text{ g}$$

The molar mass of Fe_2O_3 is 159.6 g/mol. Notice that we used the molar mass of oxygen atoms (16.0 g) rather than oxygen molecules (32.0 g). Even though the element occurs naturally as molecules of oxygen, it is atoms of oxygen that are combined in compounds. The following examples further illustrate the calculation of molar mass.

EXAMPLE EXERCISE 7.4

Calculate the molar mass for each of the following substances.

(a) silver atoms, Ag
(b) ammonia molecules, NH_3
(c) magnesium nitrate formula units, $Mg(NO_3)_2$

Solution: Begin by referring to the periodic table for atomic mass values. Express the atomic mass value in grams to obtain the mass of 1 mole. Round off the molar masses to 0.1 unit in order to have at least three significant digits in the molar mass.

(a) The atomic mass of silver is 107.868 amu. Thus, the molar mass is 107.9 g/mol.
(b) The molar masses of nitrogen and hydrogen are 14.0 g and 1.0 g, respectively. The sum for NH_3 is 14.0 g + 1.0 g + 1.0 g + 1.0 g = 17.0 g. Hence, the molar mass of ammonia, NH_3, is 17.0 g/mol.
(c) The sum of the molar masses for $Mg(NO_3)_2$ is as follows:

$$24.3\text{ g} + 2(14.0 + 16.0 + 16.0 + 16.0)\text{ g} = 24.3\text{ g} + 2(62.0)\text{ g} = 148.3\text{ g}.$$

The molar mass is therefore 148.3 g/mol.

SELF-TEST EXERCISE

Calculate the molar mass for each of the following substances.

(a) titanium, Ti
(b) sulfur hexafluoride, SF_6
(c) barium chloride, $BaCl_2$
(d) strontium acetate, $Sr(C_2H_3O_2)_2$

Answers: **(a)** 47.9 g/mol; **(b)** 146.1 g/mol; **(c)** 208.3 g/mol; **(d)** 205.6 g/mol

7.4 Mole Calculations I

OBJECTIVES

To perform calculations that relate the mass of a substance to its number of atoms, molecules, or formula units.

As you might have realized from the previous discussion, the mole is the central quantitative unit in chemistry. It is an amount of substance and relates the number of entities to the mass of the substance; that is,

$$6.02 \times 10^{23} \text{ entities} = \textit{1 mole} = \text{molar mass of substance}$$

The relationship of entities to moles to molar mass enables us to convert the number of particles of a substance to its mass, and vice versa. To do so, first we find the number of moles of substance. Given the number of particles, we use Avogadro's number to find the number of moles. Given the mass, we use the molar mass to find the moles of substance. After finding the moles of substance, we can find either remaining quantity in one step. Figure 7.2 illustrates the conversion from one quantity to another.

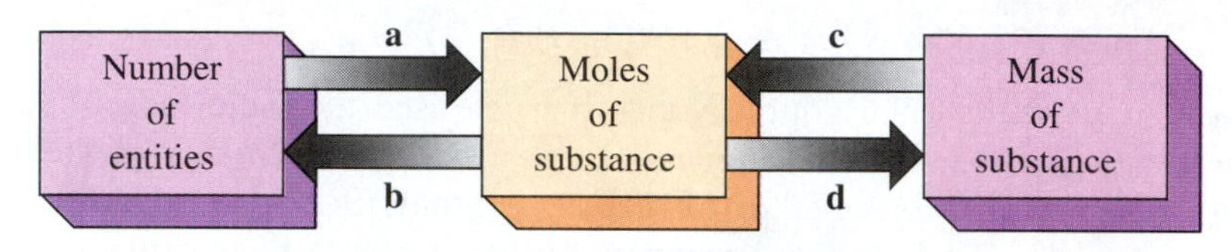

(a) Use N as a unit factor: multiply by 1 mol/6.02 x 10^{23}
(b) Use N as a unit factor: multiply by 6.02 x 10^{23}/1 mol
(c) Use molar mass as a unit factor: multiply by 1 mol/g
(d) Use molar mass as a unit factor: multiply by g/1 mol

Figure 7.2 Mole Calculations Steps a, b, c, and d outline the unit analysis conversions involved in performing a mole calculation. Notice that the mole is the central unit in the calculation.

Now we are ready to try some mole calculations. Remember to use the unit analysis method of problem solving; that is,

Step 1: Identify the units of the unknown.

Step 2: Select a given value that is related to the answer.

Step 3: Supply one or more unit factors to provide the conversion. Apply each unit factor systematically so as to cancel the previous units.

The following example exercises illustrate calculations involving moles of a substance. We will proceed from a given value to an unknown as before, only in these calculations we will use the concepts of Avogadro's number and molar mass.

EXAMPLE EXERCISE 7.5

What is the mass in grams of 2.01×10^{22} atoms of mercury?

Solution: In mole calculations the first step is to find the number of moles. In this case, we want mol Hg. We must use the relationship of 1 mole and Avogadro's number in the unit factor: 1 mol = 6.02×10^{23} atoms.

$$2.01 \times 10^{22} \text{ \sout{atoms Hg}} \times \frac{1 \text{ mol Hg}}{6.02 \times 10^{23} \text{ \sout{atoms Hg}}} = 0.0334 \text{ mol Hg}$$

To calculate the mass of mercury in grams, we multiply by the molar mass of mercury. From the periodic table, we find the molar mass of mercury is 200.6 g/mol.

$$0.0334 \text{ \sout{mol Hg}} \times \frac{200.6 \text{ g Hg}}{1 \text{ \sout{mol Hg}}} = 6.70 \text{ g Hg}$$

SELF-TEST EXERCISE

What is the mass of 2.66×10^{22} molecules of iodine, I_2?

Answer: 11.2 g I_2

EXAMPLE EXERCISE 7.6

How many O_2 molecules are present in 0.470 g of oxygen gas?

Solution: To find the number of oxygen molecules, we first calculate the mol O_2. The atomic mass of oxygen is 16.0 amu; the molecular mass of O_2 is 32.0 amu. Thus, the molar mass of O_2 is 32.0 g/mol.

$$0.470 \text{ \sout{g O}}_2 \times \frac{1 \text{ mol } O_2}{32.0 \text{ \sout{g O}}_2} = 0.0147 \text{ mol } O_2$$

We convert the moles of oxygen into molecules of O_2 by multiplying times Avogadro's number.

$$0.0147 \text{ \sout{mol O}}_2 \times \frac{6.02 \times 10^{23} \text{ molecules } O_2}{1 \text{ \sout{mol O}}_2} = 8.84 \times 10^{21} \text{ molecules } O_2$$

With practice, we can perform mole conversions in a single operation:

$$0.470 \text{ \sout{g O}}_2 \times \frac{1 \text{ \sout{mol O}}_2}{32.0 \text{ \sout{g O}}_2} \times \frac{6.02 \times 10^{23} \text{ molecules } O_2}{1 \text{ \sout{mol O}}_2} = 8.84 \times 10^{21} \text{ molecules } O_2$$

SELF-TEST EXERCISE

How many formula units of lithium fluoride are found in 0.175 g of LiF?

Answer: 4.07×10^{21} formula units LiF

Now, let's try a problem that is a little more difficult. If the molar mass of water is 18.0 g/mol, what is the mass in grams of a single molecule of water? This problem requires compound units in the answer, that is, g H_2O/molecule. To arrive at an answer with compound units, we must start with a ratio of two units; that is

$$\frac{18.0 \text{ g } H_2O}{1 \text{ mol } H_2O} \times \frac{\text{unit}}{\text{factor}} = \frac{\text{g } H_2O}{\text{molecule } H_2O}$$

The unit factor for the conversion is provided by Avogadro's number:
1 mol = 6.02×10^{23} molecules.

$$\frac{18.0 \text{ g } H_2O}{1 \text{ \sout{mol H}}_2\text{\sout{O}}} \times \frac{1 \text{ \sout{mol H}}_2\text{\sout{O}}}{6.02 \times 10^{23} \text{ molecules}} = \frac{\text{g } H_2O}{\text{molecule } H_2O}$$

$$= 2.99 \times 10^{-23} \text{ g/molecule}$$

The following example exercise further illustrates applying the mole concept to determine the mass of an individual atom or molecule.

EXAMPLE EXERCISE 7.7

Sodium metal occurs naturally as a single isotope, Na-23. The atomic mass is 23.0 amu. What is the mass of a single sodium atom in grams?

Solution: If the atomic mass of sodium is 23.0 amu, the molar mass is 23.0 g/mol. We set up the problem as follows:

$$\frac{23.0 \text{ g Na}}{1 \text{ mol Na}} \times \frac{\text{unit}}{\text{factor}} = \frac{\text{g Na}}{\text{atom Na}}$$

Once again, the unit factor for the conversion is provided by the equation $1 \text{ mol} = 6.02 \times 10^{23}$ atoms.

$$\frac{23.0 \text{ g Na}}{1 \cancel{\text{mol Na}}} \times \frac{1 \cancel{\text{mol Na}}}{6.02 \times 10^{23} \text{ atoms Na}} = \frac{\text{g Na}}{\text{atom Na}}$$

$$= 3.82 \times 10^{-23} \text{ g/atom}$$

SELF-TEST EXERCISE

Calculate the mass in grams for a single molecule of hydrogen, H_2.

Answer: 3.36×10^{-24} g/molecule of H_2

7.5 Percentage Composition

OBJECTIVES

To calculate the percentage composition of a compound, given the chemical formula of the compound.

In Section 1.11, we defined percentage as parts per hundred parts. Another way to define percentage is to say that it is the ratio of one quantity in a sample compared to everything in the entire sample, all multiplied by 100. Here we will expand our discussion of percentage and discuss percentage composition.

percentage composition An expression for the ratio of the mass of a single element compared to the mass of a given compound containing the element, all times 100.

The **percentage composition** of a compound is the mass of one element compared to the entire mass of the compound. The percentage composition of water, H_2O, for example, is 11% hydrogen and 89% oxygen. To expand this concept, recall that in Section 3.6 we introduced the law of constant composition. This law states that the elements in a compound are always present in the same proportion by mass. Therefore, water always contains 11% hydrogen and 89% oxygen, regardless of the amount. A drop of water, a liter of water, and a mole of water are always 11% hydrogen and 89% oxygen.

We are given the percentages of each element in water above, but how are the values obtained? The calculation for the percentage composition of water, H_2O, is as follows. We begin by assuming that we have 1 mole of H_2O. One mole of water contains 2 moles of hydrogen atoms and 1 mole of oxygen atoms. Thus,

$$2(\text{mol H}) + 1 \text{ mol O} = 1 \text{ mol } H_2O$$

$$2(\text{molar mass H}) + 1 \text{ molar mass O} = 1 \text{ molar mass } H_2O$$

$$2(1.01 \text{ g H}) + 16.0 \text{ g O} = \text{ g } H_2O$$

$$2.02 \text{ g H} + 16.0 \text{ g O} = 18.0 \text{ g } H_2O$$

Next, we find the percentage composition of water. We do so by comparing the separate masses of hydrogen and oxygen to the entire compound.

$$\frac{2.02 \text{ g H}}{18.0 \text{ g } H_2O} \times 100 = 11.2\% \text{ H}$$

$$\frac{16.0 \text{ g O}}{18.0 \text{ g } H_2O} \times 100 = 88.9\% \text{ O}$$

Notice that the sum of the two percentages is 100.1%. The discrepancy from 100.0% is the result of rounding off nonsignificant digits. The following examples further illustrate the calculation of percentage composition.

EXAMPLE EXERCISE 7.8

Trinitrotoluene, TNT, is a white crystalline substance that explodes at 240°C. Calculate the percentage composition of TNT, $C_7H_5(NO_2)_3$.

Solution: In calculating the percentage composition of TNT, we will assume there is 1 mole of the compound. For complex compounds that require the use of parentheses, it is necessary to count the total number of atoms of each element carefully. That is, 1 mole of $C_7H_5(NO_2)_3$ contains 7 moles of C atoms, 5 moles of H atoms, 3 moles of N atoms, and 6 moles of O atoms. We begin the calculation by finding the molar mass of TNT. First, find the molar mass of $C_7H_5(NO_2)_3$ as follows:

$$7(12.0 \text{ g C}) + 5(1.01 \text{ g H}) + 3(14.0 \text{ g N} + 32.0 \text{ g O}) = \text{ g } C_7H_5(NO_2)_3$$

$$84.0 \text{ g C} + 5.05 \text{ g H} + 42.0 \text{ g N} + 96.0 \text{ g O} = 227.1 \text{ g } C_7H_5(NO_2)_3$$

Then find the percentage composition. Compare the mass of each element to the molar mass of the compound, 227.1 g.

$$\frac{84.0 \text{ g C}}{227.1 \text{ g } C_7H_5(NO_2)_3} \times 100 = 37.0\% \text{ C}$$

$$\frac{5.05 \text{ g H}}{227.1 \text{ g } C_7H_5(NO_2)_3} \times 100 = 2.22\% \text{ H}$$

$$\frac{42.0 \text{ g N}}{227.1 \text{ g } C_7H_5(NO_2)_3} \times 100 = 18.5\% \text{ N}$$

$$\frac{96.0 \text{ g O}}{227.1 \text{ g } C_7H_5(NO_2)_3} \times 100 = 42.3\% \text{ O}$$

SELF-TEST EXERCISE

Calculate the percentage composition of iron(III) hydroxide, $Fe(OH)_3$.

Answer: 52.2% Fe, 44.9% O, and 2.84% H

CHEMISTRY CONNECTION □ HISTORICAL

Antoine Lavoisier

▶ **Can you guess the person who perhaps most influenced the enormous body of work published by Lavoisier?**

Antoine Laurent Lavoisier (1743–1794) was born into an affluent French family and received an excellent education. Initially, he followed his father into law, but later turned to science. Lavoisier was a gifted experimenter and, for his numerous contributions, is considered to be the founder of modern chemistry. One of Lavoisier's most important contributions was a magnificent laboratory that attracted scientists from all over the world. Even Benjamin Franklin and Thomas Jefferson visited his laboratory.

In 1787, Lavoisier published a book entitled *Methods of Chemical Nomenclature*. The book designated systematic principles for naming chemical substances and was so logical it became the basis for our present rules of nomenclature. In 1789, he published the landmark textbook, *Elementary Treatise on Chemistry*, which offered the first modern view of chemistry.

Before Lavoisier, it was commonly believed that in order to burn, a substance must contain the element phlogiston. Lavoisier suggested that it was not phlogiston, but an element in air, that was responsible for combustion. Lavoisier named the element oxygen, but it was the English chemist Joseph Priestly who first discovered the element. Lavoisier went on to study the respiration of animals in air and oxygen. He measured the amount of heat they produced and showed that biological reactions were similar to ordinary combustion reactions.

In 1771, Lavoisier married Marie-Anne Pierrette, who at the time was only fourteen years old. Despite her youth, she proved invaluable to his work and joined alongside him in the laboratory. Unfortunately, she was also the daughter of an important official in a firm that collected taxes. Although Lavoisier, himself, did not collect taxes, he received royalties from investments in the same firm. When the French Revolution broke out, revolutionaries threw tax collectors into jail. Lavoisier was first barred from his laboratory and then arrested. After a farce trial, he was sentenced to the guillotine, along with his father-in-law and other tax collectors.

On May 8, 1794, Lavoisier was executed and buried in an unmarked grave. The French court declared that France had no need for scientists, but the famous mathematician Lagrange lamented: "A moment was all that was necessary to strike off his head, and probably a hundred years will not be sufficient to produce another like it." Within two years, the French people realized their mistake and began unveiling statues of Lavoisier.

Antoine Lavoisier and his wife, Marie

▶ **Lavoisier's wife, Marie, deserves great credit for recording the laboratory work, translating scientific papers, and illustrating textbooks.**

7.6 Empirical Formula

To calculate the empirical formula of a compound, given the experimental data from its synthesis.

To calculate the empirical formula of a compound, given its percentage composition.

Historically, much of our understanding of the elements was based on their reactions. During the late 1700s, chemists experimented with elements to see how they reacted to form compounds. In particular, they were interested in the reactions of the elements as they combined with oxygen. By measuring the mass of the element before reaction and the mass of the oxide after reaction, chemists could determine the formula of the compound. The formula with the simplest whole-number ratio of atoms of the elements is referred to as the **empirical formula**.

empirical formula The chemical formula of a compound that expresses the simplest ratio of the atoms in a molecule or ions in a formula unit.

Elements in the periodic table are placed into groups based on similar empirical formulas as determined from chemical reactions. For example, magnesium, calcium, and barium are placed in Group IIA/2 because they react with oxygen to give similar empirical formulas. The empirical formulas of magnesium oxide, calcium oxide, and barium oxide all have a metal to oxygen ratio of 1 : 1, that is, MgO, CaO, and BaO. Other families of elements in the periodic table also show similarities in empirical formula.

Example Exercise 7.9 illustrates the placement of an element in the periodic table based on an empirical formula experiment.

EXAMPLE EXERCISE 7.9

A 1.64-g sample of radioactive radium was heated in air and found to react with 0.115 g of oxygen. Find the empirical formula for radium oxide, Ra_xO_y. (The atomic mass of radium is 226.0 amu.)

Solution: The empirical formula is the whole-number ratio of radium to oxygen in the compound radium oxide. This ratio is determined from the moles of each reactant. The moles of radium are calculated as follows.

$$1.64 \text{ g Ra} \times \frac{1 \text{ mol Ra}}{226.0 \text{ g Ra}} = 0.00726 \text{ mol Ra}$$

We use the atomic mass of oxygen to find the mol O:

$$0.115 \text{ g O} \times \frac{1 \text{ mol O}}{16.0 \text{ g O}} = 0.00719 \text{ mol O}$$

The mole ratio of the elements in radium oxide is $Ra_{0.00726}O_{0.00719}$. We simplify the mole ratio by dividing by the smaller number; that is,

$$\frac{Ra_{0.00726}O_{0.00719}}{0.00719 \quad 0.00719} = Ra_{1.01}O_{1.00}$$

The empirical formula is the smallest whole-number ratio of the elements. The ratio $Ra_{1.01}O_1$ can be rounded off to Ra_1O_1. The slight discrepancy from whole numbers is due to experimental error. The empirical formula for radium oxide is RaO. The radioactive element radium has an empirical formula similar to the other Group IIA/2 elements. Thus, radium is placed correctly in Group IIA/2.

SELF-TEST EXERCISE

A 1.751 g sample of powdered sulfur was ignited with oxygen in a chamber containing platinum gauze. If 2.619 g of oxygen reacted, what is the empirical formula of the sulfur oxide gas produced?

Answer: SO_3

EXAMPLE EXERCISE 7.10

In a laboratory experiment, 0.500 g of scandium was heated and allowed to react with oxygen from the air. The resulting oxide had a mass of 0.767 g. What is the empirical formula for scandium oxide, Sc_xO_y?

Solution: The empirical formula is the whole-number ratio of scandium and oxygen in the compound scandium oxide. This ratio is experimentally determined from the moles of each reactant. The moles of scandium are calculated as follows.

$$0.500 \text{ g Sc} \times \frac{1 \text{ mol Sc}}{45.0 \text{ g Sc}} = 0.0111 \text{ mol Sc}$$

The moles of oxygen are calculated after first subtracting the mass of Sc from the product, Sc_xO_y.

$$0.767 \text{ g } Sc_xO_y - 0.500 \text{ g Sc} = 0.267 \text{ g O}$$

The moles of oxygen are calculated as follows.

$$0.267 \text{ g O} \times \frac{1 \text{ mol O}}{16.0 \text{ g O}} = 0.0167 \text{ mol O}$$

The mole ratio of the elements in scandium oxide is $Sc_{0.0111}O_{0.0167}$. To simplify and obtain small whole-number ratios, we always divide by the smaller number; for example,

$$\frac{Sc_{0.0111}O_{0.0167}}{0.0111 \quad 0.0111} = Sc_{1.00}O_{1.50}$$

We cannot round off the experimental ratio, $Sc_1O_{1.50}$, but we can double the ratio to obtain $Sc_2O_{3.00}$. Thus, the empirical formula is Sc_2O_3.

SELF-TEST EXERCISE

Iron can react with chlorine gas to give two different compounds, $FeCl_2$ and $FeCl_3$. Under one set of conditions, 0.558 g of metallic iron reacts with chlorine gas to yield 1.621 g of iron chloride. Which iron compound is produced in the experiment?

Answer: $FeCl_3$

In 1871, Mendeleev predicted an undiscovered element and named it ekaboron. He placed ekaboron in the same group as aluminum. In 1879, ekaboron was discovered in Scandinavia and given the name scandium and the symbol Sc.

In Example Exercise 7.10, we found that the empirical formula of scandium oxide is Sc_2O_3. It is similar to the formula for aluminum oxide, Al_2O_3. But rather than being placed in Group IIIA/13 with aluminum, scandium is placed in Group IIIB/3. The reason becomes clear if we compare our current periodic table to

Mendeleev's original periodic table. The current periodic table separates representative and transition elements, but Mendeleev's periodic table of 1871 did not (see Figure 5.5).

Empirical Formulas from Percentage Composition

We briefly described an experiment that provided the empirical formula for a compound. That experiment used the classical method of igniting a substance in oxygen and analyzing the products formed; in fact, this method has been used for three centuries. Although it does ultimately provide data for the calculation of empirical formulas, it is tedious and time-consuming.

Today, modern instruments can be used to provide empirical formulas of compounds. Analysis of complex biochemical compounds has only been possible using very sophisticated instrumental methods. Amazingly, these instruments provide analyses of complicated biochemical compounds within minutes. One such instrument, the C/H/O/N analyzer, is used routinely. A lab technician injects a sample compound into the instrument and a short time later records the percentage of each element from a digital display.

Consider the compound benzene, which was used as a common solvent until 1981, when the Environmental Protection Agency (EPA) labeled benzene a carcinogen. An instrumental analysis gave the composition of benzene as 92.3% carbon and 7.77% hydrogen. Calculate the empirical formula of benzene.

The empirical formula is the formula that expresses the simplest whole-number ratio of carbon and hydrogen in benzene. To calculate the moles of each element, we need the mass of each. To convert percentage composition into a mass in grams, assume we have 100 g of benzene. In 100 g of benzene there are 92.3 g of carbon and 7.77 g of hydrogen. That is, the percentage of each element corresponds to its mass in 100 g of compound. We find the moles of C and H as we did in the previous examples.

$$92.3 \cancel{\text{g C}} \times \frac{1 \text{ mol C}}{12.0 \cancel{\text{g C}}} = 7.69 \text{ mol C}$$

$$7.77 \cancel{\text{g H}} \times \frac{1 \text{ mol H}}{1.01 \cancel{\text{g H}}} = 7.69 \text{ mol H}$$

The mole ratio of the elements in benzene is $C_{7.69}H_{7.69}$. We simplify the mole ratio by dividing both by the smaller number; that is,

$$\frac{C_{7.69}H_{7.69}}{7.69 \quad 7.69} = C_{1.00}H_{1.00}$$

The ratio $C_{1.00}H_{1.00}$ is rounded to C_1H_1. The empirical formula for benzene is CH. Since benzene is a molecular compound, the actual formula of the compound may be different from CH. In Section 7.7 we will find the actual molecular formula.

The following example illustrates another calculation of an empirical formula from percentage composition data.

EXAMPLE EXERCISE 7.11

The amino acid glycine is found in protein. An analysis of the amino acid gave the following data: 32.0% carbon, 6.7% hydrogen, 18.6% nitrogen, and 42.6% oxygen. Calculate the empirical formula of glycine.

Solution: The empirical formula is the whole-number ratio of C, H, N, and O. We can calculate the moles of each element by assuming a 100-g sample. The percentage of each element equals its mass in 100 g of compound. We find the moles of C, H, N, and O by dividing by the atomic mass.

$$32.0 \text{ g C} \times \frac{1 \text{ mol C}}{12.0 \text{ g C}} = 2.67 \text{ mol C}$$

$$6.7 \text{ g H} \times \frac{1 \text{ mol H}}{1.0 \text{ g H}} = 6.7 \text{ mol H}$$

$$18.6 \text{ g N} \times \frac{1 \text{ mol N}}{14.0 \text{ g N}} = 1.33 \text{ mol N}$$

$$42.6 \text{ g O} \times \frac{1 \text{ mol O}}{16.0 \text{ g O}} = 2.66 \text{ mol O}$$

The mole ratio of the elements in the amino acid is $C_{2.67}H_{6.7}N_{1.33}O_{2.66}$. Although this compound is more complex, we can simplify the ratio as before by dividing using the smallest number:

$$\frac{C_{2.67}H_{6.7}N_{1.33}O_{2.66}}{\;1.33\quad 1.33\quad 1.33\quad 1.33\;} = C_{2.01}H_{5.0}N_{1.00}O_{2.00}$$

The empirical formula for the amino acid glycine is $C_2H_5NO_2$.

SELF-TEST EXERCISE

Calculate the empirical formula for caffeine, given the following percentage composition: 49.5% C, 5.15% H, 28.9% N, and 16.5% O.

Answer: $C_4H_5N_2O$

7.7 Molecular Formula

To calculate the molecular formula for a molecular compound, given its empirical formula and molar mass.

molecular formula The chemical formula of a compound that expresses the actual number of atoms present in one molecule.

Molecular compounds are represented by molecules that contain two or more nonmetallic elements. To see what this means in terms of their empirical formulas, recall in the previous section where we found that the empirical formula for benzene is CH. The formula for benzene, one C atom and one H atom, does not represent a stable molecule. The actual **molecular formula** for benzene must therefore be some multiple of the empirical formula. That formula can be represented as $(CH)_n$, where n indicates that the multiple of (CH) is two or more.

Now consider acetylene, which is used in oxyacetylene welding as the fuel gas that reacts with oxygen. The empirical formula for acetylene is also CH. Even though gaseous acetylene gas and liquid benzene have totally unrelated properties, each shares the same empirical formula, CH. In addition, the compound styrene, which is used to make Styrofoam cups and insulation, also shares the same empirical formula, CH.

If benzene, acetylene, and styrene are different molecular compounds, their molecular formulas must be different. Experiments have provided the molar masses for each of these compounds. The molar mass of benzene is 78.0 g/mol, for acetylene it is 26.0 g/mol, and for styrene it is 104.0 g/mol. We can indicate the number of multiples of the empirical formula for each compound as follows:

$$\text{benzene: } (CH)_n = 78.0 \text{ g/mol}$$

$$\text{acetylene: } (CH)_n = 26.0 \text{ g/mol}$$

$$\text{styrene: } (CH)_n = 104.0 \text{ g/mol}$$

The molar mass of the empirical formula, CH, is 12.0 g C + 1.01 g H = 13.0 g/mol. Now we can compute how many multiples of the empirical formula are in the compound benzene. We have

$$\text{benzene: } \frac{(CH)_n}{CH} = \frac{78.0 \text{ g/mol}}{13.0 \text{ g/mol}}$$

$$n = 6$$

Therefore, the molecular formula for benzene is $(CH)_6$, or C_6H_6.

Similarly, we can compute the number of multiples of the empirical formula in acetylene.

$$\text{acetylene: } \frac{(CH)_n}{CH} = \frac{26.0 \text{ g/mol}}{13.0 \text{ g/mol}}$$

$$n = 2$$

The empirical formula is a twofold multiple, and therefore the molecular formula of acetylene is $(CH)_2$, or C_2H_2.

In a similar fashion, we can computer the number of multiples of the empirical formula in styrene.

$$\text{styrene: } \frac{(CH)_n}{CH} = \frac{104.0 \text{ g/mol}}{13.0 \text{ g/mol}}$$

$$n = 8$$

The empirical formula is an eightfold multiple, and the molecular formula of styrene is $(CH)_8$, or C_8H_8. Even though benzene, acetylene, and styrene share a common empirical formula, their molecular formulas are different, indicating that they are different compounds. The following example will further illustrate the determination of molecular formula.

EXAMPLE EXERCISE 7.12

The empirical formula for fructose, or fruit sugar, is CH_2O. If the molar mass of fructose is 180.0 g/mol, find the actual molecular formula for the sugar.

Solution: We can indicate the molecular formula of fructose as $(CH_2O)_n$. The molar mass of the empirical formula CH_2O is 12.0 g C + 2(1.01 g H) + 16.0 g O = 30.0 g/mol. Thus, the number of multiples of the empirical formula is

$$\text{fructose: } \frac{(CH_2O)_n}{CH_2O} = \frac{180.0 \text{ g/mol}}{30.0 \text{ g/mol}}$$

$$n = 6$$

The molecular formula is a sixfold multiple of the empirical formula. Thus, the molecular formula of fructose is $(CH_2O)_6$, or $C_6H_{12}O_6$.

SELF-TEST EXERCISE

Ethylene dibromide (EDB) was used as a grain pesticide in the United States until it was banned. Calculate the (a) empirical formula and (b) molecular formula for ethylene dibromide, given its approximate molar mass of 190 g/mol and its percentage composition: 12.7% C, 2.1% H, and 85.1% Br.

Answers: **(a)** CH_2Br; **(b)** $C_2H_4Br_2$

For molecular compounds, the empirical and molecular formulas are often different. For ionic compounds, the actual formulas are almost always identical to the empirical formulas of the compound. For example, the actual formula for sodium chloride is NaCl, not Na_2Cl_2 or some other multiple. There are a few exceptions, however, such as mercury(I) compounds, including mercury(I) chloride, Hg_2Cl_2. Notice that the actual formula is twice the empirical formula.

7.8 Molar Volume

OBJECTIVES

To state the value for the molar volume of any gas at standard temperature and pressure.

To perform calculations that relate molar mass, molar volume, and the density of a gas at standard temperature and pressure.

As you may recall, 1 mole of a gaseous substance contains Avogadro's number of molecules, that is, 6.02×10^{23} molecules. The gas can be hydrogen, oxygen, carbon dioxide, or any other gas. In 1811, Avogadro correctly proposed that two gases, under similar conditions, containing equal numbers of molecules occupy equal volumes. This statement is known as **Avogadro's law**. From this law it follows, for example, that 1 mole of hydrogen gas and 1 mole of oxygen gas occupy the same volume. That is, 6.02×10^{23} molecules of hydrogen gas occupy the same volume as 6.02×10^{23} molecules of oxygen or any other gas. In fact, 6.02×10^{23} molecules of any gas occupy the same volume.

Avogadro's law The principle that equal volumes of gases, at the same temperature and pressure, contain equal numbers of molecules.

What is the volume of gas that contains Avogadro's number of molecules? At **standard temperature and pressure (STP)**, the volume is 22.4 liters. Standard temperature has been chosen to be 0°C. Standard pressure has been chosen to be 1 atmosphere, which is the atmospheric pressure at sea level. The volume occupied by 1 mole of any gas at STP is called the **molar volume** (Figure 7.3).

standard temperature and pressure (STP) Refers to a gas at 0°C and 1.00 atmosphere pressure which have been arbitrarily chosen as standard conditions.

molar volume The volume occupied by one mole of any gas at standard conditions of temperature and pressure; at 0°C and 1.00 atm the volume of one mole of any gas is 22.4 L.

Now when we describe the quantity of a gas, we can state the number of molecules, the mass, or the volume. Table 7.1 compares 1 mole of five different gases. Notice that the number of molecules and the volume are constant, whereas the mass varies.

Gas Density

As you may recall, the density of a gas is much less than the density of a liquid or solid. In Section 2.8, we found that at 3.98°C the density of water is 1.00 g/mL. The density of air at 3.98°C is only 0.00129 g/mL. The density of water is almost 1000 times greater than that of air.

The definition of density is mass divided by volume. For any given gas, such as hydrogen or oxygen, we can calculate the density at STP. This is possible because 1 mole of any gas has a mass equal to its molar mass and a volume equal to its molar volume. The formula for **gas density** is

gas density The ratio of mass per unit volume for a gas; usually expressed in grams per liter.

Table 7.1 Mole Relationships for Selected Gases

Gas	Moles	Molecules	Mass	Volume at STP‡
hydrogen, H_2	1.00	6.02×10^{23}	2.02 g	22.4 L
oxygen, O_2	1.00	6.02×10^{23}	32.0 g	22.4 L
carbon dioxide, CO_2	1.00	6.02×10^{23}	44.0 g	22.4 L
ammonia, NH_3	1.00	6.02×10^{23}	17.0 g	22.4 L
argon, Ar*	1.00	6.02×10^{23}	39.9 g	22.4 L

* Argon gas is composed of atoms rather than molecules.
‡ Experimentally, gases demonstrate slight differences in molar volume.

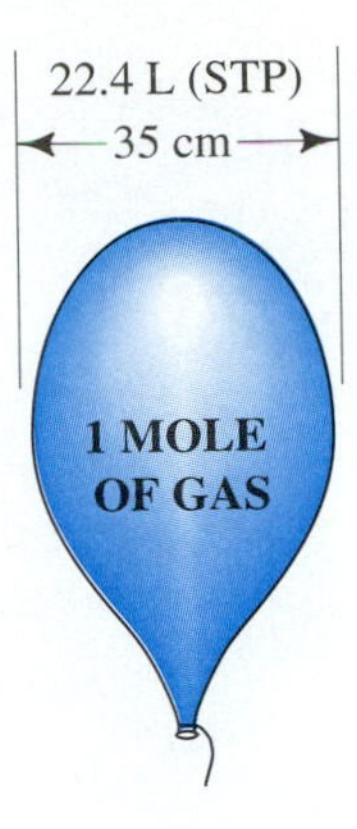

Figure 7.3 Molar Volume of a Gas A balloon containing one mole of gas has a diameter of about 35 cm. One mole of any gas at STP occupies a volume of 22.4 liters. Its mass is equal to its molar mass and contains Avogadro's number of molecules.

$$\frac{\text{molar mass in grams}}{\text{molar volume in liters}} = \text{density in g/L of a gas at STP}$$

The following examples illustrate the relationship of density, molar mass, and molar volume.

EXAMPLE EXERCISE 7.13

Calculate the density of ammonia gas, NH_3, at STP.

Solution: To find the density of a gas at STP, we divide the molar mass by the molar volume.

$$\frac{\text{molar mass, } NH_3}{\text{molar volume, } NH_3} = \text{density, g/L}$$

From the periodic table, we find the mass of an ammonia molecule is 17.0 amu. The molar mass of NH_3 is 17.0 g/mol. The molar volume is 22.4 L at STP.

$$\frac{17.0 \text{ g}}{22.4 \text{ L}} = 0.759 \text{ g/L}$$

The density of ammonia, 0.759 g/L, is less than that of air, 1.29 g/L. If we filled a balloon with ammonia gas, it would float in the air. Fluids that are less dense float on fluids that are more dense. The ammonia balloon floating in air is analogous to a less dense liquid, such as gasoline, floating on water.

SELF-TEST EXERCISE

Calculate the density of ozone, O_3, at STP.

Answer: 2.14 g/L

A variation of the above problem is to calculate the molar mass of a gas, given its density at standard conditions. If the density of an unknown gas is determined in a laboratory, this information can be used to identify the gas.

EXAMPLE EXERCISE 7.14

If 1.96 g of an unknown gas occupies a volume of 1.00 L at STP, what is the molar mass of the unknown gas?

Solution: Let's first write down the units of molar mass, g/mol. In this problem, we are given the density of the unknown gas: 1.96 g/1.00 L. We can outline the calculation as follows:

$$\frac{1.96\text{ g}}{1.00\text{ L}} \times \frac{\text{unit}}{\text{factor}} = \frac{\text{g}}{\text{mol}}$$

To convert the units of L to mol, we apply the unit factor 22.4 L/1 mol.

$$\frac{1.96\text{ g}}{1.00\ \cancel{\text{L}}} \times \frac{22.4\ \cancel{\text{L}}}{1\text{ mol}} = 43.9\text{ g/mol}$$

Suppose the unknown gas is found in a fire extinguisher. We might immediately suspect that the gas is carbon dioxide because we know that CO_2 smothers fires. We help confirm the gas is CO_2 by adding up its molar mass (44.0 g/mol) and find that it agrees with the molar mass of the unknown gas, 43.9 g/mol.

SELF-TEST EXERCISE

What is the molar mass of boron trifluoride gas, given that 1.51 g of the gas occupies a volume of 500.0 mL at STP?

Answer: 67.6 g/mol

7.9 Mole Calculations II

OBJECTIVES

To perform calculations that relate the STP volume of a gas to its mass and number of molecules.

As we mentioned previously, a mole is at the center of chemical calculations. A mole has three interpretations.

1. A mole is Avogadro's number of entities.
2. A mole of a substance has a mass equal to its formula mass expressed in grams.
3. A mole of any gas at STP occupies a volume of 22.4 liters.

We can show the interpretation of a mole in the form of equations as follows:

$$\begin{aligned} \textit{1 mole} &= 6.02 \times 10^{23}\text{ entities} \\ &= \text{molar mass} \\ &= 22.4\text{ L at STP} \end{aligned}$$

From the formulas, we see that, using the unit mole, we can relate the number of entities of substance to its mass or to its gaseous volume at STP. The process of interchanging these three quantities is referred to as mole calculations. Figure 7.4 illustrates the different mole calculation conversions.

The following example exercises illustrate calculations involving moles of substance. In these calculations, we will use the concepts of Avogadro's number, molar mass, and molar volume.

EXAMPLE EXERCISE 7.15

What is the mass of 3.36 L of ozone, O_3, at STP?

Solution: To calculate the mass of O_3, we must first find the number of moles. We use the molar volume concept. The relevant unit equation is 1 mol O_3 = 22.4 L at STP. Applying the unit factor so as to cancel units, we have

$$3.36 \cancel{\text{L } O_3} \times \frac{1 \text{ mol } O_3}{22.4 \cancel{\text{L } O_3}} = 0.150 \text{ mol } O_3$$

Next, to calculate the mass of ozone gas, we multiply by the molar mass. From the periodic table, we find that the atomic mass of oxygen is 16.0 amu. The molar mass of O_3 is 48.0 g/mol; therefore, 1 mol O_3 = 48.0 g O_3.

$$0.150 \cancel{\text{mol } O_3} \times \frac{48.0 \text{ g } O_3}{1 \cancel{\text{mol } O_3}} = 7.20 \text{ g } O_3$$

SELF-TEST EXERCISE

What is the volume occupied by 0.125 g of hydrogen sulfide gas, H_2S, at STP?

Answer: 0.0821 L H_2S (82.1 mL)

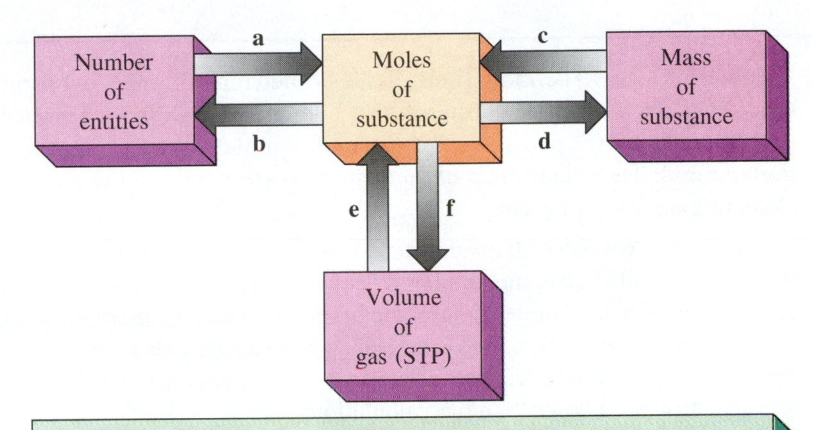

(a) Use N as a unit factor: multiply by 1 mol/6.02×10^{23}
(b) Use N as a unit factor: multiply by 6.02×10^{23}/1 mol
(c) Use MM as a unit factor: multiply by 1 mol/g
(d) Use MM as a unit factor: multiply by g/1 mol
(e) Use molar volume as a unit factor: multiply by 1 mol/22.4 L
(f) Use molar volume as a unit factor: multiply by 22.4 L/1 mol

Figure 7.4 Mole Calculations The six possible steps (a, b, c, d, e, f) outline the relationships between mole quantities. The mole is the central unit for these conversions.

EXAMPLE EXERCISE 7.16

How many molecules of hydrogen chloride gas, HCl, occupy a volume of 50.0 mL at STP?

Solution: To find the number of HCl molecules, we must first convert the volume into moles. The molar volume of hydrogen chloride is the same for all gases: 22.4 L at STP. The unit equation is 1 mol HCl = 22.4 L. Applying the unit factor to cancel units, we have

$$50.0 \cancel{\text{mL HCl}} \times \frac{1 \cancel{\text{L HCl}}}{1000 \cancel{\text{mL HCl}}} \times \frac{1 \text{ mol HCl}}{22.4 \cancel{\text{L HCl}}} = 0.00223 \text{ mol HCl}$$

To calculate the number of HCl molecules, we multiply by Avogadro's number. The unit factor is 6.02×10^{23} molecules/1 mol HCl.

$$0.00223 \cancel{\text{mol HCl}} \times \frac{6.02 \times 10^{23} \text{ molecules HCl}}{1 \cancel{\text{mol HCl}}} = 1.34 \times 10^{21} \text{ molecules HCl}$$

With practice, we can perform mole conversions in a single operation:

$$50.0 \cancel{\text{mL HCl}} \times \frac{1 \cancel{\text{mol HCl}}}{22{,}400 \cancel{\text{mL HCl}}} \times \frac{6.02 \times 10^{23} \text{ molecules HCl}}{1 \cancel{\text{mol HCl}}}$$

$$= \text{molecules HCl} = 1.34 \times 10^{21} \text{ molecules HCl}$$

SELF-TEST EXERCISE

What is the volume occupied by 3.33×10^{21} atoms of helium gas, He, at STP?

Answer: 0.124 L He (124 mL)

Summary

Sections 7.1–7.3 The **mole (mol)** is the cornerstone of chemical formula calculations. It is the amount of substance that contains **Avogadro's number** of entities, that is, 6.02×10^{23} particles. The mass of 1 mole of any substance is called the **molar mass**. The molar mass of an element corresponds to the atomic mass of the element expressed in grams.

Section 7.4 A mole is defined in terms of Avogadro's number of particles. Given the moles of substance, the number of entities (atoms, molecules, or formula units) can be calculated. A mole of substance is also related to its mass in grams. Given moles of substance, the mass of the substance can be calculated. Furthermore, the mass of the substance can be related to the number of entities. These various relationships are known as mole calculations.

Section 7.5 The **percentage composition** of a compound is computed after first finding the molar mass of the compound. The mass of each element is then divided by the mass of the entire compound. The resulting ratio, multiplied by 100, is the percentage compositon.

Section 7.6 The **empirical formula** is the simplest whole-number ratio of atoms of the elements in a compound. The empirical formula can be calculated from the experimental synthesis data. Moles of each element are calculated, and the ratio is simplified to give small whole numbers. The empirical formula can also be calculated from the percentage composition of the compound. The trick is to assume that there is a 100-g sample of the compound. The percentage of each element thus corresponds to the mass of the element in grams. Moles of each element are then calculated and simplified to a small whole-number ratio.

Section 7.7 For molecular compounds, we often find that the empirical formula is not the actual molecular formula. That is, the empirical formula is repeated a number of times to give the total number of atoms in the actual molecule. We can calculate the **molecular formula** by dividing the molar mass of the compound by the empirical formula mass. For example, the empirical formula CH_2O is shared by lactic acid and glucose. The actual formula of lactic acid is three times the empirical formula, that is, $C_3H_6O_3$. In glucose, the empirical formula repeats six times to give $C_6H_{12}O_6$.

Section 7.8 The volume of 1 mole of any gaseous substance at standard conditions is called the **molar volume**. In 1811, Avogadro stated the following law, called

Avogadro's law: Equal volumes of gases, at the same conditions, contain equal numbers of molecules. The molar volume for any gas at **standard temperature and pressure (STP)** is 22.4 L/mol. Standard conditions have been chosen as 0°C and 1 atmosphere pressure.

A mole of H_2O has 6.02×10^{23} molecules; a mole of NaCl weighs 58.5 g; and a mole of O_2 occupies 22.4 L at STP.

Section 7.9 Considering Avogadro's number, molar mass, and molar volume, mole calculations can relate three different quantities: (1) the number of entities of a substance, (2) the mass of the substance, and (3) the volume of a gaseous substance at STP. We can outline the conversion from one quantity to another.

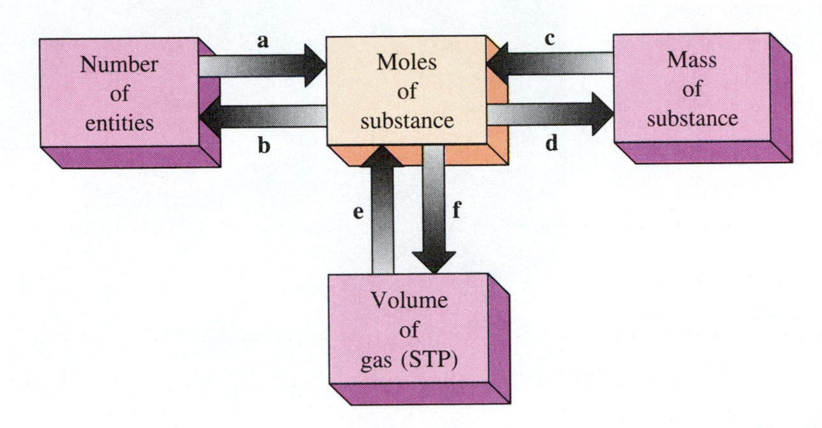

For step a, we multiply by the unit factor 1 mol/6.02×10^{23} entities; for step b, we multiply by the unit factor 6.02×10^{23} entities/1 mol. To apply step c, we multiply by the unit factor 1 mol/molar mass; for step d, we multiply by the unit factor molar mass/1 mol. To convert in step e, we need to multiply by the unit factor 1 mol/22.4 L, and in step f, multiply by the unit factor 22.4 L/1 mol.

Key Terms

Select the key term that corresponds to the following definitions.

____ **1.** the number of atoms, molecules, or formula units that constitute 1 mole of a given substance; 6.02×10^{23} individual entities.

____ **2.** the amount of substance containing the same number of entities as there are atoms in exactly 12 g of carbon-12

____ **3.** the mass of 1 mole of pure substance expressed in grams; the mass of Avogadro's number of atoms, molecules, or formula units

____ **4.** an expression for the ratio of the mass of an element compared to the mass of the compound containing the element, all times 100

____ **5.** the chemical formula of a compound that expresses the simplest ratio of the atoms in a molecule or ions in a formula unit

____ **6.** the chemical formula of a compound that expresses the actual number of atoms present in one molecule

____ **7.** the volume occupied by 1 mole of any gas at standard conditions

____ **8.** a temperature of 0°C and 1 atmosphere pressure arbitrarily chosen as standard conditions

____ **9.** equal volumes of gases, under the same conditions of temperature and pressure, contain equal numbers of molecules

____ **10.** the ratio of mass per unit volume for a gas; expressed in grams per liter

(a) Avogadro's law *(Sec. 7.8)*
(b) Avogadro's number (*N*) *(Sec. 7.1)*
(c) empirical formula *(Sec. 7.6)*
(d) gas density *(Sec. 7.8)*
(e) molar mass (MM) *(Sec. 7.3)*
(f) molar volume *(Sec. 7.8)*
(g) mole (mol) *(Sec. 7.2)*
(h) molecular formula *(Sec. 7.7)*
(i) percentage composition *(Sec. 7.5)*
(j) standard temperature and pressure (STP) *(Sec. 7.8)*

Exercises

Avogadro's Number (Sec. 7.1)

1. Refer to the periodic table and state the average mass of one atom (in amu) for each of the following elements.
   (a) H (b) Li (c) C
   (d) P (e) Ca (f) Zn
   (g) As (h) U

2. Refer to the periodic table and state the average mass (in g) of 6.02×10^{23} atoms of each of the following elements.
   (a) H (b) Li (c) C
   (d) P (e) Ca (f) Zn
   (g) As (h) U

The Mole Concept (Sec. 7.2)

3. State the number of entities in 1 mole of each of the following.
   (a) manganese atoms, Mn
   (b) nitrogen dioxide molecules, NO_2
   (c) manganese nitrate formula units, $Mn(NO_3)_2$
   (d) permanganate ions, MnO_4^-

4. State the number of moles represented by each of the following.
   (a) 6.02×10^{23} atoms of copper, Cu
   (b) 6.02×10^{23} molecules of sulfur dioxide, SO_2
   (c) 6.02×10^{23} formula units of copper(II) sulfate, $CuSO_4$
   (d) 6.02×10^{23} ions of hydrogen sulfate, HSO_4^-

5. Calculate the number of entities in each of the following.
   (a) 0.335 mol titanium atoms, Ti
   (b) 0.112 mol carbon dioxide molecules, CO_2
   (c) 1.94 mol zinc chloride formula units, $ZnCl_2$
   (d) 0.335 mol nitrite ions, NO_2^-

6. Calculate the number of entities in each of the following.
   (a) 2.12 mol argon atoms, Ar
   (b) 7.10 mol nitrogen trifluoride molecules, NF_3
   (c) 0.552 mol silver sulfate formula units, Ag_2SO_4
   (d) 0.112 mol carbonate ions, CO_3^{2-}

7. Calculate the number of moles containing each of the following.
   (a) 4.15×10^{22} atoms of iron, Fe
   (b) 3.31×10^{21} molecules of bromine, Br_2
   (c) 4.19×10^{20} formula units of cadmium nitrate, $Cd(NO_3)_2$
   (d) 8.12×10^{23} acetate ions, $C_2H_3O_2^-$

8. Calculate the number of moles containing each of the following.
   (a) 7.88×10^{24} atoms of selenium, Se
   (b) 5.55×10^{25} molecules of hydrogen sulfide, H_2S
   (c) 2.25×10^{22} formula units of strontium carbonate, $SrCO_3$
   (d) 6.55×10^{24} chromate ions, CrO_4^{2-}

Molar Mass (Sec. 7.3)

9. State the molar mass of each of the following elements.
   (a) mercury, Hg (b) silicon, Si
   (c) copper, Cu (d) selenium, Se

Each of the above samples represents one mole of substance. (Clockwise from top: water, sulfur, table sugar, mercury, and coppe

10. Calculate the molar mass for each of the following elements.
    (a) fluorine, F_2 (b) iodine, I_2
    (c) phosphorus, P_4 (d) sulfur, S_8

11. Calculate the molar mass for each of the following ionic compounds.
    (a) barium fluoride, BaF_2
    (b) potassium sulfide, K_2S
    (c) iron(III) acetate, $Fe(C_2H_3O_2)_3$
    (d) strontium phosphate, $Sr_3(PO_4)_2$

12. Calculate the molar mass for each of the following molecular compounds.
    (a) methane, CH_4
    (b) phosphorus triiodide, PI_3
    (c) diarsenic pentaoxide, As_2O_5
    (d) glycerine, $C_3H_5(OH)_3$

Mole Calculations I (Sec. 7.4)

13. Calculate the mass in grams for each of the following.
    (a) 2.95×10^{23} atoms of mercury, Hg
    (b) 1.16×10^{22} molecules of nitrogen, N_2
    (c) 5.05×10^{21} formula units of barium chloride, $BaCl_2$

14. Calculate the mass in grams for each of the following.
    (a) 1.21×10^{24} atoms of krypton, Kr
    (b) 6.33×10^{22} molecules of dinitrogen oxide, N_2O
    (c) 4.17×10^{21} formula units of magnesium perchlorate, $Mg(ClO_4)_2$

15. Calculate the number of entities in each of the following.
(a) 1.50 g potassium, K
(b) 0.470 g oxygen, O_2
(c) 20.36 g cesium chlorate, $CsClO_3$

16. Calculate the number of entities in each of the following.
(a) 7.57 g platinum, Pt
(b) 3.88 g ethane, C_2H_6
(c) 0.152 g aluminum chloride, $AlCl_3$

17. Calculate the mass in grams for a single atom of each of the following elements, which occur naturally as a single isotope.
(a) beryllium, Be
(b) sodium, Na
(c) cobalt, Co
(d) arsenic, As

18. Calculate the mass in grams for a single molecule of each of the following gaseous compounds.
(a) methane, CH_4
(b) ammonia, NH_3
(c) sulfur trioxide, SO_3
(d) dinitrogen pentaoxide, N_2O_5

Percentage Composition (Sec. 7.5)

19. Benzoyl peroxide is the active ingredient in a popular acne cream. Calculate the percentage composition for benzoyl peroxide, $C_7H_6O_3$.

20. Dynamite is nitroglycerin in a porous material such as cellulose. Calculate the percentage composition for nitroglycerin, $C_3H_5O_3(NO_2)_3$.

21. The illegal drug cocaine has the chemical formula $C_{17}H_{21}NO_4$. Calculate the percentage composition for the compound.

22. The amino acid methionine has the chemical formula $C_5H_{11}NSO_2$. Calculate the percentage composition for the compound.

23. Mustard gas was used in chemical warfare during World War I. Find the percentage composition for the active ingredient, $C_4H_8SCl_2$, in mustard gas.

24. Chlorophyll is a dark green plant pigment. Calculate the percentage composition for chlorophyll, $C_{55}H_{70}MgN_4O_6$.

25. Monosodium glutamate (MSG) is added to food to enhance the flavor. Find the percentage composition for MSG given its formula, $NaC_5H_8NO_4$.

26. Mercurochrome has the formula $HgNa_2C_{20}H_8Br_2O_6$. Calculate the percentage composition for the medicinal compound.

Empirical Formula (Sec. 7.6)

27. In a laboratory experiment, 0.500 g of tin was oxidized with nitric acid to give tin oxide. If the oxide had a mass of 0.635 g, what is the empirical formula of tin oxide?

28. In a laboratory experiment, 0.500 g of nickel was oxidized in air to give 0.704 g of nickel oxide. What is the empirical formula of the oxide?

29. In a laboratory experiment, 1.550 g of mercury oxide was decomposed by heat to give oxygen gas and 1.435 g of liquid mercury. What is the empirical formula of mercury oxide?

30. In a laboratory experiment, 2.410 g of copper oxide was reduced to 1.925 g of copper metal after heating with hydrogen gas. What is the empirical formula of the oxide?

31. A 1.115-g sample of cobalt was heated with sulfur to give 2.025 g of cobalt sulfide. Calculate the empirical formula for the sulfide.

32. A 0.715-g sample of titanium was heated with chlorine gas to give 2.836 g of titanium chloride. Calculate the empirical formula of the titanium compound.

33. Calculate the empirical formula for the following ionic compounds, given their percentage composition.
(a) manganese fluoride, 59.1% Mn and 40.9% F
(b) copper chloride, 64.1% Cu and 35.9% Cl
(c) tin bromide, 42.6% Sn and 57.4% Br
(d) thallium iodide, 61.7% Tl and 38.3% I

34. Calculate the empirical formula for the following ionic compounds, given their percentage composition.
(a) potassium superoxide, 55.0% K and 45.0% O
(b) vanadium oxide, 68.0% V and 32.0% O
(c) zirconium oxide, 74.0% Zr and 26.0% O
(d) bismuth oxide, 89.7% Bi and 10.3% O

35. Trichloroethylene (TCE) is a common organic solvent used to degrease machine parts. Calculate the empirical formula for TCE if the percentage composition is 18.25% C, 0.77% H, and 80.99% Cl.

36. Dimethyl sulfoxide (DMSO) is a powerful solvent and has been used in the treatment of arthritis. Calculate the empirical formula for DMSO, given that the percentage composition is 30.7% C, 7.74% H, 20.5% O, and 41.0% S.

Molecular Formula (Sec. 7.7)

37. Aspirin has a molar mass of approximately 180 g/mol. If the empirical formula is $C_9H_8O_4$, what is the molecular formula of aspirin?

38. Quinine is used to treat malaria. If the molar mass is approximately 325 g/mol and the empirical formula is $C_{10}H_{12}NO$, what is the molecular formula of the compound?

39. Adipic acid is used to manufacture nylon. If the molar mass is approximately 147 g/mol and the empirical formula is $C_3H_5O_2$, what is the molecular formula of adipic acid?

40. Hexamethylene diamine is used to manufacture nylon. If the molar mass is approximately 115 g/mol and the empirical formula is C_3H_8N, what is the molecular formula of the compound?

41. Calculate the molecular formula for ethylene glycol (antifreeze). The molar mass is approximately 62 g/mol and the percentage composition is 38.7% C, 9.74% H, and 51.6% O.

42. Calculate the molecular formula for dioxane, a plastics solvent discovered to be carcinogenic. The molar mass is approximately 88 g/mol, and the percentage composition is 54.5% C, 9.15% H, and 36.3% O.

43. The compound lindane has been used as an insecticide. It has an approximate molar mass of 290 g/mol, and the percentage composition is 24.8% C, 2.08% H, and 73.1% Cl. Calculate the empirical and molecular formulas of lindane.

44. Mercurous chloride is used in calomel electrodes and as a fungicide. The molar mass is approximately 470 g/mol and the percentage composition is 85.0% Hg and 15.0% Cl. Calculate the molecular formula of mercurous chloride.

45. Nicotine, found in tobacco, has an approximate molar mass of 160 g/mol. If the percentage composition is 74.0% C, 8.70% H, and 17.3% N, what is the molecular formula of nicotine?

46. Galactose is sometimes referred to as cerebrose or brain sugar. Its molar mass is approximately 180 g/mol, and its percentage composition is 40.0% C, 6.72% H, and 53.3% O. What is the molecular formula of galactose?

Molar Volume (Sec. 7.8)

47. State standard conditions in degrees Celsius and atmospheres.

48. State STP conditions in Kelvin units and millimeters Hg. (*Hint:* 1 atm = 76 cm Hg.)

49. Given 1 mole of each of the following gases, complete the table.

Gas	Molecules	Mass	Volume at STP
fluorine, F_2			
hydrogen fluoride, HF			
silicon tetrafluoride, SiF_4			
oxygen difluoride, OF_2			

50. Given 1 mole of each of the following gases, complete the table.

Gas	Molecules	Mass	Volume at STP
radon, Rn			
hydrogen sulfide, H_2S			
phosphine, PH_3			
butane, C_4H_{10}			

51. Calculate the density for each of the following gases at STP.
(**a**) neon, Ne (**b**) chlorine, Cl_2
(**c**) nitrogen dioxide, NO_2 (**d**) hydrogen iodide, HI

52. Calculate the density for each of the following gases at STP.
(**a**) xenon, Xe (**b**) fluorine, F_2
(**c**) propane, C_3H_8 (**d**) sulfur trioxide, SO_3

53. Calculate the molar mass for each of the following gases, given the STP density.
(**a**) ethane, 1.34 g/L (**b**) diborane, 1.23 g/L
(**c**) Freon-12, 5.40 g/L (**d**) nitrous oxide, 2.05 g/L

54. Calculate the molar mass for each of the following gases, given the STP density.
(**a**) isobutane, 2.59 g/L (**b**) silane, 1.43 g/L
(**c**) Freon-22, 3.86 g/L (**d**) nitric oxide, 1.34 g/L

Mole Calculations II (Sec. 7.9)

55. Calculate the mass in grams for each of the following gases at STP.
(**a**) 1.05 L of hydrogen sulfide, H_2S
(**b**) 5.33 L of dinitrogen trioxide, N_2O_3
(**c**) 75.0 mL of dichlorine monoxide, Cl_2O

56. Calculate the volume in liters for each of the following gases at STP.
(**a**) 0.250 g of helium, He
(**b**) 5.05 g of nitrogen, N_2
(**c**) 0.885 g of carbon monoxide, CO

57. Calculate the number of molecules in each of the following gases at STP.
(**a**) 10.0 mL of hydrogen, H_2
(**b**) 70.5 mL of ammonia, NH_3
(**c**) 1.00 L of carbon dioxide, CO_2

58. Calculate the volume in liters for each of the following gases at STP.
(**a**) 2.22×10^{22} molecules of methane, CH_4
(**b**) 4.18×10^{24} molecules of ethane, C_2H_6
(**c**) 5.09×10^{21} molecules of propane, C_3H_8

59. Given 0.225 mol of each of the following gases, complete the table by calculating the number of molecules, total number of atoms, mass in grams, and volume in liters at STP.

Gas	Molecules	Atoms	Mass	Volume at STP
N_2				
NO_2				
NO				
N_2O_4				

60. Use the given quantity of each gas listed in the table and calculate the related quantities. That is, calculate the number of molecules, total number of atoms, mass in grams, and volume in liters at STP for each gas.

Gas	Molecules	Atoms	Mass	Volume at STP
HCl	1.15×10^{22}			
Cl_2		4.27×10^{24}		
Cl_2O			10.0 g	
ClO_2				0.282 L

General Exercises

61. A mole of electrons is called a faraday after the great English scientist Michael Faraday. A mole of photons is called an einstein after the German-American physicist Albert Einstein. How many electrons are in a farday? How many photons are in an einstein?

62. A 1-carat diamond has a mass of 0.200 g. If diamond is composed exclusively of carbon, does the diamond contain more than 1 trillion carbon atoms?

63. A quadrillion is approximately the number of red blood cells in 50,000 people. Which is greater: a quadrillion, 1×10^{15} red blood cells, or the number of nickel atoms in a 5.0-g nickel coin?

64. By rubbing the Lincoln profile on a one-cent coin, a student reduced the mass just enough to be detected by an analytical balance, 0.001 g. How many copper atoms were rubbed off the coin?

65. Which weighs more: 1 mole of furry moles or the earth? Assume the average rodent mole has a mass of 100 grams. The mass of the earth is 6×10^{24} kilograms.

66. Which is longer: 1 mole of furry moles (head to tail) or 10 light years? Assume the average rodent mole has a length of 17 cm. A light-year is the distance light travels in 1 year, 9.5×10^{12} kilometers.

67. In 1871, Mendeleev predicted the undiscovered element *ekaaluminum*. In 1875, the element was discovered in Gaul (France) and given the name gallium. If 0.500 g of gallium reacts with oxygen gas to give 0.672 g of gallium oxide, what is its empirical formula?

68. In 1871, Mendeleev predicted the undiscovered element *ekasilicon*. In 1886, the element was discovered in Germany and given the name germanium. If 0.500 g of germanium reacts with chlorine gas to give 0.978 g of germanium chloride, what is its empirical formula?

69. Calculate the cubic centimeter volume occupied by one molecule of water in a mole of water. Recall that the density of water is 1.00 g/cm^3.

70. Calculate the milliliter volume occupied by one molecule of ethyl alcohol, C_2H_5OH, in a beaker of alcohol. The density of ethyl alcohol is 0.789 g/mL.

71. Calculate the number of carbon atoms in 1.00 g of table sugar, $C_{12}H_{22}O_{11}$.

72. Calculate the number of carbon atoms in 1.00 g of blood sugar, $C_6H_{12}O_6$.

73. Vitamin K has a molar mass of 173 g/mol and is 76.3% carbon by mass. How many carbon atoms are in one molecule of vitamin K?

74. What is the mass of rust, Fe_2O_3, that contains 10.0 g of iron?

75. The volume occupied by each copper atom in a 1-mol crystal is 0.0118 nm^3. If the density of the copper crystal is 8.92 g/cm^3, what is the experimental value of Avogadro's number?

76. Each atom in a crystal of aluminum metal occupies a theoretical cube that is 0.255 nm on a side. If the density of the aluminum crystal is 2.70 g/cm^3, what is the experimental value for Avogadro's number?

CHAPTER

8

Writing Chemical Equations

A seasonal change to fall colors, such as these shown in the Butchert Garden on Vancouver Island, British Columbia, are produced by chemical reactions.

In Section 3.7, we distinguished between a physical change and a chemical change. Recall that a physical change, such as melting or boiling, does not change the composition of the substance. A chemical change, on the other hand, produces new substances that have different chemical formulas and properties. One way to describe a chemical change is in terms of a chemical reaction. For example, baking soda reacts with vinegar while releasing bubbles of carbon dioxide gas. This reaction is shown in Figure 8.1.

In this chapter, we will use our formula writing skills acquired in Chapter 6 to describe chemical reactions. For example, vinegar contains acetic acid, which reacts with baking soda to produce sodium acetate, water, and carbon dioxide gas. If we translate the names of substances into chemical formulas, we have

$HC_2H_3O_2$ and $NaHCO_3$ *react to produce* $NaC_2H_3O_2$, H_2O, and CO_2

Figure 8.1 Observing a Chemical Reaction When vinegar is poured into a solution of baking soda, bubbles are produced. The bubbles are carbon dioxide gas and indicate a chemical reaction.

8.1 Evidence for Chemical Reactions

OBJECTIVES

To state four experimental observations that provide evidence for a chemical reaction.

We know that when we combine two substances the change can be physical or chemical. We now need to know how we can distinguish between a physical change and a chemical reaction. One way is to observe what has occurred. For example, each of the following is a *strong indication* that a chemical reaction has taken place.

1. *A gas is produced.* We can observe the experimental evidence for a gas in a number of different ways. After combining two solutions, we may see bubbles. Bubbles are evidence for a gas being released. For example, Alka-Seltzer gives evidence for reaction after being added to water because it fizzes. The citric acid and baking soda in the tablet react to give a gas.

   Suppose we heat a solid substance in a test tube for a couple of minutes and then insert a flaming splint into the test tube. If we observe that the flame is immediately extinguished, we can conclude that the substance produced a gas, such as CO_2, that does not support combustion.

   Now suppose that, in a different experiment, we heat a substance in a flask. If we insert a glowing splint into the flask and the splint bursts into flames, we can conclude that the substance produced a gas, such as O_2, that does support combustion.
2. *A precipitate is formed.* After adding two solutions together, we observe that an insoluble substance is formed. This insoluble particulate substance is called a **precipitate**. The formation of a precipitate is evidence for a chemical reaction.

precipitate An insoluble substance produced by a reaction in aqueous solution.

3. *A permanent color change is observed.* Many chemical reactions involve a permanent change in color. Adding aqueous ammonia to a solution of copper(II) sulfate, for example, changes the color from light blue to deep royal blue. To observe an acid reacting with a base, we use an indicator that changes color. The indicator enables us to follow a reaction directly that would otherwise not be visible.

   Heating a substance often causes a change in color. If the color change is permanent, a chemical reaction has occurred. If we heat copper wire to red heat, for example, it turns black upon cooling. In this experiment, copper has been oxidized to copper(II) oxide. The color change is therefore evidence for a chemical reaction. If we heat platinum wire to red heat and then cool it, it returns to its original metallic luster. This experiment shows that platinum has not undergone a permanent color change. Therefore, a chemical reaction did not take place.
4. *A heat energy change is noted.* In chemical reactions there is often a detectable change in heat energy. A reaction that releases heat is said to be an **exothermic reaction**. For example, if a highway flare is ignited, heat and light are observed. Heat and light are two forms of energy that indicate a chemical reaction is taking place.

   A reaction that absorbs heat energy is said to be an **endothermic reaction**. For example, if a reaction takes place between two solutions and there is a detectable lowering of temperature, the process is an endothermic reaction.

exothermic reaction A reaction that releases heat energy.

endothermic reaction A reaction that absorbs heat energy.

In summary, each of the above four criteria—the production of gas, the formation of a precipitate, a change of color, or a change in energy—is an indication that a **chemical reaction** has occurred (Figure 8.2). A chemical reaction, however, may have occurred even though no gas, precipitate, color change, or heat change is observed. In some reactions, the energy change may be too subtle to notice.

chemical reaction The process of undergoing a chemical change.

EXAMPLE EXERCISE 8.1

Which of the following is experimental evidence for a chemical reaction?
(a) pouring vinegar on baking soda causes foamy bubbles
(b) mixing two solutions produces yellow insoluble particles
(c) mixing two colorless solutions gives a pink solution
(d) mixing two solutions produces a temperature increase

Solution: We can analyze each of these observations based on the stated criteria.
(a) The bubbles produced are a strong indication that a chemical reaction is occurring.
(b) The insoluble particles formed, that is a precipitate, are evidence for a chemical reaction.
(c) The pink color produced from two colorless solutions is an indication of a chemical reaction.
(d) The temperature increase indicates that heat energy is being released; thus, we have evidence that suggests an exothermic chemical reaction is taking place.

SELF-TEST EXERCISE

What are four indications that a chemical reaction has occurred?

Answers:
(1) A gas is evolved; **(2)** A precipitate is produced; **(3)** A color change is noted; **(4)** A heat change is observed.

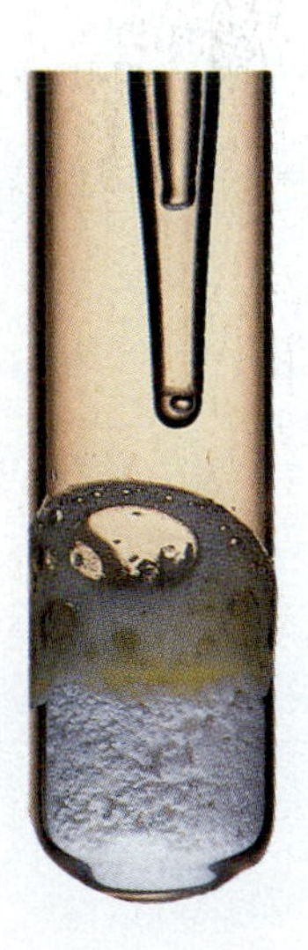

Figure 8.2 Evidence for a Chemical Reaction Each of the following is strong evidence that a chemical reaction is occurring: (a) the production of a gas, (b) the formation of a precipitate, (c) a change of color, or (d) the release of heat energy.

Figure 8.2 Evidence for a Chemical Reaction Each of the following is strong evidence that a chemical reaction is occurring: (a) the production of a gas, (b) the formation of a precipitate, (c) a change of color, or (d) the release of heat energy. (*Continued*)

(b)

(c)

(d)

Each of the aforementioned four criteria is an indication of a chemical reaction. There are some exceptions, however, when an indication is erroneous. For example, if we hold a platinum wire over a low burner flame, the wire may turn black, but the change is physical. The platinum wire is not undergoing a chemical reaction; it is simply collecting unburnt carbon from the burner flame. In the laboratory, each of the above criteria strongly suggests a chemical reaction although, in some cases, additional testing may be necessary.

8.2 Writing Chemical Equations

OBJECTIVES

To write a chemical equation using formulas and symbols when the description of the chemical reaction is given.

In Chapter 6, we learned to use formulas to describe chemical compounds. In this chapter, we will learn to use formulas and symbols to describe a chemical reaction; that is, we will learn to write a **chemical equation**. Consider the following general description of a chemical reaction: substance A added to substance B causes a chemical change creating substances C and D. This statement is described in symbols as follows:

chemical equation A shorthand representation using formulas and symbols to describe a chemical change.

$$A + B \longrightarrow C + D$$

In this general chemical equation, A and B are called the **reactants,** and C and D are called the **products.**

reactant A substance undergoing a chemical reaction.

product A substance resulting from a chemical reaction.

We can provide more information about the reaction if we specify the physical state of each substance, that is, solid, liquid, or gas. The physical state is specified using the abbreviations (s), (l), or (g). By convention, we indicate an **aqueous solution** by the symbol (aq), and an insoluble substance by the symbol (s). For example, an aqueous solution of A reacts with a gaseous substance B to yield a precipitate of C and an aqueous solution D. This statement can be written as follows:

$$A(aq) + B(g) \longrightarrow C(s) + D(aq)$$

A **catalyst** is a substance that speeds up a reaction without being consumed or permanently altered. For example, a catalytic converter in an automobile contains metallic granules that speed up the conversion of unburnt fuel to carbon dioxide and water. A catalyst is indicated by placing its formula above the arrow. Table 8.1 lists the symbols used to describe a chemical reaction.

catalyst A substance that increases the rate of reaction but can be recovered without being permanently changed.

Let's interpret the symbols used in the following chemical equation for the reaction of hydrochloric acid with baking soda:

$$HCl(aq) + NaHCO_3(s) \longrightarrow NaCl(aq) + H_2O(l) + CO_2(g)$$

The formulas and symbols can be read as follows: hydrochloric acid is added to solid sodium hydrogen carbonate, which produces aqueous sodium chloride, liquid water, and carbon dioxide gas. Example Exercise 8.2 further illustrates writing chemical equations.

Table 8.1 Chemical Equation Symbols

Symbol	Translation of Symbol
$\longrightarrow$	produces, yields, gives (points from reactants to products)
+	reacts with, added to, plus (separates two or more reactants or two or more products)
NR	no reaction
(s)	solid substance or precipitate
(l)	liquid substance
(g)	gaseous substance
(aq)	aqueous solution

EXAMPLE EXERCISE 8.2

Write a chemical equation for each of the following chemical reactions.

(a) Iron metal is heated with sulfur powder to produce solid iron(II) sulfide.

(b) Zinc metal reacts with sulfuric acid to give aqueous zinc sulfate and hydrogen gas.

(c) Aqueous solutions of sodium iodide and silver nitrate yield silver iodide precipitate and aqueous sodium nitrate.

(d) Acetic acid reacts with aqueous potassium hydroxide to give aqueous potassium acetate plus water.

Solution: To write the chemical equation, we must provide formulas and symbols for each substance. We can describe each of the above chemical reactions as follows.

(a) $Fe(s) + S(s) \rightarrow FeS(s)$

(b) $Zn(s) + H_2SO_4(aq) \rightarrow ZnSO_4(aq) + H_2(g)$

(c) $NaI(aq) + AgNO_3(aq) \rightarrow AgI(s) + NaNO_3(aq)$

(d) $HC_2H_3O_2(aq) + KOH(aq) \rightarrow KC_2H_3O_2(aq) + H_2O(l)$

SELF-TEST EXERCISE

Write a chemical equation for each of the following chemical reactions.

(a) Metallic aluminum is heated with red phosphorus powder to produce solid aluminum phosphide.

(b) Solid barium carbonate is decomposed with heat to give solid barium oxide and carbon dioxide gas.

(c) Magnesium metal reacts with aqueous tin(II) nitrate to produce metallic tin and an aqueous solution of magnesium nitrate.

(d) Aqueous solutions of lithium chloride and silver nitrate react to give a silver chloride precipitate and aqueous lithium nitrate.

Answers:

(a) $Al(s) + P(s) \rightarrow AlP(s)$; **(b)** $BaCO_3(s) \rightarrow BaO(s) + CO_2(g)$;
(c) $Mg(s) + Sn(NO_3)_2(aq) \rightarrow Sn(s) + Mg(NO_3)_2(aq)$;
(d) $LiCl(aq) + AgNO_3(aq) \rightarrow AgCl(s) + LiNO_3(aq)$

Note: In this section, we learned to write chemical equations based on a description of the chemical reaction. In each of the examples chosen, the equation has equal numbers of atoms on each side and is said to be balanced. In Section 8.3, we will learn how to balance a chemical equation where the numbers of atoms are not the same.

8.3 Balancing Chemical Equations

To become familiar with the general directions for balancing chemical equations by inspection.

To gain practice in balancing chemical equations by inspection.

In the previous section, we translated the description of a chemical reaction into a chemical equation. Because the number of atoms of each element was the same on both sides of the arrow, the chemical equation is said to be a balanced chemical equation.

More often, the number of atoms of each element in the reactants and products is not the same when the equation is first written. In those cases, it becomes necessary to balance the number of atoms of each element on each side of the chemical equation. We can balance a chemical equation by placing a whole number **coefficient** in front of each substance. As an example, hydrogen gas reacts with chlorine gas to give hydrogen chloride, a gas with a sharp odor. We can write the chemical equation as follows:

coefficient A digit placed in front of a chemical formula in order to balance a chemical equation.

$$H_2(g) + Cl_2(g) \longrightarrow HCl(g)$$

Notice that the formula **subscript** for H_2 and Cl_2 is 2, but only one H atom and Cl atom appear in the product. The equation is not balanced. To balance the number of H atoms on both sides of the equation, we place the coefficient 2 in front of the HCl. This gives us

subscript A digit in a chemical formula that represents the number of atoms or ions occurring in a substance.

$$H_2(g) + Cl_2(g) \longrightarrow 2\ HCl(g)$$

By using the reaction coefficient 2, we represent two molecules of HCl. Now we have two H and two Cl atoms on each side of the equation. This equation is now balanced.

Let's try a more difficult example: Aluminum metal is heated with oxygen gas to give solid aluminum oxide. The formula for aluminum oxide is Al_2O_3. The chemical equation is

$$Al(s) + O_2(g) \longrightarrow Al_2O_3(s)$$

Notice that two O atoms appear in the reactants, but three O atoms appear in the product. This reaction is not balanced. To balance the numbers of O atoms, we will use the lowest common multiple of 2 and 3; the number is 6. We place the coefficient 3 in front of the O_2 to give six oxygen atoms in the reactants and the coefficient 2 in front of Al_2O_3 to give six oxygens in the products. This gives us

$$Al(s) + 3\ O_2(g) \longrightarrow 2\ Al_2O_3(s)$$

Now there are equal numbers of O atoms on each side of the equation.

The numbers of Al atoms, however, are not balanced. On the reactant side we have one Al atom. On the product side, we have 2 units of Al_2O_3, for a total of 4 Al atoms. We can place the coefficient 4 in front of the reacting Al metal. This gives a balanced chemical equation:

$$4\ Al(s) + 3\ O_2(g) \longrightarrow 2\ Al_2O_3(s)$$

This method of placing coefficients by systematically analyzing each side of an equation is called *balancing by inspection*. Although there is no formal prescription for balancing an equation by inspection, the following general guidelines will be helpful in balancing an equation.

1. Before placing a coefficient in an equation, check the formula of each substance to make sure the formula subscripts are correct. (*Note:* Subscripts of correct formulas are never changed in order to balance an equation.)
2. Balance each element in the equation by placing a coefficient in front of each substance. Coefficients of 1 are assumed and do not appear in the balanced chemical equation.
   - **(a)** Begin balancing the equation by starting with the most complex chemical formula. Usually, the formula containing the greatest number of atoms is considered the most complex.
   - **(b)** Balance polyatomic ions as a unit rather than treat the elements separately. Occasionally, this is not possible because the polyatomic ion changes, and you will have to balance each element separately.
   - **(c)** The coefficients must be whole numbers. Occasionally, a fractional coefficient is helpful in obtaining a balance. If a fraction is used, multiply all coefficients to clear the denominator. For example, if the coefficient is 1/2, multiply all coefficients in the equation by 2.
3. After balancing the equation, check (√) each symbol of every element (or polyatomic ion) to verify that the coefficients are correct. Proceed back and forth between reactants and products. The procedure for verification is to multiply the coefficient times the subscript of each element; the total should be the same on both sides of the equation.

The following example exercises illustrate these guidelines to balance chemical equations.

EXAMPLE EXERCISE 8.3

Aqueous solutions of calcium acetate and potassium phosphate react to yield an insoluble white precipitate of calcium phosphate and aqueous potassium acetate. Write a balanced chemical equation, given

$$Ca(C_2H_3O_2)_2(aq) + K_3PO_4(aq) \longrightarrow Ca_3(PO_4)_2(s) + KC_2H_3O_2(aq)$$

Solution: Both calcium compounds appear complex. However, $Ca_3(PO_4)_2$ is the more complicated compound because each ion in the formula is followed by a subscript. Let's start with $Ca_3(PO_4)_2$. There are three Ca on the right side of the equation and only one Ca on the left side. Placing the coefficient 3, we have

$$3\ Ca(C_2H_3O_2)_2(aq) + K_3PO_4(aq) \longrightarrow Ca_3(PO_4)_2(s) + KC_2H_3O_2(aq)$$

We now have six $C_2H_3O_2$ on the left side of the equation and only one on the right. Thus, we need the coefficient 6 in front of $KC_2H_3O_2$.

$$3\ Ca(C_2H_3O_2)_2(aq) + K_3PO_4(aq) \longrightarrow Ca_3(PO_4)_2(s) + 6\ KC_2H_3O_2(aq)$$

The coefficient 6 in front of $KC_2H_3O_2$ generates 6 K. We now need 6 K on the left side. Let's place a 2 in front of K_3PO_4. The coefficient 2 makes two PO_4 on the left side, which equals the number in $Ca_3(PO_4)_2$. Finally, we can check off each element and polyatomic ion to verify that the chemical equation is balanced.

$$\overset{\surd}{3}\ \overset{\surd}{Ca}(C_2H_3O_2)_2(aq) + \overset{\surd}{2}\ \overset{\surd}{K_3}PO_4(aq) \longrightarrow \overset{\surd}{Ca_3}(\overset{\surd}{PO_4})_2(s) + \overset{\surd}{6}\ \overset{\surd}{K}C_2H_3O_2(aq)$$

SELF-TEST EXERCISE

Aqueous solutions of aluminum sulfate and barium nitrate react to give a white precipitate of barium sulfate and aqueous aluminum nitrate. Write a balanced chemical equation, given

$$Al_2(SO_4)_3(aq) + Ba(NO_3)_2(aq) \longrightarrow BaSO_4(s) + Al(NO_3)_3(aq)$$

Answer:

$$\overset{\checkmark\ \checkmark}{Al_2(SO_4)_3(aq)} + 3\ \overset{\checkmark\ \checkmark}{Ba(NO_3)_2(aq)} \longrightarrow 3\ \overset{\checkmark\ \checkmark}{BaSO_4(s)} + 2\ \overset{\checkmark\ \checkmark}{Al(NO_3)_3(aq)}$$

EXAMPLE EXERCISE 8.4

Sulfuric acid reacts with aqueous sodium hydroxide to give aqueous sodium sulfate and water. Write a balanced chemical equation given

$$H_2SO_4(aq) + NaOH(aq) \longrightarrow Na_2SO_4(aq) + H_2O(l)$$

Solution: We can start with Na_2SO_4, which contains the same number of atoms as H_2SO_4. There is one SO_4 on each side of the equation, so it is balanced. However, there are two Na on the right side of the equation and one Na on the left. Let's place the coefficient 2 in front of NaOH.

$$H_2SO_4(aq) + 2\ NaOH(aq) \longrightarrow Na_2SO_4(aq) + H_2O(l)$$

Notice that we now have two OH on the left side of the equation and only H_2O on the right side. After reacting, the OH is now part of the H_2O. By placing a 2 in front of the H_2O we can balance both the OH and the two H on the left side. Finally, we check off Na, the polyatomic ion, and the O and H to verify that we have a balanced chemical equation.

$$\overset{\checkmark\ \checkmark}{H_2SO_4(aq)} + 2\ \overset{\checkmark\ \checkmark}{NaOH(aq)} \longrightarrow \overset{\checkmark\ \checkmark}{Na_2SO_4(aq)} + 2\ \overset{\checkmark\ \checkmark}{H_2O(l)}$$

SELF-TEST EXERCISE

Carbonic acid reacts with aqueous ammonium hydroxide to give aqueous ammonium carbonate and water. Write a balanced chemical equation, given

$$H_2CO_3(aq) + NH_4OH(aq) \longrightarrow (NH_4)_2CO_3(aq) + H_2O(l)$$

Answer:

$$\overset{\checkmark\ \checkmark}{H_2CO_3(aq)} + 2\ \overset{\checkmark\ \checkmark}{NH_4OH(aq)} \longrightarrow \overset{\checkmark\ \checkmark}{(NH_4)_2CO_3(aq)} + 2\ \overset{\checkmark\ \checkmark}{H_2O(l)}$$

note Balancing an equation can be a straightforward task. Start with the most complex formula. Then balance each element or polyatomic ion. Proceed systematically, back and forth, between reactants and products. Once you arrive at the final substance to be balanced, the coefficient will be obvious. If you encounter difficulty in the final step, it is probably because the original equation had an incorrect subscript in a chemical formula.

8.4 Classifying Chemical Reactions

To classify a chemical reaction as one of the following types: combination, decomposition, single replacement, double replacement, or neutralization.

There are hundreds of thousands of chemical reactions. How can we attempt to master such a large number of reactions and even try to predict their products? The answer is that we will use the same technique we learned in Chapter 6 to master chemical formulas. We will *classify chemical reactions* and put them into categories. In Chapter 17, we will study oxidation–reduction reactions. In Chapter 20, we will study the reactions of organic compounds. In this chapter, we will study five introductory types of reactions: combination, decomposition, single replacement, double replacement, and neutralization.

combination reaction A type of reaction in which two substances react to produce a single compound.

The first type of reaction, a **combination reaction**, involves two or more substances being combined into a single compound. It is also referred to as a *synthesis reaction*. In the general case, the element or compound A combines with element or compound Z to produce the compound AZ. The chemical equation is

$$A + Z \longrightarrow AZ$$

decomposition reaction A type of reaction in which two or more substances are produced from a single compound.

In the second type of reaction, a **decomposition reaction**, a single compound is broken down into two or more substances. In this case, heat or light is usually applied in order to decompose the compound. In the general case, compound AZ decomposes into the substances A and Z. The chemical equation is

$$AZ \longrightarrow A + Z$$

single-replacement reaction A type of reaction in which a more active element displaces a less active element from a solution or compound.

The third type of reaction, a **single-replacement reaction**, occurs when one element displaces another from a compound or aqueous solution. It is also referred to as a *displacement reaction*, *a substitution reaction*, or simply a *replacement reaction*. The element that is replaced shows less tendency to undergo reaction. It is less active. In the general case, element A replaces element B in the compound BZ to give the compound AZ and element B. The chemical equation is

$$A + BZ \longrightarrow AZ + B$$

double-replacement reaction A type of reaction in which two cations in different compounds exchange anions.

In a **double-replacement reaction**, two compounds exchange anions. It is also referred to as a *metathesis reaction*. The compound AX reacts with BZ to yield the products AZ and BX. We can write the general form of the reaction as

$$AX + BZ \longrightarrow AZ + BX$$

neutralization reaction A type of reaction in which an acid and base react to produce a salt and water.

A fifth type of reaction involves an acid and a base and is called a **neutralization reaction**. An acid, HX, reacts with a base, BOH, to give an ionic compound, BX, and water. The chemical reaction can be written as

$$HX + BOH \longrightarrow BX + HOH$$

If we examine the general form of a neutralization reaction, we see it is actually a special type of double-replacement reaction. The acid and base are simply switching anion partners.

Notice that we wrote water as HOH rather than H_2O. The formula HOH more clearly shows the double-replacement nature of a neutralization reaction. In addition, we will find it is easier to balance neutralization equations if we write water as HOH.

Before studying each reaction type in detail, let's categorize some example reactions.

EXAMPLE EXERCISE 8.5

Classify the following chemical reactions as combination, decomposition, single replacement, double replacement, or neutralization.

(a) Copper metal heated with oxygen gas produces solid copper(II) oxide.

$$2\ Cu(s) + O_2(g) \longrightarrow 2\ CuO(s)$$

(b) Aqueous sodium chromate reacts with aqueous barium chloride to give insoluble barium chromate and aqueous sodium chloride.

$$Na_2CrO_4(aq) + BaCl_2(aq) \longrightarrow BaCrO_4(s) + 2\ NaCl(aq)$$

(c) Nitric acid neutralizes aqueous potassium hydroxide to give aqueous potassium nitrate and water.

$$HNO_3(aq) + KOH(aq) \longrightarrow KNO_3(aq) + H_2O(l)$$

(d) Heating powdered iron(III) carbonate produces solid iron(III) oxide and carbon dioxide gas.

$$Fe_2(CO_3)_3(s) \longrightarrow Fe_2O_3(s) + 3\ CO_2(g)$$

(e) Aluminum metal reacts with aqueous manganese(II) sulfate to give aqueous aluminum sulfate and manganese metal.

$$2\ Al(s) + 3\ MnSO_4(aq) \longrightarrow Al_2(SO_4)_3(aq) + 3\ Mn(s)$$

Solution:

(a) The two elements Cu and O_2 are combining into a single compound; this is an example of a *combination reaction.*

(b) The two compounds Na_2CrO_4 and $BaCl_2$ are exchanging anions; this is an example of a *double-replacement reaction.*

(c) The acid HNO_3 is reacting with the base KOH to form water; this is an example of a *neutralization reaction.*

(d) The compound $Fe_2(CO_3)_3$ is heated and breaks down into a simpler compound and a gas; this is an example of a *decomposition reaction.*

(e) The element Al is displacing Mn from aqueous $MnSO_4$; this is an example of a *single-replacement reaction.*

SELF-TEST EXERCISE

Classify the following chemical equations as one of the following types: combination, decomposition, single replacement, double replacement, or neutralization.

(a) $Zn(s) + CuSO_4(aq) \rightarrow ZnSO_4(aq) + Cu(s)$

(b) $2\ Sr(s) + O_2(g) \rightarrow 2\ SrO(s)$

(c) $Cd(HCO_3)_2(s) \rightarrow CdCO_3(s) + H_2O(g) + CO_2(g)$

(d) $HC_2H_3O_2(aq) + NaOH(aq) \rightarrow NaC_2H_3O_2(aq) + H_2O(l)$

(e) $AgNO_3(aq) + KCl(aq) \rightarrow AgCl(s) + KNO_3(aq)$

Answers:

(a) single-replacement reaction; **(b)** combination reaction;
(c) decomposition reaction; **(d)** neutralization reaction;
(e) double-replacement reaction

8.5 Combination Reactions

OBJECTIVES

To write balanced chemical equations for combination reactions of a metal and oxygen gas.

To write balanced chemical equations for combination reactions of a nonmetal and oxygen gas.

To write balanced chemical equations for combination reactions of a metal and a nonmetal.

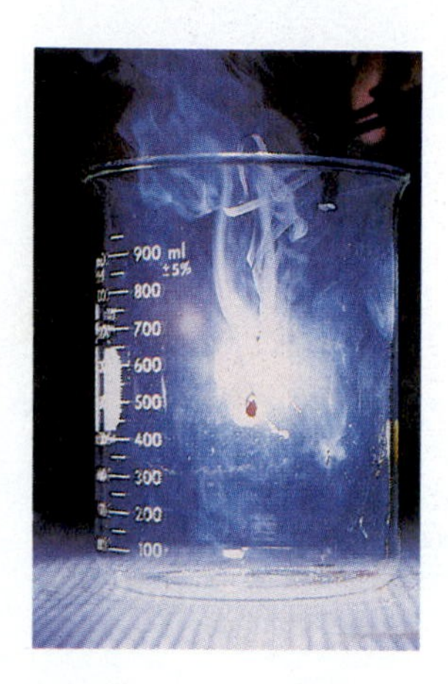

Igniting a strip of magnesium in oxygen gives a white smoke of MgO particles and a bright white light.

There are many examples of combination reactions. Heating two substances often causes the substances to combine into a more complex compound. We will study a few important combination reactions in this section.

Metal and Oxygen Gas

One combination reaction is that of a metal and oxygen gas. Here is the general reaction for a metal combining with oxygen gas. In the example, the metal is magnesium.

1. metal + oxygen gas $\longrightarrow$ metal oxide

$$2\,Mg(s) + O_2(g) \longrightarrow 2\,MgO(s)$$

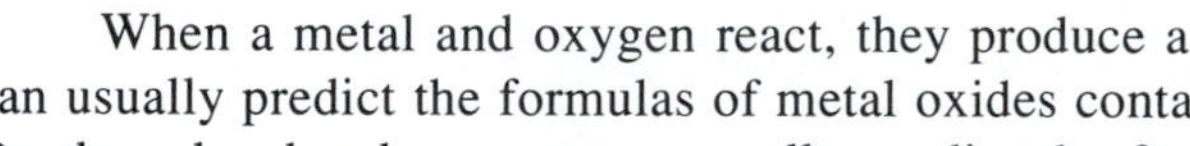

When a metal and oxygen react, they produce a metal oxide compound. We can usually predict the formulas of metal oxides containing a representative metal. On the other hand, we cannot usually predict the formulas of most metal oxides containing a transition metal. For transition metal oxides, we must be given the charge of the metal ion before writing the chemical formula of the metal oxide. Historically, it is interesting to note that Mendeleev relied on the formula of metal oxides for the correct placement of metals within a group in his periodic table.

EXAMPLE EXERCISE 8.6

Write a balanced chemical equation for the following combination reactions.

(a) Zinc metal is heated with oxygen gas in air and combines to form solid zinc oxide.

(b) Bismuth metal is heated with oxygen in air to form solid bismuth(III) oxide.

Solution: A metal and oxygen react to produce a metal oxide.

(a) The formula of zinc oxide is predictable. That is, the zinc ion is Zn^{2+}, so zinc oxide is ZnO. The balanced equation for the reaction is

$$2\,Zn(s) + O_2(g) \longrightarrow 2\,ZnO(s)$$

(b) Bismuth can have a variable charge, but the name bismuth(III) oxide indicates that bismuth has a 3+ charge in the metal oxide product. Therefore, the balanced equation for the reaction is

$$4\,Bi(s) + 3\,O_2(g) \longrightarrow 2\,Bi_2O_3(s)$$

SELF-TEST EXERCISE

Write a balanced chemical equation for the following combination reactions.

(a) Metallic lead shot is heated with oxygen in air and combines to yield solid lead(IV) oxide.

(b) Cobalt metal is heated with oxygen in air to give solid cobalt(III) oxide.

Answers:
(a) $Pb(s) + O_2(g) \longrightarrow PbO_2(s)$; **(b)** $4\ Co(s) + 3\ O_2(g) \longrightarrow 2\ Co_2O_3(s)$

Nonmetal and Oxygen Gas

Another combination reaction is that of a nonmetal and oxygen gas. Here is the general reaction for a nonmetal combining with oxygen gas. In the example, the nonmetal is sulfur.

$$\textbf{2.} \quad \text{nonmetal} + \text{oxygen gas} \longrightarrow \text{nonmetal oxide}$$

$$S(s) \quad + \quad O_2(g) \quad \longrightarrow \quad SO_2(g)$$

Burning powdered sulfur in oxygen gives a colorless gas of SO_2 and an intense blue flame.

Nonmetal oxides are molecular compounds that demonstrate multiple combining capacities. In general, the formula of a nonmetal oxide product is unpredictable. It varies greatly with conditions of temperature and pressure. For example, all the following are formulas for stable oxides of nitrogen: NO, NO_2, N_2O, N_2O_3, N_2O_4, and N_2O_5. To complete and balance an equation of a nonmetal and oxygen gas, the formula or name of the nonmetal oxide must be given.

EXAMPLE EXERCISE 8.7

Write a balanced chemical equation for the following combination reactions.
(a) Carbon is heated with oxygen to produce carbon dioxide gas.
(b) Phosphorus is heated with oxygen and combines to form solid diphosphorus pentaoxide.

Solution: A nonmetal and oxygen react to produce a nonmetal oxide.
(a) The formula of the nonmetal oxide is unpredictable. We are given that the product is carbon dioxide and not carbon monoxide. The balanced equation for the reaction is

$$C(s) + O_2(g) \longrightarrow CO_2(g)$$

(b) The formula for the oxide of phosphorus is not predictable. Again, we must have the name of the nonmetal oxide product. The formula for diphosphorus pentaoxide is P_2O_5. The balanced equation is

$$4\ P(s) + 5\ O_2(g) \longrightarrow 2\ P_2O_5(s)$$

SELF-TEST EXERCISE

Write a balanced chemical equation for the following combination reactions.
(a) Nitrogen gas is heated with oxygen to give dinitrogen trioxide gas.
(b) Chlorine gas is heated with oxygen to give dichlorine oxide gas.

Answers:
(a) $2\ N_2(g) + 3\ O_2(g) \rightarrow 2\ N_2O_3(g)$; **(b)** $2\ Cl_2(g) + O_2(g) \rightarrow 2\ Cl_2O(g)$

Metal and Nonmetal

Another combination reaction is that of a metal and a nonmetal. Here is the general reaction for a metal combining with a nonmetal. In the example the metal is sodium and the nonmetal is chlorine.

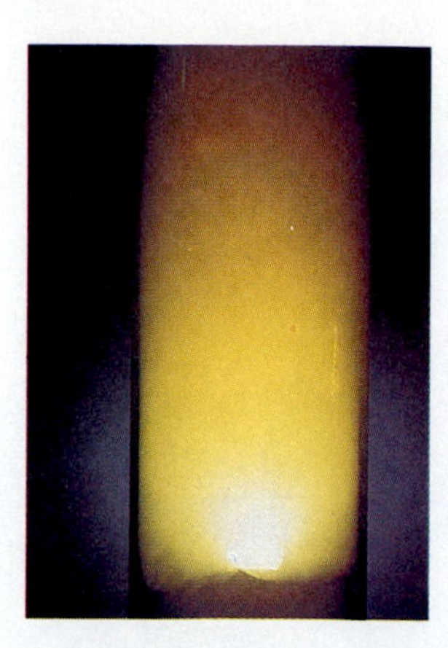

Igniting sodium metal in chlorine gas produces white NaCl smoke and a colorful yellow light.

3. metal + nonmetal $\longrightarrow$ ionic compound

$$2\ Na(s) + Cl_2(g) \longrightarrow 2\ NaCl(s)$$

When a metal and a nonmetal react, they produce an ionic compound. The formula of an ionic compound containing a representative metal is often predictable. The formulas of compounds containing a transition metal are not usually predictable.

EXAMPLE EXERCISE 8.8

Write a balanced chemical equation for the following combination reactions.

(a) Aluminum metal is heated with sulfur and gives a solid product.

(b) Chromium metal is heated with iodine gas and produces powered chromium(III) iodide.

Solution: A metal and nonmetal react to produce an ionic compound.

(a) The formula of the product is predictable. Aluminum combines with sulfur to give aluminum sulfide, Al_2S_3. The balanced equation for the reaction is

$$2\ Al(s) + 3\ S(s) \longrightarrow Al_2S_3(s)$$

(b) Chromium is a transition metal, so we cannot predict the formula for the product. We are given, however, the name of the compound, chromium(III) iodide; the formula is CrI_3. The equation is

$$2\ Cr(s) + 3\ I_2(g) \longrightarrow 2\ CrI_3(s)$$

SELF-TEST EXERCISE

Write a balanced chemical equation for the following combination reactions.

(a) Calcium is heated with fluorine gas to yield solid calcium fluoride.

(b) Manganese metal reacts with bromine vapor to give crystalline manganese(IV) bromide.

Answers:

(a) $Ca(s) + F_2(g) \rightarrow CaF_2(s)$; **(b)** $Mn(s) + 2\ Br_2(g) \rightarrow MnBr_4(s)$

8.6 Decomposition Reactions

OBJECTIVES

To write balanced chemical equations for the decomposition reactions of metal hydrogen carbonate compounds.

To write balanced chemical equations for the decomposition reactions of metal carbonate compounds.

To write balanced chemical equations for the decomposition reactions of miscellaneous compounds that evolve oxygen gas.

Heating a compound often causes the compound to decompose into simpler substances. We will study a few of the most important decomposition reactions in this section.

Metal Hydrogen Carbonates

A metal hydrogen carbonate undergoes a decomposition reaction when it is heated. For example, you may be aware that baking soda is a natural fire extinguisher. That is, heat from a fire decomposes baking soda (sodium bicarbonate) and releases carbon dioxide gas. Since carbon dioxide is more dense than air, it can smother a fire by excluding oxygen.

Here is the general reaction for the decomposition of a metal hydrogen carbonate. In the example, sodium hydrogen carbonate is decomposed by heating.

1. metal hydrogen carbonate $\longrightarrow$ metal carbonate + water (steam) + carbon dioxide

$$2\ NaHCO_3(s) \longrightarrow Na_2CO_3(s) + H_2O(g) + CO_2(g)$$

During the decomposition reaction of a metal hydrogen carbonate, the ionic charge of the metal does not change. Therefore, the formulas of the metal carbonate products are predictable. Even transition metal hydrogen carbonates are predictable. If nickel(II) hydrogen carbonate decomposes, one of the products is nickel(II) carbonate.

EXAMPLE EXERCISE 8.9

Write a balanced chemical equation for the following decomposition reactions.

(a) Lithium hydrogen carbonate decomposes upon heating.

(b) Lead(II) hydrogen carbonate decomposes upon heating.

Solution: A metal hydrogen carbonate decomposes with heat to give a metal carbonate, water, and carbon dioxide gas.

(a) All the formulas are predictable, including the product, Li_2CO_3. The balanced equation for the reaction is

$$2\ LiHCO_3(s) \longrightarrow Li_2CO_3(s) + H_2O(g) + CO_2(g)$$

(b) Although the ionic charge on lead is variable, we are given that lead(II) hydrogen carbonate is the reactant. Therefore, the product is lead(II) carbonate, $PbCO_3$. The balanced equation for the reaction is

$$Pb(HCO_3)_2(s) \longrightarrow PbCO_3(s) + H_2O(g) + CO_2(g)$$

SELF-TEST EXERCISE

Write a balanced chemical equation for the following decomposition reactions.

(a) Barium hydrogen carbonate is decomposed by heating the compound.

(b) Copper(I) hydrogen carbonate is decomposed with heat.

Answers:

(a) $Ba(HCO_3)_2(s) \rightarrow BaCO_3(s) + H_2O(g) + CO_2(g)$;

(b) $2\ CuHCO_3(s) \rightarrow Cu_2CO_3(s) + H_2O(g) + CO_2(g)$

Metal Carbonates

After a metal hydrogen carbonate decomposes into a metal carbonate, the metal carbonate can further decompose with prolonged heating. The carbonate decomposes into a metal oxide while releasing more carbon dioxide gas.

Here is the general reaction for the decomposition of a metal carbonate. In this example, calcium carbonate is decomposed by heating.

2. metal carbonate $\longrightarrow$ metal oxide + carbon dioxide

$$CaCO_3(s) \longrightarrow CaO(s) + CO_2(g)$$

During the decomposition reaction of a metal carbonate compound, the ionic charge on the metal does not change. Therefore, we can predict the formula for the products from a carbonate decomposition. If cobalt(II) carbonate is decomposed, cobalt(II) oxide is produced along with CO_2.

EXAMPLE EXERCISE 8.10

Write a balanced chemical equation for the following decomposition reactions.

(a) Magnesium carbonate decomposes upon heating.

(b) Copper(I) carbonate decomposes upon heating.

Solution: A metal carbonate decomposes with heat to a metal carbonate and carbon dioxide gas.

(a) All the formulas are predictable, including the metal oxide, MgO. The balanced equation for the reaction is

$$MgCO_3(s) \longrightarrow MgO(s) + CO_2(g)$$

(b) The ionic charge on copper can either be 1+ or 2+. Since the reactant is copper(I) carbonate, the product is copper(I) oxide, Cu_2O. Thus, the balanced equation for the reaction is

$$Cu_2CO_3(s) \longrightarrow Cu_2O(s) + CO_2(g)$$

SELF-TEST EXERCISE

Write a balanced chemical equation for the following decomposition reactions.

(a) Aluminum carbonate is decomposed by heating the compound.

(b) Iron(II) carbonate is decomposed with heat.

Answers:

(a) $Al_2(CO_3)_3(s) \rightarrow Al_2O_3(s) + 3\ CO_2(g)$; **(b)** $FeCO_3(s) \rightarrow FeO(s) + CO_2(g)$

Miscellaneous Oxygen-Containing Compounds

Under the proper experimental conditions, most compounds containing oxygen will decompose and release oxygen gas.

3. oxygen-containing compounds $\longrightarrow$ oxygen gas

These decomposition reactions, however, are not generally predictable. For compounds that decompose and liberate oxygen gas, you must be given specific information about the reactants and products. For example, given that potassium chlorate decomposes to give potassium chloride and oxygen gas, we can write

$$2\ KClO_3(s) \longrightarrow 2\ KCl(s) + 3\ O_2(g)$$

Example Exercise 8.11 gives additional practice in illustrating this type of reaction.

EXAMPLE EXERCISE 8.11

Write a balanced chemical equation for the following decomposition reactions.

(a) Sodium nitrate is a white crystalline substance that decomposes with heat to give solid sodium nitrite and oxygen gas.

(b) Manganese(II) sulfate is a pink powder that decomposes with heat to give solid manganese(II) oxide and sulfur trioxide gas.

Solution: There is no general format for these reactions; however, we are given the names of the products.

(a) We can write the formula for each substance and then balance the chemical equation. The balanced equation for the reaction is

$$2\ NaNO_3(s) \longrightarrow 2\ NaNO_2(s) + O_2(g)$$

(b) The decomposition of manganese(II) sulfate gives MnO and SO_3. Thus, the balanced equation for the reaction is

$$MnSO_4(s) \longrightarrow MnO(s) + SO_3(g)$$

Heating orange HgO powder produces colorless oxygen gas and beads of silvery mercury.

SELF-TEST EXERCISE

Write a balanced chemical equation for the following decomposition reactions.

(a) Mercuric oxide, an orange powder, is decomposed with heat to give small droplets of liquid mercury and oxygen gas.

(b) An aqueous solution of hydrogen peroxide, H_2O_2, decomposes with heat to give water and oxygen gas.

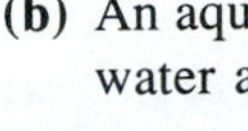

Answers:

(a) $2\ HgO(s) \rightarrow 2\ Hg(l) + O_2(g)$; **(b)** $2\ H_2O_2(aq) \rightarrow 2\ H_2O(l) + O_2(g)$

8.7 Activity Series

OBJECTIVES

To predict whether or not a replacement reaction occurs by referring to the activity series for metals.

When a metal undergoes a replacement reaction, it displaces another element from a compound or aqueous solution. The metal displaces the other element because it has a greater tendency to undergo a reaction. In other words, it is more chemically active than the element it replaces. Thus, if element A is more active than element B in the compound BZ, element A will replace B from the compound. We can show the equation as

$$A + BZ \longrightarrow AZ + B$$

The activity of an element is measured as it competes with another element in a replacement reaction. Therefore, if we compare several metals, we can establish an activity series. An **activity series**, sometimes called an electromotive series, is a sequence of elements arranged according to their ability to undergo reaction. Metals that are more active appear first in the activity series. Metals that are less active appear later in the series. A listing of the relative activity of several metals is as follows:

activity series A relative order of elements arranged by their ability to undergo reaction; also called an electromotive series.

Activity Series

Li > K > Ba > Sr > Ca > Na > Mg > Al > Mn > Zn > Fe > Cd > Co > Ni > Sn > Pb > (H) > Cu > Ag > Hg > Au

As you see, lithium metal precedes potassium in the series, which precedes barium, and so on. Although hydrogen is not a metal, it is included in the series as a reference (H). Metals that precede (H) in the series react with an acid by displacing hydrogen gas from the acid solution. Metals that follow (H) in the series do not react with acids under the conditions we will study in this chapter. That is, the metals Cu, Ag, Au, and Hg do not react with acids.

Consider the reaction of zinc metal and an aqueous solution of iron(II) sulfate:

$$Zn(s) + FeSO_4(aq) \longrightarrow ZnSO_4(aq) + Fe(s)$$

Zinc precedes iron in the activity series, so Zn displaces Fe from aqueous solution. Now consider the reaction of iron metal and copper(II) sulfate solution.

$$Fe(s) + CuSO_4(aq) \longrightarrow FeSO_4(aq) + Cu(s)$$

Dipping an iron nail into a $CuSO_4$ solution produces a deposit of reddish-brown copper metal on the nail.

Iron precedes copper in the series, so Fe displaces Cu from aqueous solution. If you perform the reaction in the laboratory, you see a reddish-brown copper deposit on the iron metal. It is this type of experiment that establishes the order of metals in the electromotive series; that is, Zn > Fe > Cu.

Now consider the reaction of copper metal and an aqueous solution of iron(II) sulfate. In this case, since copper follows iron in the activity series, Cu cannot displace Fe from solution. If we put copper wire into a solution of iron(II) sulfate, we observe

$$Cu(s) + FeSO_4(aq) \longrightarrow \text{NR (no reaction)}$$

Consider another experiment. Let's drop zinc granules into hydrochloric acid. The equation is

$$Zn(s) + 2\ HCl(aq) \longrightarrow ZnCl_2(aq) + H_2(g)$$

Zinc precedes (H) in the activity series, so Zn displaces H_2 gas from an acid solution. If we observe the reaction, we see tiny bubbles being released from the zinc granules. Now consider the reaction of copper metal and hydrochloric acid.

$$Cu(s) + HCl(aq) \longrightarrow \text{NR}$$

Copper is after (H) in the activity series, so Cu cannot displace H_2 gas from an acid solution. If we observe the copper metal, there is no evidence for reaction.

active metal A metal that is sufficiently active to react with water at 25°C.

There are a few metals that are so reactive that they react directly with water at 25°C. These metals, called the **active metals**, include most of the metals in Groups IA/1 and IIA/2. Specifically, Li, Na, K, Rb, Cs, Ca, Sr, and Ba react with water. From the activity series, we have

Active Metals:

Li > K > Ba > Sr > Ca > Na

Consider an experiment where calcium metal is dropped into water at 25°C. Since calcium is an active metal, it reacts with cold water according to the following equation.

$$Ca(s) + 2\ H_2O(l) \longrightarrow Ca(OH)_2(aq) + H_2(g)$$

Example Exercise 8.12 further illustrates the predictability of metals to undergo reaction with aqueous solutions, aqueous acids, and water.

Dropping a small piece of calcium metal into water gives bubbles of hydrogen gas.

EXAMPLE EXERCISE 8.12

Predict whether or not a reaction occurs for each of the following.
(a) Aluminum foil is put into iron(II) sulfate solution.
(b) Lead shot is dropped into sodium nitrate solution.
(c) Magnesium ribbon is placed in acetic acid.
(d) Lithium metal is dropped into water.

Solution: We will refer to the activity series for the relative positions of each metal.
(a) Aluminum precedes iron in the activity series: Al > Fe. Therefore, a reaction occurs, and Al displaces Fe from solution.
(b) Lead follows sodium in the series: Na > Pb. Therefore, there is no reaction.
(c) Magnesium precedes hydrogen in the series: Mg > (H). Therefore, a reaction occurs, and Mg produces H_2 gas that evolves from solution.
(d) Lithium is an active metal in Group IA/1. Therefore, lithium reacts with water and produces hydrogen gas.

SELF-TEST EXERCISE

Predict whether or not a reaction occurs for each of the following.
(a) Cadmium metal is put into lead(II) nitrate solution.
(b) Gold foil is dropped into dilute nitric acid.
(c) Zinc granules are added to hydrochloric acid.
(d) Chromium metal is dropped into water.

Answer:
(a) Yes; cadmium precedes lead in the activity series.
(b) No; gold follows (H) in the activity series.
(c) Yes; zinc precedes (H) in the activity series.
(d) No; chromium is not an active metal.

In this brief introduction to the concept of an activity series, we have only considered the activity of metals. However, the ability of nonmetals to undergo reaction also gives rise to an activity series. For example, within the halogens the order of activity is F > Cl > Br > I. Consider the reaction

$$Cl_2(g) + 2\ NaBr(aq) \longrightarrow 2\ NaCl(aq) + Br_2(l)$$

This replacement reaction occurs because Cl precedes Br in the series; thus, the more active nonmetal displaces the less active one from solution. Conversely, Cl follows F in the series, so it is less reactive and we have

$$Cl_2(g) + NaF(aq) \longrightarrow NR$$

8.8 Single-Replacement Reactions

OBJECTIVES

To complete and balance single-replacement reactions of a metal in an aqueous solution of an ionic compound.

To complete and balance single-replacement reactions of a metal in an acid.

To complete and balance single-replacement reactions of an active metal in water.

There are many examples of single-replacement reactions in which a more active element replaces a less active element from solution. In general, the activity of an element follows the trend in metallic character. Metals on the far left of the periodic table are most active. The coinage metals in Group IB/11 (Cu, Ag, Au) are least active. To find the exact position of an element, however, we need to refer to the activity series.

Metal and Aqueous Solution

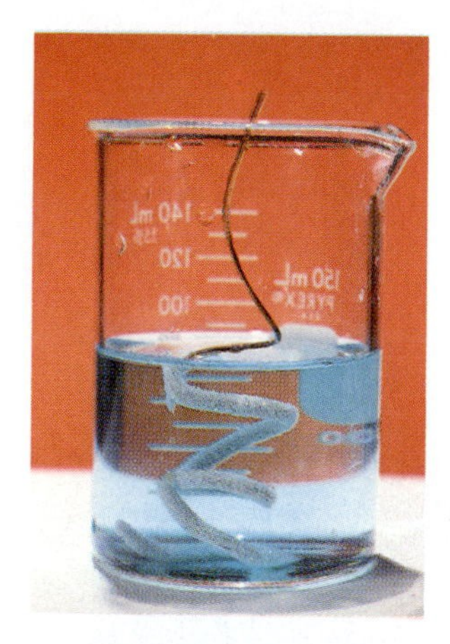

Inserting a coil of copper wire into a $AgNO_3$ solution produces white crystals of silver metal on the wire.

One type of single-replacement reaction is that of a metal and aqueous solution of an ionic compound. Here is the general reaction for a metal replacing an element from aqueous solution. In the example, zinc reacts with silver nitrate.

1. $$\text{metal}_1 + \text{aqueous solution}_1 \longrightarrow \text{metal}_2 + \text{aqueous solution}_2$$

$$Zn(s) + 2\ AgNO_3(aq) \longrightarrow 2\ Ag(s) + Zn(NO_3)_2(aq)$$

First, refer to the activity series to see which metal is more active. Zinc is above silver in the series: Zn > Ag. Therefore, in the equation, Zn metal displaces Ag metal from aqueous solution.

Now let's consider the reverse reaction, that is, the reaction of zinc nitrate and silver metal. Since Ag metal is less active than Zn metal, we can correctly predict that there is no reaction (NR).

$$Ag(s) + Zn(NO_3)_2(aq) \longrightarrow \text{NR}$$

Metal and Aqueous Acid Solution

Now look at a second type of single-replacement reaction in which a metal and an aqueous acid react to produce hydrogen gas and an aqueous solution of an ionic compound.

2. $$\text{metal} + \text{aqueous acid} \longrightarrow \text{aqueous solution} + \text{hydrogen gas}$$

Consider the reaction of nickel metal and hydrochloric acid. First, refer to the activity series to check the position of nickel. Nickel is above hydrogen in the series: Ni > (H). Therefore, Ni displaces hydrogen gas from the acid. If we assume that the reaction produces nickel(II) chloride, we can write the balanced chemical equation.

$$Ni(s) + 2\ HCl(aq) \longrightarrow NiCl_2(aq) + H_2(g)$$

Active Metal and Water

The third type of single-replacement reaction involves an active metal and water. An active metal reacts with water to produce a metal hydroxide and hydrogen gas. The general equation for the reaction is

3. active metal + water $\longrightarrow$ metal hydroxide + hydrogen gas

Consider the reaction of sodium metal in water. The balanced equation for the reaction is

$$2\ Na(s) + 2\ H_2O(l) \longrightarrow 2\ NaOH(aq) + H_2(g)$$

Example Exercise 8.13 illustrates writing balanced chemical equations for the three types of single-replacement reactions.

EXAMPLE EXERCISE 8.13

Complete and balance the following single-replacement reactions.
(a) Nickel metal is placed in a tin(II) sulfate solution.
(b) Cobalt metal is put in a cadmium(II) nitrate solution.
(c) Manganese chips are added to sulfuric acid.
(d) A small, soft chunk of potassium is dropped into water.

Solution: First, we will refer to the activity series for the relative positions of each metal. Then we will write the equations.

(a) Nickel is above tin in the activity series: Ni > Sn. Therefore, a reaction occurs, and Ni displaces Sn from solution.

$$Ni(s) + SnSO_4(aq) \longrightarrow NiSO_4(aq) + Sn(s)$$

(b) Cobalt is below cadmium in the series: Cd > Co. Therefore, there is no reaction.

$$Co(s) + Cd(NO_3)_2(aq) \longrightarrow NR$$

(c) Manganese is above hydrogen in the series: Mn > (H). Therefore, a reaction occurs, and Mn produces H_2 gas that bubbles from solution.

$$Mn(s) + H_2SO_4(aq) \longrightarrow MnSO_4(aq) + H_2(g)$$

(d) Potassium is a Group IA/1 active metal. Therefore, a reaction occurs, and K reacts to evolve H_2 gas from solution.

$$2\ K(s) + 2\ H_2O(l) \longrightarrow 2\ KOH(aq) + H_2(g)$$

SELF-TEST EXERCISE

Complete and balance the following single-replacement reactions (refer to the activity series).
(a) Nickel metal is placed in a silver nitrate solution.
(b) Gold metal is placed in a silver nitrate solution.
(c) A chunk of cadmium metal is dropped into hydrochloric acid.
(d) A small piece of strontium metal is dropped into water.

Answers:
(a) $Ni(s) + 2\ AgNO_3(aq) \rightarrow 2\ Ag(s) + Ni(NO_3)_2(aq)$;
(b) $Au(s) + AgNO_3(aq) \rightarrow NR$; **(c)** $Cd(s) + 2\ HCl(aq) \rightarrow CdCl_2(aq) + H_2(g)$;
(d) $Sr(s) + 2\ H_2O(l) \rightarrow Sr(OH)_2(aq) + H_2(g)$

Placing an iron nail into a sulfuric acid solution yields bubbles of hydrogen gas.

When a metal undergoes a single-replacement reaction, it is not always possible to predict the resulting ionic charge on the metallic cation. For example, when copper metal reacts with aqueous silver nitrate solution, the resulting copper compound may contain Cu^{+} or Cu^{2+}. That is, the product may be $CuNO_3$ or $Cu(NO_3)_2$. In the absence of specific information, you can assume that either compound is possible.

8.9 Solubility Rules

OBJECTIVES

To predict whether or not a compound dissolves in water given the solubility rules for ionic compounds.

Before studying the next type of chemical reaction, double-replacement reactions, let's consider the driving force for reactants to become products. There is a fundamental principle that a reaction proceeds spontaneously if the products have less energy (that is, are more stable) than the reactants. In other words, a reaction occurs because the products are more stable than the reactants. In an aqueous solution, a substance that does not dissolve is very stable. An insoluble substance formed in solution is called a precipitate. Thus, the formation of a precipitate drives a reaction toward completion.

In a double-replacement reaction, two ionic compounds in aqueous solution may react to form a precipitate. If we know that an insoluble compound is produced, we can predict that the reaction has a driving force for completion. Even though *every* insoluble compound is slightly soluble in aqueous solution, we will consider a slightly soluble compound as insoluble.

Table 8.2 lists the general rules for the solubility of solid ionic compounds. There are numerous exceptions to the general rules, a few of which are indicated in the table.

Now let's apply the solubility rules to selected compounds and state whether or not the compound is soluble in water.

Table 8.2 Solubility Rules for Ionic Compounds

Compounds containing the following ions are generally *soluble* in water:

1. alkali metal ions and ammonium ions, Li^{+}, Na^{+}, K^{+}, NH_4^{+}
2. acetate ion, $C_2H_3O_2^{-}$
3. nitrate ion, NO_3^{-}
4. halide ions (X), Cl^{-}, Br^{-}, I^{-} (AgX, Hg_2X_2, and PbX_2 are insoluble exceptions)
5. sulfate ion, SO_4^{2-} ($SrSO_4$, $BaSO_4$, and $PbSO_4$ are insoluble exceptions)

Compounds containing the following ions are generally *insoluble* in water:

6. carbonate ion, CO_3^{2-} (see rule 1 exceptions, which are soluble)
7. chromate ion, CrO_4^{2-} (see rule 1 exceptions, which are soluble)
8. phosphate ion, PO_4^{3-} (see rule 1 exceptions, which are soluble)
9. sulfide ion, S^{2-} (CaS, SrS, BaS, and rule 1 exceptions are soluble)
10. hydroxide ion, OH^{-} [$Ca(OH)_2$, $Sr(OH)_2$, $Ba(OH)_2$, and rule 1 exceptions are soluble]

EXAMPLE EXERCISE 8.14

State whether the following compounds are soluble or insoluble in water.

(a) sodium sulfate, Na_2SO_4
(b) lead(II) acetate, $Pb(C_2H_3O_2)_2$
(c) aluminum nitrate, $Al(NO_3)_3$
(d) mercury(I) chloride, Hg_2Cl_2
(e) barium sulfate, $BaSO_4$
(f) calcium carbonate, $CaCO_3$
(g) potassium chromate, K_2CrO_4
(h) zinc phosphate, $Zn_3(PO_4)_2$
(i) ammonium sulfide, $(NH_4)_2S$
(j) iron(III) hydroxide, $Fe(OH)_3$

Solution: We will refer to the solubility rules for ionic compounds (Table 8.2).

(a) Sodium sulfate contains the alkali metal ion, Na^+. According to rule 1, Na_2SO_4 is *soluble*.

(b) Lead(II) acetate contains the acetate ion, $C_2H_3O_2^-$. According to rule 2, $Pb(C_2H_3O_2)_2$ is *soluble*.

(c) Aluminum nitrate contains the nitrate ion, NO_3^-. According to rule 3, $Al(NO_3)_3$ is *soluble*.

(d) Mercury(I) chloride contains the chloride ion, Cl^-. However, according to a rule 4 exception, Hg_2Cl_2 is *insoluble*.

(e) Barium sulfate contains the sulfate ion, SO_4^{2-}. However, according to a rule 5 exception, $BaSO_4$ is *insoluble*.

(f) Calcium carbonate contains the carbonate ion, CO_3^{2-}. According to rule 6, $CaCO_3$ is *insoluble*.

(g) Potassium chromate contains the chromate ion, CrO_4^{2-}. However, according to the rule 7 exception, K_2CrO_4 is *soluble*.

(h) Zinc phosphate contains the phosphate ion, PO_4^{3-}. According to rule 8, $Zn_3(PO_4)_2$ is *insoluble*.

(i) Ammonium sulfide contains the ammonium ion, NH_4^+. According to rule 1, $(NH_4)_2S$ is *soluble*.

(j) Iron(III) hydroxide contains the hydroxide ion, OH^-. According to rule 10, $Fe(OH)_3$ is *insoluble*.

SELF-TEST EXERCISE

State whether the following compounds are soluble or insoluble in water. (Refer to the solubility rules for ionic compounds.)

(a) iron(II) acetate, $Fe(C_2H_3O_2)_2$
(b) magnesium carbonate, $MgCO_3$
(c) lead(II) sulfide, PbS
(d) mercury(II) bromide, $HgBr_2$

Answers:
(a) soluble (rule 2); **(b)** insoluble (rule 6); **(c)** insoluble (rule 9); **(d)** soluble (rule 4)

8.10 Double-Replacement Reactions

To complete and balance double-replacement reactions of two aqueous solutions of ionic compounds.

In a double-replacement reaction, two ionic compounds in aqueous solution switch anions to produce two new compounds. The general form of the reaction is

$$AX(aq) + BZ(aq) \longrightarrow AZ(aq) + BX(s)$$

As we mentioned in Section 8.9, the formation of an insoluble substance drives a reaction toward completion. In the reaction of AX and BZ, we can usually assume that either AZ or BX is a precipitate. If both AZ and BX are soluble, there is no driving force and therefore no reaction.

The reaction of an aqueous acid and aqueous base is also an example of a double-replacement reaction. The reactions of acids and bases are so important, however, that we treat them as a special case in Section 8.11.

Consider the double-replacement reaction of two aqueous solutions of ionic compounds.

$$\text{aqueous solution}_1 + \text{aqueous solution}_2 \longrightarrow \text{precipitate} + \text{aqueous solution}_3$$

In the laboratory, when we add a few drops of sodium hydroxide solution to aqueous copper(II) sulfate, we get a light-blue precipitate in a blue solution. The two ionic compounds switch anions to give copper(II) hydroxide and sodium sulfate. If we refer to solubility rule 10, we find that copper(II) hydroxide is insoluble. We can write the balanced chemical equation as follows:

$$CuSO_4(aq) + 2\ NaOH(aq) \longrightarrow Cu(OH)_2(s) + Na_2SO_4(aq)$$

Now, consider the reaction of aqueous solutions of calcium iodide and silver nitrate. If we add the two solutions together, we get a yellow precipitate in solution. The two aqueous ionic compounds switch anions to give calcium nitrate and silver iodide. If we refer to the solubility rule 4 exception, we find that silver iodide is insoluble. We can write the balanced chemical equation as follows:

$$CaI_2(aq) + 2\ AgNO_3(aq) \longrightarrow Ca(NO_3)_2(aq) + 2\ AgI(s)$$

Notice that silver iodide is written as $AgI(s)$; the (s) indicates that AgI is insoluble in water. Example Exercise 8.15 illustrates writing balanced chemical equations for double-replacement reactions.

EXAMPLE EXERCISE 8.15

Complete and balance the following double-replacement reactions.

(a) An aqueous solution of barium chloride is added dropwise into a potassium chromate solution.

(b) An aqueous solution of lead(II) nitrate is added dropwise into an iron(III) sulfate solution.

(c) An aqueous solution of strontium acetate is added dropwise into a lithium hydroxide solution.

Solution: For double-replacement reactions, we switch anions for the two compounds and check the solubility rules for an insoluble compound.

(a) Barium chloride and potassium chromate give barium chromate and potassium chloride. According to solubility rule 7, we find that barium chromate is insoluble. The balanced equation is

$$BaCl_2(aq) + K_2CrO_4(aq) \longrightarrow BaCrO_4(s) + 2\ KCl(aq)$$

(b) Lead(II) nitrate and iron(III) sulfate give lead(II) sulfate and iron(III) nitrate. According to solubility rule 5 exception, we find that lead(II) sulfate is insoluble. The balanced chemical equation is

$$3\ Pb(NO_3)_2(aq) + Fe_2(SO_4)_3(aq) \longrightarrow 3\ PbSO_4(s) + 2\ Fe(NO_3)_3(aq)$$

(c) Strontium acetate and lithium hydroxide give strontium hydroxide and lithium acetate. The solubility rules indicate all the compounds are soluble. Therefore, the equation is written

$$Sr(C_2H_3O_2)_2(aq) + LiOH(aq) \longrightarrow NR$$

SELF-TEST EXERCISE

Complete and balance the following double-replacement reactions (refer to the solubility rules for ionic compounds).

(a) Aqueous solutions of zinc sulfate and sodium carbonate are added together.

(b) Aqueous solutions of manganese(II) nitrate and potassium sulfide are added together.

Answers:

(a) $ZnSO_4(aq) + Na_2CO_3(aq) \rightarrow ZnCO_3(s) + Na_2SO_4(aq)$;

(b) $Mn(NO_3)_2(aq) + K_2S(aq) \rightarrow MnS(s) + 2\ KNO_3(aq)$

When a double-replacement reaction takes place, the ionic charge of a metal cation does not change. For example, if a reactant compound contains Fe^{2+}, the product compound contains Fe^{2+}. In the reaction of iron(II) sulfate to give iron(II) carbonate, the iron has a 2+ charge in both the reactant ($FeSO_4$) and in the product ($FeCO_3$). Even though transition metals have cations with charges that vary, you can assume that the ionic charge does not change during a double-replacement reaction.

8.11 Neutralization Reactions

OBJECTIVES

To complete and balance neutralization reactions of an acid and a base.

A neutralization reaction is a special case of a double-replacement reaction. In a neutralization reaction, an acid and a base react to give a salt and water. An **acid** is any substance that releases hydrogen ions in solution, and a **base** is a substance that releases hydroxide ions. The resulting **salt** is composed of the cation from the base and the anion from the acid. The general form of the reaction is

salt an ionic compound resulting from a neutralization reaction.

$$\mathbf{HX}(aq) + \mathbf{BOH}(aq) \longrightarrow \mathbf{BX}(aq) + \mathbf{HOH}(l)$$

Notice that we have written water as HOH to emphasize the switching of ions. We will also discover that writing water in this form is helpful in balancing equations for neutralization reactions.

In Section 8.9, we stated that reactions need a driving force to convert reactants to products. In Section 8.10, we saw that in a double-replacement reaction the driving force is usually the formation of a precipitate. In a neutralization reaction, the driving force for the reaction is the formation of water molecules.

Consider the neutralization reaction of an aqueous acid by an aqueous base.

$$\text{aqueous acid} + \text{aqueous base} \longrightarrow \text{aqueous salt} + \text{water}$$

In the laboratory, when we add hydrochloric acid to sodium hydroxide solution, we get sodium chloride and water. Other than a slight warming, there is no evidence of reaction. We can write the balanced chemical equation as follows:

$$HCl(aq) + NaOH(aq) \longrightarrow NaCl(aq) + HOH(l)$$

Now consider the reaction of sulfuric acid and potassium hydroxide solution. To provide evidence for reaction, we can add a drop of phenolphthalein indicator to the acid. The indicator is colorless in acid and pink in base. As we add the potassium hydroxide base, we see flashes of pink and eventually the solution turns pink. With an indicator, we can see a color change, which is evidence for a reaction. The balanced chemical equation for the reaction is

$$H_2SO_4(aq) + 2\ KOH(aq) \longrightarrow K_2SO_4(aq) + 2\ HOH(l)$$

It is also possible to neutralize an acid with a carbonate, hydrogen carbonate, or oxide. Since these neutralization reactions do not fit the above format, these special cases have not been included in this section. Example Exercise 8.16 provides additional practice in writing balanced chemical equations for ordinary neutralization reactions.

EXAMPLE EXERCISE 8.16

Complete and balance the following neutralization reactions.

(a) Barium hydroxide solution is added to nitric acid containing a drop of phenolphthalein indicator.

(b) Sodium hydroxide solution is added to phosphoric acid until the acid is completely neutralized.

Solution: Each of these neutralization reactions produce a salt and water.

(a) Barium hydroxide completely neutralizes the nitric acid when the solution turns pink. The products are barium nitrate and water. The balanced equation requires two HNO_3 for every $Ba(OH)_2$:

$$Ba(OH)_2(aq) + 2\ HNO_3(aq) \longrightarrow Ba(NO_3)_2(aq) + 2\ HOH(l)$$

(b) Sodium hydroxide completely neutralizes phosphoric acid to give sodium phosphate and water. The balanced equation requires three NaOH for every one H_3PO_4:

$$H_3PO_4(aq) + 3\ NaOH(aq) \longrightarrow Na_3PO_4(aq) + 3\ HOH(l)$$

SELF-TEST EXERCISE

Complete and balance the following neutralization reactions.

(a) Chloric acid is added into strontium hydroxide solution.

(b) Sulfuric acid is added into ammonium hydroxide solution.

Answers:

(a) $2\ HClO_3(aq) + Sr(OH)_2(aq) \rightarrow Sr(ClO_3)_2(aq) + 2\ HOH(l)$;

(b) $H_2SO_4(aq) + 2\ NH_4OH(aq) \rightarrow (NH_4)_2SO_4(aq) + 2\ HOH(l)$

Although we have generally assumed that chemical reactions are irreversible, frequently this is not the case. Many reactions are reversible. That is, they show a tendency to at least partially revert back to reactants. After a while, the rate of the reverse reaction takes place as fast as the original forward reaction. When this happens, a reaction is said to be in a state of *chemical equilibrium*.

For those reactions that have a strong driving force, such as the formation of a precipitate or gas, the equilibrium favors the products. In Chapter 16, we will study chemical equilibrium and learn experimental techniques for shifting the equilibrium forward or backward.

Summary

Sections 8.1–8.2 In this chapter, we learned that the evidence for a **chemical reaction** is any of the following: a gas is detected, a **precipitate** is formed, a color change is observed, or any energy change is noted. A **chemical equation** is a description of a reaction using formulas and symbols.

Section 8.3 In a balanced chemical equation, the reactants and products have the same number of atoms of each element. To balance an equation by inspection, a **coefficient** is systematically placed in front of each substance until there are the same number of atoms of each element on both sides of the equation. If the coefficient is omitted, it is understood to be 1.

Section 8.4 Although there are many thousands of different chemical reactions, most can be classified as one of five basic types (see Table 8.3). In a **combination reaction**, two elements combine to form a single compound. In a **decomposition reaction**, a compound decomposes into two or more substances. In a **single replacement reaction**, a more active element replaces another element in a compound. In a **double-replacement reaction**, two ionic compounds switch anions in aqueous solution. In a **neutralization reaction**, an acid and a base react to give a salt and water.

Section 8.5 There are three types of combination reactions: a metal plus oxygen, a nonmetal plus oxygen, and a metal plus a nonmetal. Often, we can predict the products from the reaction. Nonmetals, however, combine with oxygen to give a variety of oxides, so the product is not predictable.

1. metal + oxygen gas ⟶ metal oxide
2. nonmetal + oxygen gas ⟶ nonmetal oxide
3. metal + nonmetal ⟶ ionic compound

Section 8.6 There are three types of decomposition reactions. In the first, we can predictably decompose a metal hydrogen carbonate with heat to a metal carbonate product. In the second, we can further decompose the metal carbonate to a metal

Table 8.3 Summary of Five Basic Reaction Types

Reaction Type	General Format
Combination reaction	$A + Z \longrightarrow AZ$
Decomposition reaction	$AZ \longrightarrow A + Z$
Single-replacement reaction	$A + BZ \longrightarrow AZ + B$
Double-replacement reaction	$AX + BZ \longrightarrow AZ + BX$
Neutralization reaction	$HX + BOH \longrightarrow BZ + HOH$

oxide. In the third, we can predict compounds containing a polyatomic ion, such as chlorate or nitrate, decompose to give oxygen gas.

1. metal hydrogen carbonate $\longrightarrow$ metal carbonate + water + carbon dioxide
2. metal carbonate $\longrightarrow$ metal oxide + carbon dioxide
3. oxygen-containing compounds $\longrightarrow$ oxygen gas

Sections 8.7–8.8 There are three types of single-replacement reactions: a metal and aqueous salt, a metal and a dilute acid, and an active metal and water. The **activity series** lists, in order, the relative reactivity of each metal. We can use this list to predict whether or not a reaction occurs. An element higher in the series always displaces an element lower in the series. Metals that are above (H) will displace hydrogen gas from an acid. The **active metals**, which react directly with water at 25°C, include all the Group IA/1 and IIA/2 elements except Be and Mg.

1. metal_1 + $\text{aqueous solution}_1$ $\longrightarrow$ metal_2 + $\text{aqueous solution}_2$
2. metal + aqueous acid $\longrightarrow$ aqueous solution + hydrogen gas
3. active metal + water $\longrightarrow$ metal hydroxide + hydrogen gas

Sections 8.9–8.10 In a double-replacement reaction, two ionic compounds react in aqueous solution. An insoluble substance called a precipitate is usually one of the products and drives the reaction. The substance that is insoluble is determined by referring to the solubility rules for ionic compounds.

$\text{aqueous solution}_1$ + $\text{aqueous solution}_2$ $\longrightarrow$ $\text{aqueous solution}_3$ + precipitate

Section 8.11 In a neutralization reaction, an **acid** and a **base** react to yield a **salt** and water. A neutralization reaction is a special type of double-replacement reaction. It is advisable to write the formula for water as HOH. Not only does this more clearly show the exchange of anions, but it is helpful in balancing neutralization reactions.

aqueous acid + aqueous base $\longrightarrow$ aqueous salt + water

Key Terms

Select the key term that corresponds to the following definitions.

____ 1. the process of undergoing a chemical change
____ 2. a shorthand representation using formulas and symbols to describe a chemical change
____ 3. a substance undergoing a chemical change
____ 4. a substance resulting from a chemical change
____ 5. a substance that increases the rate of reaction but that can be recovered without being permanently changed
____ 6. a number in a chemical formula that represents the number of atoms or ions appearing in a substance
____ 7. a number placed in front of a chemical formula in order to balance a chemical equation

(a) acid *(Sec. 8.11)*
(b) active metal *(Sec. 8.7)*
(c) activity series *(Sec. 8.7)*
(d) aqueous solution (aq) *(Sec. 8.2)*
(e) base *(Sec. 8.11)*
(f) catalyst *(Sec. 8.2)*
(g) chemical equation *(Sec. 8.2)*
(h) chemical reaction *(Sec. 8.1)*
(i) coefficient *(Sec. 8.3)*

____ **8.** a reaction that evolves heat energy
____ **9.** a reaction that absorbs heat energy
____ **10.** a homogeneous mixture of a substance dissolved in water
____ **11.** a relative order of elements arranged by their ability to undergo reaction; also called an electromotive series
____ **12.** a metal that is sufficiently active to react with water at 25°C
____ **13.** an insoluble substance in solution produced by a reaction
____ **14.** a reaction in which two substances produce a single compound
____ **15.** a reaction in which two or more substances are produced from a single compound
____ **16.** a reaction in which a more active element displaces a less active element from a solution or compound
____ **17.** a reaction in which two cations in different compounds exchange anions
____ **18.** a reaction of an acid and base to produce a salt and water
____ **19.** a substance that releases hydrogen ions in aqueous solution
____ **20.** a substance that releases hydroxide ions in aqueous solution
____ **21.** an ionic compound produced from an acid-base reaction; an ionic compound that does not contain the hydroxide ion

(j) combination reaction *(Sec. 8.4)*
(k) decomposition reaction *(Sec. 8.4)*
(l) double-replacement reaction *(Sec. 8.4)*
(m) endothermic reaction *(Sec. 8.1)*
(n) exothermic reaction *(Sec. 8.1)*
(o) neutralization reaction *(Sec. 8.4)*
(p) precipitate *(Sec. 8.1)*
(q) product *(Sec. 8.2)*
(r) reactant *(Sec. 8.2)*
(s) salt *(Sec. 8.11)*
(t) single-replacement reaction *(Sec. 8.4)*
(u) subscript *(Sec. 8.3)*

Exercises

Evidence for Chemical Reactions (Sec. 8.1)

1. Indicate any of the following that are evidence for a chemical reaction.
   (a) mixing two solutions produces gas bubbles
   (b) heating a compound extinguishes a flaming splint
   (c) mixing two solutions produces a precipitate
   (d) mixing two colorless solutions results in a pink solution
   (e) mixing two solutions raises the temperature 5°C
2. Indicate any of the following that are evidence for a chemical reaction.
   (a) heating a compound gives a sharp odor
   (b) heating a compound causes a glowing splint to burst into flames
   (c) mixing two colorless solutions results in a colorless aqueous solution
   (d) heating gold wire in a vacuum causes the metal to explode into a vapor
   (e) mixing two solutions lowers the temperature from 20° to 15°C

Writing Chemical Equations (Sec. 8.2)

3. Write a chemical equation to describe each of the following chemical changes.
   (a) Tin metal heated with oxygen gas yields solid tin(IV) oxide.
   (b) Solid zinc carbonate decomposes to give zinc oxide powder and carbon dioxide gas.
   (c) Magnesium metal reacts with aqueous cadmium nitrate to produce aqueous magnesium nitrate and cadmium metal.
   (d) Aqueous solutions of lithium bromide and silver nitrate yield insoluble silver bromide and aqueous lithium nitrate.
   (e) Acetic acid reacts with potassium hydroxide solution to give aqueous potassium acetate plus water.
4. Write a chemical equation to describe each of the following chemical changes.
   (a) Copper metal heated with fluorine gas yields solid cupric fluoride.
   (b) Solid iron(II) hydrogen carbonate decomposes to give iron(II) carbonate powder, steam, and carbon dioxide gas.
   (c) Manganese metal reacts with sulfuric acid to produce aqueous manganese(II) sulfate and hydrogen gas.
   (d) Aqueous solutions of sodium chromate and calcium sulfide yield a calcium chromate precipitate and aqueous sodium sulfide.
   (e) Nitric acid neutralizes ammonium hydroxide solution to give aqueous ammonium nitrate and water.

Balancing Chemical Equations (Sec. 8.3)

5. Which of the following is not a general guideline for balancing a chemical equation by inspection?
   (a) Use coefficients to balance the elements in each substance.

(b) Begin balancing the equation by starting with the most complex formula.
(c) Balance polyatomic ions as a unit rather than the elements separately.
(d) After balancing the equation, verify the coefficients.
(e) If an equation coefficient is a fraction, change a subscript in the chemical formula to give whole number coefficients.

6. If a chemical equation is impossible to balance, what is most likely the problem?

7. Balance each of the following chemical equations by inspection.
(a) $Co(s) + O_2(g) \rightarrow Co_2O_3(s)$
(b) $CsClO_3(s) \rightarrow CsCl(s) + O_2(g)$
(c) $Cu(s) + AgC_2H_3O_2(aq) \rightarrow Cu(C_2H_3O_2)_2(aq) + Ag(s)$
(d) $Pb(NO_3)_2(aq) + LiCl(aq) \rightarrow PbCl_2(s) + LiNO_3(aq)$
(e) $H_2SO_4(aq) + Al(OH)_3(aq) \rightarrow Al_2(SO_4)_3(aq) + H_2O(l)$

8. Balance each of the following chemical equations by inspection.
(a) $Sr(s) + H_2O(l) \rightarrow Sr(OH)_2(aq) + H_2(g)$
(b) $H_3PO_4(aq) + Mn(OH)_2(aq) \rightarrow Mn_3(PO_4)_2(s) + H_2O(l)$
(c) $H_2(g) + N_2(g) \rightarrow NH_3(g)$
(d) $Al(HCO_3)_3(s) \rightarrow Al_2(CO_3)_3(s) + CO_2(g) + H_2O(g)$
(e) $K_2SO_4(aq) + Ba(OH)_2(aq) \rightarrow BaSO_4(s) + KOH(aq)$

9. Balance each of the following chemical equations by inspection.
(a) $Pb(s) + O_2(g) \rightarrow PbO(s)$
(b) $LiNO_3(s) \rightarrow LiNO_2(s) + O_2(g)$
(c) $Mg(s) + HC_2H_3O_2(aq) \rightarrow Mg(C_2H_3O_2)_2(aq) + H_2(g)$
(d) $Hg_2(NO_3)_2(aq) + NaBr(aq) \rightarrow Hg_2Br_2(s) + NaNO_3(aq)$
(e) $H_2CO_3(aq) + NH_4OH(aq) \rightarrow (NH_4)_2CO_3(aq) + HOH(l)$

10. Balance each of the following chemical equations by inspection.
(a) $Fe(s) + Cd(NO_3)_2(aq) \rightarrow Fe(NO_3)_3(aq) + Cd(s)$
(b) $HClO_4(aq) + Ba(OH)_2(aq) \rightarrow Ba(ClO_4)_2(aq) + HOH(l)$
(c) $Sn(s) + P(s) \rightarrow Sn_3P_2(s)$
(d) $Fe_2(CO_3)_3(s) \rightarrow Fe_2O_3(s) + CO_2(g)$
(e) $Co(NO_3)_2(aq) + H_2S(g) \rightarrow CoS(s) + HNO_3(aq)$

Classifying Chemical Reactions (Sec. 8.4)

11. Classify each reaction in Exercise 7 as one of the following: combination, decomposition, single replacement, double replacement, or neutralization.

12. Classify each reaction in Exercise 8 as one of the following: combination, decomposition, single replacement, double replacement, or neutralization.

13. Classify each reaction in Exercise 9 as one of the following: combination, decomposition, single replacement, double replacement, or neutralization.

14. Classify each reaction in Exercise 10 as one of the following: combination, decomposition, single replacement, double replacement, or neutralization.

Combination Reactions (Sec. 8.5)

15. Write a balanced equation for each of the following.
(a) Copper metal is heated with oxygen to give cuprous oxide.
(b) Tin metal is heated with oxygen to produce stannous oxide.
(c) Iron metal is heated with oxygen to form iron(III) oxide.
(d) Titanium metal is heated with oxygen to yield titanium(IV) oxide.

Igniting steel wool and plunging it into pure oxygen gas gives a shower of Fe_2O_3 sparks and a bright white light.

16. Write a balanced chemical equation for each of the following.
(a) Carbon is heated with oxygen to yield carbon monoxide gas.
(b) Phosphorus is heated with oxygen to form diphosphorus trioxide.
(c) Nitrogen is heated with oxygen to produce dinitrogen pentaoxide gas.
(d) Chlorine is heated with oxygen to give dichlorine trioxide gas.

17. Write a balanced chemical equation for each of the following.
(a) Mercury is heated with chlorine to give solid mercuric chloride.
(b) Lead is heated with phosphorus to yield plumbous phosphide.
(c) Iron is heated with fluorine to produce ferric fluoride.
(d) Cobalt is heated with sulfur to form cobalt(II) sulfide.

18. Write a balanced chemical equation for each of the following.
(a) Chromium metal is heated with oxygen to give chromium(III) oxide.

(b) Chromium is heated with nitrogen to yield chromium(III) nitride.
(c) Sulfur is heated with oxygen to produce sulfur dioxide gas.
(d) Sulfur is heated with oxygen and platinum catalyst to form sulfur trioxide gas.

19. Complete and balance the following combination reactions.
(a) $Cd + O_2 \rightarrow$ (b) $Al + Cl_2 \rightarrow$
(c) $Sr + S \rightarrow$ (d) $Ba + N_2 \rightarrow$

20. Complete and balance the following combination reactions.
(a) $Na + I_2 \rightarrow$ (b) $Ca + O_2 \rightarrow$
(c) $Zn + P \rightarrow$ (d) $K + F_2 \rightarrow$

21. Complete and balance the following combination reactions.
(a) $Sr + O_2 \rightarrow$ (b) $Bi + S \rightarrow$
(c) $Li + Br_2 \rightarrow$ (d) $Mg + Cl_2 \rightarrow$

22. Complete and balance the following combination reactions.
(a) $Al + O_2 \rightarrow$ (b) $Zn + F_2 \rightarrow$
(c) $K + S \rightarrow$ (d) $Ba + N_2 \rightarrow$

Decomposition Reactions (Sec. 8.6)

23. Write a balanced chemical equation for each of the following.
(a) Silver hydrogen carbonate decomposes upon heating.
(b) Cupric hydrogen carbonate decomposes upon heating.

24. Write a balanced chemical equation for each of the following.
(a) Gold(III) hydrogen carbonate decomposes upon heating.
(b) Tin(IV) hydrogen carbonate decomposes upon heating.

25. Write a balanced chemical equation for each of the following.
(a) Copper(II) carbonate decomposes upon heating.
(b) Manganese(II) carbonate decomposes upon heating.

26. Write a balanced chemical equation for each of the following.
(a) Chromium(III) carbonate decomposes upon heating.
(b) Lead(IV) carbonate decomposes upon heating.

27. Write a balanced chemical equation for each of the following.
(a) Plumbic oxide is decomposed with heat to give plumbous oxide and oxygen gas.
(b) Silver nitrate is decomposed with heat to give silver nitrite and oxygen.

28. Write a balanced chemical equation for each of the following.
(a) Calcium nitrate is decomposed with heat to give calcium nitrite and oxygen gas.
(b) Silver chlorate is decomposed with heat to give silver chloride and oxygen gas.

29. Complete and balance the following decomposition reactions.
(a) $KHCO_3 \rightarrow$ (b) $Zn(HCO_3)_2 \rightarrow$
(c) $Li_2CO_3 \rightarrow$ (d) $CdCO_3 \rightarrow$

30. Complete and balance the following decomposition reactions.
(a) $Sr(ClO_3)_2 \rightarrow SrCl_2 + ?$
(b) $NaClO_3 \rightarrow NaCl + ?$
(c) $Al(NO_3)_3 \rightarrow Al(NO_2)_3 + ?$
(d) $CsNO_3 \rightarrow CsNO_2 + ?$

Activity Series (Sec. 8.7)

31. Predict whether or not a reaction occurs for each of the following metals when dropped into an aqueous solution of iron(II) nitrate (refer to the activity series in Section 8.7).
(a) Hg (b) Zn (c) Cd (d) Mg

32. Predict whether or not a reaction occurs for each of the following metals when dropped into an aqueous solution of nickel(II) nitrate (refer to the activity series in Section 8.7).
(a) Ag (b) Sn (c) Co (d) Mn

33. Predict whether or not a reaction occurs for each of the following metals when dropped into an aqueous solution of dilute hydrochloric acid.
(a) Ni (b) Zn (c) Ca (d) Al

34. Predict whether or not a reaction occurs for each of the following metals when dropped into an aqueous solution of dilute sulfuric acid.
(a) Fe (b) Mn (c) Cd (d) Au

35. Predict whether or not a reaction occurs for each of the following metals when dropped into cold water.
(a) Li (b) Mg (c) Ca (d) Al

36. Predict whether or not a reaction occurs for each of the following metals when dropped into cold water.
(a) Ba (b) Mn (c) Sn (d) K

37. Predict whether or not a reaction occurs for each of the following (refer to the activity series in Section 8.7).
(a) Manganese is dropped into an aluminum nitrate solution.
(b) Iron metal is put into a tin(II) acetate solution.
(c) Cobalt shot is placed in chloric acid.
(d) Silver metal is added to sulfuric acid.

38. Predict whether or not a reaction occurs for each of the following (refer to the activity series in Section 8.7).
(a) Lead shot is dropped into a zinc chloride solution.
(b) Magnesium metal is put into an iron(II) sulfate solution.
(c) Copper metal is placed in dilute sulfuric acid.
(d) Tin strips are added to hydrochloric acid.

Single-Replacement Reactions (Sec 8.8)

39. Complete and balance an equation for each of the following reactions (refer to the activity series in Section 8.7).
(a) Copper wire is placed in an aluminum nitrate solution.

(b) Aluminum wire is placed in a copper(II) nitrate solution.
(c) Cadmium metal is put into an iron(II) sulfate solution.
(d) Iron metal is put into a cadmium sulfate solution.

40. Complete and balance an equation for each of the following reactions (refer to the activity series in Section 8.7).
(a) Nickel metal is put into a lead(II) acetate solution.
(b) Lead metal is put into a nickel(II) acetate solution.
(c) Iron filings are added to a mercury(II) sulfate solution.
(d) Drops of liquid mercury are added to a ferrous sulfate solution.

41. Complete and balance an equation for each of the following reactions (refer to the activity series in Section 8.7).
(a) Cadmium metal is added to dilute hydrochloric acid.
(b) Manganese chips are added to acetic acid.
(c) Zinc granules are placed in dilute sulfuric acid.
(d) Magnesium ribbon is added to carbonic acid.

42. Complete and balance an equation for each of the following reactions (refer to the activity series in Section 8.7).
(a) Cobalt metal is added to dilute hydrochloric acid.
(b) Drops of mercury are added to dilute sulfuric acid.
(c) Aluminum granules are added to acetic acid.
(d) Metallic mossy zinc is added to phosphoric acid.

43. Complete and balance an equation for each of the following reactions.
(a) Grayish-white lithium beads are added to water.
(b) A small soft chunk of lead is added to water.
(c) Metallic cobalt shot is added to water.
(d) Gray calcium turnings are added to water.

44. Complete and balance an equation for each of the following reactions.
(a) A small piece of radioactive radium is dropped into water.
(b) A small piece of radioactive gold is dropped into water.
(c) A small piece of radioactive technetium is dropped into water.
(d) A small piece of radioactive francium is dropped into water.

Placing a strip of zinc metal into $Pb(NO_3)_2$ solution produces a gray layer of spongy lead.

45. Complete and balance the following single-replacement reactions (refer to the activity series in Section 8.7).
(a) $Zn(s) + Pb(NO_3)_2(aq) \rightarrow$
(b) $Co(s) + NiSO_4(aq) \rightarrow$
(c) $Ni(s) + SnSO_4(aq) \rightarrow$
(d) $Sn(s) + LiC_2H_3O_2(aq) \rightarrow$
(e) $Zn(s) + HCl(aq) \rightarrow$

46. Complete and balance the following single-replacement reactions (refer to the activity series in Section 8.7).
(a) $Ni(s) + HNO_3(aq) \rightarrow$
(b) $Mg(s) + H_2SO_4(aq) \rightarrow$
(c) $Na(s) + H_2O(l) \rightarrow$
(d) $Mg(s) + H_2O(l) \rightarrow$
(e) $Ba(s) + H_2O(l) \rightarrow$

Solubility Rules (Sec. 8.9)

47. State whether the following compounds are soluble or insoluble in water (refer to the solubility rules for ionic compounds in Section 8.9).
(a) ammonium hydroxide, NH_4OH
(b) iron(II) sulfate, $FeSO_4$
(c) potassium chromate, K_2CrO_4
(d) lead(II) acetate, $Pb(C_2H_3O_2)_2$
(e) aluminum nitrate, $Al(NO_3)_3$
(f) mercury(I) chloride, Hg_2Cl_2
(g) manganese(II) iodide, MnI_2
(h) mercury(II) bromide, $HgBr_2$

48. State whether the following compounds are soluble or insoluble in water (refer to the solubility rules for ionic compounds in Section 8.9).
(a) silver bromide
(b) lead(II) sulfate
(c) sodium sulfide
(d) iron(III) phosphate
(e) strontium carbonate
(f) cobalt(II) hydroxide
(g) nickel(II) sulfide
(h) silver chloride

Double-Replacement Reactions (Sec. 8.10)

49. Complete and balance the following double-replacement reactions (refer to the solubility rules for ionic compounds in Section 8.9).
(a) $MgSO_4(aq) + BaCl_2(aq) \rightarrow$
(b) $AlBr_3(aq) + Na_2CO_3(aq) \rightarrow$
(c) $NiSO_4(aq) + Li_3PO_4(aq) \rightarrow$

50. Complete and balance the following double-replacement reactions (refer to the solubility rules for ionic compounds in Section 8.9).
(a) $AgC_2H_3O_2(aq) + SrI_2(aq) \rightarrow$
(b) $FeSO_4(aq) + Ca(OH)_2(aq) \rightarrow$
(c) $Sn(NO_3)_2(aq) + K_2CrO_4(aq) \rightarrow$

51. Complete and balance the following double-replacement reactions (refer to the solubility rules for ionic compounds in Section 8.9).
(a) Aqueous solutions of manganese(II) sulfate and ammonium hydroxide are added together.
(b) Aqueous solutions of zinc chloride and mercury(I) nitrate are added together.

52. Complete and balance the following double-replacement reactions (refer to the solubility rules for ionic compounds in Section 8.9).
(a) Aqueous solutions of tin(II) chloride and sodium sulfide are added together.
(b) Aqueous solutions of cobalt(II) nitrate and potassium chromate are added together.

Neutralization Reactions (Sec. 8.11)

53. Complete and balance each of the following neutralization reactions.
(a) $HCl(aq) + Ca(OH)_2(aq) \rightarrow$
(b) $H_2SO_4(aq) + LiOH(aq) \rightarrow$
(c) $HNO_2(aq) + Sr(OH)_2(aq) \rightarrow$

54. Complete and balance each of the following neutralization reactions.
(a) $HBr(aq) + Al(OH)_3(s) \rightarrow$
(b) $HClO_2(aq) + Mg(OH)_2(aq) \rightarrow$
(c) $H_3PO_4(aq) + KOH(aq) \rightarrow$

55. Complete and balance each of the following neutralization reactions.
(a) Sodium hydroxide solution is added to nitric acid.
(b) Barium hydroxide solution is added to phosphoric acid.

56. Complete and balance each of the following neutralization reactions.
(a) Potassium hydroxide solution is added to carbonic acid.
(b) Strontium hydroxide solution is added to acetic acid.

General Exercises

57. Balance the following chemical equations by inspection.
(a) $SO_2(g) + O_2(g) \rightarrow SO_3(g)$
(b) $F_2(g) + NaBr(aq) \rightarrow Br_2(l) + NaF(aq)$
(c) $Cl_2(g) + H_2O(l) \rightarrow HCl(aq) + HClO(aq)$
(d) $PCl_5(s) + H_2O(l) \rightarrow H_3PO_4(aq) + HCl(aq)$
(e) $Sb_2S_3(s) + HCl(aq) \rightarrow SbCl_3(aq) + H_2S(aq)$

58 Balance the following chemical equations by inspection.
(a) $TiCl_4(s) + H_2O(g) \rightarrow TiO_2(s) + HCl(g)$
(b) $MnO_2(l) + Al(l) \rightarrow Al_2O_3(l) + Mn(l)$
(c) $Fe(s) + H_2O(g) \rightarrow Fe_3O_4(s) + H_2(g)$
(d) $FeO(l) + Al(l) \rightarrow Al_2O_3(l) + Fe(l)$
(e) $FeS(s) + O_2(g) \rightarrow Fe_2O_3(s) + SO_2(g)$

59. A spark is provided for the following combustion reactions. Balance each of the chemical equations by inspection.
(a) $CH_4(g) + O_2(g) \rightarrow CO_2(g) + H_2O(g)$
(b) $CH_4O(l) + O_2(g) \rightarrow CO_2(g) + H_2O(g)$
(c) $C_3H_8(g) + O_2(g) \rightarrow CO_2(g) + H_2O(g)$
(d) $C_3H_8O(l) + O_2(g) \rightarrow CO_2(g) + H_2O(g)$

60. A spark is provided for the following combustion reactions. Complete and balance each of the chemical equations by inspection.
(a) $C_2H_6(g) + O_2(g) \rightarrow$
(b) $C_2H_6O(l) + O_2(g) \rightarrow$
(c) $C_6H_6(g) + O_2(g) \rightarrow$
(d) $C_6H_6O(l) + O_2(g) \rightarrow$

61. Chlorine gas is prepared industrially by the Deacon method. Hydrogen chloride gas is heated with oxygen to give steam and chlorine gas. Write a balanced chemical equation for the Deacon method.

62. An industrial blast furnace is used to convert iron ore to the element iron. Iron(III) oxide is heated to 1500°C, and carbon monoxide gas is passed through the molten liquid. In addition to iron, the reaction produces carbon dioxide gas. Write a balanced chemical equation for the blast furnace reaction.

63. The Ostwald process is an important industrial method for making nitric acid. In the first step, ammonia and oxygen gases are heated to produce nitrogen monoxide gas and steam. In the second step, nitrogen monoxide and oxygen gases react to give nitrogen dioxide. In the third step, nitrogen dioxide gas is passed through water to yield nitric acid and nitrogen monoxide gas. Write three balanced chemical equations for the Ostwald process.

CHAPTER

9 Chemical Equation Calculations

Roller-blades are usually manufactured with four wheels for recreational use and five wheels for racing competition. By analogy, sulfur oxides are manufactured with two oxygen atoms (SO_2) for making sulfurous acid and three oxygen atoms (SO_3) for making sulfuric acid.

In Chapter 7, chemical formula calculations and the mole concept were introduced. As you may recall, a mole is the amount of substance containing Avogadro's number of particles. The mass of 1 mole of any substance is the **molar mass**. *A molar mass has the units of grams per mole. The volume of 1 mole of any gaseous substance at* **standard conditions** *of temperature and pressure is the* **molar volume**. *At 0°C and 1 atmosphere pressure, the molar volume of any gas is 22.4 liters per mole.*

So far, we have performed mole calculations on chemical formulas. In this chapter, we will perform mole calculations on chemical equations. That is, we will apply the mole concept to relate quantities of reactants and products. In chemistry, calculations that relate quantities of substance are a topic known as **stoichiometry**. *The term is derived from the Greek words* stoicheion, *meaning element, and* metron, *meaning measure. In this chapter, you can think of stoichiometry as chemical equation calculations.*

stoichiometry The relationship of quantities (mass of substance or volume of gas) in a chemical change according to the balanced chemical equation.

Stoichiometry is one of the most important quantitative topics in chemistry. It involves the use of chemical formulas, mole calculations, and chemical equations. Stoichiometry is also essential in industry. There, it is used to do the cost analysis for manufacturing chemicals. In fact, manufacturing processes are financed according to the cost of reactants and the value of products as determined by stoichiometric calculations.

9.1 Interpreting a Chemical Equation

To relate the coefficients in a balanced chemical equation to:

(a) molecules or formula units of reactants and products

(b) moles of reactants and products

(c) liters of gaseous reactants and products (at the same conditions)

To verify the conservation of mass law using the molar mass of reactants and products in a balanced chemical equation.

law of combining volumes The principle that volumes of gases that combine in a chemical reaction, at the same temperature and pressure, are in the ratio of small whole numbers; also called Gay-Lussac's law of combining volumes.

In 1804, the French chemist Joseph Gay-Lussac (1778–1850) ascended to 23,000 feet in a hydrogen-filled balloon. During his record-setting ascent, he took samples of the atmosphere and later studied their composition. Four years later, in 1808 Gay-Lussac proposed the **law of combining volumes**. This law states that volumes of gases, under similar conditions, combine in small, whole-number ratios. In one experiment, Gay-Lussac studied the reaction of hydrogen and chlorine and found that one volume of gaseous hydrogen combined with one volume of gaseous chlorine to give two volumes of hydrogen chloride gas. That is,

	hydrogen gas +	chlorine gas $\longrightarrow$	hydrogen chloride gas
	1 volume	*1 volume*	*2 volumes*
for example,	*50.0 mL*	*50.0 mL*	*100.0 mL*

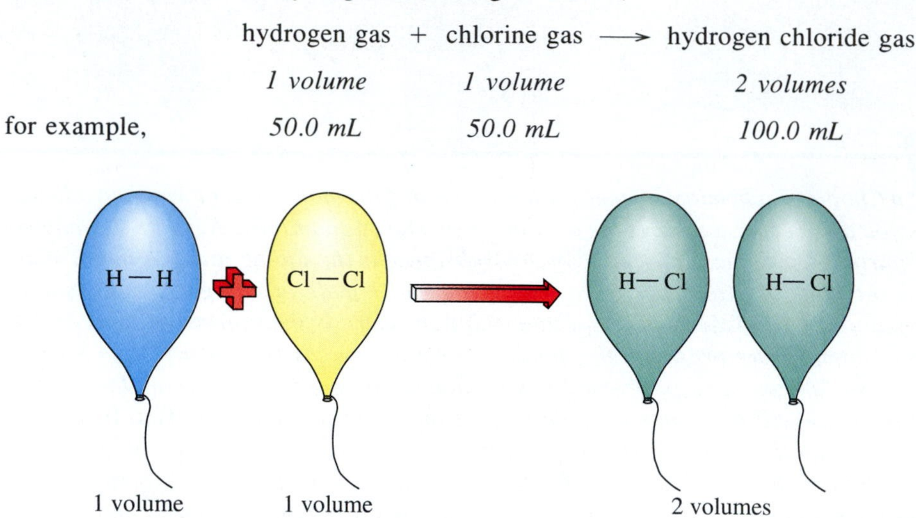

In 1811, the Italian physicist Amedeo Avogadro (1776–1856) explained the law of combining volumes. Avogadro proposed that equal volumes of gas, at the same temperature and pressure, contain the same number of molecules. That is, any two gases containing the same number of molecules will occupy equal volumes. This statement is known as **Avogadro's law**. From the law, it was clear that gases react in small, whole-number ratios because molecules of two gases react in small, whole-number ratios.

Let's reconsider the reaction of hydrogen and chlorine gases at the molecular level. We have

$$H_2(g) \quad + \quad Cl_2(g) \quad \longrightarrow \quad 2\ HCl(g)$$

1 molecule   *1 molecule*   *2 molecules*

Avogadro's law claims that equal numbers of H_2 molecules and Cl_2 molecules would have equal volumes. Since we have twice the number of HCl molecules as H_2 or Cl_2, we would have twice the volume. Thus,

In another experiment, Gay-Lussac studied the reaction of hydrogen and oxygen. He found that two volumes of hydrogen gas combined with one volume of oxygen gas. In this case, the gases combine in the small, whole-number ratio of 2:1. That is,

	hydrogen gas +	oxygen gas $\longrightarrow$	water vapor
	2 volumes	*1 volume*	*2 volumes*
for example,	*100.0 mL*	*50.0 mL*	*100.0 mL*

We can also look at the reaction of hydrogen and oxygen molecules.

$$2\ H_2(g) + O_2(g) \longrightarrow 2\ H_2O(g)$$

Once again applying Avogadro's law, there are twice the number of H_2 molecules as O_2 molecules; therefore, the volume of H_2 would be twice the volume of O_2. Since the coefficients of H_2O and H_2 are the same, they would have equal volumes. Illustrated as above, we have

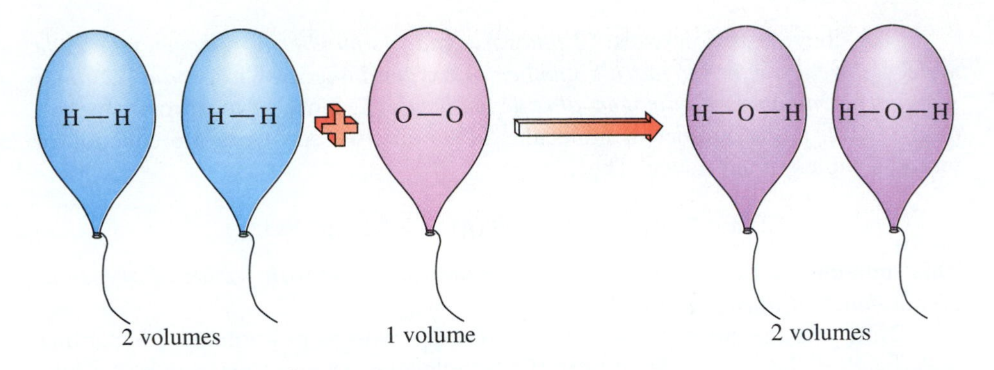

Thus, Avogadro's law neatly explained Gay-Lussac's law of combining volumes. That is, volumes of gases combine in small, whole-number ratios because equal volumes of gases contain the same number of molecules.

Interpreting Balanced Chemical Equations

Now let's examine the information we can obtain from a balanced chemical equation. Consider nitrogen monoxide gas, which is present in automobile emissions. In the atmosphere, nitrogen monoxide reacts with oxygen to produce reddish-brown nitrogen dioxide smog. Ultraviolet light (UV) initiates the reaction. The balanced equation is

$$2\ NO(g) + O_2(g) \xrightarrow{UV} 2\ NO_2(g)$$

The coefficients in the balanced equation indicate that 2 molecules of NO react with 1 molecule of O_2 to produce 2 molecules of NO_2. It therefore follows that multiples of these coefficients will be in the same ratios. For example, 2000 molecules of NO react with 1000 molecules of O_2 to give 2000 molecules of NO_2. Since molecules are so small, let's consider an even larger number than 2000. Let's consider Avogadro's number (N) of molecules. We can substitute this very large number of molecules for the coefficients and the equation remains balanced. This gives us

$$2\ N\ NO(g) + 1\ N\ O_2(g) \xrightarrow{UV} 2\ N\ NO_2(g)$$

A commemorative postage stamp honoring Amedeo Avogadro (1776-1856). Avogadro's law is stated on the stamp in Italian.

The above equation reads: "*2 times Avogadro's number of nitrogen monoxide molecules react with Avogadro's number of oxygen molecules to produce 2 times Avogadro's number of nitrogen dioxide molecules.*" Since Avogadro's number, 6.02×10^{23}, is the number of molecules in 1 mole, we can write the equation in terms of moles of substance, Thus,

$$2 \text{ mol NO(g)} + 1 \text{ mol } O_2(g) \xrightarrow{UV} 2 \text{ mol } NO_2(g)$$

This equation reads, "*2 moles of nitrogen monoxide react with 1 mole of oxygen to give 2 moles of nitrogen dioxide.*"

Note that the coefficients also represent the ratio of gas volumes. According to Avogadro's law, the same number of gas molecules occupy equal volumes. Thus, 2 volumes of nitrogen monoxide gas react with 1 volume of oxygen gas to produce 2 volumes of nitrogen dioxide gas, conditions of temperature and pressure remaining constant. We can now write a balanced equation in terms of gas volumes.

$$2 \text{ volumes NO(g)} + 1 \text{ volume } O_2(g) \xrightarrow{UV} 2 \text{ volumes } NO_2(g)$$

We can use any units for the gas volume that are convenient. The smallest whole-number ratio of the coefficients will be 2:1:2. For example, 20.0 mL of nitrogen monoxide react with 10.0 mL of oxygen to produce 20.0 mL of nitrogen dioxide. The smallest whole-number ratio of gaseous volumes (20.0:10.0:20.0) corresponds to the coefficients in the balanced chemical equation: 2:1:2.

Table 9.1 summarizes the information that can be obtained from the coefficients of a balanced chemical equation.

Table 9.1 Summary of Coefficient Interpretations

For the General Equation:	$2\ A + 3\ B \longrightarrow C + 2\ D$
The ratio of molecules is:	2:3:1:2
The ratio of moles is:	2:3:1:2
The ratio of volumes of gas is:	2:3:1:2

The following example exercise illustrates the information that can be obtained by interpreting the coefficients in a balanced chemical equation.

EXAMPLE EXERCISE 9.1

After balancing the following equation for the combustion of propane and oxygen, interpret the coefficients in terms of (a) moles and (b) liters.

$$C_3H_8(g) + O_2(g) \longrightarrow CO_2(g) + H_2O(g)$$

Solution: We can supply the following coefficients to obtain a balanced chemical equation.

$$C_3H_8(g) + 5\ O_2(g) \longrightarrow 3\ CO_2(g) + 4\ H_2O(g)$$

(a) The coefficients in the equation (1:5:3:4) indicate the ratio of moles as well as molecules. Thus,

$$1 \text{ mole} + 5 \text{ moles} \longrightarrow 3 \text{ moles} + 4 \text{ moles}$$

(b) The coefficients in the equation (1:5:3:4) indicate the ratio of volumes of gases as well as molecules. Expressing the volume in liters, we have

$$1 \text{ liter} + 5 \text{ liters} \longrightarrow 3 \text{ liters} + 4 \text{ liters}$$

SELF-TEST EXERCISE

After balancing the following equation for the combustion of butane and oxygen, interpret the coefficients in terms of (a) moles and (b) liters.

$$C_4H_{10}(g) + O_2(g) \longrightarrow CO_2(g) + H_2O(g)$$

Answers: $2\ C_4H_{10}(g) + 13\ O_2(g) \longrightarrow 8\ CO_2(g) + 10\ H_2O(g)$

(a) 2 moles + 13 moles → 8 moles + 10 moles;
(b) 2 liters + 13 liters → 8 liters + 10 liters

Verifying the Conservation of Mass Law

In Section 3.9, we introduced the **law of conservation of mass**. In 1787, after conducting numerous experiments, Antoine Lavoisier stated that the combined masses of reactants before reaction was equal to the combined masses of the products. In other words, for any given chemical reaction, the mass remains constant.

law of conservation of mass The statement that the total mass of products from a chemical reaction must equal the sum of the masses of reactants.

We can verify the conservation of mass law by examining a chemical equation. To do so, let's heat potassium chlorate, which decomposes to give potassium chloride and oxygen gas. Manganese(IV) oxide is used as a catalyst for the reaction. The balanced chemical equation for the decomposition is

$$2\ KClO_3(s) \xrightarrow{MnO_2} 2\ KCl(s) + 3\ O_2(g)$$

According to the coefficients in the balanced equation, the ratio of the substances is 2:2:3. Therefore, the mole ratio of each substance is 2:2:3.

$$2\ \text{mol}\ KClO_3 \longrightarrow 2\ \text{mol}\ KCl + 3\ \text{mol}\ O_2$$

Since the mass of a mole of a given substance is constant, the coefficients also express the ratios of the molar mass (MM). Thus,

$$2\ \text{MM}\ KClO_3 \longrightarrow 2\ \text{MM}\ KCl + 3\ \text{MM}\ O_2$$

By summing the atomic masses for each substance, we compute the molar mass for $KClO_3$ is 122.6 g/mol, KCl is 74.6 g/mol, and O_2 is 32.0 g/mol. Thus,

$$2(122.6\ \text{g}) \longrightarrow 2(74.6\ \text{g}) + 3(32.0\ \text{g})$$

Simplifying,

$$245.2\ \text{g} \longrightarrow 149.2\ \text{g} + 96.0\ \text{g}$$

and

$$245.2\ \text{g} \longrightarrow 245.2\ \text{g}$$

In this example, the total mass of the reactants is 245.2 g, the same as the total mass of the products. From our calculation, we have verified, theoretically, the conservation of mass law. Example Exercise 9.2 further illustrates the conservation of mass law.

EXAMPLE EXERCISE 9.2

Verify the conservation of mass law by calculating the masses of reactants and products, given the following chemical equation.

$$NaNO_3(s) \longrightarrow NaNO_2(s) + O_2(g)$$

Solution: Before we can verify the conservation of mass law, we must write a balanced chemical equation for the reaction:

$$2\ NaNO_3(s) \longrightarrow 2\ NaNO_2(s) + O_2(g)$$

Next, we must determine the molar masses of each substance by summing the atomic masses. We find that the molar mass of $NaNO_3$ is 85.0 g/mol, $NaNO_2$ is 69.0 g/mol, and O_2 is 32.0 g/mol. Thus,

$$2(85.0\text{ g}) \longrightarrow 2(69.0\text{ g}) + 1(32.0\text{ g})$$

$$170.0\text{ g} \longrightarrow 138.0\text{ g} + 32.0\text{ g}$$

$$170.0\text{ g} \longrightarrow 170.0\text{ g}$$

The mass of the reactant is 170.0 g, which is the same as the total mass of the products. Thus, we have verified the conservation of mass law.

SELF-TEST EXERCISE

Verify the conservation of mass law by calculating the masses of reactants and products for the following unbalanced chemical equation.

$$C_7H_{16}(g) + O_2(g) \longrightarrow CO_2(g) + H_2O(g)$$

Answer: 452.0 g → 452.0 g

9.2 Mole–Mole Problems

To relate moles of any two substances participating in a balanced chemical equation.

As we demonstrated in Section 9.1, the coefficients in the chemical equation also indicate the ratio of moles of reactants and products. Consider the complete combustion of natural gas. Methane, CH_4, reacts with oxygen to give carbon dioxide and water. The equation is

$$CH_4(g) + 2\ O_2(g) \xrightarrow{\text{spark}} CO_2(g) + 2\ H_2O(g)$$

The coefficients indicate that 1 mol of methane reacts with 2 mol of oxygen.

Suppose we wanted to find out how many moles of oxygen react with 2.25 mol of CH_4. We can set up the problem as follows:

$$2.25\text{ mol }CH_4 \times \frac{\text{unit}}{\text{factor}} = \text{mol }O_2$$

From the balanced equation, we have 1 mol CH_4 gives 2 mol O_2. Applying the unit factor to cancel units, we have

$$2.25\ \cancel{\text{mol }CH_4} \times \frac{2\text{ mol }O_2}{1\ \cancel{\text{mol }CH_4}} = 4.50\text{ mol }O_2$$

We could also calculate the moles of carbon dioxide produced from the reaction. In the balanced equation, we have 1 mol CH_4 yields 1 mol CO_2. Applying the unit factor as before, we have

$$2.25\ \cancel{\text{mol }CH_4} \times \frac{1\text{ mol }CO_2}{1\ \cancel{\text{mol }CH_4}} = 2.25\text{ mol }CO_2$$

In this type of problem, we can convert from moles of one substance to moles of another using a single unit factor. This type of conversion is sometimes referred to as a **mole–mole problem**. Example Exercise 9.3 further illustrates a mole–mole problem calculation.

mole–mole problem A type of calculation that relates the moles of two substances participating in a balanced chemical equation.

EXAMPLE EXERCISE 9.3

Calculate the number of moles of hydrogen that react with 0.125 mol of nitrogen gas. Also calculate the number of moles of ammonia, NH_3, produced from the reaction.

$$N_2(g) + 3\ H_2(g) \longrightarrow 2\ NH_3(g)$$

Solution: The mole ratio comes from the coefficients in the balanced chemical equation, that is, 1 mol N_2 = 3 mol H_2.

$$0.125\ \cancel{\text{mol } N_2} \times \frac{3\text{ mol } H_2}{1\ \cancel{\text{mol } N_2}} = 0.375\text{ mol } H_2$$

The mole ratio of nitrogen to ammonia is the same as the coefficients. Thus, 1 mole N_2 = 2 mol NH_3.

$$0.125\ \cancel{\text{mol } N_2} \times \frac{2\text{ mol } NH_3}{1\ \cancel{\text{mol } N_2}} = 0.250\text{ mol } NH_3$$

SELF-TEST EXERCISE

Iron is produced in an industrial blast furnace by passing carbon monoxide gas through molten iron(III) oxide at 1500°C. The unbalanced equation is

$$Fe_2O_3(l) + CO(g) \longrightarrow Fe(l) + CO_2(g)$$

(a) How many moles of carbon monoxide react with 2.50 mol of iron(III) oxide?
(b) How many moles of iron are produced from 2.50 mol Fe_2O_3?

Answers: **(a)** 7.50 mol CO; **(b)** 5.00 mol Fe

9.3 Types of Stoichiometry Problems

OBJECTIVES

To classify the three basic types of stoichiometry problems: mass–mass, mass–volume, and volume–volume.

To state the three-step procedure for solving a stoichiometry problem, given the balanced chemical equation.

Stoichiometry, as we have stated, is the quantitative relationship of substances, that is, reactants and products, in a balanced chemical equation. The quantities can be masses of substance or volumes of gases.

Some stoichiometric calculations relate a given amount of reactant to an unknown amount of product. Other stoichiometric calculations relate a given amount of product back to an unknown amount of reactant. Whether we relate a given reactant to another reactant or a given reactant to a product, the calculations use the same format. Similarly, the format applies when we relate a given product to a reactant or a product to another product.

We can classify types of stoichiometry problems based on the given quantity and the quantity to be calculated. Consider the general equation for any reaction:

$$aA + bB \longrightarrow cC + dD$$

mass–mass problem A type of stoichiometry calculation that relates the masses of substances in a balanced chemical equation.

mass–volume problem A type of stoichiometry calculation that relates the mass of a substance to the volume of a gaseous substance in a balanced chemical equation.

volume–volume problem A type of stoichiometry calculation that relates the volumes of two gases (at the same temperature and pressure) according to a balanced chemical equation.

If we are given grams of A and asked to calculate how many grams of C are produced, the type of calculation is a **mass–mass problem**. If we are given grams of A and asked to calculate the milliliters of a gas D, it is a **mass–volume problem**. If we are given milliliters of a gas A and asked to find the liters of a gas D, the type is a **volume–volume problem**. To summarize, stoichiometry problems are of three basic types:

Type 1: mass–mass problem

Type 2: mass–volume problem

Type 3: volume–volume problem

Once we classify a problem, we can apply the unit analysis method of problem solving to its solution. Each type of problem has a solution that is characteristic of the unknown and given quantities. The following example exercise offers practice in applying this procedure.

EXAMPLE EXERCISE 9.4

Classify the type of stoichiometry problem for each of the following reactions.

(a) How many grams of zinc react with sulfuric acid to give 0.500 g of $ZnCl_2$?

(b) How many liters of hydrogen react with chlorine to yield 50.0 cm^3 of HCl gas?

(c) How many kilograms of iron react with hydrochloric acid to produce 50.0 mL of H_2 gas?

Solution: After analyzing the problem for the unknown quantity and the relevant given quantity, we can classify the type of problem.

(a) The problem asks for grams (mass) and we are given 0.500 g of $ZnCl_2$ (mass). This calculation is a *mass–mass* problem.

(b) The problem asks us to find liters (volume) and we are given 50.0 cm^3 of HCl gas (volume). This calculation is a *volume–volume* problem.

(c) The problem asks us to find kilograms (mass) and we are given 50.0 mL of H_2 gas (volume). This calculation relates mass to volume, so it is classified as a *mass–volume* type of problem.

SELF-TEST EXERCISE

Classify the type of stoichiometry problem for each of the following.

(a) How many grams of mercuric oxide must be decomposed to produce 0.500 L of oxygen at STP?

(b) How many grams of silver chloride are produced from the reaction of 0.500 g of sodium chloride and excess silver nitrate?

(c) Calculate the number of milliliters of hydrogen that are necessary to react with nitrogen to produce 1.00 L of ammonia gas.

Answers: **(a)** mass–volume problem; **(b)** mass–mass problem; **(c)** volume–volume problem

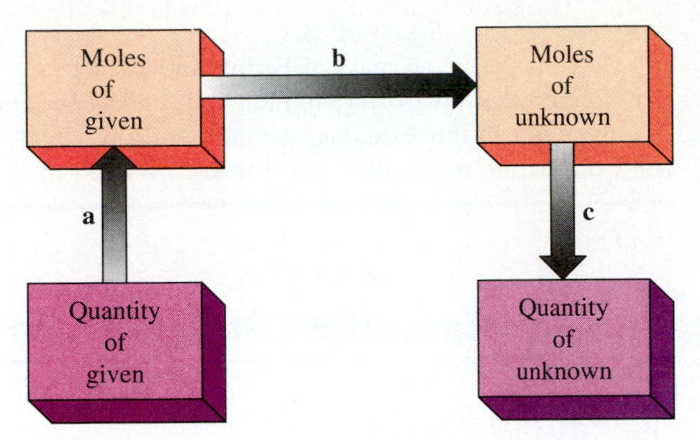

Figure 9.1 General Stoichiometry Format The quantity of given can be either a mass of substance or volume of gas. Similarly, the quantity of unknown can be either a mass or volume of gaseous substance. The quantity of given and quantity of unknown are related by applying the unit analysis method of problem solving.

In Chapter 7, we applied the unit analysis method of problem solving to chemical formula calculations. In this chapter, we are tackling more difficult stoichiometry problems. Although the problems are more complex, we can still apply the unit analysis method of problem solving. To perform stoichiometry calculations, we must always start with a balanced chemical equation. After balancing the equation, we proceed as follows:

(a) Convert the given quantity of substance (that is, a mass of substance or volume of gas) to moles of the given substance.

(b) Convert the number of moles of the given substance to the number of moles of the unknown substance using the coefficients from the balanced chemical equation.

(c) Convert the moles of the unknown to a quantity of substance (that is, a mass of substance or volume of gas).

Figure 9.1 illustrates the general format for the stoichiometry calculation process.

The following example exercise further illustrates the relationship of given and unknown quantities using the general stoichiometry format.

EXAMPLE EXERCISE 9.5

Heating iron(III) carbonate decomposes the solid to give a red iron(III) oxide powder and carbon dioxide gas. State the procedure for finding the unknown volume of CO_2 gas produced at STP from a given mass of $Fe_2(CO_3)_3$.

Solution: First, it is necessary to write a balanced chemical equation for the reaction; that is,

$$Fe_2(CO_3)_3(s) \longrightarrow Fe_2O_3(s) + 3\ CO_2(g)$$

Next, refer to the general stoichiometry format in Figure 9.1 and proceed as follows.

(a) Convert the given mass of $Fe_2(CO_3)_3$ into moles of $Fe_2(CO_3)_3$.

(b) Convert the moles of $Fe_2(CO_3)_3$ to moles of CO_2 using the coefficients from the balanced equation. In this example, the mole ratio is 1 to 3.

(c) Convert the moles of CO_2 to the unknown volume of CO_2 at STP.

SELF-TEST EXERCISE

Metallic filings of iron are heated with yellow sulfur powder to give iron(II) sulfide. Write the procedure for finding the mass of sulfur that reacts with a given mass of iron.

Answer: $Fe(s) + S(s) \rightarrow FeS(s)$.

(a) Convert the given mass of Fe into moles of Fe.

(b) Convert the moles of Fe to moles of S using the coefficients from the balanced equation. In this example, the mole ratio is 1 to 1.

(c) Convert the moles of S to the unknown mass of S.

9.4 Mass–Mass Stoichiometry Problems

To perform mass–mass stoichiometry calculations.

A mass–mass stoichiometry problem is so named because an *unknown mass* of a substance is calculated from a *given mass* of a reactant or product in a chemical equation. The process for solving mass–mass stoichiometry problems is outlined in Figure 9.2.

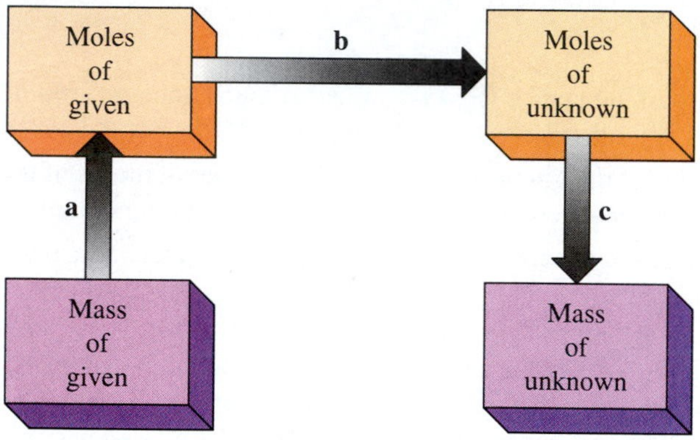

Moles of given —b→ Moles of unknown

Mass of given —a→ Moles of given

Moles of unknown —c→ Mass of unknown

Figure 9.2 Mass-Mass Stoichiometry Format We can see from the diagram that the mass of a given substance is related to the mass of the unknown. In performing mass-mass stoichiometry calculations, the unit factor method can be used for each step of the calculation.

After balancing the chemical equations, proceed as follows:

(a) Convert the relevant given mass of substance into moles using the molar mass as a unit factor.

(b) Convert the number of moles of the given substance into moles of unknown substance using the coefficients in the balanced chemical equation.

(c) Calculate the mass of the unknown substance by using the molar mass of the unknown as a unit factor.

To see how to solve a mass–mass problem, consider the high-temperature reduction of 14.4 g iron(II) oxide to elemental iron with aluminum metal. The balanced equation for the reaction is

$$3\ FeO(l) +, 2\ Al(l) \longrightarrow 3\ Fe(l) + Al_2O_3(l)$$

To calculate the mass of aluminum necessary for the reaction requires three steps. The first step is to calculate the moles of iron(II) oxide. Since the molar mass of FeO is 71.8 g/mol, the unit factor conversion is as follows:

$$14.4\ \text{g}\ \cancel{\text{FeO}} \times \frac{1\ \text{mol FeO}}{71.8\ \text{g}\ \cancel{\text{FeO}}} = 0.201\ \text{mol FeO}$$

In the second step, we convert the number of moles of FeO to moles of Al by applying the coefficients from the balanced equation. The relationship of the two substances is 3 mol FeO = 2 mol Al. The unit factor conversion is,

$$0.201 \cancel{\text{mol FeO}} \times \frac{2 \text{ mol Al}}{3 \cancel{\text{mol FeO}}} = 0.134 \text{ mol Al}$$

In the third step, we use the molar mass of aluminum as a unit factor to obtain mass of Al reacting with 14.4 g of FeO. Thus

$$0.134 \cancel{\text{mol Al}} \times \frac{27.0 \text{ g Al}}{1 \cancel{\text{mol Al}}} = 3.61 \text{ g Al}$$

The reaction of iron oxide and powdered aluminum is highly exothermic and produces a shower of metallic iron sparks.

After you gain confidence in the three-step method for stoichiometry problems, it will be more convenient to perform one continuous calculation. For example, the solution to the above problem can be shown as

$$14.4 \cancel{\text{g FeO}} \times \frac{1 \cancel{\text{mol FeO}}}{71.8 \cancel{\text{g FeO}}} \times \frac{2 \cancel{\text{mol Al}}}{3 \cancel{\text{mol FeO}}} \times \frac{27.0 \text{ g Al}}{1 \cancel{\text{mol Al}}} = 3.61 \text{ g Al}$$

The following example exercises give additional illustrations of mass–mass stoichiometry calculations using unit factors.

EXAMPLE EXERCISE 9.6

Calculate the mass of titanium (IV) chloride produced from the reaction of 1.25 g of titanium metal and excess chlorine gas.

$$\text{Ti}(s) + 2 \text{ Cl}_2(g) \longrightarrow \text{TiCl}_4(s)$$

Solution: In the first step, we calculate the number of moles of titanium. The molar mass of Ti is obtained from the periodic table, 47.9 g/mol.

$$1.25 \cancel{\text{g Ti}} \times \frac{1 \text{ mol Ti}}{47.9 \cancel{\text{g Ti}}} = 0.0261 \text{ mol Ti}$$

In the second step, we find the number of moles of $TiCl_4$ from the coefficients in the balanced equation. In this example, 1 mol Ti = 1 mol $TiCl_4$.

$$0.0261 \cancel{\text{mol Ti}} \times \frac{1 \text{ mol TiCl}_4}{1 \cancel{\text{mol Ti}}} = 0.0261 \text{ mol TiCl}_4$$

In the third step, we calculate the mass of product. The molar mass of $TiCl_4$ is 189.9 g/mol (47.9 + 35.5 + 35.5 + 35.5 + 35.5 = 189.9). Thus,

$$0.0261 \cancel{\text{mol TiCl}_4} \times \frac{189.9 \text{ g TiCl}_4}{1 \cancel{\text{mol TiCl}_4}} = 4.96 \text{ g TiCl}_4$$

SELF-TEST EXERCISE

Calculate the mass of carbon dioxide released from the decomposition of 10.0 g of cobalt(III) carbonate given the unbalanced chemical equation.

$$\text{Co}_2(\text{CO}_3)_3(s) \longrightarrow \text{Co}_2\text{O}_3(s) + \text{CO}_2(g)$$

Answer: 4.43 g CO_2

EXAMPLE EXERCISE 9.7

Calculate the mass of solid potassium iodide (166.0 g/mol) required to yield 1.78 g of yellow lead(II) iodide precipitate (461.0 g/mol).

$$2\ KI(s) + Pb(NO_3)_2(aq) \longrightarrow PbI_2(s) + 2\ KNO_3(aq)$$

Solution: First, calculate the number of moles of PbI_2. The molar mass of PbI_2 (461.0 g/mol) is given. Thus,

$$1.78\ \cancel{g\ PbI_2} \times \frac{1\ mol\ PbI_2}{461.0\ \cancel{g\ PbI_2}} = 0.00386\ mol\ PbI_2$$

Second, calculate the moles of potassium iodide required for the reaction. From the coefficients in the balanced equation, we have 2 mol KI = 1 mol PbI_2. Therefore, the unit factor is

$$0.00386\ \cancel{mol\ PbI_2} \times \frac{2\ mol\ KI}{1\ \cancel{mol\ PbI_2}} = 0.00772\ mol\ KI$$

Third, find the mass of potassium iodide. Since the molar mass of KI is 166.0 g/mol, we have

$$0.00772\ \cancel{mol\ KI} \times \frac{166.0\ g\ KI}{1\ \cancel{mol\ KI}} = 1.28\ g\ KI$$

In summary, 1.28 g of KI is necessary to yield 1.78 g of PbI_2 precipitate. Notice that the stoichiometry format is the same basic three-step process.

SELF-TEST EXERCISE

Calculate the mass of iron filings required to produce 0.455 g of silver metal, given the unbalanced equation.

$$Fe(s) + AgNO_3(aq) \longrightarrow Fe(NO_3)_3(aq) + Ag(s)$$

Answer: 0.0784 g Fe

note

It is not often that the exact amounts of chemicals, as calculated from the chemical equation, undergo reaction. More often, one of the reactants is more important (for example, more expensive or less available), and an excess quantity of other reactants is used to ensure that all the important reactant is converted to product. The more important reactant is called the *limiting reactant* because the amount of product formed is limited by its amount.

All chemical reactions stop when a reactant runs out. For example, the combustion reaction in an automobile stops when the gasoline runs out. In this example, the oxygen in the atmosphere that supports combustion is the excess reactant. Since there is usually an excess of less important reactants, a reaction stops when the limiting reactant is consumed.

The limiting reactant concept and associated calculations are covered in detail in Section 18.4. After finishing this chapter, you can go directly to this section with no break in continuity. Alternatively, the limiting reactant concept can be taken up at a later time after you have had a chance to assimilate the introductory aspects of stoichiometry presented in this chapter.

9.5 Mass–Volume Stoichiometry Problems

To perform mass–volume stoichiometry calculations.

In a mass–volume problem, we first convert the mass of a substance to moles. We then convert the moles of a given substance to moles of an unknown substance using the coefficients of the balanced equation. Finally, we multiply the moles of the unknown gaseous substance by the molar volume to obtain the volume in liters. Figure 9.3 shows this three-step process.

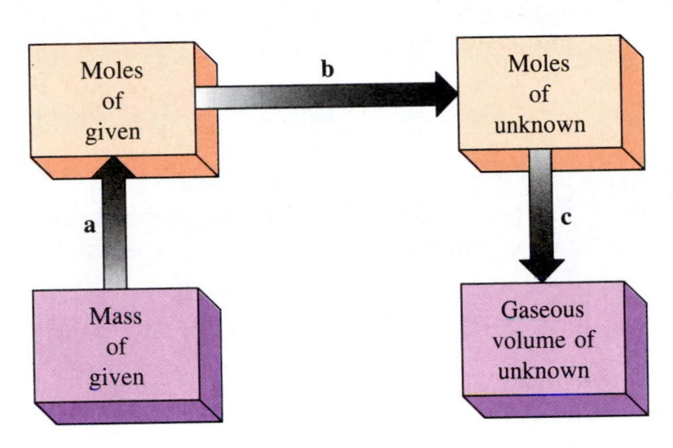

Figure 9.3 Mass-Volume Stoichiometry Format We can see from the diagram that the mass of given substance is related to the gaseous volume of the unknown. In performing mass-volume stoichiometry calculations, the unit analysis method can be used for each step of the calculation.

We must always begin with a balanced chemical equation. After balancing the equation, proceed as follows:

(a) Convert the relevant given mass of substance into moles using the molar mass as a unit factor.
(b) Convert the number of moles of given substance into moles of unknown substance using the coefficients in the balanced chemical equation.
(c) Calculate the volume of the unknown substance by using the molar volume of the gas (22.4 L/mol at STP) as a unit factor.

Of course, it is also possible to perform the reverse calculation; that is, we can find a mass of unknown substance given the volume of gas. Figure 9.4 shows the three-step process for a volume to mass calculation. For simplicity, this reverse process is also referred to as a mass–volume type of stoichiometry problem.

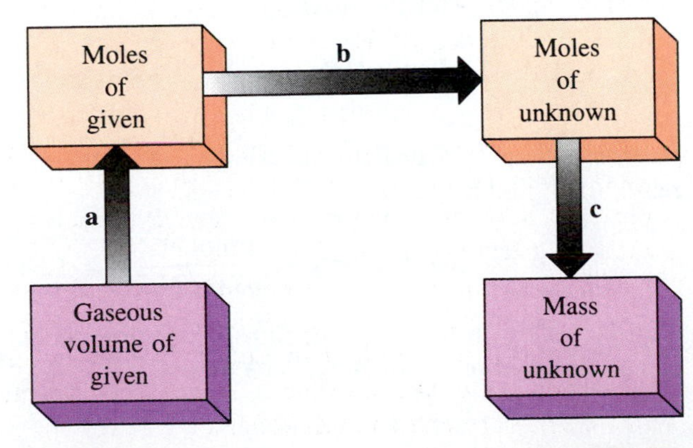

Figure 9.4 Volume-Mass Stoichiometry Format We can see from the diagram how the volume and moles of a given gaseous substance are related to the moles and mass of an unknown substance.

To see how to solve a mass–volume stoichiometry problem, consider the reaction of 0.165 g of aluminum metal with dilute hydrochloric acid. The balanced equation for the reaction is

$$2\ Al(s) + 6\ HCl(aq) \longrightarrow 2\ AlCl_3(aq) + 3\ H_2(g)$$

Let's calculate the volume, in milliliters, of hydrogen gas at STP produced from the reaction. This is a mass–volume stoichiometry problem and requires three steps. In the first step, we calculate the moles of aluminum. The molar mass of Al is 27.0 g/mol. Thus,

$$0.165\ \cancel{g\ Al} \times \frac{1\ mol\ Al}{27.0\ \cancel{g\ Al}} = 0.00611\ mol\ Al$$

In the second step, we use the coefficients in the balanced equation to find the moles of hydrogen gas. The relationship of the two substances is 2 mol Al = 3 mol H_2. Applying the unit factor, we have

$$0.00611\ \cancel{mol\ Al} \times \frac{3\ mol\ H_2}{2\ \cancel{mol\ Al}} = 0.00917\ mol\ H_2$$

In the third step, we multiply by the molar volume, 22.4 L/mol, to obtain the volume of H_2 gas produced at STP. Since the problem asked for mL units, we have one additional unit factor.

$$0.00917\ \cancel{mol\ H_2} \times \frac{22.4\ \cancel{L\ H_2}}{1\ \cancel{mol\ H_2}} \times \frac{1000\ mL\ H_2}{1\ \cancel{L\ H_2}} = 205\ mL\ H_2$$

Alternatively, the solution to the above problem can be performed as a continuous calculation:

$$0.165\ \cancel{g\ Al} \times \frac{1\ \cancel{mol\ Al}}{27.0\ \cancel{g\ Al}} \times \frac{3\ \cancel{mol\ H_2}}{2\ \cancel{mol\ Al}} \times \frac{22.4\ \cancel{L\ H_2}}{1\ \cancel{mol\ H_2}} \times \frac{1000\ mL\ H_2}{1\ \cancel{L\ H_2}} = 205\ mL\ H_2$$

In mass–volume stoichiometry, the problems usually require three unit factors to obtain a solution. In our problem, four unit factors were necessary because the units of volume were converted from liters to milliliters. The following example exercises provide additional practice in solving mass–volume stoichiometry problems.

EXAMPLE EXERCISE 9.8

Inflatable air-bags are a safety feature in many automobiles. In the event of a collision, a motion sensor sets off a spark, causing an unstable compound to decompose explosively. Assume that an air-bag contains 100.0 g of sodium azide NaN_3 (65.0 g/mol). Find the volume of nitrogen gas produced at STP, given the balanced chemical equation for the reaction.

$$2\ NaN_3(s) \xrightarrow{spark} 2\ Na(s) + 3\ N_2(g)$$

Solution: Step 1: Calculate the number of moles of NaN_3 using the given molar mass, 65.0 g/mol.

$$100.0\ \cancel{g\ NaN_3} \times \frac{1\ mol\ NaN_3}{65.0\ \cancel{g\ NaN_3}} = 1.54\ mol\ NaN_3$$

Step 2: Calculate the moles of nitrogen gas produced. According to the balanced equation, we have 2 mol NaN_3 = 3 mol N_2.

$$1.54 \cancel{\text{mol NaN}_3} \times \frac{3 \text{ mol N}_2}{2 \cancel{\text{mol NaN}_3}} = 2.31 \text{ mol N}_2$$

Step 3: Find the volume of nitrogen gas produced at STP. The molar volume is 22.4 L per mole.

$$2.31 \cancel{\text{mol N}_2} \times \frac{22.4 \text{ L N}_2}{1 \cancel{\text{mol N}_2}} = 51.7 \text{ L N}_2 \text{ (STP)}$$

In about 40 thousandths of a second, the air-bag fills with 51.7 L of nitrogen gas. This volume of gas is somewhat less than 2 cubic feet.

SELF-TEST EXERCISE

Calculate the volume of hydrogen gas produced at STP, given the unbalanced chemical equation for the reaction of 1.55 g of sodium metal.

$$\text{Na(s)} + \text{H}_2\text{O(l)} \longrightarrow \text{NaOH(aq)} + \text{H}_2\text{(g)}$$

Answer: 0.755 L H_2

EXAMPLE EXERCISE 9.9

Sodium bicarbonate can be used as a fire extinguisher. When heated, it decomposes to give carbon dioxide gas, which smothers a fire. If a sample of $NaHCO_3$ (84.0 g/mol) produces 0.500 L of carbon dioxide gas at STP, what is the mass of the sample? The balanced equation for the reaction is

$$2\ \text{NaHCO}_3\text{(s)} \longrightarrow \text{Na}_2\text{CO}_3\text{(s)} + \text{H}_2\text{O(g)} + \text{CO}_2\text{(g)}$$

Solution: In this problem, we will work from products to reactants. In step 1, we find the moles of CO_2. The volume occupied by 1 mole of CO_2 is 22.4 L at STP. Thus,

$$0.500 \cancel{\text{L CO}_2} \times \frac{1 \text{ mol CO}_2}{22.4 \cancel{\text{L CO}_2}} = 0.0223 \text{ mol CO}_2 \text{ (STP)}$$

In step 2, we find the moles of sodium bicarbonate gas used. According to the balanced equation, 2 moles of $NaHCO_3$ produce 1 mole of CO_2.

$$0.0223 \cancel{\text{mol CO}_2} \times \frac{2 \text{ mol NaHCO}_3}{1 \cancel{\text{mol CO}_2}} = 0.0446 \text{ mol NaHCO}_3$$

In step 3, we find the mass of sodium bicarbonate. The molar mass is given, 84.0 g/mol.

$$0.0446 \cancel{\text{mol NaHCO}_3} \times \frac{84.0 \text{ g NaHCO}_3}{1 \cancel{\text{mol NaHCO}_3}} = 3.75 \text{ g NaHCO}_3$$

SELF-TEST EXERCISE

Find the mass of aluminum required to produce 2160 mL of hydrogen gas at STP from the reaction with sulfuric acid given the unbalanced equation.

$$\text{Al(s)} + \text{H}_2\text{SO}_4\text{(aq)} \longrightarrow \text{Al}_2(\text{SO}_4)_3\text{(aq)} + \text{H}_2\text{(g)}$$

Answer: 1.74 g Al

9.6 Volume–Volume Stoichiometry Problems

To perform volume–volume stoichiometry calculations.

We can solve volume–volume stoichiometry problems using a simplified, one-step approach. According to Gay-Lussac's law of combining volumes, gases combine in small, whole-number ratios. Avogadro explained that equal volumes of gases, under similar conditions of temperature and pressure, contain the same number of molecules.

As we learned in Section 9.1, the coefficients of a balanced equation are a statement of the ratios of reacting volumes of gas. Therefore, volume–volume problems require a single step. Figure 9.5 shows this one-step process.

To see how to solve a volume–volume problem, let's look at an important industrial reaction. Sulfur dioxide gas is converted to sulfur trioxide using heat and platinum catalyst. The sulfur trioxide gas is then passed through water to produce sulfuric acid. This process is extremely important, since each year more sulfuric acid is used than any other single chemical. The balanced equation for the conversion of SO_2 to SO_3 is

$$2\ SO_2(g) + O_2(g) \xrightarrow{Pt} 2\ SO_3(g)$$

Let's calculate the volume of oxygen gas that reacts with 37.5 L of sulfur dioxide. This is a volume–volume problem and requires only one step. From the balanced equation we have 2 volumes of SO_2 = 1 volume of O_2. This relationship is valid regardless of the units of volume. Therefore, 2 L of SO_2 = 1 L of O_2.

$$37.5\ \cancel{L\ SO_2} \times \frac{1\ L\ O_2}{2\ \cancel{L\ SO_2}} = 18.8\ L\ O_2$$

Now let's calculate the volume, in liters, of sulfur trioxide produced. From the balanced equation we have 2 volumes of SO_2 = 2 volumes of SO_3. Therefore,

$$37.5\ \cancel{L\ SO_2} \times \frac{2\ L\ SO_3}{2\ \cancel{L\ SO_2}} = 37.5\ L\ SO_3$$

Example Exercise 9.10 further illustrates the application of stoichiometry principles to volume–volume type problems.

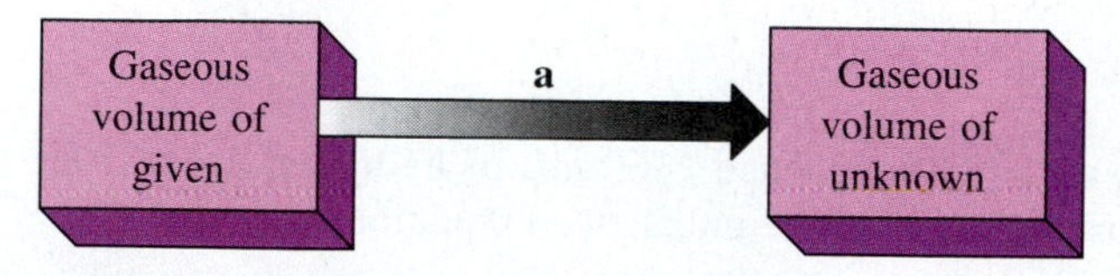

Figure 9.5 Diagram for Volume-Volume Stoichiometry Volume-volume stoichiometry problems involve a single step to relate the gaseous volume of the given to the gaseous volume of the unknown. Simply apply the coefficients in the balanced equation.

CHEMISTRY CONNECTION □ APPLIED

The Haber Process

► Can you name the common household chemical that is one of the ten most important industrial chemicals?

In 1990, over 15 million tons of ammonia were produced in the United States. The chief use of ammonia is for agriculture where it is used to fertilize and return nitrogen to the soil that crops consume. Although our atmosphere is a vast potential source of nitrogen for agriculture, the gas is in its elemental form and is essentially unreactive. With the exception of certain bacteria that can fix nitrogen from the atmosphere, the gas is inert.

In 1905, the German chemist Fritz Haber (1868–1934) successfully prepared ammonia for the first time in the laboratory. Haber discovered that nitrogen and hydrogen gases combine directly, at high temperatures and pressures, in the presence of metal oxide catalysts. The Haber process thus provides a method for converting unreactive atmospheric nitrogen into the versatile compound ammonia, NH_3. Ammonia, in turn, is easily converted into important compounds such as NH_4NO_3, $(NH_4)_2SO_4$, and $(NH_4)_2HPO_4$. Today, the Haber process is the main source for manufacturing nitrogen compounds throughout the world.

Until the beginning of the 1900s, the bulk of the world's nitrate was supplied from the rich saltpeter deposits in South America. Although nitrates were used principally as fertilizer, they were also used to make explosives. Indirectly, the Haber process had a pronounced effect on World War I. Soon after the outbreak of the war, the British Blockade halted the supply of Chilean saltpeter to Germany. Without a source of nitrate, Germany would have soon run out of ammunition. But by late 1914, Germany had built factories that applied the Haber process for the manufacture of ammonia and nitrate compounds.

It is often noted that Haber was very patriotic and sympathetic to Germany's war effort. He actively contributed to chemical warfare by helping to develop poisonous chlorine gas and the even more lethal mustard gas. In 1918, the same year World War I ended, Haber was awarded the Nobel prize in chemistry. Following the announcement of the Nobel prize, he was denounced by American, English, and French scientists for his involvement in the war.

It is an ironic twist of fate that Haber was later condemned by the Nazis because he was Jewish. In 1933, he was forced to resign as director of the Kaiser Wilhelm Institute and accepted an invitation to join Cambridge University. Only a few months later, he decided to travel through Europe and then visit Palestine. While in Switzerland, however, he became very ill and died.

Liquid ammonia is used in agricultural fields where it is injected into irrigation water or directly into the soil.

► Ammonia, NH_3, occurs in household products, such as window cleaner, and in 1990 was the sixth most important industrial chemical.

EXAMPLE EXERCISE 9.10

The Haber process combines elemental nitrogen and hydrogen gases to give ammonia gas. Given 5.55 L of nitrogen, calculate the volume of hydrogen that reacts and the volume of ammonia produced. Assume all gas volumes are measured at the same conditions.

$$N_2(g) + 3\ H_2(g) \xrightarrow[500°C/300\ atm]{Fe/Al_2O_3} 2\ NH_3(g)$$

Solution: The coefficients of the balanced chemical equation indicate that 1 volume of N_2 = 3 volumes of H_2. We calculate the volume of H_2 as follows:

$$5.55\ L\ N_2 \times \frac{3\ L\ H_2}{1\ \cancel{L\ N_2}} = 16.7\ L\ H_2$$

The equation also indicates that 1 volume of N_2 = 2 volumes of NH_3. Therefore,

$$5.55\ \cancel{L\ N_2} \times \frac{2\ L\ NH_3}{1\ \cancel{L\ N_2}} = 11.1\ L\ NH_3$$

SELF-TEST EXERCISE

Calculate the volumes of (a) nitrogen monoxide and (b) bromine gas that react to yield 50.0 mL of nitrosyl bromide gas, NOBr, given the following unbalanced equation. Assume all gases are at similar conditions.

$$NO(g) + Br_2(g) \longrightarrow NOBr(g)$$

Answers: **(a)** 50.0 mL NO; **(b)** 25.0 mL Br_2

Mass–mass and mass–volume stoichiometry problems require three unit factors for their solution. Volume–volume problems require only one unit factor. Since the volume of a gas depends on temperature and pressure, in volume–volume problems the experimental conditions must be the same before and after reaction. The conditions need not be STP. It is only necessary that the conditions remain constant.

9.7 Percent Yield

To calculate the percent yield for a reaction, given the actual yield and theoretical yield.

To understand the concept of yield, consider the following laboratory experiment. First, a student weighs a 2.000-g sample of barium chloride and dissolves it in water. Then, the student adds aqueous potassium chromate solution to obtain a precipitate of barium chromate. The balanced equation for the reaction is

$$BaCl_2(aq) + K_2CrO_4(aq) \longrightarrow BaCrO_4(s) + 2\ KCl(aq)$$

The student collects the insoluble barium chromate in filter paper and weighs its mass on a balance. The mass of precipitate is called the experimental yield or **actual yield**.

The mass of barium chromate obtained, starting with the 2.000-g sample of barium chloride, can be predicted using a stoichiometry calculation. The calculated amount of precipitate that can be obtained is called the **theoretical yield**. We assume all of the sample is converted to product when we calculate the theoretical yield. In practice, this is usually not the case as there are factors that affect yield in every experiment. We can express the lab results in terms of percent yield. The **percent yield** is the actual yield compared to the theoretical yield expressed as a percent. The equation is

actual yield The amount of substance experimentally measured in a laboratory procedure.

theoretical yield The amount of product that is calculated from given amounts of reactants.

percent yield The ratio of the actual yield compared to the theoretical yield, all times 100.

$$\frac{\text{actual yield}}{\text{theoretical yield}} \times 100 = \text{percent yield}$$

To find percent yield, we start with the 2.000-g sample of barium chloride. Applying stoichiometry to the balanced equation, we can calculate the amount of barium chromate precipitate, which in this example is 2.433 g. Suppose the student weighs a precipitate and finds it has a mass of 2.399 g. To calculate the percent yield we compare the actual yield (2.399 g) to the theoretical yield (2.433 g) as follows:

$$\frac{2.399 \text{ g}}{2.433 \text{ g}} \times 100 = 98.60\%$$

In this trial the percent yield is 98.60%.

Suppose the student performed a second trial for the same experiment. Again, the student used a 2.000-g sample of barium chloride. In this second trial, the precipitate had a mass of 2.555 g. Since the student started with 2.000 g of barium chloride, the theoretical yield for the two trials is the same, that is, 2.433 g. As before, to find the percent yield we compare the actual and theoretical yields.

$$\frac{2.555 \text{ g}}{2.433 \text{ g}} \times 100 = 105.0\%$$

Notice that the percent yield for this second trial is greater than 100%. Although surprising, this result is possible. When performing an experiment, some errors lead to high results, while other errors lead to low results. In the two trials there were different types of experimental errors. Example Exercise 9.11 provides additional practice in calculating percent yield.

EXAMPLE EXERCISE 9.11

A chemistry student prepared aspirin in the laboratory. Starting with 1.500 g of salicylic acid, the student collected 1.183 g of aspirin. If the theoretical yield is 1.956 g, what is the percent yield?

Solution: The percent yield is the ratio of the actual yield compared to the theoretical yield. In this experiment, the actual yield is 1.183 g and the theoretical yield is 1.956 g. The percent yield is, therefore,

$$\frac{1.183 \text{ g}}{1.956 \text{ g}} \times 100 = \text{percent yield}$$

$$= 60.48\%$$

Although the percent yield is only about 60%, this is a typical student result for an aspirin synthesis experiment.

SELF-TEST EXERCISE

Ammonium nitrate is used in explosives and is produced from the reaction of ammonia, NH_3, and nitric acid. The unbalanced equation is

$$NH_3(g) + HNO_3(aq) \longrightarrow NH_4NO_3(s)$$

If 15.0 kg of ammonia give an actual yield of 65.3 kg of ammonium nitrate, what is the percent yield? The calculated yield of ammonium nitrate for the experiment is 70.6 kg.

Answer: 92.5%

9.8 Experimental Accuracy and Precision

To express accuracy and precision for an analysis, given the experimental results and the theoretical value.

In Chapter 1 we discussed the uncertainty in measurements obtained from instruments. In addition to instrumental error, there is analyst error associated with every laboratory technique. Frequently, a laboratory analysis involves three or more experimental trials. This is because an analysis with three trials provides a check on the experimental results. For example, consider the following three-trial analysis. A student performed a three-trial analysis for the sodium chloride content in an unknown sample and obtained the following results:

Trial 1 = 90.5% NaCl

Trial 2 = 86.2% NaCl

Trial 3 = 88.0% NaCl

The student reported the average value of 88.2% for the three trials. The average was obtained by finding the sum of the results (90.5% + 86.2% + 88.0%) and dividing by the number of trials (3). The laboratory instructor graded the result based on student error. To find the absolute error, or simply **error**, the instructor subtracted the theoretical result from the experimental result. That is,

error Expresses the difference between the experimental result and the theoretical value for an analysis.

$$\text{experimental result} - \text{theoretical value} = \text{error}$$

Thus, if the theoretical value for the unknown is 90.0%, the error is calculated as follows:

$$\text{experimental result} - \text{theoretical value} = \text{error}$$

$$88.2\% - 90.0\% = -1.8\%$$

Notice that the error is −1.8%. A negative error indictes that the experimental result was low. A positive error indicates that the experimental result was high. An experienced chemist knows the type of errors that lead to high or low results. If the error is too great to be acceptable, the experimental procedure is changed.

The **accuracy** of the results obtained from an experiment is a measure of student, or instrument, error. The **range** of results is the difference between the high and low values. In the above example, the results spread from a high of 90.5% to a low of 86.2%. We can express the difference as follows:

accuracy Refers to the error in the result obtained from an experiment.

range Expresses the spread of results for an experiment having two or more trials.

$$\text{highest result} - \text{lowest result} = \text{range}$$

$$90.5\% - 86.2\% = 4.3\%$$

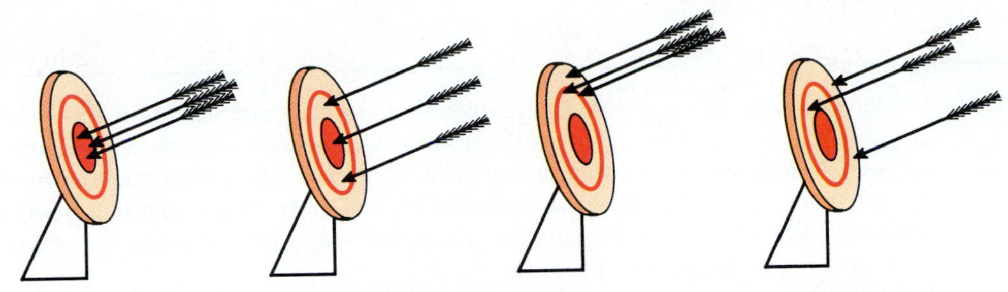

1. Accurate/precise 2. Accurate/not precise 3. Precise/not accurate 4. Not precise/not accurate

Figure 9.6 Bullseye Analogy for Accuracy and Precision The results of three trial analyses can be likened to three arrows shot at an archery target. The results may all be accurate (target 1) or the average may be accurate (target 2). The results may be precise but not accurate (target 3), or neither precise nor accurate (target 4).

The **precision** of an analysis measures consistency from trial to trial. If an experimental procedure has good precision, the number of trials can be reduced. If a method lacks precision, more trials are required. The best results are obtained using an experimental method that is both accurate and precise.

precision Refers to the range in the results obtained from an experiment.

Figure 9.6 illustrates, by analogy, the possible combinations of accuracy and precision.

The following example exercise further illustrates the principles of accuracy and precision for a set of experimental data.

EXAMPLE EXERCISE 9.12

A mixture containing barium chloride was analyzed by precipitation. Three trials were performed, which gave the following results: 45.0%, 46.6%, and 43.9%. Assuming that the theoretical value for barium chloride is 43.5%, express the (a) accuracy and (b) precision of the analysis.

Solution: Accuracy is expressed as experimental error. Precision is expressed as the range of results. Thus,

(a) The experimental result minus the theoretical value equals the error. The best experimental result is the average of the three trials: (45.0% + 46.6% + 43.9%) divided by 3 = 45.2%. Since the theoretical value is 43.5%, we can express the accuracy as

$$45.2\% - 43.5\% = 1.7\% \text{ error}$$

(b) The high result minus the low result equals the range. The high value is 46.6% and the low value is 43.9%; therefore, we can express the precision as

$$46.6\% - 43.9\% = 2.7\% \text{ range}$$

SELF-TEST EXERCISE

A mixture containing potassium chlorate was analyzed by decomposition. Three trials were performed, which gave the following results: 72.5%, 70.5%, and 69.3%. Assuming that the theoretical value for potassium chlorate is 70.0%, express the (a) accuracy and (b) precision of the analysis.

Answers: **(a)** 0.8% error; **(b)** 3.2% range

Summary

Sections 9.1–9.2 In this chapter, we learned to interpret the coefficients in a balanced chemical equation in terms of molecules of substance, moles of substance, or volumes of gaseous substance. Furthermore, we verified the **law of conservation of mass** by substituting the molar mass for amounts of each mole of reactant and product. Given a balanced chemical equation for a reaction, we can relate the moles of each substance in the reactants or products.

Section 9.3 We also learned to classify three types of stoichiometry problems (see Table 9.2). In an **mass–mass problem**, the masses of two substances are related according to a balanced chemical equation. The two substances can be reactants, products, or a reactant and a product. In a **mass–volume problem**, the mass and volume of gaseous reactants and products are related. In a **volume–volume problem**, the volumes of two gases are related. The first two types of stoichiometry problems require a three-step process. A volume–volume problem can be solved in one step.

Sections 9.4–9.6 In stoichiometry problems, we most often relate the amount of a reactant to the amount of product. However, it is also possible to relate the amounts of two reactants or two products. The following diagram will help you visualize the overall process.

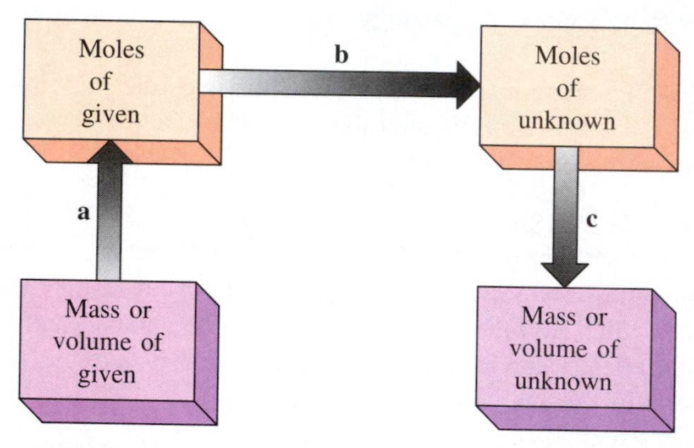

To convert from a mass of substance to the number of moles of substance, we use molar mass. The molar mass is either given in the problem or found from adding up the atomic masses found in the periodic table. To convert from volume of gas to moles of gas, we use the molar volume. The volume occupied by 1 mole of any gas is 22.4 liters at STP. Standard conditions are 0°C and 1 atmosphere pressure.

Table 9.2 Summary of Stoichiometry Format

Type of Problem*	Steps in Conversion
mass–mass	(a) mass of given to moles of given (b) moles of given to moles of unknown (c) moles of unknown to mass of unknown
mass–volume	(a) mass of given to moles of given (b) moles of given to moles of unknown (c) moles of unknown to gaseous volume of unknown
volume–volume	(a) gaseous volume of given to gaseous volume of unknown

*Every type of stoichiometry problem must always begin with a balanced chemical equation.

Section 9.7 The percent yield is an evaluation of a laboratory method and the skill of the analyst. **Percent yield** is defined as the ratio of the **actual yield** compared to a **theoretical yield**, all times 100.

Section 9.8 **Accuracy** is a measure of the **error** in an analysis. Error is calculated by subtracting the theoretical value from the experimental result. Error may either be positive or negative, depending on whether or not the result is greater than or less than the theoretical value. **Precision** is a measure of the **range** in results for an analysis involving more than one experimental trial. The best results from a laboratory analysis are both accurate and precise.

Key Terms

Select the key term that corresponds to the following definitions.

____ **1.** the principle that, at the same temperature and pressure, the volumes of gases that combine in a chemical reaction are in the ratio of small whole numbers

____ **2.** the principle that, at the same temperature and pressure, equal volumes of gases contain equal numbers of molecules

____ **3.** the relationship of quantities (for example, mass of substance or volume of gas) in a chemical change according to the balanced chemical equation

____ **4.** the statement that the total mass of reactants in a chemical reaction is equal to the total mass of products

____ **5.** the mass in grams of 1 mole of a substance

____ **6.** a type of calculation that relates the moles of two substances participating in a balanced chemical equation

____ **7.** a type of stoichiometry calculation that relates the masses of substances in a balanced chemical equation

____ **8.** a type of stoichiometry calculation that relates the mass of a substance to the volume of a gaseous substance in a balanced chemical equation

____ **9.** a type of stoichiometry calculation that relates the volumes of two gaseous substances (at the same temperature and pressure) in a balanced chemical equation

____ **10.** the volume occupied by 1 mole of any gas at standard conditions; 22.4 L/mol

____ **11.** a gas at 0°C and 1.00 atmosphere pressure

____ **12.** the amount of substance experimentally measured in a laboratory procedure

____ **13.** the amount of product that is calculated to be obtained from a given amount of reactant

____ **14.** the ratio of the actual yield compared to the theoretical yield, all times 100

____ **15.** expresses the difference between the experimental result and the theoretical value for an analysis

____ **16.** expresses the spread of results for an experiment having two or more trials

____ **17.** refers to the error in the results obtained from an experiment

____ **18.** refers to the range in the results obtained from an experiment

(a) accuracy *(Sec. 9.8)*
(b) actual yield *(Sec. 9.7)*
(c) Avogadro's Law *(Sec. 9.1)*
(d) error *(Sec. 9.8)*
(e) law of combining volumes *(Sec. 9.1)*
(f) law of conservation of mass *(Sec. 9.1)*
(g) mass–mass problem *(Sec. 9.3)*
(h) mass–volume problem *(Sec. 9.3)*
(i) molar mass *(Intro.)*
(j) molar volume *(Intro.)*
(k) mole–mole problem *(Sec. 9.2)*
(l) percent yield *(Sec. 9.7)*
(m) precision *(Sec. 9.8)*
(n) range *(Sec. 9.8)*
(o) standard conditions *(Intro.)*
(p) stoichiometry *(Intro.)*
(q) theoretical yield *(Sec. 9.7)*
(r) volume–volume problem *(Sec. 9.3)*

Exercises

Interpreting a Chemical Equation (Sec. 9.1)

1. State three quantities that are proportional to the coefficients in the balanced equation for a chemical reaction.
2. Consider the general chemical equation:

$$2\ A + 3\ B \longrightarrow C + 2\ D$$

   (a) How many molecules of B react with two molecules of A?
   (b) How many moles of C are produced from two moles of A?
   (c) How many volumes of gas D are produced from two volumes of gas A?
   (d) How many molar masses of A react to give one molar mass of C?

3. Consider the general chemical equation:

$$A + 2\ B \longrightarrow C + 3\ D$$

   (a) How many formula units of B react with 5 formula units of A?
   (b) How many moles of C are produced from 4 moles of A?
   (c) How many liters of gas D are produced from 2 liters of gas A?
   (d) How many molar masses of B react to give 6 molar masses of D?

4. Consider the general chemical equation: $A + 3\ B \longrightarrow 2\ C$.
   (a) If 10.0 g of A react with 15.0 g of B, what is the mass of the product C?
   (b) If 50.0 g of A react to produce 75.0 g of C, what is the mass of B that reacted?
5. Verify the conservation of mass law by calculating the molar masses of reactants and products for each of the following balanced equations.
   (a) $2\ KNO_3(s) \rightarrow 2\ KNO_2(s) + O_2(g)$
   (b) $2\ Al(s) + 3\ S(s) \rightarrow Al_2S_3(s)$
6. Verify the conservation of mass law by calculating the molar masses of reactants and products for each of the following unbalanced equations.
   (a) $CH_4(g) + O_2(g) \xrightarrow{\text{spark}} CO_2(g) + H_2O(g)$
   (b) $C_2H_6(g) + O_2(g) \xrightarrow{\text{spark}} CO_2(g) + H_2O(g)$

Mole-Mole Problems (Sec. 9.2)

7. Given a balanced chemical equation, how many moles of oxygen react with 0.500 mol of hydrogen gas? How many moles of water vapor are produced?

$$2\ H_2(g) + O_2(g) \longrightarrow 2\ H_2O(g)$$

8. How many moles of oxygen gas react with 2.50 mol of phosphorus? How many moles of diphosphorus pentaoxide are produced?

$$P(s) + O_2(g) \longrightarrow P_2O_5(s)$$

9. Given a balanced chemical equation, how many moles of chlorine react with 0.333 mol of metallic iron? How many moles of iron(III) chloride are produced?

$$2\ Fe(s) + 3\ Cl_2(g) \longrightarrow 2\ FeCl_3(s)$$

10. How many moles of barium undergo reaction to produce 0.333 mol of barium nitride? How many moles of nitrogen gas react?

$$Ba(s) + N_2(g) \longrightarrow Ba_3N_2(s)$$

11. Given a balanced chemical equation, how many moles of ammonia react to give 1.50 mol of nitrogen monoxide gas? How many moles of water are produced? How many moles of oxygen react?

$$4\ NH_3(g) + 5\ O_2(g) \longrightarrow 4\ NO(g) + 6\ H_2O(g)$$

12. How many moles of propane react with 1.75 mol of oxygen gas? How many moles of water are produced? How many moles of carbon dioxide are produced?

$$C_3H_8(g) + O_2(g) \longrightarrow CO_2(g) + H_2O(g)$$

Types of Stoichiometry Problems (Sec. 9.3)

13. Classify the type of stoichiometry problem for each of the following.
    (a) How many kilograms of iron oxide are produced from the reaction of excess steam and 256.4 g of iron?
    (b) How many milliliters of nitrogen dioxide are produced from the reaction of excess oxygen and 1.00 L of nitrogen monoxide gas?
    (c) How many milliliters of hydrogen gas are produced from the reaction of excess sulfuric acid and 1.414 g of cadmium metal?
14. Classify the type of stoichiometry problem for each of the following.
    (a) How many liters of hydrogen gas react with bromine to yield 25.0 mL of HBr gas?
    (b) How many grams of of aluminum react with nitric acid to give 0.403 g of $Al(NO_3)_3$?
    (c) How many grams of potassium chlorate are decomposed by heating to yield 45.5 cm^3 of O_2 gas?

Mass-Mass Stoichiometry Problems (Sec. 9.4)

15. Given the balanced chemical equation, calculate the mass in grams of zinc oxide that can be prepared from 2.36 g of zinc metal.

$$2\ Zn(s) + O_2(g) \longrightarrow 2\ ZnO(s)$$

16. How many grams of oxygen must react to give 1.28 g of ZnO? (See equation in exercise 15.)

17. Given the balanced chemical equation, find the mass in grams of bismuth chloride that can be produced from 3.45 g of bismuth metal.

$$2\ Bi(s) + 3\ Cl_2(g) \longrightarrow 2\ BiCl_3(s)$$

18. How many grams of chlorine gas must react to give 3.52 g of $BiCl_3$? (See equation in Exercise 17.)

19. Given the balanced chemical equation, what is the mass in grams of silver that are prepared from 0.615 g of copper metal?

$$Cu(s) + 2\ AgNO_3(aq) \longrightarrow Cu(NO_3)_2(aq) + 2\ Ag(s)$$

20. How many grams of silver nitrate must react to give 1.00 g of Ag? (See equation in Exercise 19.)

21. Balance the following chemical equation and calculate the grams of mercury that are produced from 1.25 g of cobalt metal.

$$Co(s) + HgCl_2(aq) \longrightarrow CoCl_3\ (aq) + Hg(l)$$

22. How many grams of mercuric chloride must react to give 5.11 g of Hg? (See equation in exercise 21.)

23. Balance the following chemical equation and determine the mass of calcium phosphate that is precipitated, assuming 1.78 g of Na_3PO_4 reacts completely.

$$Na_3PO_4(aq) + Ca(OH)_2(aq) \longrightarrow Ca_3(PO_4)_2(s) + NaOH(aq)$$

24. What mass of calcium hydroxide must react to give 2.39 g of $Ca_3(PO_4)_2$? (See equation in exercise 23.)

Mass-Volume Stoichiometry Problems (Sec. 9.5)

25. Given the balanced chemical equation, how many milliliters of carbon dioxide gas at STP are produced from the decomposition of 1.59 g of ferric carbonate?

$$Fe_2(CO_3)_3(s) \longrightarrow Fe_2O_3(s) + 3\ CO_2(g)$$

26. How many milliliters of oxygen gas at STP are released from the decomposition of 2.57 g of calcium chlorate?

$$Ca(ClO_3)_2(s) \longrightarrow CaCl_2(s) + O_2(g)$$

27. Given the balanced chemical equation, how many milliliters of carbon dioxide gas at STP are liberated from the decomposition of 1.59 g of lithium hydrogen carbonate?

$$2\ LiHCO_3(s) \longrightarrow Li_2CO_3(s) + H_2O(l) + CO_2(g)$$

28. How many milliliters of oxygen gas at STP are released from the decomposition of 2.50 g of mercuric oxide?

$$HgO(s) \longrightarrow Hg(l) + O_2(g)$$

29. Given the balanced chemical equation, calculate the mass of magnesium metal that reacts with sulfuric acid to evolve 225 mL of hydrogen gas at STP.

$$Mg(s) + H_2SO_4(aq) \longrightarrow MgSO_4(aq) + H_2(g)$$

30. Calculate the grams of sodium metal that react with water to give 75.0 mL of hydrogen gas at STP.

$$Na(s) + H_2O(l) \longrightarrow NaOH(aq) + H_2(g)$$

31. Given the balanced chemical equation, how many grams of hydrogen peroxide must decompose to give 55.0 mL of oxygen gas at STP?

$$2\ H_2O_2(l) \longrightarrow 2\ H_2O(l) + O_2(g)$$

32. How many grams of manganese(II) chloride must react with sulfuric acid to release 49.5 mL of hydrogen chloride gas at STP?

$$MnCl_2(s) + H_2SO_4(aq) \longrightarrow MnSO_4(aq) + HCl(g)$$

Volume-Volume Stoichiometry Problems (Sec. 9.6)

33. Given the balanced chemical equation, how many liters of oxygen gas react with 2.00 L of carbon monoxide? (Assume all gas volumes are measured at the same temperature and pressure.)

$$2\ CO(g) + O_2(g) \longrightarrow 2\ CO_2(g)$$

34. What is the volume of hydrogen iodide gas that is produced from 125 mL of hydrogen? (Assume all gas volumes are measured at the same conditions.)

$$H_2(g) + I_2(g) \longrightarrow HI(g)$$

35. Given the balanced chemical equation, calculate the volumes of (a) nitrogen gas and (b) hydrogen gas that react to give 38.5 mL of ammonia gas. (Assume all gases are at the same conditions.)

$$3\ H_2(g) + N_2(g) \longrightarrow 2\ NH_3(g)$$

36. Calculate the volumes of (a) chlorine gas and (b) oxygen gas that react to yield 1.75 L of dichlorine trioxide. (Assume all gases are at the same conditions.)

$$Cl_2(g) + O_2(g) \longrightarrow Cl_2O_3(g)$$

37. Given the balanced chemical equation, calculate the volume of sulfur trioxide gas produced from 25.0 L of oxygen gas. (Assume all gases are at the same conditions.)

$$2\ SO_2(g) + O_2(g) \longrightarrow 2\ SO_3(g)$$

38. Calculate the volumes of (a) nitrogen gas and (b) oxygen gas that react to yield 500.0 mL of dinitrogen pentaoxide. (Assume all gases are at the same conditions.)

$$N_2(g) + O_2(g) \longrightarrow N_2O_5(g)$$

Percent Yield (Sec. 9.7)

39. A 1.55-g sample of lead(II) nitrate reacts with aqueous potassium iodide to give insoluble lead(II) iodide weighing 2.01 g. If the theoretical yield of precipitate is 2.16 g, what is the percent yield?

40. A beginning chemistry student made acetone by decomposing calcium acetate. Starting with 31.6 g of calcium acetate, the student collected 10.4 g of acetone. If the calculated yield is 11.6 g, what is the percent yield?

41. A 1.50-g sample of sodium nitrate is decomposed by heat. The resulting sodium nitrite weighs 1.29 g. If the theoretical yield of sodium nitrite is 1.22 g, what is the percent yield?

42. A 1.000-g sample of potassium bicarbonate is decomposed by heat. The resulting potassium carbonate has a mass of 0.725 g. If the calculated yield of potassium carbonate is 0.716 g, what is the percent yield?

Experimental Accuracy and Precision (Sec. 9.8)

43. Dolomite tablets are taken as a dietary supplement for calcium and magnesium. A sample of dolomite was analyzed for its calcium content and gave the following results: 19.5%, 20.4%, and 19.1%. Assuming the theoretical value for calcium is 21.7%, express the (a) error and (b) range for the analysis.

44. When dolomite was analyzed for its magnesium content, it gave the following results: 15.5%, 14.2%, and 14.1%. Assuming the theoretical value is 13.2% Mg, express the (a) error and (b) range for the analysis.

45. The amount of copper in a copper penny was determined by a weighing method. Three pennies were analyzed, which gave the following percentages of copper: 93.5%, 97.0%, and 96.9%. Assuming the theoretical value for copper in a copper penny is 95.0%, calculate the (a) accuracy and (b) precision for the analysis.

46. The amount of copper in a "zinc penny" was determined by a weighing method. Three pennies were analyzed, which gave the following percentages of copper: 2.55%, 2.41%, and 2.35%. Assuming the theoretical value for copper in a "zinc penny" is 2.50%, calculate the (a) accuracy and (b) precision for the analysis.

General Exercises

47. The volcano reaction produces green chromium(III) oxide, water, and nitrogen gas from the decomposition of orange ammonium dichromate. Calculate the (a) mass in grams of chromium(III) oxide and (b) STP volume in milliliters of nitrogen gas from the decomposition of 1.54 g of ammonium dichromate.

48. Manganese metal and aluminum oxide are produced from the reaction of manganese(IV) oxide and aluminum. How many grams of aluminum are necessary to give 1.00 kg of manganese metal?

49. The mineral stibnite, antimony(III) sulfide, is treated with hydrochloric acid to give antimony(III) chloride and hydrogen sulfide gas. What is the STP volume of H_2S evolved from a 3.00-g sample of stibnite?

50. When an electric current is passed through water, hydrogen gas and oxygen gas are liberated. Calculate the STP milliliter volume of each gas from the electrolysis of 100.0 mL of water. (*Hint:* The density of water is 1.00 g/mL.)

51. Propane tanks are commonplace on recreational vehicles. The combustion of propane and oxygen produces carbon dioxide and water. Calculate the (a) mass in grams of water and (b) milliliter volume of carbon dioxide gas at STP from the combustion of 10.0 g of propane, C_3H_8.

CHAPTER

10

Modern Atomic Theory

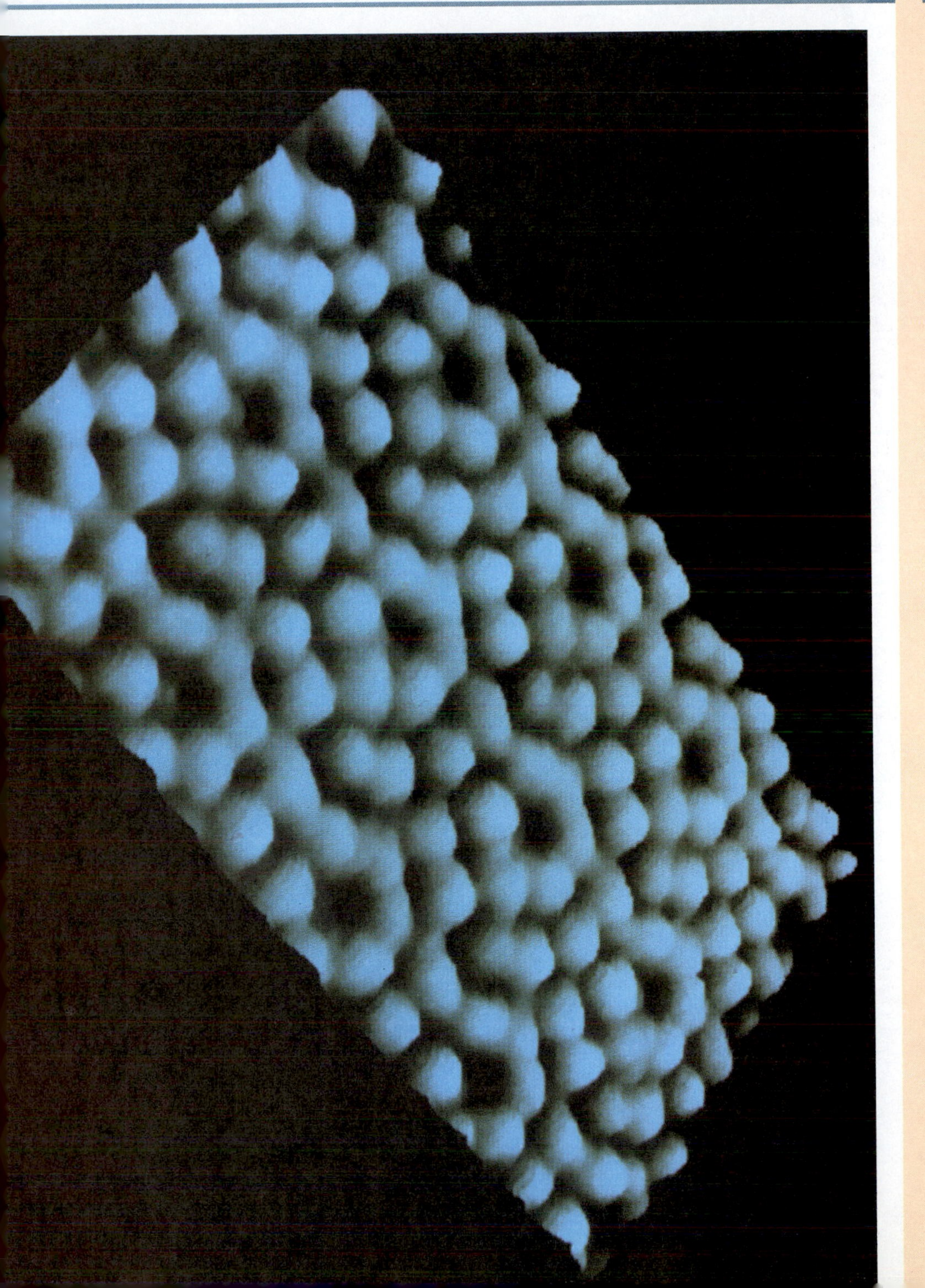

The blue spheres in the photograph are images of silicon atoms magnified over a hundred million times. Using an instrument called a scanning tunneling microscope (STM), it is possible to obtain images of individual atoms.

In the sixteenth century, the Polish astronomer Nicholas Copernicus (1473–1543) presented a revolutionary view of the universe. Since the time of the Greek astronomer Ptolemy, it had been thought that the earth was at the center of the universe and all heavenly bodies revolved about our planet. In the face of great opposition, Copernicus suggested that all the planets in our solar system, including the earth, circled about the sun.

Although it was heresy to believe that the earth was not the center of the universe, the German astronomer Johannes Kepler (1571–1630) agreed with Copernicus. Kepler was a gifted mathematician and observer of planetary motion. After meticulous calculation, he found that he could not accurately predict the path of the planets assuming they traveled in circular orbits about the sun. Later, he modified Copernicus' proposal and stated that the planets travel in orbits that are ovals, or ellipses—and not circles. Describing planetary motion as elliptical orbits about the sun agreed with mathematical calculations (Figure 10.1). Thus, the journey began to scientifically understand the outer universe.

In the seventeenth century, the microscope was invented by the Dutch scientist Anthony van Leeuwenhoek (1632–1723). Although the first microscopes were not very powerful, Leeuwenhoek was able to observe bacteria and even red blood cells. Thus, the journey began to scientifically understand the inner world (Figure 10.2). Today, using a scanning tunneling microscope, individual atoms can be observed.

Historically, astronomy was probing further into the universe while physics was penetrating deeper into the atom. It is not surprising, therefore, that astronomers

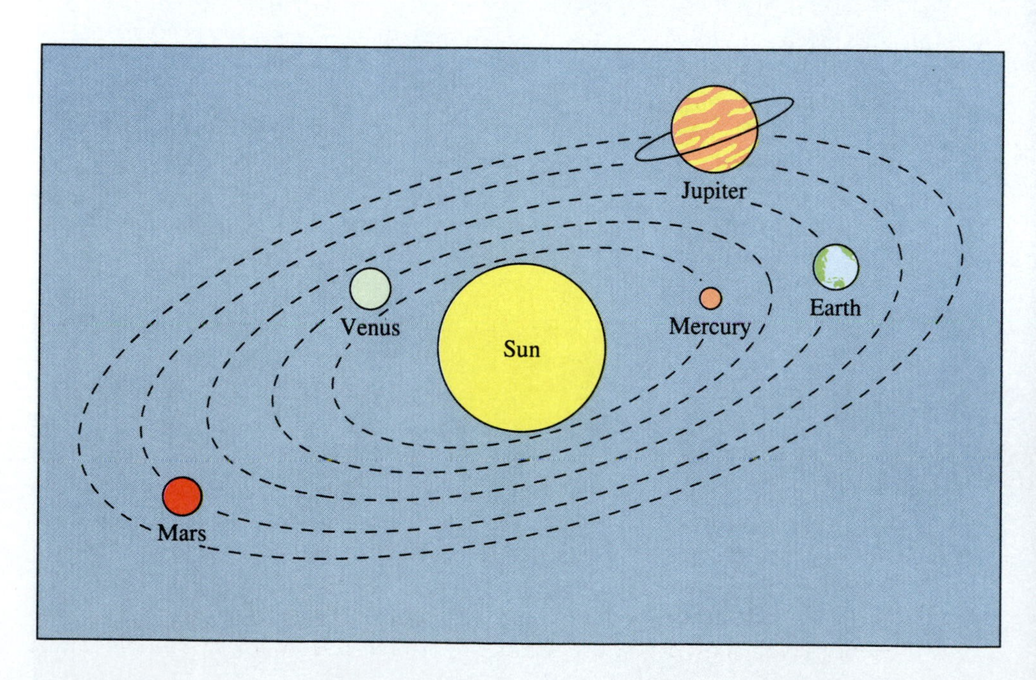

Figure 10.1 Early Model of the Solar System The planets travel in elliptical orbits about the sun. Similarly, the planetary model of the atom was proposed with our solar system in mind. That is, electrons orbit about the nucleus as the planets orbit about the sun.

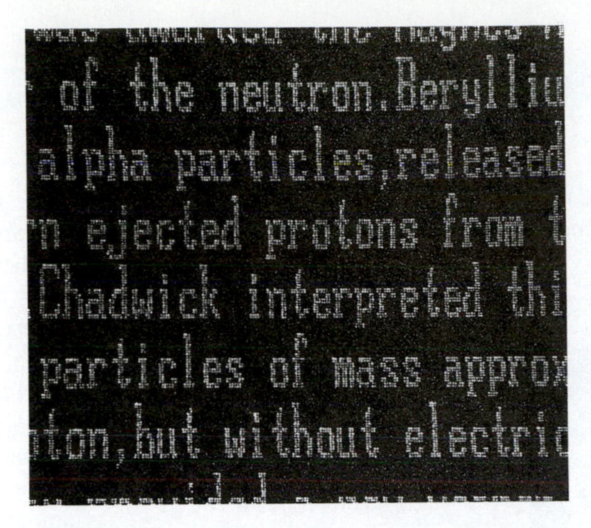

Figure 10.2 Writing with an Electron Beam A beam of high-energy electrons has been used to etch the surface of an aluminum fluoride crystal with a few lines from the *Encyclopedia Brittanica*. The letters are so small that the entire contents of the *Encyclopedia Brittanica* would occupy an area no larger than the size of the period at the end of this sentence.

and physicists used similar models to describe their journey. In 1913, the planetary model of the atom described electrons in circular orbits about the nucleus. Later, the electron orbits were changed from circles to ellipses, just as Kepler had modified the planetary orbits suggested by Copernicus.

10.1 Bohr Model of the Atom

OBJECTIVES

To describe the Bohr model of the atom.

In Chapter 4, we introduced a model of the atom proposed in 1913 by the Danish physicist Niels Bohr. According to the **Bohr atom**, electrons orbit in circles about the nucleus in much the same way the planets orbit the sun. One important flaw in this analogy is that the planets in our solar system are slowly losing energy and spiraling toward the sun. (Eventually, in perhaps billions of years, the earth will spiral into the sun.) Conversely, in the Bohr model of the atom, the electron does not lose energy and occupies a fixed energy state.

In the Bohr model, electrons are found only in circular orbits that possess a discrete amount of energy. The electron can gain energy, however, and jump to an orbit farther from the nucleus. Orbits farther from the nucleus are considered to be of higher energy. Conversely, an excited electron can lose energy and drop to an orbit closer to the nucleus that is of lower energy. The electron must, however, occupy a discrete orbit of specific energy. The electron is said to occupy an allowed energy state and cannot occupy just any position of unspecified energy.

The greatest triumph of the Bohr model of the atom was that it explained the lines in the emission spectrum of hydrogen. That is, each line in the emission spectrum was the result of excited electrons dropping from one given energy level to another. You may recall from Section 4.7 that, when excited electrons drop to the second energy level, they produce lines in the visible spectrum. The red line corresponds to electrons dropping from the third energy level to the second; the blue-green line corresponds to electrons dropping from the fourth level to the second; the violet line results from electrons dropping from the fifth to second level.

When electrons drop to the first energy level, they produce a series of powerful ultraviolet lines that are not visible to the human eye. When electrons drop to the third energy level, they produce a series of infrared lines that are also invisible.

CHEMISTRY CONNECTION □ APPLIED

Fireworks

► **Can you name the element that is responsible for the red color in the fireworks display?**

Have you ever watched a Fourth of July fireworks display and wondered how the colors were produced? The color of fireworks is explained as follows. A rocket shell is packed with chemicals, fitted with a fuse, and fired into the air. When the fuse burns down, gunpowder ignites and sets off an explosion that propels chemicals throughout the sky. If the rocket shell is packed with a sodium compound, a yellow color is produced. If the shell is packed with a barium compound, a green color is observed.

Not all chemicals produce a colored display. You may recall having seen fireworks that simply produce a shower of white sparks. Aluminum or magnesium metals can be used to produce this effect. The following table lists the chemicals and metals used to produce various colors in fireworks.

Chemical/ Metal	Fireworks Color
Na compounds	yellow
Ba compounds	green
Ca compounds	orange
Sr compounds	red
Li compounds	crimson
Cu compounds	blue
Al or Mg metals	white sparks
Fe filings	gold sparks

Interestingly, the colors of fireworks and the colors of "neon" advertising signs are explained using the same principle. In a fireworks display an element is energized by heat, whereas in a "neon" light, an element is energized by electricity. In either case, electrons are first excited and then immediately lose energy by emitting light. The observed color corresponds to the wavelength of the light emitted. For example, if the emitted light has a wavelength around 650 nm, it is observed as red. If the light has a wavelength around 450 nm, it appears to be blue.

Chemists can often identify an element by the color of its flame test. For example, sodium gives a yellow flame test and barium a green flame test. Flame-test colors are identical to fireworks colors containing the same element. In the laboratory, a flame test is performed by placing a small amount of chemical on the tip of a wire and placing the wire in a hot flame. The flame test for lithium is illustrated below.

► **The red color suggests that the element is either strontium or lithium.**

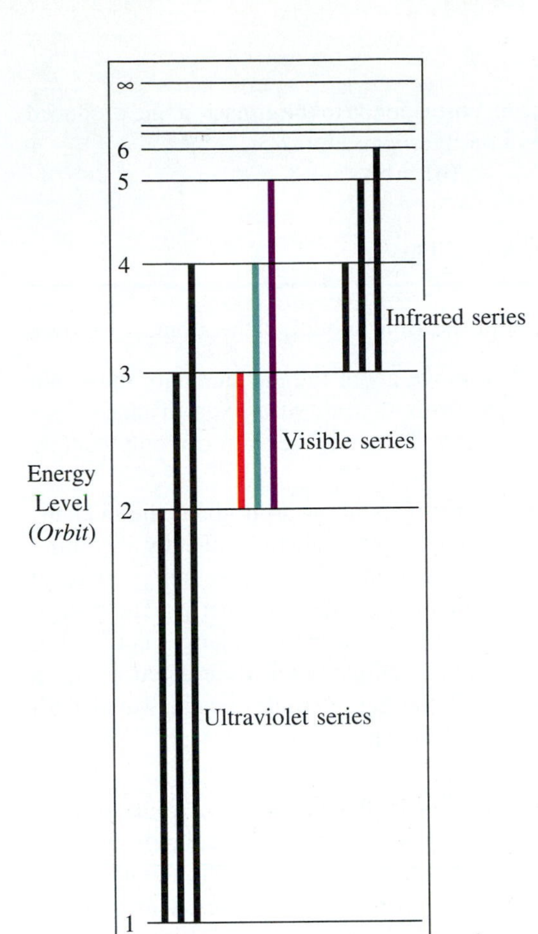

Figure 10.3 Emission Lines and Energy Level in the Hydrogen Atom The emission spectrum of hydrogen produces several lines. However, only when electrons drop to the second level is the radiation in the visible spectrum. When electrons drop to the first energy level, they produce lines in the ultraviolet spectrum. When electrons drop to the third energy level, they produce lines in the infrared spectrum.

Figure 10.3 illustrates electron transitions between orbits in the hydrogen atom.

In the Bohr atom, the energy levels are closer together as the electron orbits are farther from the nucleus. In Figure 10.3, we notice that the energy difference between the second and third orbits is less than between the first and second orbits. And the energy difference between the third and fourth orbits is less than between the second and third orbits. The following example exercise relates energy levels in the hydrogen atom to the observed spectral lines.

EXAMPLE EXERCISE 10.1

Which of the following lines in the emission spectrum of hydrogen is most energetic: red, blue-green, or violet?

Solution: Referring to Figure 10.3, we see that the emission lines are the result of the following electron transitions in the hydrogen atom:

red line: electrons drop from energy level 3 to 2
blue-green line: electrons drop from energy level 4 to 2
violet line: electrons drop from energy level 5 to 2

Since the energy difference is greatest when electrons drop from 5 to 2, the violet line is most energetic. Conversely, the energy difference is least for the 3 to 2 transition, so the red line is the least energetic.

SELF-TEST EXERCISE

What region of the radiant energy spectrum corresponds to the emission line produced from the following electron transitions in a hydrogen atom?

(**a**) orbit 3 to 1 (**b**) orbit 4 to 3
(**c**) orbit 6 to 2

Answers: (**a**) ultraviolet; (**b**) infrared; (**c**) visible

We can summarize the description of the Bohr model of the atom as follows:

1. The electron in a hydrogen atom can move about the nucleus only in certain circular orbits. The electrons occupy orbits that are at a fixed distance from the nucleus and possess a definite energy. Thus, the electron can only occupy certain allowed energy states.
2. The electron does not lose energy as it travels in an orbit about the nucleus. However, the electron can change from one energy state to another by absorbing or emitting radiant energy. When electrons drop from one energy state to another, particles of light are emitted that have a specific energy. The energy of a light particle is related to its wavelength, and the wavelength is found in a particular region of the spectrum. For example, each time electrons drop from the third energy level to the second, the light particles have a wavelength that is found in the red region of the visible spectrum.

Note: In the early part of the twentieth century, physicists were rapidly gaining a deeper understanding of atomic structure. Almost immediately after the Bohr atom was introduced, flaws began to develop in the model. Upon careful examination, the energy state of the electron was not consistent with the electron moving in a circular orbit. Similar to Kepler's suggestion regarding planetary motion, physicists theorized that the electron moved in elliptical orbits. Using an elliptical orbit to describe the energy state of an electron agreed more closely with experimental data.

The greatest limitation of the Bohr model, however, is that it could not adequately explain the energy states of electrons in atoms other than hydrogen. Moreover, it could not be extended to explain the behavior of electrons in molecules. Little by little, evidence accumulated that the Bohr atom was limited at best, and perhaps the model was incorrect.

Max Planck was a professor of physics at the University of Berlin. In 1900, he proposed the quantum theory and in 1918 he was awarded the Nobel prize for his revolutionary idea.

10.2 Quantum Theory

OBJECTIVES

To apply the quantum theory to the description of electrons in atoms.

The study of physics previous to 1900 is called classical physics and is based on the laws of motion stated by the English physicist Isaac Newton (1642–1727). Classical physics adequately explains the movement of large objects, including the planets. However, classical physics does not adequately explain the behavior of small particles in the atomic world. The study of physics after 1900 is called modern physics and is based on the quantum concept introduced by the German physicist Max Planck (1858–1947). Modern physics, or quantum physics, explains the behavior of very small particles such as electrons.

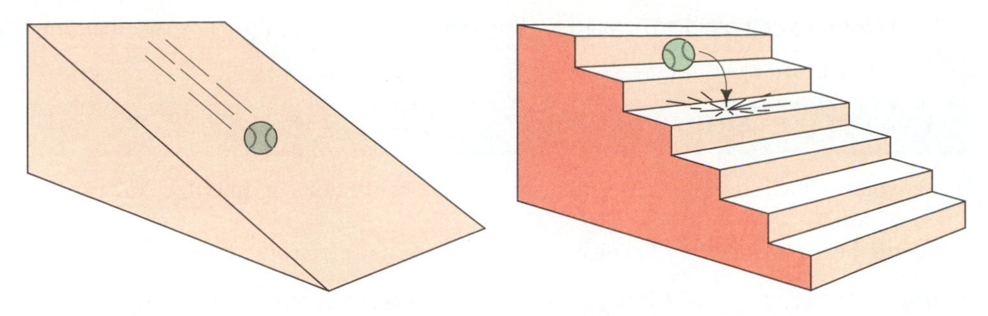

Figure 10.4 Stair Analogy for the Quantum Principle A ball rolling down a ramp loses potential energy continuously. Conversely, a ball rolling down a flight of stairs loses potential energy in specific amounts each time it drops from one step to another. In this analogy, the loss of potential energy on the ramp is continuous and on the stairs is quantized.

In Section 4.7, we briefly mentioned the quantum concept to explain the emission spectra of electrically excited gases. When an excited gas loses energy, we might expect it to lose energy gradually and continuously. This is not the case. If an excited gas lost energy gradually, the emitted radiation would be a continuous band of light energy. Recall, however, that the emission spectrum of an excited gas in a discharge tube is not continuous. Rather, the emission spectrum contains a series of narrow colorful lines.

In 1900 Planck proposed that when a body radiates light energy, the energy exists in tiny bundles. He called a tiny bundle of radiant energy a **quantum** (*pl.* quanta). According to Planck's **quantum theory**, a single red line in an emission spectrum is actually composed of many individual quanta. Niels Bohr combined quantum theory and line spectra experiments to establish the idea of discrete energy levels for electrons in atoms. Specifically, Bohr stated that the electron in a hydrogen atom may be found only in discrete levels where its energy is fixed.

quantum theory A theory that specifically explains the gain or loss of energy by electrons in atoms. A theory that generally explains energy changes in objects by small specific increments.

If the quantum idea seems difficult, consider the following analogy. A heavy ball rolling down a ramp continuously loses potential energy while gaining kinetic energy. Conversely, a heavy ball rolling down a flight of stairs loses potential energy in specific amounts each time it drops from one step to another. In this analogy, the loss of potential energy is continuous as the ball rolls down the ramp and quantized as it rolls down the stairs. Figure 10.4 contrasts the continuous change in energy versus the quantized change in energy.

Another analogy for a quantized event versus a continuous change involves an electric guitar and keyboard. Since the guitar is a string instrument, it is possible to alter its pitch continuously. On the other hand, the keyboard has only specific, discrete notes that can be played. Thus, the pitch of a guitar can be changed continuously, while the keyboard represents quantized changes. Figure 10.5 offers another analogy for a quantized change.

Figure 10.5 Musical Analogy for the Quantum Principle Although it is possible to continuously vary the pitch of a violin, it is only possible to vary the notes on a piano in steps. In this analogy, the piano notes are quantized, while the violin notes are not.

Albert Einstein and Niels Bohr, who were close friends, are shown walking through the streets of Brussels, Belgium in 1933.

The following example exercise illustrates practical applications of quantum theory.

EXAMPLE EXERCISE 10.2

State whether the following instruments produce a continuous or quantized measurement of mass.

(a) triple-beam mechanical balance **(b)** digital electronic balance

Solution: Refer to Figure 1.2 if you have not used these balances in the laboratory.

(a) On a mechanical balance, a small metal rider is moved along the most sensitive beam until balance is achieved. Since the rider can be moved to any position on the beam, a mechanical balance gives a *continuous* mass measurement.

(b) On a digital electronic balance, the display indicates the mass of an object to a particular decimal place. For example, a nickel coin may have a mass of 5.015 g. Since the display is limited to three decimal places, a digital electronic balance produces a *quantized* mass measurement.

SELF-TEST EXERCISE

State whether the following instruments produce a continuous or quantized measurement of time.

(a) watch with digital display **(b)** clock with sweep-second hand

Answers: **(a)** quantized; **(b)** continuous

The Photoelectric Effect

In 1900, Max Planck announced the quantum theory and subsequently received the Nobel prize in 1918 for his contribution. In 1905, Albert Einstein (1879–1955) used quantum theory to explain the photoelectric effect. It was for his explanation of the photoelectric effect that Einstein won the Nobel prize in 1921, and not for the theory of relativity (Section 3.11) for which he is more famous.

What, then, is the photoelectric effect? When light shines on a metallic surface, the metal can emit electrons. This is called the **photoelectric effect**. However, the light must be sufficiently energetic; that is, the light must have a minimum frequency to dislodge electrons from a metal. Moreover, each metal has a characteristic minimum frequency that is required to produce the photoelectric effect. This phenomenon only occurs when the light has sufficient energy, regardless of its intensity. If a certain energy of violet light is required for a metal to emit electrons, the phenomenon does not occur with red light no matter how intensely or how long the red light shines on the metal. Red light has a lower frequency than violet light and simply does not have the energy required to "knock" electrons out of the metal. Figure 10.6 illustrates this phenomenon.

photoelectric effect The phenomenon of a metal ejecting an electron when struck by a photon of sufficient energy.

Einstein explained the photoelectric effect using quantum theory as follows. Light is a wave of radiant energy composed of individual particles called **photons**. When light shines on a metallic surface, an individual photon of light must possess sufficient energy to knock an electron out of an atom on the surface of the metal. As shown in Figure 10.6, a photon of violet light has sufficient energy to dislodge an electron in a metal atom. A photon of red light, however, is not powerful enough to knock out an electron. No matter how many red photons strike the metal, red light cannot produce the photoelectric effect. On the other hand, a violet photon striking the metal gives the photoelectric effect because violet light has sufficient energy.

photon An individual unit of radiant energy that corresponds to the particle nature of light.

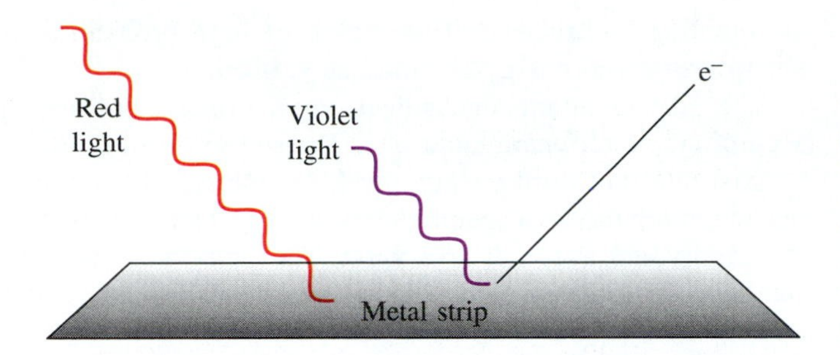

Figure 10.6 Photoelectric Effect In this illustration the more energetic violet light knocks electrons out of the metal. The red light, which has a lower frequency and is not as energetic as the violet light, cannot produce the photoelectric effect.

There are many practical applications of the photoelectric effect. Scientific instruments called spectrophotometers are used to measure subtle energy changes in molecules when the molecules are radiated with light. These instruments have detectors that trigger an electric current when struck by light energy. The detector relies on the photoelectric effect to function. Similarly, remote controls for stereos, televisions, and garage doors operate using this principle.

10.3 Quantum Mechanical Model of the Atom

OBJECTIVES

To describe the quantum mechanical model of the atom in terms of the probability of finding an electron with a given energy.

To describe the relative sizes and shapes of *s* and *p* orbitals.

In the mid-1920s, an entirely new model of the atom began to emerge. Specifically, the behavior of electrons in atoms could not be explained using the Bohr model of the atom. The German physicist Werner Heisenberg (1901–1976) utilized a type of mathematics called matrix algebra to calculate the energy of electrons in atoms. His calculations were based on the wavelengths of emitted light in the emission spectra of the elements. He further concluded that it was not possible to determine accurately both the position and energy of an electron. In his **uncertainty principle**, Heisenberg stated that it is impossible to measure both the location and **momentum** of a small particle simultaneously. In fact, the more precisely the position of an electron in an atom is known, the less precisely can its momentum be calculated.

uncertainty principle the statement that it is impossible to precisely measure both the location and momentum of a particle at the same time.

momentum The quantity that expresses the mass of a particle times its velocity. The momentum of an electron is equal to its mass times its velocity.

In 1932, Heisenberg won the Nobel prize in physics for his uncertainty principle. Not everyone, however, subscribed to the principle of uncertainty. Some physicists found it unsettling to consider that they might live in a universe ruled by chance. Even Einstein was sufficiently troubled by uncertainty that he offered the famous quote, "*It seems hard to look in God's cards. But I cannot for a moment believe He plays dice as the current quantum theory alleges He does.*" Although the uncertainty principle was initially controversial, it was an essential contribution to the new view of the atom.

Gradually, the deeper nature of the atom came into focus. The new model retained the idea of quantized energy levels but incorporated the concept of uncertainty. The new model that emerged became known as the **quantum mechanical atom**. Recall that in the Bohr model the energy of an electron is defined in terms of a fixed orbit about the nucleus. In the quantum mechanical model, the energy of an electron is described in terms of its probability of being within a spatial volume

quantum mechanical atom A sophisticated model of the atom that describes the energy of an electron in terms of its probability of being found in a particular location about the nucleus.

orbital A region in space surrounding the nucleus of an atom in which there is a high probabilty (~95%) of finding an electron with a given energy.

quantum energy level A general term that refers to an allowed energy state for an electron in an atom.

principal quantum number (*n*) The quantum number that describes the size and energy of a particular orbital. The allowed values of *n* are 1, 2, 3, ...

second quantum number (*l*) The quantum number that describes the shape of an orbital; also referred to as the subsidiary quantum number. The allowed values of *l* are 0, 1, 2, 3, . . . , $n-1$.

surrounding the nucleus. This region of high probability (~95%) for finding an electron of given energy is called an **orbital**.

In the quantum mechanical atom, orbitals are arranged about the nucleus according to their energy and size. The energy of an orbital is quantized and is said to exist in a **quantum energy level**, or simply, a quantum level. Thus, an electron in a given orbital has a specific, fixed energy. Quantum numbers are used to describe the energy and shape of an orbital. The **principal quantum number** (symbol *n*) can have the integer values 1, 2, 3, 4, ... As the value of *n* increases, the energy and size of a particular orbital also increase.

Principal Quantum Number:

Number designation:	1	2	3	4	5	6	. . .
Orbital energy:	*increasing* →						
Orbital size:	*increasing* →						

The **second quantum number** (*l*) can have the integer values 0, 1, 2, . . . up to a maximum of $n-1$. The value of *l* describes the shape of a particular orbital. In lieu of quantum numbers, the letters *s*, *p*, *d*, and *f* are often used to designate the shapes of orbitals. For example, the notations 4*s*, 4*p*, 4*d*, and 4*f* designate the shapes of different orbitals having a principal quantum number of 4. The letters *s*, *p*, *d*, and *f* correspond to the *l* values of 0, 1, 2, and 3, respectively. When *l* equals 0, the shape of the orbital is that of a sphere. When *l* equals 1, the shape of the orbital is that of a dumbbell. For higher values of *l*, orbital shapes are too complex for our discussion.

Second Quantum Number:

Number designation:	0	1	2	3
Letter designation:	*s*	*p*	*d*	*f*
Orbital shape:	sphere	dumbbell	complex	complex

Sizes and Shapes of Orbitals

The size and shape of an orbital are usually designated by combining the principal quantum number and the letter of the second quantum number. For example, the designations 1*s*, 2*p*, and 3*d* indicate three orbitals that differ in size, energy, and shape. All *s* orbitals are spherically shaped, but they are not all the same size. A 3*s* orbital is a larger sphere than a 2*s*; a 2*s* orbital is larger than a 1*s*. Figure 10.7 illustrates the relationship of *s* orbitals about the nucleus.

All *p* orbitals have the same shape, but they are not all equal in size. As with *s* orbitals, the size of *p* orbitals increases as the value of *n* becomes greater. Therefore,

Figure 10.7 Relative Size and Relationship of *s* Orbitals
The relative size of 1*s*, 2*s*, and 3*s* orbitals are shown. As the principal quantum number becomes greater, the size and energy of the orbital increases. The atomic nucleus is considered to be where the axes cross. The sketch on the far right demonstrates the interrelationship of the 1*s*, 2*s*, and 3*s* orbitals.

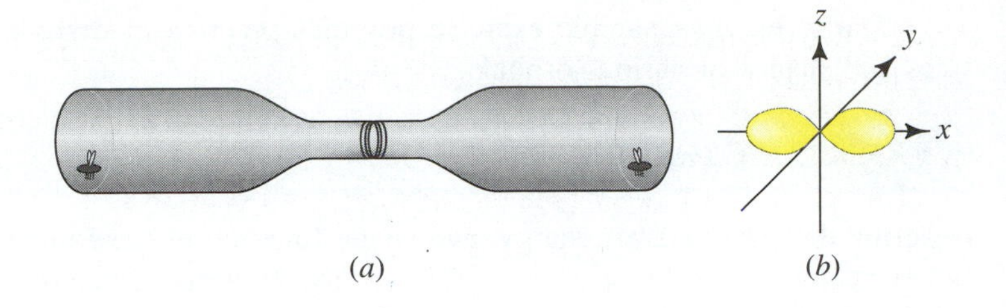

Figure 10.8 Analogy for a *p* Orbital (a) Notice the two insects trapped within the bottles held end-to-end. The two insects can both be in the left bottle, the right bottle, or one insect can be in the left bottle and the other in the right. (b) Similarly, two electrons have a high probability of being found anywhere within the two lobes of the *p* orbital.

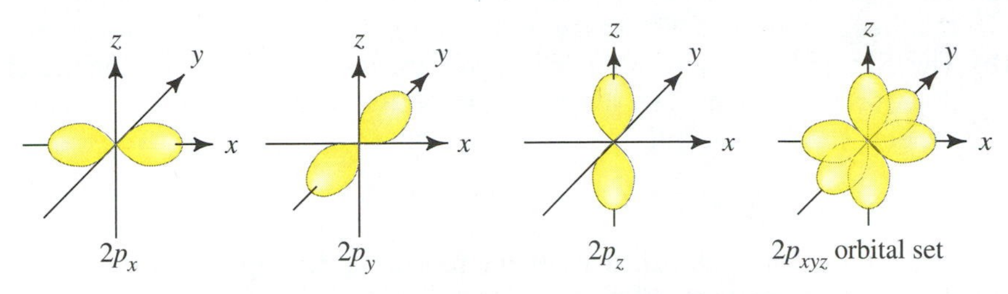

Figure 10.9 Orientation and Relationship of 2*p* Orbitals The size and shape of the three 2*p* orbitals are identical. The 2*p* orbitals do not have an absolute orientation, but rather, they are mutually perpendicular to each other. An electron in a $2p_x$ orbital has the identical probability of occupying the $2p_y$ or $2p_z$ orbital.

a 3*p* orbital is larger than a 2*p* orbital. Every *p* orbital has two lobes and is said to resemble a dumbbell. Electrons in a *p* orbital may occupy any portion of the orbital volume. By analogy, try to visualize a flying insect trapped inside two bottles with their open ends held together. The insect would be free to fly about the entire inner volume generated by the two bottles. In this analogy, the insect represents an electron and the two bottles correspond to the two lobes of a single *p* orbital. Thus, there is a high probability of finding the electron anywhere within the volume of the entire *p* orbital. Figure 10.8 illustrates this analogy for electrons in orbitals.

There are three different *p* orbitals in each principal quantum level. Although the three 2*p* orbitals are identical in size and shape, they differ in their orientation to each other. That is, the three 2*p* orbitals intersect at the nucleus, but they are oriented in different directions. Figure 10.9 illustrates the relationship of the $2p_x$, $2p_y$, and $2p_z$ orbitals. The p_x orbital is oriented along the *x*-axis of a three-dimensional axes system. The p_y and p_z orbitals are oriented along the *y*-axis and *z*-axis, respectively.

Let's proceed one step further in our discussion and illustrate the relationship of the 1*s*, 2*s*, and 2*p* orbitals to each other. In Figure 10.10, the 1*s*, 2*s*, and $2p_x$ orbitals are shown individually and collectively. Notice that the $2p_x$ orbital is shown superimposed onto the *s* orbitals. In these orbital diagrams, we have simulated a three-dimensional model by the intersection of *x*-, *y*-, and *z*-axes.

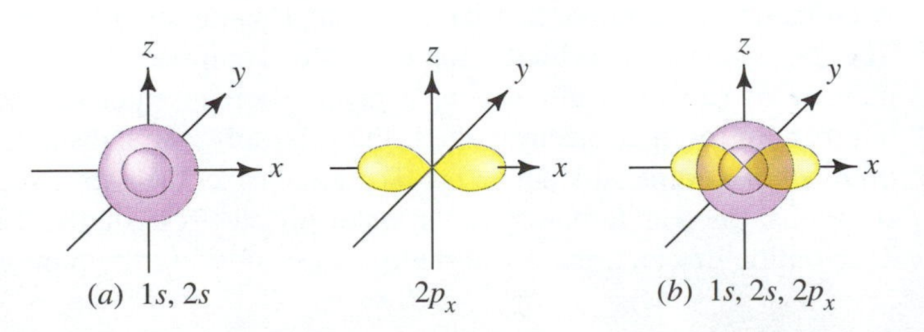

Figure 10.10 Relationship of *s* and *p* Orbitals (a) The relative size and shape of the 1*s*, 2*s*, and $2p_x$ orbitals are shown. (b) Notice that the three orbitals overlap. Thus, there are regions where there is a small probability of finding electrons of different energy.

The following example exercise provides practice in describing the relative sizes and shapes of various orbitals.

EXAMPLE EXERCISE 10.3

Describe the relative size, energy, and shape for each of the following orbitals.

(a) 4*s* versus 3*s* and 5*s* **(b)** 4*p* versus 3*p* and 5*p*

Solution: The size and energy of an orbital are indicated by the principal quantum number. A letter designation following the number indicates the shape of the orbital.

(a) The size and energy of a 4*s* orbital are greater than for a 3*s* orbital, but are less than for a 5*s* orbital. The shape of a 4*s* orbital, and all *s* orbitals, is spherical.

(b) The size and energy of a 4*p* orbital are greater than for a 3*p* orbital and less than for a 5*p* orbital. The shape of a 4*p* orbital, and all *p* orbitals, is similar to the shape of a dumbbell.

SELF-TEST EXERCISE

Select the orbital in each pair that fits the following descriptions.

(a) the higher-energy orbital: 3*p* or 4*p*

(b) the larger-size orbital: 4*d* or 5*d*

Answers: **(a)** 4*p*; **(b)** 5*d*

10.4 Distribution of Electrons by Orbital

OBJECTIVES

To designate the maximum number of electrons in a given orbital, subshell, or shell.

The rules of quantum mechanics limit orbitals to a maximum of two electrons. It makes no difference what type, *s*, *p*, *d*, or *f*; a single orbital can hold no more than two electrons. However, there can be more than one orbital with the same energy and shape. For example, the $2p_x$, $2p_y$, and $2p_z$ orbitals are identical except for their orientation. A collection of orbitals that have the same value of *n* and *l* is sometimes called an **electron subshell**.

electron subshell A collection of orbitals that have the same value of *n* and *l*; for example, the $2p_x$, $2p_y$, and $2p_z$ orbitals comprise the 2*p* subshell.

It therefore follows that the *s* subshell can hold a maximum of 2 electrons because there is only one *s* orbital. There are three orbitals in a *p* subshell. Thus, a *p* subshell can hold a maximum of 6 electrons. The *d* subshell is composed of five orbitals and has a maximum of 10 electrons. The *f* subshell has seven orbitals and can have a maximum of 14 electrons.

The principal quantum number (*n*) designates the size and energy of an orbital. A collection of orbitals that have the same value of *n* is called an **electron shell**. The 3*s*, 3*p*, and 3*d* orbitals, for example, comprise the third electron shell. The number of subshells allowed in a given shell is equal to the principal quantum number. In the first electron shell, there is only one subshell; in the second shell, there are two subshells; an so on. To summarize, orbitals of equal energy make up subshells, and, in turn, subshells make up electron shells. Table 10.1 relates the distribution of electrons in subshells by considering the number of orbitals.

electron shell A collection of orbitals that have the same value of *n*; for example, the 3*s*, 3*p*, and 3*d* orbitals comprise the third electron shell.

UPDATE

Electrons: Particles or Waves?

► **Can you explain the cobweb lines that are observed if you hold your thumbs slightly apart at the tip of your nose and look directly between them into a bright light?**

Atomic physics was progressing rapidly at the turn of the twentieth century. Thomson had determined the mass and charge for the electron, Rutherford had discovered the atomic nucleus, and Bohr had established that electrons circled the nucleus in fixed orbits. At the time, it appeared that the controversy over atoms, going all the way back to Aristotle, had reached a final conclusion. Ironically, this was not the case. Another controversy was brewing that made atomic theory a great deal more complex.

In 1900, Max Planck had proposed the quantum concept and light was given a dual nature. That is, light could behave as a wave of energy although it is composed of particles. While still a graduate student, the French physicist Louis De Broglie wrote two papers on the wave-particle nature of light. He began with the principle that light has both a wave and particle nature. Since an electron is a particle, De Broglie reasoned that perhaps an electron also has a wave nature. In 1923, De Broglie defended his doctoral thesis and proposed that an electron has wave-like properties. Although his research was a little too original for the doctoral committee at the University of Paris, they did him the courtesy of asking for Einstein's opinion. Einstein came to his defense and said: "*It may look crazy, but it is really sound.*" De Broglie was awarded his doctorate and the wave nature of the electron had thus gained credibility.

However, to firmly establish a scientific idea, experimental evidence is necessary. (Recall that Bohr needed the emission line spectrum of hydrogen to support his idea of electrons in fixed energy levels.) The idea that an electron may be a wave, just as light is a wave, was based on theory and not on experiment. Therefore, experimental evidence was necessary for the electron wave theory to gain full acceptance. Within two years, however, an American physicist discovered the necessary evidence.

An interesting property of light is that light waves can interfere with one another. This phenomenon is called an interference pattern. In 1925, Clinton Davisson, working at the Bell Telephone Laboratories, was reflecting electrons off the surface of nickel crystals. He noticed that the reflection patterns appeared to resemble light interference patterns. Davisson correctly concluded that the electrons bouncing off the nickel atoms were creating a wave interference pattern. Thus, there was experimental evidence for electron waves. (See figures.)

Today, we speak of electrons as having a wave-particle nature. The Bohr model of the atom is consistent with the particle nature of the electron. However, the quantum mechanical atom is necessary to explain the wave nature of the electron and the concept of atomic orbitals.

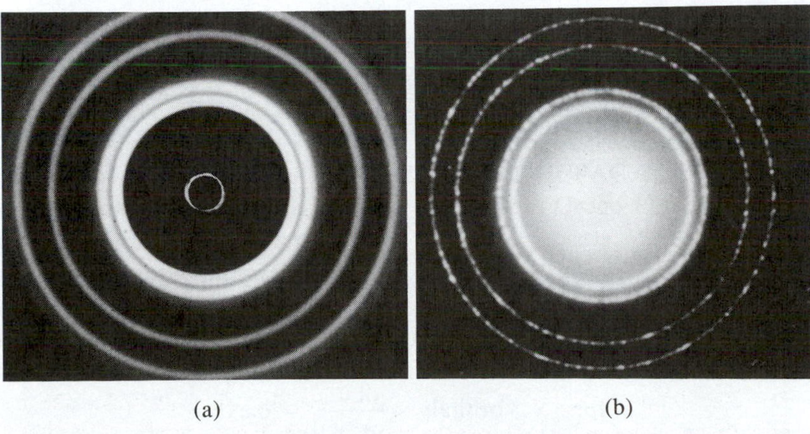

Diffraction patterns produced by (a) a beam of fast-moving electrons and (b) a beam of X-rays passing through a thin aluminum foil. The similarity of the patterns shows that electrons behave like X-rays and have wavelike properties.

► **If you look at light through the narrow slit formed by your thumbs, you will see thin black lines that are evidence for an interference pattern.**

Table 10.1 Distribution of Electrons

Electron Shell	Electron Subshells	Orbitals in Subshells	Maximum Electrons in Subshell
1	1*s*	1	2 e^-
2	2*s*	1	2 e^-
	2*p*	3	6 e^-
3	3*s*	1	2 e^-
	3*p*	3	6 e^-
	3*d*	5	10 e^-
4	4*s*	1	2 e^-
	4*p*	3	6 e^-
	4*d*	5	10 e^-
	4*f*	7	14 e^-

The following example exercise provides additional practice in determining the distribution of electrons.

EXAMPLE EXERCISE 10.4

State the number of orbitals in each of the following electron subshells.

(a) 5*p* **(b)** 5*f*

Solution: Remember that the number of orbitals in a subshell depends only on the type of subshell. All *s* subshells consist of one orbital, all *p* subshells consist of three orbitals, all *d* subshells consist of five orbitals, and all *f* subshells are made up of seven orbitals.

(a) In a 5*p* subshell, there are three *p* orbitals (p_x, p_y, and p_z).

(b) In a 5*f* subshell, there are seven *f* orbitals.

SELF-TEST EXERCISE

State the number of orbitals in each of the following electron subshells.

(a) 5*s* **(b)** 5*d*

Answers: **(a)** 1; **(b)** 5

EXAMPLE EXERCISE 10.5

Calculate the maximum number of electrons that can occupy each of the following subshells.

(a) 4*p* **(b)** 4*f*

Solution: First, let's recall the number of orbitals in each given subshell.

(a) In the 4*p* subshell, there are three *p* orbitals and each orbital can hold 2 electrons. The maximum number of electrons is therefore

$$4p: \quad 3 \text{ orbitals} \times \frac{2\ e^-}{\text{orbital}} = 6\ e^-$$

(b) In the 4*f* sublevel, there are seven *f* orbitals. Each orbital can hold two electrons; so the maximum number of electrons is

$$4f: \quad 7 \text{ orbitals} \times \frac{2\ e^-}{\text{orbital}} = 14\ e^-$$

SELF-TEST EXERCISE

Calculate the maximum number of electrons in each of the following.

(a) 4*s* subshell **(b)** 4*d* subshell

Answers: **(a)** 2 e^-; **(b)** 10 e^-

EXAMPLE EXERCISE 10.6

Determine the maximum number of electrons that can occupy the third electron shell.

Solution: The third shell is composed of three subshells: 3*s*, 3*p*, and 3*d*. The *s* subshell has one orbital, the *p* subshell has three, and the *d* subshell has five orbitals. We can find the total number of electrons by adding up the three subshells as follows:

$$3s: \quad 1 \text{ orbital} \times \frac{2\ e^-}{\text{orbital}} = 2\ e^-$$

$$3p: \quad 3 \text{ orbitals} \times \frac{2\ e^-}{\text{orbital}} = 6\ e^-$$

$$3d: \quad 5 \text{ orbitals} \times \frac{2\ e^-}{\text{orbital}} = \underline{10\ e^-}$$

$$\text{Total maximum number of electrons} = 18\ e^-$$

SELF-TEST EXERCISE

Determine the maximum number of electrons that can occupy the second electron shell.

Answer: 8 e^- (2 e^- + 6 e^-)

Note: The quantum mechanical model of an atom is much more powerful than the Bohr model. While the Bohr atom adequately explains energy levels in hydrogen, the quantum mechanical atom explains energy levels for all of the elements in the periodic table. In Section 5.6, we discussed the blocks of elements in the periodic table in terms of energy sublevels. The *s* block of elements includes Groups 1 and 2 and the atoms are filling *s* sublevels; the elements in Groups 3 through 12 are filling *d* sublevels Groups 13 through 18 are filling *p* sublevels, and the inner transition elements are filling *f* sublevels.

In the quantum mechanical model of the atom, electrons exist in orbitals. The rules of quantum mechanics dictate the number of orbitals in a given subshell. Using an orbital description, we can predict the number of groups of elements that comprise a given subshell. An *s* subshell is allowed only one orbital and can hold a maximum of two electrons. An *s* subshell corresponds to two groups in the periodic table, that is, Groups 1 and 2. A *p* subshell is allowed three orbitals that can hold a total of 6 electrons. The 6 electrons correspond to the elements in Groups 13 through 18. A *d* subshell is allowed five orbitals and can hold a total of 10 electrons. The 10 electrons correspond to elements in Groups 3 through 12 in the periodic table. An *f* subshell has seven orbitals that can hold a total of 14 electrons. The 14 electrons

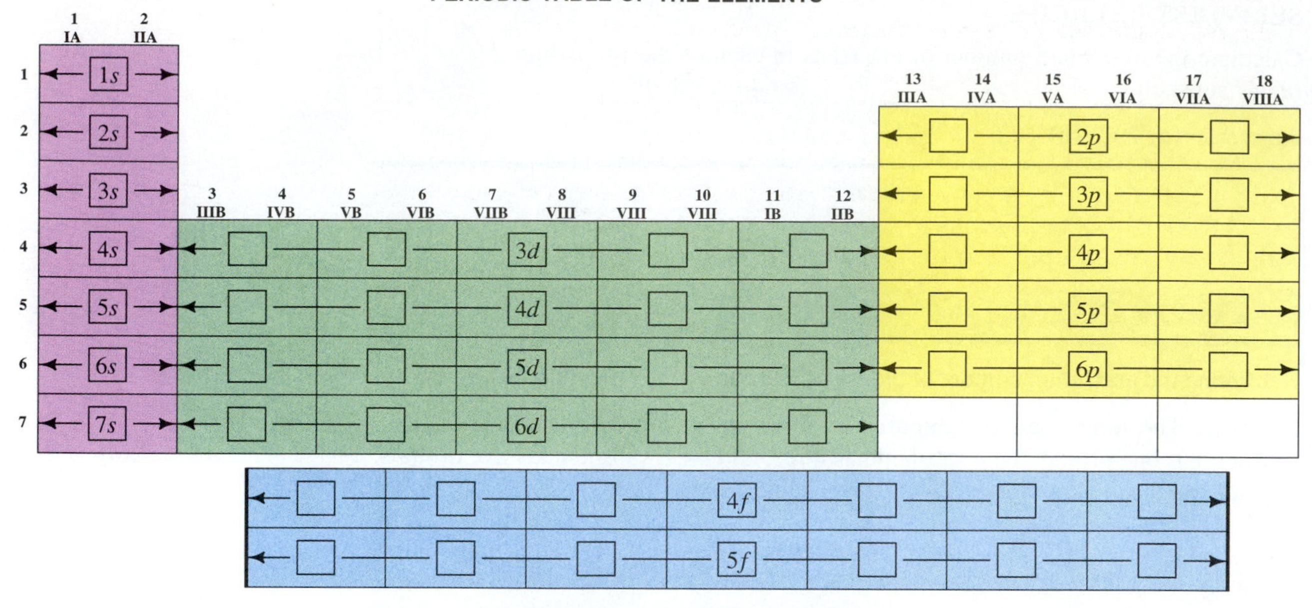

Figure 10.11 Orbitals and Subshells in the Periodic Table The *s*, *p*, *d*, and *f* blocks of elements in the periodic table. There is one *s* orbital in each *s* subshell, three *p* orbitals in each *p* subshell, five *d* orbitals in each *d* subshell, and seven *f* orbitals in each *f* subshell.

correspond to elements in the lanthanide series and the actinide series. These relationships are summarized in the periodic table shown in Figure 10.11.

10.5 The Four Quantum Numbers

OBJECTIVES

To state the allowed values for the four quantum numbers that indicate the energy and location of an electron in an atom.

In Section 10.3, we introduced the idea of a quantum number. A quantum number is used to define the energy and location of an electron in the quantum mechanical atom. A total of four quantum numbers is necessary to completely define the energy and location of an electron. Each of the four quantum numbers has only certain allowed values. The first number, called the principal quantum number (n), describes the size and energy of an orbital. The principal quantum number can have only integer values, that is, 1, 2, 3, 4, ... As the value of n increases, so do the size and energy of the orbital.

Allowed Values for the Principal Quantum Number

$$n = 1, 2, 3, 4, 5, 6, 7, \ldots$$

The second quantum number (l) describes the shape of the orbital in which the electron is found. It can have the integer values 0, 1, 2, . . . up to $n - 1$. When l equals 0, the shape of the orbital is that of a sphere. When l equals 1, the shape of the orbital is that of a dumbbell. For higher values of l, the orbital shapes become very complex.

Allowed Values for the Second Quantum Number

$$l = 0(s), 1(p), 2(d), 3(f), \ldots, n-1$$

The **third quantum number** (symbol $\boldsymbol{m}$) is sometimes called the magnetic quantum number. This quantum number describes the relative orientation of a particular orbital, for example, the relative orientation of the three p orbitals (p_x, p_y, p_z) to each other. The third quantum number can have integer values that correspond to $+l, \ldots 0, \ldots -l$.

third quantum number (m) The quantum number that describes the orientation of orbitals having the same energy (for example, $2p_x$, $2p_y$, $2p_z$); also referred to as the magnetic quantum number. The allowed values of m are $+l, \ldots 0, \ldots -l$.

Allowed Values for the Third Quantum Number

$$m = +l, \ldots, +1, 0, -1, \ldots -l$$

The **fourth quantum number** (s) is sometimes called the electron spin quantum number. There is experimental evidence that an electron in an orbital acts like a tiny charged particle spinning on an axis through its center. This **electron spin** creates a slight magnetic field that can either align with or against a strong surrounding magnetic field. Figure 10.12 illustrates this behavior of the electron in a magnetic field.

Thus, the fourth quantum number describes the electron spin in a particular orbital. Since the electron has only two orientations, up or down, there are only two possible values for the spin quantum number. The two allowed values of s are designated $+\frac{1}{2}$ and $-\frac{1}{2}$.

fourth quantum number (s) The quantum number that describes the spin of an electron in an orbital. The allowed values of s are $+\frac{1}{2}$ or $-\frac{1}{2}$.

electron spin A property of an electron that simulates a magnet. The electron spin creates a small magnetic field that either aligns with or against an external magnetic field.

Allowed Values for the Fourth Quantum Number

$$s = +\tfrac{1}{2}, -\tfrac{1}{2}$$

Using quantum numbers, we can designate the location and energy of an electron in a particular orbital. For example, an electron in a $2s$ orbital can be described as $n = 2$, $l = 0$, $m = 0$, and s $+\frac{1}{2}$ or $-\frac{1}{2}$. An electron in a $2p$ orbital can be described as $n = 2$, $l = 1$, $m = +1, 0$, or -1, and $s = +\frac{1}{2}$ or $-\frac{1}{2}$. Notice that in a $2p$ orbital an electron has three possible values for the third quantum number, as well as two possible spin values. The following example exercise provides practice in designating allowed quantum number values for a particular electron.

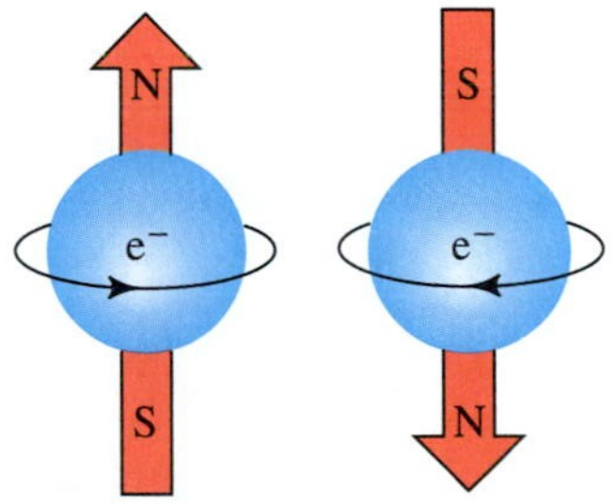

Figure 10.12 The Fourth Quantum Number The electron behaves like a spinning top. Since the electron is a charged particle, it creates a magnetic field as it rotates. If the electron is placed into a strong magnetic field, its weak magnetic field aligns with the strong field. The electron may also 'flip' upside down and be aligned against the strong field.

EXAMPLE EXERCISE 10.7

State the allowed values for the four quantum numbers that correspond to an electron in each of the following orbitals.

(a) $3p$ orbital **(b)** $4d$ orbital

Solution: If an orbital is specified, there is only one value for the principal and second quantum numbers. The third and fourth quantum numbers, however, may have more than one allowed value.

(a) For an electron in a $3p$ orbital, $n = 3$ and $l = 1$. The allowed values for the magnetic quantum number range from $+l, \ldots 0, \ldots -l$. Since $l = 1$, the values for m can be $+1$, 0, or -1. There are two spin states for the fourth quantum number. The allowed values for s are $+\frac{1}{2}$ and $-\frac{1}{2}$.

(b) The principal and second quantum numbers for an electron in a $4d$ orbital are $n = 4$ and $l = 2$. The allowed values for the third quantum number when $l = 2$ are $+2$, $+1$, 0, -1 or -2. The allowed values for the spin quantum number are $+\frac{1}{2}$ or $-\frac{1}{2}$.

SELF-TEST EXERCISE

State the allowed values for the four quantum numbers that correspond to an electron in each of the following orbitals.

(a) $5s$ orbital

(b) $4f$ orbital

Answers: **(a)** $n = 5$, $l = 0$, $m = 0$, $s = +\frac{1}{2}, -\frac{1}{2}$;
(b) $n = 4$, $l = 3$, $m = +3, +2, +1, 0, -1, -2, -3$, $s = +\frac{1}{2}, -\frac{1}{2}$

10.6 Orbital Energy Diagrams

OBJECTIVES

To write out the electron configuration and orbital diagram for the distribution of electrons in atoms of selected elements.

To draw an orbital energy diagram for the electrons in hydrogen atoms and nonhydrogen atoms.

In Section 4.9, we learned to write the electron configuration for most of the elements in the periodic table. You may recall that we can use the periodic table to arrange electrons into sublevels according to increasing energy; that is,

$$1s < 2s < 2p < 3s < 3p < 4s < 3d < 4p < 5s \ldots.$$

We systematically add electrons into the lowest-energy sublevel that is available. For example, the electron configuration for a neon atom is $1s^2\,2s^2\,2p^6$. The superscripts indicate the number of electrons in each sublevel. The total of the superscript numbers is ten, which equals the number of electrons in a neon atom. (The atomic number of neon is 10.)

In considering the quantum mechanical atom, we know that electrons are distributed in orbitals that make up subshells. We can, therefore, write a more complete description of the electrons in an atom by indicating the location of electrons in orbitals rather than in sublevels. The electron configuration of a hydrogen atom is $1s^1$. If we draw a small box to symbolize an orbital and an arrow to indicate the electron, we have the following orbital diagram:

		$1s$ orbital
H:	$1s^1$	[↑]

The electron configuration for a helium atom is $1s^2$. Thus, there are two electrons in a $1s$ orbital and the two electrons must have an opposite spin. We usually represent the first electron in an orbital with an upward arrow and the second electron with a downward arrow. The two electrons are said to be spin paired in the orbital. Similar to the orbital diagram for hydrogen, we draw two arrows in a small box to represent the spin-paired electrons.

		1*s* orbital
He:	$1s^2$	↑↓

In Section 4.9, we wrote electron configurations of elements by systematically adding electrons into sublevels. Now we can systematically add electrons into orbitals represented by small boxes. In Section 10.4, we stated that the rules of quantum mechanics limit every type of orbital to a maximum of 2 electrons. It makes no difference whether it is an *s*, *p*, *d*, or *f* orbital. Furthermore, in a given subshell, electrons tend to occupy different orbitals before two electrons fill an available orbital. This principle is known as **Hund's rule**. Alternately, we can state Hund's rule as follows: electrons in a given subshell remain unpaired as long as orbitals of the same energy are available.

Hund's rule The principle which states that electrons in a given subshell occupy different orbitals of the same energy. That is, electrons in a given subshell remain unpaired as long as orbitals of the same energy are available.

Now let's apply Hund's rule to an orbital diagram. Consider the electron configuration for a nitrogen atom, which is $1s^2\ 2s^2\ 2p^3$. There are three electrons in the 2*p* subshell, which is comprised of three orbitals: p_x, p_y, and p_z. According to Hund's rule, the three electrons in the 2*p* subshell will be unpaired. The three 2*p* electrons will occupy the separate p_x, p_y, and p_z orbitals and resist spin pairing. Thus, the orbital diagram for a nitrogen atom is

		1*s*	2*s*	2*p*		
N:	$1s^2\ 2s^2\ 2p^3$	↑↓	↑↓	↑	↑	↑

Orbitals of higher energy are available in all atoms even if they are unoccupied. Table 10.2 shows the electron distribution in orbitals for selected elements.

The following example exercise provides practice in writing the orbital diagram for the distribution of electrons.

Table 10.2 Electron Distribution of Selected Elements

Element	Number of Electrons	Electron Configuration	Orbital Diagram 1*s*	2*s*	2*p*			3*s*
Li	3	$1s^2\ 2s^1$	↑↓	↑				
Be	4	$1s^2\ 2s^2$	↑↓	↑↓				
B	5	$1s^2\ 2s^2\ 2p^1$	↑↓	↑↓	↑			
C	6	$1s^2\ 2s^2\ 2p^2$	↑↓	↑↓	↑	↑		
N	7	$1s^2\ 2s^2\ 2p^3$	↑↓	↑↓	↑	↑	↑	
O	8	$1s^2\ 2s^2\ 2p^4$	↑↓	↑↓	↑↓	↑	↑	
F	9	$1s^2\ 2s^2\ 2p^5$	↑↓	↑↓	↑↓	↑↓	↑	
Ne	10	$1s^2\ 2s^2\ 2p^6$	↑↓	↑↓	↑↓	↑↓	↑↓	
Na	11	$1s^2\ 2s^2\ 2p^6\ 3s^1$	↑↓	↑↓	↑↓	↑↓	↑↓	↑
Mg	12	$1s^2\ 2s^2\ 2p^6\ 3s^2$	↑↓	↑↓	↑↓	↑↓	↑↓	↑↓

EXAMPLE EXERCISE 10.8

Write out the electron configuration for aluminum and draw the orbital diagram.

Solution: From the periodic table, we find the atomic number for Al is 13. Thus, aluminum has 13 electrons and the electron configuration is $1s^2\ 2s^2\ 2p^6\ 3s^2\ 3p^1$. We construct the orbital diagram by representing each orbital using a small box and indicating electrons with upward or downward arrows.

	$1s$	$2s$	$2p$	$3s$	$3p$
Al	↑↓	↑↓	↑↓ ↑↓ ↑↓	↑↓	↑ ◻ ◻

SELF-TEST EXERCISE

Write out the electron configuration for silicon and draw the orbital diagram.

Answer: $1s^2\ 2s^2\ 2p^6\ 3s^2\ 3p^2$

$1s$	$2s$	$2p$	$3s$	$3p$
↑↓	↑↓	↑↓ ↑↓ ↑↓	↑↓	↑ ↑ ◻

The Hydrogen Atom

In the quantum mechanical model of the hydrogen atom, all the orbitals in a given subshell have the same energy. If we consider the simplest case, the hydrogen atom, even the orbitals in a given shell have the same energy. The $2s$ orbital, for instance, has the same energy as each of the three $2p$ orbitals. These orbitals are said to be degenerate. We can show the energy relationship of the orbitals in a given subshell relative to the orbitals in other subshells. By symbolizing orbitals using small boxes, we can profile the relative energy of different subshells. This type of arrangement is called an **orbital energy diagram**. Figure 10.13 illustrates the orbital energy diagram for the hydrogen atom.

orbital energy diagram An energy profile of the orbitals in various subshells

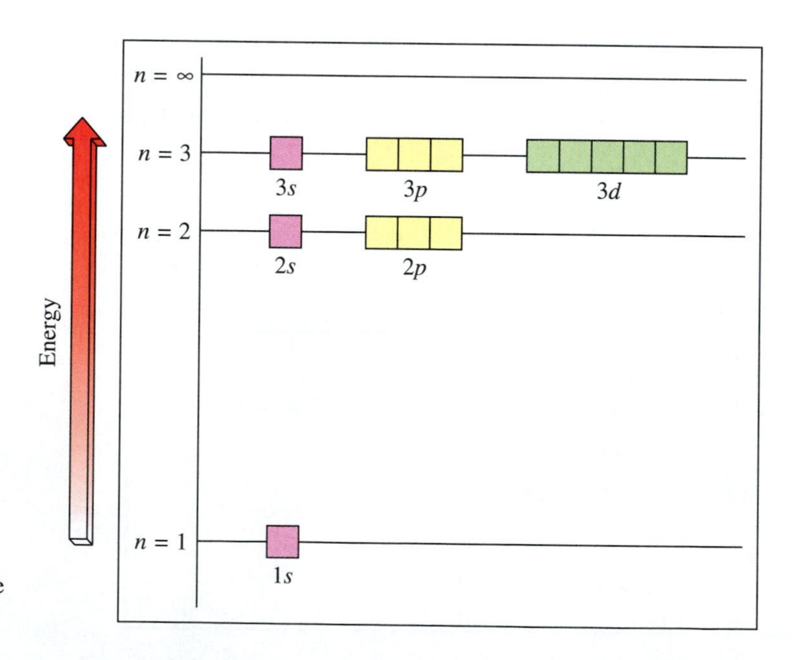

Figure 10.13 Orbital Energy Diagram for the Hydrogen Atom In the hydrogen atom the orbitals in the same electron shell possess equal energy. Notice that the $3s$, $3p$, and $3d$ orbitals have the same energy and are said to be degenerate.

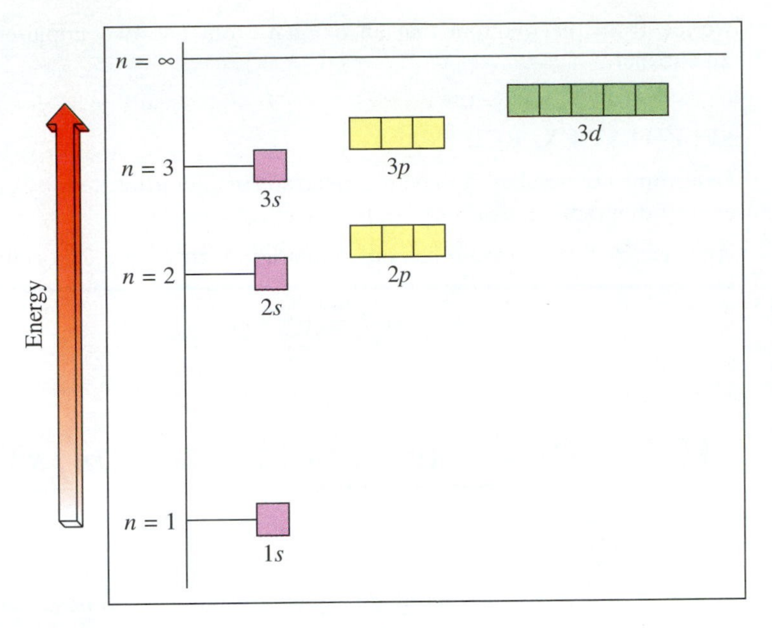

Figure 10.14 Orbital Energy Diagram for Nonhydrogen Atoms In atoms having two or more electrons, orbitals in different subshells do not have the same energy. Notice that within the second electron shell the $2p$ orbitals are higher in energy than the $2s$ orbitals. Similarly, the $3s$, $3p$, and $3d$ orbitals do not have the same energy.

Nonhydrogen Atoms

Unlike the hydrogen atom, atoms with two or more electrons are referred to as many-electron atoms. In atoms having two or more electrons, subshells within a given electron shell are at a different energy. For example, in the second shell the $2p$ orbitals are at a higher energy than the $2s$ orbital.

The energy of orbitals is arranged in the same sequence as the energy sublevels we previously studied in Section 4.9. Orbitals can be arranged in order of increasing energy as follows: $1s < 2s < 2p < 3s < 3p < 4s < 3d < 4p < 5s$ . . . . The corresponding orbital energy diagram reflects the splitting of shells into subshells. Figure 10.14 illustrates the general orbital energy diagram for many-electron atoms.

The following example exercise provides practice in assigning the distribution of electrons to an orbital energy diagram.

EXAMPLE EXERCISE 10.9

Draw the orbital energy diagram for an atom of oxygen.

Solution: From the periodic table, we find the atomic number for O is 8. Oxygen has eight electrons and the electron configuration is $1s^2\ 2s^2\ 2p^4$. After drawing the orbital energy diagram, we assign the eight electrons to the orbitals with the lowest available energy. Thus,

		↑↓ ↑ ↑
$n = 2$ ———	↑↓	$2p$
	$2s$	
$n = 1$ ———	↑↓	
	$1s$	

Notice from the diagram that an oxygen atom has two unpaired electrons in its $2p$ subshell.

SELF-TEST EXERCISE

Determine the number of unpaired electrons in a fluorine atom by drawing the orbital energy diagram.

Answer: A fluorine atom has one unpaired electron in a $2p$ orbital.

10.7 Quantum Notation: (n, l, m, s)

To write a set of four quantum numbers to describe the highest-energy electron in an atom of an element.

In Section 10.5, we introduced the four quantum numbers that describe the energy and location of an electron in an atom. The four quantum numbers represent the following.

1. The first quantum number, n describes the size and energy of the given orbital.
2. The second quantum number, l, describes the shape of the given orbital.
3. The third quantum number, m, describes the orientation of the orbital.
4. The fourth quantum number, s, describes the electron spin.

Pauli exclusion principle The statement that no two electrons in an atom can have the same set of four quantum numbers.

In 1925 the Austrian-American physicist Wolfgang Pauli (1900–1958) stated the following rule: *no two electrons in an atom can have the same set of quantum numbers*. This rule is known as the **Pauli exclusion principle**. The exclusion principle is responsible for the spin quantum number. That is, if two electrons occupy the same orbital, the electrons must have opposite spin states.

The concept that an electron can spin along its central axis gives rise to the notion that two electrons can be spin paired. In any given orbital, s, p, d, or f, one electron spins clockwise and a second electron spins counterclockwise. Since two electrons cannot have the same set of quantum numbers, and the orbital values for n, l, and m are fixed, a fourth quantum number is required to distinguish between electron spins. The values for the electron spin quantum number are designated as $+\frac{1}{2}$ or $-\frac{1}{2}$.

Quantum Notation

quantum notation A method for describing the energy and location of an electron in an atom using a set of four quantum numbers, (n, l, m, s).

Let's now apply the Pauli exclusion principle to a particular electron in a given atom. To describe the electron completely, it is necessary to use a set of four quantum numbers. We can use **quantum notation** (n, l, m, s) to indicate the four quantum numbers that describe an orbital having a certain energy, shape, and orientation, in addition to the spin of the electron. Table 10.3 summarizes the allowed values for each quantum number.

Table 10.3 Summary of Allowed Quantum Number Values

Orbital	Orbital Energy, n	Orbital Shape, l	Orbital Orientation, m	Electron Spin,* s
1s	1	0	0	$+\frac{1}{2}, -\frac{1}{2}$
2s	2	0	0	$+\frac{1}{2}, -\frac{1}{2}$
2p	2	1	$+1, 0, -1$	$+\frac{1}{2}, -\frac{1}{2}$
3s	3	0	0	$+\frac{1}{2}, -\frac{1}{2}$
3p	3	1	$+1, 0, -1$	$+\frac{1}{2}, -\frac{1}{2}$
3d	3	2	$+2, +1, 0, -1, -2$	$+\frac{1}{2}, -\frac{1}{2}$
4s	4	0	0	$+\frac{1}{2}, -\frac{1}{2}$
4p	4	1	$+1, 0, -1$	$+\frac{1}{2}, -\frac{1}{2}$
4d	4	2	$+2, +1, 0, -1, -2$	$+\frac{1}{2}, -\frac{1}{2}$
4f	4	3	$+3, +2, +1, 0, -1, -2, -3$	$+\frac{1}{2}, -\frac{1}{2}$

* An electron spin value of $+\frac{1}{2}$ or $-\frac{1}{2}$ is available for each value of m; that is, for $m = +1$, the spin can either be $+\frac{1}{2}$ or $-\frac{1}{2}$. The order of assigning the fourth quantum, however, is specific. That is, a value of $+\frac{1}{2}$ is assigned to the first electron, which is unpaired. A value of $-\frac{1}{2}$ is assigned to the second electron, which is spin paired in the orbital.

Let's use quantum notation to indicate the electron in a hydrogen atom. Since the electron configuration is $1s^1$, the value of n is 1. The quantum number for an s orbital is $l = 0$. There is only one allowed value for the third quantum number when $l = 0$; that is, $m = 0$. By convention, the first electron in an orbital is usually assigned $s = +\frac{1}{2}$. We can write the complete set of quantum numbers as follows:

$$\text{H: } n = 1, l = 0, m = 0, s = +\tfrac{1}{2} \qquad (1, 0, 0, +\tfrac{1}{2})$$

Next, let's use quantum notation to describe the second electron in a helium atom. Since the electron configuration is $1s^2$, the value of $n = 1$. For the s orbital, $l = 0$. The only permissible value for the third quantum number when $l = 0$ is $m = 0$. By convention, the second electron in an orbital is usually assigned $s = -\frac{1}{2}$. We can designate the second electron in helium using four quantum numbers as follows:

$$\text{He: } n = 1, l = 0, m = 0, s = -\tfrac{1}{2} \qquad (1, 0, 0, -\tfrac{1}{2})$$

Next, let's try a more difficult example. Consider the highest-energy electron in a nitrogen atom. A nitrogen atom has seven electrons, so the electron configuration is $1s^2\ 2s^2\ 2p^3$. The value of n is 2 and l is 1 for a 2p orbital. There are three allowed values for the magnetic quantum number when $l = 1$, that is, $+1, 0, -1$. According to Hund's rule, the three electrons in the 2p subshell will be unpaired and in different orbitals. By convention, the magnetic quantum numbers are usually assigned in the order of $+l, 0, -l$. Thus, the value for m is -1. The electron is unpaired, so $s = +\frac{1}{2}$. The quantum notation for the highest-energy electron in a nitrogen atom is

$$\text{N: } n = 2, l = 1, m = -1, s = +\tfrac{1}{2} \qquad (2, 1, -1, +\tfrac{1}{2})$$

Table 10.4 lists sets of quantum numbers for the highest-energy electron in atoms of elements in the third row of the periodic table. By using elements all from

Table 10.4 Quantum Notation for the Highest-Energy Electron in an Atom of the Elements in Period 3

Element	Electron Configuration	Highest-energy Orbital	Quantum Notation (n, l, m, s)
Na	[Ne] $3s^1$	$3s$	$(3, 0, 0, +\frac{1}{2})$
Mg	[Ne] $3s^2$	$3s$	$(3, 0, 0, -\frac{1}{2})$
Al	[Ne] $3s^2\ 3p^1$	$3p$	$(3, 1, +1, +\frac{1}{2})$
Si	[Ne] $3s^2\ 3p^2$	$3p$	$(3, 1, 0, +\frac{1}{2})$
P	[Ne] $3s^2\ 3p^3$	$3p$	$(3, 1, -1, +\frac{1}{2})$
S	[Ne] $3s^2\ 3p^4$	$3p$	$(3, 1, +1, -\frac{1}{2})$
Cl	[Ne] $3s^2\ 3p^5$	$3p$	$(3, 1, 0, -\frac{1}{2})$
Ar	[Ne] $3s^2\ 3p^6$	$3p$	$(3, 1, -1, -\frac{1}{2})$

the same period, we can more easily see the systematic way the third and fourth quantum numbers are assigned.

As a final illustration, let's write a set of quantum numbers for the highest-energy electron in an iron atom. An iron atom has 26 electrons, so the electron configuration is $1s^2\ 2s^2\ 2p^6\ 3s^2\ 3p^6\ 4s^2\ 3d^6$. The value of n is 3, and l is 2 for a $3d$ orbital. There are five allowed values for the third quantum number when $l = 2$, that is, $+2, +1, 0, -1, -2$. According to Hund's rule, the first 5 electrons in the $3d$ subshell will be unpaired and in separate orbitals. The sixth electron will spin pair with the electron occupying the first $3d$ orbital. Thus, the magnetic quantum number will have a value of $+2$. The electron is spin paired, so s is $-\frac{1}{2}$. The quantum notation for the highest-energy electron in an iron atom is

$$\text{Fe: } n = 3, l = 2, m = +2, s = -\tfrac{1}{2} \qquad (3, 2, +2, -\tfrac{1}{2})$$

The following example exercises provide additional practice in writing quantum notation for the highest-energy electron in atoms of selected elements.

EXAMPLE EXERCISE 10.10

Write a set of four quantum numbers for the highest-energy electron in an atom of the following elements.

(a) beryllium **(b)** oxygen

Solution: We begin by finding the atomic number of the element in the periodic table. The atomic number equals the number of electrons in an atom. The atomic numbers of Be and O are 4 and 8, respectively.

(a) We can write the electron configuration for Be as $1s^2\ 2s^2$. An electron in a $2s$ orbital has $n = 2$ and $l = 0$. The only allowed value for m is 0. The two electrons in the $2s$ orbital are spin paired, so $s = -\frac{1}{2}$. The quantum notation for Be is $(2, 0, 0, -\frac{1}{2})$.

(b) The electron configuration for O is $1s^2\ 2s^2\ 2p^4$. The last electron in a $2p$ orbital has $n = 2$ and $l = 1$. The allowed values for m are $+1, 0, -1$. The fourth electron in the $2p$ orbital has a value of $+1$ for the third quantum number. It is the second electron in the orbital, so the spin is paired and $s = -\frac{1}{2}$. The quantum notation for O is $(2, 1, +1, -\frac{1}{2})$.

SELF-TEST EXERCISE

Write a set of four quantum numbers for the highest-energy electron in an atom of titanium.

Answers: $(3, 2, +1, +\frac{1}{2})$

Quantum Notation and the Periodic Table

In Section 10.4, we related the types of orbitals to the positions of subshells in the periodic table. That is, *s* orbitals correspond to Groups 1 and 2, *p* orbitals correspond to Groups 13 through 18, and *d* orbitals correspond to Groups 3 through 12. We can now further illustrate the relationship of quantum numbers to the periodic table. Figure 10.15 indicates the quantum numbers that correspond to the blocks of elements in the periodic table.

If we understand the arrangement of orbitals from the positions of subshells in the periodic table, we have a powerful tool for predicting quantum numbers directly. For example, a zinc atom has a filled 3*d* subshell. Therefore, the quantum numbers are $n = 3$ and $l = 2$. The values for the third quantum number are in the sequence $+2, +1, 0, -1, -2$; for zinc, $m = -2$. The last electron is spin paired,

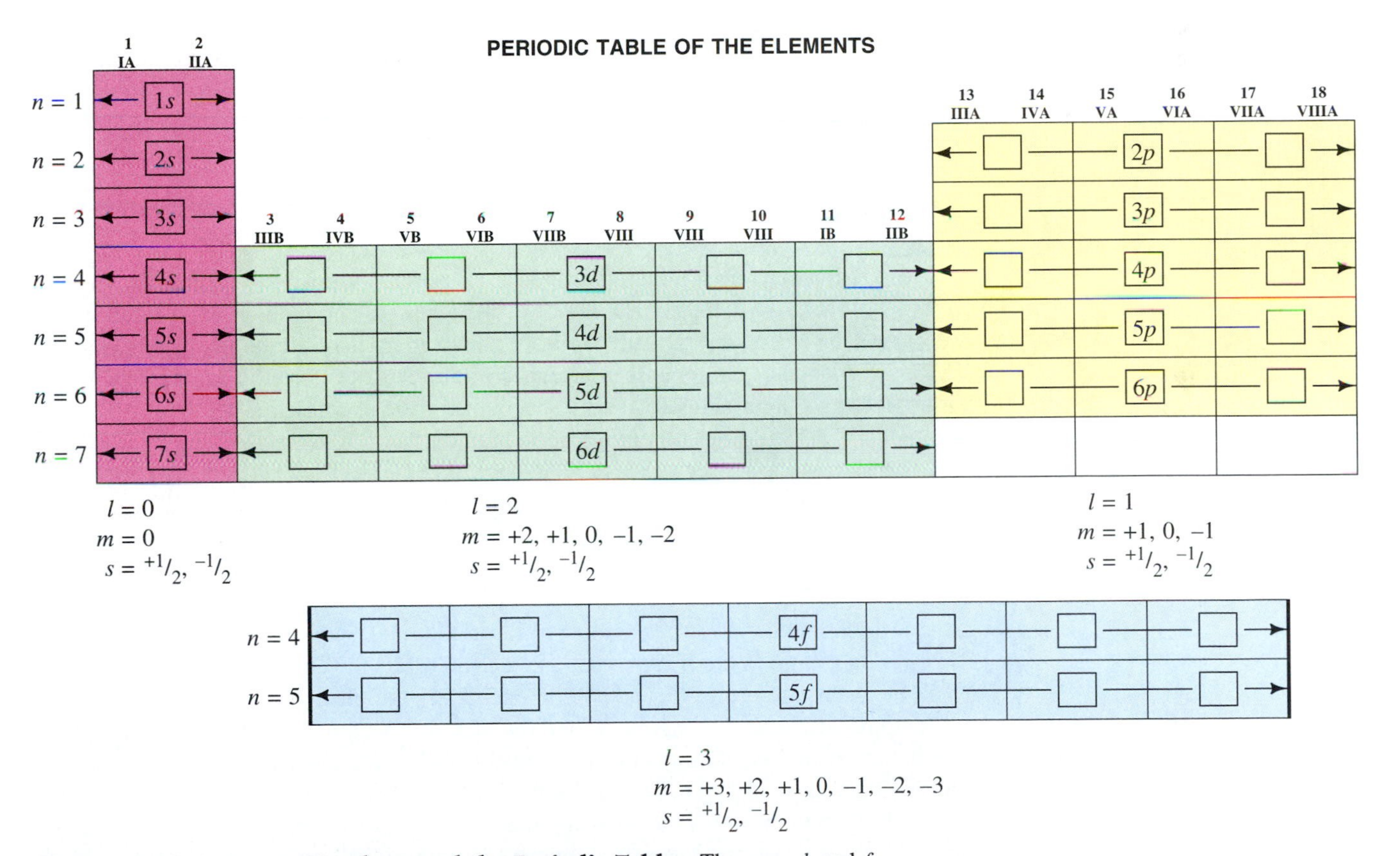

Figure 10.15 Quantum Numbers and the Periodic Table The *s*, *p*, *d*, and *f* subshells are listed in the periodic table. The principal quantum number corresponds to rows in the periodic table. The second quantum number corresponds to the *s*, *p*, *d*, and *f* blocks of elements in the table. The sequence of allowed values for the third and fourth quantum number are indicated below each block of elements.

so $s = -\frac{1}{2}$. The quantum notation for the highest-energy electron in zinc is (3, 2, −2, $-\frac{1}{2}$).

The following example exercise provides practice in correlating quantum notation with the positions of elements in the periodic table.

EXAMPLE EXERCISE 10.11

Write a set of four quantum numbers of the highest-energy electron in an atom of the following elements.

(a) lithium **(b)** chlorine

Solution: Let's use our knowledge of the positions of subshells in the periodic table to find each of the quantum numbers.

(a) Since the third electron in Li is in a 2*s* subshell, $n = 2$ and $l = 0$. The only allowed value for m is equal to 0. There is one electron in the 2*s* orbital, so $s = +\frac{1}{2}$. Thus, the quantum notation for Li is (2, 0, 0, $+\frac{1}{2}$).

(b) We find from the periodic table that highest-energy electron in Cl is in a 3*p* subshell; thus, $n = 3$ and $l = 1$. The allowed values for the third quantum number are +1, 0, −1. There are five electrons in the 3*p* subshell, so $m = 0$. The last electron is spin paired in the orbital; thus, $s = -\frac{1}{2}$. The quantum notation for Cl is (3, 1, 0, $-\frac{1}{2}$).

SELF-TEST EXERCISE

Refer to the periodic table and predict the element whose highest-energy electron has the following quantum notation.

(a) (5, 0, 0, $-\frac{1}{2}$) **(b)** (4, 3, +3, $+\frac{1}{2}$)

Answers: **(a)** Sr; **(b)** Ce

In Chapters 4 and 5, we related simplified atomic theory and the periodic table. In the early chapters, we were able to show the general relationship between electron sublevels and the properties of elements. In this chapter, we examined the relationship between orbitals and the periodic table. This latter, more rigorous approach allows us to account for the exceptional behavior of certain elements that could not otherwise be explained.

Summary

Section 10.1 In 1913, Niels Bohr proposed that electrons travel about the nucleus in circular orbits. In the **Bohr** atom, an electron orbits the nucleus at a fixed distance. The electron, therefore, has a specific energy. If the electron changes orbits, there is a stepwise change in energy. When the electron becomes excited by gaining heat or electrical energy, it jumps to a higher orbit. When the electron becomes unexcited by losing heat or light energy, it drops to a lower orbit.

Bohr supported his theory using the line spectra from gas discharge tubes. The emission spectrum for hydrogen contained three vivid lines: a violet line, a blue-green line, and a red line. Bohr argued that, when an electron drops from a higher energy level to a lower level, a quantum of light is released. The energy of the quantum of light is equal to the difference in energy between the two orbits. Unfortunately, Bohr was only able to explain the energy states of the electron in hydrogen atoms.

Section 10.2 In 1900, Max Planck introduced the quantum concept. Bohr applied the concept to the hydrogen atom, as **quantum theory** suggested particles of energy were released whenever any matter underwent a change in energy. The particle of light energy that is emitted is a **quantum** of light. This was a novel concept because, previous to Planck, light had been considered to be a continuous wave of radiant energy. A few years later, Einstein used the quantum concept to explain the **photoelectric effect**. This is the phenomenon of a metal ejecting an electron after being struck by a **photon** of light. The photon, however, must have sufficient energy. Einstein proposed that a single photon of light could cause a metal to emit a single electron.

Section 10.3 In the 1920s, our understanding of electrons in atoms became very sophisticated. In 1925, Werner Heisenberg suggested the **uncertainty principle**, that is, it was impossible to simultaneously know both the precise location and momentum of an electron. Instead, the momentum of an electron could only be known in terms of its probability of being located somewhere within the atom. This description gave rise to the **quantum mechanical atom**. The locations within the atom where there is a high probability of finding an electron having a certain energy are called **orbitals**. An orbital is a region about the nucleus having a given energy, size, and shape. The shape of an *s* orbital is spherical, a *p* orbital resembles the shape of a dumbbell, and the shapes of *d* and *f* orbitals are rather complex.

Section 10.4 The properties of an electron in an orbital are designated by a set of quantum numbers. The **principal quantum number** (n) describes the size and energy of a particular orbital. The principal quantum number has integer values of 1, 2, 3, ... The **second quantum number** (l) describes the shape of an orbital. The allowed values of l are 0, 1, 2, 3, . . . , $n - 1$. A collection of orbitals having the same value of n and l is called an **electron subshell**. For example, $2p_x$, $2p_y$, and $2p_z$ make up the $2p$ subshell. A collection of subshells having the same value of n, for example, $3s$, $3p$, $3d$, is referred to as an **electron shell**.

Section 10.5 The **third quantum number** (m) describes the orientation of orbitals in a subshell, for example, the orientation of $2p_x$, $2p_y$, $2p_z$ orbitals. The allowed values of m are $+l, \ldots 0, \ldots -l$. The **fourth quantum number** (s) describes the **electron spin** in an orbital. A spinning electron is similar to a magnet spinning on its axis. A spinning electron creates a small magnetic field that can either align with or against an external magnetic field. In an orbital, two electrons are said to be spin paired or to have opposite spins. By convention, the first electron in an orbit is given a value of $+\frac{1}{2}$; the second electron is given a value of $-\frac{1}{2}$.

Section 10.6 **Hund's rule** states that electrons in a given subshell tend to occupy different orbitals. That is, electrons remain unpaired as long as orbitals of the same energy are available. We can draw an energy profile of the orbitals in a given subshell relative to the orbitals in other subshells. This is called an **orbital energy diagram**. The orbital energy diagram uses small boxes to represent orbitals. Orbital boxes are grouped together by subshell and then arranged according to increasing energy.

Section 10.7 The **Pauli exclusion principle** states that no two electrons in an atom can have four identical quantum numbers. This rule supports the idea that two electrons in the same orbital have opposite spins. Moreover, four quantum numbers are necessary to completely describe an electron in an orbital. We use **quantum notation** to designate the orbital energy, orbital shape, orbital orientation, and electron spin in an atom, that is, (n, l, m, s).

Key Terms

Select the key term that corresponds to the following definitions.

____ **1.** a simplistic model of the atom that describes the energy of an electron in terms of a particular orbit about the nucleus

____ **2.** a bundle of radiant energy that is emitted when an electron in an excited atom drops to a lower-energy level

____ **3.** a theory that specifically explains the gain or loss of energy by electrons in atoms; a theory that generally explains energy changes in objects by small specific increments

____ **4.** the phenomenon of a metal ejecting an electron when struck by a photon of sufficient energy

____ **5.** an individual unit of radiant energy that corresponds to the particle nature of light

____ **6.** the statement that it is impossible to precisely measure both the location and momentum of a particle at the same time

____ **7.** the quantity that expresses the mass of a particle times its velocity

____ **8.** a sophisticated model of the atom that describes an electron in terms of its probability of being found in a particular location about the nucleus

____ **9.** a region in space surrounding the nucleus of an atom in which there is a high probability (~95%) of finding an electron with a given energy

____ **10.** a general term that refers to an allowed energy state for an electron in an atom

____ **11.** the quantum number that describes the size and energy of a particular orbital; the allowed values are 1, 2, 3, . . .

____ **12.** the quantum number that describes the shape of an orbital; the allowed values are 0, 1, 2, 3, . . . , $n - 1$

____ **13.** a collection of orbitals that have the same energy and shape, for example, the $2p_x$, $2p_y$, and $2p_z$ orbitals

____ **14.** a collection of orbitals that have the same principal quantum number, for example, the 3*s*, 3*p*, and 3*d* orbitals

____ **15.** the quantum number that describes the orientation of orbitals having the same value of *n* and *l* (for example, $2p_x$, $2p_y$, $2p_z$); the allowed values are $+l, \ldots 0, \ldots -l$

____ **16.** the quantum number that describes the spin of an electron in an orbital; the allowed values of *s* are $+\frac{1}{2}$ or $-\frac{1}{2}$

____ **17.** a property of an electron that simulates a charged particle spinning on its axis; the electron spin creates a small magnetic field that either aligns with or against an external magnetic field

____ **18.** the statement that electrons in a given subshell tend to occupy different orbitals; that is, electrons remain unpaired as long as orbitals of the same energy are available

____ **19.** an energy profile of the orbitals in a given subshell relative to the orbitals in other subshells

____ **20.** the statement that no two electrons in an atom can have the same set of four quantum numbers

____ **21.** a method for describing the energy and location of an electron in an atom using a set of four quantum numbers, that is, *n*, *l*, *m*, *s*

(a) Bohr atom *(Sec. 10.1)*
(b) electron shell *(Sec. 10.4)*
(c) electron spin *(Sec. 10.5)*
(d) electron subshell *(Sec. 10.4)*
(e) fourth quantum number (*s*) *(Sec. 10.5)*
(f) Hund's rule *(Sec. 10.6)*
(g) momentum *(Sec. 10.3)*
(h) orbital *(Sec. 10.3)*
(i) orbital energy diagram *(Sec. 10.6)*
(j) Pauli exclusion principle *(Sec. 10.7)*
(k) photoelectric effect *(Sec. 10.2)*
(l) photon *(Sec. 10.2)*
(m) principal quantum number (*n*) *(Sec. 10.3)*
(n) quantum *(Sec. 10.2)*
(o) quantum energy level *(Sec. 10.3)*
(p) quantum mechanical atom *(Sec. 10.3)*
(q) quantum notation *(Sec. 10.7)*
(r) quantum theory *(Sec. 10.2)*
(s) second quantum number (*l*) *(Sec. 10.3)*
(t) third quantum number (*m*) *(Sec. 10.5)*
(u) uncertainty principle *(Sec. 10.3)*

Exercises

Bohr Model of the Atom (Sec. 10.1)

1. Which of the following statements is false according to the Bohr model of the atom?
   (a) Electrons circle the nucleus similar to the way the planets circle the sun.
   (b) Electrons lose energy as they orbit the nucleus.
2. Which of the following statements is false according to the Bohr model of the atom?
   (a) Electrons become excited in atoms that receive heat or electrical energy.
   (b) Electrons gain energy if they drop to an orbit closer to the nucleus.
3. Which of the following lines in the emission spectrum of hydrogen is most energetic: red, blue-green, or violet?
4. Which of the following series of lines in the emission spectrum of hydrogen is most energetic: ultraviolet, visible, or infrared?
5. How many quanta of light are emitted for the following electron energy changes in hydrogen atoms?
   (a) 1 e^- drops from energy level 3 to 1
   (b) 1 e^- drops from energy level 3 to 2
   (c) 100 e^- drop from energy level 3 to 2
   (d) 100 e^- drop from energy level 4 to 2
   (e) 500 e^- drop from energy level 5 to 2
   (f) 500 e^- drop from energy level 5 to 3
6. What is the color of the spectral line emitted for the following electron energy changes in excited hydrogen gas?
   (a) electrons drop from energy level 2 to 1
   (b) electrons drop from energy level 3 to 2
   (c) electrons drop from energy level 4 to 2
   (d) electrons drop from energy level 5 to 1
   (e) electrons drop from energy level 5 to 2
   (f) electrons drop from energy level 5 to 3
7. What is the greatest success for the Bohr model of the atom?
8. What is the greatest limitation for the Bohr model of the atom?

Quantum Theory (Sec. 10.2)

9. Indicate whether the following correspond to a continuous or a quantized spectrum.
   (a) a rainbow
   (b) an emission line spectrum
10. Indicate whether the following musical instruments give a continuous or a quantized pitch.
    (a) piano (b) violin
11. Indicate whether the following scientific instruments give a continuous or a quantized measurement.
    (a) 10-mL graduated cylinder
    (b) 10-mL volumetric pipet
12. Indicate whether the following gauges in an automobile give a continuous or a quantized measurement.
    (a) digital odometer (b) analog speedometer
13. Explain the following observation. A metal is irradiated for 2 seconds with infrared light and no electrons are ejected. The same metal is irradiated for 2 seconds with ultraviolet light and thousands of electrons are emitted.
14. How many photons are responsible for ejecting 100 electrons from a sheet of aluminum foil?

Quantum Mechanical Model of the Atom (Sec. 10.3)

15. What is the distinction between an orbit and an orbital?
16. What are two important differences between the Bohr model of the atom and the quantum mechanical model?
17. Describe the uncertainty principle.
18. Describe the concept of electron probability.
19. Given the principal quantum number, which of the following orbitals has the higher energy?
    (a) $n = 1$ or $n = 2$ (b) $n = 1$ or $n = 3$
    (c) $n = 2$ or $n = 3$ (d) $n = 3$ or $n = 5$
20. Given the principal quantum number, which of the following orbitals has the larger size?
    (a) $n = 1$ or $n = 2$ (b) $n = 1$ or $n = 3$
    (c) $n = 2$ or $n = 3$ (d) $n = 3$ or $n = 5$
21. Which of the following pairs of orbital designations are identical?
    (a) s and $l = 1$ (b) p and $l = 1$
    (c) d and $l = 2$ (d) f and $l = 4$
22. Which of the following designations are not allowed for a given orbital?
    (a) $n = 0, l = 0$ (b) $n = 1, l = 0$
    (c) $n = 2, l = 2$ (d) $n = 3, l = 2$
    (e) $n = 4, l = 3$ (f) $n = 5, l = 0$
23. Sketch a three-dimensional representation for each of the following orbitals. Label the x-, y-, and z-axes.
    (a) $1s$ (b) $4s$
    (c) $2p_x$ (d) $3p_x$
    (e) $3p_y$ (f) $4p_z$
24. Sketch a three-dimensional representation for each of the following collections of orbitals. Label the x-, y-, and z-axes.
    (a) $1s$, $2s$, $3s$, $4s$ (b) $3p_x$, $3p_y$, $3p_z$
    (c) $1s$, $2s$, $2p_x$ (d) $2s$, $2p_x$, $2p_y$, $2p_z$
25. Which of the following orbitals has the higher energy?
    (a) $2s$ or $3s$ (b) $2p_x$ or $3p_x$
    (c) $2p_x$ or $2p_y$ (d) $4p_y$ or $4p_z$
26. Which of the following orbitals has the larger size?
    (a) $2s$ or $3s$ (b) $2p_x$ or $3p_x$
    (c) $2p_x$ or $2p_y$ (d) $4p_y$ or $4p_z$

27. Designate the orbital that fits each of the following descriptions.
(a) spherical orbital in the fifth shell
(b) dumbbell-shaped orbital in the fourth shell

28. Designate the orbital that fits each of the following descriptions.
(a) spherical orbital in the sixth shell
(b) dumbbell-shaped orbital in the third shell

Distribution of Electrons by Orbital (Sec. 10.4)

29. State the maximum number of electrons that can occupy each of the following types of orbitals.
(a) s (b) p
(c) d (d) f

30. State the maximum number of electrons that can occupy each of the following orbitals.
(a) $7s$ (b) $6p$
(c) $6d$ (d) $5f$

31. State the maximum number of electrons that can occupy each of the following orbitals.
(a) $l = 0$ (b) $l = 1$
(c) $l = 2$ (d) $l = 3$

32. State the maximum number of electrons that can occupy each of the following orbitals.
(a) $n = 1, l = 0$ (b) $n = 2, l = 0$
(c) $n = 3, l = 2$ (d) $n = 4, l = 1$

33. State the number of orbitals within each of the following electron subshells.
(a) $3s$ (b) $4p$
(c) $3d$ (d) $4f$

34. State the number of orbitals within each of the following electron subshells.
(a) $n = 3, l = 1$ (b) $n = 4, l = 0$
(c) $n = 5, l = 3$ (d) $n = 6, l = 2$

35. How many subshells exist within each of the following electron shells?
(a) $n = 1$ (b) $n = 2$
(c) $n = 3$ (d) $n = 4$

36. How many subshells theoretically exist within each of the following shells?
(a) $n = 5$ (b) $n = 6$
(c) $n = 7$ (d) $n = 8$

37. Calculate the maximum number of electrons that can occupy each of the following subshells.
(a) $4s$ (b) $4p$
(c) $4d$ (d) $4f$

38. Calculate the maximum number of electrons that can occupy each of the following subshells.
(a) $7s$ (b) $6p$
(c) $5d$ (d) $5f$

39. Calculate the maximum number of electrons that can occupy each of the following electron shells.
(a) $n = 1$ (b) $n = 2$
(c) $n = 3$ (d) $n = 4$

40. Calculate the maximum number of electrons that can theoretically occupy the fifth electron shell. (*Hint:* A $5g$ subshell would have nine orbitals.)

The Four Quantum Numbers (Sec. 10.5)

41. Indicate all the allowed values for the principal quantum number.

42. Which of the following values are not allowed for the principal quantum number?
(a) $n = 0$ (b) $n = 1$
(c) $n = 2.5$ (d) $n = 3$

43. Indicate all the allowed values for the second quantum number, given the following principal quantum numbers.
(a) $n = 1$ (b) $n = 2$
(c) $n = 3$ (d) $n = 4$

44. Which of the following values are not allowed for the second quantum number?
(a) $l = 0$ (b) $l = 1$
(c) $l = 3.5$ (d) $l = 5$

45. Indicate all the allowed values for the third quantum number, given the following second quantum numbers.
(a) $l = 0$ (b) $l = 1$
(c) $l = 2$ (d) $l = 3$

46. Which of the following values are not allowed for the third quantum number?
(a) $m = 0$ (b) $m = +3$
(c) $m = 3.5$ (d) $m = -3$

47. Indicate all the allowed values for the third quantum number, given the following orbitals.
(a) $4s$ (b) $4p$
(c) $4d$ (d) $4f$

48. Indicate all the allowed values for the third quantum number, given the following orbitals.
(a) $6s$ (b) $2p$
(c) $3d$ (d) $5f$

49. Indicate all the allowed values for the fourth quantum number, given the following third quantum numbers.
(a) $m = 0$ (b) $m = +1$
(c) $m = +2$ (d) $m = -3$

50. Which of the following values are not allowed for the fourth quantum number?
(a) $s = 0$ (b) $s = -\frac{1}{2}$
(c) $s = +\frac{1}{2}$ (d) $s = +1$

51. What property of an electron in an atom is specified by the principal quantum number?

52. What property of an electron in an atom is specified by the second quantum number?

53. What property of an electron in an atom is specified by the third quantum number?

54. What property of an electron in an atom is specified by the fourth quantum number?

Orbital Energy Diagrams (Sec. 10.6)

55. Write out the electron configuration by subshell for atoms of the following elements.
(a) C (b) P
(c) K (d) Co

56. Write out the electron configuration by subshell for atoms of the following elements.
(a) Na (b) Ti
(c) Ca (d) Se

57. Write out the orbital diagram for the following elements (refer to Exercise 55).
(a) C (b) P
(c) K (d) Co

58. Write out the orbital diagram by subshell for the following elements (refer to Exercise 56).
(a) Na (b) Ti
(c) Ca (d) Se

59. Determine the number of unpaired electrons in an atom of sulfur. (*Hint:* Draw the orbital diagram.)

60. Determine the number of unpaired electrons in an atom of nickel. (*Hint:* Draw the orbital diagram.)

61. Draw the orbital energy diagram for a helium atom.

62. Draw the orbital energy diagram for a hydrogen atom.

63. Draw the orbital energy diagram for each of the following.
(a) Li (b) C
(c) Mg (d) Cl

64. Draw the orbital energy diagram for each of the following.
(a) Na (b) F
(c) Mn (d) As

Quantum Notation: (n, l, m, s) (Sec. 10.7)

65. Write a set of four quantum numbers for the highest-energy electron in an atom of the following elements.
(a) boron (b) neon
(c) magnesium (d) oxygen

66. Write a set of four quantum numbers for the highest-energy electron in an atom of the following elements.
(a) potassium (b) sulfur
(c) nickel (d) bromine

67. Refer to the periodic table and write a set of four quantum numbers for the highest-energy electron in an atom of the following elements.
(a) Na (b) F
(c) Sc (d) Kr

68. Refer to the periodic table and write a set of four quantum numbers for the highest-energy electron in an atom of the following elements.
(a) Mg (b) P
(c) V (d) Ga

69. Refer to the periodic table and determine the element whose highest-energy electron has the following quantum notation.
(a) $(2, 0, 0, +\frac{1}{2})$ (b) $(3, 1, +1, -\frac{1}{2})$
(c) $(4, 2, 0, +\frac{1}{2})$ (d) $(5, 1, 0, -\frac{1}{2})$

70. Refer to the periodic table and determine the element whose highest-energy electron has the following quantum notation.
(a) $(3, 2, 0, +\frac{1}{2})$ (b) $(3, 1, -1, -\frac{1}{2})$
(c) $(4, 3, +3, +\frac{1}{2})$ (d) $(5, 3, +1, -\frac{1}{2})$

General Exercises

71. Matter and energy can be viewed as being either continuous or quantized. What is the particle corresponding to the quantized nature of each of the following?
(a) element (b) compound
(c) light energy (d) electrical energy

72. Distinguish between the terms quantum and photon.

73. List some practical applications that utilize the photoelectric effect.

74. Explain the significance of the spatial overlap for $1s$ and $2s$ orbitals.

75. Explain the significance of the spatial overlap for $2s$ and $2p$ orbitals.

76. The actual electron configuration for copper is $1s^2\ 2s^2\ 2p^6\ 3s^2\ 3p^6\ 4s^1\ 3d^{10}$ and not $1s^2\ 2s^2\ 2p^6\ 3s^2\ 3p^6\ 4s^2\ 3d^9$. Propose a reasonable explanation for this exceptional behavior.

77. The actual electron configuration for chromium is $1s^2\ 2s^2\ 2p^6\ 3s^2\ 3p^6\ 4s^1\ 3d^5$ and not $1s^2\ 2s^2\ 2p^6\ 3s^2\ 3p^6\ 4s^2\ 3d^4$. Using the orbital concept and Hund's rule, give a reasonable explanation for this exceptional behavior.

78. Explain the concept of paired electron spins.

79. The modern view of atomic theory is supported by the magnetic properties of certain elements. Elements that have unpaired electrons are attracted when placed into a magnetic field. These elements are said to be paramagnetic. Which of the following elements are paramagnetic? (*Hint:* Draw the orbital diagram.)
(a) Ti (b) Zn
(c) Ni (d) Ba

CHAPTER

11

Chemical Bonding

This color-enhanced photograph shows the intricate and detailed geometry of a snowflake. The shape of the snowflake is the result of countless water molecules bonded together in a regular, repeating pattern.

The late nineteenth and early twentieth centuries were an active time for physical scientists. J. J. Thomson discovered the electron in 1897. Ernest Rutherford unveiled the atomic nucleus in 1911. Using emission spectra from the elements, Niels Bohr established the planetary model of the atom in 1913. During this same period of time, scientists began to theorize that electrons were responsible for "gluing" atoms together in compounds. They proposed that electrons provided the necessary "glue" for the bonding of one atom to another. Ever since then, the bonding of atoms in compounds has been an area of great interest to chemists.

By studying the noble gases, the American chemist G. N. Lewis (1875–1946) formulated one of the first theories of chemical bonding. He noted that the noble gases were unusually stable and, except for helium, all had eight electrons in their outer shell. Lewis suggested that atoms react in such a way as to attain a stable noble gas electron structure and proposed the octet rule in 1916. The octet rule states that atoms bond in such a way that each atom attains eight valence electrons (an octet) in its outer shell. Lewis initially pictured each of the valence electrons around the nucleus at the corners of a hypothetical cube, as shown in Figure 11.1.

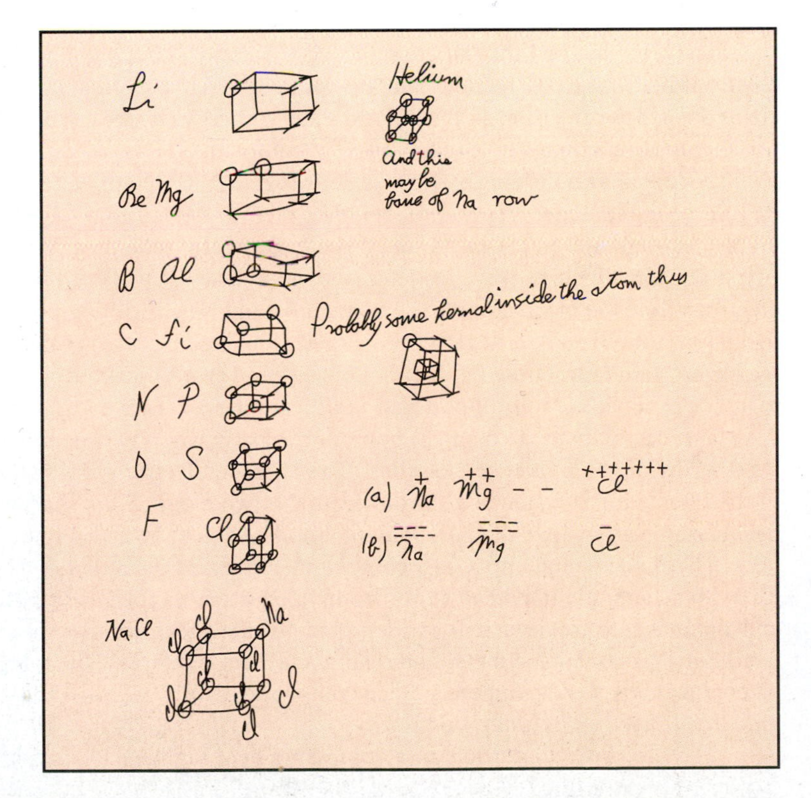

Figure 11.1 A Page from the Notebook of G. N. Lewis An original sketch by Lewis of the octet rule in which 8 electrons are located at the corners of a hypothetical cube. The idea occurred to Lewis in 1902 while lecturing to an introductory chemistry class.

Lewis eventually realized that his cube model was not quite correct. He later suggested that atoms become stable when surrounded by four pairs of electrons. The Lewis concept of a filled valence shell, however, was still valid. The octet rule remains today as a fundamental principle of chemical bonding.

In his own words Lewis described the chemical bond concept as follows:

> *Two atoms may conform to the rule of eight, or the octet rule, not only by the transfer of electrons from one atom to another, but also by sharing one or more pairs of electrons. These electrons which are held in common by two atoms may be considered to belong to the outer shells of both atoms.*
>
> *The simplest explanation of the predominant occurrence of an even number of electrons in the valence shells of molecules is that the electrons are definitely paired with one another.*
>
> *Two electrons thus coupled together, when lying between two atomic centers, and held jointly in the shells of the two atoms, I have considered to be the chemical bond.*

11.1 Valence Electrons and Chemical Bonds

To describe valence electron changes for the formation of ionic and covalent bonds.

To predict whether a compound is held together by ionic or covalent bonds.

valence electrons The electrons in the highest *s* and *p* subshells in an atom that undergo reactions and form chemical bonds.

chemical bond The attraction between two atoms or two ions.

octet rule The statement that each atom in a compound must attain eight valence electrons in order to be energetically stable. Hydrogen is an exception to the rule and requires only two electrons to be stable.

ionic bond A chemical bond characterized by the attraction of positive metal ions and negative nonmetal ions.

covalent bond A chemical bond characterized by the sharing of one or more pairs of valence electrons.

formula unit The smallest representative entity in a compound held together by ionic bonds.

Previously, we learned that an atom has core electrons and valence electrons. Core electrons lie close to the nucleus and are not involved in chemical reactions. **Valence electrons** are the electrons in the highest *s* and *p* energy subshells and are responsible for chemical reactions and producing new compounds.

A **chemical bond** holds atoms or ions together in a compound. The **octet rule**, or rule of eight, states that atoms bond in such a way so that each atom acquires eight electrons (an octet) in its outer shell. Basically, chemical bonds develop in two ways. In the first way, a metal and nonmetal react by transferring valence electrons from the metal atom to the nonmetal atom. The resulting positively charged metal ion (cation) and the negatively charged nonmetal ion (anion) are drawn together by electrostatic attraction. The attraction holding two oppositely charged particles together is called an **ionic bond** and is illustrated in Figure 11.2.

The second way a chemical bond develops is when two nonmetal atoms join together by sharing valence electrons. The valence electrons are shared as a result of the interaction of valence electrons from the two atoms. This sharing of electrons between two atoms is called a **covalent bond** and is shown in Figure 11.2.

The fundamental entity in a compound composed of ionic bonds is a **formula unit**. Since an ionic bond can result from valence electron transfer between a metal and nonmetal, a formula unit is identified by the fact that the chemical formula nearly always contains a metal and a nonmetal. For example, NaCl, MgO, and AlF_3 all contain metals and nonmetals. Each compound, therefore, is composed of formula units held together by ionic bonds.

The fundamental particle in a compound held together by covalent bonds is a

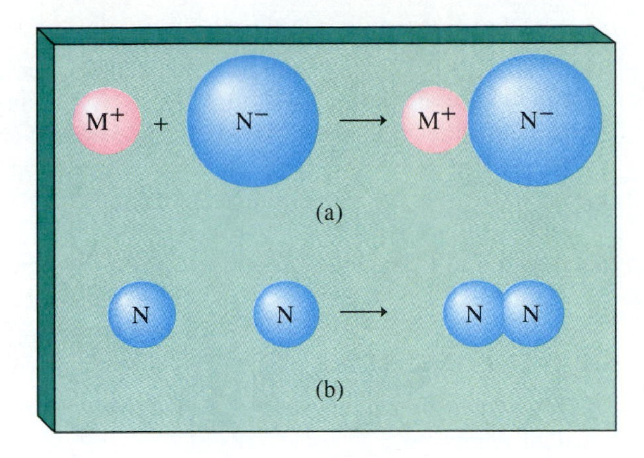

Figure 11.2 Chemical Bonds (a) An ionic bond is formed when a metal cation (M^+) is attracted to a nonmetal anion (N^-). (b) A covalent bond results from the sharing of valence electrons between two nonmetal atoms (N).

molecule. Since a covalent bond is the result of sharing valence electrons between two nonmetals, a molecule is easily identified by the fact that it contains two nonmetals. For example, HCl, H_2O, and Cl_2O_5 all contain two nonmetals. Each compound, therefore, is composed of molecules held together by covalent bonds. Figure 11.3 summarizes the relationship between formula units and molecules.

molecule The smallest representative entity in a compound held together by covalent bonds.

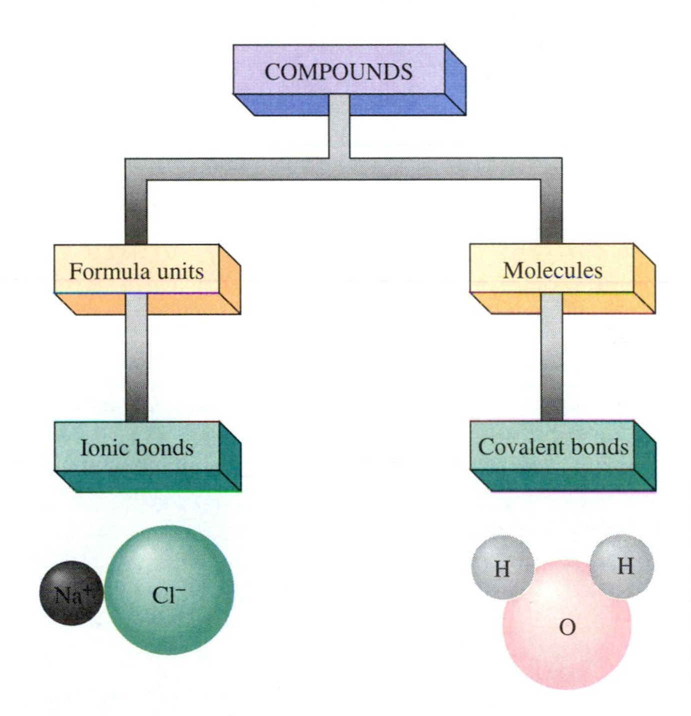

Figure 11.3 Classification of Representative Particles in Compounds Compounds are composed of formula units (for example, NaCl) held together by ionic bonds or molecules (for example, H_2O) held together by covalent bonds.

The following example exercise further illustrates the relationship of bonding and the type of fundamental particles in a compound.

EXAMPLE EXERCISE 11.1

Predict whether each of the following compounds is held together by ionic or covalent bonds.

(a) zinc oxide, ZnO
(b) nitrogen monoxide, NO
(c) ammonia, NH_3
(d) iron pyrite, FeS_2

Ionic crystals of iron pyrite, FeS_2.

Solution: Refer to Figure 11.3.

(a) Zinc oxide is a compound containing a metal (Zn) and a nonmetal (O). Therefore, ZnO is a formula unit held together by *ionic* bonds.

(b) Nitrogen monoxide is a compound composed of two nonmetals (N and O). Thus, NO is a molecule held together by *covalent* bonds.

(c) Ammonia contains the nonmetals nitrogen and hydrogen. It follows that NH_3 molecules have *covalent* bonds.

(d) Iron pyrite (commonly referred to as "fool's gold") contains a metal (Fe) and a nonmetal (S). It follows that FeS_2 has *ionic* bonds.

SELF-TEST EXERCISE

Predict whether each of the following compounds is held together by ionic or covalent bonds.

(a) aluminum oxide, Al_2O_3 **(b)** sulfur hexafluoride, SF_6

Answers: **(a)** ionic; **(b)** covalent

11.2 The Ionic Bond

To describe the formation of an ionic bond between a metal atom and a nonmetal atom.

As was shown in Figure 11.2(a), an ionic bond is the result of the attraction between a positively charged ion and a negatively charged ion. This is termed electrostatic attraction and is similar to the attraction between opposite ends of two magnets. In

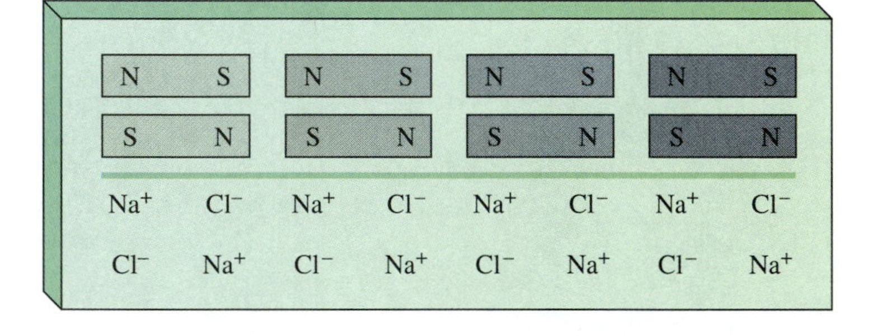

Figure 11.4 Electrostatic Attraction Opposite ends of ordinary magnets are attracted to each other in much the same way that oppositely charged cations and anions are attracted. Notice that attraction can take place vertically as well as horizontally.

sodium chloride, NaCl, the attraction is between Na^+ and Cl^-. Figure 11.4 illustrates electrostatic attraction.

An ionic compound results from the combination of cations and anions. In the ionic compound table salt, NaCl, the strong ionic bonds between sodium ions and chloride ions create a rigid crystalline structure (see Figure 11.5).

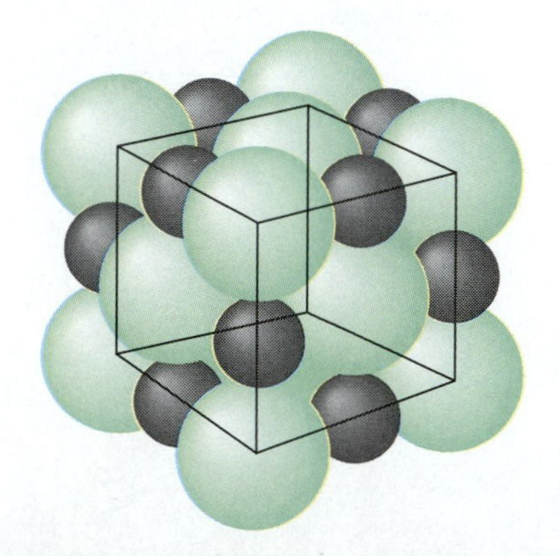

Figure 11.5 Crystalline Structure of Sodium Chloride The compound sodium chloride is composed of NaCl formula units that repeat regularly. Notice, however, that each sodium ion is attracted to several nearby chloride ions in the crystal. Conversely, each Cl^- is attracted to several adjacent Na^+. The alternating pattern of cations and anions gives rise to a crystalline structure.

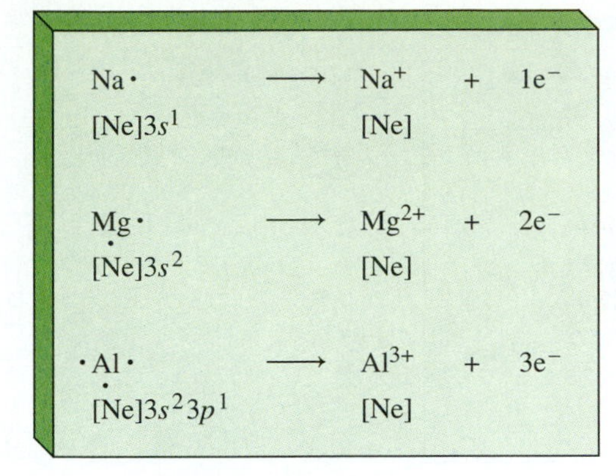

Figure 11.6 Formation of Metal Ions The metals in period 3 lose 1, 2, and 3 electrons, respectively, to form cations. In each case the metal becomes isoelectronic with the noble gas neon, [Ne].

Formation of Cations

In general, metal atoms lose valence electrons and become positively charged. Recall that the number of valence electrons corresponds to the group number of the element in the periodic table. For example, a magnesium atom (Group IIA/2) loses two valence electrons to form Mg^{2+}. If we refer to the periodic table, we notice that Mg^{2+} has 10 $(12 - 2)$ electrons, the same number of electrons as the noble gas neon. Although there are numerous exceptions, metals lose valence electrons in order to achieve a stable noble gas electron configuration. That is, metal ions attempt to become isoelectronic with a noble gas.

Similarly, an aluminum atom (Group IIIA/13) loses 3 electrons to form Al^{3+}. An aluminum ion has 10 $(13 - 3)$ electrons and is isoelectronic with the noble gas neon. In Section 5.8, we learned to write electron dot formulas for atoms of the elements. We can now diagram electron dot formulas to clarify the formation of ions. Electron dot diagrams allow us to focus our attention on only those electrons involved in this process of forming ions. Figure 11.6 illustrates the formation of ions by the third-period metals.

Notice that the electron dot diagrams of the ions in Figure 11.6 do not have any dots. Each of these ions now has the electron configuration of neon, which is [He] $2s^2\ 2p^6$. Each of these ions, as well as neon, has a stable electron configuration consisting of 8 electrons, that is, a complete octet. By convention, chemists do not draw the underlying octet of electrons as they are considered core electrons. According to Lewis, when atoms form cations they tend to lose electrons until a stable noble gas electron configuration is achieved. Electron dot formulas indicate this by showing either eight dots or zero dots.

Formation of Anions

In contrast to the metals, nonmetal atoms gain electrons to fill their valence shell and thus become negatively charged. A nonmetal atom with 6 valence electrons, for example, gains 2 electrons to achieve an octet. The nonmetal atom would then have a negative charge of 2−. Thus, sulfur (Group VIA/16) gains 2 valence electrons to form S^{2-}. By referring to the periodic table, we can calculate that S^{2-} has 18 $(16 + 2)$ electrons. Therefore, S^{2-} has the same number of electrons as the noble gas argon (18). Similarly, a phosphorus atom (Group VA/15) gains 3 electrons to form P^{3-}. Since P^{3-} has 18 $(15 + 3)$ electrons, it is also isoelectronic with the noble gas argon. Figure 11.7 illustrates the formation of ions by the third-period nonmetals.

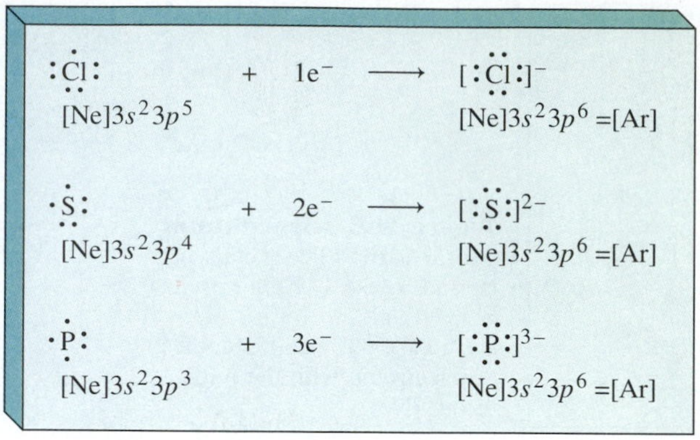

:Cl:	+	1e⁻	⟶	[:Cl:]⁻
$[Ne]3s^2 3p^5$				$[Ne]3s^2 3p^6 = [Ar]$
·S:	+	2e⁻	⟶	[:S:]²⁻
$[Ne]3s^2 3p^4$				$[Ne]3s^2 3p^6 = [Ar]$
·P:	+	3e⁻	⟶	[:P:]³⁻
$[Ne]3s^2 3p^3$				$[Ne]3s^2 3p^6 = [Ar]$

Figure 11.7 Formation of Nonmetal Ions The nonmetals in period 3 gain 1, 2, and 3 electrons, respectively. In each case the nonmetal ion becomes isoelectronic with the noble gas argon, [Ar].

Notice that the electron dot formulas of the ions in Figure 11.7 show eight dots. Each of these ions now has the electron configuration of argon, which is [Ne] $3s^2\ 3p^6$. Each of these ions, as well as argon, has a stable electron configuration consisting of eight electrons; that is, each has a complete octet. When atoms form anions, they tend to gain electrons until a stable noble gas electron configuration is achieved. Electron dot diagrams indicate this by showing eight dots surrounding the symbol of the element.

The following example exercise will further illustrate the formation of cations and anions.

EXAMPLE EXERCISE 11.2

Which noble gas has an electron configuration identical to each of the following ions?

(a) lithium ion **(b)** oxide ion
(c) calcium ion **(d)** bromide ion

Solution: Refer to the group number in the periodic table for the number of valence electrons.

(a) Lithium (Group IA/1) forms an ion by losing 1 valence electron; Li^+ has only 2 electrons remaining and is the same as *He*.
(b) Oxygen (Group VIA/16) forms an ion by gaining 2 electrons; thus, O^{2-} has 10 electrons and is the same as *Ne*.
(c) Calcium (Group IIA/2) forms an ion by losing 2 electrons; Ca^{2+} has 18 electrons remaining and is isoelectronic with *Ar*.
(d) Bromine (Group VIIA/17) forms an ion by gaining 1 electron; Br^- has 36 electrons and is isoelectronic with *Kr*.

SELF-TEST EXERCISE

Which noble gas has an electron configuration identical to each of the following ions?

(a) potassium ion **(b)** nitride ion
(c) strontium ion **(d)** iodide ion

Answers:
(a) K^+ is isoelectronic with Ar; **(b)** N^{3-} is isoelectronic with Ne;
(c) Sr^{2+} is isoelectronic with Kr; **(d)** I^- is isoelectronic with Xe.

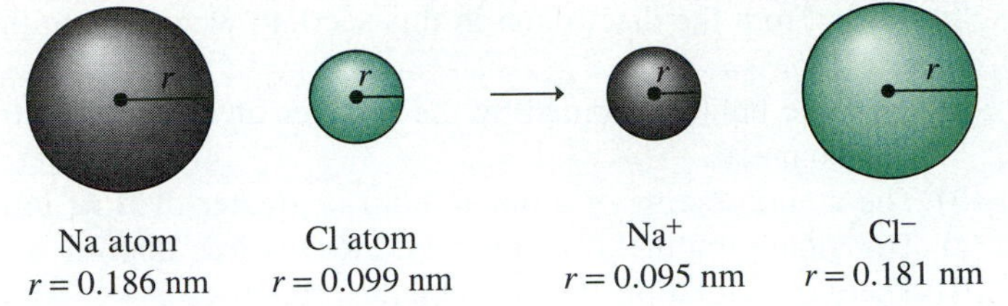

Figure 11.8 Formation of Sodium and Chloride Ions The atomic radius (r) of the sodium atom decreases as a result of losing an electron while the atomic radius of the chlorine atom increases as a result of gaining an electron.

Ionic Radii

To visualize an ionic bond more clearly, picture a sodium atom becoming smaller after losing its valence electron. The atomic radius of a sodium atom is 0.186 nm, while the radius of the ion is only 0.095 nm. The reason for this decrease in radius is twofold. First, the entire 3*s* energy sublevel has been lost by the sodium atom. Second, the nuclear charge is still 11+, but the total electron charge is only 10−. The positive nuclear charge decreases the radius by drawing fewer electrons closer to the nucleus.

Now visualize a chloride atom becoming larger after gaining another valence electron. The atomic radius of a chlorine atom is 0.099 nm, while the radius of the ion increases to 0.181 nm. The reason for the increase in radius is that the electron gained by the chloride ion repels other negative electrons that are already present. The nuclear charge is 17+, but the total electron charge is now 18−. The net effect is that the ion becomes larger than the neutral atom. In fact, the radius of the charged ion is almost twice as big as the neutral atom.

Regarding atomic radius, the behaviors of metallic sodium and nonmetallic chlorine show the same general trend observed for all metals and nonmetals. That is, the radius of a *cation is smaller* than that of the corresponding metal atom; the radius of an *anion is larger* than that of the corresponding nonmetal atom. The specific changes in atomic radius for sodium and chlorine atoms are shown in Figure 11.8.

Heat of Reaction

Chemical changes involve a change in energy, and heat is always released when two atoms in the gaseous state react to produce an ionic bond. That is, the reaction is always exothermic. In the previous example, the individual sodium and chlorine atoms have more potential energy than the compound sodium chloride. Thus, the formation of NaCl from sodium and chlorine atoms releases heat energy.

$$\text{Na(g)} + \text{Cl(g)} \longrightarrow \text{Na} \qquad \text{Cl(g)} + \text{heat energy}$$

The following example exercise will summarize the characteristics associated with the formation of an ionic bond.

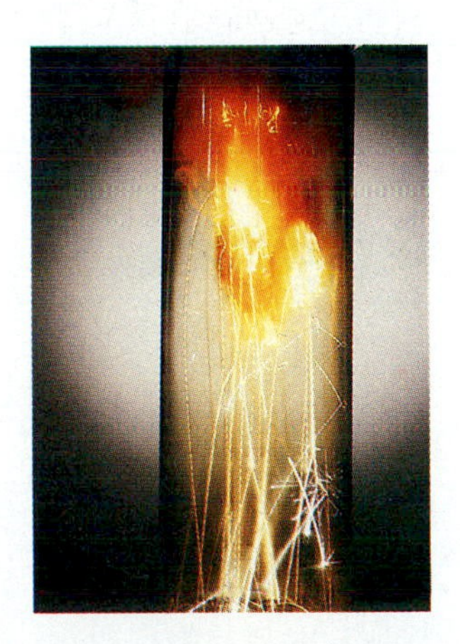

The reaction of steel wool and chlorine gas releases heat energy and white-hot sparks of iron. The product is iron(III) chloride, $FeCl_3$, the reddish-brown deposit at the top of the cylinder.

EXAMPLE EXERCISE 11.3

Which of the following statements is correct regarding the formation of an ionic bond between calcium and oxygen?

(a) Valence electrons are transferred from a calcium atom to an oxygen atom.
(b) The calcium atom is larger in radius than the calcium ion.
(c) The oxygen atom is smaller in radius than the oxide ion.
(d) The properties of calcium and calcium oxide are not related.
(e) Energy is released when calcium and oxygen atoms combine by forming Ca—O ionic bonds.

Solution: From the discussion in this section, we can state that all of the above are correct. That is,

(a) An ionic bond is formed by the transfer of valence electrons from the metal to the nonmetal.
(b) The atomic radius of a metal atom is greater than its ionic radius.
(c) The atomic radius of a nonmetal atom is less than its ionic radius.
(d) There is no relationship between the properties of a compound and the properties of its constituent elements.
(e) Energy is always released when atoms of a metal and nonmetal combine by forming ionic bonds.

SELF-TEST EXERCISE

Which of the following statements is *not* correct regarding the formation of an ionic bond between zinc and sulfur?

(a) Valence electrons are transferred from a zinc atom to a sulfur atom.
(b) The zinc atom is larger in radius than the zinc ion.
(c) The sulfur atom is smaller in radius than the sulfide ion.
(d) Zinc and zinc sulfide have similar properties.
(e) The formation of an ionic bond between a zinc atom and a sulfur atom releases energy.

Answer: **(d)** There is no similarity in properties.

Piles of pure salt, NaCl.

Sodium chloride is an inexpensive chemical found in nature. Sodium metal and chlorine gas are relatively expensive and are not naturally occurring. Why is this? The answer is that NaCl is produced spontaneously from its elements while releasing energy in the process. Conversely, it is necessary to use energy to decompose NaCl in order to produce the metal and nonmetal. Sodium metal and chlorine gas are expensive because *energy costs money*.

11.3 The Covalent Bond

To describe the formation of a covalent bond between two nonmetal atoms.

Recall that a covalent bond is formed by the sharing of a pair of electrons between two nonmetal atoms; refer to Figure 11.2(b). You should understand that both of the electrons are distributed over each atom. Since the electron pair belongs to both nonmetal atoms, each atom can use the electrons to complete their valence shells. A filled valence shell is very stable, and this accounts for the stability of the resulting particle.

Now let's see what happens during the formation of a covalent bond. As an example, consider that hydrogen and chlorine react to yield hydrogen chloride, HCl. During bond formation, the hydrogen atom shares its 1 valence electron with the chlorine atom. This additional electron gives chlorine 8 electrons in its valence shell. This process completes an octet in the valence shell of chlorine. Chlorine thus becomes isoelectronic with argon and is stable.

In the same process, the chlorine atom shares one of its valence electrons with the hydrogen atom. The additional electron gives the hydrogen atom 2 electrons in its valence shell. Hydrogen becomes isoelectronic with the noble gas helium and is therefore stable. Both the hydrogen atom and chlorine atom become stable in the process of sharing an electron pair and forming a covalent bond.

Bond Length

To understand the change more clearly, picture the valence shells of the two atoms overlapping each other. The overlapping occurs because the 1*s* energy subshell of the hydrogen atom mixes with the 3*p* subshell of the chlorine atom. This mixing of subshells draws the two nuclei closer together. The radius of the hydrogen atom is 0.037 nm and the chlorine atom is 0.099 nm. If the two atoms were simply next to each other, the distance from the hydrogen nucleus to the chlorine nucleus would be (0.037 + 0.099) nm or 0.136 nm. But experiments reveal that the distance between the two nuclei is actually 0.127 nm. Thus, the shells overlap as shown in Figure 11.9. The actual distance from one nucleus to another is referred to as the **bond length**.

bond length The distance between the nuclei of two atoms that are covalently bonded.

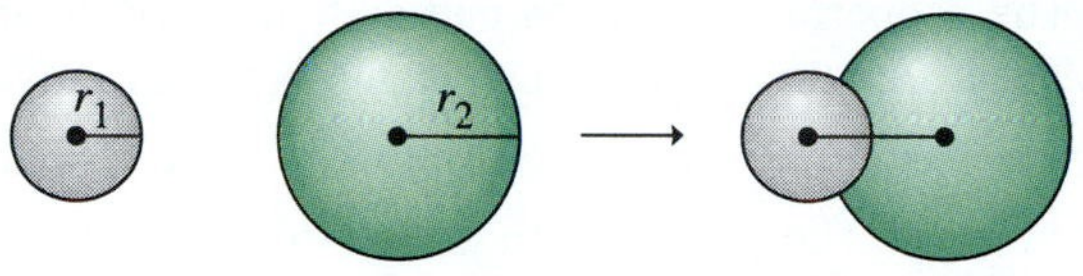

Figure 11.9 Formation of a Covalent Bond The sharing of an electron pair between a hydrogen atom and a chlorine atom produces a covalent bond. The two atoms are held together in a molecule of HCl. The bond length (0.127 nm) is less than the sum of the two atomic radii (0.037 + 0.099 = 0.136 nm).

Bond Energy

When discussing the ionic bond, we stated that energy is released when two neutral atoms react to produce a chemical bond. The same is true for covalent bonds. When hydrogen atoms and chlorine atoms combine to form HCl, 103 kilocalories (or 431 kilojoules) of heat energy are released for every mole of HCl produced.

Conversely, energy is needed to break the H—Cl bond. The amount of energy required to break a given bond in a mole of gaseous substance is called the **bond energy**. The bond energy for HCl is 103 kcal/mol (431 kJ/mol).

bond energy The amount of energy required to break a given bond in a mole of substance in the gaseous state.

$$HCl(g) + 103 \text{ kcal} \longrightarrow H(g) + Cl(g)$$

The following example exercise will summarize the characteristics associated with the formation of a covalent bond.

EXAMPLE EXERCISE 11.4

Which of the following statements is correct regarding the formation of a covalent bond between hydrogen and oxygen to form water?

(a) Valence electrons are shared between a hydrogen atom and an oxygen atom.
(b) The pair of bonding electrons is loosely distributed about the hydrogen atom and the oxygen atom.
(c) The H—O bond length is less than the sum of the atomic radii of a hydrogen atom and oxygen atom.
(d) The properties of hydrogen and water are not related.
(e) Energy is required to break the H—O covalent bond.

Solution: From the discussion in this section, we can state that all of the above are correct. That is,

(a) A covalent bond is formed by sharing valence electrons between the two non-metals.

(b) The shared bonding electrons are in motion and move about both atoms in the covalent bond.

(c) The bond length in a covalent bond is less than the sum of the two atomic radii.

(d) There is no direct relationship between the properties of a compound and the properties of its constituent elements.

(e) An amount of energy equal to the bond energy is required to break any covalent bond regardless of the constitutent elements.

SELF-TEST EXERCISE

Which of the following statements is *not* correct regarding the formation of a covalent bond between carbon and sulfur to form carbon disulfide?

(a) Valence electrons are shared between the carbon and sulfur atoms.

(b) Bonding electrons are distributed over both the carbon and sulfur atoms.

(c) The bond length between carbon and sulfur atoms is equal to the sum of the two atomic radii.

(d) The properties of carbon and carbon disulfide are not related.

(e) The formation of covalent bonds between carbon atoms and sulfur atoms releases energy.

Answer: **(c)** The bond length is less than the sum of the atomic radii.

It is always true that energy is released when *two neutral aotms* combine to form a chemical bond. Most reactions, however, do not involve neutral atoms. For example, consider the formation of hydrogen iodide. It is *molecules* of H_2 and I_2 that react to produce HI. Thus, energy is required to break the H—H and I—I bonds before neutral atoms of H and I are available to combine. In fact, the formation of HI is endothermic because the energy required to break H—H and I—I bonds is greater than the energy released in the formation of H—I bonds.

11.4 Electron Dot Formulas of Molecules

To apply the octet rule to the electron dot formula of a molecule, given the arrangement of atoms.

To write the structural formula for a molecule, given the electron dot formula.

In Section 5.8, we wrote the electron dot formula for a representative atom of an element. We found the element in the periodic table, observed its group number, and recorded the number of valence electrons. For example, referring to the second series in the periodic table, we find a Li atom has 1 valence electron (Group IA/1), Be has 2 (Group IIA/2), B has 3 (Group IIIA/13), C has 4 (Group IVA/14), N has

5 (Group VA/15), O has 6 (Group VIA/16), F has 7 (Group VIIA/17), and the noble gas Ne has 8 (Group VIIIA/18) valence electrons. We represented each valence electron by a dot surrounding the symbol of the element.

Now we will write the **electron dot formulas** for molecules. The guidelines we will follow are simple and direct.

electron dot formula A representation of a molecule or polyatomic ion that shows the chemical symbol of each atom surrounded by a dot for each valence electron; also called a Lewis diagram.

1. Calculate the total number of valence electrons in the molecule by adding together all the valence electrons for each atom in the molecule. The total is nearly always an even number because of the octet rule. If the total is an odd number, check your calculations.
2. Divide the total number of valence electrons by 2 to find the total number of electron pairs.
3. Surround the central atom with four electron pairs. Use the remaining electron pairs to complete an octet around each of the other atoms. Hydrogen is the sole exception as H requires only 2 e^-, that is, one electron pair. The electron pairs between bonded atoms are called **bonding electrons**. The other electron pairs simply complete the octet and are called **nonbonding electrons**.
4. If there are not enough electron pairs to provide an octet for each atom, move a nonbonding electron pair between two atoms. The nonbonding electron pair thus becomes a bonding pair. A shared pair of electrons between two atoms is called a **single bond**. Two electron pairs between two atoms are referred to as a **double bond**. In some instances, three electron pairs are shared between two atoms; this is called a **triple bond**.

bonding electrons The valence electrons in a molecule that are shared between two atoms.

nonbonding electrons The valence electrons in a molecule that are not shared.

single bond A bond between two atoms composed of one electron pair. A single bond can be represented as a dash between the symbols of two atoms.

double bond A bond between two atoms composed of two electron pairs. A double bond can be represented as two dashes between the symbols of two atoms.

triple bond A bond between two atoms composed of three electron pairs. A triple bond can be represented as three dashes between the symbols of two atoms.

Electron Dot Formula For Water

To see how to write electron dot formulas for molecules, let's write the electron dot formula for water, H_2O. The total number of valence electrons in the molecule is $2(1\ e^-) + 6\ e^- = 8\ e^-$. The number of electron pairs is four ($8/2 = 4$).

In the formula H_2O, oxygen is the central atom. We can place the four pairs of electrons around the oxygen to provide the necessary octet:

:Ö: (with a pair of dots below)

We can place the two hydrogen atoms at any of the four electron pair positions. For example.

H:Ö:H (with a pair of dots below O)

bonding electrons → (pair between H and O); *nonbonding electrons* → (pair below O)

Notice that there are two bonding and two nonbonding electron pairs. We can also place the two hydrogen atoms at angles to each other; for example,

H:Ö: (with a pair of dots below O)
H

The electron dot formula can be simplified. Each bonding pair of electrons can be represented by a single dash and the nonbonding electrons can be omitted. The result is called the **structural formula** of the molecule.

structural formula A representation of a molecule or polyatomic ion that shows each atom connected by a dash for each pair of bonding electrons.

We can write the structural formula for water in more than one way; for example,

H—O—H   H—O
               |
               H

bond angle The angle formed by two atoms bonded to the central atom in a molecule.

At this point, either of the electron dot formulas for water is correct. In Section 13.7, however, we will learn that the properties of water are best explained if the two hydrogens are at an angle to each other. The angle formed by two bonds to a central atom is referred to as the **bond angle**. For example, for water the bond angle is 104.5° and looks like

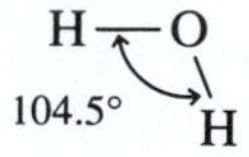

```
H—O
   \
104.5°  H
```

Electron Dot Formula For Sulfur Trioxide

Now we'll try a more difficult example. Let's write the electron dot formula for sulfur trioxide, SO_3. Sulfur is the central atom in the molecule and is indicated in bold. The total number of valence electrons in the molecule is $6\ e^- + 3(6\ e^-) = 24\ e^-$. The number of electron pairs is $24/2 = 12$ pairs of electrons.

Since sulfur is the central atom, we can begin by placing four pairs of electrons around the sulfur and then attaching the three oxygen atoms. This gives us

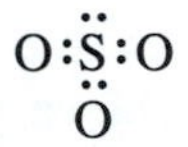

```
  ..
O:S:O
  ..
  O
```

We started with 12 electron pairs, so we have 8 pairs remaining. Let's place the remaining pairs around the oxygen atoms.

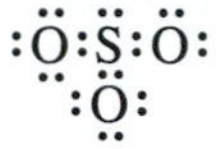

```
 ..  ..  ..
:O:S:O:
 ..  ..  ..
    :O:
     ..
```

Notice that one of the oxygen atoms does not have an octet; it has only three electron pairs. According to our guidelines, we should therefore move a nonbonding electron pair to provide two bonding pairs between the S and the O atoms. For example,

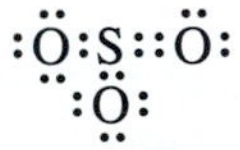

```
 ..      ..
:O:S::O:
 ..  ..
    :O:
     ..
```

The two electron pairs constitute a double bond. All four electrons in the double bond are shared between the S and O atoms. Therefore, the central sulfur atom still has an octet. The oxygen atom has now gained 2 electrons to complete its octet. We could have just as easily moved the nonbonding electron pairs on the sulfur atom to give any of these electron dot formulas.

```
 ..      ..       ..    ..      ..      ..
:O:S::O:       :O:S:O:       :O::S:O:
 ..  ..           ..  ::  ..          ..  ..
    :O:              O:             :O:
     ..              ..              ..
```

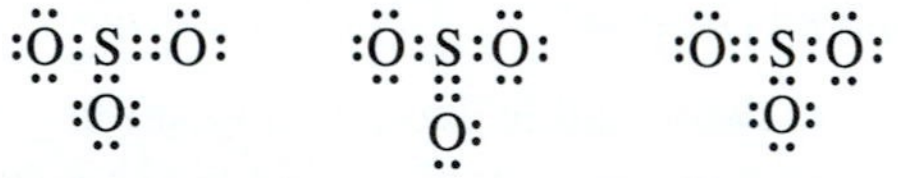

The molecule is free to rotate, so all three electron dot formulas are identical. No matter which way we represent SO_3, it has one double bond and two single bonds. In the structural formula we represent the double bond by a double dash. Therefore, the structural formula for SO can be shown by any one of the following:

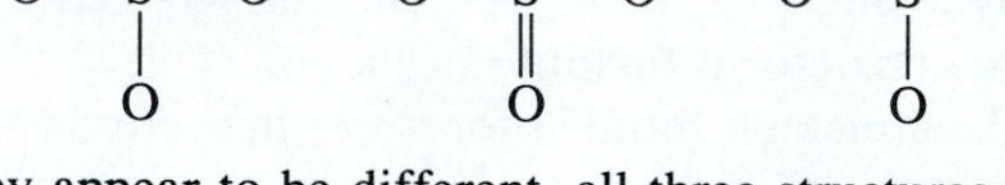

```
O—S=O     O—S—O     O=S—O
  |         ||        |
  O         O         O
```

Although they appear to be different, all three structures are identical. If we construct the molecule using molecular models, we can easily verify this statement. Each model can be rotated to look identical to the other two.

Electron Dot Formula For Hydrogen Cyanide

Let's try one more example. Let's write the electron dot formula for hydrogen cyanide, HCN. Carbon is the central atom in the molecule and is indicated in bold. The total number of valence electrons in the molecule is 1 e^- + 4 e^- + 5 e^- = 10 e^-. The number of electron pairs is 10/2 or 5 pairs of electrons. Since carbon is the central atom, we can begin by placing four pairs of electrons around the carbon and attaching the hydrogen and nitrogen atoms. This gives us

$$H:\underset{\cdot\cdot}{\overset{\cdot\cdot}{C}}:N$$

We started with five electron pairs and there is only one pair remaining. Let's place the remaining pair next to the nitrogen atom. That gives us

$$H:\underset{\cdot\cdot}{\overset{\cdot\cdot}{C}}:N:$$

The hydrogen atom has its 2 required electrons. The nitrogen atom, however, only has four electrons, 4 e^- less than an octet. According to the guidelines, we should move nonbonding electron pairs between the C and N atoms. For example,

$$H:\underset{\cdot\cdot}{C}::N:$$

The octet around the carbon is still intact, but the nitrogen has only six electrons. Let's move the carbon's other nonbonding electron pair. We now have

$$H:C:::N:$$

The three electron pairs produce a triple bond. All 6 electrons in the triple bond are shared between the C and N atoms. Therefore, both the carbon atom and nitrogen atom have obtained an octet. In the structural formula, we represent a triple bond using a triple dash. The structural formula for HCN is therefore

$$H—C\equiv N$$

EXAMPLE EXERCISE 11.5

Write the electron dot formula and structural formula for a chloroform molecule, $CHCl_3$.

Solution: Carbon is the central atom in the chloroform molecule and is indicated in bold. The total number of valence electrons is 4 e^- + 1 e^- + 3(7 e^-) = 26 e^-. The number of electron pairs is 26/2, or 13. We can begin by placing four pairs of electrons around the carbon and adding the H and three Cl atoms.

$$\begin{array}{c} H \\ Cl:\underset{\cdot\cdot}{\overset{\cdot\cdot}{C}}:Cl \\ Cl \end{array}$$

We have 13 electron pairs minus the 4 pairs we used for carbon. With the 9 remaining electron pairs, let's place 3 pairs around each chlorine.

$$\begin{array}{c} H \\ :\underset{\cdot\cdot}{\overset{\cdot\cdot}{Cl}}:\underset{\cdot\cdot}{\overset{\cdot\cdot}{C}}:\underset{\cdot\cdot}{\overset{\cdot\cdot}{Cl}}: \\ :\underset{\cdot\cdot}{Cl}: \end{array}$$

Each atom is surrounded by an octet of electrons, except hydrogen, which requires only two. This is the correct electron dot formula. The corresponding structural formula replaces each bonding electron pair with a single dash. The structural formula is

```
       H
       |
  Cl — C — Cl
       |
       Cl
```

Once again, remember that molecules are free to rotate. We could have also written the hydrogen atom on either side or below the carbon atom.

SELF-TEST EXERCISE

Write the electron dot formula and structural formula for a molecule of **SiHClBrI**.

Answers:

```
      H                      H
  ··  ··  ··                 |
 :I :Si:Cl:            I — Si — Cl
  ··  ··  ··                 |
     :Br:                    Br
      ··
```

EXAMPLE EXERCISE 11.6

Write the electron dot formula and structural formula for a carbon dioxide molecule, $\mathbf{CO_2}$.

Solution: Carbon is the central atom in the carbon dioxide molecule, as indicated in bold. The total number of valence electrons is 4 e^- + 2(6 e^-) = 16 e^-. The number of electron pairs is 16/2 or 8. We can begin by placing four pairs of electrons around the carbon atom and adding the two O atoms.

```
  ··
O:C:O
  ··
```

There are four remaining electron pairs. Since the carbon atom already has 8 electrons, let's add two pairs to each oxygen.

```
··  ··  ··
O : C : O
··  ··  ··
```

Each oxygen atom has 6 electrons, 2 less than an octet. Let's use the nonbonding electron pairs around carbon. Move one pair to the oxygen on the left and the other pair to the oxygen on the right. Each carbon-oxygen bond shares two electron pairs.

```
··      ··
O :: C :: O
··      ··
```

Now the octet rule is satisfied for each atom. This is a correct electron dot formula. There are two electron pairs between each oxygen and carbon. Thus, there are two double bonds in the carbon dioxide molecule. The structural formula represents each double bond with a double dash.

O═C═O

SELF-TEST EXERCISE

Write the electron dot formula and structural formula for a molecule of $\mathbf{SiO_2}$.

Answers:

```
 ··       ··
:O :: Si :: O:      O═Si═O
 ··       ··
```

11.5 Polar Covalent Bonds

OBJECTIVES

To state the general electronegativity trends for elements in the periodic table.

To define and give an example of a polar covalent bond.

To apply delta notation (δ^+ and δ^-) to a polar bond.

Ionic bonds result from the attraction of ions. Covalent bonds result from the sharing of valence electrons. To this point in our discussion we have assumed electrons are shared equally in a covalent bond. What about a covalent bond in which the two atoms do not share electrons equally? In many instances one of the two atoms holds the electron pair more tightly. When the electrons are drawn more closely to one of the atoms, the bond is said to be polarized. This type of bond is called a **polar covalent bond**. In other words, a polar covalent bond results when one of the two bonded atoms has a greater attraction for the shared electron pair.

polar covalent bond A bond in which one or more pairs of electrons are shared unequally.

Electronegativity Trends

electronegativity The ability of an atom to attract a shared pair of electrons in a chemical bond.

Each element in the periodic table has an inherent ability to attract valence electrons. This ability to attract valence electrons is related to the proximity of the valence shell to the nucleus. It is also related to the number of positive charges in the nucleus of an atom. The ability of an atom to attract electrons in a chemical bond is referred to as its **electronegativity**. Atoms of elements that strongly attract bonding electrons are said to be highly electronegative.

In the 1950s, Linus Pauling devised a method for measuring the relative electronegativity values of each of the elements. He arbitrarily assigned carbon a value of 2.5. He then determined the ability of other elements to attract bonding electrons relative to carbon. Pauling found fluorine to be the most electronegative

Figure 11.10 Electronegativity Values for the Elements The Pauling electronegativity values for the elements are shown below the symbol. In general, the electronegativity trends increase across a period and up a group. The most electronegative elements are the nonmetals on the far right of the periodic table. Since the noble gases form compounds that tend to be unstable, are electronegativity values not given.

PERIODIC TABLE OF THE ELEMENTS

ELECTRONEGATIVE VALUE - H 2.1

	1 IA	2 IIA	3 IIIB	4 IVB	5 VB	6 VIB	7 VIIB	8 VIII	9 VIII	10 VIII	11 IB	12 IIB	13 IIIA	14 IVA	15 VA	16 VIA	17 VIIA	18 VIIIA
1																		He ——
2	Li 1.0	Be 1.5											B 2.0	C 2.5	N 3.0	O 3.5	F 4.0	Ne ——
3	Na 0.9	Mg 1.2											Al 1.5	Si 1.8	P 2.1	S 2.5	Cl 3.0	Ar ——
4	K 0.8	Ca 1.0	Sc 1.3	Ti 1.5	V 1.6	Cr 1.6	Mn 1.8	Fe 1.8	Co 1.8	Ni 1.8	Cu 1.9	Zn 1.6	Ga 1.6	Ge 1.8	As 2.0	Se 2.4	Br 2.8	Kr ——
5	Rb 0.8	Sr 1.0	Y 1.2	Zr 1.4	Nb 1.6	Mo 1.8	Tc 1.9	Ru 2.2	Rh 2.2	Pd 2.2	Ag 1.9	Cd 1.7	In 1.7	Sn 1.8	Sb 1.9	Te 2.1	I 2.5	Xe ——
6	Cs 0.7	Ba 0.9	La 1.1	Hf 1.3	Ta 1.5	W 1.7	Re 1.9	Os 2.2	Ir 2.2	Pt 2.2	Au 2.4	Hg 1.9	Tl 1.8	Pb 1.8	Bi 1.9	Po 2.0	At 2.2	Rn ——
7	Fr	Ra	Ac	Unq	Unp	Unh	Uns	Uno	Une									

Electronegativity increases

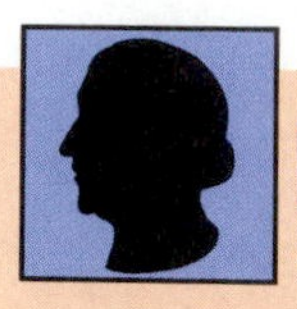

CHEMISTRY CONNECTION □ HISTORICAL

G. N. Lewis

► **Can you name the famous 20th century chemist who won a Nobel peace prize as well as the Nobel prize for chemistry?**

Gilbert Newton Lewis (1875–1946) had a strong voice in the development of the chemical bond concept. After completing his Ph.D. at Harvard in 1899, he traveled abroad for postdoctoral study. In Europe, he began his study of the structure of atoms and molecules. When Lewis returned to the United States, he joined the faculty at the Massachusetts Institute of Technology. After a visit to the University of California at Berkeley in 1912, he accepted a position there as professor of chemistry. It was at the University of California that he first presented his concept of the rule of eight and proposed the electron dot formulas that bear his name. In addition, he helped build an international reputation for the UC Berkeley chemistry department.

Although Lewis is perhaps best known for the octet rule, he had an enormous influence on the entire field of chemical bonding. He clarified the role of energy changes that occur during chemical reactions and was a coauthor of the definitive book on the subject. Despite his long career and many contributions, Lewis never received a Nobel prize. However, he was acknowledged by Linus Pauling, perhaps the most famous chemist of the twentieth century. In *The Nature of the Chemical Bond,* Pauling dedicated his classic textbook to G. N. Lewis.

Among his many credits, Linus Pauling invented the concept of electronegativity and provided an ingenious method for calculating the values of elements. One of his most brilliant accomplishments was his explanation for the unusual stability of the benzene molecule (C_6H_6). During the late 1940s and early 1950s, Pauling investigated proteins and found that their structures resembled a helix. He also established the relationship between molecular abnormality and disease. His work in this area earned him the 1954 Nobel prize in chemistry.

In the 1960s, Pauling turned his attention to nuclear test bans and pointed out the serious genetic dangers of radiation. He spoke passionately against war and encouraged disarmament. For his efforts, Pauling was awarded the 1962 Nobel peace prize. In the 1970s, he stirred controversy when he publicly advocated massive doses of vitamin C to combat the common cold. In addition to two Nobel prizes, Pauling has received numerous medals and awards, honorary memberships in several scientific societies in various countries, and published more than 400 research papers.

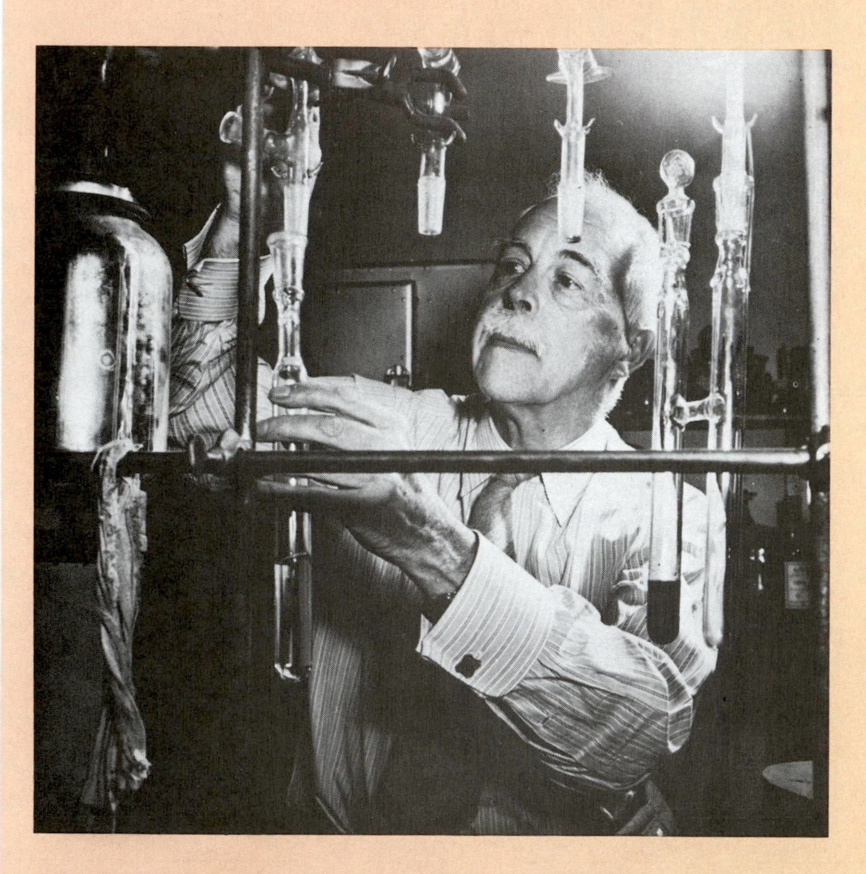

G. N. Lewis working in his laboratory.

► **Linus Pauling (1901–) won the Nobel peace prize in 1962 for his efforts to ban above-ground testing of nuclear weapons.**

element. It has a value of 4.0 compared to carbon. Other highly electronegative elements are oxygen, 3.5; nitrogen, 3.0; and chlorine, 3.0. The most electronegative elements are the nonmetals on the far right of the periodic table. Figure 11.10 shows selected elements in the periodic table and their Pauling electronegativity value.

Notice the electronegativity trend in Figure 11.10. First, the elements generally become more electronegative from left to right in the periodic table. This trend is consistent with the chemical properties of the elements. That is, nonmetals react by gaining electrons and metals react by losing electrons. Second, the elements become more electronegative moving up within a group. This follows the trend in nonmetallic character. As an example, examine Group VA/15: Bi is a metal, Sb and As are semimetals, P and N are nonmetals. Therefore, the general trend in electronegativity increases as the trend in nonmetallic character increases. As you see in Figure 11.10, the nonmetals are more electronegative than the semimetals, which are more electronegative than the metals.

Electronegativity Trends

The general trend in electronegativity in the periodic table *increases* across a period from left to right and *increases* within a group of elements from bottom to top.

The following example exercise will further illustrate the prediction of electronegativity trends from the periodic table.

EXAMPLE EXERCISE 11.7

Predict which of the following elements is more electronegative according to the general electronegativity trends in the periodic table.

(a) F or Cl
(b) O or N
(c) Si or C
(d) Se or Br

Solution: According to the electronegativity trends, elements that lie to the right in a series or at the top of a group are more electronegative. Thus,

(a) F is more electronegative than Cl.
(b) O is more electronegative than N.
(c) C is more electronegative than Si.
(d) Br is more electronegative than Se.

H—F

H—Cl

H—Br

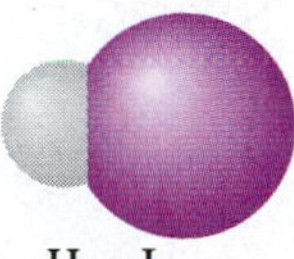
H—I

Molecular models of the hydrogen halides, which all have polar covalent bonds.

SELF-TEST EXERCISE

Predict which of the following elements is more electronegative according to the general trends in the periodic table.

(a) Br or I
(b) P or S

Answers: **(a)** Br; **(b)** S

Delta Notation For Polar Bonds

In Section 11.4, we introduced the concept of the covalent bond, using H—Cl as an example. In Figure 11.10, we saw that the electronegativity value of H is 2.1 and Cl is 3.0. Since there exists a difference in electronegativity between the two elements, $3.0 - 2.1 = 0.9$, the bond in an HCl molecule is polar covalent. Moreover, since Cl is the more electronegative element, the bonding electron pair is drawn away from the H atom and toward the Cl atom. The Cl atom thus becomes slightly negatively charged, whereas the H atom becomes slightly positively charged.

delta (δ) notation A method of indicating the partial positive charge (δ^+) and partial negative charge (δ^-) in a polar covalent bond.

We can identify a polar bond using a special label. We indicate the atom having a partial negative charge with the label δ^-. Similarly, we indicate the atom having a partially positive charge with the label δ^+. These labels, δ^- and δ^+, use the Greek letter delta (δ) and are referred to as **delta notation**. We can use delta notation to illustrate the polar bond in an HCl molecule as follows:

$$\delta^+ \text{ H—Cl } \delta^-$$

The chlorine atom is more electronegative so it draws the negative electron pair closer. As a result, the hydrogen atom becomes slightly positive. Example Exercise 11.8 further illustrates the application of delta notation.

EXAMPLE EXERCISE 11.8

Calculate the electronegativity difference and apply delta notation to the bond between carbon and oxygen, C—O.

Solution: From Figure 11.10, we find that the electronegativity value of C is 2.5 and of O is 3.5. By convention, we always subtract the lesser value from the greater. The difference between the two elements is $3.5 - 2.5 = 1.0$. This difference indicates that the C—O bond is polarized. Since O is the more electronegative element, the bonding electron pair is drawn away from the C atom and toward the O atom. Thus, the O atom is slightly negatively charged, whereas the C atom is slightly positively charged. Applying the delta convention, we have

$$\delta^+ \text{ C—O } \delta^-$$

SELF-TEST EXERCISE

Using delta notation (δ^+ and δ^-), label each atom in the following polar covalent bonds.

(a) N—O **(b)** H—F

Answer: **(a)** δ^+ N—O δ^-; **(b)** δ^+ H—F δ^-

Keep in mind that the properties of polar molecules are very different from the properties of ionic formula units. The delta convention indicates the partially negative and positive atoms in a polar covalent bond. A polar bond is, however, very different than an ionic bond.

11.6 Nonpolar Covalent Bonds

To define and give an example of a nonpolar covalent bond.

To identify the seven nonmetallic elements (H, N, O, F, Cl, Br, and I) that occur naturally as diatomic molecules.

In Section 11.5, we learned that polar covalent bonds result from the unequal sharing of bonding electrons. In that case, one of the two bonded atoms generally has a greater tendency to attract the bonded electron pair. The atom having the greater electronegativity attracts the bonded electron pair and becomes partially negatively charged. Now we learn what happens to covalent bonds between two atoms having the same electronegativity.

Let's begin by reexamining the Pauling electronegativity values shown in Figure 11.10. In general, the trend in electronegativity increases across a period and up a group. Notice, however, that several elements have the same Pauling electronegativity values. For example, N and Cl both have a value of 3.0; C, S, and I all have a value of 2.5.

How then would we describe a N—Cl bond? Since the electronegativity of each atom is the same, the bond is not polarized. Covalent bonds between two atoms having the same electronegativity are referred to as **nonpolar covalent bonds**. In terms of properties, molecular compounds having nonpolar bonds behave differently from compounds with polar bonds. Example Exercise 11.9 further illustrates polar and nonpolar covalent bonds.

nonpolar covalent bond A bond in which one or more pairs of electrons are shared equally.

EXAMPLE EXERCISE 11.9

Classify each of the following covalent bonds in molecules as polar or nonpolar based on the electronegativity values found in Figure 11.10.

(a) C—S
(b) S—Cl
(c) Cl—P
(d) P—H

Solution: From Figure 11.10, we find the following electronegativity values: C = 2.5, S = 2.5, Cl = 3.0, P = 2.1, and H = 2.1.

(a) The C—S bond is nonpolar (2.5 − 2.5 = 0).
(b) The S—Cl bond is polar (3.0 − 2.5 = 0.5).
(c) The Cl—P bond is polar (3.0 − 2.1 = 0.9).
(d) The P—H bond is nonpolar (2.1 − 2.1 = 0).

SELF-TEST EXERCISE

Refer to Figure 11.10 and indicate which of the following are nonpolar covalent bonds.

(a) F—F
(b) Cl—I

Answers:

(a) The F—F bond is nonpolar (4.0 − 4.0 = 0).
(b) The Cl—I bond is polar (3.0 − 2.5 = 0.5).

Diatomic Molecules

A **diatomic molecule** consists of two nonmetal atoms joined by a covalent bond. The best example of a nonpolar covalent bond is found between two identical atoms. For example, the element oxygen occurs naturally in the atmosphere as a diatomic gas. In fact, the air we breathe is about 21% O_2 molecules. Even though oxygen is the second most electronegative element, the bond in an O_2 molecule is nonpolar covalent. This is because each oxygen atom draws the electron pair with the same attractive force.

diatomic molecule A particle composed of two nonmetal atoms that are covalently bonded.

The elements nitrogen and hydrogen also occur naturally in the atmosphere as diatomic gases. Air is approximately 78% N_2 molecules. There is, however, very little hydrogen in the atmosphere. Hydrogen, in the elemental state, is found as H_2 molecules.

In Section 5.3, we briefly studied the elements in Group VIIA/17. Recall that,

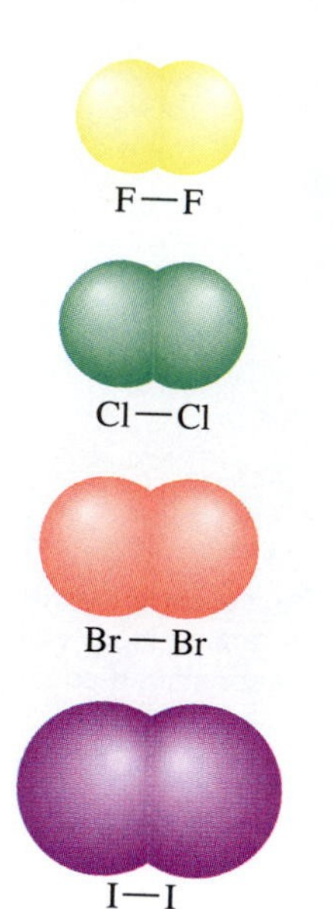

Molecular models of the halogens, which all have nonpolar covalent bonds.

collectively, these elements are called the halogens. The halogens (F, Cl, Br, and I) are a family and have similar properties. All four elements occur as diatomic molecules: F_2, Cl_2, Br_2, and I_2. Fluorine is a yellowish-green gas, chlorine is a greenish-yellow gas, bromine is a reddish-brown liquid, and iodine is a dark violet crystal solid.

There are seven elements that exist as diatomic molecules: H_2, N_2, O_2, F_2, Cl_2, Br_2, and I_2. Although each molecule is distinctly different, each exhibits a nonpolar bond that is perfectly covalent.

Note For our discussion, we classified a bond as nonpolar when there is no electronegativity difference between the two bonded atoms. In reality, slight differences in electronegativity between two bonded atoms do not markedly alter the properties. In fact, a more sophisticated understanding of bonding views electronegativity differences on a continuous spectrum. That is, the electrons in a covalent bond may be shared equally, slightly polarized, polarized, or highly polarized. Even the distinction between an ionic bond and a highly polar covalent bond becomes blurred. For example, the electronegativity difference in an ionic bond can be similar to the electronegativity difference in a highly polarized covalent bond.

11.7 Coordinate Covalent Bonds

OBJECTIVES

To define and identify a coordinate covalent bond.

As we said, a covalent bond is composed of an electron pair that is shared between two nonmetal atoms. Additionally, each nonmetal is usually surrounded by nonbonding electron pairs to complete its octet. An atom such as oxygen with six valence electrons can join with a nonbonding electron pair on another atom to attain its octet. The resulting bond involves a special type of electron sharing. A covalent bond resulting from the donation of an electron pair is referred to as a **coordinate covalent bond**.

coordinate covalent bond A bond in which an electron pair is shared but both electrons have been donated by a single atom.

A good example of a coordinate covalent bond is found in a molecule of ozone, O_3. Here, an oxygen atom coordinates a nonbonding electron pair on an oxygen molecule to produce ozone.

$$:\ddot{O}::\ddot{O}: + \ddot{\underset{..}{O}}: \longrightarrow :\ddot{O}::\ddot{O}:\ddot{\underset{..}{O}}:$$

↑ *coordinate covalent bond*

Although a coordinate covalent bond is similar to any other covalent bond, it is valuable to be able to identify it. To identify a coordinate covalent bond, we first draw the electron dot formula for a molecule. The following example further illustrates the concept of a coordinate covalent bond.

EXAMPLE EXERCISE 11.10

Burning yellow sulfur powder produces sulfur dioxide, SO_2. Sulfur dioxide is a colorless gas with a suffocating odor. It is used to kill insect larva in dried fruit and is the odor released when a sealed package of dried fruit iş opened. Draw the electron dot formula for SO_2. Then show the formation of a coordinate covalent bond in SO_3.

Solution: The total number of valence electrons in one molecule of SO_2 is $6\ e^- + 2(6\ e^-) = 18\ e^-$. After experimenting, we find that the electron dot formula for SO_2 requires a double bond. The sulfur has a nonbonding pair of electrons that can coordinate with the additional oxygen atom. A diagram of the formation of the coordinate covalent bond is as follows:

:Ö::S: + Ö: ⟶ :Ö::S:Ö:
   :Ö   :Ö

nonbonding electron pair *coordinate covalent bond*

SELF-TEST EXERCISE

A nitrogen molecule can form a coordinate covalent bond with an oxygen atom to give nitrous oxide, N_2O. First, draw the electron dot formula for a nitrogen molecule; than attach an oxygen atom.

Answers: :N:::N: :N:::N:Ö:

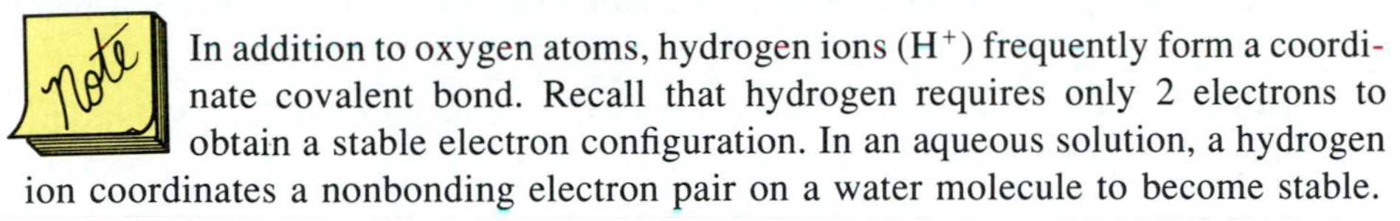

In addition to oxygen atoms, hydrogen ions (H^+) frequently form a coordinate covalent bond. Recall that hydrogen requires only 2 electrons to obtain a stable electron configuration. In an aqueous solution, a hydrogen ion coordinates a nonbonding electron pair on a water molecule to become stable.

H:Ö: + H^+ ⟶ [H:Ö:H]$^+$
   H   H

Another interesting example of the formation of a coordinate covalent bond is found in the conversion of ammonia gas, NH_3, to ammonium compounds. The nitrogen in NH_3 has a nonbonding electron pair capable of coordinating a hydrogen ion. The resulting ammonium ion, NH_4^+, is found in compounds that are important industrially, such as explosives and agricultural products.

11.8 Polyatomic Ions

OBJECTIVES

To calculate the total number of valence electrons for a polyatomic ion and write the electron dot formula, given the arrangement of atoms.

To write the structural formula for a polyatomic ion, given the electron dot formula.

So far in our discussion of ions, we have only considered single atoms that have gained or lost electrons. These are referred to as **monoatomic ions**. But there are other ions that contain two or more atoms. A collection of two or more atoms that

monoatomic ion A single atom that has lost one or more valence electrons to become positively charged or gained electrons to become negatively charged.

polyatomic ion A unit containing two or more atoms covalently bonded together and bearing a positive or negative charge as the result of losing or gaining valence electrons.

bears a positive or negative charge is called a **polyatomic ion**. Some common polyatomic ions are the ammonium ion, NH_4^+, found in fertilizer; the hydroxide ion, OH^-, found in caustic lye; and the hydrogen carbonate ion, HCO_3^-, found in baking soda.

In Section 11.4 we learned to write the electron dot formula for molecules. Now let's write the electron dot formula for polyatomic ions. These are the guidelines we will follow.

1. Calculate the total number of valence electrons in the polyatomic ion. Begin by adding up the valence electrons for all the atoms in the polyatomic ion. If the ion is negatively charged, add the number of electrons equal to the negative charge. If the ion is positively charged, subtract the number of electrons equal to the positive charge. In each case, the resulting total number of valence electrons should be an even number.
2. Divide the total number of valence electrons by 2 to find the number of electron pairs in the polyatomic ion.
3. Surround the central atom with four electron pairs. Use the remaining electron pairs to complete an octet around each of the other atoms. Hydrogen is the only exception; H requires only 2 e^- (one electron pair).
4. If there are not enough electron pairs to provide an octet for each atom, move a nonbonding electron pair between two atoms that already share a bonding pair.

Now let's write the electron dot formula for the ammonium ion, NH_4^+. The total number of valence electrons in the polyatomic ion is $5\ e^- + 4(1\ e^-) - 1\ e^- = 8\ e^-$. The number of electron pairs is 8/2 or 4 pairs of electrons. In the formula NH_4^+, nitrogen is the central atom. We can place the four pairs of electrons around the nitrogen atom to provide the necessary octet. After adding the four H atoms, we have

$$\left[\begin{array}{c} \text{H} \\ \text{H}:\ddot{\underset{..}{\text{N}}}:\text{H} \\ \text{H} \end{array}\right]^+$$

The electron dot formula is written correctly for the polyatomic ammonium ion. The N has an octet and each hydrogen shares two electrons.

We can simplify the electron dot formula to give the structural formula of the polyatomic ion. In that case, we represent each bonding pair of electrons by a single dash. We can write the structural formula for the ammonium ion as follows:

$$\left[\begin{array}{c} \text{H} \\ | \\ \text{H}-\text{N}-\text{H} \\ | \\ \text{H} \end{array}\right]^+$$

The usual way to write electron dot and structural formulas of polyatomic ions is to enclose them in brackets. The overall charge on the polyatomic ion is indicated outside the brackets. By placing the charge outside the brackets, we are emphasizing that the positive charge from the missing electron is distributed over the entire ion.

Let's write the electron dot formula for the negative chlorate ion, ClO_3^-. The total number of valence electrons in the ion is $7\ e- + 3(6\ e^-) + 1\ e^- = 26\ e^-$. The number of electron pairs is 26/2 or 13 pairs. Since chlorine is the central atom, we can begin by placing four pairs of electrons around the Cl and add the three O atoms. This gives us

```
  ..
O:Cl:O
  ..
  O
```

We started with 13 electron pairs, so let's place the remaining 9 pairs around the oxygen atoms to complete each octet. We then have

```
 ..  ..  ..  -
[:O:Cl:O:]
 ..  ..  ..
    :O:
     ..
```

In the structural formula, we represent each bonding electron pair with a single dash. The nonbonding electrons are usually omitted. The structural formula for ClO_3^- is as follows:

```
[O—Cl—O]⁻
    |
    O
```

Let's try a more difficult example. Let's write the electron dot formula for the carbonate ion, CO_3^{2-}. The total number of valence electrons in the polyatomic ion is $4\ e^- + 3(6\ e^-) + 2\ e^- = 24\ e^-$. The total number of electron pairs is 24/2 or 12 pairs.

Carbon is the central atom. We can begin by placing four pairs of electrons around the carbon atom and adding the three oxygen atoms. We then have

```
  ..
O:C:O
  ..
  O
```

We started with 12 electron pairs, so we have 8 pairs remaining. Let's place the remaining pairs around the oxygen atoms to give octets as follows:

```
 ..  ..
:O:C:O:
 .. .. ..
   :O:
    ..
```

Notice that one of the oxygen atoms does not have an octet; it has only three electron pairs. According to the guidelines above, we can move the nonbonding electron pair on the carbon. Let's place the nonbonding pair between the C and the O atoms; for example,

```
 ..     ..  2-
[:O:C::O:]
 .. ..
   :O:
    ..
```

The two electron pairs constitute a double bond. All four electrons in the double bond are shared between the C and O atoms. Therefore, the central carbon atom still has an octet. The oxygen atom has now gained 2 electrons to complete its octet.

We use brackets to convey the idea that bonding electrons are free to move about the polyatomic ion. In this example, there appear to be three different structural formulas for the carbonate ion. That is,

```
[C—C=O]²⁻     [O—C—O]²⁻     [O=C—O]²⁻
   |             ‖             |
   O             O             O
```

Although they appear to be different, all three structures for the carbonate ion are equivalent. No matter which way we represent CO_3^{2-}, there is one double bond and two single bonds. Example Exercise 11.11 further illustrates writing electron dot formulas of polyatomic ions.

EXAMPLE EXERCISE 11.11

Write the electron dot formula and structural formula for the sulfate ion, SO_4^{2-}.

Solution: The total number of valence electrons is the sum of each atom plus two for the negative charge: $6\ e^- + 4(6\ e^-) + 2\ e^- = 32\ e^-$. The number of electron pairs is 32/2 or 16. We can begin by placing 4 pairs of electrons around the central sulfur atom and attaching the four oxygen atoms as follows:

```
   O
   ..
O :S: O
   ..
   O
```

We have 12 remaining electron pairs, so let's place 3 pairs around each oxygen atom:

```
[     ..      ]2-
[    :O:      ]
[  ..  ..  .. ]
[ :O: S :O:   ]
[  ..  ..  .. ]
[    :O:      ]
[     ..      ]
```

Notice that each atom is surrounded by an octet of electrons. This is the correct electron dot formula. The structural formula replaces each bonding electron pair with a single dash. The structural formula is

```
[      O      ]2-
[      |      ]
[ O —  S  — O ]
[      |      ]
[      O      ]
```

SELF-TEST EXERCISE

Write the electron dot formula and structural formula for the sulfite ion, SO_3^{2-}.

Answers:

```
[  ..  ..  .. ]2-     [ O — S — O ]2-
[ :O: S :O:   ]       [     |     ]
[  ..  ..  .. ]       [     O     ]
[    :O:      ]
[     ..      ]
```

EXAMPLE EXERCISE 11.12

Write the electron dot formula and structural formula for the nitrate ion, NO_3^-.

Solution: Nitrogen is the central atom, and the total number of valence electrons is $5\ e^- + 3(6\ e^-) + 1\ e^- = 24\ e^-$. The number of electron pairs is $24/2 = 12$. We begin by placing four pairs of electrons around the nitrogen and attaching three oxygen atoms to give

```
   ..
O :N: O
   ..
   O
```

There are eight remaining electron pairs $(12 - 4 = 8)$. Let's add the remaining electron pairs to each oxygen:

```
 ..  ..  ..
:O: N :O:
 ..  ..  ..
   :O:
    ..
```

One oxygen atom shares only 6 electrons, 2 less than an octet. Let's use the nonbonding electron pairs around nitrogen. Move the nonbonding pair toward the oxygen on the right as follows:

$$\left[\begin{array}{c} :\ddot{O}:\ddot{N}::\ddot{O}: \\ :\ddot{\underset{..}{O}}: \end{array}\right]^-$$

Now the octet rule is satisfied for each atom. The N atom shares two electron pairs with one of the O atoms. This means that N and O share a double bond. The structural formula represents the double bond using a double dash:

$$\left[\begin{array}{c} O—N{=}O \\ | \\ O \end{array}\right]^-$$

Remember, the brackets indicate that the negative charge is dispersed over the entire polyatomic ion.

SELF-TEST EXERCISE

Write the electron dot formula and structural formula for the nitrite ion, NO_2^-.

Answers: $[:\ddot{O}::\ddot{N}:\underset{..}{\ddot{O}}:]^-$ $[O{=}N—O]^-$

11.9 Properties of Ionic and Molecular Compounds

OBJECTIVES

To compare the properties of ionic compounds to those of molecular compounds.

To relate the properties of compounds to ionic or covalent bonds.

As we have learned, an ionic compound is composed of positive and negative ions that form ionic bonds. The strong attraction between oppositely charged ions throughout an ionic compound produces a unique set of properties. For example, ionic compounds are typically crystalline solids with a definite geometric shape. Owing to the strong attraction between ions, these compounds usually have high melting and boiling points. In addition, aqueous solutions of ionic compounds are usually good conductors of electricity and undergo reaction instantaneously.

Molecular compounds are composed of individual molecules that are held together by covalent bonds. The attractive forces between molecules are much less than the electrostatic forces between ions. Thus, the typical properties of molecular

Table 11.1 General Properties of Compounds

Property*	Ionic Compounds	Molecular Compounds
physical state	solid	solid, liquid, gas
structure	crystalline	often noncrystalline
melting point	usually above 500°C	usually below 300°C
boiling point	very high	low
solution conductivity	conductor	nonconductor
rate of reaction	very fast	usually slow

*There are many exceptions to these general guidelines.

compounds are different from those of ionic compounds. Molecular compounds are found as gases and liquids, as well as solids. They usually have low melting points (below 300°C). Aqueous solutions of molecular compounds are poor conductors of electricity and usually undergo reaction very slowly. Table 11.1 contrasts the properties of ionic and molecular compounds.

The following example exercise provides practice in comparing the properties of ionic and molecular compounds.

EXAMPLE EXERCISE 11.13

Predict whether the following properties most likely indicate an ionic or molecular compound.

(a) solid, deep-blue crystal
(b) colorless liquid
(c) melts at 2775°C
(d) boils at −61°C
(e) nonconducting aqueous solution
(f) reacts instantly with $AgNO_3$

Solution: Refer to the properties listed in Table 11.1.

(a) A blue crystal is probably an *ionic* compound because it is a solid and has a crystalline structure.
(b) A colorless liquid is most likely a *molecular* compound because of its physical state.
(c) The extremely high melting point suggests an *ionic* compound.
(d) The low boiling point indicates a *molecular* compound.
(e) An aqueous solution that does not conduct an electric current is probably a solution of a *molecular* compound.
(f) A substance that reacts rapidly is most often an *ionic* compound.

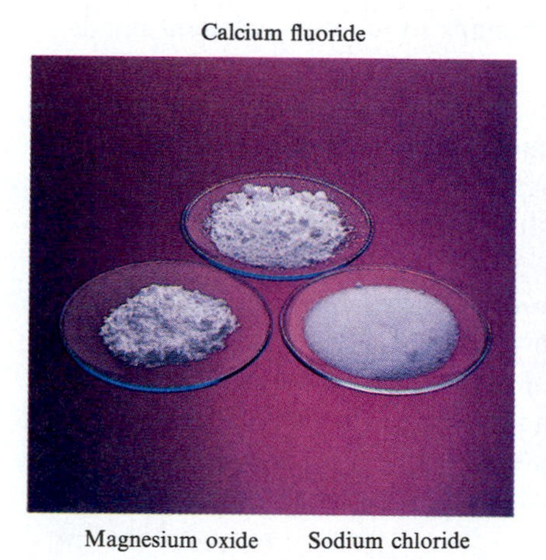

Examples of (a) ionic compounds—CaF_2, MgO, NaCl—held together by ionic bonds and (b) molecular substances—Br_2, $C_{12}H_{22}O_{11}$, S_8—held together by covalent bonds.

SELF-TEST EXERCISE

Predict whether the following properties most likely indicate a compound having ionic or covalent bonds.

(a) liquid physical state
(b) highly crystalline structure
(c) low melting point
(d) very high boiling point
(e) conducts an electric current in aqueous solution
(f) reacts slowly

Answers:
(**a**) covalent bonds; (**b**) ionic bonds;
(**c**) covalent bonds; (**d**) ionic bonds;
(**e**) ionic bonds; (**f**) covalent bonds

Summary

Section 11.1 In 1916, G. N. Lewis proposed the octet rule to describe the **chemical bond**. Based on the stability of the noble gases, Lewis theorized that the **valence electrons** in different atoms interact in such a way that each atom completes its valence shell. Since eight electrons are found in the valence shell of all the noble gases, except helium, the Lewis theory became known as the rule of eight, or the **octet rule**. The only exception to the octet rule is hydrogen, which requires two electrons.

Chemical bonds "glue" atoms together in compounds. The glue is composed of electrons found in the valence shells of the bonded atoms. If valence electrons are transferred from a metal to a nonmetal, an **ionic bond** results. If valence electrons are shared between two nonmetals, a **covalent bond** results. The smallest representative particle in a compound held together by ionic bonds is a **formula unit**. The smallest representative particle in a compound held together by covalent bonds is a **molecule**.

Section 11.2 In the formation of ionic bonds, metal atoms lose valence electrons and are often left with a noble gas structure. Simultaneously, nonmetal atoms gain electrons to complete their valence shells. In the process of forming ions, the radius of metal atoms becomes smaller, while the radius of nonmetal atoms becomes larger. The formation of ionic bonds from metal and nonmetal atoms always releases heat energy.

Section 11.3 In the formation of covalent bonds, the outer valence shells of two nonmetal atoms overlap and share electrons. Each atom in the bond shares the number of electrons necessary to complete their individual octets. The **bond length** between the two atoms is always less than the sum of their atomic radii. When atoms of different elements react to produce a compound, energy is always released from the reaction. The amount of energy released during the formation of a bond is exactly equal to the amount of energy required to break that bond. The amount of energy required to separate two bonded atoms is called the **bond energy**.

Section 11.4 There is more than one method for writing **electron dot formulas**. In the method we used, we first calculate the total number of electrons available from the valence shells of all the atoms bonded in the molecule. We then divide the valence electron total by 2 to find the number of electron pairs. The electron pairs are first placed around the central atom and then the remaining atoms in the molecule so as to provide an octet for each atom. The position of the electron dots is only a device for keeping track of octets. Actually, electrons are free to move about the entire molecule. Molecules may contain one, two, or three bonding electron pairs between two atoms. These bonds are described as **single bonds, double bonds**, and **triple bonds**, respectively. Multiple bonds result from an insufficient number of valence electrons in the molecule to provide single bonds with complete octets.

Sections 11.5–11.7 Covalent bonds can be further classified as **polar covalent** or **nonpolar covalent**. A nonpolar covalent bond shares the bonded electron pair equally. In a polar covalent bond, one of the atoms has a greater attraction for the bonded electron pair. The atom that more strongly attracts the bonding electrons is said to have a higher **electronegativity**. The more electronegative atom becomes partially negatively charged. Using **delta notation**, this atom is labeled δ^-. The other atom in the polar bond is partially positive and is labeled δ^+. Furthermore, one of the atoms in a covalent bond may donate the bonding electron pair. If both electrons are donated by one of the atoms, the bond is a **coordinate covalent bond**.

Section 11.8 Ionic compounds may be composed of **polyatomic ions** as well as **monoatomic ions**. Ammonium chloride, NH_4Cl, and copper sulfate, $CuSO_4$, are ionic compounds containing a polyatomic ion. In summing the total number of valence electrons for a polyatomic cation, we subtract the number of electrons equal to its ionic charge. For the ammonium cation, NH_4^+, the total number of valence electrons is 8 e^- (9 e^- minus 1 e^-). In summing the valence electrons for a polyatomic anion, we add the number of electrons equal to its ionic charge. For the sulfate ion, SO_4^{2-}, the total number of valence electrons is 32 e^- (30 e^- plus 2 e^-).

Section 11.9 The properties of ionic compounds are observed to be different from the properties of molecular compounds. Ionic compounds are crystalline solids having high melting and boiling points. Their aqueous solutions strongly or weakly conduct an electric current. In addition, ionic compounds usually undergo reaction at a rapid rate. The properties of molecular compounds, in general, are just the opposite.

Key Terms

Select the key term that corresponds to the following definitions.

____ **1.** the attraction between two atoms in a molecule or two ions in a formula unit
____ **2.** the electrons in the outermost energy shell of an atom that are available for bonding
____ **3.** the statement that each atom in a compound must attain 8 valence electrons in order to be energetically stable
____ **4.** a chemical bond characterized by the attraction of a cation and an anion
____ **5.** a chemical bond characterized by the sharing of one or more pairs of valence electrons
____ **6.** the smallest representative entity in a compound held together by ionic bonds
____ **7.** the smallest representative entity in a compound held together by covalent bonds
____ **8.** a particle composed of two covalently bonded nonmetal atoms
____ **9.** the distance between the nuclei of two covalently bonded atoms
____ **10.** the amount of energy required to break a given bond in 1 mole of substance in the gaseous state
____ **11.** the valence electrons in a molecule that are shared
____ **12.** the valence electrons in a molecule that are not shared

(a) bond angle *(Sec. 11.4)*
(b) bond energy *(Sec. 11.3)*
(c) bond length *(Sec. 11.3)*
(d) bonding electrons *(Sec. 11.4)*
(e) chemical bond *(Sec. 11.1)*
(f) coordinate covalent bond *(Sec. 11.7)*
(g) covalent bond *(Sec. 11.1)*
(h) delta (δ) notation *(Sec. 11.5)*
(i) diatomic molecule *(Sec. 11.6)*
(j) double bond *(Sec. 11.4)*
(k) electron dot formula *(Sec. 11.4)*
(l) electronegativity *(Sec. 11.5)*
(m) formula unit *(Sec. 11.1)*

____ **13.** a diagram of a molecule or polyatomic ion in which each atom is represented by its chemical symbol surrounded by a dot for each bonding or nonbonding electron; also referred to as a Lewis structure

____ **14.** a diagram of a molecule or polyatomic ion in which each atom is represented by its chemical symbol connected by a dash for each pair of bonding electrons

____ **15.** the angle formed by two atoms attached to the central atom in a molecule

____ **16.** a bond composed of an electron pair shown as a single dash between two atoms

____ **17.** a bond composed of two electron pairs shown as two dashes between two atoms

____ **18.** a bond composed of three electron pairs shown as three dashes between two atoms

____ **19.** the ability of an atom to attract a shared pair of electrons

____ **20.** a bond in which one or more pairs of electrons are shared equally

____ **21.** a bond in which one or more pairs of electrons are shared unequally

____ **22.** a method of indicating the partial positive charge (δ^+) and partial negative charge (δ^-) in a polar covalent bond

____ **23.** a bond in which an electron pair is shared but both electrons have been donated by a single atom

____ **24.** a single atom that bears a charge as the result of gaining or losing electrons

____ **25.** a collection of two or more atoms covalently bonded that bears a charge as the result of gaining or losing electrons

(n) ionic bond *(Sec. 11.1)*
(o) molecule *(Sec. 11.1)*
(p) monoatomic ion *(Sec. 11.8)*
(q) nonbonding electrons *(Sec. 11.4)*
(r) nonpolar covalent bond *(Sec. 11.6)*
(s) octet rule *(Sec. 11.1)*
(t) polar covalent bond *(Sec. 11.5)*
(u) polyatomic ion *(Sec. 11.8)*
(v) single dashes *(Sec. 11.4)*
(w) structural formula *(Sec. 11.4)*
(x) triple bond *(Sec. 11.4)*
(y) valence electrons *(Sec. 11.1)*

Exercises

Valence Electrons and Chemical Bonds (Sec. 11.1)

1. Describe the formation of an ionic bond in terms of valence electrons.

2. Describe the formation of a covalent bond in terms of valence electrons.

3. State the number of electrons in the valence shells of a magnesium atom and a sulfur atom before and after reacting to produce MgS.

4. State the number of electrons in the valence shells of a hydrogen atom and an iodine atom before and after reacting to produce HI.

5. Predict whether each of the following compounds is held together by ionic or covalent bonds.
(a) water, H_2O
(b) sodium chloride, NaCl
(c) methane, CH_4
(d) cadmium oxide, CdO

6. Predict whether each of the following compounds is held together by ionic or covalent bonds.
(a) dinitrogen tetraoxide, N_2O_4
(b) lithium chlorate, $LiClO_3$
(c) iron(II) sulfate, $FeSO_4$
(d) iodine heptafluoride, IF_7

7. State whether the representative entity for each of the following substances is a molecule or formula unit.
(a) ethyl alcohol, C_2H_5OH
(b) gold(III) sulfide, Au_2S_3
(c) phosphine, PH_3
(d) dinitrogen tetraoxide, N_2O_4
(e) oxygen difluoride, OF_2
(f) potassium carbonate, K_2CO_3

8. State whether the representative entity for each of the following substances is a molecule or formula unit.
(a) butane, C_4H_{10}
(b) strontium iodide, SrI_2
(c) potassium bromide, KBr
(d) lead(IV) oxide, PbO_2
(e) acetone, C_3H_6O
(f) uranium(IV) fluoride, UF_4

9. State whether the representative entity for each of the following substances is an atom, molecule, or formula unit.
(a) neon, Ne
(b) chlorine, Cl_2
(c) cocaine, $C_{17}H_{21}NO_4$
(d) iron, Fe

(e) phosphorus, P_4
(f) plutonium bromide, $PuBr_3$

10. State whether the representative entity for each of the following substances is an atom, molecule, or formula unit.
(a) ethane, C_2H_6 (b) copper, Cu
(c) magnetite, Fe_3O_4 (d) soda ash, Na_2CO_3
(e) sulfur, S_8 (f) bromine, Br_2

The Ionic Bond (Sec. 11.2)

11. Use the periodic table to predict a charge for each of the following metal ions.
(a) Na ion (b) Cs ion
(c) Sn ion (d) Pb ion

12. Use the periodic table to predict a charge for each of the following metal ions.
(a) Be ion (b) Sr ion
(c) Ga ion (d) Cs ion

13. Use the periodic table to predict the charge on each of the following nonmetal ions.
(a) F ion (b) Br ion
(c) S ion (d) N ion

14. Use the periodic table to predict the charge on each of the following nonmetal ions.
(a) I ion (b) S ion
(c) Se ion (d) P ion

15. Write out the electron configuration for each of the following metal ions.
(a) Li^+ (b) K^+
(c) Ca^{2+} (d) Ra^{2+}

16. Write out the electron configuration for each of the following metal ions.
(a) Sc^{3+} (b) Y^{3+}
(c) Ti^{4+} (d) Zr^{4+}

17. Write out the electron configuration for each of the following nonmetal ions.
(a) Cl^- (b) I^-
(c) S^{2-} (d) P^{3-}

18. Write out the electron configuration for each of the following nonmetal ions.
(a) Br^- (b) O^{2-}
(c) Se^{2-} (d) N^{3-}

19. Which noble gas has an electron configuration identical to each of the following metal ions?
(a) Li^+ (b) K^+
(c) Ca^{2+} (d) Ra^{2+}

20. Which noble gas has an electron configuration identical to each of the following nonmetal ions?
(a) Cl^- (b) I^-
(c) O^{2-} (d) P^{3-}

21. Which noble gas is isoelectronic with each of the following metal ions?
(a) Sc^{3+} (b) Y^{3+}
(c) Ti^{4+} (d) Zr^{4+}

22. Which noble gas is isoelectronic with each of the following nonmetal ions?
(a) F^- (b) Br^-
(c) Se^{2-} (d) N^{3-}

23. Which of the following has a larger radius?
(a) Li atom or Li ion (b) Mg atom or Mg ion
(c) F atom or F ion (d) O atom or O ion

24. Which of the following has a larger radius?
(a) Al atom or Al ion (b) Pb atom or Pb ion
(c) Se atom or Se ion (d) N atom or N ion

25. Which of the following statements is correct regarding the formation of an ionic bond between a metal atom and a nonmetal atom?
(a) An ionic bond can be formed by the transfer of valence electrons from the nonmetal atom to the metal atom.
(b) The ionic radius of a metal atom is greater than its atomic radius.
(c) The ionic radius of a nonmetal atom is less than its atomic radius.
(d) The properties of a compound are usually the average of the properties of the constituent elements.
(e) Energy is released when an ionic bond is dissociated into ions.

26. Which of the following statements is *not* true regarding the formation of an ionic bond between cobalt and bromine?
(a) Valence electrons are transferred from a cobalt atom to a bromine atom.
(b) The cobalt atom is larger in radius than the cobalt ion.
(c) The bromine atom is smaller in radius than the bromide ion.
(d) Cobalt and cobalt bromide have similar properties.
(e) Energy is released when an ionic bond is formed between a cobalt atom and a bromine atom.

The Covalent Bond (Sec. 11.3)

27. Which of the following is greater?
(a) the sum of the H and I atomic radii or the bond length in H—I
(b) the sum of the N and O atomic radii or the bond length in N—O

28. Which of the following is greater?
(a) the sum of the C and Cl atomic radii or the bond length in C—Cl
(b) the sum of the S and F atomic radii or the bond length in S—F

29. Which of the following statements is correct regarding the formation of a covalent bond between two nonmetal atoms?
(a) A covalent bond is formed by transferring valence electrons to the more electronegative atom.
(b) The bonding electrons are located an equal distance between the two atoms in the covalent bond.
(c) The bond length in a covalent bond is equal to the sum of the two atomic radii.

(d) The properties of the compound are about the average of the properties of the constituent elements.
(e) An amount of energy equal to the bond energy is released when a covalent bond is broken.

30. Which of the following statements is *not* true regarding the formation of a covalent bond between nitrogen and oxygen to give nitric oxide, NO?
(a) Valence electrons are shared between the nitrogen and oxygen atoms.
(b) Bonding electrons are distributed over the entire NO molecule.
(c) The bond length between nitrogen and oxygen atoms is greater than the sum of the two atomic radii.
(d) The properties of nitrogen and nitric oxide are not related.
(e) The formation of covalent bonds between nitrogen atoms and oxygen atoms releases energy.

Electron Dot Formulas of Molecules (Sec. 11.4)

31. Find the total number of valence electrons in each of the following molecules. Write the electron dot formula and corresponding structural formula. (*The central atom is indicated in* ***bold***.)
(a) $\mathbf{H}_2$ (b) $\mathbf{F}_2$
(c) $H\mathbf{Br}$ (d) $\mathbf{N}H_3$
(e) $\mathbf{C}Cl_4$ (f) $\mathbf{O}F_2$

32. Write the electron dot formula and corresponding structural formula for each of the following molecules.
(a) $\mathbf{Cl}_2$ (b) $\mathbf{O}_2$
(c) $H\mathbf{I}$ (d) $\mathbf{P}H_3$
(e) $\mathbf{C}H_4$ (f) $\mathbf{N}F_3$

33. Find the total number of valence electrons in each of the following molecules. Write the electron dot formula and corresponding structural formula.
(a) $HONO_2$ (b) SO_2
(c) C_2H_4 (d) C_2H_2
(e) $HOCl$ (f) $HONO$

34. Write the electron dot formula and corresponding structural formula for each of the following molecules.
(a) $\mathbf{N}_2$ (b) $\mathbf{P}I_3$
(c) $\mathbf{CS}_2$ (d) $\mathbf{CH_3OH}$
(e) $H_2\mathbf{O}_2$ (f) $\mathbf{HOCN}$

Polar Covalent Bonds (Sec. 11.5)

35. What is the general trend in electronegativity down a group in the periodic table?

36. What is the general trend in electronegativity across a series in the periodic table?

37. Which elements are more electronegative, metals or nonmetals?

38. In general, which elements are more electronegative, semimetals or nonmetals?

39. Predict which of the following elements is more electronegative according to the general electronegativity trends in the periodic table.
(a) Br or Cl (b) O or S
(c) Al or B (d) Se or As
(e) N or F (f) P or Sb

40. Predict which of the following elements is more electronegative according to the general electronegativity trends in the periodic table.
(a) Se or Br (b) O or H
(c) Te or S (d) C or B
(e) Co or Ca (f) Ba or Be

41. Refer to the electronegativity values in Figure 11.10 and calculate the polarity for each of the following bonds.
(a) H—Br (b) H—F
(c) I—Cl (d) Br—F
(e) C—O

42. Refer to the electronegativity values in Figure 11.10 and calculate the polarity for each of the following bonds.
(a) H—Cl (b) H—I
(c) Br—Cl (d) I—Br
(e) N—O

43. Refer to Figure 11.10 and label each atom in the following polar covalent bonds using delta notation (δ^+ and δ^-).
(a) H—S (b) O—S
(c) Cl—B (d) S—Cl
(e) N—F

44. Refer to Figure 11.10 and label each atom in the following polar covalent bonds using delta notation.
(a) C—H (b) Se—O
(c) Ge—Cl (d) H—Br
(e) P—I

Nonpolar Covalent Bonds (Sec. 11.6)

45. Refer to Figure 11.10 and indicate which of the following are nonpolar covalent bonds.
(a) Cl—Cl (b) Cl—N
(c) N—H (d) H—P
(e) P—Te

46. Refer to Figure 11.10 and indicate which of the following are nonpolar covalent bonds.
(a) I—C (b) C—S
(c) S—H (d) H—Br
(e) Te—I

47. Which of the following elements occur naturally as nonpolar diatomic molecules?
(a) hydrogen (b) nitrogen
(c) chlorine (d) phosphorus
(e) helium (f) fluorine
(g) iodine (h) selenium
(i) oxygen (j) bromine

48. Write the chemical formula for each of the seven elements that occurs naturally as diatomic molecules.

Coordinate Covalent Bonds (Sec. 11.7)

49. The chlorine atom in a molecule of HOCl can form a coordinate covalent bond with an oxygen atom. Draw the electron dot formula for HOCl; add an oxygen atom and label the coordinate covalent bond.

50. The iodine atom in a molecule of HOI can form a coordinate covalent bond with an oxygen atom. Draw the electron dot formula for HOI; add an oxygen atom and label the coordinate covalent bond.

51. An ammonia molecule, NH_3, can coordinate with a hydrogen ion, H^+, to give NH_4^+. Draw the electron dot formula for an ammonia molecule; add the hydrogen ion and label the coordinate covalent bond.

52. A phosphine molecule, PH_3, can coordinate with a hydrogen ion, H^+, to give PH_4^+. Draw the electron dot formula for a phosphine molecule; add the hydrogen ion and label the coordinate covalent bond.

53. A nitrite ion, NO_2^-, can coordinate an oxygen atom to give the nitrate ion, NO_3^-. Draw the electron dot formula for NO_2^-; add an oxygen atom and label the coordinate covalent bond in NO_3^-.

54. A phosphite ion, PO_3^{3-}, can coordinate an oxygen atom to give the phosphate ion, PO_4^{3-}. Draw the electron dot formula for PO_3^{3-}; add an oxygen atom and label the coordinate covalent bond.

55. A sulfate ion, SO_4^{2-}, can coordinate a hydrogen ion, H^+, to give the hydrogen sulfate ion, HSO_4^-. Draw the electron dot formula for SO_4^{2-}; add a hydrogen ion and label the coordinate covalent bond.

56. A hydrogen sulfate ion, HSO_4^-, can coordinate a second hydrogen ion to give sulfuric acid, H_2SO_4. Draw the electron dot formula for HSO_4^-; add a hydrogen ion and label the additional coordinate covalent bond.

Polyatomic Ions (Sec. 11.8)

57. Find the total number of valence electrons in each of the following polyatomic ions. Write the electron dot formula and corresponding structural formula. *(Central atoms are indicated in* ***bold****.)*

(a) $\mathbf{O}H^-$ (b) $\mathbf{I}O^-$
(c) $\mathbf{Cl}O_2^-$ (d) $\mathbf{P}H_4^+$
(e) $\mathbf{I}O_4^-$

58. Write the electron dot formula and corresponding structural formula for each of the following polyatomic ions.

(a) $\mathbf{Br}O^-$ (b) $\mathbf{S}O_3^{2-}$
(c) $\mathbf{B}O_3^{3-}$ (d) $H\mathbf{S}O_3^-$
(e) $\mathbf{Br}O_3^-$

59. Find the total number of valence electrons in each of the following polyatomic ions. Write the electron dot formula and corresponding structural formula.

(a) $\mathbf{Cl}O^-$ (b) $\mathbf{Br}O_2^-$
(c) $\mathbf{I}O_3^-$ (d) $H_3\mathbf{O}^+$
(e) $\mathbf{Cl}O_4^-$

60. Write the electron dot formula and corresponding structural formula for each of the following polyatomic ions.

(a) $\mathbf{C}N^-$ (b) $H\mathbf{S}^-$
(c) $\mathbf{P}O_3^{3-}$ (d) $H\mathbf{C}O_3^-$
(e) $\mathbf{Se}O_3^{2-}$

Properties of Ionic and Molecular Compounds (Sec. 11.9)

61. Describe the general properties of ionic compounds with respect to each of the following.

(a) physical state
(b) structure
(c) melting point
(d) boiling point
(e) electrical conductivity of an aqueous solution
(f) rate of reaction

62. Describe the typical properties of molecular compounds with respect to each of the following.

(a) physical state
(b) structure
(c) melting point
(d) boiling point
(e) electrical conductivity of an aqueous solution
(f) rate of reaction

63. Predict whether the following properties most likely indicate an ionic or molecular compound.

(a) emerald crystal
(b) greenish-yellow gas
(c) melts at 801°C
(d) boils at −79°C
(e) aqueous solutions conduct electric current
(f) reacts slowly in solution

64. Predict whether the following properties most likely indicate a compound having ionic or covalent bonds.

(a) colorless gas
(b) transparent crystalline solid
(c) low boiling point
(d) colorless liquid
(e) nonconductor in aqueous solution
(f) reacts rapidly in solution

The red ruby, the blue sapphire, and the green emerald are gemstones having ionic crystalline structures. The diamond is a molecular structure.

General Exercises

65. Explain why the radius of a sodium ion (0.095 nm) is about half that of a sodium atom (0.186 nm).

66. Explain why the radius of a chloride ion (0.181 nm) is about twice that of a chlorine atom (0.099 nm).

67. Iron rusts to give ferric oxide. What is the relationship between the properties of ferric oxide and the elements iron and oxygen?

68. What is the relationship between the properties of a compound composed of formula units having ionic bonds and its constituent elements?

69. Write neutral formula units by combining the following cations and anions. (Refer to Section 6.4 as necessary.)
(a) Ca^{2+} and I^- (b) Ra^{2+} and O^{2-}
(c) Ga^{3+} and F^- (d) Ba^{2+} and P^{3-}

70. Write neutral formula units by combining the following cations and anions.
(a) Bi^{3+} and S^{2-} (b) Sr^{2+} and As^{3-}
(c) Sc^{3+} and N^{3-} (d) Ti^{4+} and O^{2-}

71. Write the formula for the compound composed of the following ions.
(a) Al^{3+} and CO_3^{2-} (b) Sr^{2+} and OH^-
(c) Ag^+ and PO_4^{3-} (d) Cd^{2+} and NO_3^-

72. Write the formula for the compound composed of the following ions.
(a) Hg^{2+} and HCO_3^- (b) Bi^{3+} and BrO_3^-
(c) NH_4^+ and CO_3^{2-} (d) Hg_2^{2+} and PO_4^{3-}

73. An automobile engine produces oxides of nitrogen from nitrogen and oxygen gases. What is the relationship between the properties of nitrogen dioxide and the elements nitrogen and oxygen?

74. What is the relationship between the properties of a compound composed of molecules having covalent bonds and its constituent elements?

75. Find the total number of valence electrons in each of the following molecules containing a semimetal as the central atom. Write the electron dot formula and corresponding structural formula.
(a) SbH_3 (b) $GeCl_4$

76. Find the total number of valence electrons in each of the following polyatomic ions containing a semimetal as the central atom. Write the electron dot formula and corresponding structural formula.
(a) AsO_3^{3-} (b) SiO_3^{2-}

77. To write the electron dot formula for a molecule, we first find the total number of valence electrons from each atom. Why it is unnecessary to keep the valence electrons separate for each atom?

78. The structural formula for a polyatomic ion is usually placed within brackets followed by the ionic charge. Why is it incorrect to place the ionic charge on a single atom within the polyatomic ion?

79. There are two satisfactory electron dot formulas for sulfur dioxide SO_2. Each formula is equivalent and is referred to as a resonance form of the molecule. Draw the two equivalent electron dot formulas for SO_2.

80. There are two possible structural formulas for the nitrite ion, NO_2^-. Each structure is equivalent and the ion resonates back and forth between the two. Draw the two equivalent structural formulas for NO_2^-.

81. Some stable molecules violate the octet rule. A few examples include boron trichloride, BCl_3 (B is surrounded by 6 valence electrons); phosphorus pentabromide, PBr_5 (P is surrounded by 10 valence electrons); and sulfur hexafluoride, SF_6 (S is surrounded by 12 valence electrons). Draw the electron dot formula and structural formula for BCl_3, PBr_5, and SF_6.

Crystals of a xenon compound, XeF_2.

82. In 1962, the first compound to contain an inert gas was isolated. Within months, compounds containing the inert gases xenon and krypton were made in the laboratory. One of the first compounds to be synthesized was xenon trioxide, XeO_3. Draw the electron dot and structural formulas for XeO_3.

83. Some periodic tables place hydrogen in both Groups IA/1 and VIIA/17 because a hydrogen atom can lose or gain electrons. Write the formula for the two ions of hydrogen.

CHAPTER

12

The Gaseous State

The helium-filled balloons float up into the atmosphere after being released because they are less dense than air. The mass of a balloon filled with helium is less than the mass of nitrogen and oxygen in an equal volume of air.

Greek philosophers valued thoughtful mental exercises, but they were not interested in practical experiments. That's why, in the time of the early Greeks, logic and reason dominated natural science. The Greeks thought that matter was composed of four basic elements: air, earth, fire, and water. They came to that conclusion by theoretical speculation. In the 1600s, interest in science shifted to the laboratory. Scientists agreed that matter was composed of elements, but disagreed with the Greek view of the four basic elements. The English chemist Robert Boyle proposed that a basic element could only be discovered through practical experiments.

Robert Boyle is best known for his work with gases. He believed that a gas could be compressed because the particles could be squeezed closer together. He studied the atmosphere and discovered that air could be compressed. He thus came to the conclusion that air must be composed of discrete particles separated by a void. His experiments were remarkable because they offered the first evidence for the particle nature of matter. In 1803, John Dalton proposed the atomic theory which was supported, in part, by Boyle's experiments with gases.

12.1 Properties of Gases

To describe five observed properties of a gas.

By the end of the 1700s, the study of gases was of great interest, especially in France, where high-altitude ballooning was popular. Scientists experimented with various gases and found they had common characteristics. Even mixtures of gases, such as air, had the same general properties. We can summarize these properties as follows:

1. *Gases have an indefinite shape.* A gas takes on the shape of its container and fills it uniformly. If the volume of the container changes, so does the volume of the gas.
2. *Gases may be expanded.* A gas continuously expands and distributes itself throughout a closed container. This means that the volume of gas in a closed cylinder can be increased by moving a piston to enlarge the volume.
3. *Gases may be compressed.* The volume of gas in a closed cylinder can be decreased by moving a piston to reduce the volume. If the volume is reduced sufficiently, however, the gas may eventually liquefy.
4. *Gases have low densities.* The density of air is about 1.3 g/L. The density of water, on the other hand, is 1.0 g/mL. Air is almost 1000 times less dense than water.

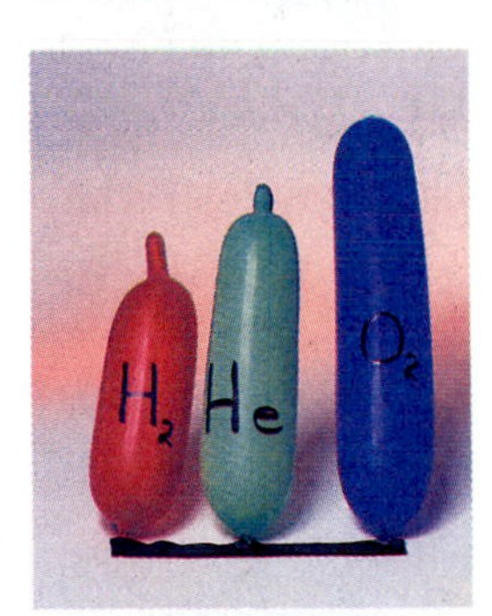

Initially, the balloons shown above have the same volume. After a few hours, the same balloons have different volumes. The explanation is that H_2 is least dense and diffuses through the balloon faster than He and O_2. The O_2 is most dense and diffuses slower than H_2 and He.

5. *Gases mix spontaneously to form homogeneous mixtures.* Air is a good example of a gaseous mixture. To see how gases mix uniformly, consider that oxygen produced by photosynthesis from green plants distributes itself evenly throughout our atmosphere. This is also true of the sulfur dioxide pollution from coal-burning industries. These gases are said to diffuse throughout the atmosphere.

EXAMPLE EXERCISE 12.1

Which of the following is *not* an observed property of a gas?
(a) varies its shape and volume to fit the container
(b) expands infinitely and distributes uniformly
(c) compresses infinitely and uniformly
(d) has a low density
(e) mixes homogeneously with other gases

Solution: According to the properties described above, the answer is (c).
(a) A gas can vary its shape and volume.
(b) A gas can expand indefinitely and uniformly.
(c) A gas *cannot* be compressed infinitely.
(d) A gas has a density that is much less than water.
(e) A gas can mix homogeneously with another gas.

SELF-TEST EXERCISE

Describe five observable properties of a gas.

Answer: Refer to the preceding list of five properties.

Gases were first understood in terms of their observable properties. Even though gases are usually invisible, it was easy to observe experimental changes in gases. Increasing or decreasing the volume of a gas brought about a change in pressure. Heating or cooling a gas brought about a change in volume. By the mid-1800s, scientists began to formulate a model for the behavior of gases based on the behavior of individual gas molecules. In Section 12.10, we will study this molecular model of gases, called the kinetic theory of gases.

12.2 Atmospheric Pressure and the Barometer

To state standard atmospheric pressure in the following units: atmospheres, millimeters of mercury, torr, centimeters of mercury, inches of mercury, pounds per square inch, and kilopascals.

To convert a given gas pressure into the following units: atm, mm Hg, torr, cm Hg, in. Hg, psi, and kPa.

gas pressure A physical quantity that measures the frequency and energy of molecular collisions. The force per unit area exerted by a gas.

Gas pressure is the result of molecules being in constant motion and striking surfaces. The pressure exerted by a gas depends on how often and how hard the molecules strike a surface (Figure 12.1).

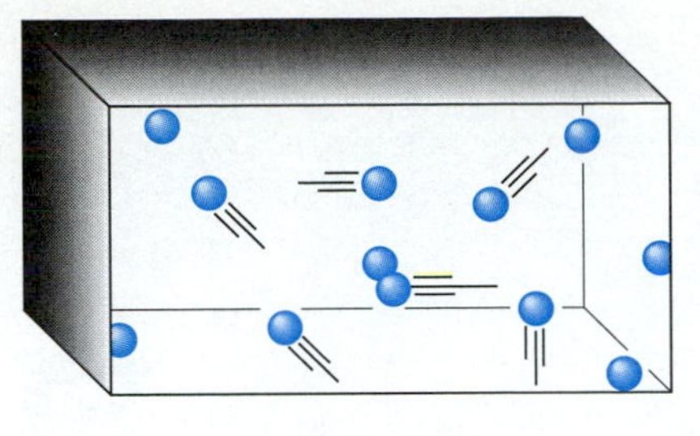
(a) Higher pressure

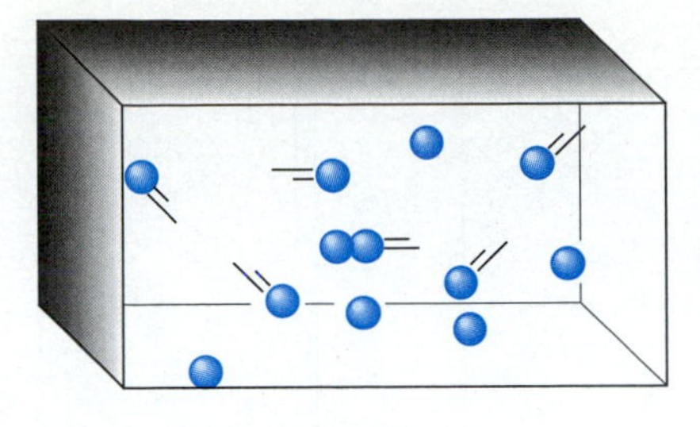
(b) Lower pressure

Figure 12.1 Pressure Depends on the Frequency of Collision (a) The molecules are moving faster and colliding more frequently with the container. In container (b) the molecules are moving more slowly and colliding less frequently.

1. If the frequency of collisions increases, the gas pressure increases.
2. If the molecules collide with more energy, the pressure increases.
3. As the temperature increases, molecules move faster and collide with more energy.

Atmospheric Pressure

From the time of the Greek civilization, it has been observed that a wine barrel with a single hole emptied slowly. Aristotle suggested the concept of a vacuum to explain the observation. He proposed that a vacuum was created as wine emptied from the barrel. He further stated that "nature abhors a vacuum" and a vacuum went against natural principles.

In 1644, the Italian physicist Evangelista Torricelli (1608–1647) published a scientific explanation of a **vacuum.** Torricelli proposed that the earth was surrounded by a sea of air. He argued that the air exerted a pressure on everything it touched. Therefore, air pressure was responsible for slowing the flow of liquid from a barrel. According to Torricelli, if we opened a second hole in the top of the barrel, air would rush in and allow the liquid out. Later, Torricelli carried out experiments that supported his theory. By the end of the 1600s, the concept of atmospheric pressure was well established. Today, we understand that the **atmospheric pressure** is the pressure exerted by air molecules colliding with surfaces in the environment. The pressure of the atmosphere is considerable, about 15 pounds on every square inch. For an average size person, the atmosphere exerts a total force of nearly 20 tons on the surface of his or her body! More effects of atmospheric pressure are illustrated in Figure 12.2.

vacuum A volume that does not contain gas molecules or any other matter.

atmospheric pressure The pressure exerted by the gas molecules in air; at sea level this pressure supports a 760-mm column of mercury.

In 1643, Torricelli invented the first **barometer**, an instrument that measures atmospheric pressure. He took a 4-foot length of glass tubing, sealed one end, and filled the tubing with mercury. He put a stopper in the open end and turned it upside down in a dish of mercury. After removing the stopper, he found the level of mercury inside the tubing was about 30 inches. Each day he measured the height of mercury and found it varied slightly. Torricelli explained that atmospheric pressure was responsible for the fluctuations in the barometer. Today, a column of mercury measuring 29.9 in. is defined as standard atmospheric pressure. Standard pressure is given the value of 1 atmosphere (atm). Figure 12.3 illustrates Torricelli's barometer.

barometer An instrument for measuring atmospheric pressure.

We can express standard pressure in many other units besides atmospheres and inches of mercury. Table 12.1 lists selected units that are frequently used to express pressure.

Since all the values in Table 12.1 are expressions for standard pressure, we can convert them from one unit to another. Given a barometer reading of 31.5 in. Hg, what is the pressure in atmospheres? Let's apply the unit analysis method

Figure 12.2 Illustrations of Atmospheric Pressure
(a) The pressure of the atmosphere supports a column of water 34 feet high. (b) Atmospheric pressure on the card keeps the water in the inverted glass.
(c) After boiling water in a can, the opening is capped. As the trapped steam cools to a liquid, the internal pressure decreases and the external atmospheric pressure crushes the can.

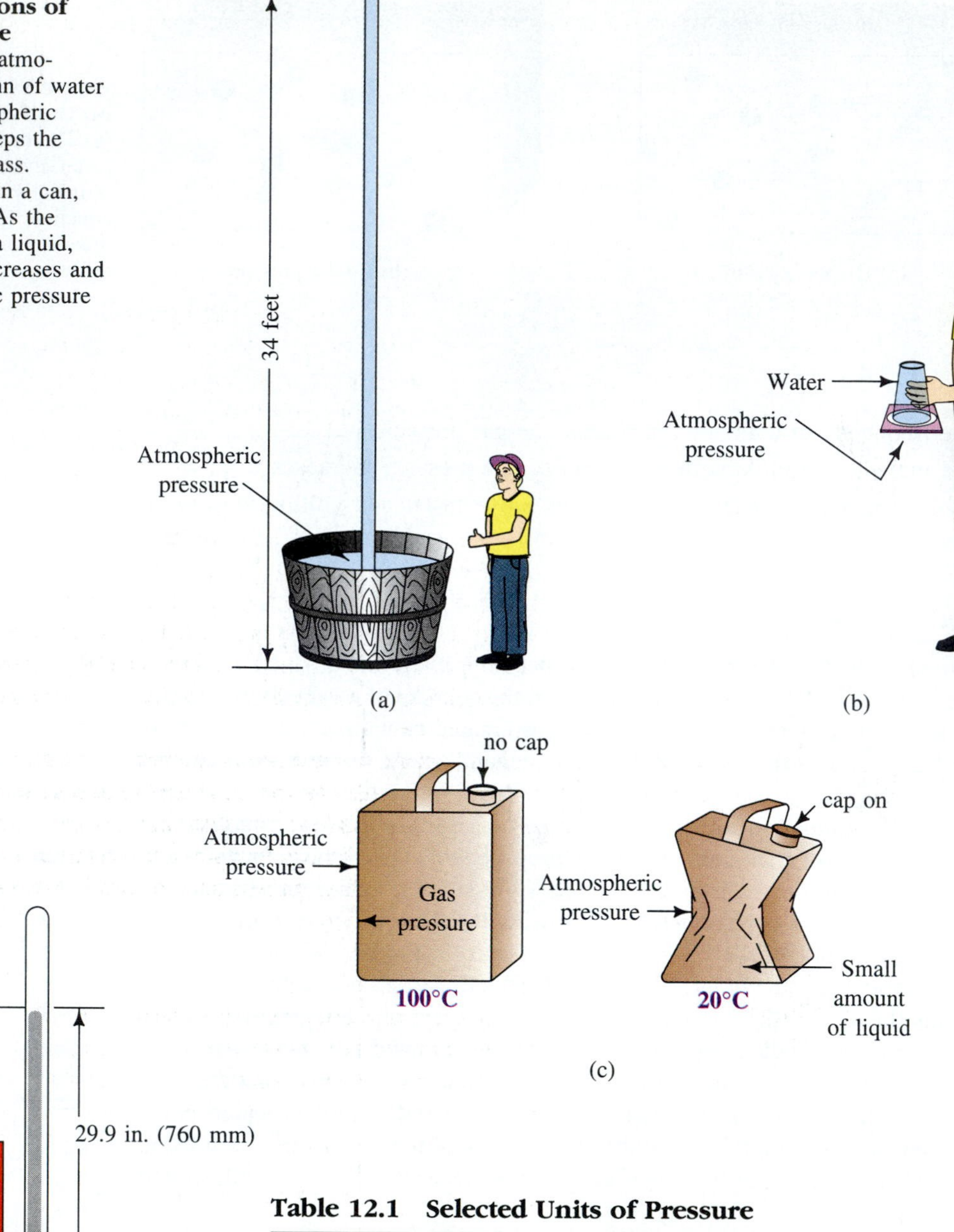

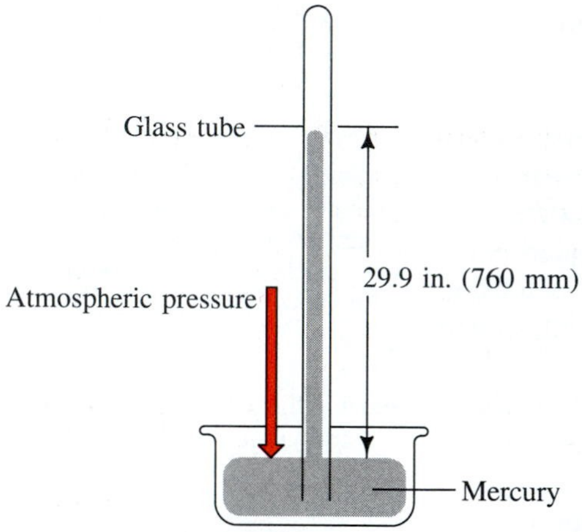

Figure 12.3 Torricelli's Mercury Barometer A sealed glass tube is filled with liquid mercury, inverted, and placed into a dish of mercury. The pressure of the atmosphere at sea level and 0°C supports a column of mercury 29.9 inches high (760 mm). There is no gas trapped above the column of mercury inside the glass tube.

Table 12.1 Selected Units of Pressure

Unit	Standard Pressure
atmosphere	1.00 atm
inches of mercury	29.9 in. Hg
centimeters of mercury	76.0 cm Hg
millimeters of mercury	760 mm Hg
torr*	760 torr
pounds per square inch	14.7 psi
kilopascal*	101 kPa

* Recently, 1 mm of Hg was redefined as 1 torr in honor of Torricelli. Thus, standard pressure can be given as 760 torr. The kilopascal, kPa, is the SI unit of standard pressure. A pascal is defined as a force of 1 newton on an area of 1 square meter (see Section 2.11).

to set up the problem. The unknown unit is atm, the given value is 31.5 in. Hg, and 1 atm = 29.9 in. Hg; thus,

$$31.5 \text{ \sout{in. Hg}} \times \frac{1 \text{ atm}}{29.9 \text{ \sout{in. Hg}}} = 1.05 \text{ atm}$$

To convert the pressure from one unit to another, we derive a unit factor from the relationship of standard pressures.

EXAMPLE EXERCISE 12.2

Given a barometer reading of 777 mm Hg, express the pressure in each of the following units.

(a) torr
(b) centimeters of Hg
(c) pounds per square inch
(d) kilopascals

torr A unit of pressure equal to 1 mm Hg.

Solution: For each conversion, we will write a unit factor based on the relationship for units of standard pressure.

(a) We can apply the unit analysis method after writing the equivalent relationship: 760 mm Hg = 760 torr.

$$777 \text{ \sout{mm Hg}} \times \frac{760 \text{ torr}}{760 \text{ \sout{mm Hg}}} = 777 \text{ torr}$$

(b) To convert to centimeters of Hg, we use the equivalent relationship: 760 mm Hg = 76.0 cm Hg.

$$777 \text{ \sout{mm Hg}} \times \frac{76.0 \text{ cm Hg}}{760 \text{ \sout{mm Hg}}} = 77.7 \text{ cm Hg}$$

(c) To express the pressure in pounds per square inch, psi, we use the equivalent relationship: 760 mm Hg = 14.7 psi.

$$777 \text{ \sout{mm Hg}} \times \frac{14.7 \text{ psi}}{760 \text{ \sout{mm Hg}}} = 15.0 \text{ psi}$$

(d) To find kilopascals, we use the relationship: 760 mm Hg = 101 kPa.

$$777 \text{ \sout{mm Hg}} \times \frac{101 \text{ kPa}}{760 \text{ \sout{mm Hg}}} = 103 \text{ kPa}$$

SELF-TEST EXERCISE

A sample of nitric oxide gas, NO, is at a pressure of 2550 mm Hg. Express the pressure of the gas in each of the following units.

(a) torr
(b) atm
(c) cm Hg
(d) in. Hg
(e) psi
(f) kPa

Answers: **(a)** 2550 torr; **(b)** 3.36 atm; **(c)** 255 cm Hg; **(d)** 100 in. Hg; **(e)** 49.3 psi; **(f)** 339 kPa

12.3 Vapor Pressure

To understand the concept of vapor pressure and the relationship of boiling point to temperature.

Everyone has observed that water in an open container evaporates. Evaporation is the result of water molecules having enough energy to escape the liquid. If the container is enclosed, the molecules cannot escape. However, molecules continuously leave the liquid. Simultaneously, water molecules above the liquid in the gaseous vapor return to the liquid. When the rate of evaporation is equal to the rate of condensation, a state of equilibrium is reached. **Vapor pressure** is defined as the pressure exerted by gaseous molecules in dynamic equilibrium with the same molecules in the liquid state (Figure 12.4).

vapor pressure The pressure exerted by gas molecules in dynamic equilibrium with its liquid state.

The vapor pressure of a liquid can be determined using a mercury barometer (Figure 12.5). A drop of liquid is introduced at the bottom of the barometer. The less dense liquid droplet floats to the top of the mercury and vaporizes. The vaporized liquid exerts a gas pressure, thus driving the mercury level down. *The decrease in the mercury level corresponds to the vapor pressure of the liquid.* If the mercury level drops from 760 to 740 mm, the vapor pressure of the liquid is recorded as 20 mm Hg.

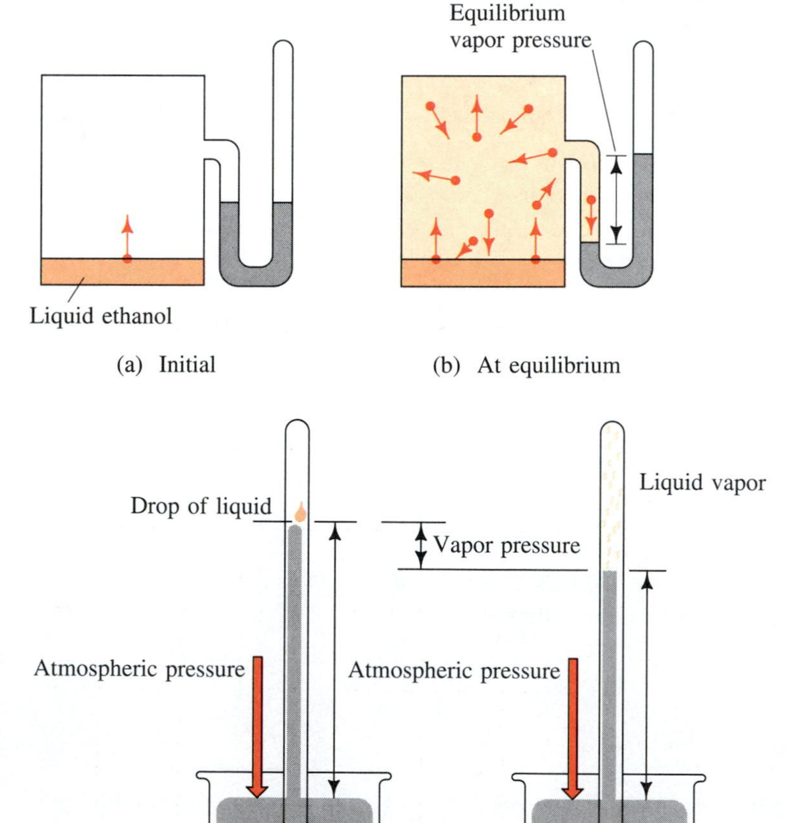

Figure 12.4 Vapor Pressure of a Liquid The vapor pressure of ethanol can be illustrated as follows. (a) Initially, ethanol is placed into a closed container and molecules begin to escape from the liquid. After awhile, some of the ethanol molecules in the vapor return to the liquid. (b) Eventually, the ethanol molecules are evaporating and condensing at the same rate. The vapor pressure is measured by the difference in the height of mercury in the side tube.

Figure 12.5 Measuring Vapor Pressure of a Liquid The vapor pressure of a liquid can be measured as follows. (a) Initially, a drop of liquid is introduced into the tube of mercury. As molecules of liquid evaporate, they begin to exert a pressure against the column of mercury. (b) Eventually, the molecules are evaporating and condensing at the same rate. At a specified temperature, the difference in the height of mercury is the vapor pressure of the liquid.

Table 12.2 Vapor Pressure of Water

Temperature (°C)	Pressure (mm Hg)	Temperature (°C)	Pressure (mm Hg)
5	6.5	55	118.0
10	9.2	60	149.4
15	12.8	65	187.5
20	17.5	70	233.7
25	23.8	75	289.1
30	31.8	80	355.1
35	41.2	85	433.6
40	55.3	90	525.8
45	71.9	95	633.9
50	92.5	100	760.0

It is an observed property of a gas that its pressure increases as its temperature increases. Similarly, vapor pressure is greater as the temperature increases. Vapor pressures, however, increase more dramatically as the temperature increases. At 25°C, the vapor pressure of water is 23.8 mm Hg. At 50°C, the vapor pressure increases to 92.5 mm Hg. Table 12.2 lists selected values for the vapor pressure of water.

Boiling Point

At 20°C the vapor pressure of water is 17.5 mm Hg; at 30°C the pressure is 31.8 mm Hg; at 100°C the vapor pressure is 760 mm Hg. When the upward pressure of the vapor equals the downward pressure of the atmosphere, the liquid begins to boil. The **boiling point** is that temperature where the two pressures are equal. Figure 12.6 shows the relationship between vapor pressure and temperature for three common liquids.

boiling point The temperature at which the vapor pressure of a liquid is equal to the atmospheric pressure.

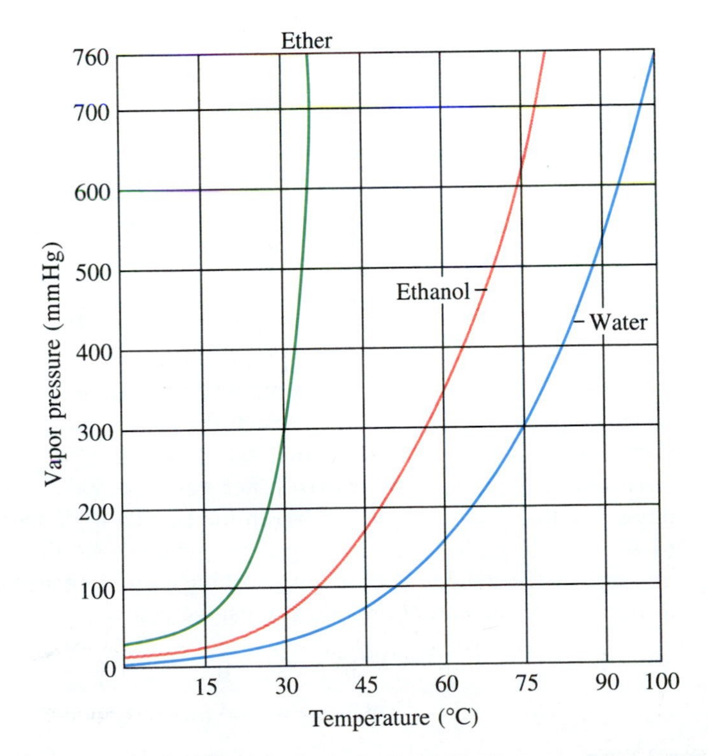

Figure 12.6 Vapor Pressure and Temperature The vapor pressure of a liquid becomes greater as the temperature increases. Notice that the vapor pressure of water equals 760 mm Hg at 100°C. Thus, at 100°C the vapor pressure of water (760 mm Hg) equals the atmospheric pressure at sea level (760 mm Hg) and water begins to boil. On top of Mt. Everest, the atmospheric pressure is only 250 mm Hg and water boils at about 70°C.

EXAMPLE EXERCISE 12.3

Refer to Figure 12.6 and state the boiling point of (a) ether and (b) ethanol.

Solution: The boiling point of a liquid is the temperature at which the vapor pressure equals standard atmospheric pressure, 760 mm Hg.

(a) Referring to Figure 12.6, we find that the vapor pressure of ether is 760 mm Hg at about 35°C. The actual boiling point of ether is 36°C.

(b) The vapor pressure of ethanol reaches 760 mm Hg at about 80°C. The actual boiling point of ethanol is 78°C.

SELF-TEST EXERCISE

The vapor pressure of methanol is 100 mm Hg at 21°C, 400 mm Hg at 50°C, and 760 mm Hg at 65°C. What is the normal boiling point of methanol?

Answer: 65°C

12.4 Dalton's Law of Partial Pressures

To apply Dalton's law of partial pressures to gases in a mixture.

To become familiar with the procedure of collecting a gas over water.

In Section 4.1, we discussed the atomic theory presented by John Dalton. The evidence Dalton offered to support the atomic theory was partially based on the behavior of gases. His knowledge of gases was related to his interest in meteorology. In fact, he kept daily records of the weather till the day he died. In 1801, Dalton proposed that each gas in a mixture exerted the same pressure that it would if it were the only gas present. This theory is known as **Dalton's law of partial pressures**, which is usually stated, *the total pressure of a gaseous system is equal to the sum of the partial pressures of each gas in the mixture*. We can write this relationship as follows:

Dalton's law of partial pressures The pressure exerted by a mixture of gases is equal to the sum of the pressures exerted by each gas in the mixture.

$$P_1 + P_2 + P_3 + \cdots = P_{total}$$

partial pressure The pressure exerted by a single gas in a mixture of two or more gases.

The pressure exerted by a gas in a mixture of gases is called the **partial pressure**. In the equation, P_1, P_2, and P_3 represent the partial pressures for each of the gases in the mixture. P_{total} symbolizes the total of all partial pressures.

Let's suppose we have two 1-liter gas cylinders. In one of the cylinders is oxygen with a pressure of 550 torr. In the second cylinder is helium at a pressure of 100 torr. If the helium and oxygen are pumped into the same vessel, what is the pressure in the cylinder? Assuming no temperature changes, the pressure is equal to the total of the partial pressures; thus, we can write

$$P_{O_2} \quad + \quad P_{He} \quad = \quad P_{total}$$
$$550 \text{ torr} + 100 \text{ torr} = 650 \text{ torr}$$

EXAMPLE EXERCISE 12.4

In 1952, a landmark experiment took place at the University of Chicago. In an attempt to prove that the molecular building blocks of life could evolve from simple atmospheric gases, Stanley Miller simulated earth's primordial atmosphere. He placed the gases hydrogen, ammonia, methane, and water vapor in a glass sphere and energized the mixture with ultraviolet light and an electric spark. Assuming the total gas pressure in the sphere was 760 torr, what was the partial pressure of the hydrogen given

$$P_{\text{ammonia}} = 125 \text{ torr}, \quad P_{\text{methane}} = 340 \text{ torr}, \quad P_{\text{water vapor}} = 20 \text{ torr}$$

Solution: Dalton's law of partial pressures states that the sum of the individual partial pressures equals the total pressure; therefore,

$$P_{\text{hydrogen}} + P_{\text{ammonia}} + P_{\text{methane}} + P_{\text{water vapor}} = P_{\text{total}}$$

Rearranging the equation in terms of the partial pressure of hydrogen,

$$P_{\text{hydrogen}} = P_{\text{total}} - P_{\text{ammonia}} - P_{\text{methane}} - P_{\text{water vapor}}$$

Substituting the given partial pressures and solving, we have

$$P_{\text{hydrogen}} = 760 \text{ torr} - 125 \text{ torr} - 340 \text{ torr} - 20 \text{ torr} = 275 \text{ torr}$$

After a week, the apparatus was disassembled and the contents analyzed. In the "primordial soup" were a few amino acids and nucleic acid residues, the basic building blocks of life! This famous experiment is frequently cited as evidence for the chemical evolution of life.

SELF-TEST EXERCISE

What is the total pressure in a steel cylinder containing the following gases at their given partial pressures? $P_{\text{nitrogen}} = 5.02$ atm, $P_{\text{oxygen}} = 1.15$ atm, $P_{\text{argon}} = 0.05$ atm, and $P_{\text{carbon dioxide}} = 0.01$ atm.

Answer: 6.23 atm

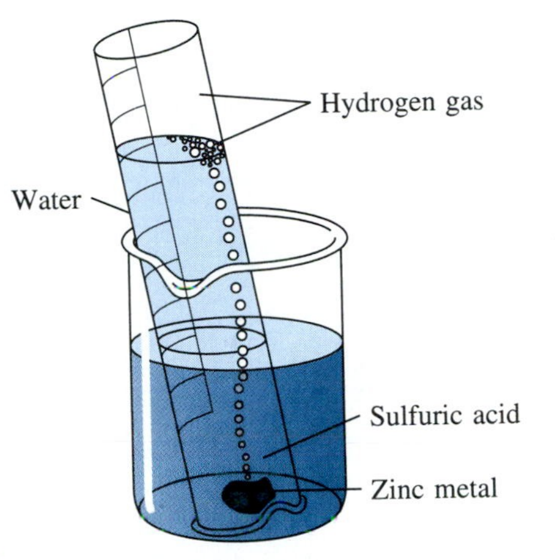

Figure 12.7 Collecting a Gas over Water Zinc metal reacts with sulfuric acid to give bubbles of hydrogen gas. A graduated cylinder full of water is placed over the metal to collect the gas bubbles. The volume of water displaced from the graduated cylinder equals the volume of hydrogen gas liberated from the acid.

Collecting a Gas Over Water

Gases expand in their containers and are usually invisible. How then do we measure the volume of a gas? We determine the volume of a gas indirectly. In Section 2.7, we discussed measuring **volume by displacement**, whereby a gas displaces a volume of water. The volume of water displaced equals the volume of the gas.

volume by displacement A technique for determining the volume of a gas by measuring the amount of water it displaces.

A laboratory experiment that evolves a gas is the reaction of zinc metal and sulfuric acid to give zinc sulfate and hydrogen gas.

$$Zn(s) + H_2SO_4(aq) \longrightarrow ZnSO_4(aq) + H_2(g)$$

If we invert a graduated cylinder filled with water over the reacting metal, we can trap the hydrogen gas. Figure 12.7 shows the experimental apparatus.

When the reaction is complete the gas pressure inside the graduated cylinder is equal to the atmospheric pressure (assuming the liquid level is the same inside and outside the cylinder). The atmosphere exerts a pressure on the surface of the water in the beaker that is equal to the pressure of the collected gas. However, there are actually two gases inside the graduated cylinder! One of the gases is hydrogen produced from the reaction. The other gas is water vapor. Since hydrogen is collected over the aqueous solution, water vapor is present. We can use Dalton's law of partial pressures to determine the pressure of the hydrogen:

$$P_{\text{hydrogen}} + P_{\text{water vapor}} = P_{\text{atmosphere}}$$

The reaction takes place at 25°C and the barometer reads 767 mm Hg. What is the partial pressure of the hydrogen gas? We must refer to Table 12.2 to find the vapor pressure of water at 25°C. The pressure is 23.8 mm Hg. Now we can substitute into the equation and calculate the hydrogen gas pressure:

$$P_{\text{hydrogen}} + 23.8 \text{ mm Hg} = 767 \text{ mm Hg}$$

rearranging,

$$\begin{aligned} P_{\text{hydrogen}} &= 767 \text{ mm Hg} - 23.8 \text{ mm Hg} \\ &= 743 \text{ mm Hg} \end{aligned}$$

EXAMPLE EXERCISE 12.5

Sodium hydrogen carbonate is decomposed by heating. The carbon dioxide gas that is liberated is collected over water at 20°C and a barometric pressure of 30.5 in. Hg. Calculate the partial pressure of carbon dioxide gas in mm Hg.

Solution: First, we can use Dalton's law of partial pressures to write the equation:

$$P_{\text{carbon dioxide}} + P_{\text{water vapor}} = P_{\text{atmosphere}}$$

Referring to Table 12.2, we find that the vapor pressure of water at 20°C is 17.5 mm Hg. Substituting into the equation, we have

$$P_{\text{carbon dioxide}} + 17.5 \text{ mm Hg} = 30.5 \text{ in. Hg}$$

We are asked to find the partial pressure in mm Hg. Since the barometric pressure is in units of in. Hg, we must convert as follows.

$$30.5 \cancel{\text{in. Hg}} \times \frac{760 \text{ mm Hg}}{29.9 \cancel{\text{in. Hg}}} = 775 \text{ mm Hg}$$

Once again using the above equation of partial pressures, we have

$$\begin{aligned} P_{\text{carbon dioxide}} + 17.5 \text{ mm Hg} &= 775 \text{ mm Hg} \\ P_{\text{carbon dioxide}} &= 775 \text{ mm Hg} - 17.5 \text{ mm Hg} \\ &= 758 \text{ mm Hg} \end{aligned}$$

SELF-TEST EXERCISE

Hydrogen gas is collected over water at 22°C and 764 torr. What is the partial pressure of the hydrogen gas in torr? The vapor pressure of water at 22°C is 19.8 mm Hg.

Answer: 744 torr

When a gas is collected over water, it is often called a wet gas. That is, the collected gas contains water vapor. A gas that does not contain water vapor is sometimes referred to as a dry gas.

12.5 Variables Affecting Gas Pressure

To identify the three variables that directly affect the pressure exerted by a gas.

To state whether gas pressure increases or decreases with a given change in volume, temperature, or moles of gas.

We learned in Section 12.2 that gas pressure is related to the frequency and energy of molecular collisions. Experimentally, we cannot directly change the frequency or energy of collision. Therefore, we must indirectly affect collisions to change the pressure. In any gaseous system, there are only three ways to change the pressure.

1. *Increase or decrease the volume of the container:* If we increase the volume, gas molecules are farther apart and collide less frequently, and the pressure decreases. If we decrease the volume, gas molecules are closer together and collide more frequently, and the pressure increases. The pressure is said to be *inversely* related to the volume. That is, if the volume increases, the pressure decreases. If the volume decreases, the pressure increases.
2. *Increase or decrease the temperature of the gas:* If we increase the temperature, gas molecules move faster and collide with a greater frequency and energy. If we decrease the temperature, the gas molecules move slower and collide with less frequency and energy. The pressure is said to be *directly* related to the temperature. If the temperature increases, the pressure increases. If the temperature decreases, the pressure decreases.
3. *Increase or decrease the number of molecules in the container:* If we increase the number of gas molecules, there are more collisions and the pressure increases. If we decrease the number of gas molecules, there are fewer collisions and the pressure decreases. The pressure is said to be *directly* related to the number of gas molecules. If the number of molecules increases, the pressure increases. If the number of molecules decreases, the pressure decreases.

The effects of these three variables, volume, temperature, and the number of molecules, on pressure are illustrated in Figure 12.8.

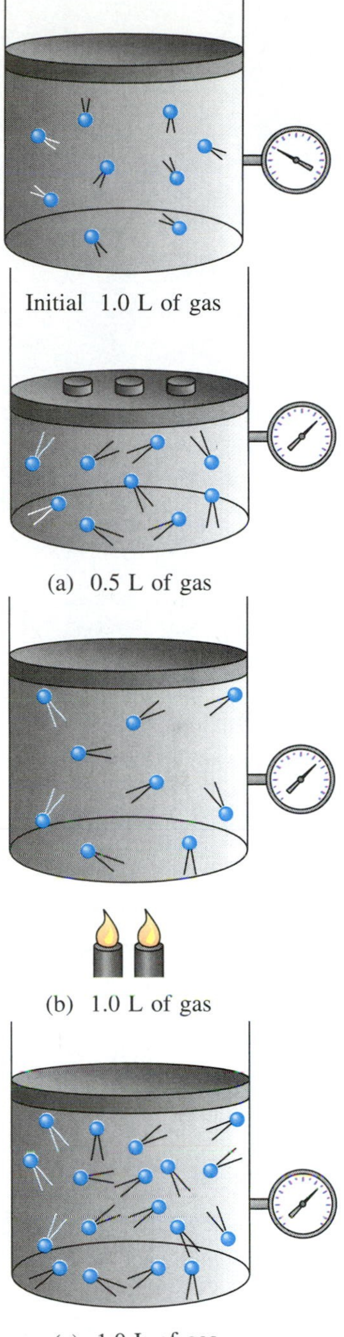

Figure 12.8 Variables Affecting Pressure A gas is contained in a cylinder with a moving piston. The pressure of the gas is affected by (a) changing the volume, (b) changing the temperature, or (c) changing the number of gas molecules in the cylinder. (Note the pressure gauge in each case.)

EXAMPLE EXERCISE 12.6

State whether the pressure of a gas in a closed system increases or decreases with the following changes.

(a) volume changes from 250 to 500 mL

(b) temperature changes from 20° to −80°C

(c) moles of gas change from 1.00 to 1.50 mol

Solution: For each of the changes, we must consider whether the number of molecular collisions increases or decreases.

(a) The volume increases, so the number of collisions decreases; the pressure *decreases.*

(b) The temperature decreases, so the number of collisions decreases; the pressure *decreases.*

(c) If the moles of gas increase, the number of molecules increases. With more molecules, there are more collisions and the pressure increases.

SELF-TEST EXERCISE

Indicate whether gas pressure increases or decreases for each of the following changes in a closed gaseous system.
(a) increasing the temperature
(b) increasing the volume
(c) increasing the number of gas molecules

Answers: **(a)** gas pressure increases; **(b)** gas pressure decreases; **(c)** gas pressure increases

In the following sections, we will consider pressure and only one other variable at a time to simplify our discussion. First, we will assume that we can change the volume of a gas and only affect its pressure; that is, the temperature remains constant. Second, we will assume that we can change the temperature of a gas while the volume remains constant. Third, we will assume that we can add or remove gas molecules and affect only pressure. In real gas systems, these assumptions are not usually valid.

12.6 Boyle's Law: Pressure/Volume Changes

To sketch a graph of the pressure/volume relationship for a gas.

To calculate the new pressure or volume for a gas after a change in one of these conditions using either (a) modified unit analysis or (b) an algebraic method.

We mentioned in the chapter introduction that Robert Boyle was the founder of the scientific method. In one of his experiments Boyle trapped air in a J-shaped tube with liquid mercury. As he added mercury into the tube, he found that the volume of trapped air decreased (see Figure 12.9). If he doubled the mercury pressure, the volume of air halved. If he tripled the pressure, the volume of air was cut to a third. Conversely, when the pressure was halved the volume doubled. Thus, in 1662, Boyle discovered the relationship between volume and pressure for a sample of air.

Boyle's law At constant temperature, the pressure and volume of a gas are inversely proportional.

inversely proportional Refers to two related variables; if one variable doubles, the other variable is reduced to one-half; if one quadruples, the other is reduced to one-fourth; and so on.

The results of Boyle's experiments are known as **Boyle's law**. Formally, Boyle's law states *the volume of a fixed mass of gas is inversely proportional to the pressure, temperature remaining constant*. We can express the relationship as

$$V \propto \frac{1}{P} \qquad (T \text{ constant})$$

In other words, the volume (V) is proportional to ($\propto$) the reciprocal of pressure ($1/P$); temperature (T) remains constant. **Inversely proportional** means that one variable is

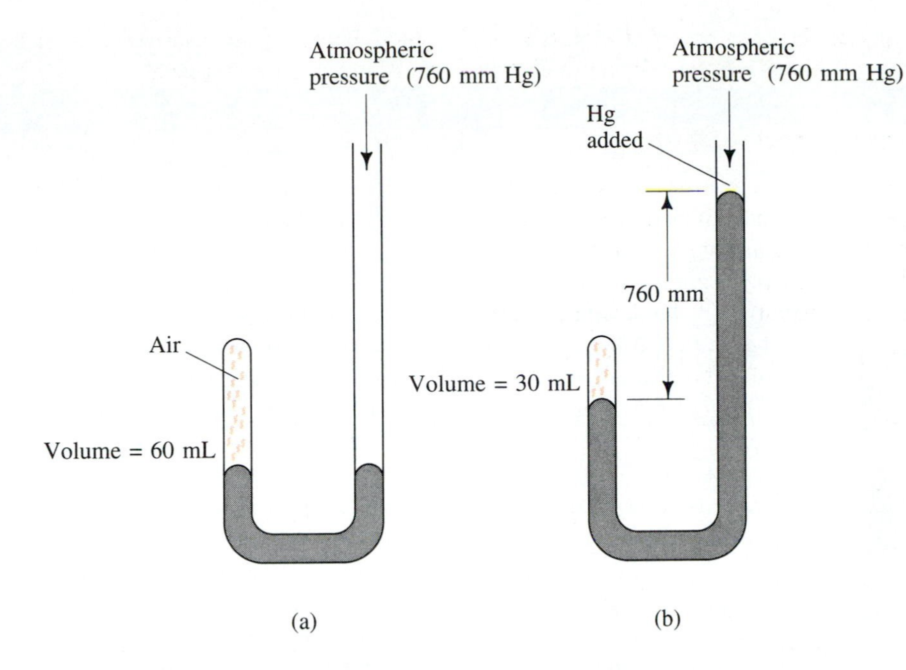

Figure 12.9 Illustration of Boyle's Experiment (a) The initial volume of gas at atmospheric pressure (760 mm Hg) is 60 mL. (b) After adding more mercury to increase the pressure, the volume of gas is 30 mL. Note that the pressure is doubled when the gas volume is halved. In (b) the pressure is 2 atmospheres pressure; that is, atmospheric pressure (760 mm Hg) plus an additional 760 mm column of Hg.

proportional to the reciprocal of another. That is, as one variable gets larger, the other variable gets smaller. As Figure 12.10 shows, when we plot experimental data for the relationship, we get a graph which is not a straight line.

The relationship of pressure to volume can be written as an equation by using a proportionality constant (k). We can write

$$V = k \times \frac{1}{P}$$

By multiplying both sides of the equation by P, we see that the product of pressure and volume is equal to the constant k. The new equation is

$$PV = k$$

We can consider a sample of a gas at different conditions. Let's indicate the initial conditions of pressure and volume as P_1 and V_1, respectively. After a change in conditions, we can indicate the final pressure and volume as P_2 and V_2. Since the product of pressure and volume is a constant, we can write

$$P_1V_1 = k = P_2V_2$$

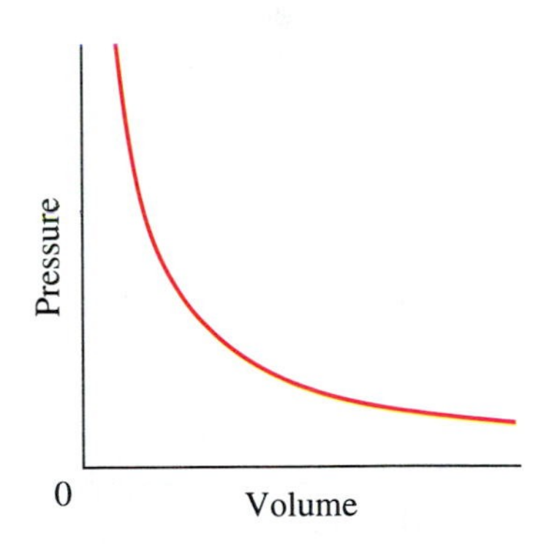

Figure 12.10 Graph of Pressure versus Volume for a Gas As the volume of a gas increases, the gas pressure decreases. As the volume of a gas decreases, the gas pressure increases. The volume and pressure of a gas are inversely proportional.

Solving Boyle's Law Problems

There are two methods for solving gas law problems. One, we can apply algebra to the equation $P_1V_1 = k = P_2V_2$ and solve for an unknown variable. For example, if we know P_1, V_1, and V_2, we can solve the equation for P_2.

Two, we can use a reasoning method. Since the pressure and volume are proportional, we can apply a proportionality factor to find the unknown variable. For example, to find the pressure after a change in conditions, we apply a volume factor to the initial pressure. The equation looks like this:

$$P_1 \times V_{\text{factor}} = P_2$$

CHEMISTRY CONNECTION □ HISTORICAL

Robert Boyle

Robert Boyle (1627–1691)

► **Can you name the instrument that Robert Boyle invented and then used to demonstrate that a feather and a lump of lead are affected identically by gravity?**

Robert Boyle was a child prodigy who was speaking Latin and Greek by the age of eight. Boyle was the son of wealthy English aristocrats and traveled throughout Europe with a private tutor to gain a broad education. At age 18, his father died and left him with a lifetime income. Boyle continued his studies and at age 27 attended Oxford. There he became interested in experimentation, although at the time laboratory work was considered to be of minor importance. Most influential scientists believed that reason was far superior to experiment.

In 1657, Boyle designed a vacuum pump which was a great improvement over the first one developed a few years earlier. In one experiment, he evacuated most of the air from a sealed chamber and showed that a ticking clock could not be heard in a vacuum. He then correctly concluded that sound does not exist in the absence of air. Boyle was especially fascinated by the behavior of gases and formulated the law that is associated with his name; that is, the pressure and volume of a gas are inversely related.

In 1661, Boyle published *The Sceptical Chemist,* which argued that theories were no better than the experiments on which they were based. Gradually, this point of view was accepted and his text marked a turning point for the importance of experimentation. For numerous contributions to chemistry and physics, he is generally regarded as the founder of the modern scientific method.

Boyle felt strongly about keeping meticulous notes and reporting experimental results. He thought that everyone should be able to learn from published accounts of laboratory inquiry. He felt that experiments should be held up to scientific scrutiny and that others should have the opportunity to confirm or refute the results. This practice has become a cornerstone of science and—except for military and industrial secrets—research experiments are published and available to the scientific community, as well as the public at large.

Historians are quick to point out that Boyle was devoutly religious, studied the Bible, wrote essays on religion, and personally financed Christian missionary work. After his death, funds from his will supported the Boyle Lectures which publicly defended Christianity.

► **Boyle invented a vacuum pump which he used to remove air from a cylinder. He then released a feather and a lump of lead from the same height, and found they landed simultaneously.**

The reasoning method is actually just a slight variation of the unit analysis method of problem solving.

Step 1: Write down the unknown units.

Step 2: Find the relevant given value.

Step 3: Apply a factor to obtain the answer.

In Boyle's law problems, the factor is derived from the relationship of pressure and volume. If the volume increases, the final pressure decreases. Thus, the V_{factor}

must be a ratio that is less than 1 for the pressure to be less. If the volume decreases, the final pressure increases, and the V_{factor} is a ratio greater than 1.

Consider the following example of a Boyle's gas law problem. Suppose a 250-mL helium sample at 1.20 atm is compressed to 125 mL. To calculate the new pressure, we can write

$$1.20 \text{ atm} \times V_{factor} = P_2$$

Since the volume of the helium gas *decreases* from 250 to 125 mL, the final pressure must *increase*. Therefore, the V_{factor} must be greater than 1. That is, the larger volume appears in the numerator, and the smaller volume appears in the denominator. The equation is

$$1.20 \text{ atm} \times \frac{250 \cancel{\text{mL}}}{125 \cancel{\text{mL}}} = 2.40 \text{ atm}$$

Notice that unit cancellation takes place in the applied factor (V_{factor}). Although the numerator and denominator are not equal, this reasoning method uses the same format as unit analysis problem solving.

In another experiment, a 3.00-L sample of oxygen gas was changed from 200 to 600 torr. To calculate the new volume, V_2, we can write

$$V_1 \times P_{factor} = V_2$$

Since the pressure of the oxygen gas increases from 200 to 600 torr, the final volume must decrease. Therefore, the P_{factor} must be less than 1. That is, the smaller pressure value appears in the numerator. Thus,

$$3.00 \text{ L} \times \frac{200 \cancel{\text{torr}}}{600 \cancel{\text{torr}}} = 1.00 \text{ L}$$

In the following example exercises, you may use either the modified unit analysis method or algebra to obtain an answer.

EXAMPLE EXERCISE 12.7

A 1.50-L sample of methane gas exerted a pressure of 1650 mm Hg. Calculate the new pressure if the volume was changed to 7.00 L. Assume temperature remains constant.

Solution: Applying Boyle's law, we can find the new pressure, P_2, using the relationship

$$P_1 \times V_{factor} = P_2$$

The volume increases from 1.50 to 7.00 L. Thus, the pressure decreases. The V_{factor} must be less than 1. Hence,

$$1650 \text{ mm Hg} \times \frac{1.50 \cancel{\text{L}}}{7.00 \cancel{\text{L}}} = 354 \text{ mm Hg}$$

Alternatively, we can solve this problem using algebra. We have the equation

$$P_1V_1 = P_2V_2$$

Solving for P_2 gives

$$\frac{P_1V_1}{V_2} = P_2$$

Substituting for each variable and simplifying, we obtain

$$\frac{1650 \text{ mm Hg} \times 1.50 \cancel{\text{L}}}{7.00 \cancel{\text{L}}} = 354 \text{ mm Hg}$$

SELF-TEST EXERCISE

A hydrogen sulfide gas sample has a volume of 500.0 mL at 25°C and 225 torr. What is the volume of the gas at 25°C and 755 torr?

Answer: 149 mL

The reasoning method for solving gas law problems is valuable because it reinforces the relationship between variables. We tend to become more involved with how the changes in pressure, volume, or temperature affect a gas in a closed system. This reasoning approach allows us to apply a modified unit analysis method of problem solving that we have practiced in several previous topics. The factor, however, is different from before because the same units are found in both the numerator and denominator.

12.7 Charles' Law: Volume/Temperature Changes

OBJECTIVES

To sketch a graph of the volume/temperature relationship for a gas.

To calculate the new volume or temperature for a gas after a change in one of these conditions using either (a) modified unit analysis or (b) an algebraic method.

In 1783, the French physicist Jacques Charles (1746–1823) was one of the first to try high-altitude ballooning. In a balloon filled with hydrogen, he rose to 10,000 feet—nearly two miles! Four years later, he discovered the effect of temperature on the volume of a gas. In his experiments, he found that heating a gas 1°C increased the volume by 1/273 of the original volume. Heating a gas 10°C increased the volume by 10/273. He also found the opposite to be true. Cooling the gas decreased the volume by 1/273 of the original volume. This means that as the temperature of a gas approaches −273°C the volume of the gas approaches zero. Theoretically, the volume of a gas reaches zero at −273°C. Actually, real gases liquefy before reaching this temperature.

The mathematical relationship between volume and temperature is based on the absolute, or Kelvin, temperature scale. The relationship is known as **Charles' law**. Formally, Charles' law states that *the volume of a fixed mass of gas is directly proportional to the absolute temperature, pressure remaining constant.* **Directly proportional** means that one variable changes in relation to another. If we double the absolute temperature, we double the volume of the gas. If we halve the Kelvin temperature, we decrease volume to one-half. Plotting experimental data for the relationship stated in Charles' law gives us a graph that is a straight line (Figure 12.11).

Charles' law At constant pressure, the volume and Kelvin temperature of a gas are directly proportional.

directly proportional Refers to two related variables; if one variable doubles, the other variable doubles; if one variable triples, the other triples; and so on.

We can express the relationship of volume to Kelvin temperature as

$$V \propto T \qquad (P \text{ constant})$$

The relationship of volume and Kelvin temperature can also be written as an equation using a proportionality constant (k). We can write

$$V = kT$$

By dividing both sides of the equation by T, we see that the ratio of volume to temperature equals the constant k. The new equation is

$$\frac{V}{T} = k$$

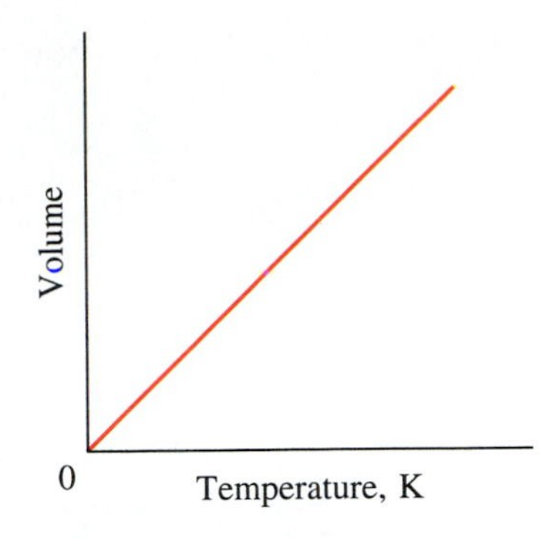

Figure 12.11 Graph of Volume versus Temperature for a Gas As the Kelvin temperature decreases, the volume of a gas decreases. Before the temperature reaches 0 K, however, all gases liquefy. The Kelvin temperature and volume of a gas are directly proportional.

Let's consider a sample of gas at different conditions. We will indicate the initial conditions of volume and temperature as V_1 and T_1, respectively. After a change in conditions, we can indicate the final volume and temperature as V_2 and T_2. Since the ratio of volume to temperature equals the constant k, we can write

$$\frac{V_1}{T_1} = k = \frac{V_2}{T_2}$$

Solving Charles' Law Problems

Since the volume and temperature of a gas are directly proportional, we can solve Charles' law problems using the modified unit analysis approach. To find the Kelvin temperature after a change in volume, for example, we apply a volume factor. Thus,

$$T_1 \times V_{\text{factor}} = T_2$$

To find a new volume after a change in temperature, we apply a temperature factor. Thus,

$$V_1 \times T_{\text{factor}} = V_2$$

In Charles' law problems, the factor is determined by the relationship of volume and Kelvin temperature. If the temperature increases, the final volume (V_2) increases. Thus, the T_{factor} must be a ratio that is greater than 1. If the temperature decreases, the final volume (V_2) decreases. Thus, the T_{factor} is a ratio that is less than 1.

To see how to solve Charles' gas law problems, let's suppose 25.0 mL of hydrogen gas are heated and the temperature rises from 200 to 650 K. To calculate the new volume, we can apply the above relationship, $V_1 \times T_{\text{factor}} = V_2$.

$$25.0 \text{ mL} \times T_{\text{factor}} = V_2$$

Since the temperature of the hydrogen gas increases from 200 to 650 K, the final volume must increase. Therefore, the T_{factor} must be greater than 1. That is, we place the larger value in the numerator and smaller value in the denominator. The equation is

$$25.0 \text{ mL} \times \frac{650 \cancel{\text{K}}}{200 \cancel{\text{K}}} = 81.3 \text{ mL}$$

In another experiment, suppose 3.50 L of argon gas were heated to 500 K. If the final volume was 12.5 L, what was the original temperature?

To calculate the original temperature, T_1, we can write

$$T_2 \times V_{\text{factor}} = T_1$$

Since the volume of the argon gas increases from 3.50 to 12.5 L, the initial temperature must be less. Therefore, the V_{factor} must be less than 1. The equation is

$$500 \text{ K} \times \frac{3.50 \cancel{\text{L}}}{12.5 \cancel{\text{L}}} = 140 \text{ K}$$

The initial temperature is 140 K. Now let's try some example exercises to reinforce our understanding of Charles' law.

EXAMPLE EXERCISE 12.8

A 275-L hydrogen balloon was cooled from 20° to −20°C. Calculate the new volume. Assume that the pressure remains constant.

Solution: We first must convert the Celsius temperature to Kelvin temperature. In Chapter 2, we learned to convert to Kelvin by adding 273 units. Thus,

$$°\text{C} + 273 = \text{K}$$

$$20°\text{C} + 273 = 293 \text{ K}$$

$$-20°\text{C} + 273 = 253 \text{ K}$$

Applying Charles' law, we can find the new volume, V_2:

$$V_1 \times T_{factor} = V_2$$

The temperature decreases from 293 to 253 K. Therefore, the volume decreases. The T_{factor} must be less than 1. Hence,

$$275 \text{ L} \times \frac{253 \cancel{\text{K}}}{293 \cancel{\text{K}}} = 237 \text{ L}$$

Alternatively we can solve this problem using algebra. We have the equation

$$\frac{V_1}{T_1} = \frac{V_2}{T_2}$$

Solving for V_2 gives us

$$\frac{V_1 T_2}{T_1} = V_2$$

Substituting for each variable and simplifying, we obtain

$$\frac{275 \text{ L} \times 253 \cancel{\text{K}}}{293 \cancel{\text{K}}} = 237 \text{ L}$$

SELF-TEST EXERCISE

A hydrogen balloon has a volume of 555 mL at 21°C. If the balloon is cooled and the volume decreases to 475 mL, what is the new temperature? Assume the pressure remains constant.

Answer: 252 K (−21°C)

In Charles' law problems, we can state the volume in any units. For example, we can give the volume in liters or cubic centimeters. Temperature, however, must *always* be expressed in Kelvin. For example, if we increase the temperature from 50 to 100 K, the volume doubles. If we increase the Celsius temperature from 50° to 100°C, the volume increases only 15%.

12.8 Gay-Lussac's Law: Pressure/Temperature Changes

OBJECTIVES

To sketch a graph of the pressure/temperature relationship for a gas.

To calculate the new pressure or temperature for a gas after a change in one of these conditions using either (a) modified unit analysis or (b) an algebraic method.

In 1802, the French scientist Joseph Gay-Lussac (1778–1850) investigated the effect of temperature on the volume and pressure of a gas. His work confirmed Charles's discovery that the volume and temperature of a gas are directly related. Furthermore, Gay-Lussac found that the pressure exerted by a gas increased or decreased proportionally with a change in temperature. For example, raising the temperature of a gas 1°C caused a pressure increase of 1/273. Lowering the temperature 10°C caused a pressure decrease of 10/273.

The mathematical relationship between pressure and temperature is based on the absolute, or Kelvin, temperature scale. This relationship is usually referred to as **Gay-Lussac's law**. Formally, Gay-Lussac's law states that *the pressure of a fixed mass of gas is directly proportional to the absolute temperature, volume remaining constant*. Recall that directly proportional means that a variable changes proportionally with another. Thus, if we triple the absolute temperature, we triple the pressure of the gas. If the Kelvin temperature is reduced to one-third, the pressure decreases to one-third. When we plot experimental data for the relationship, we obtain a graph that is a straight line; see Figure 12.12.

Gay-Lussac's Law At constant volume, the pressure exerted by a gas is directly proportional to its Kelvin temperature.

We can express the relationship of pressure to Kelvin temperature as

$$P \propto T \qquad (V \text{ constant})$$

The relationship of pressure and temperature can also be written as an equation using the proportionality constant (k). We can write

$$P = kT$$

By dividing both sides of the equation by T, we see that the ratio of pressure to temperature equals the constant k. The new equation is

$$\frac{P}{T} = k$$

Let's consider a sample of gas at different conditions. We will indicate the initial conditions of pressure and temperature as P_1 and T_1, respectively. After a change in conditions, we can indicate the final pressure and temperature as P_2 and T_2. Since the ratio of pressure to temperature equals the constant k, we can write

$$\frac{P_1}{T_1} = k = \frac{P_2}{T_2}$$

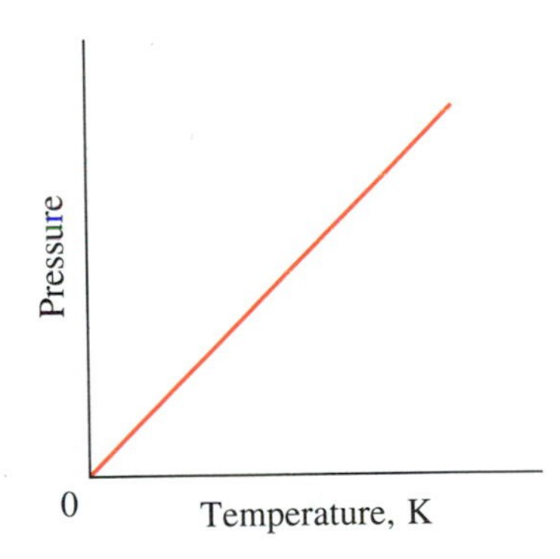

Figure 12.12 Graph of Pressure versus Temperature for a Gas As the Kelvin temperature decreases, the pressure exerted by a gas decreases. Before the temperature reaches 0 K, all gases condense to a liquid. The Kelvin temperature and pressure of a gas are directly proportional.

Pressure and temperature are directly proportional. Therefore, we can solve Gay-Lussac's law problems using the unit analysis approach. For example, to find the Kelvin temperature accompanying a pressure change, we apply a pressure factor. Thus,

$$T_1 \times P_{\text{factor}} = T_2$$

To find a new pressure after a change in temperature, we have

$$P_1 \times T_{\text{factor}} = P_2$$

In Gay-Lussac's law problems, the factor comes from the relationship of pressure and Kelvin temperature. If the temperature increases, the final pressure (P_2) increases. Thus, the T_{factor} must be a ratio that is greater than 1. If the temperature decreases, the final pressure (P_2) decreases. Here the T_{factor} is a ratio that is less than 1.

To see how to solve Gay-Lussac's law, let's suppose a sample of carbon dioxide gas at 2.50 atm is heated from 293 to 425 K. To calculate the new pressure, we use the above relationship, $P_1 \times T_{\text{factor}} = P_2$. This gives us

$$2.50 \text{ atm} \times T_{\text{factor}} = P_2$$

Since the temperature of the carbon dioxide gas increases from 293 to 425 K, the final pressure must increase. Therefore, the T_{factor} must be greater than 1. That is, we place the larger value in the numerator. The equation is

$$2.50 \text{ atm} \times \frac{425 \cancel{\text{K}}}{293 \cancel{\text{K}}} = 3.63 \text{ atm}$$

Thus, the new pressure is 3.63 atm.

In another experiment, cooling neon gas with liquid nitrogen reduced the pressure from 1200 to 320 torr. If the initial temperature is 290 K, what is the final temperature? To calculate the final temperature, T_2, we can write

$$T_1 \times P_{\text{factor}} = T_2$$

Since the pressure of the neon gas decreases from 1200 to 320 torr, the final temperature must also decrease. Therefore, the P_{factor} must be less than 1. Thus,

$$290 \text{ K} \times \frac{320 \cancel{\text{torr}}}{1200 \cancel{\text{torr}}} = 77 \text{ K}$$

The new temperature is 77 K. Now let's try some example exercises to reinforce our understanding of Gay-Lussac's law.

EXAMPLE EXERCISE 12.9

An automobile tire has a pressure of 21.0 psi at a temperature of 18°C. After traveling at high speed, the tire pressure reads 27.0 psi. What is the new tire temperature in degrees Celsius? Assume the tire volume remains constant.

Solution: We first must convert the Celsius temperature to Kelvin by adding 273 units. Thus, 18°C + 273 = 291 K. The volume is constant, so we will apply Gay-Lussac's law to find the new temperature, T_2. This gives us

$$T_1 \times P_{\text{factor}} = T_2$$

Since the pressure increases from 21.0 to 27.0 psi, the temperature must increase. Hence, the P_{factor} must be greater than 1.

$$291 \text{ K} \times \frac{27.0 \text{ psi}}{21.0 \text{ psi}} = 374 \text{ K}$$

In Celsius, the tire temperature is 374 − 273 = 101°C.

Alternatively, we can solve this problem using algebra. We have the equation

$$\frac{P_1}{T_1} = \frac{P_2}{T_2}$$

Solving for T_2 gives us

$$\frac{P_2 T_1}{P_1} = T_2$$

Substituting for each variable and simplifying gives

$$\frac{27.0 \text{ psi} \times 291 \text{ K}}{21.0 \text{ psi}} = 374 \text{ K}$$

SELF-TEST EXERCISE

A 125-mL sample of gaseous bromine vapor is heated from 250° to 450°C. If the initial pressure is 14.7 psi and the volume remains constant, what is the new pressure?

Answer: 20.3 psi

In Gay-Lussac's law problems, the pressure can be stated in any units, for example, atm, torr, or psi. The temperature, however, must always be expressed using the Kelvin temperature scale.

12.9 Combined Gas Law

To calculate a new pressure, volume, or temperature for a gas after a change in conditions using either (a) modified unit analysis or (b) an algebraic method.

We began our discussion of gas laws with a simplifying assumption. In our treatment of Boyle's, Charles's, and Gay-Lussac's laws, we assumed that we could study two variables at a time. Experimentally, all three variables—pressure, volume, and temperature—usually change. Therefore, we will now incorporate all three variables into a single expression. This expression is called the **combined gas law.** The equation is

combined gas law The pressure exerted by a gas is inversely proportional to its volume and directly proportional to its Kelvin temperature.

$$\frac{P_1 V_1}{T_1} = \frac{P_2 V_2}{T_2}$$

Since the variables P, V, and T are proportional, we can solve combined gas law problems using the unit analysis approach. To calculate a new pressure, P_2, we apply a volume factor and a temperature factor to the initial pressure, P_1. That is,

$$P_1 \times V_{\text{factor}} \times T_{\text{factor}} = P_2$$

CHEMISTRY CONNECTION □ ENVIRONMENTAL

The Greenhouse Effect

► Can you explain why planting millions of trees has been suggested as a means of combatting global warming?

When we read about the greenhouse effect, we may think that it is life-threatening to our planet. However, it is actually beneficial. In fact, it is responsible for making the earth habitable and helps to maintain the climatic temperature of our planet. The current fear is that our atmosphere has an abundance of so-called greenhouse gases, and there will be drastic global warming. Although carbon dioxide is usually singled out as the chief offender, other gases such as chlorofluorocarbons (CFCs), methane (CH_4), and the oxides of nitrogen (NO_x), also contribute to global warming.

Scientists explain the greenhouse effect as follows. The earth's atmosphere is nearly transparent to short-wavelength radiation from the sun. After striking the earth's surface, the short-wavelength radiation from the sun is emitted back into space as long-wavelength heat energy. This heat energy can be absorbed by carbon dioxide and other trace gases in the atmosphere. Subsequently, these trace gas molecules release heat energy in all directions and some of this energy is radiated back toward the earth. The net effect is that carbon dioxide and other gases in our atmosphere act as a giant canopy to trap heat energy. These gases act similarly to glass panels of an ordinary greenhouse, hence the name, greenhouse effect.

Our present concern is that trace gases in the atmosphere are increasing the temperature of our planet at an alarming rate. Global warming can result from high concentrations of carbon dioxide in our atmosphere. Carbon dioxide is produced by burning fossil fuels (coal, oil, and gas) in automobiles, homes, and power plants. The problem is further accelerated by massive deforestation, especially in South America. The loss of forests is harmful because trees remove carbon dioxide from the atmosphere.

One prediction, regarding the greenhouse effect, is that the earth's climate may be substantially different by the middle of the 21st century. Using computer climate models, meteorologists forecast that the average increase in global warming may be 2° to 5°C by the year 2050. The effect is predicted to be greater at extreme latitudes. That is, global warming will be greater at the poles than at the equator. In fact, the North and South Poles may warm as much as 10°C. As a result, a portion of the polar ice may melt, causing sea levels to rise. This, in turn, could redefine the coastlines of nations.

While some scientists do not believe global warming is as severe as the dire predictions, other scientists believe we must do something immediately. One futuristic suggestion has been to add chemicals to the oceans that will help grow plankton. The plankton will remove carbon dioxide from the air through the process of photosynthesis. Currently, our efforts are directed at planting millions of new trees and reducing the emission of greenhouse gases released into the atmosphere.

Greenhouse gases can trap heat energy from the sun much the same way as a plant greenhouse.

► The greenhouse gas that most contributes to global warming is carbon dioxide. By the process of photosynthesis, trees convert carbon dioxide to oxygen.

To calculate a new volume, V_2, we apply a pressure factor and a temperature factor to the initial volume, V_1.

$$V_1 \times P_{\text{factor}} \times T_{\text{factor}} = V_2$$

To calculate a new temperature, T_2, we apply a pressure factor and a volume factor to the initial temperature, T_1. That is,

$$T_1 \times P_{\text{factor}} \times V_{\text{factor}} = T_2$$

In combined gas law problems, we have three variables at initial and final conditions. This gives us six pieces of data and it is helpful to use a table. We can place the six pieces of data in a table that is organized as follows:

	P	***V***	***T***
initial	P_1	V_1	T_1
final	P_2	V_2	T_2

Let's apply the combined gas law by considering an experiment in which 10.0 L of carbon dioxide gas at 273 K and 1.00 atm is heated. If both the volume and Kelvin temperature double, what is the final pressure in atm? The conditions are 273 K and 1.00 atm. Doubling the temperature gives us 546 K; doubling the volume gives 20.0 L. Now let's place all the data in the table:

	P	***V***	***T***
initial	1.00 atm	10.0 L	273 K
final	P_2	20.0 L	546 K

We can calculate the final pressure, P_2, by applying a V_{factor} and T_{factor} to the initial pressure. Thus,

$$1.00 \text{ atm} \times V_{\text{factor}} \times T_{\text{factor}} = P_2$$

Since the volume of the gas increases from 10.0 to 20.0 L, P_2 must decrease. The V_{factor} is less than 1 and the smaller value is placed in the numerator. Thus,

$$1.00 \text{ atm} \times \frac{10.0 \not{L}}{20.0 \not{L}} \times T_{\text{factor}} = P_2$$

The absolute temperature of the gas increases from 273 to 546 K; therefore, P_2 increases. The T_{factor} is greater than 1, and the larger value is placed in the numerator:

$$1.00 \text{ atm} \times \frac{10.0 \not{L}}{20.0 \not{L}} \times \frac{546 \not{K}}{273 \not{K}} = 1.00 \text{ atm}$$

Notice that the pressure did not change even though the volume and temperature both doubled. The pressure remained unchanged, because pressure and volume are inversely proportional and pressure and temperature are directly proportional.

Standard Conditions

In Section 7.8, we defined **standard conditions** for a gas as 0°C and 1 atmosphere pressure (STP). Alternatively, we can express standard temperature as 273 K.

standard conditions Refers to a temperature of 0°C and a pressure of 1.00 atmosphere for a gas; also referred to as STP conditions.

We can also express standard pressure in different units. Most frequently, the value is 760 mm Hg; however, standard pressure may be expressed as 760 torr, 76.0 cm Hg, 29.9 in. Hg, 14.7 psi, or 101 kPa. The following example exercise illustrates STP conditions.

EXAMPLE EXERCISE 12.10

If a krypton gas sample at −80°C and 1250 torr occupies 50.5 mL, what is the STP volume?

Solution: The initial temperature of −80°C must be converted to Kelvins. Although the final conditions are not given, we know that STP is 273 K and 760 torr. Placing the information into the table, we have

	P	***V***	***T***
initial	1250 torr	50.5 mL	$-80 + 273 = 193$ K
final	760 torr	V_2	273 K

We can calculate the final volume by applying a P_{factor} and T_{factor} to the initial volume:

$$50.5 \text{ mL} \times P_{factor} \times T_{factor} = V_2$$

The pressure decreases, so the volume increases. The P_{factor} is greater than 1. The temperature increases, so the volume increases. The T_{factor} is also greater than 1.

$$50.5 \text{ mL} \times \frac{1250 \cancel{\text{torr}}}{760 \cancel{\text{torr}}} \times \frac{273 \cancel{\text{K}}}{193 \cancel{\text{K}}} = 117 \text{ mL}$$

Alternatively, we could solve this problem using algebra. We have the equation

$$\frac{P_1V_1}{T_1} = \frac{P_2V_2}{T_2}$$

Solving for V_2,

$$\frac{P_1V_1T_2}{T_1P_2} = V_2$$

Substituting for each variable and simplifying, we have

$$\frac{1250 \cancel{\text{torr}} \times 50.5 \text{ mL} \times 273 \cancel{\text{K}}}{193 \cancel{\text{K}} \times 760 \cancel{\text{torr}}} = 117 \text{ mL}$$

SELF-TEST EXERCISE

An oxygen sample has a volume of 70.0 cm^3 at 20°C and 765 torr. What is the resulting pressure if the sample is heated to 75°C and the volume increases to 95.5 cm^3?

Answer: 666 torr

12.10 Ideal Gas Behavior

To describe five characteristics of an ideal gas according to the kinetic theory of gases.

To determine the value of absolute zero from a graph of volume versus temperature, or pressure versus temperature.

By the early 1800s, experiments provided much information about the behavior of gases. There was, however, no clear way to interpret that information. Then, about 1850, a theory began to emerge. The British physicist James Joule (1818–1889) suggested that *heat energy is related to the molecular motion of a gas.* In other words, Joule proposed that gas temperatures could be understood in terms of the motion of molecules. At higher temperatures, gas molecules move faster. At lower temperatures, gas molecules move slower.

Kinetic Theory of Gases

For two decades, from 1850 to 1870, scientists continued to work on a model for an ideal gas. An **ideal gas** is a theoretical gas that behaves in a consistent and predictable manner. **Real gases** do not behave ideally under all conditions. A real gas, however, may approach ideal gas behavior at high temperature and low pressure. The model describing ideal gas behavior is called the **kinetic theory of gases.**

ideal gas A hypothetical gas that perfectly obeys the kinetic theory.

real gas A gas such as hydrogen or oxygen that does not strictly obey ideal behavior.

kinetic theory of gases A set of assumptions that describes gas molecules demonstrating perfectly consistent behavior.

According to the kinetic theory of gases, an ideal gas has the following characteristics:

1. *Gases are made up of very tiny molecules.* The distance between molecules is very large. Therefore, gases are mostly empty space. For an ideal gas, the molecules occupy zero volume.
2. *Gas molecules demonstrate rapid motion, move in straight lines, and travel in random directions.*
3. *Gas molecules show no attraction for one another.* After colliding with each other, molecules simply bounce off in different directions.
4. *Gas molecules have* ***elastic collisions****.* That is, gas molecules do not lose kinetic energy after colliding. If a high-energy molecule strikes a less energetic molecule, part of the energy can be transferred. The total energy of both molecules, before and after collision, does not change.
5. *The average kinetic energy of the gas molecules is proportional to the Kelvin temperature of the gas.* At the same temperature, all gases have the same average molecular kinetic energy. At higher temperatures, molecules move faster and collide more frequently. At lower temperatures, molecules move slower and collide less frequently. Recall that the kinetic energy of a molecule is proportional to its mass and square of velocity. As a consequence, smaller molecules move faster and larger molecules move slower at the same temperature.

elastic collision Refers to two gas molecules which collide without losing energy.

The average molecular kinetic energy of a given gas is equal to the average molecular kinetic energy of any other gas at the same temperature. For example, at the same temperature, the average kinetic energy of hydrogen molecules is equal to the average kinetic energy of oxygen molecules. Even though the average kinetic energy of hydrogen and oxygen is the same at any given temperature, the hydrogen molecules move faster because they are lighter.

EXAMPLE EXERCISE 12.11

Suppose we have two 5.00-L samples of gas, each at 25°C. One sample is ammonia, NH_3, and the other nitrogen dioxide, NO_2. Which has the greater kinetic energy? Which has the faster molecules?

Solution: According to the kinetic theory, kinetic energy is proportional to the temperature. Since the temperature of each gas is 25°C, the kinetic energy is the same for NH_3 and NO_2. From the kinetic theory, we know that smaller molecules move faster than larger molecules at the same temperature. The molecular mass of NH_3 is 17 amu and NO_2 is 46 amu. Since NH_3 has a lower mass than NO_2, the ammonia molecules have a higher velocity than nitrogen dioxide molecules.

SELF-TEST EXERCISE

Which of the following statements is *not* true of molecules in the gaseous state according to the kinetic theory of gases?

(a) Molecules have a negligible volume.
(b) Molecules show rapid, random motion.
(c) Molecules move in straight-line paths.
(d) Molecules show no attraction for one another.
(e) Molecules have elastic collisions.
(f) Molecules of different gases have the same average kinetic energy at the same temperature.
(g) The kinetic energy of a gas is proportional to the Celsius temperature.
(h) Smaller molecules have greater velocities than larger molecules at the same temperature.

Answer: All statements are true except (g). The kinetic energy of a gas molecule is proportional to its Kelvin temperature.

Absolute Zero

In Section 12.8, we introduced Gay-Lussac's law and the relationship of pressure and temperature. Recall that pressure and temperature are directly proportional. That is, as the temperature decreases, molecules slow down and the pressure decreases. The pressure of an ideal gas will eventually reach zero.

absolute zero The theoretical temperature at which the kinetic energy of a gas is zero.

The temperature at which the pressure reaches zero is defined as **absolute zero**. Absolute zero is the coldest possible temperature. It corresponds to a temperature of −273°C, or 0 K. An ideal gas at absolute zero has no kinetic energy.

Absolute zero is impossible to attain experimentally. How then do we determine the value for absolute zero? We do so by obtaining data for the pressure and temperature of a real gas. We then plot that pressure and temperature data and extend the graph to zero pressure. Figure 12.13 illustrates how we determine a value for absolute zero.

We can also use Charles's law to find absolute zero. From that law, we know that volume and temperature are directly proportional. Therefore, as the temperature decreases, the volume decreases. Theoretically, the volume of an ideal gas will eventually reach zero. When the volume is zero, the temperature is 0 K. Obviously, no real gases have zero volume. In fact, long before reaching 0 K, all real gases

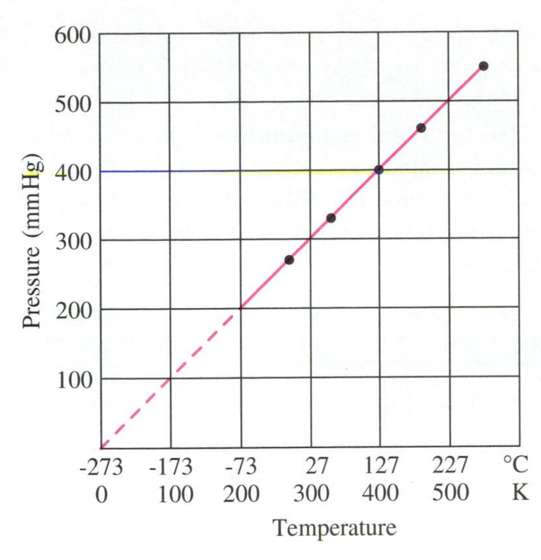

Figure 12.13 Graph of Pressure versus Temperature As the temperature of the gas decreases, the pressure decreases. When the gas has zero pressure, the temperature is absolute zero.

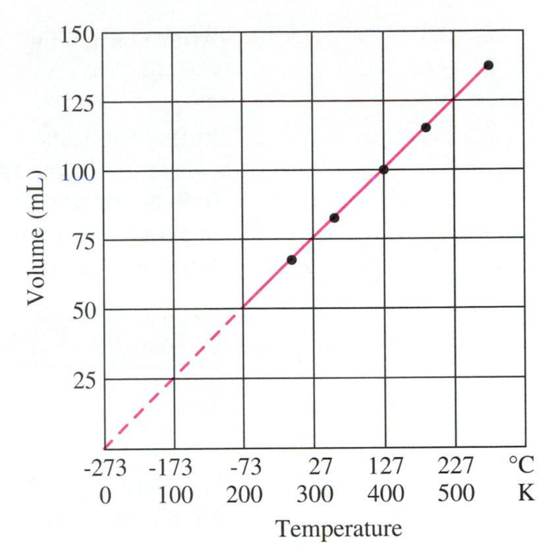

Figure 12.14 Graph of Volume versus Temperature As the temperature of the gas decreases, the volume decreases. By extending the graph to zero volume, we obtain absolute zero.

condense to a liquid. We can, however, plot a graph of volume and temperature and determine absolute zero. Figure 12.14 shows how we determine absolute zero from volume and temperature data.

The atomic theory and the kinetic theory stand as two of the greatest scientific achievements of the nineteenth century. Dalton used evidence from experiments with gases to establish the atomic theory. Later, scientists used the atomic theory to support the idea of the molecular motion of gases. Ultimately, each theory gained strength from the other. As the twentieth century began, the atomic and kinetic theories were on solid foundations and scientists turned their attention to the subatomic nature of matter.

12.11 Ideal Gas Equation

OBJECTIVES

To use the ideal gas equation to find the pressure, volume, temperature, or number of moles of gas, given any three of the four variables.

To calculate the molar mass of a gas using the ideal gas equation.

In Section 12.5, we learned that the pressure (P) of a gas is inversely proportional to the volume (V). We also learned that pressure is directly proportional to the number of moles (n) and the Kelvin temperature (T) of the gas. We can therefore write

$$P \propto \frac{nT}{V}$$

By introducing the proportionality constant R, we can write this relationship as an equation. That is,

$$P = \frac{RnT}{V}$$

After rearranging, we have

$$PV = nRT$$

ideal gas equation The relationship that $PV = nRT$ for an ideal gas.

ideal gas constant The proportionality constant R in the equation $PV = nRT$.

In this form, the relationship is called the **ideal gas equation**. The proportionality constant R is the **ideal gas constant**. Its value can be expressed as 0.0821 atm·L/mol·K. To use R, the variables must be expressed in the following units: pressure in atmospheres, volume in liters, and temperature in kelvins. The following examples illustrate the application of the ideal gas law.

EXAMPLE EXERCISE 12.12

What is the pressure in atmospheres if 2.50 mol of nitrogen gas occupies 10.2 L at 25°C?

Solution: Let's begin by rearranging the ideal gas equation, $PV = nRT$, to solve for the pressure, P. We then have

$$\frac{nRT}{V} = P$$

Next, convert the temperature to Kelvin (25°C + 273), which gives us 298 K. Now, substitute for each variable and simplify the equation. For R, we can use the value 0.0821 atm·L/mol·K. This gives

$$\frac{2.50\ \cancel{\text{mol}}}{10.2\ \cancel{\text{L}}} \times \frac{0.0821\ \text{atm}\cdot\cancel{\text{L}}}{1\ \cancel{\text{mol}}\cdot\cancel{\text{K}}} \times 298\ \cancel{\text{K}} = 6.00\ \text{atm}$$

SELF-TEST EXERCISE

What is the volume in liters of 1.10 mol of hydrogen gas at 30°C and 30.0 psi?

Answer: 13.4 L

EXAMPLE EXERCISE 12.13

Find the number of moles of hydrogen gas that occupy a volume of 500.0 mL at STP.

Solution: We begin by rearranging the ideal gas equation, $PV = nRT$, and solving for n. We have

$$\frac{PV}{RT} = n$$

Since R has the units of atm·L/mol·K, we will express STP conditions as P = 1.00 atm and T = 273 K. We must convert the given volume to units of liters. This gives us 500.0 mL = 0.5000 L. Now let's substitute for each variable and simplify the equation. We now have

$$\frac{1.00\ \cancel{\text{atm}} \times 0.5000\ \cancel{\text{L}}}{273\ \cancel{\text{K}}} \times \frac{1\ \text{mol}\cdot\cancel{\text{K}}}{0.0821\ \cancel{\text{atm}}\cdot\cancel{\text{L}}} = 0.0223\ \text{mol}$$

The ideal gas constant, R, is displayed separately to show the cancellation of units.

SELF-TEST EXERCISE

What is the temperature of 0.250 mol of chlorine gas at 655 torr if the volume is 3.50 L?

Answer: 147 K (−126°C)

Molar Mass of a Gas

In Section 7.3, we defined molar mass (MM) as the mass of 1 mole of any substance. Using the ideal gas equation, we can compute the molar mass of a gas. The moles of gas equal its mass divided by its molar mass; that is, g/MM. After substituting for moles, n, in the ideal gas equation, $PV = nRT$, we have

$$PV = \frac{\text{g}\,RT}{\text{MM}}$$

If we are given the mass of a gas and values for P, V, and T, we can calculate the molar mass from the ideal gas equation. The following example will illustrate.

EXAMPLE EXERCISE 12.14

An unknown gas having a mass of 2.041 g occupies a volume of 1.15 L at 740 torr and 20°C. Calculate the molar mass of the unknown gas.

Solution: We begin by rearranging the equation $PV = \text{g}\,RT/\text{MM}$ and we solve for MM.

$$\text{MM} = \frac{\text{g}\,RT}{PV}$$

We must convert the temperature to K and the pressure to atm. The temperature is 20°C + 273 = 293 K. The pressure is calculated from the ratio 740 torr/760 torr = 0.974 atm. Substituting for each variable and simplifying the equation, we have

$$\frac{2.041\ \text{g}}{1.15\ \cancel{\text{L}} \times 0.974\ \cancel{\text{atm}}} \times \frac{0.0821\ \cancel{\text{atm}}\cdot\cancel{\text{L}}}{1\ \text{mol}\cdot\cancel{\text{K}}} \times 293\ \cancel{\text{K}} = 43.8\ \text{g/mol}$$

SELF-TEST EXERCISE

Find the molar mass of an unknown gas, given that 0.320 g has a volume of 275 mL at 35°C and 745 mm Hg.

Answer: 30.0 g/mol

The identification of a substance can be confirmed by determining its molar mass. Let's suppose a gas sample is obtained from a recreational vehicle and its molar mass is 43.8 g/mol. If the fuel may be either propane, C_3H_8, or butane, C_4H_{10}, which gas does the RV use? Since the molar mass of C_3H_8 is 44.0 g/mol, the unknown gas is propane rather than butane.

Summary

Section 12.1 In this chapter, we studied the behavior of gases. Gases expand and can be compressed. In either case, they fill their containers uniformly. This is because the shape and volume of a gas are variable.

Section 12.2 The pressure exerted by the atmosphere is measured with a **barometer**. The **atmospheric pressure** can be expressed in any of the following units: atm, mm Hg, torr, cm Hg, in. Hg, psi, or kPa. Standard atmospheric pressure, or standard pressure, is usually given as 1.00 atm or 760 mm Hg (760 torr). The SI value for standard pressure is 101 kPa.

Section 12.3 **Vapor pressure** is the pressure exerted by vapor molecules in equilibrium with the liquid state, for example, water vapor above liquid water. Vapor pressure increases as the temperature increases. When the vapor pressure equals the atmospheric pressure, we've reached the **boiling point** and a liquid boils. The normal boiling point of a liquid is that temperature at which the vapor pressure is 760 mm Hg.

Section 12.4 **Dalton's law of partial pressures** states that the total pressure in a gaseous system is equal to the sum of the partial pressures of each gas in the mixture. A wet gas collected over water contains water vapor. Using Dalton's law, we can find the partial pressure exerted by the gas. The partial pressure is obtained by subtracting the vapor pressure of water from the total pressure.

Sections 12.4–12.8 The pressure of a gas is related to the energy and frequency of collisions of the gas molecules. We can change the pressure of a gas by changing one of three variables. If we increase the volume, molecules collide less frequently and the pressure decreases. If we increase the temperature, molecules collide more frequently and the pressure increases. If we increase the number of molecules, there are more collisions and the pressure increases. The relationship between the three variables of pressure, volume, and temperature (Table 12.3) gives rise to the basic gas laws.

Section 12.9 The three variables—pressure, volume, and temperature—can be combined into a single equation. This **combined gas law** let us calculate changes for a gas involving all three variables.

We can work out gas law problems in two ways. One way is to use the unit analysis approach and apply special volume, temperature, or pressure factors. The other way is to use an algebraic approach to solve the combined gas law equation. For example, we can use the combined gas law to find a final volume, V_2, by substituting and rearranging the following equation.

$$\frac{P_1V_1}{T_1} = \frac{P_2V_2}{T_2}$$

Section 12.10 According to the **kinetic theory of gases**, a gas is made up of individual molecules distributed in empty space. Ideal gas molecules move about randomly in straight-line paths, show no attraction for one another, and have **elastic collisions**. The kinetic energy of a gas is directly proportional to its Kelvin temperature. The velocity of a gas molecule increases with temperature and decreases with greater molecular mass.

Section 12.11 A sophisticated method of solving gas law problems involves the **ideal gas equation**. In the equation $PV = nRT$, the symbol R is the **ideal gas**

Table 12.3 Summary of Gas Law Variables

Gas Law	Pressure	Volume	Temperature
Boyle's	increases	decreases	*constant*
	decreases	increases	*constant*
Charles'	*constant*	increases	increases
	constant	decreases	decreases
Gay-Lussac's	increases	*constant*	increases
	decreases	*constant*	decreases

constant and has a value of 0.0821 atm·L/mol·K. By substituting g/MM in the equation for n, we can calculate the molar mass (MM) of a gas.

Key Terms

Select the key term that corresponds to the following definitions.

____ **1.** a physical quantity that measures the frequency and energy of molecular collisions against the container

____ **2.** the pressure exerted by the gas molecules in air; at sea level this pressure supports a 760-mm column of mercury

____ **3.** an instrument for measuring atmospheric pressure

____ **4.** the arbitrarily chosen values of 0°C (273 K) and 1 atmosphere (760 mm Hg) pressure for a gas

____ **5.** a unit of pressure equal to 1 mm Hg

____ **6.** the pressure exerted by molecules in the gaseous state in dynamic equilibrium with the same type of molecules in the liquid state; for example, water molecules above liquid water

____ **7.** the temperature at which the vapor pressure of a liquid is equal to the atmospheric pressure

____ **8.** the pressure exerted by a mixture of gases is equal to the sum of the pressures exerted by each gas in the mixture

____ **9.** the pressure exerted by a single gas in a mixture of two or more gases

____ **10.** a technique for determining the volume of a gas by measuring the amount of water it displaces; the gas contains water vapor and is referred to as a wet gas

____ **11.** a volume that does not contain any gas molecules

____ **12.** refers to two related variables; if one variable doubles, the other variable doubles; if one variable triples, the other triples; and so on

____ **13.** refers to two related variables; if one variable doubles, the other variable is reduced to one-half; if one quadruples, the other is reduced to one-fourth; and so on

____ **14.** at constant temperature, the pressure and volume of a gas are inversely proportional

____ **15.** at constant pressure, the volume and Kelvin temperature of a gas are directly proportional

____ **16.** at constant volume, the pressure exerted by a gas is directly proportional to its Kelvin temperature

____ **17.** the pressure exerted by a gas is inversely proportional to its volume and directly proportional to its Kelvin temperature

____ **18.** a set of assumptions that describes gas molecules demonstrating perfectly consistent behavior

____ **19.** a hypothetical gas that perfectly obeys the kinetic theory

____ **20.** a gas that actually exists and deviates slightly from ideal behavior

____ **21.** refers to two gas molecules that collide without losing energy

____ **22.** the theoretical temperature at which the kinetic energy of a gas is zero

____ **23.** the relationship $PV = nRT$ for an ideal gas

____ **24.** the proportionality constant R in the equation $PV = nRT$

(a) absolute zero *(Sec. 12.10)*
(b) atmospheric pressure *(Sec. 12.2)*
(c) barometer *(Sec. 12.2)*
(d) boiling point *(Sec. 12.3)*
(e) Boyle's law *(Sec. 12.6)*
(f) Charles' law *(Sec. 12.7)*
(g) combined gas law *(Sec. 12.9)*
(h) Dalton's law of partial pressures *(Sec. 12.4)*
(i) directly proportional *(Sec. 12.7)*
(j) elastic collision *(Sec. 12.10)*
(k) gas pressure *(Sec. 12.2)*
(l) Gay-Lussac's law *(Sec. 12.8)*
(m) ideal gas *(Sec. 12.10)*
(n) ideal gas constant *(Sec. 12.11)*
(o) ideal gas equation *(Sec. 12.11)*
(p) inversely proportional *(Sec. 12.6)*
(q) kinetic theory of gases *(Sec. 12.10)*
(r) partial pressure *(Sec. 12.4)*
(s) real gas *(Sec. 12.10)*
(t) standard conditions *(Sec. 12.9)*
(u) torr *(Sec. 12.2)*
(v) vacuum *(Sec. 12.2)*
(w) vapor pressure *(Sec. 12.3)*
(x) volume by displacement *(Sec. 12.4)*

Exercises

Properties of Gases (Sec. 12.1)

1. State five observed properties of gases.
2. Approximately how much less dense is air than water?

Atmospheric Pressure and the Barometer (Sec. 12.2)

3. Give the value for standard atmospheric pressure in each of the following units.
   (a) atmospheres (b) millimeters of mercury
   (c) torr (d) centimeters of mercury
4. Give the value for standard atmospheric pressure in each of the following units.
   (a) inches of mercury
   (b) pounds per square inch
   (c) kilopascals
5. If oxygen gas in a steel cylinder is at a pressure of 5.25 atm, what is the pressure expressed in each of the following units?
   (a) mm Hg (b) torr
   (c) cm Hg (d) in. Hg
6. If an automobile piston compresses a fuel–air mixture to a pressure of 7555 torr, what is the pressure expressed in each of the following units?
   (a) cm Hg (b) in. Hg
   (c) psi (d) kPa
7. A newscast program states that the barometer reads 28.8 in. of Hg. Express the pressure in each of the following units.
   (a) atm (b) mm Hg
   (c) cm Hg (d) torr
8. A Canadian newscast states that the atmospheric pressure is 99.9 kPa. Express the pressure in each of the following units.
   (a) atm (b) mm Hg
   (c) psi (d) in. Hg

Vapor Pressure (Sec. 12.3)

9. How is the vapor pressure of a liquid determined?
10. Explain how a glass of water evaporates using the concept of vapor pressure.
11. What is the general relationship between the vapor pressure of a liquid and its temperature?
12. What is the general relationship between the boiling point of a liquid and its vapor pressure?
13. The vapor pressure of acetone is 1 torr at −59°C, 10 torr at −31°C, 100 torr at 8°C, 400 torr at 40°C, and 760 torr at 56°C. What is the boiling point of acetone?
14. The vapor pressure of methanol is 1 atm at 65°C, 2 atm at 84°C, 5 atm at 112°C, 10 atm at 138°C, and 20 atm at 168°C. What is the boiling point of methanol?

Dalton's Law of Partial Pressures (Sec. 12.4)

15. Define the expression "collecting a gas over water."
16. Define the term *wet gas*.
17. Oxygen is collected over water at a temperature of 25°C and 766 torr. What is the partial pressure of the oxygen if the vapor pressure of water is 24 torr at 25°C?
18. Nitrogen is collected over water at 35°C. What is the atmospheric pressure if the partial pressure of nitrogen is 731 mm Hg? Refer to Table 12.2 for the vapor pressure of water.
19. Air is composed of 78% nitrogen, 21% oxygen, 1% argon, and traces of other gases. If the partial pressure of nitrogen is 587 torr, oxygen is 158 torr, and argon is 7 torr, what is the atmospheric pressure? (Ignore trace gases.)
20. An alloy cylinder contains nitrogen, hydrogen, and ammonia gases at 500 K and 5.00 atm. If the partial pressure of nitrogen is 1850 torr and hydrogen is 1150 torr, what is the partial pressure of ammonia in torr?

Variables Affecting Gas Pressure (Sec. 12.5)

21. State the three variables that can directly affect the pressure of a gas.
22. Explain how increasing the temperature of a gas increases its pressure.
23. Indicate what happens to the pressure of a gas for the following changes.
    (a) volume increases (b) temperature increases
    (c) moles of gas increase
24. Indicate what happens to the pressure of a gas for the following changes.
    (a) volume decreases (b) temperature decreases
    (c) moles of gas decrease
25. State whether the pressure of a gas in a closed system increases or decreases with the following changes.
    (a) volume changes from 2.50 to 5.00 L
    (b) temperature changes from 20° to 100°C
    (c) moles of gas change from 0.500 to 0.250 mol
26. State whether the pressure of a gas in a closed system increases or decreases with the following changes.
    (a) volume changes from 75.0 to 50.0 mL
    (b) temperature changes from 0° to −195°C
    (c) moles of gas change from 1.00 to 5.00 mol

Boyle's Law: Pressure/Volume Changes (Sec. 12.6)

27. Sketch the graph of pressure versus volume. Assume that temperature is constant, and label pressure on the vertical axis and volume on the horizontal.

28. Sketch the graph of pressure versus the reciprocal of volume. That is, sketch the graph of P versus $1/V$. Label the vertical axis P and the horizontal axis $1/V$.

29. A sample of air at 0.750 atm is expanded from 250.0 to 655.0 mL. If the temperature remains constant, what is the new pressure in atm?

30. What is the final volume of argon gas if 2.50 L at 705 torr are compressed to a pressure of 1550 torr? Assume that the temperature remains constant.

31. A 50.0-mL sample of carbon monoxide gas at 25°C has a pressure of 15.0 psi. If the new volume is 44.0 mL at 25°C, what is the new pressure in psi?

32. Calculate the volume of chlorine gas at 20°C and 75.0 mm Hg if the volume of the gas is 1.10 L at 20°C and 95.5 cm Hg.

Charles' Law: Volume/Temperature Changes (Sec. 12.7)

33. Sketch the graph of volume versus Kelvin temperature. Assume that the pressure is constant, and label volume on the vertical axis and temperature on the horizontal axis.

34. Sketch the graph of volume versus Celsius temperature. Label volume on the vertical axis and temperature on the horizontal axis. Assume that the pressure remains constant.

35. A 335-mL sample of oxygen at 25°C is heated to 50°C. If the pressure remains constant, what is the new milliliter volume?

36. What is the final Celsius temperature if 4.50 L of nitric oxide gas at 35°C is cooled until the volume reaches 1.00 L? Assume that the pressure remains constant.

37. A 80.0-cm^3 sample of fluorine gas at 0°C has a pressure of 761 torr. If the gas is heated to 100°C at 761 torr, what is the new cubic centimeter volume?

38. Calculate the new Celsius temperature of hydrogen chloride gas if 0.500 L at 35°C and 0.950 atm is heated until the volume reaches 1.26 L at 0.950 atm.

Gay-Lussac's Law: Pressure/Temperature Changes (Sec. 12.8)

39. Sketch the graph of pressure versus Kelvin temperature. Assume that volume is constant, and label pressure on the vertical axis and temperature on the horizontal axis.

40. Sketch the graph of pressure versus Celsius temperature. Label pressure on the vertical axis and temperature on the horizontal axis. Assume that the volume remains constant.

41. A sample of air at 760 torr is heated from 20° to 200°C. If the volume remains constant, what is the new pressure in torr?

42. What is the final Celsius temperature if a sample of ammonia gas at 10°C and 0.570 atm is cooled until the pressure reaches 0.100 atm? Assume that the volume remains constant.

43. A 1.00-L sample of neon gas at 0°C has a pressure of 76.0 cm Hg. If the gas is heated to 100°C, what is the new pressure in cm Hg if the volume is constant?

44. Calculate the new Celsius temperature of sulfur dioxide if 0.500 L of the gas at 35°C and 650 mm Hg is heated until the pressure reaches 745 mm Hg. Assume that the volume remains 0.500 L.

Combined Gas Law (Sec. 12.9)

45. A 100.0-mL sample of hydrogen gas is collected at 772 mm Hg and 21°C. Calculate the volume of hydrogen at STP.

46. A 5.00-L sample of nitrogen dioxide gas is collected at 5.00 atm and 500°C. What is the volume of nitrogen dioxide at standard conditions?

47. If a sample of air occupies 2.00 L at STP, what is the volume at 75°C and 365 torr?

48. If a sample of gas occupies 25.0 mL at −25°C and 650 mm Hg, what is the volume at 25°C and 350 mm Hg?

49. A sample of hydrogen fluoride gas has a volume of 1250 mL at STP. What is the pressure in torr if the volume is 255 mL at 300°C?

50. A sample of air occupies 0.750 L at standard conditions. What is the pressure in atm if the volume is 100.0 mL at 25°C?

51. A sample of krypton gas has a volume of 500.0 mL at 225 mm Hg and −125°C. Calculate the pressure in mm Hg if the gas occupies 220.0 mL at 100°C.

52. A sample of gas has a volume of 1.00 L at STP. What is the temperature in °C if the volume is 10.0 L at 2.00 atm?

53. A sample of air occupies 50.0 mL at standard conditions. What is the Celsius temperature if the volume is 350.0 mL at 350 torr?

54. A sample of oxygen gas occupies 500.0 mL at 75.0 cm Hg and −185°C. Calculate the Celsius temperature if the gas has a volume of 225.0 mL at 55.0 cm Hg.

Ideal Gas Behavior (Sec. 12.10)

55. State the five characteristics of an ideal gas according to the kinetic theory.

56. Distinguish between a real gas and an ideal gas.

57. What are the conditions of temperature and pressure for a real gas to behave most like an ideal gas?

58. At what Celsius temperature does an ideal gas possess zero kinetic energy?

59. A stainless steel cylinder contains three noble gases: He, Ne, and Ar. State which of the gases fits the following description.
    (a) most kinetic energy
    (b) least kinetic energy
    (c) highest velocity
    (d) lowest velocity

60. Given cylinders of hydrogen gas and oxygen gas, each at 20°C, which gas has the greater kinetic energy? Which gas has the faster molecular velocity?

61. What is the pressure exerted by an ideal gas at 0 K?

62. What is the volume occupied by an ideal gas at absolute zero?

Ideal Gas Equation (Sec. 12.11)

63. What is the pressure in atm if 0.500 mol of hydrogen gas occupies 50.0 mL at 25°C?

64. If 1.25 mol of oxygen gas at 25°C exert a pressure of 1200 torr, what is the volume of gas in liters?

65. How many moles of nitrous oxide gas occupy a volume of 10.0 L at 373 K and 125 psi?

66. What is the Celsius temperature of 0.100 mol of argon gas if the volume is 2.15 L at 725 torr?

67. The density of ozone at STP is 2.14 g/L. What is the molar mass of ozone?

68. An unknown gas occupies a volume of 3.00 L at 20°C and 760 torr. If the mass of unknown gas is 1.95 g, what is the molar mass of the unknown gas?

69. A sample of chlorine gas occupies a volume of 1550 mL at 20°C and 710 torr. What is the mass of the chlorine gas sample?

70. Freon-12 is used as a refrigerant and aerosol propellent. If the density is 5.40 g/L at 0°C and 760 torr, what is the molar mass of Freon-12?

71. Express the ideal gas constant in the following units: torr · L/mol · K.

72. Express the ideal gas constant in the following SI units: J/mol · K. (Given: 1 atm · L = 101.27 J.)

General Exercises

73. If the surface area of a human is 2500 square inches, what is the force of the atmosphere on the body? Assume standard pressure and express the weight in pounds.

74. If the atmospheric pressure is 760 torr, what is the height in feet of a barometer filled with water? (Given: Mercury is 13.6 times more dense than water.)

75. A beaker of water at 25°C is placed under a bell jar, and a vacuum pump is used to evacuate the air. Explain why the water begins to boil.

76. Refer to Figure 12.6 and determine the boiling point of water at 650 torr.

77. The normal boiling point of ethyl alcohol is 78°C. Refer to Figure 12.6 and determine the boiling point of alcohol at an elevation where the atmospheric pressure is 0.5 atm.

78. A steel cylinder contains hydrogen, chlorine, and hydrogen chloride gases. If the partial pressures of the three gases are 3.15 atm, 50.0 psi, and 2500 torr, what is the total pressure (in atmospheres) in the cylinder?

79. The decomposition of potassium chlorate produces oxygen gas. If 42.5 mL of wet oxygen gas are collected over water at 22°C and 764 mm Hg, what is the volume of dry oxygen gas at STP conditions? (The vapor pressure of water at 22°C is 19.8 mm Hg.)

80. Zinc metal reacts with hydrochloric acid to produce hydrogen gas. If the volume of hydrogen gas, collected over water, is 79.9 mL at 16°C and 758 mm Hg, what is the volume of gas at STP conditions? (The vapor pressure of water at 16°C is 13.6 mm Hg.)

81. A 5.00-L sample of krypton gas at 25°C contains 1.51×10^{24} atoms. What is the pressure of the gas in atmospheres?

82. How many molecules of carbon monoxide, CO, are present in 1.00 cm^3 of gas at STP?

83. What is the volume occupied by 3.38×10^{22} molecules of nitrogen monoxide, NO, at 100°C and 255 torr?

84. Given samples of propane gas, C_3H_8, and butane gas, C_4H_{10}, each at 100°C, which gas has the greater kinetic energy? Which gas has the faster molecular velocity?

85. Which of the following remains constant in a closed gaseous system: pressure, volume, temperature, number of molecules?

86. Ammonia gas, NH_3, makes you cry and nitrous oxide, N_2O, makes you laugh. Which would you do first if both gases are released at the same time and at the same distance?

87. Deep-sea divers breathe a special gas mixture of oxygen and helium. Explain why deep-sea divers have high squeaky voices while breathing the mixture. (*Hint:* The pitch of your voice rises as breath molecules move faster across your vocal cords.)

CHAPTER

13

Liquids, Solids, and Water

This eruption of Old Faithful in Yellowstone National Park illustrates the different physical states of H_2O. In winter, snow is observed on the ground while hot steam forces a spray of water from the geyser.

Water is the most important liquid on earth and covers about 75% of the surface of our planet (see Figure 13.1). It is necessary for all the chemical reactions that support plant and animal life. Although the percentage of water in animals varies, about 65% of human body mass is water. Blood is more than 80% water. With the exception of oxygen, water is our most critical substance. We can survive quite some time without food but only a few days without water.

The physical properties of water are unusual in several ways. For one thing, the density of a substance in the solid state is almost always greater than in the liquid state. Water, however, is one of the few exceptions. The density of solid ice is less than that of liquid water. Because the density of ice is less than water, it floats. If ice were more dense than water, marine life would not survive. Rivers and lakes would freeze into solid chunks of ice. Initially, ice would form on the surface and then sink, and eventually the entire body of water would freeze.

The melting point and boiling point of water are unusually high for a compound with such a small molar mass. Again, this is fortunate since it provides a range of temperatures in the environment that support plant and animal life. Also, the amount of heat required to melt ice and to evaporate water are unusually high. Once again this is fortunate. To a large extent, the climate of our planet is buffered by the freezing of the polar regions and the evaporation of the oceans.

Figure 13.1 The Earth's Water Resources This NASA photograph of earth shows the abundance of our water resources; however, 97% is salt water. Of the remaining 3% fresh water, most of it is found in glaciers and the polar icecaps.

13.1 The Liquid State

To describe five observed properties of a liquid.

To describe the forces of intermolecular attraction in a liquid, that is, dispersion forces, dipole forces, and hydrogen bonding.

Unlike gases, liquids do not respond dramatically to temperature and pressure changes. Also, the mathematical relationships we applied to gases, such as the combined gas law, do not apply to liquids. Rather, when we study the liquid state, we observe the following general properties.

1. *Liquids have an indefinite shape but a fixed volume.* The shape of a liquid conforms to the shape of its container.
2. *Liquids usually flow readily.* Different liquids flow at different rates. Oil flows less readily than water. Glass is actually a liquid that flows very slowly. Because of this property, panes of glass in the Sistine Chapel (built in 1473) are thicker at the bottom than at the top.
3. *Liquids do not compress or expand to any degree.* The volume of a liquid changes very little with changes in temperature or pressure. There are no mathematical relationships, similar to the combined gas law, that apply to liquids.
4. *Liquids have high densities compared to gases.* Gases and liquids are both fluids, but the liquid state is about a 1000 times more dense. For example, the density of water is 1.00 g/mL, whereas the density of air is about 0.001 g/mL.
5. *Liquids that are soluble in one another mix uniformly.* Liquids diffuse more slowly than gases. Soluble liquids, however, eventually form a homogeneous mixture. If alcohol is added to water, for example, the liquids will slowly diffuse and mix uniformly.

13.2 Properties of Liquids

To relate the properties of compounds in the liquid state to the strength of the intermolecular forces.

In Chapter 12, we introduced the kinetic theory of gases to explain the behavior of gases. Here we will extend that theory to explain the properties of liquids. According to the theory as it applies to gases, the forces of attraction between molecules are negligible. This is not the case for liquids. The molecules are in contact with each other, and the forces of attraction restrict molecular movement. The individual molecules, however, have sufficient kinetic energy to overcome the forces of attraction. Thus, molecules are more or less free to move about one another. By analogy, the distinction between the liquid state and gaseous state is likened to honey bees swarming in the hive (liquid state) as compared to individual bees gathering pollen (gaseous state).

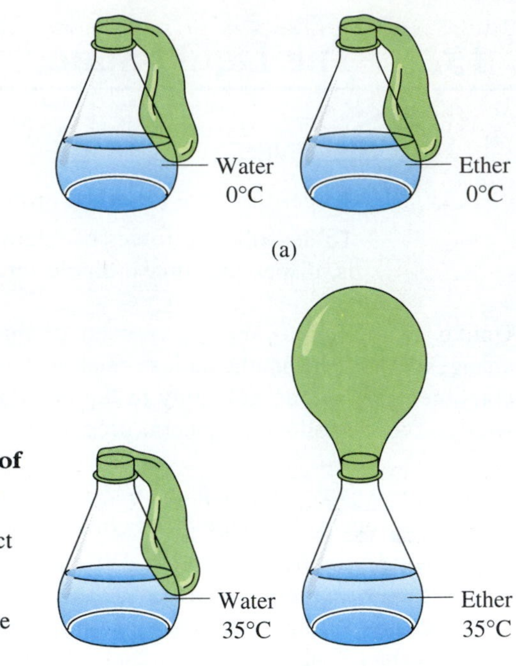

Figure 13.2 Vapor Pressure of Water and Ether (a) At 0°C, neither water nor ether has sufficient vapor pressure to affect the balloons. (b) At 35°C, the vapor pressure of water is still low; however, the vapor pressure of ether is considerable and inflates the balloon.

Vapor Pressure

The kinetic theory relates the average energy of molecules to temperature. The higher the temperature, the greater the kinetic molecular energy. However, not all molecules at the same temperature have identical energies. Some are more energetic than others. At the surface of a liquid, for example, some molecules have enough energy to completely escape the attraction of neighboring molecules. The molecules that escape are in the vapor state. The process is called *evaporation*. In the reverse process, some molecules in the vapor state return to the liquid. This process is called *condensation*. When the rates of evaporation and condensation are equal, the liquid and vapor are said to be in equilibrium.

The pressure exerted by the gas molecules above the liquid is called **vapor pressure** (Section 12.3). The attraction between molecules varies with the liquid. Water has strong attractive forces between molecules. Ethyl ether has weak forces. Because of the lesser molecular attraction, ether molecules escape the liquid state more readily than water molecules. The vapor pressure of ether is therefore greater than that of water at the same temperature (Figure 13.2).

Table 13.1 compares the molecular attraction and vapor pressure of some liquids including water and ethyl ether.

Let's consider molecules of similar mass and size. In general, as the attraction

Table 13.1 Vapor Pressure of Selected Liquids

Liquid	Molar Mass (g/mol)	Molecular Attraction	Vapor Pressure at 20°C (mm Hg)
water	18.0	strong	17.5
propionic acid, C_2H_5COOH	74.0	strong	5.0
butyl alcohol, C_4H_9OH	74.0	strong	6.3
propyl chloride, C_3H_7Cl	78.5	weak	300
ethyl ether, $C_2H_5OC_2H_5$	74.0	weak	450

Table 13.2 Viscosity of Selected Liquids

Liquid	Molar Mass (g/mol)	Molecular Attraction	Viscosity* at 20°C
water	18.0	strong	1.00
propionic acid, C_2H_5COOH	74.0	strong	1.10
butyl alcohol, C_4H_9OH	74.0	strong	2.95
propyl chloride, C_3H_7Cl	78.5	weak	0.35
ethyl ether, $C_2H_5OC_2H_5$	74.0	weak	0.23

* Viscosity is often expressed in centipoise units (g/cm × 100 s), as it is here.

between molecules increases, the vapor pressure decreases. In Table 13.1, we see that propionic acid and butyl alcohol have strong attractive forces. Their vapor pressures at 20°C are similarly low. Conversely, propyl chloride and ethyl ether have weak molecular attraction. Their vapor pressures are considerably higher. Notice that the molar mass of water is much less than that of the other liquids. Water has strong attractive forces between molecules and thus a low vapor pressure. Recall from Section 12.3 that the boiling point is the temperature at which the vapor pressure of the liquid equals the atmospheric pressure. Therefore, liquids with low vapor pressures have high boiling points.

Viscosity

Some liquids are easier to pour than others. Water pours easily, whereas honey does not. The resistance of a liquid to flow is a property called **viscosity**. Viscosity is the result of attractive forces between molecules. It is also affected by factors such as the size and shape of the molecule. In principle, the greater the molecular attraction, the higher the viscosity. Table 13.2 compares the molecular attraction and viscosity of selected liquids.

viscosity The resistance of a liquid to flow as a result of intermolecular attraction.

In Table 13.2, we see propionic acid and butyl alcohol each have strong molecular attraction. Their viscosities are greater than water. We also see that propyl chloride and ethyl ether have weak molecular attraction. Their viscosities are much lower than water. Other examples, such as honey and molasses, are much higher. We can therefore predict considerable attraction between the molecules in honey and molasses.

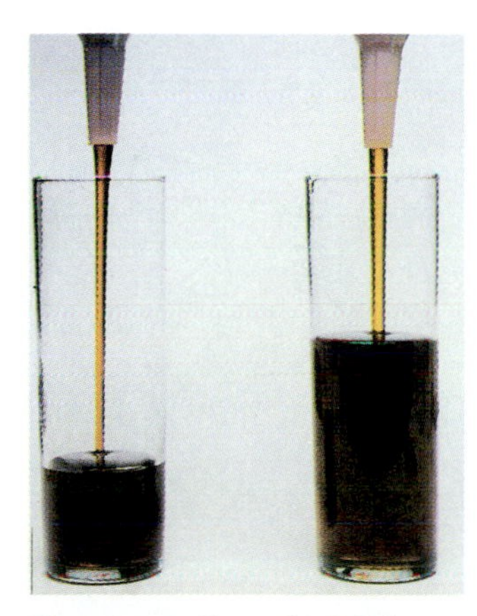

The motor oil on the left is more viscous and flows more slowly than the oil on the right.

Surface Tension

At some time you have probably noticed a small insect or some other object floating on water. For any object to sink in a liquid, it has to break through the surface. But

Water has a high surface tension, and this allows the water strider to "walk" on water.

Table 13.3 Surface Tension of Selected Liquids

Liquid	Molar Mass (g/mol)	Molecular Attraction	Surface Tension* at 20°C
water	18.0	strong	70
propionic acid, C_2H_5COOH	74.0	strong	27
butyl alcohol, C_4H_9OH	74.0	strong	25
propyl chloride, C_3H_7Cl	78.5	weak	18
ethyl ether, $C_2H_5OC_2H_5$	74.0	weak	17

* Surface tension is expressed in terms of energy per unit area. The units here are dynes ($g \times cm/s^2$) per cm^2.

surface tension The intermolecular attraction that causes a liquid to have a minimum surface area.

the molecules on the surface of liquid resist being pushed apart. The attraction between the surface molecules in a liquid is called **surface tension**. There are other factors to consider, but, in general, the greater the attraction between molecules, the higher the surface tension. Table 13.3 compares the molecular attraction and surface tension of liquids.

We all know rain forms drops as it falls. In fact, when we spray any liquid, it forms drops. Moreover, each drop assumes the shape of a small sphere. The reason drops of liquid are spherical is that surface tension causes drops to have the smallest possible surface area. The smallest surface area corresponds to a compact, spherical droplet of liquid. Table 13.3 shows that water has an unusually high surface tension. We can therefore predict that under similar conditions water forms larger droplets than other liquids. Because other liquids have a lower surface tension than water, the sizes of their drops are smaller.

EXAMPLE EXERCISE 13.1

Consider the following properties of liquids. State whether the value for the property will be high or low for a liquid with strong intermolecular forces.

(a) vapor pressure
(b) boiling point
(c) viscosity
(d) surface tension

Solution: Properties (b), (c), and (d) are generally high; (a) is low.

(a) Molecular attraction slows evaporation. Therefore, vapor pressure is *lower* for liquids with strong intermolecular forces.
(b) Attraction between molecules inhibits boiling. Thus, the boiling point is *higher* for liquids with strong intermolecular forces.
(c) Molecular attraction increases the resistance for a liquid to flow. The viscosity is *higher* for liquids with strong intermolecular forces.
(d) Attraction between molecules draws a drop of liquid into a sphere. Surface tension is *higher* for liquids with strong intermolecular forces.

SELF-TEST EXERCISE

In pentane, C_5H_{12}, the intermolecular attraction is less than in isopropyl alcohol, C_3H_7OH. Predict which of the compounds has the higher value for each of the following.

(a) vapor pressure
(b) boiling point
(c) viscosity
(d) surface tension

Answers:
(a) C_5H_{12}; **(b)** C_3H_7OH; **(c)** C_3H_7OH; **(d)** C_3H_7OH

13.3 Intermolecular Forces

To state and distinguish the three attractive forces operating between molecules in the liquid state.

The attractive forces between molecules that explain vapor pressure, viscosity, and surface tension are the result of three different types of phenomena. The three types of intermolecular forces are temporary dipole attraction, permanent dipole attraction, and hydrogen bonds. Each of these three forces involves what is called a dipole. In Section 11.7, we learned about polar covalent bonds. The combined effect of polar covalent bonds in a molecule is to concentrate positive and negative charge in different regions of the molecule. A molecule with two such regions, one positive and one negative, is called a **dipole**. These two charged regions are created by uneven distribution of electrons about the molecule.

dipole A directional shift of an electron pair in a polar covalent bond to give a molecule with regions of partial positive and negative charge.

Let's first consider temporary dipole attraction. Temporary dipole attraction is found between nonpolar molecules. Even though atoms in a nonpolar molecule share electrons equally, the electrons are constantly shifting about. This shifting about of electrons produces temporary regions in the molecule that are electron rich and hence slightly negative. Consequently, another region of the molecule becomes electron poor and hence slightly positive. These temporary charges create a weak force of attraction between molecules. This temporary attraction between molecules is referred to as **dispersion forces**, or London forces. Although the dispersion forces last for only brief periods of time, they happen frequently. At any one time, so many of these temporary dipoles exist that they represent a significant force between large nonpolar molecules. Figure 13.3 illustrates the temporary nature of London forces.

dispersion forces Intermolecular attraction based on temporary dipoles in molecules; also termed London forces.

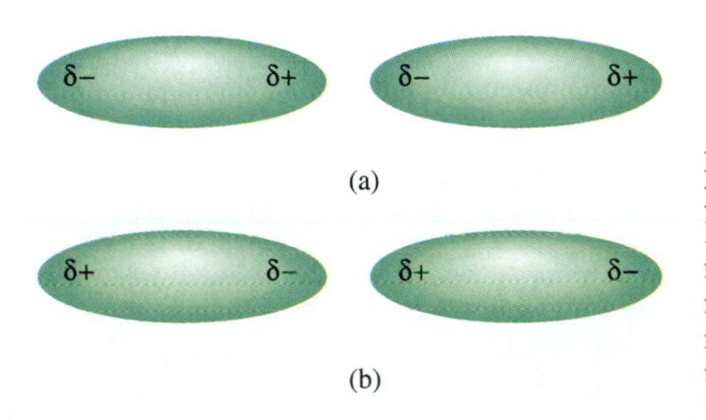

Figure 13.3 Intermolecular Dispersion Forces (a) The nonpolar molecule on the left forms a temporary dipole. This molecule can induce a dipole in the molecule on the right. (b) The two nonpolar molecules are attracted to each other after the temporary dipole reverses.

In contrast to temporary dipole attractions, permanent dipole attractions, also called dipole forces, operate continuously. Permanent dipole attractions occur between polar molecules (see Figure 13.4). Because of the electronegativity differences of two atoms, a polar molecule shares bonding electrons unequally. The result is that the molecule has a permanent negative end and a permanent positive end. Similar to miniature magnets, molecules are attracted to each other. Since dipole forces are permanent and involve greater charge concentrations, the molecular attraction is stronger than temporary London dispersion forces. This assumes that the molecules are of similar size.

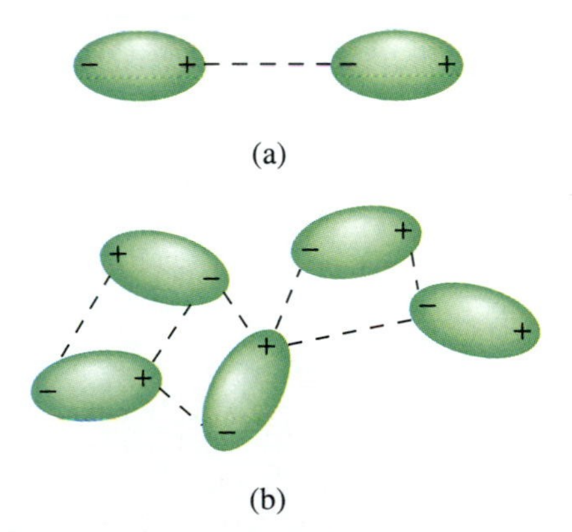

Figure 13.4 Intermolecular Dipole Forces (a) Dipole–dipole attraction between two polar molecules. (b) Dipole forces between several polar molecules in a liquid.

The dipole forces of attraction between certain molecules can be especially strong. For instance, when a hydrogen atom is bonded to a highly electronegative element such as fluorine, oxygen, or nitrogen, a very polar bond is produced. It is

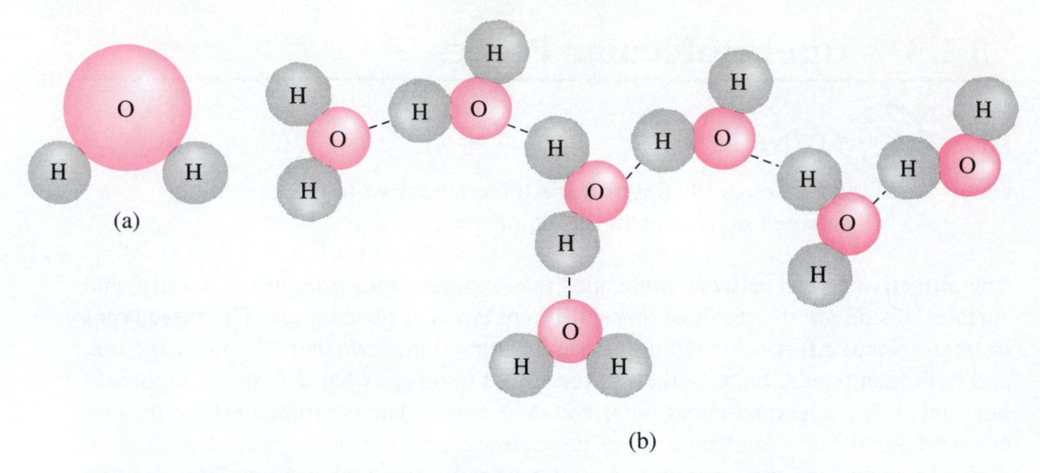

Figure 13.5 Intermolecular Hydrogen Bonds (a) A water molecule has two polar bonds. (b) Polar water molecules form intermolecular hydrogen bonds. Notice that the hydrogen atom is attracted to the nonbonded electrons on the highly electronegative oxygen atom.

hydrogen bond An intermolecular bond such as tha between hydrogen and oxygen i two molecules of water.

the large difference in electronegativity that creates a strong dipole in the molecule. In addition, the small hydrogen atom allows a close intermolecular approach. This special type of dipole attraction is called a **hydrogen bond** (see Figure 13.5). Although a hydrogen bond strongly attracts two molecules, it is typically less than 10% the strength of a normal covalent bond.

EXAMPLE EXERCISE 13.2

Indicate which of the following statements are true of liquids.

(a) Nonpolar molecules are attracted only by London dispersion forces.
(b) Polar molecules are attracted only by temporary–dipole forces.
(c) Molecules having H—F, H—O, or H—N bonds produce a strong dipole force called a hydrogen bond.

Solution: Statements (a) and (c) are true, (b) is false.

(a) Nonpolar molecules can only have temporary dipoles. Thus, intermolecular attraction is limited to London dispersion forces.
(b) Polar molecules can have both permanent and temporary dipoles. Thus, dispersion and strong dipole forces operate in the liquid.
(c) Molecules such as HF, H_2O, and NH_3 produce a hydrogen bond and have strong intermolecular forces.

SELF-TEST EXERCISE

Describe the three forces of intermolecular attraction in a liquid.

Answers:

(a) Dispersion forces between molecules result from temporary dipoles.
(b) Permanent dipole attractions between molecules result when a polar covalent bond is present.
(c) Hydrogen bonds between molecules result when strong dipoles such as O—H or N—H are present.

13.4 The Solid State

OBJECTIVES

To describe five observed properties of a solid.

Unlike liquids, solids have a fixed shape. The reason is that the individual particles of the solid are not free to move. Moreover, the volume of a solid shows very little response to changes in temperature or pressure. Here are some of the observed properties of the solid state.

1. *Solids have a definite shape and a fixed volume.* Unlike liquids, solids are rigid and their shape is fixed.
2. *Solids are either crystalline or noncrystalline.* **Crystalline solids**, or crystals, contain particles arranged in a regular repeating pattern. Each particle occupies a fixed position in the crystal. The high degree of order of the particles can produce beautiful and valuable crystals. For example, crystals, such as diamond and ruby, reflect light brilliantly and are indeed valuable. Other common examples of crystalline solids are table salt and sugar.
   Solids in which particles are arranged randomly are *noncrystalline*. Noncrystalline solids such as pearls do not have a regular geometric pattern. Rubber, plastic, and wood are noncrystalline solids. Their particles are distributed more or less randomly in the material. This type of noncrystalline solid is said to be an amorphous solid. Technically, glass is a viscous liquid, although it is considered an amorphous solid. The particles in glass are not permanently fixed in position and do not have a highly defined geometric order. Figure 13.6 shows the arrangement of particles in crystalline and amorphous solids.
3. *Solids do not compress or expand to any degree.* Assuming there is not a change of physical state, temperature and pressure have a negligible effect on the volume of a solid.
4. *Solids usually have a slightly higher density than their corresponding liquids.* For example, solid chunks of iron sink in a high-temperature furnace containing molten iron. An important exception to this rule is water; that is, ice is less dense than the liquid. As a result, ice floats on water. Similarly, solid ammonia floats on liquid ammonia.

crystalline solid A substance in the solid state that contains particles that repeat in a regular geometric pattern.

(a)

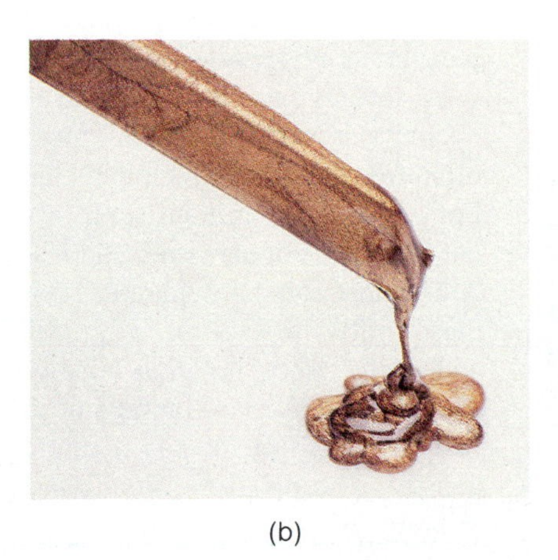

(b)

Figure 13.6 Crystalline Solid versus Noncrystalline Solid
(a) Crystalline solid in which the particles have a regular geometric pattern. (b) Amorphous solid in which the particles have no order and are distributed randomly.

5. *Solids do not mix by diffusion.* If a mixture is heterogeneous when it solidifies, the particles are not free to diffuse and form a homogeneous mixture. Alloys are homogeneous because they mix uniformly in the molten liquid state before cooling to a solid.

CHEMISTRY CONNECTION □ ENVIRONMENTAL

Recycling Aluminum

▶ **Can you guess how much energy is saved by making a can from recycled aluminum rather than manufacturing it from aluminum ore?**

Aluminum is a very abundant metal, in fact, it is the most common metal in the earth's crust. Why, then, is there so much concern about recycling aluminum? One reason is that it takes a great deal more energy to extract aluminum from the earth than to recycle it. In order to produce one ton of aluminum from its raw ore, it takes about 8 tons of coal.

% Al cans recycled

~0% ~20% ~30% ~45% ~60% ~75%

1970 1975 1980 1985 1990 1995

Notice the dramatic increase in the percentage of recycled aluminum cans.

However, to produce one ton of aluminum from recycled scrap, it requires only 0.4 ton of coal. This is a 95% savings of energy. By comparison, recycling iron is a 75% savings of energy, and recycling paper is about a 70% savings.

In a recent publication, the Environmental Protection Agency (EPA) estimated more than one million tons of aluminum cans are produced annually in the United States. Although we are using more aluminum each year, we are also recycling more. From 1970 to 1988, we increased our use of aluminum from 0.10 to 1.25 million tons for beverage cans alone. But during that same period of time, we increased the amount of aluminum recycled from essentially zero to 0.8 million tons. The bar graph (left) shows the rising percentage of recycled aluminum.

A serious problem with the recycling process is that aluminum is often found with other scrap metals. For example, aluminum scrap may contain iron, copper, zinc, and lead. Aluminum can be separated from these other metals in a process called the froth flotation method. This method takes advantage of the fact that each metal has a unique density. During froth flotation, a mixture of scrap metals is first melted and then systematically separated as the least dense metal floats to the surface. Since aluminum has a low density, it is removed before other metals, such as iron, which are more dense.

A modern aluminum recycling center.

Hopefully, we can recycle at least 75% of the aluminum used for disposable containers in the United States. This means that about 1.2 million tons of aluminum metal will be recycled, rather than being processed from an aluminum ore.

▶ **It requires about twenty times less energy to make a can from recycled aluminum. Recycling 24 aluminum cans is an energy saving equivalent to a gallon of gasoline.**

13.5 Crystalline Solids

OBJECTIVES

To describe three types of crystalline solids, that is, ionic, molecular, and metallic.

We now know that the particles of a crystalline solid are arranged in a regular geometric pattern (Figure 13.7). The particles, however, can be of different types. They can be ionic, molecular, or metallic. In ionic solids, the crystals are composed of ions in a regular pattern. In molecular solids, the molecules form a repeating pattern. In metallic solids, the atoms have a definite arrangement.

Ionic Solids

Crystalline **ionic solids** are made up of positive and negative ions. Table salt, for example, is a crystalline solid of NaCl (Figure 13.8). Here, sodium ions, Na^+, and chloride ions, Cl^-, are arranged in a regular three-dimensional structure. Other ionic compounds, such as NaF, CaF_2, and $CaCO_3$ that occur as crystals have different geometric shapes.

ionic solid A crystalline solid composed of ions that repeat in a regular pattern.

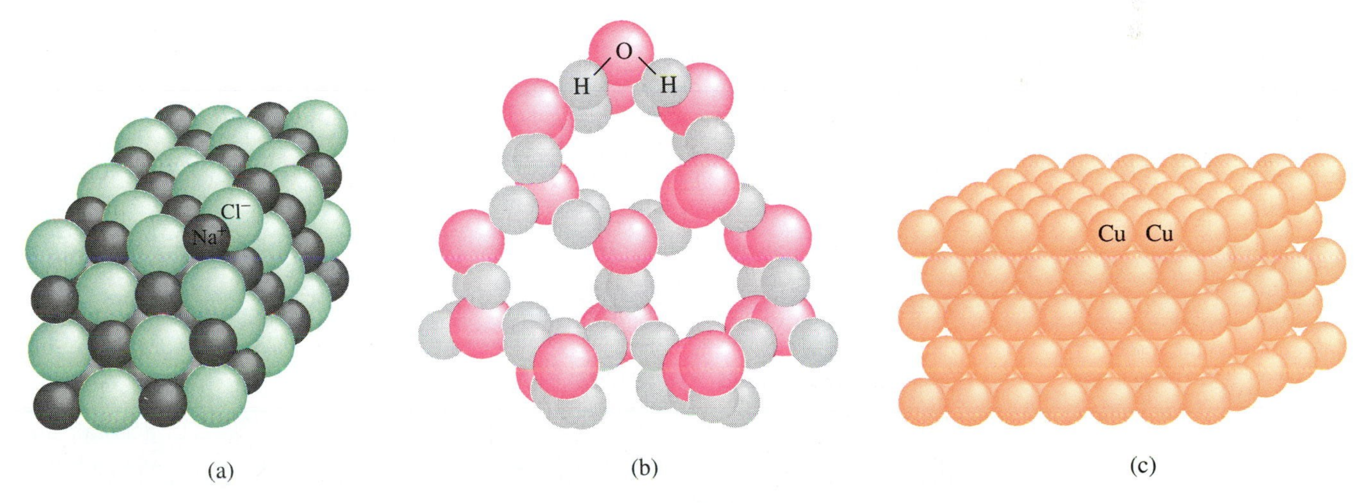

Figure 13.7 Ionic, Molecular, and Metallic Solids (a) The arrangement of **ions** in salt, NaCl. (b) The pattern of water **molecules** in ice. (c) The geometry of copper **atoms** in a metallic crystal.

Figure 13.8 A Crystalline Ionic Solid Sodium chloride crystals are composed of NaCl formula units which repeat to give a definite pattern.

Figure 13.9 A Crystalline Molecular Solid Sulfur crystals are composed of S_8 molecules arranged in a definite structure.

Molecular Solids

molecular solid A solid, often crystalline, composed of molecules that repeat in a regular pattern.

Crystalline **molecular solids** have molecules arranged in a particular configuration. Crystalline sucrose (table sugar), for example, is composed of $C_{12}H_{22}O_{11}$ molecules. The sucrose molecules are arranged in a regular order that allows light to pass through the crystal. Therefore, a large crystal of sucrose may appear transparent. Other molecular solids are sulfur and white phosphorus. Sulfur crystals are made up of S_8 molecules. Phosphorus contains P_4 molecules. Owing to different crystalline structures, phosphorus may appear as either a white or red solid. Figure 13.9 shows crystalline sulfur.

Metallic Solids

metallic solid A crystalline solid composed of metal atoms that repeat in a regular pattern.

electron sea model A theory for metallic crystals that explains the electrical conductivity of metals.

Crystalline **metallic solids** have atoms of metals arranged in a definite pattern. That is, a metallic crystal is made up of positive metal ions surrounded by valence electrons. Metals are good conductors of electricity because their valence electrons are free to move about the crystal. Specifically, metal cations are arranged in a highly ordered pattern, while the valence electrons are free to move about. This pattern is referred to as the **electron sea model** (Figure 13.10).

Classifying Crystalline Solids

It is helpful to classify crystalline solids to predict their properties. In general, the properties of ionic solids such as melting point, hardness, conductivity, and solubility

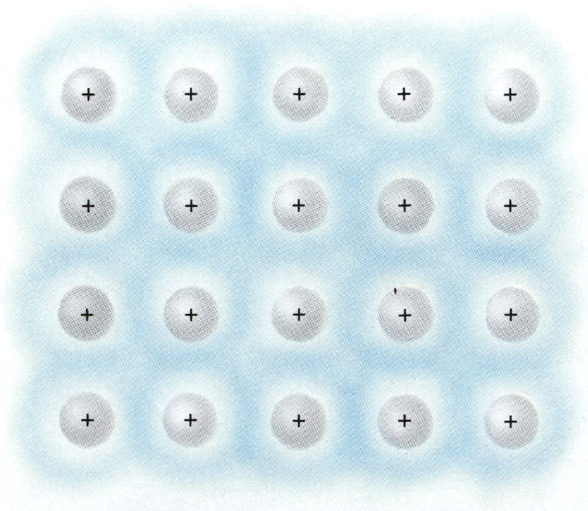

Figure 13.10 Electron Sea Model of a Metallic Crystal The electron sea model shows metal cations arranged in a fixed pattern surrounded by electrons in constant motion. The "sea" of electrons holds the crystal together.

Table 13.4 General Properties of Crystalline Solids

Type of Solid	General Properties	Examples
ionic	high mp, hard, brittle, at least partially soluble in water, conducts electricity when melted or in solution	NaCl, $CaCO_3$, $MgSO_4$
molecular	low mp, generally insoluble in water, nonconductor of electricity	S_8, $C_{10}H_8$, $C_6H_{12}O_6$
metallic	low to high mp, malleable, ductile, electrical conductor, insoluble in most solvents	Fe, Ag, Au

(a)

(b)

Samples of (a) gold and (b) silver in naturally occurring crystalline forms.

are similar. The properties of molecular solids are usually similar as well. Metals show a range of physical properties, but all are malleable, ductile, and good conductors of electricity. Table 13.4 lists the general properties for each of the three types of crystalline solids.

Dry Ice is a commercial product that sublimes at low temperatures. Dry Ice contains CO_2 molecules; it is therefore a molecular solid. The following example exercise further illustrates ionic, molecular, and metallic types of crystalline solids.

EXAMPLE EXERCISE 13.3

Classify the following crystalline solids as ionic, molecular, or metallic.

(a) nickel, Ni
(b) nickel oxide, NiO
(c) iodine I_2
(d) silver iodide, AgI

Solution: The type of crystalline solid is dictated by the type of particle.

(a) Nickel is a metal composed of atoms; thus, Ni is a *metallic* solid.
(b) Nickel oxide contains ions and is therefore an *ionic* solid.
(c) Dark violet iodine crystals contain I_2 molecules; thus, iodine is a *molecular* solid.
(d) Silver iodide crystals are used to seed rain clouds. Since AgI is composed of metal and nonmetal ions, we will classify it as an *ionic* solid.

SELF-TEST EXERCISE

Classify the following crystalline solids as ionic, molecular, or metallic.

(a) urea $CO(NH_2)_2$
(b) cobalt, Co
(c) cobalt(II) sulfide, CoS

Answers: **(a)** molecular; **(b)** metallic; **(c)** ionic

We have classified crystalline solids as ionic, molecular, or metallic. There are, however, special crystalline solids that do not fit any of these categories. For example, diamond is crystalline carbon and quartz crystalline silicon dioxide, SiO_2. In diamonds, each carbon is part of a large, continuous network. Therefore, a diamond does not exist as individual atoms. Similarly, quartz has no individual particles. Silicon and oxygen are covalently bonded continuously throughout the crystal. Diamond, quartz, and other such solids are sometimes referred to as *network solids*.

13.6 Changes of Physical State

OBJECTIVES

To calculate heat changes for a substance that involve heat of fusion, specific heat, and heat of vaporization.

Heat can be used to change the physical state of a substance. To understand heat changes, first recall that we discussed specific heat in Section 2.10. Specific heat is the amount of heat required to raise 1.00 g of a substance 1°C. Every substance has a unique value for its specific heat. Water is considered a reference and its specific heat is 1.00 calorie per gram per degree Celsius (1.00 cal/g·°C). It is interesting to note that the specific heats for ice and steam are approximately half that of water.

heat of fusion The heat required to convert a solid to a liquid at its melting point.

heat of vaporization The heat required to convert a liquid to a gas at its boiling point.

Next, consider that a substance changes state from a solid to a liquid at its melting point. The amount of heat required to melt 1.00 g of substance is called the **heat of fusion** (H_{fusion}). For water, the heat of fusion is 80.0 cal/g. Water releases the same amount of heat energy, 80.0 cal/g, when it changes from the liquid to the solid state. This heat change is called the *heat of solidification* (H_{solid}). The heats of fusion and solidification are equal for all substances.

A substance rapidly changes state from a liquid to a vapor at its boiling point. The amount of heat used to vaporize 1.00 g of a substance is called the **heat of vaporization** (H_{vapor}). For water, it is 540 cal/g. Conversely, water releases the same amount of heat energy, 540 cal/g, when it condenses from a gas to a liquid. This heat change is called the *heat of condensation* (H_{cond}). The heats of vaporization and condensation are equal for all substances. Table 13.5 lists the heat values for water, ice, and steam.

To be able to see the change in temperature with a constant application of heat, we can draw a temperature–energy graph, sometimes called a heating graph. The heating graph for water is shown in Figure 13.11.

Let's combine the concept of the temperature–energy graph with the heat values in Table 13.5. For example, let's find the amount of heat energy necessary to convert 25.0 g of ice at −5.0°C to steam at 100°C. This problem requires four steps: (1) heat the ice from −5.0° to 0.0°C; (2) convert the ice to water at 0.0°C; (3) heat the water from 0.0° to 100.0°C; and (4) convert the water to steam at 100.0°C.

1. To calculate the amount of energy to heat the ice, we must use its mass (25.0 g), the temperature change (−5.0° to 0.0°C), and the specific heat of ice (0.50 cal/g·°C); therefore, we have

$$25.0\ \cancel{g} \times [0.0 - (-5.0)]°\cancel{C} \times \frac{0.50\ \text{cal}}{1\ \cancel{g}\cdot°\cancel{C}} = \text{cal}$$

$$= 63\ \text{cal}$$

Table 13.5 Heat Values for Water

Substance	Specific heat (cal/g·°C)	H_{fusion} (cal/g)	H_{solid} (cal/g)	H_{vapor} (cal/g)	H_{cond} (cal/g)
ice, $H_2O(s)$	0.50	80.0			
water, $H_2O(l)$	1.00		80.0	540	
steam, $H_2O(g)$	0.48				540

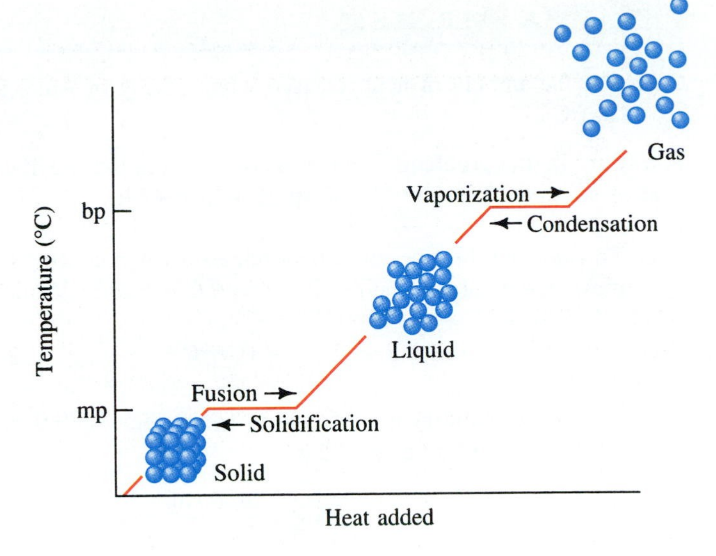

Figure 13.11 Temperature–Energy Graph As heat is continuously added to a substance, the substance eventually changes its physical state. Notice that the temperature remains constant during the change of state from solid to liquid and liquid to gas.

2. The heat of fusion for ice, 80.0 cal/g, is found in Table 13.5. The energy required to melt 25.0 g of ice is

$$25.0\ \cancel{g} \times \frac{80.0\ \text{cal}}{1\ \cancel{g}} = \text{cal}$$

$$= 2000\ \text{cal}$$

3. To calculate the amount of energy to heat the water, we must know its mass, temperature change (0.0° to 100.0°C), and the specific heat of water; therefore, we have

$$25.0\ \cancel{g} \times (100.0 - 0.0)°\cancel{C} \times \frac{1.00\ \text{cal}}{1\ \cancel{g}{\cdot}\cancel{C}} = \text{cal}$$

$$= 2500\ \text{cal}$$

4. The heat of vaporization, 540 cal/g, is found in Table 13.5. The energy required to vaporize the water to steam is

$$25.0\ \cancel{g} \times \frac{540\ \text{cal}}{1\ \cancel{g}} = \text{cal}$$

$$= 13{,}500\ \text{cal}$$

The total heat energy required to heat and vaporize the water is equal to the sum of steps 1 through 4. That is,

$$63\ \text{cal} + 2000\ \text{cal} + 2500\ \text{cal} + 13{,}500\ \text{cal} = \text{total heat}$$

$$= 18{,}100\ \text{cal}$$

The heat required to raise the temperature of the ice at −5.0°C to steam at 100.0°C is 18,100 cal, or 18.1 kcal.

The following example exercise illustrates the heat changes associated with the cooling of water and its solidification to ice.

EXAMPLE EXERCISE 13.4

Calculate the amount of heat released when 15.5 g of water at 22.5°C cools to ice at −10.0°C.

Solution: In this problem we have to consider (1) the specific heat of water, (2) the heat of solidification, and (3) the specific heat of ice.

1. To calculate the amount of heat released when cooling the water, consider the mass, temperature change (22.5° to 0.0°C), and specific heat. Thus,

$$15.5\ \not{g} \times (22.5 - 0.0)°\not{C} \times \frac{1.00\ \text{cal}}{1\ \not{g}\cdot°\not{C}} = 349\ \text{cal}$$

2. The heat of solidification, found in Table 13.5, is 80.0 cal/g. The heat released when water solidifies to ice is

$$15.5\ \not{g} \times \frac{80.0\ \text{cal}}{1\ \not{g}} = 1240\ \text{cal}$$

3. The specific heat of ice is 0.50 cal/g·°C. The heat released as the ice cools to −10.0°C is found as follows:

$$15.5\ \not{g} \times [0.0 - (-10.0)]°\not{C} \times \frac{0.50\ \text{cal}}{1\ \not{g}\cdot°\not{C}} = 78\ \text{cal}$$

The total heat energy released when the water cools to ice at −10.0°C equals the sum of steps 1 through 3. It is

$$349\ \text{cal} + 1240\ \text{cal} + 78\ \text{cal} = 1670\ \text{cal}$$

Thus, the heat released when the water cools from 22.5°C to ice at −10.0°C is 1670 cal, or 1.67 kcal.

SELF-TEST EXERCISE

Calculate the amount of heat released when 50.0 g of steam at 100.0°C cools to ice at 0.0°C.

Answer: 36,000 cal (36.0 kcal)

13.7 The Water Molecule

To illustrate the following information for a water molecule.

(a) electron dot formula
(b) structural formula
(c) observed bond angle
(d) polar covalent bonds (delta notation)
(e) net dipole (arrow notation)

We learned in Section 13.1 that the boiling point and surface tension of water are unusually high. We also saw that water has strong intermolecular forces because of hydrogen bonding. To understand these concepts more completely, let's review the covalent bonding in a water molecule. Specifically, let's consider the electron dot formula, the structural formula, the observed bond angle, polar covalent bonds and the net dipole of water.

Electron Dot Formula

We begin by writing the electron dot formula of water, H_2O. The total number of valence electrons in one molecule is $2(1\ e^-) + 6\ e^- = 8\ e^-$. Thus, there are four electron pairs. We can place the four pairs of electrons around the oxygen to provide the necessary octet. This gives

H:Ö:H

bonding electrons *nonbonding electrons*

Notice that there are two bonding and two nonboding electron pairs. Also, it seems that H—O—H is linear and the hydrogen atoms are separated by 180°. This is not the case. Actually, each of the four electron pairs is at a corner of a tetrahedron. Moreover, the nonbonding electron pairs exert a greater repelling force than the bonding electron pairs. Figure 13.12 shows the tetrahedral arrangement of electron pairs.

Structural Formula and Bond Angle

The structural formula for a molecule uses dashes to represent covalent bonds. Using this notation, the unusual properties of water are best explained if the two hydrogen atoms are at an angle to each other. Experimental evidence shows that the angle between the two hydrogen atoms is 104.5°. The angle formed by the H—O—H bonds is referred to as the **bond angle**.

H—O
104.5°
H

Polar Covalent Bonds

In a water molecule the two covalent bonds are polarized. That is, the more electronegative oxygen atom draws the bonding electrons closer to the central atom. In

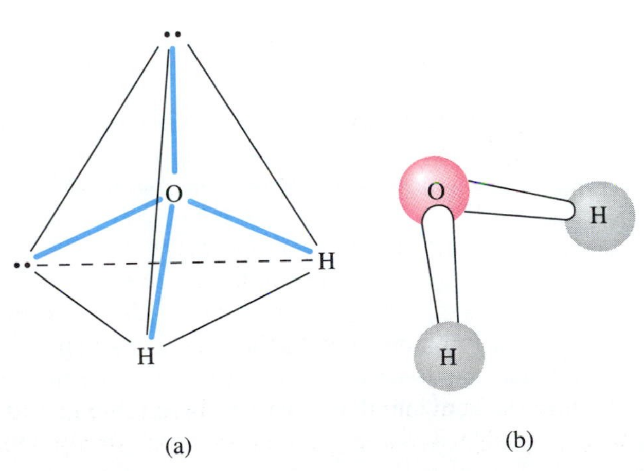

Figure 13.12 Bonding in a Water Molecule (a) The arrangement of the electron pairs in a water molecule forms the four corners of a three-dimensional tetrahedron. (b) A water molecule is described as V-shaped with the hydrogen atoms separated by 104.5°. In fact, the two hydrogen atoms are repelled slightly toward each other by the two nonbonding electron pairs.

turn, the hydrogen atoms become slightly positive. In Chapter 11, we used delta notation (δ^+ and δ^-) to indicate bond polarity. Since the bond between H and O has a partially positive and partially negative aspect, it is referred to as a dipole.

δ^+ H—O δ^-, O—H δ^+

Net Dipole

net dipole The overall effect of one or more dipoles operating in a molecule.

Notice that the water molecule has two dipoles, each pulling an electron pair toward the central atom. The resulting overall force is a dipole through the center of the molecule. This overall force is called the net molecular dipole, or simply the **net dipole**.

H—O—H

The net dipole produces a negative and a positive end in the water molecule. The negative end of the molecule is indicated by the arrow. The positive end is indicated by the plus sign on the opposite end of the net dipole.

13.8 Physical Properties of Water

OBJECTIVES

To explain the unusual properties of water.
- **(a)** density compared to ice
- **(b)** melting point compared to Group VIA/16 hydrogen compounds
- **(c)** boiling point compared to Group VIA/16 hydrogen compounds
- **(d)** heat of fusion compared to Group VIA/16 hydrogen compounds
- **(e)** heat of vaporization compared to Group VIA/16 hydrogen compounds

To compare the properties of ordinary water and heavy water.

Water is a colorless, odorless, and tasteless liquid. It is also a powerful solvent. At room temperature, water has the highest specific heat, highest heat of fusion (except for ammonia), and highest heat of vaporization of any comparable liquid substance.

Density

Generally, a substance in the solid state has a higher density than one in the liquid state. Therefore, we would predict that the density of ice is greater than that of water. But it's obvious that this is not correct. We know ice floats in water. The reason

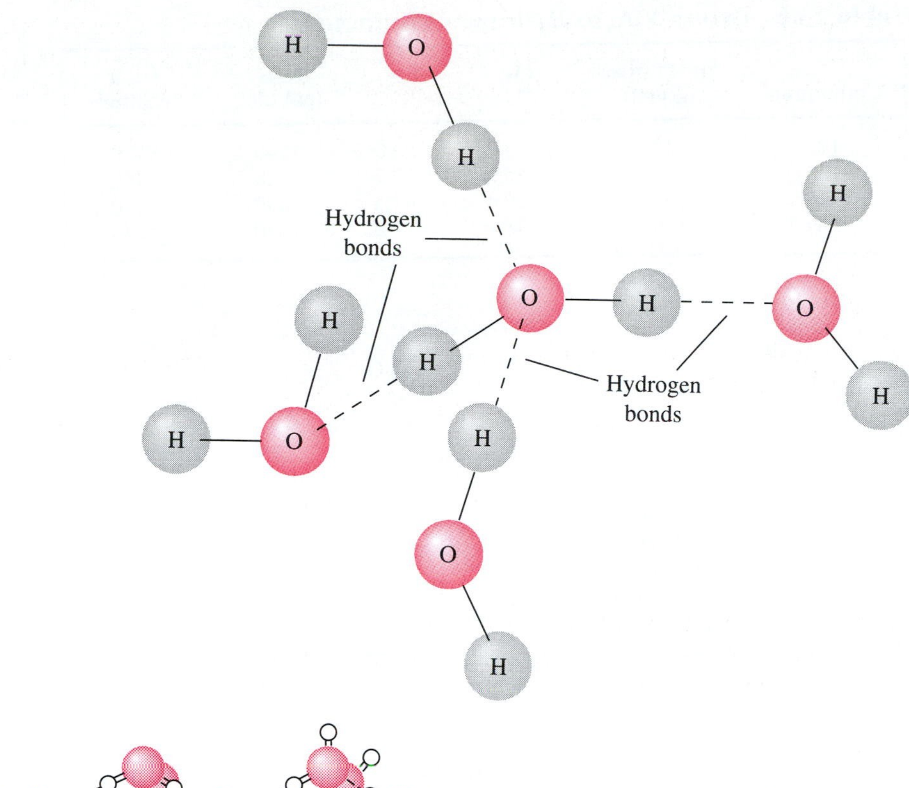

Figure 13.13 Hydrogen Bonding Each water molecule is attached to four other molecules. The intermolecular hydrogen bond is about 50% longer than the intramolecular covalent bonds. Since a hydrogen bond is longer, it is weaker and requires much less energy to break.

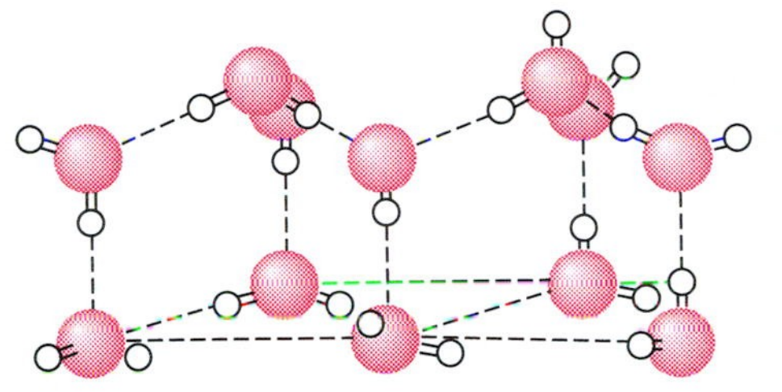

Figure 13.14 Structure of Ice Crystals Water molecules hydrogen bond forming six-member rings. The rings, in turn, hydrogen bond to other rings producing large three-dimensional crystalline structures.

ice is less dense than water relates to hydrogen bonding. Figure 13.13 illustrates three-dimensional hydrogen bonding in water.

When water freezes to solid ice, the hydrogen bonds produce a three-dimensional crystal. Figure 13.14 illustrates the structure of an ice crystal. Because of the arrangement of water molecules, the crystal has empty spaces. These holes create a volume for ice greater than that for liquid water. Furthermore, since the volume of ice is greater than that of water, its density is less than that of water. At 0°C, the density of ice is 0.917 g/mL. The density of water is 1.00 g/mL.

Melting and Boiling Points

Water has an unusually high melting point and boiling point for a substance with such small molecules. To see how unusual these properties are, compare them with some hydrogen compounds of Group VIA/16 as outlined in Table 13.6.

If we ignore water, we see a clear trend in the melting and boiling points of Group VIA/16 hydrogen compounds. First, note the increase in molar mass for H_2S through H_2Te. Next, note that the values for the melting point and boiling point increase simultaneously. As the density, the properties of water are unusual because of hydrogen bonding. Hydrogen bonding produces a strong intermolecular force that resists the movement of molecules. Therefore, a higher temperature is needed to

Table 13.6 Group VIA/16 Hydrogen Compounds

Compound	Molar Mass (g/mol)	Mp (°C)	Bp (°C)	H_{fusion} (cal/mol)	H_{vapor} (cal/mol)
H_2O	18.0	0.0	100.0	1440	9720
H_2S	34.0	−85.5	−60.7	568	4450
H_2Se	81.0	−60.4	−41.5	899	4620
H_2Te	129.6	−48.9	−2.2	1670	5570

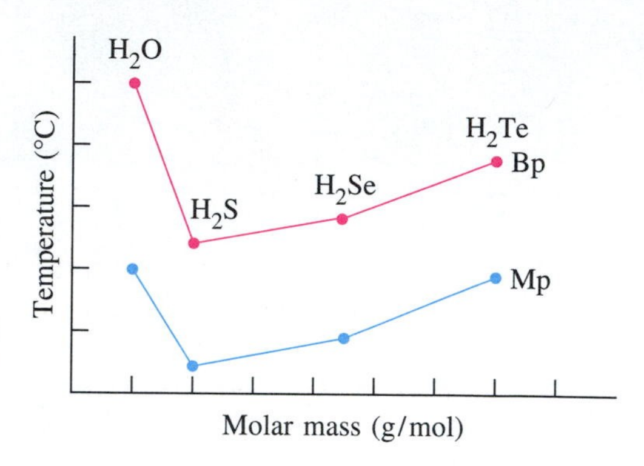

Figure 13.15 Melting and Boiling Points of Group VIA/16 Hydrogen Compounds Notice the systematic trend in melting point and boiling point as the molar mass increases. Water is a striking exception because of strong intermolecular hydrogen bonds.

melt ice and to boil water. Figure 13.15 illustrates the melting and boiling points of the Group VIA/16 hydrogen compounds.

Heats of Fusion and Vaporization

Water also has surprisingly high heats of fusion and vaporization. To see this, again look at Table 13.6. As before, ignore water. Note that as the molar mass of H_2S through H_2Te increases, the values for the heat of fusion and heat of vaporization increase. The explanation for this trend is that, as the molecular size increases, the attractive forces increase slightly. Therefore, it requires more energy to melt a solid or vaporize a liquid. The unusually high values of water are again because of hydrogen bonding.

Heavy Water

Most hydrogen atoms have one proton as their nuclei. However, about 1 in 6000 hydrogen atoms has both a neutron and a proton in its nucleus. This isotope of hydrogen is called deuterium (symbol, D). The mass of deuterium is about twice that of hydrogen. When deuterium atoms replace the hydrogen atoms in water, the

Table 13.7 Properties of Water versus Heavy Water

Property	Water, H_2O	Heavy Water, D_2O
appearance	colorless liquid	colorless liquid
molar mass	18.0 g/mol	20.0 g/mol
density	1.000 g/mL at 4°C	1.105 g/mL at 4°C
melting point	0.00°C	3.82°C
boiling point	100.00°C	101.42°C
heat of fusion	1440 cal/mol	1516 cal/mol

resulting compound is called **heavy water**. The systematic name of heavy water is deuterium oxide. The formula is D_2O. Heavy water is colorless, odorless, and tasteless, but animals find it toxic. Table 13.7 lists the properties of heavy water and ordinary water.

heavy water A molecule of water in which the hydrogen atoms are replaced by deuterium atoms, D_2O.

13.9 Chemical Properties of Water

To complete and balance equations that illustrate chemical reactions of water.

- **(a)** electrolysis of water
- **(b)** reaction of water with active metals
- **(c)** reaction of water with metal oxides
- **(d)** reaction of water with nonmetal oxides

To complete and balance equations for chemical reactions that produce water.

- **(a)** combination of hydrogen and oxygen
- **(b)** oxidation of a hydrocarbon
- **(c)** neutralization of an acid with a base
- **(d)** decomposition of a hydrate

Reactions of Water

Recall that in Chapter 8 we studied five basic types of chemical reactions. Water is often a solvent for these chemical reactions, although it does react under selected conditions. One such condition is the **electrolysis** of water. For example, passing an electric current through an aqueous solution decomposes H_2O into hydrogen and oxygen gases. From the balanced chemical equation, we note that two volumes of hydrogen are produced for every volume of oxygen. That is,

electrolysis A nonspontaneous chemical reaction produced from the passage of electricity through an aqueous solution.

$$2\ H_2O(l) \xrightarrow{\text{electricity}} 2\ H_2(g) + O_2(g)$$

One of the five basic types of reactions is a replacement reaction. In this reaction an active metal (Li, Na, K, Ca, Sr, or Ba) reacts directly with water to give a metal hydroxide and hydrogen gas. These reactions occur rapidly at room temperature. At 25°C, calcium metal reacts as follows:

$$Ca(s) + 2\ H_2O(l) \longrightarrow Ca(OH)_2(aq) + H_2(g)$$

The oxides of most metals react with water to yield a metal hydroxide. Hydroxide compounds are said to be basic or alkaline. Since a **metal oxide** reacts with water to yield a basic solution, a metal oxide is referred to as a basic oxide. For example, magnesium oxide reacts with water as follows:

metal oxide A compound that forms an alkaline solution when it dissolves in water; also termed a basic oxide.

$$MgO(s) + H_2O(l) \longrightarrow Mg(OH)_2(aq) \qquad (\textit{a base})$$

The oxides of most nonmetals react with water to yield an acidic solution. Since a **nonmetal oxide** reacts with water to yield an acid, a nonmetal oxide is referred to as an acidic oxide. For example, carbon dioxide reacts with water as follows:

nonmetal oxide A compound that forms an acidic solution when it dissolves in water; also termed an acidic oxide.

$$CO_2(g) + H_2O(l) \longrightarrow H_2CO_3(aq) \qquad (\textit{an acid})$$

Reactions That Produce Water

Water is produced by several types of reactions. The simplest reaction is the formation of water directly from hydrogen and oxygen. In this reaction, hydrogen and oxygen gases react to give H_2O. The reaction takes place very slowly at room temperature, but explosively if exposed to a flame. From the balanced chemical equation, we note that two volumes of hydrogen react with one volume of oxygen. That is,

$$2\ H_2(g) + O_2(g) \xrightarrow{\text{spark}} 2\ H_2O(l)$$

Another reaction that produces water is the oxidation of hydrocarbons. Hydrocarbons are organic compounds that contain hydrogen and carbon. They burn in oxygen to give carbon dioxide and water. For example, propane, C_3H_8, undergoes oxidation as follows:

$$C_3H_8(g) + 5\ O_2(g) \xrightarrow{\text{spark}} 3\ CO_2(g) + 4\ H_2O(g)$$

Hydrocarbon derivatives that contain oxygen also undergo combustion to give carbon dioxide and water. Ethanol, C_2H_5OH, for example, is currently blended with gasoline to give gasohol. It undergoes combustion to give carbon dioxide and water as follows:

$$C_2H_5OH(g) + 3\ O_2(g) \xrightarrow{\text{spark}} 2\ CO_2(g) + 3\ H_2O(g)$$

Recall the neutralization reactions we studied in Section 8.11. These reactions also produce water. An acid neutralizes a base to produce an aqueous salt and water. For example, battery acid, H_2SO_4, reacts with aqueous lye, NaOH, to produce sodium sulfate and water. That is,

$$H_2SO_4(aq) + 2\ NaOH(aq) \longrightarrow Na_2SO_4(aq) + 2\ H_2O(l)$$

hydrate A compound that contains a specific number of water molecules per formula unit in a crystalline compound.

Another reaction that produces water is the decomposition of a hydrate compound. A **hydrate** is a crystalline compound that contains a specific number of water molecules per formula unit. Gypsum is a hydrate of calcium sulfate. The formula $CaSO_4{\cdot}2\ H_2O$ indicates that two water molecules are attached to each formula unit. Heating a hydrate releases water from the compound. For example, heat (Δ) decomposes gypsum to give $CaSO_4$ and two molecules of water. Thus,

$$CaSO_4{\cdot}2\ H_2O(s) \xrightarrow{\Delta} CaSO_4(s) + 2\ H_2O(g)$$

13.10 Hydrates

To write the chemical formula and IUPAC systematic name for hydrates.

To calculate the percentage of water in a hydrate, given the chemical formula.

To calculate the water of hydration for a hydrate, given the anhydrous compound and the percentage water.

As we said in Section 13.9, a hydrate is a crystalline ionic compound containing water. Each formula unit in the hydrate has a specific number of water molecules

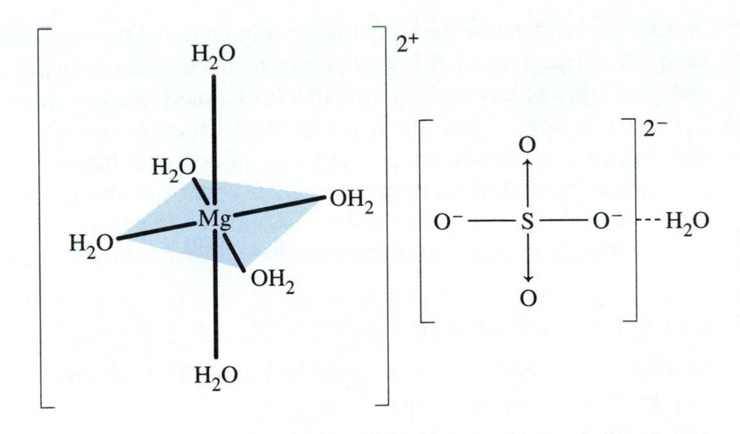

Figure 13.16 Crystal Structure of Epsom Salt Epsom salts crystals, $MgSO_4{\cdot}7\ H_2O$, are composed of six water molecules attached to the magnesium ion and a seventh water molecule attached to the sulfate ion.

Figure 13.17 Hydrate and Anhydrous Forms of Copper Sulfate Copper(II) sulfate pentahydrate, $CuSO_4{\cdot}5\ H_2O$, is a deep-blue crystalline substance; whereas, anhydrous copper(II) sulfate, $CuSO_4$, is a white powder.

attached to it. Common examples of hydrates include borax, $Na_2B_4O_7{\cdot}10\ H_2O$ and Epsom salts, $MgSO_4{\cdot}7\ H_2O$. The dot (·) in the hydrate formula indicates that water molecules are bonded directly to each unit of crystalline hydrate. In Epsom salts, $MgSO_4{\cdot}7\ H_2O$, for example, seven molecules of water are associated with each formula unit of $MgSO_4$ (see Figure 13.16).

Heating a hydrate produces an **anhydrous** compound and water. For example, when we heat a hydrate of copper(II) sulfate, it decomposes to give anhydrous copper (II) sulfate and water. The equation for the reaction is

anhydrous Refers to a compound that does not contain water.

$$CuSO_4{\cdot}5\ H_2O(s) \xrightarrow{\Delta} CuSO_4(s) + 5\ H_2O(g)$$

The water molecules in the hydrate are referred to as the **water of hydration** (or water of crystallization). Thus, the number of waters of hydration for copper(II) sulfate is five. In other words, copper(II) sulfate has five waters of crystallization. Figure 13.17 shows the hydrate and anhydrous form of copper(II) sulfate.

water of hydration The water molecules bound to a formula unit in a hydrate; also called the water of crystallization.

IUPAC prescribes rules for the nomenclature of hydrate compounds. According to the rules first name the anhydrous compound and then indicate the water of hydration by a Greek prefix and attach the word hydrate. For example, gypsum, $CaSO_4{\cdot}2\ H_2O$, is systematically named calcium sulfate dihydrate; the di indicates two waters of hydration. The following example exercise further illustrates naming hydrate compounds.

EXAMPLE EXERCISE 13.5

Supply a systematic name for each of the following hydrate compounds.

(a) $CaCl_2{\cdot}6\ H_2O$ **(b)** $FeSO_4{\cdot}7\ H_2O$

Solution: First, name the anhydrous compound and then indicate the water of hydration. (Refer to Table 6.5 if you do not recall Greek prefixes.)

(a) $CaCl_2$ is a binary ionic compound. It is named calcium chloride. The Greek prefix for 6 is hexa. Thus, the name of the hydrate is *calcium chloride hexahydrate*.

(b) $FeSO_4$ is a ternary ionic compound. Since iron has a variable ionic charge, it can be named using either the Stock system or the Latin system. Thus, $FeSO_4$ is named iron(II) sulfate or ferrous sulfate. The Greek prefix for 7 is hepta. The hydrate is named *iron(II) sulfate heptahydrate* or ferrous sulfate heptahydrate.

SELF-TEST EXERCISE

Provide the formula for each of the following hydrate compounds.

(a) zinc sulfate heptahydrate

(b) calcium nitrate tetrahydrate

Answers: **(a)** $ZnSO_4{\cdot}7\ H_2O$; **(b)** $Ca(NO_3)_2{\cdot}4\ H_2O$

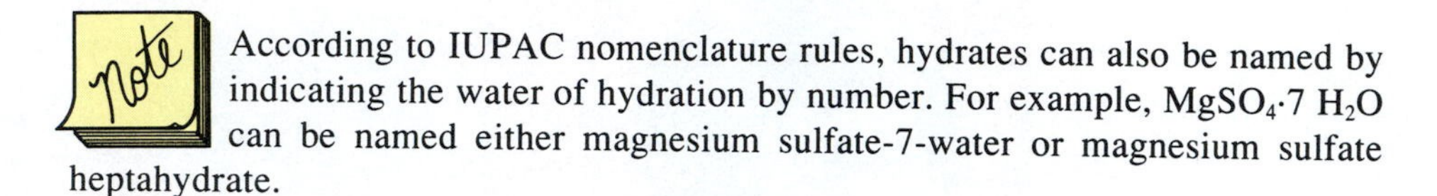

According to IUPAC nomenclature rules, hydrates can also be named by indicating the water of hydration by number. For example, $MgSO_4{\cdot}7\ H_2O$ can be named either magnesium sulfate-7-water or magnesium sulfate heptahydrate.

Percentage Composition of a Hydrate

In Section 7.5, we studied the percentage composition of compounds. Here we see that the percentage composition of a hydrate is the ratio of the mass of water compared to the mass of the hydrate times 100. The formula is

$$\frac{\text{mass of water}}{\text{mass of hydrate}} \times 100 = \%\ H_2O$$

As an example, we'll find the percentage of water in gypsum, $CaSO_4{\cdot}2\ H_2O$. Using the periodic table, we find the molar mass of H_2O is 18.0 g and of $CaSO_4$ is 136.1 g. We find the percentage of water from the following ratio:

$$\frac{2\ (18.0\ \text{g})}{136.1\ \text{g} + 2\ (18.0\ \text{g})} \times 100 = 20.9\%\ H_2O$$

Notice that in the numerator we multiplied the molar mass of water by 2 because the compound is a *di*hydrate. In the denominator, we added twice the molar mass of water to that of the anhydrous compound for the same reason. Example Exercise 13.6 provides additional illustrations of calculating the percentage composition of a hydrate.

EXAMPLE EXERCISE 13.6

Calculate the percentage of water in each of the following hydrates.

(a) $CuSO_4{\cdot}5\ H_2O$ **(b)** $Na_2B_4O_7{\cdot}10\ H_2O$

Solution: In each example, first obtain the molar mass of the anhydrous compound using the periodic table. The molar mass of water is 18.0 g.

(a) The molar mass of $CuSO_4$ is 159.6 g (63.5 g + 32.1 g + 64.0 g). Since the hydrate has 5 waters of hydration, we have

$$\frac{5\,(18.0\text{ g})}{159.6\text{ g} + 5\,(18.0\text{ g})} \times 100 = \%\ H_2O$$

$$= 36.1\%\ H_2O$$

(b) The molar mass of $Na_2B_4O_7$ is 201.2 g (46.0 g + 43.2 g + 112.0 g). The hydrate has 10 waters of crystallization. Therefore, the ratio is

$$\frac{10\,(18.0\text{ g})}{201.2\text{ g} + 10\,(18.0\text{ g})} \times 100 = \%\ H_2O$$

$$= 47.2\%\ H_2O$$

SELF-TEST EXERCISE

Calculate the percentage of water in the following hydrates.

(a) $NaC_2H_3O_2 \cdot 3\ H_2O$ **(b)** $Na_2S_2O_3 \cdot 5\ H_2O$

Answers: **(a)** 39.7%; **(b)** 36.3%

Determining the Formula of a Hydrate

In Section 7.6, we calculated the empirical formula for a compound from its percentage composition. To determine the water of hydration for a hydrate, we will proceed in a similar fashion.

The empirical formula of a compound is the simplest whole number ratio of its elements. The empirical formula of a hydrate is the simplest whole number ratio of water molecules to the anhydrous compound. In the formula for washing soda, $Na_3PO_4 \cdot X\ H_2O$, the X represents the water of hydration. To determine the value of X, we must be given the percentage composition. Let's assume $Na_3PO_4 \cdot X\ H_2O$ is found by experiment to contain 52.3% water. The equation for the decomposition reaction is as follows:

$$Na_3PO_4 \cdot X\ H_2O(s) \xrightarrow{\Delta} Na_3PO_4(s) + X\ H_2O(g)$$

Since the hydrate contains 52.3% water, the percentage of Na_3PO_4 is 47.7% (100.0 − 52.3 = 47.7%). As with an empirical formula, let's assume we have a 100.0-g sample of hydrate. Therefore, our sample has 52.3 g of water and 47.7 g of anhydrous compound.

The next step is to calculate the number of moles of water. That is,

$$52.3\text{ g }\cancel{H_2O} \times \frac{1\text{ mol } H_2O}{18.0\text{ g }\cancel{H_2O}} = 2.91\text{ mol } H_2O$$

From the periodic table, we find the molar mass of Na_3PO_4 to be 164.0 g. The moles of anhydrous compound are

$$47.7\text{ g }\cancel{Na_3PO_4} \times \frac{1\text{ mol } Na_3PO_4}{164.0\text{ g }\cancel{Na_3PO_4}} = 0.291\text{ mol } Na_3PO_4$$

We can write the mole ratio of the hydrate as $Na_3PO_4 \cdot (2.91/0.291) H_2O$, where the water of hydration is 2.91/0.291. This ratio simplifies to 10/1. Thus, the water of hydration is ten, and the formula of the hydrate is $Na_3PO_4 \cdot 10\ H_2O$.

EXAMPLE EXERCISE 13.7

Determine the water of crystallization for the hydrate of magnesium iodide. In an experiment, $MgI_2 \cdot X\ H_2O$ was found to contain 34.0% water.

Solution: Begin by writing an equation for the decomposition. That is,

$$MgI_2 \cdot X\ H_2O(s) \xrightarrow{\Delta} MgI_2(s) + X\ H_2O(g)$$

Since the hydrate contains 34.0% water, the percentage of MgI_2 is 66.0% (100.0 − 34.0 = 66.0%). Assume we have a 100.0-g sample of hydrate. Therefore, we have 34.0 g of water and 66.0 g of anhydrous compound. The moles of water are

$$34.0\ \text{g}\ \cancel{H_2O} \times \frac{1\ \text{mol}\ H_2O}{18.0\ \text{g}\ \cancel{H_2O}} = 1.89\ \text{mol}\ H_2O$$

The molar mass of MgI_2 is 278.3 g (24.3 g + 254.0 g). We find the moles of anhydrous compound as follows:

$$66.0\ \text{g}\ \cancel{MgI_2} \times \frac{1\ \text{mol}\ MgI_2}{278.3\ \text{g}\ \cancel{MgI_2}} = 0.237\ \text{mol}\ MgI_2$$

We can write the mole ratio of the hydrate as $MgI_2 \cdot (1.89/0.237)H_2O$. The ratio 1.89/0.237 reduces to 7.97 and rounds off to the whole number 8. The water of crystallization is eight and the formula is $MgI_2 \cdot 8\ H_2O$. The name of the hydrate is magnesium iodide octahydrate.

SELF-TEST EXERCISE

Determine the water of crystallization for the hydrate of copper(II) fluoride. In an experiment, $CuF_2 \cdot X\ H_2O$ was found to contain 26.2% water.

Answer: $CuF_2 \cdot 2\ H_2O$.

13.11 Water Purification

To explain the treatment of hard water to produce:

- **(a)** soft water
- **(b)** deionized water
- **(c)** distilled water

Air and water are two of our most critical resources. Unfortunately, both have suffered pollution as a result of twentieth-century technology. Automobiles have put harmful oxides of nitrogen into our atmosphere, while coal-burning plants have contributed sulfur oxides. We have polluted our lakes, rivers, and oceans with everything from toxic chemicals to nuclear waste. Since three-fourths of the earth is covered with water, we often forget to appreciate this precious substance.

Hard Water

In some regions, the minerals dissolved in water give it a high concentration of sodium, calcium, magnesium, and iron. In addition, the water has negative ions, including chloride, carbonate, sulfate, and phosphate. Such water is called **hard water**. Sometimes the mineral content in hard water is so abundant it is not healthy to drink. In some areas, hard water can cause plumbing and corrosion problems.

hard water Water containing a variety of cations and anions such as Ca^{2+}, Mg^{2+}, Fe^{3+}, CO_3^{2-}, SO_4^{2-}, and PO_4^{3-}.

Soft Water

The minerals in water can be removed to eliminate the problems caused by hard water. The resulting water is called **soft water**. As an example, consider the soap ring we sometimes see in bathtubs. The soap ring is formed by the cations in hard water (Ca^{2+}, Mg^{2+}, and Fe^{3+}) reacting with soap to create a compound that deposits as an insoluble film. The reaction is as follows:

soft water Water containing sodium ions and a variety of anions.

$$Ca^{2+}(aq) + 2\ Na(soap)(aq) \longrightarrow Ca(soap)_2(s) + 2\ Na^+(aq)$$

Magnesium ions and iron ions each give a similar reaction.

We can eliminate this problem by removing Ca^{2+}, Mg^{2+}, and Fe^{3+} from hard water. To do so, we pass hard water through a water softener. This process replaces Ca^{2+}, Mg^{2+}, and Fe^{3+} with Na^+. Compounds containing Na^+ are generally soluble.

A commercially available ion-exchange resin that softens water is Zeolite. A typical reaction of Zeolite in a water softening tank is as follows:

$$Ca^{2+}(aq) + Na_2Zeolite \longrightarrow CaZeolite + 2\ Na^+(aq)$$

Eventually, the ions in hard water saturate the Zeolite resin. To regenerate the resin, a concentrated solution of sodium chloride is passed through the water softening tank. The Ca^{2+}, Mg^{2+}, and Fe^{3+} are displaced from the resin by the sodium ions as follows:

$$CaZeolite + 2\ Na^+(aq) \longrightarrow Na_2Zeolite + Ca^{2+}(aq)$$

The hard ions that are displaced pass out of the water softening tank. The resin in the tank is once again ready to soften the hard water. Note, however, that soft water still has a high concentration of other ions. Specifically, soft water has Na^+, Cl^-, CO_3^{2-}, SO_4^{2-}, and PO_4^{3-}.

Deionized Water

In a laboratory, even the sodium ions in soft water can interfere with a chemical analysis. Therefore, chemists routinely use water that does not contain any ions. They purify water by removing the minerals using an ion-exchange system. Water purified by this method is called **deionized water**, or demineralized water.

deionized water Water purified by removing ions using an ion-exchange method; also termed demineralized water.

Deionizing water is accomplished using a special type of ion-exchange resin. The resin has both a cation and anion exchange component. First, the cations in the water are exchanged for hydrogen ions on the resin. A typical cation exchange reaction is as follows:

$$Na^+(aq) + H(resin) \longrightarrow Na(resin) + H^+(aq)$$

Second, the anions in the water are exchanged for hydroxide ions. A typical anion exchange reaction is as follows.

$$Cl^-(aq) + (resin)OH \longrightarrow (resin)Cl + OH^-(aq)$$

Notice that the ion-exchange resin produces both hydrogen ions and hydroxide ions. However, H^+ and OH^- readily combine to give deionized water. That is,

$$H^+(aq) + OH^-(aq) \longrightarrow H_2O(l)$$

The net result is that the resin removes *all ions* from the water passing through the deionizing system.

Distilled Water

Sometimes even deionized water is not pure enough. Chemists then put water through a process called *distillation*. First, hard water is heated to 100°C. As the water begins to boil, steam is produced. The steam is then passed through a cooling column, which condenses the vapor to a liquid. The water that is collected is called **distilled water**. It is pure and free of ions.

distilled water Water purified by boiling hard water and collecting the condensed vapor.

In some areas of the country, water is becoming a scarce resource. Southern California has access to the Pacific Ocean, but salt water is not suitable for drinking or agriculture. Political fights have arisen over the control of water for southern California and Arizona. This problem has been addressed in several ways, but unsuccessfully. Distilling large quantities of seawater requires an enormous amount of energy which is very expensive. Although probably not practical, some scientists have suggested hauling large icebergs from polar regions. Icebergs do not contain salt, and pure water would be provided as they melted.

Summary

Sections 13.1–13.3 We began this chapter by studying the liquid and solid states. The liquid state has a variable shape but a fixed volume. Liquids have a resistance to flow, and this property is called **viscosity**. Liquids do not compress or expand as do gases. The densities of liquids vary but are approximately 1000 times greater than the densities of gases. Liquids mix and diffuse uniformly.

Surface tension is a measure of the attractive force between molecules, or intermolecular attraction, at the surface of a liquid. There are three basic types of intermolecular attraction. They are **dispersion forces**, **dipole attractions**, and **hydrogen bonds**. Dispersion forces have the weakest attraction and result from a temporary dipole attraction. Polar molecules with permanent dipoles are attracted more strongly. Molecules having polar —OH or —NH groups are attracted to each other by a very strong dipole force that is called a hydrogen bond.

Section 13.4 The solid state has a definite shape and volume. Solids that have highly regular structures are called **crystalline solids**. If the arrangement of particles is not regular, the solid is said to be a noncrystalline solid. Solids do not compress or expand to any large degree. The density of a substance in the solid state is usually higher than that in the liquid state. Water and ammonia are two interesting exceptions to the general rule. Solid ice and solid ammonia float on their respective liquids. Particles in a solid do not diffuse.

Section 13.5 There are three basic types of crystalline solids. **Ionic solids** are made up of ions. The ions are attracted to each other and form repeating geometric patterns. **Molecular solids** form crystals made up of molecules. **Metallic solids** are made up of metal atoms arranged in a definite pattern. The properties of ionic, molecular, and metallic solids differ. Ionic solids conduct electricity only when they are melted or in an aqueous solution. As a rule, molecular solids do not conduct electricity. Metallic solids are good conductors of electricity.

Section 13.6 The heat required to melt a substance is called the **heat of fusion**. Heat released when a substance freezes to a solid is called the *heat of solidification*. The heat of solidification of a substance is equal to its heat of fusion. The heat required to convert a liquid to a gas at its boiling point is the **heat of vaporization**. When a vapor condenses to a liquid, it releases the same amount of heat energy that was necessary to vaporize the liquid and is termed the *heat of condensation*.

Section 13.7 Water has unusual properties because of the **net dipole** in the water molecule. The net dipole creates hydrogen bonds that give water a high value for viscosity and surface tension. Hydrogen bonding also explains why the density of ice is less than water. For a small molecule, water has an unusually high melting point, boiling point, heat of fusion, and heat of vaporization.

Section 13.8 Under certain conditions, water undergoes a chemical reaction. (1) An electric current decomposes water into hydrogen and oxygen gases. (2) The active metals of Groups IA/1 and IIA/2 react with water to give a metal hydroxide and hydrogen gas. (3) A metal oxide reacts with water to give a metal hydroxide. (4) A nonmetal oxide combines with water to yield an acidic solution (Table 13.8).

Section 13.9 Water is produced from (1) the reaction of hydrogen and oxygen gases, (2) burning hydrocarbons, (3) neutralizing an acid with a base, and (4) heating a hydrate (Table 13.9).

Section 13.10 A **hydrate** is named by stating the anhydrous compound followed by a Greek prefix and the word hydrate. For example, $BaCl_2 \cdot 2\ H_2O$ is named barium chloride dihydrate. The percentage of water in a hydrate is calculated by dividing the mass of water by the mass of the entire hydrate. The formula of a hydrate is determined in a similar fashion to finding an empirical formula. From the percentage composition of a hydrate, calculate the moles of water and anhydrous compound. The ratio of water to anhydrous compound gives the **water of hydration**.

Section 13.11 Water can be purified by a number of methods. **Soft water** is obtained by passing **hard water** through a water softener containing an ion-exchange resin. **Deionized water** is obtained by passing hard water through an ion-exchange system that removes both cations and anions. **Distilled water** is collected from a process called distillation. During distillation hard water is boiled. The steam produced is then cooled and collected.

Table 13.8 Summary of the Reactions of Water

Reactants		Products
water	$\xrightarrow{\text{electricity}}$	hydrogen + oxygen
active metal + water	$\longrightarrow$	metal hydroxide + hydrogen
metal oxide + water	$\longrightarrow$	basic solution
nonmetal oxide + water	$\longrightarrow$	acidic solution

Table 13.9 Summary of Reactions Producing Water

Reactants		Products
hydrogen + oxygen	$\xrightarrow{\text{spark}}$	water
hydrocarbon + oxygen	$\xrightarrow{\text{spark}}$	carbon dioxide + water
acid + base	$\longrightarrow$	salt + water
hydrate	$\xrightarrow{\Delta}$	anhydrous compound + water

Key Terms

Select the key term that corresponds to the following definitions.

____ 1. the pressure exerted by molecules in the gaseous state in equilibrium with the same molecules in the liquid state

____ 2. the resistance of a liquid to flow as a result of intermolecular attraction

____ 3. the attraction between molecules of a liquid that cause minimum-size drops

____ 4. a directional shift of an electron pair in a polar covalent bond to give a molecule with regions of positive charge and negative charge

____ 5. an intermolecular attraction based on temporary dipoles in molecules; also termed London forces

____ 6. an intermolecular bond such as that between hydrogen and oxygen in two molecules having a polar —OH group

____ 7. a substance in the solid state that contains particles that repeat in a regular geometric pattern

____ 8. a crystalline solid composed of ions that repeat in a regular pattern

____ 9. a crystalline solid composed of molecules that repeat in a regular pattern

____ 10. a crystalline solid composed of metal atoms that repeat in a regular pattern

____ 11. a theory for metallic crystals that explains the electrical conductivity of metals

____ 12. the heat required to convert a solid to a liquid at its meling point

____ 13. the heat required to convert a liquid to a gas at its boiling point

____ 14. the angle formed by two atoms bonded to a central atom in a molecule

____ 15. the overall effect resulting from one or more dipoles in a molecule

____ 16. a molecule of water in which the hydrogen atoms are replaced by deuterium atoms, D_2O

____ 17. a chemical reaction produced from the passage of electric current through an aqueous solution

____ 18. a compound that reacts with water to form a basic solution

____ 19. a compound that reacts with water to form an acidic solution

____ 20. a compound that contains a specific number of water molecules per formula unit in a crystalline compound

____ 21. the water molecules bound in a hydrate

____ 22. refers to a compound that does not contain water

____ 23. water containing a variety of cations and anions, such as Ca^{2+}, Mg^{2+}, Fe^{3+}, CO_3^{2-}, SO_4^{2-}, and PO_4^{3-}

____ 24. water containing sodium ions and a variety of anions

____ 25. water purified by removing ions using an ion-exchange method; also termed demineralized water

____ 26. water purified by boiling hard water and collecting the condensed vapor

(a) anhydrous *(Sec. 13.10)*
(b) bond angle *(Sec. 13.7)*
(c) crystalline solid *(Sec. 13.4)*
(d) deionized water *(Sec. 13.11)*
(e) dipole *(Sec. 13.3)*
(f) dispersion forces *(Sec. 13.3)*
(g) distilled water *(Sec. 13.11)*
(h) electrolysis *(Sec. 13.9)*
(i) electron sea model *(Sec. 13.5)*
(j) hard water *(Sec. 13.11)*
(k) heat of fusion *(Sec. 13.6)*
(l) heat of vaporization *(Sec. 13.6)*
(m) heavy water *(Sec. 13.8)*
(n) hydrate *(Sec. 13.9)*
(o) hydrogen bond *(Sec. 13.3)*
(p) ionic solid *(Sec. 13.5)*
(q) metal oxide *(Sec. 13.9)*
(r) metallic solid *(Sec. 13.5)*
(s) molecular solid *(Sec. 13.5)*
(t) net dipole *(Sec. 13.7)*
(u) nonmetal oxide *(Sec. 13.9)*
(v) soft water *(Sec. 13.11)*
(w) surface tension *(Sec. 13.2)*
(x) vapor pressure *(Sec. 13.2)*
(y) viscosity *(Sec. 13.2)*
(z) water of hydration *(Sec. 13.10)*

Exercises

The Liquid State (Sec. 13.1)

1. List five general properties of the liquid state.
2. Distinguish between a liquid and a gas at the molecular level.
3. Indicate the physical state (solid, liquid, gas) for each of the following at the designated temperature.
   (a) H_2O at −20.0°C (b) H_2O at 120.0°C
   (c) NH_3 at −195.0°C (d) NH_3 at 0.0°C
   (e) $CHCl_3$ at −55.5°C (f) $CHCl_3$ at 100.0°C
   The melting points and boiling points for water, ammonia, and chloroform are:

Substance	Melting Point (°C)	Boiling Point (°C)
water, H_2O	0.0	100.0
ammonia, NH_3	−77.7	−33.4
chloroform, $CHCl_3$	−63.5	61.7

4. Indicate the physical state (solid, liquid, gas) for each of the following noble gases at the designated temperature.
   (a) Ne at −248°C (b) Ne at −225°C
   (c) Ar at −187°C (d) Ar at −212°C
   (e) Kr at −100°C (f) Kr at −195°C
   The melting points and boiling points for neon, argon, and krypton are:

Element	Melting Point (°C)	Boiling Point (°C)
neon, Ne	−248.7	−245.9
argon, Ar	−189.2	−185.7
krypton, Kr	−156.6	−152.3

Properties of Liquids (Sec. 13.2)

5. Using water as an example, define and illustrate the concept of vapor pressure.
6. Using water as an example, define and illustrate the concept of boiling point.
7. Using water as an example, define and illustrate the concept of viscosity.
8. Using water as an example, define and illustrate the concept of surface tension.
9. If the molecules in a liquid have strong intermolecular forces, which of the following properties, in general, have a high value?
   (a) vapor pressure (b) boiling point
   (c) viscosity (d) surface tension
10. If the molecules in a liquid have weak intermolecular forces, which of the following properties, in general, have a high value?
   (a) vapor pressure (b) boiling point
   (c) viscosity (d) surface tension

Intermolecular Forces (Sec. 13.3)

11. Indicate which of the three intermolecular forces (dispersion, dipole, hydrogen bond) is the strongest for a liquid containing the following molecules.
   (a) $CH_3CH_2CH_2CH_2CH_2CH_3$
   (b) CH_3OH
   (c) CH_3Cl
   (d) CH_3OCH_3
12. Compare the following intermolecular forces in terms of the strength of attraction between molecules.
   (a) dispersion versus dipole
   (b) dipole versus hydrogen bond
13. Predict which liquid in each pair has the highest vapor pressure.
   (a) CH_3COOH or C_2H_5Cl
   (b) C_2H_5OH or CH_3OCH_3
14. Predict which liquid in each pair has the higher boiling point.
   (a) CH_3COOH or C_2H_5Cl
   (b) C_2H_5OH or CH_3OCH_3
15. Predict which liquid in each pair has the higher viscosity.
   (a) CH_3COOH or C_2H_5Cl
   (b) C_2H_5OH or CH_3OCH_3
16. Predict which liquid in each pair has the highest surface tension.
   (a) CH_3COOH or C_2H_5Cl
   (b) C_2H_5OH or CH_3OCH_3

The Solid State (Sec. 13.4)

17. List five general properties of the solid state.
18. Distinguish between a solid and a liquid at the molecular level.
19. Indicate the physical state (solid, liquid, gas) for each of the following metals after being placed in ice water or boiling water.
   (a) Ga in ice water (b) Ga in boiling water
   (c) Sn in ice water (d) Sn in boiling water
   (e) Hg in ice water (f) Hg in boiling water
   The melting points and boiling points for the three metals are:

Element	Melting Point (°C)	Boiling Point (°C)
gallium, Ga	29.8	2403
tin, Sn	232.0	2270
mercury, Hg	−38.9	357

20. Indicate the physical state (solid, liquid, gas) at each designated temperature for the following Group VIA/16 hydro-

gen compounds. Refer to Table 13.6 for melting point and boiling point data.

(a) H_2S at −75.0°C
(b) H_2S at −50.0°C
(c) H_2Se at −50.0°C
(d) H_2Se at −25.0°C
(e) H_2Te at −51.5°C
(f) H_2Te at 0.0°C

Crystalline Solids (Sec. 13.5)

21. List three examples of crystalline solids.

22. List three types of crystalline solids.

23. State the type of particles that compose each of the following.
(a) ionic solid
(b) molecular solid
(c) metallic solid

24. State whether the following list of properties is most descriptive of an ionic, molecular, or metallic solid.
(a) wide mp range, malleable, ductile, electrical conductor
(b) high mp, hard, soluble in water, conducts electricity when melted
(c) low mp, generally insoluble in water, nonconductor of electricity

25. Classify each of the following crystalline solids as ionic, molecular, or metallic.
(a) zinc, Zn
(b) zinc oxide, ZnO
(c) phosphorus, P_4
(d) iodine monobromide, IBr

26. Classify each of the following crystalline solids as ionic, molecular, or metallic.
(a) sulfur, S_8
(b) sulfur dioxide, SO_2
(c) silver, Ag
(d) silver nitrate, $AgNO_3$

Sulfur can occur in two crystalline forms. In this photo, S_8 molecules are shown as crystalline needles.

Changes of Physical State (Sec. 13.6)

27. Draw the general shape of the temperature–energy graph for the heating of ethanol from −120° to 120°C. (Given: mp = −117.3°C; bp = 78.5°C.)

28. Draw the general shape of the temperature–energy graph for the cooling of acetone from 100 to −100°C. (Given: mp = −95.4°C; bp = 56.2°C.)

29. Calculate the amount of heat required to melt 125 g of ice at 0°C.

30. Calculate the amount of heat released when 75.5 g of steam condenses to a liquid at 100°C.

31. Calculate the amount of heat required to convert 25.0 g of water at 25.0°C to steam at 100.0°C.

32. Calculate the amount of heat released when 65.5 g of water at 55.5°C cools and freezes into ice at 0.0°C.

33. Calculate the amount of heat required to convert 115 g of ice at 0.0°C to steam at 100.0°C.

34. Calculate the amount of heat released when 155 g of steam at 100.0°C cools and freezes into ice at 0.0°C.

35. Calculate the amount of heat required to convert 38.5 g of ice at −20.0°C to steam at 100.0°C.

36. Calculate the amount of heat released when 90.5 g of steam at 110.0°C cools to ice at 0.0°C.

37. Calculate the amount of heat required to convert 100.0 g of ice at −40.0°C to steam at 125.0°C.

38. Calculate the amount of heat released when 0.500 kg of steam at 150.0°C cools to ice at −50.0°C.

The Water Molecule (Sec. 13.7)

39. Draw the electron dot and structural formulas for a molecule of water.

40. How many bonding and nonbonding electron pairs are in a water molecule?

41. What is the observed bond angle in a water molecule?

42. The center of a tetrahedron forms an angle of 109° with its corners. If oxygen is at the center of the tetrahedron and hydrogens are at two corners, explain why the observed bond angle is less than 109°.

43. Indicate the two dipoles in a water molecule using delta notation.

44. Draw the net dipole in a molecule of water.

45. Draw two molecules of hydrogen fluoride, HF, and diagram an intermolecular hydrogen bond.

46. Draw two molecules of ammonia, NH_3, and diagram an intermolecular hydrogen bond.

Physical Properties of Water (Sec. 13.8)

47. The density of solid ammonia is less than its liquid. Does an "ammonia ice cube" float or sink in liquid ammonia?

48. A soft drink bottle is accidentally filled completely and then capped. What will happen if the soft drink is frozen solid?

49. Without referring to Table 13.6, predict which of the following has the higher melting point.
(a) H_2O or H_2S
(b) H_2S or H_2Se

50. Without referring to Table 13.6, predict which of the following has the higher boiling point.
(a) H_2O or H_2Se
(b) H_2S or H_2Te

51. Without referring to Table 13.6, predict which of the following has the higher heat of fusion (cal/mol).
(a) H_2O or H_2S (b) H_2S or H_2Se

52. Without referring to Table 13.6, predict which of the following has the higher heat of vaporization (cal/mol).
(a) H_2O or H_2Se (b) H_2S or H_2Te

53. In general, as the molar mass of Group VIA/16 hydrogen compounds increases, do the following increase or decrease?
(a) melting point (b) boiling point
(c) heat of fusion (d) heat of vaporization

54. Refer to the trends in Table 13.6 and estimate the predicted values for the properties of radioactive H_2Po. Find the molar mass of H_2Po from the periodic table and predict a value for the mp, bp, H_{fusion}, and H_{vapor}.

55. Indicate the physiological similarities and differences between ordinary water and heavy water.

56. Predict whether light water or heavy water has the higher value for each of the following properties.
(a) molar mass (b) density
(c) melting point (d) boiling point
(e) heat of fusion (f) heat of vaporization

Chemical Properties of Water (Sec. 13.9)

57. Write a balanced chemical equation for the electrolysis of water.

58. Write a balanced equation for the reaction of hydrogen and oxygen gases.

59. Complete and balance the following equations.
(a) $Li(s) + H_2O(l) \rightarrow$ (b) $Rb(s) + H_2O(l) \rightarrow$
(c) $Na_2O(s) + H_2O(l) \rightarrow$ (d) $Cs_2O(s) + H_2O(l) \rightarrow$
(e) $CO_2(g) + H_2O(l) \rightarrow$ (f) $P_2O_5(s) + H_2O(l) \rightarrow$

60. Complete and balance the following equations.
(a) $Ba(s) + H_2O(l) \rightarrow$ (b) $Mg(s) + H_2O(l) \rightarrow$
(c) $CaO(s) + H_2O(l) \rightarrow$ (d) $SrO(s) + H_2O(l) \rightarrow$
(e) $N_2O_3(g) + H_2O(l) \rightarrow$ (f) $N_2O_5(g) + H_2O(l) \rightarrow$

61. Complete and balance the following equations.
(a) $C_3H_6(g) + O_2(g) \xrightarrow{\text{spark}}$
(b) $C_3H_6O(g) + O_2(g) \xrightarrow{\text{spark}}$
(c) $HF(aq) + Ca(OH)_2(aq) \longrightarrow$
(d) $H_2CO_3(aq) + KOH(aq) \longrightarrow$
(e) $Na_2Cr_2O_7 \cdot 2\ H_2O(s) \xrightarrow{\Delta}$
(f) $Ca(NO_3)_2 \cdot 4\ H_2O(s) \xrightarrow{\Delta}$

62. Complete and balance the following equations.
(a) $C_4H_{10}(g) + O_2(g) \xrightarrow{\text{spark}}$
(b) $C_4H_{10}O(g) + O_2(g) \xrightarrow{\text{spark}}$
(c) $HNO_3(aq) + Ba(OH)_2(aq) \longrightarrow$
(d) $H_3PO_4(aq) + NaOH(aq) \longrightarrow$
(e) $Co(C_2H_3O_2)_2 \cdot 4\ H_2O(s) \xrightarrow{\Delta}$
(f) $KAl(SO_4)_2 \cdot 12\ H_2O(s) \xrightarrow{\Delta}$

Hydrates (Sec. 13.10)

63. Supply a systematic name for each of the following hydrate compounds.
(a) $MgSO_4 \cdot 7\ H_2O$ (b) $Co(CN)_3 \cdot 3\ H_2O$
(c) $MnSO_4 \cdot H_2O$ (d) $Na_2Cr_2O_7 \cdot 2\ H_2O$
(e) $Sr(NO_3)_2 \cdot 6\ H_2O$ (f) $Co(C_2H_3O_2)_2 \cdot 4\ H_2O$
(g) $CuSO_4 \cdot 5\ H_2O$ (h) $Cr(NO_3)_3 \cdot 9\ H_2O$

64. Provide the formula for each of the following hydrate compounds.
(a) sodium acetate trihydrate
(b) calcium sulfate dihydrate
(c) potassium chromate tetrahydrate
(d) zinc sulfate heptahydrate
(e) sodium carbonate decahydrate
(f) nickel(II) nitrate hexahydrate
(g) cobalt(III) iodide octahydrate
(h) chromium(III) acetate monohydrate

65. Calculate the percentage of water in each of the following hydrates.
(a) $SrCl_2 \cdot 6\ H_2O$ (b) $K_2Cr_2O_7 \cdot 2\ H_2O$
(c) $MgSO_4 \cdot 7\ H_2O$ (d) $Co(CN)_3 \cdot 3\ H_2O$
(e) $MnSO_4 \cdot H_2O$ (f) $Na_2CrO_4 \cdot 4\ H_2O$

66. Calculate the percentage of water in each of the following hydrates.
(a) $NiCl_2 \cdot 2\ H_2O$ (b) $Sr(NO_3)_2 \cdot 6\ H_2O$
(c) $Co(C_2H_3O_2)_2 \cdot 4\ H_2O$ (d) $ZnSO_4 \cdot 7\ H_2O$
(e) $Cr(NO_3)_3 \cdot 9\ H_2O$ (f) $KAl(SO_4)_2 \cdot 12\ H_2O$

67. Determine the water of hydration for the following hydrates and write the chemical formula.
(a) $NiCl_2 \cdot X\ H_2O$ is found to contain 21.7% water.
(b) $Sr(NO_3)_2 \cdot X\ H_2O$ is found to contain 33.8% water.
(c) $CrI_3 \cdot X\ H_2O$ is found to contain 27.2% water.
(d) $Ca(NO_3)_2 \cdot X\ H_2O$ is found to contain 30.5% water.

68. Determine the water of hydration for the following hydrates and write the chemical formula.
(a) $SrCl_2 \cdot X\ H_2O$ is found to contain 18.5% water.
(b) $Ni(NO_3)_2 \cdot X\ H_2O$ is found to contain 37.2% water.
(c) $CoSO_4 \cdot X\ H_2O$ is found to contain 10.4% water.
(d) $Na_2B_4O_7 \cdot X\ H_2O$ is found to contain 30.9% water.

Water Purification (Sec. 13.11)

69. List the cations commonly found in each of the following.
(a) hard water (b) soft water
(c) deionized water (d) distilled water

70. List the anions commonly found in each of the following.
(a) hard water (b) soft water
(c) deionized water (d) distilled water

71. How is a bathtub "soap ring" produced?

72. Water that contains soluble calcium bicarbonate is said to have temporary hardness. Heating water with temporary hardness produces a scaly precipitate of insoluble calcium carbonate. Write a balanced equation for this reaction.

General Exercises

73. State the approximate percent of water on earth for each of the following.
    (a) salt water (b) nonfrozen fresh water

74. State the approximate percent of water in humans for each of the following.
    (a) human body (b) blood

75. Define the term *hemogeneous liquid mixture* and give an example.

76. Define the term *network solid* and give an example.

77. The atmosphere on Venus has clouds of sulfuric acid. Explain why it is reasonable to predict that the "rain-drops" on Venus are small spheres.

78. Describe the electron sea model for metallic solids.

79. Ethylene glycol is a permanent antifreeze. Calculate the amount of heat released when 1250 g of liquid at 25.0°C cools to a solid at its melting point. (For ethylene glycol: specific heat = 0.561 cal/g·°C; mp = −11.5°C; bp = 197.6°C; H_{fusion} = 43.3 cal/g; H_{vapor} = 293 cal/g.)

80. Methanol is considered a temporary antifreeze because it evaporates from solution. Calculate the amount of heat necessary to convert 1250 g of liquid at 25.0°C to a vapor at its boiling point. (For methanol: specific heat = 0.610 cal/g·°C; mp = −97.9°C; bp = 65.2°C; H_{fusion} = 23.7 cal/g; H_{vapor} = 293 cal/g.)

81. Refer to the trends in Table 13.6 and estimate the predicted values for the properties for water. That is, predict the mp, bp, H_{fusion}, and H_{vapor} in the absence of hydrogen bonding.

82. Hydrogen peroxide, H_2O_2, decomposes upon heating to give water and oxygen gas. Write a balanced chemical equation for the reaction.

83. Using hard water, it is difficult to get soap to lather. In fact, hard water often produces a soapy film on our skin. Write an equation for the formation of a magnesium soap film.

CHAPTER

14

Solutions

The ocean is a concentrated solution containing sodium chloride and other dissolved salts. Sea water is about 10% more dense than pure water and its greater density helps support the surfer.

In Section 3.3 we learned that a solution is a homogeneous mixture. This means that a solution is the same throughout, and every sample of the solution has the same properties and chemical composition. In this chapter, we will learn more about solutions. For one thing, solutions consist of a solute, such as sugar, dissolved in a solvent such as water. Other examples are carbon dioxide dissolved in a soft drink, and alcohol dissolved in liquor. The relative proportions of the solute and solvent can vary, and we will calculate the concentration of the solute in solution.

We usually think of solutions in the liquid state. Solutes scattered throughout solvents exist, however, in any physical state. That is, there are gaseous, liquid, and solid solutions. Table 14.1 lists many such common solutions.

14.1 Gases in a Liquid Solution

OBJECTIVES

To indicate how the solubility of a gas in a liquid is affected by temperature and pressure.

Muriatic acid, available from the supermarket, is used to increase the acidity of swimming pools. The systematic name for muriatic acid is hydrochloric acid. It is produced by dissolving hydrogen chloride gas in water. Like muriatic acid, household ammonia is another common example of a gas dissolved in a liquid. Ammonia solutions have gaseous NH_3 dissolved in water. Champagne and soft drinks are also liquids containing dissolved gases. The dissolved gas in carbonated beverages is carbon dioxide, CO_2.

Temperature Effects

Solutions of gases in liquids are greatly affected by changes in temperature. As the temperature increases, the kinetic energy of the solute gas becomes greater. The gas molecules acquire more of a tendency to escape from the solvent. Thus, as the temperature increases, the solubility of a gas in a liquid decreases.

A practical example of this principle is illustrated by a carbonated beverage. You have probably taken a beverage from the refrigerator and let it warm to room temperature. When you opened the beverage, it foamed as the gas escaped from the solution. Compare that to what happens when you open a beverage directly from the refrigerator. At colder temperatures, the carbon dioxide is more soluble and foaming is minimal. Table 14.2 indicates the effect of temperature on the solubility of carbon dioxide in water.

Table 14.1 Types of Solutions

Solute	Solvent	Solution Example
		Gaseous Solutions
gas	gas	deep-sea diving mixture, $O_2(g)$ in $He(g)$
liquid	gas	fog
solid	gas	smoke
		Liquid Solutions
gas	liquid	carbonated drinks, $CO_2(g)$ in water
		household ammonia, $NH_3(g)$ in water
liquid	liquid	vinegar, $HC_2H_3O_2(l)$ in water
		antifreeze, $C_2H_4(OH)_2(l)$ in water
solid	liquid	salt water, $NaCl(s)$ in water
		sugar solution, $C_{12}H_{22}O_{11}(s)$ in water
		Solid Solutions
gas	solid	sponge, pumice stone
liquid	solid	dental fillings, $Hg(l)$ in $Ag(s)$
solid	solid	sterling silver, $Cu(s)$ in $Ag(s)$ alloy

Table 14.2 Solubility of Carbon Dioxide in Water

Temperature (°C)	Pressure (atm)	Solubility of CO_2 (g/100 mL water)*
Temperature Effect		
0	1	0.348
20	1	0.176
40	1	0.097
60	1	0.058
Pressure Effect		
0	1	0.348
0	2	0.696
0	3	1.044

* Notice higher temperatures decrease the solubility; higher pressures increase the solubility.

Pressure Effects

The solubility of a gas in a liquid is strongly influenced by pressure. In 1803, the English chemist William Henry (1774–1836) conducted experiments on the solubility of gases in liquids. Henry found that the solubility of a gas was proportional to the partial pressure of the gas above the liquid. This is known as **Henry's law**. As an example, consider carbonated beverages. If we double the partial pressure of CO_2, we double the solubility. If we halve the partial pressure of CO_2, we cut the carbonation in half. Table 14.2 indicates the effect of pressure on the solubility of carbon dioxide in water.

Henry's law The principle that the solubility of a gas in a liquid is proportional to the partial pressure of the gas above the liquid.

If we examine the effect of pressure on the solubility of carbon dioxide as shown in Table 14.2, we see that the solubility doubles when the pressure increases from 1 to 2 atm, and the solubility triples when the pressure increases from 1 to 3 atm. Let's apply Henry's law and calculate the solubility of carbon dioxide when the pressure is 2.50 atm. The effect of increasing the pressure is to proportionally increase the solubility, which is given in the table as 0.348 g/100 mL water at 0°C and 1 atm. Thus,

$$0.348 \text{ g/100 mL} \times \frac{2.50 \text{ atm}}{1 \text{ atm}} = 0.870 \text{ g/100 mL}$$

Notice that the solubility of the carbon dioxide in water is given in g/100 mL. Gas solubilities are also expressed as the volume of gas dissolved in solution. For example, the solubility of oxygen in water is listed in the *Handbook of Chemistry and Physics* as 4.78 cm^3/100 mL at 25°C and 1 atm. The following exercise further illustrates the application of Henry's law.

EXAMPLE EXERCISE 14.1

Calculate the solubility of oxygen gas in water at 25°C and a partial pressure of 1150 torr. A reference book lists the solubility of oxygen in water at 25°C and 760 torr as 0.00414 g/100 mL.

Solution: Henry's law states that the solubility of the oxygen gas is proportional to the partial pressure of the gas above the liquid. Since 1150 torr is greater than 760 torr, the solubility increases. We can write

$$\text{standard solubility} \times \text{pressure factor} = \text{new solubility}$$

$$0.00414 \text{ g/100 mL} \times \frac{1150 \cancel{\text{torr}}}{760 \cancel{\text{torr}}} = 0.00626 \text{ g/100 mL}$$

SELF-TEST EXERCISE

Apply the solubility principles for a gas in a liquid to each of the following examples.

(a) One lake is at sea level and another at an elevation of 7500 feet. Assuming that the two lakes are at the same temperature, which has the lesser concentration of dissolved oxygen from the air?

(b) A nuclear energy plant is located next to a lake in order to dissipate the heat that is produced. A neighboring lake at the same elevation is not heated by the nuclear power plant. Which lake has the lesser oxygen concentration?

Answers: **(a)** the lake at 7500 feet (atmospheric pressure is less at higher elevations); **(b)** the lake heated by the power plant

14.2 Liquids in a Liquid Solution

To predict whether two liquids are miscible or immiscible by applying the *like dissolves like rule.*

solution A general term for a solute dissolved in a solvent.

solute The component of a solution that is the lesser quantity.

solvent The component of a solution that is the greater quantity.

A **solution** is composed of a solute dissolved in a solvent. The **solute** is the lesser quantity and the **solvent** is the greater. Let's consider the factors that affect the solubility of a liquid solute in a liquid solvent.

Dipoles

In Section 11.5, we learned that polar covalent bonds are the result of a difference in electronegativity between two bonded atoms. That is, a polar bond in a molecule creates areas of partial positive and partial negative charge. In a water molecule, for example, the more electronegative oxygen atom has a partial negative charge (δ^-). The two hydrogen atoms are less electronegative than the oxygen atom and have a partial positive charge (δ^+). The separation of charge between the hydrogen atom and oxygen atom is called a ***dipole***. There are two dipoles in a water molecule. If we resolve the two dipoles into one force in a single direction, we have a ***net dipole*** for the molecule. The symbol for a net dipole is an arrow pointing to the negative end of the molecule, as shown in Figure 14.1.

Polar and Nonpolar Solvents

polar solvent A dissolving liquid composed of polar molecules.

A liquid made up of polar molecules is called a **polar solvent**. Water is the most common polar solvent. Although there are many exceptions, solvent molecules

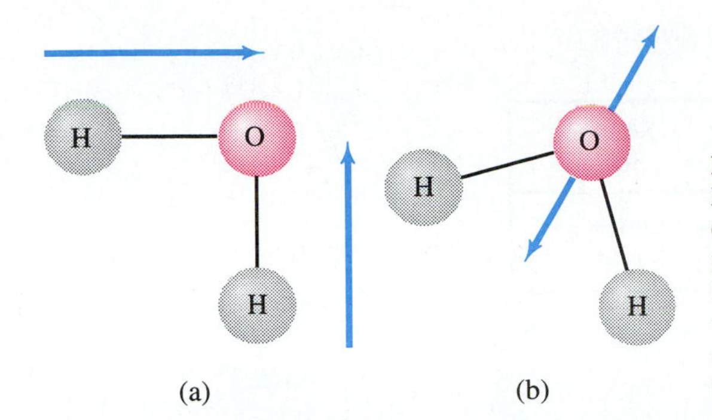

Figure 14.1 The Polar Water Molecule (a) The more electronegative oxygen atom polarizes the O—H bond, which in turn, creates two dipoles in a water molecule. (b) The two dipoles produce a net dipole for the entire water molecule.

Table 14.3 Selected Polar and Nonpolar Solvents

Polar Solvents	Nonpolar Solvents
water, H_2O	hexane, C_6H_{14}
methyl alcohol, CH_3OH	heptane, C_7H_{16}
ethyl alcohol, C_2H_5OH	octane, C_8H_{18}
isopropyl alcohol, C_3H_7OH	toluene, C_7H_8
acetone, C_3H_6O	carbon tetrachloride, CCl_4
methyl ethyl ketone, C_4H_8O	chloroform, $CHCl_3$
formic acid, $HCHO_2$	methylene chloride, CH_2Cl_2
acetic acid, $HC_2H_3O_2$	trichloroethylene, C_2HCl_3
	ethyl ether, $C_4H_{10}O$*

* The general rule that oxygen-containing solvents are polar has many exceptions. For example, ethyl ether contains oxygen and yet is a nonpolar solvent.

Oil layer

Water

Oil and water are immiscible because oil molecules are nonpolar and water molecules are polar.

containing oxygen atoms, such as water, are usually polar. Ethyl alcohol, CH_3CH_2OH, and acetic acid, CH_3COOH, are polar solvents.

Nonpolar molecules have elements with comparable electronegativities. A liquid made up of nonpolar molecules is a **nonpolar solvent**. Common nonpolar solvents are carbon tetrachloride, CCl_4, hexane, C_6H_{14}, and other compounds of carbon and hydrogen, C_xH_y. Table 14.3 lists common polar and nonpolar solvents.

nonpolar solvent A dissolving liquid composed of nonpolar molecules.

"Like Dissolves Like" Rule

Polar solvents such as water and ethyl alcohol dissolve in one another. Nonpolar solvents such as hexane and chloroform also dissolve in one another. From these observations, we see that when two solvents are similar they interact by dissolving in each other. This principle is the **like dissolves like rule**. The rule states that two liquids dissolve in one another because their molecules are alike in polarity.

Two liquids that are soluble in one another are said to be **miscible**. For example, water and ethyl alcohol are miscible. From the rule it also follows that a polar solvent and a nonpolar solvent are not miscible. These liquids are said to be **immiscible**. Experimentally, mixtures of immiscible liquids separate into layers. For example, water and gasoline are immiscible and if mixed will separate into two layers, with gasoline floating on water. The *like dissolves like* rule is summarized in Table 14.4.

Example Exercise 14.2 illustrates the *like dissolves like* rule to predict the miscibility for two liquids.

like dissolves like rule The general principle that solubility is greatest when the polarity of the solute is similar to that of the solvent.

miscible Refers to two or more liquids that are infinitely soluble in one another.

immiscible Refers to two liquids that are not soluble in one another, and if mixed, separate into two layers.

Oil and water are immiscible liquids and the less-dense oil floats on top. Light reflecting off the film of oil produces a colorful pattern.

Table 14.4 Summary of *Like Dissolves Like* Rule for Two Liquids

Solute	Polar Solvent	Nonpolar Solvent
polar	miscible	immiscible
nonpolar	immiscible	miscible

EXAMPLE EXERCISE 14.2

Predict whether the following solvents are miscible or immiscible with water.

(a) methanol, CH_3OH **(b)** toluene, C_7H_8
(c) acetone, C_3H_6O **(d)** chloroform, $CHCl_3$

Solution: Let's use the simplifying assumption that most solvents containing oxygen are polar. Thus, methanol and acetone are polar solvents, and toluene and chloroform are nonpolar. Applying the *like dissolves like* rule gives the following:

(a) CH_3OH is *miscible* with H_2O because both solvents are polar.
(b) C_7H_8 is nonpolar and therefore *immiscible* with H_2O.
(c) C_3H_6O is *miscible* with H_2O because both liquids are polar.
(d) $CHCl_3$ is a nonpolar solvent and thus *immiscible* with water.

We can verify our predictions by referring to Table 14.3. We see that the polar solvents are CH_3OH, C_3H_6O, and H_2O. Hence, each of the above predictions is correct.

SELF-TEST EXERCISE

Predict whether the following solvents are miscible or immiscible with water.

(a) trichloroethane, $C_2H_3Cl_3$ **(b)** glycerine, $C_3H_5(OH)_3$
(c) isopropyl alcohol, C_3H_7OH **(d)** decane, $C_{10}H_{22}$

Answers: **(a)** immiscible (nonpolar); **(b)** miscible (polar); **(c)** miscible (polar); **(d)** immiscible (nonpolar)

Note At this level of our discussion, we will consider two solvents to be either miscible or immiscible, that is, either soluble or insoluble in each other. At a more sophisticated level, two solvents can be partially soluble in one another. That is, a slightly polar solvent can partially dissolve in a polar solvent. Although ethyl ether is generally considered a nonpolar solvent, it is actually slightly polar and partially dissolves (~7%) in water.

14.3 Solids in a Liquid Solution

To predict whether a solid compound is soluble or insoluble in water by applying the *like dissolves like* rule.

A solid substance, such as sugar, dissolves in a liquid, such as water, because the solute (sugar) molecules and solvent (water) molecules are attracted to each other.

Table 14.5 Summary of *Like Dissolves Like* Rule for a Solid in a Liquid

Solid Solute	Polar Solvent	Nonpolar Solvent
polar	soluble	insoluble
nonpolar	insoluble	soluble
ionic	soluble*	insoluble

* Many ionic compounds are only slightly soluble in water.

In general, a polar molecular compound is likely to dissolve in a polar solvent such as water (see Table 14.5). Table sugar, $C_{12}H_{22}O_{11}$, contains several oxygen atoms and is therefore a polar compound. Applying the *like dissolves like* rule, we can predict that polar $C_{12}H_{22}O_{11}$ is soluble in water.

Similarly, molecules in a nonpolar molecular compound are attracted by molecules in a nonpolar solvent. Thus, nonpolar molecular compounds dissolve in nonpolar solvents. We are aware that water is not a good solvent to remove grease from our hands. Grease is a nonpolar compound, and water is a polar solvent. Turpentine, however, can dissolve the grease. This is because turpentine is a nonpolar solvent. Dissolving grease with turpentine is an illustration of the *like dissolves like* rule.

Since ionic compounds are made up of charged ions, they are similar to polar compounds. In general, ionic compounds are more soluble in a polar solvent than in a nonpolar solvent. Consider ordinary table salt, which is the ionic compound sodium chloride, NaCl. Table salt dissolves readily in the polar solvent water. It does not dissolve readily in a nonpolar solvent such as gasoline. This is because the solute (salt) is ionic and the solvent (gasoline) is nonpolar. It is important to note, however, that many ionic compounds are only slightly soluble in water (refer to the solubility rules in Section 8.9).

Example Exercise 14.3 illustrates the *like dissolves like* rule for a solid compound in water.

EXAMPLE EXERCISE 14.3

Predict whether the following solid compounds are soluble or insoluble in water.

(a) fructose, $C_6H_{12}O_6$
(b) iron(II) fluoride, FeF_2
(c) lithium carbonate, Li_2CO_3
(d) paradichlorobenzene, $C_6H_4Cl_2$

Solution: Generally, we can apply the *like dissolves like* rule to determine if a compound is soluble. Since water is a polar solvent, we can predict that water dissolves polar and many ionic compounds.

(a) Fructose has six oxygen atoms and is a polar compound. We predict that $C_6H_{12}O_6$ is *soluble* in water.

(b) Iron(II) fluoride contains the metal iron and the nonmetal fluorine; it is thus an ionic compound. We predict that FeF_2 is *soluble* in water.

(c) Lithium carbonate contains the metal lithium and nonmetals; it is therefore an ionic compound. We predict that Li_2CO_3 is *soluble* in water.

(d) Paradichlorobenzene does not contain oxygen and is a nonpolar compound. Thus, $C_6H_4Cl_2$ is *insoluble* in water.

SELF-TEST EXERCISE

Predict whether the following solid compounds are soluble or insoluble in water.

(a) naphthalene, $C_{10}H_8$ **(b)** cobalt(III) chloride, $CoCl_3$

(c) cupric sulfate, $CuSO_4$ **(d)** lactic acid, $HC_3H_5O_3$

Answers: **(a)** insoluble; **(b)** soluble; **(c)** soluble; **(d)** soluble

Some compounds contain both polar and nonpolar components. Although cholesterol, $C_{27}H_{46}O$, for example, contains oxygen, it is considered nonpolar. It is nonpolar because it contains such a large number of carbon and hydrogen atoms. Using the *like dissolves like* rule, we can correctly predict that cholesterol is insoluble in water and soluble in a nonpolar substance. In the human body, cholesterol deposits in fat tissue, which is nonpolar. The following exercise further illustrates this principle.

EXAMPLE EXERCISE 14.4

Predict whether the following vitamins are water soluble or fat soluble.

(a) vitamin A, $C_{20}H_{30}O$ **(b)** vitamin B_2, $C_{17}H_{20}N_4O_6$

(c) vitamin B_{12}, $C_{63}H_{88}CoN_{14}O_{14}P$ **(d)** vitamin E, $C_{29}H_{50}O_2$

Solution: Applying the *like dissolves like* rule, we predict that polar compounds are water soluble and nonpolar compounds are fat soluble.

(a) Vitamin A has one oxygen atom but is mostly nonpolar. We would predict that $C_{20}H_{30}O$ is a *fat-soluble* vitamin.

(b) Vitamin B_2, riboflavin, has six oxygen atoms and is therefore polar. We predict that $C_{17}H_{20}N_4O_6$ is a *water-soluble* vitamin.

(c) Vitamin B_{12} contains a metallic cobalt atom, which makes it an ionic compound. We predict that $C_{63}H_{88}CoN_{14}O_{14}P$ is a *water-soluble* vitamin.

(d) Vitamin E has two oxygen atoms but is overwhelmingly nonpolar. We can predict that $C_{29}H_{50}O_2$ is a *fat-soluble* vitamin.

Nutritionists teach that the B vitamins are water soluble and eliminated in the urine. They also teach that large doses of vitamins A, D, E, and K can be toxic because they are fat soluble and accumulate in the body.

SELF-TEST EXERCISE

Predict whether the following vitamins are soluble or insoluble in water.

(a) vitamin C, $C_6H_8O_6$ **(b)** vitamin D, $C_{27}H_{44}O$

Answers: **(a)** soluble; **(b)** insoluble

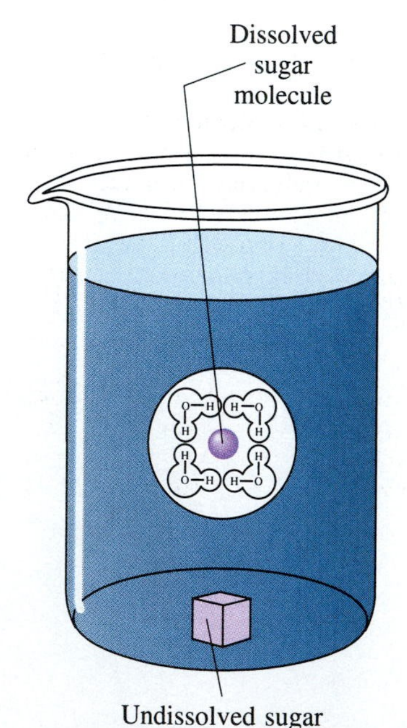

Figure 14.2 Glucose Sugar Dissolving in Water The water solvent molecules attack the glucose solute along the edges of the crystal. Each glucose molecule, $C_6H_{12}O_6$, in solution is surrounded by several water molecules.

14.4 The Dissolving Process

To describe and diagram the process of dissolving a solid compound composed of:

(a) molecules

(b) ions

Let's try to visualize the dissolving process. When a solute crystal is dropped into a solution, the crystal begins to dissolve. This is because water molecules attack the crystal and begin pulling away part of it. Specifically, water molecules attack the edges and corners of the crystal.

As an example, suppose we drop a glucose crystal into water. Water molecules are attracted to the polar glucose, $C_6H_{12}O_6$, and pull glucose molecules into the solution. Several water molecules surround each glucose molecule in the solution. The glucose molecules, which are held within a cluster of water molecules, are said to be in a **solvent cage**. The number of water molecules varies, depending on the concentration of the solute (Figure 14.2).

solvent cage Refers to a cluster of water molecules surrounding a solute molecule or ion in solution.

As another example, consider what happens when a crystal of table salt dissolves in water. By obeying the *like dissolves like* rule, water molecules are attracted to the ionic salt compound. Once again, the water molecules attack the edges of the crystal and begin pulling away part of it. For an ionic compound such as table salt, NaCl, the water molecules pull away positive and negative ions during the dissolving process. The negatively charged oxygen atom in a water molecule is attracted to the positively charged sodium ion, Na^+. The negatively charged chloride ion, Cl^-, is pulled into solution by the more positively charged hydrogen atoms in the water molecule. Figure 14.3 illustrates this process.

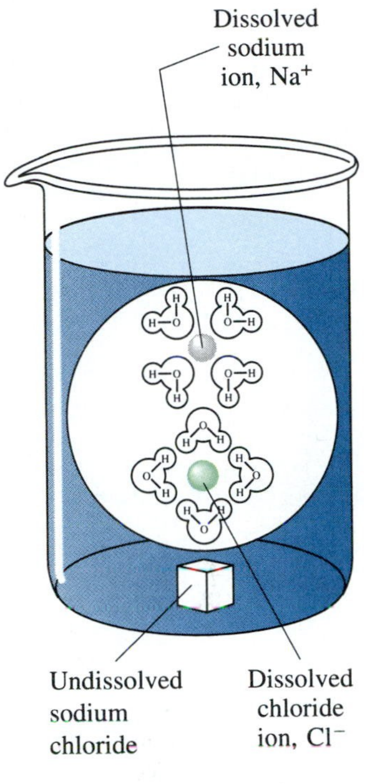

Figure 14.3 Table Salt Dissolving in Water The water solvent molecules attack the sodium chloride solute along the edges of the crystal. Carefully note the orientation of the water molecules as they surround each ion in solution. The Na^+ is attracted to the more electronegative oxygen atom in a water molecule, while the Cl^- is attracted to the more electropositive hydrogen atoms.

14.5 Rate of Dissolving

OBJECTIVES

To state the effects of temperature, stirring, and particle size on the rate of dissolving of a solid compound in water.

The rate at which a solid compound, such as table sugar, dissolves in a solution depends on three factors. That is, we can increase the rate of dissolving by:

1. heating the solution
2. stirring the solution
3. grinding the solid solute

Figure 14.4 Rate of Dissolving of Table Sugar A sugar cube in water at 20°C dissolves slowly. An equal amount of powdered sugar stirred in water at 100°C dissolves immediately.

By heating and stirring the solution, or grinding the solute, we increase the rate at which solvent molecules attack the solute. *Heating the solution* increases the kinetic energy of the solution, and the solvent molecules move faster. In an aqueous solution, water molecules attack the solute more frequently. Solute molecules are pulled into the solution faster, thus increasing the rate of dissolving. *Stirring a solution* increases the interaction between water molecules and the solute. As fresh solvent contacts the solute more often, the rate of dissolution is faster.

Grinding the solute into smaller crystals creates more surface area. As the solute crystals become smaller, the total surface area increases. We know that solvent molecules attack the surface of crystals along the edges. By creating smaller crystals and greater surface area, more solute is exposed to attack by water molecules. The water molecules attack the solute more frequently, thus increasing the rate of dissolving (Figure 14.4).

14.6 Solubility and Temperature

To state the effect of temperature on the solubility of solid substances in water.

If we heat a solution, the interaction between the solute and solvent increases. In general, the solubility of the solute is greater as the temperature increases. A few compounds are exceptions, however, and become less soluble as the temperature increases. We will define the **solubility** of a compound as the maximum amount of solute that can be dissolved in 100 g of water at a specified temperature. Figure 14.5 graphs the solubility of various compounds at different temperatures.

solubility Refers to the maximum amount of solute that dissolves in a solvent at a specified temperature; usually expressed in grams of solute per 100 g of water.

People who enjoy sugar in a hot beverage may know an interesting illustration of the effect of temperature on solubility. Let's suppose a person dissolves a lot of sugar in hot coffee and allows the coffee to cool in a cup. The solubility of the sugar decreases as the coffee cools. If the person stirs the cold coffee with a spoon,

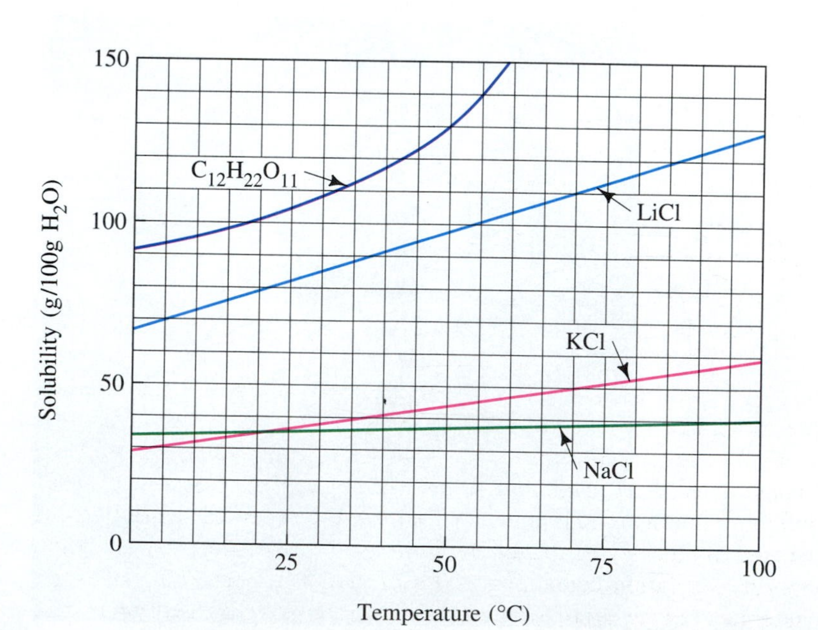

Figure 14.5 Solubility of Various Solid Compounds Although there are exceptions, solid compounds are usually more soluble as the temperature increases.

sugar can be detected that has crystallized from solution at the lower temperature.

Figure 14.5 illustrates that the effect of temperature on solubility varies with the compound. The solubility of table salt, NaCl, in water is affected very little by temperature. Its solubility is about 35 g of NaCl per 100 g of water at 20°C and increases to 40 g per 100 g of water at 100°C. Interestingly, the solubility of table sugar, $C_{12}H_{22}O_{11}$, is greatly affected by an increase in temperature. At 20°C, the solubility of table sugar is about 100 g per 100 g of water. At 60°C, the solubility is about 150 g per 100 g of water and increasing rapidly.

EXAMPLE EXERCISE 14.5

Determine the solubility of the following compounds at 50°C as shown in Figure 14.5.

(a) NaCl
(b) KCl
(c) LiCl
(d) $C_{12}H_{22}O_{11}$

Solution: From Figure 14.5, let's find the point at which the solubility curve of the compound intersects a line drawn at 50°C.

(a) The solubility of NaCl at 50°C is about 38 g per 100 g of water.
(b) The solubility of KCl at 50°C is about 45 g per 100 g of water.
(c) The solubility of LiCl at 50°C is about 98 g per 100 g of water.
(d) The solubility of $C_{12}H_{22}O_{11}$ at 50°C is about 130 g per 100 g of water.

SELF-TEST EXERCISE

Refer to the solubility behavior in Figure 14.5 and determine the minimum temperature to obtain the following solutions.

(a) 35 g of NaCl per 100 g of water
(b) 40 g of KCl per 100 g of water
(c) 100 g of LiCl per 100 g of water
(d) 140 g of $C_{12}H_{22}O_{11}$ per 100 g of water

Answers: **(a)** 5°C; **(b)** 35°C; **(c)** 55°C; **(d)** 55°C

14.7 Unsaturated, Saturated, and Supersaturated Solutions

OBJECTIVES

To state whether a solution is saturated, unsaturated, or supersaturated, given its concentration, temperature, and solubility.

In Section 14.6, we stated that the solubility of a compound usually increases as the temperature increases. To further illustrate, consider the solubility of sodium acetate, $NaC_2H_3O_2$. The solubility of the sodium acetate at 55°C is 100 g of $NaC_2H_3O_2$ per 100 g of water. This is the maximum amount of solute that dissolves at 55°C. A solution containing the maximum amount of solute at a specified temperature is said to be **saturated**. By increasing the temperature to 75°C, the solubility is higher. At that temperature, a solution containing 100 g of $NaC_2H_3O_2$ per 100 g of water is no longer saturated. Since more solute can be dissolved, the solution is said

saturated solution A solution that contains the maximum amount of dissolved solute at a given temperature.

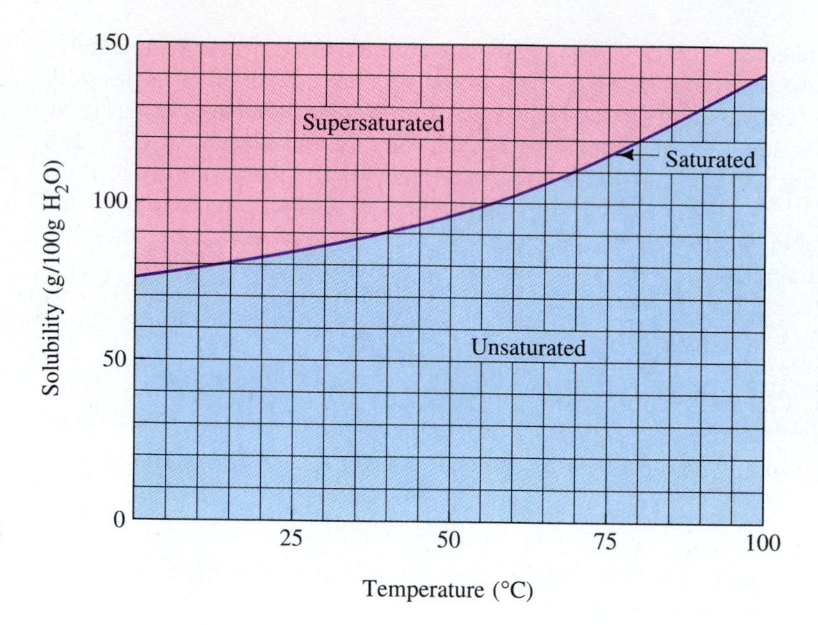

Figure 14.6 Solubility of Sodium Acetate The curve on the graph represents a saturated solution of $NaC_2H_3O_2$ at various temperatures. The region below the curve represents solutions that are unsaturated while those above are supersaturated.

unsaturated solution A solution that contains less than the maximum amount of dissolved solute at a given temperature.

supersaturated solution A solution that contains more dissolved solute than will ordinarily dissolve at a given temperature.

to be **unsaturated**. Figure 14.6 graphs the solubility of $NaC_2H_3O_2$ as a function of temperature.

Under special circumstances it is possible to exceed the usual maximum solubility of a compound. Solutions that contain more solute than ordinarily dissolves at a given temperature are said to be **supersaturated**. For example, recall that 100 g of $NaC_2H_3O_2$ can be dissolved in 100 g of water at 55°C. This solution is saturated. If we allow the solution to cool, while being careful not to disturb it, the excess solute will remain in solution. It is possible to cool the solution to 20°C and still have 100 g of $NaC_2H_3O_2$ remain in solution. At 20°C, the maximum solubility is only about 82 g per 100 g of water. Since the excess solute remains in the solution at 20°C, the solution is supersaturated.

Supersaturated solutions are unstable. In our example, one tiny crystal can cause 18 g of $NaC_2H_3O_2$ per 100 g H_2O to crystallize from solution. As another example, consider rock candy, which is made by suspending a small sugar crystal in a supersaturated sugar solution. The small crystal turns into a large rock of sugar as excess solute deposits from the supersaturated solution. Example Exercise 14.6 provides practice in obtaining information from the graph of solubility.

EXAMPLE EXERCISE 14.6

A sodium acetate solution contains 110 g of $NaC_2H_3O_2$ per 100 g of water. Refer to Figure 14.6 and determine whether the solution is unsaturated, saturated, or supersaturated at the following temperatures.

(a) 50°C **(b)** 70°C
(c) 90°C

Solution:

(a) At 50°C the solubility of $NaC_2H_3O_2$ is about 95 g per 100 g of water. Since the solution contains more solute, 110 g per 100 g water, the solution is *supersaturated.*

(b) At 70°C the solubility of $NaC_2H_3O_2$ is about 110 g per 100 g of water. Since the solution contains the same amount of solute, 110 g per 100 g water, the solution is *saturated.*

(c) At 90°C the solubility of $NaC_2H_3O_2$ is about 130 g per 100 g of water. Since the solution has only 110 g per 100 g water, the solution is *unsaturated.*

(a)

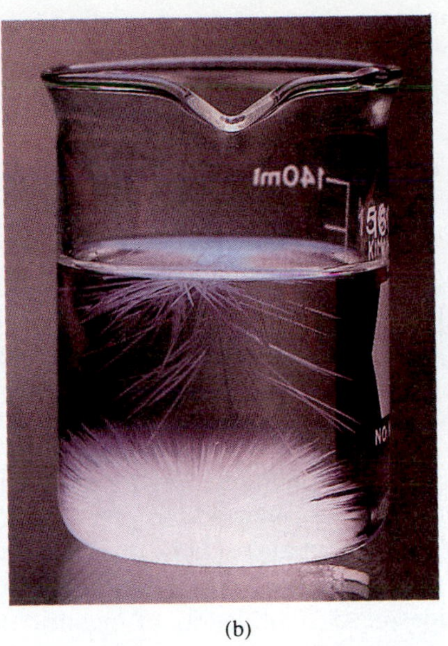

(b)

(c)

(a) A single crystal of sodium acetate is about to be dropped into a supersaturated solution of $NaC_2H_3O_2$. (b) A small crystal can cause gross crystallization, and (c) eventually a solid mass.

SELF-TEST EXERCISE

A sodium acetate solution contains 80 g of $NaC_2H_3O_2$ per 100 g of water. Refer to Figure 14.6 and determine whether the solution is unsaturated, saturated, or supersaturated at the following temperatures.

(a) 0°C **(b)** 15°C

(c) 45°C

Answers: **(a)** supersaturated; **(b)** saturated; **(c)** unsaturated

14.8 Mass Percent Concentration

OBJECTIVES

To write three pairs of unit factors given the mass/mass percent concentration of a solution.

To calculate the unknown quantity for a solution given any two of the following: mass of solute, mass of solvent, mass percent concentration.

The concentration of a solution tells us how much dissolved solute is in a given volume of the solution. We use the terms dilute and concentrated to describe the relative concentration of a solution. But since those terms are not specific, they are not used for quantitative laboratory procedures. The concentration of a solution can, however, be described precisely by comparing the mass of the solute to the mass of the solution. This **mass/mass percent** (m/m%) concentration is the gram mass of solute dissolved in 100 grams of solution. We can express the ratio as follows:

mass/mass percent (m/m %) A solution concentration expression that relates the mass of solute in grams dissolved in each 100 grams of solution.

$$\frac{\text{mass of solute}}{\text{mass of solution}} \times 100 = \text{m/m\%}$$

If a chemist prepares a standard solution from 5.00 g of NaF dissolved in 95.0 g of water, what is the mass/mass percent concentration?

Let's substitute the data into our equation for percent concentration thus,

CHEMISTRY CONNECTION □ ENVIRONMENTAL

Water Fluoridation

► **Can you guess what percentage of the people in the United States drink fluoridated water?**

As early as the 1930s, studies by the U.S. Public Health Service indicated that fluoride can help prevent dental cavities. The action of fluoride is to make the enamel surface of teeth more resistant to decay.

Tooth enamel is made mainly of hydroxyapatite, $Ca_{10}(PO_4)_6(OH)_2$, the hardest substance in our bodies. However, bacteria in the mouth react with food to produce acids that attack tooth enamel. These acids can react with hydroxyapatite causing enamel to dissolve and form pits. Fluoride ions can prevent cavities by converting $Ca_{10}(PO_4)_6(OH)_2$ to $Ca_{10}(PO_4)_6F_2$, which is more resistant to acid attack.

In 1950, the Public Health Service officially endorsed the practice of adding fluoride to public drinking water. Since then, there has been widespread fluoridation of public water supplies. Typically, fluoridated water has a concentration of fluoride ion that is less than 1 part per million (1 mg/L). A U.S. Department of Health census revealed that every state has drinking water with natural or controlled fluoridation. However, the number of people that drink fluoridated water varies widely: 2% in Utah, 22% in California, 36% in Florida, 40% in Washington, 59% in Texas, 66% in New York, 86% in Illinois. In Canada, over 50% of the people drink water that has fluoride.

The benefits of fluoride for preventing tooth decay have been proven repeatedly. In one study, children who grew up with fluoridation had an average of three teeth with cavities. Conversely, children who grew up without fluoridation had an average of ten teeth with cavities. In other studies, it has been shown that fluoride is most effective when it is available while teeth are still developing.

Despite proven benefits, not everyone agrees that fluoride should be added to the public water supply. One reason is that fluoride may have adverse side effects. For example, it has been reported that fluoridation has caused brown mottling in the teeth of some children. Adding fluoride to the public water supply also raises the issue of free choice, thus sparking a political debate.

As an alternative to fluoridation, people can choose to take inexpensive fluoride tablets. In addition, many toothpastes and mouthwashs are readily available sources of fluoride. Stannous fluoride, SnF_2, and sodium fluoride, NaF, are commonly found in toothpaste. Another alternative is in-office fluoride treatments by a dentist, but this procedure is often expensive.

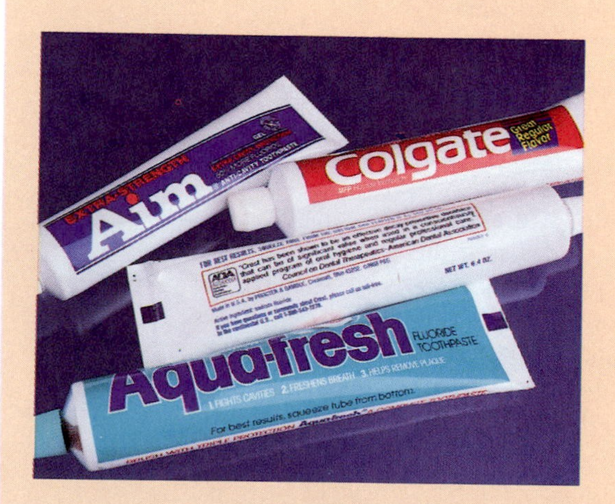

Toothpastes often contain fluoride

► **Currently, about 50% of the U.S. population drinks water that is naturally or artificially fluoridated.**

$$\frac{5.00 \text{ g NaF}}{5.00 \text{ g NaF} + 95.0 \text{ g } H_2O} \times 100 = \text{m/m\%}$$

$$\frac{5.00 \text{ g NaF}}{100.0 \text{ g solution}} \times 100 = 5.00\%$$

Notice that in the denominator we added the mass of the solute to the mass of solvent to obtain the mass of solution.

Writing Solution Concentration Unit Factors

The unit analysis method of problem solving involves three steps: unknown units, relevant given value, and the application of a unit factor. We can solve mass/mass percent calculations by using the solution concentration as a unit factor. For example, a 5.00% solution of NaF contains 5.00 g of solute in each 100.0 g of solution. We can write a pair of unit factors as

$$\frac{5.00 \text{ g NaF}}{100.0 \text{ g solution}} \quad \text{and} \quad \frac{100.0 \text{ g solution}}{5.00 \text{ g NaF}}$$

Since the solution is made up of solute and solvent, we can write a second pair of unit factors. The solution contains 5.00 g of solute in 95.0 g of solvent. Thus,

$$\frac{5.00 \text{ g NaF}}{95.0 \text{ g water}} \quad \text{and} \quad \frac{95.0 \text{ g water}}{5.00 \text{ g NaF}}$$

Furthermore, we can express the ratio of the mass of solvent to the mass of solution in a third pair of unit factors. The solution contains 95.0 g of solvent in 100.0 g of solution. Thus,

$$\frac{95.0 \text{ g water}}{100.0 \text{ g solution}} \quad \text{and} \quad \frac{100.0 \text{ g solution}}{95.0 \text{ g water}}$$

The following example exercise further illustrates writing unit factors associated with various solution concentrations.

EXAMPLE EXERCISE 14.7

Write three pairs of unit factors related to a concentrated hydrochloric acid solution, which is 36.0% HCl.

Solution: A 36% solution of HCl contains 36.0 g of the solute in each 100.0 g of acid solution. The first pair of unit factors is

$$\frac{36.0 \text{ g HCl}}{100.0 \text{ g solution}} \quad \text{and} \quad \frac{100.0 \text{ g solution}}{36.0 \text{ g HCl}}$$

Since the solution is made up of HCl solute in water, the solution contains 36.0 g of HCl in 64.0 g of water. Thus, a second pair of unit factors is

$$\frac{36.0 \text{ g HCl}}{64.0 \text{ g water}} \quad \text{and} \quad \frac{64.0 \text{ g water}}{36.0 \text{ g HCl}}$$

We can also express the ratio of the mass of water to the mass of solution. The solution contains 64.0 g of water in 100.0 g of solution. Thus, a third pair of unit factors is

$$\frac{64.0 \text{ g water}}{100.0 \text{ g solution}} \quad \text{and} \quad \frac{100.0 \text{ solution}}{64.0 \text{ g water}}$$

SELF-TEST EXERCISE

Write three pairs of unit factors that are related to a 12.5% aqueous solution of methanol, CH_3OH.

Answers:

$$\frac{12.5\ \text{g}\ CH_3OH}{100.0\ \text{g solution}} \quad \text{and} \quad \frac{100.0\ \text{g solution}}{12.5\ \text{g}\ CH_3OH}$$

$$\frac{12.5\ \text{g}\ CH_3OH}{87.5\ \text{g water}} \quad \text{and} \quad \frac{87.5\ \text{g water}}{12.5\ \text{g}\ CH_3OH}$$

$$\frac{87.5\ \text{g water}}{100.0\ \text{g solution}} \quad \text{and} \quad \frac{100.0\ \text{g solution}}{87.5\ \text{g water}}$$

Calculating Unknown Quantities

We can now apply unit factors derived from mass/mass percent concentration to solution calculations. For example, intravenous injections of glucose are sometimes administered to patients with low blood sugar. If a normal glucose solution is 5.00%, what is the mass of solution that contains 25.2 g of glucose sugar?

Let's use the unit analysis method of problem solving. Step 1: The unknown quantity is grams of solution. Step 2: The relevant given value is 25.2 g sugar. Thus,

$$25.2\ \text{g sugar} \times \frac{\text{unit}}{\text{factor}} = \text{g solution}$$

The percent concentration provides the unit factor. Since the solution concentration is 5.00%, there are 5.00 g of solute in 100.0 g of solution. We can write the unit factors as

$$\frac{5.00\ \text{g sugar}}{100.0\ \text{g solution}} \quad \text{and} \quad \frac{100.0\ \text{g solution}}{5.00\ \text{g sugar}}$$

We should select the second unit factor to cancel units properly; that is,

$$25.2\ \cancel{\text{g sugar}} \times \frac{100.0\ \text{g solution}}{5.00\ \cancel{\text{g sugar}}} = 504\ \text{g solution}$$

The following example exercises further illustrate calculations based on the mass/mass percent concentration of a solution.

EXAMPLE EXERCISE 14.8

A glucose tolerance test is given to patients to diagnose diabetes or hypoglycemia (high or low blood sugar). If a patient is given a 30.0% glucose solution containing 250.0 g of water, what is the mass of glucose?

Solution: Let's use the three-step unit analysis method of problem solving. Step 1: The unknown quantity is grams of sugar. Step 2: The relevant given value is 250.0 g of water.

$$250.0\ \text{g water} \times \frac{\text{unit}}{\text{factor}} = \text{g sugar}$$

The percent concentration provides the unit factor. Since the solution is 30.0%, there are 30.0 g of solute in 100.0 g of solution. Therefore, there are 30.0 g of sugar in 70.0 g of water. We can write the unit factors as

$$\frac{30.0 \text{ g sugar}}{70.0 \text{ g water}} \quad \text{and} \quad \frac{70.0 \text{ g water}}{30.0 \text{ g sugar}}$$

Selecting the first unit factor to cancel units properly, we have

$$250.0 \cancel{\text{g water}} \times \frac{30.0 \text{ g sugar}}{70.0 \cancel{\text{g water}}} = 107 \text{ g sugar}$$

In a typical glucose tolerance test, a patient is given 107 g of glucose in a 30.0% solution. A blood sample is taken at the start of the test and again every hour for up to 6 hours.

SELF-TEST EXERCISE

Given 125 g of a 15.0% sucrose sugar solution, what is the mass of sucrose?

Answer: 18.8 g sucrose

EXAMPLE EXERCISE 14.9

Intravenous saline injections are sometimes given to restore the electrolyte balance in trauma patients. What is the mass of water required to dissolve 2.00 g of NaCl for a 0.90% saline solution?

Solution: Using the unit analysis method of problem solving, the unknown quantity is grams of water and the relevant given value is 2.00 g NaCl.

$$2.00 \text{ g NaCl} \times \frac{\text{unit}}{\text{factor}} = \text{g water}$$

Since the solution concentration is 0.90%, there is 0.90 g of solute in 100.0 g of solution. Therefore, there is 0.90 g of NaCl in 99.1 g of water. We can write the unit factors as

$$\frac{0.90 \text{ g NaCl}}{99.1 \text{ g water}} \quad \text{and} \quad \frac{99.1 \text{ g water}}{0.90 \text{ g NaCl}}$$

Selecting the second unit factor to cancel units, we have

$$2.00 \cancel{\text{g NaCl}} \times \frac{99.1 \text{ g water}}{0.90 \cancel{\text{g NaCl}}} = 220 \text{ g water}$$

SELF-TEST EXERCISE

A 7.50% potassium chloride solution is prepared by dissolving enough of the salt to give 100.0 g of solution. What is the mass of water required?

Answer: 92.5 g water

The term percent concentration is ambiguous. It is used to express the mass of solute in 100 g of solution, the mass of solute in 100 mL of solution, and sometimes the volume of solute in 100 mL of solution. Therefore, there are three interpretations of percent concentration expressions, that is, mass/mass, mass/volume, and volume/volume. Unless specifically told otherwise, we will assume that percent concentration refers to the mass/mass expression.

14.9 Molar Concentration

OBJECTIVES

To write two pairs of unit factors given the molar concentration of a solution.

To calculate the unknown quantity for a solution given the molar mass of the solute and any two of the following: mass of solute, volume of solution, molar concentration.

molarity (M) A solution concentration expression that relates the moles of solute dissolved in each liter of solution.

In Section 7.3, we discussed the concept of moles as it applies to the mass of a substance. In Section 7.8, we introduced the mole concept as it applies to the volume of a gas. Not surprisingly, the mole is used to express the concentration of a solution. The most common expression of solution concentration is **molarity**. Molarity (symbol $\boldsymbol{M}$) is the number of moles of a solute dissolved in 1 liter of a solution. We can express this ratio as follows:

$$\frac{\text{moles of solute}}{\text{liter of solution}} = M$$

To see how to determine molarity, consider a household drain cleaner that is a solution of caustic sodium hydroxide, NaOH. If a manufacturer prepares a solution from 12.0 g of NaOH dissolved in 0.100 L of solution, what is the molarity? Let's calculate the molarity of NaOH (40.0 g/mol) as follows:

$$\frac{12.0 \cancel{\text{g NaOH}}}{0.100 \text{ L of solution}} \times \frac{1 \text{ mol NaOH}}{40.0 \cancel{\text{g NaOH}}} = \frac{\text{mol NaOH}}{1 \text{ L of solution}}$$

$$= 3.00\ M \text{ NaOH}$$

Notice that we started the calculation with a ratio of units (g/L) in order to obtain an answer that is also a ratio of two units (mol/L).

Writing Unit Factors

The unit analysis method of problem solving involves three steps: unknown units, relevant given value, and the application of a unit factor. We can solve molarity calculations by using the solution concentration as a unit factor. For example, a 3.00 M solution of NaOH contains 3.00 mol of solute in each liter of solution. We can write this pair of unit factors as

$$\frac{3.00 \text{ mol NaOH}}{1 \text{ L solution}} \quad \text{and} \quad \frac{1 \text{ L solution}}{3.00 \text{ mol NaOH}}$$

In the laboratory, the volume of a solution is usually measured in milliliters. Since there are 1000 mL in a liter, we can write a second pair of unit factors for the 3.00 M solution.

$$\frac{3.00 \text{ mol NaOH}}{1000 \text{ mL solution}} \quad \text{and} \quad \frac{1000 \text{ mL solution}}{3.00 \text{ mol NaOH}}$$

Calculating Unknown Quantities

We can now apply unit factors derived from molar concentration to solution calculations. Consider a solution prepared from dissolving cupric sulfate in water. Let's calculate the mass of $CuSO_4$ dissolved in 0.250 L of 0.100 M $CuSO_4$ solution.

We begin with the unit analysis method of problem solving. Step 1: The unknown quantity is grams of $CuSO_4$. Step 2: The relevant given value is 0.250 L of solution.

$$0.250 \text{ L solution} \times \frac{\text{unit}}{\text{factors}} = \text{g } CuSO_4$$

Since the solution concentration is 0.100 *M*, there is 0.100 mol of solute in each liter of solution. We can write two unit factors:

$$\frac{0.100 \text{ mol } CuSO_4}{1 \text{ L solution}} \quad \text{and} \quad \frac{1 \text{ L solution}}{0.100 \text{ mol } CuSO_4}$$

We choose the first unit factor to cancel units properly. In addition, we must use the molar mass of cupric sulfate (159.6 g/mol).

$$0.250 \cancel{\text{ L solution}} \times \frac{0.100 \cancel{\text{ mol } CuSO_4}}{1 \cancel{\text{ L solution}}} \times \frac{159.6 \text{ g } CuSO_4}{1 \cancel{\text{ mol } CuSO_4}} = 3.99 \text{ g } CuSO_4$$

The following example exercises further illustrate calculations based on the molar concentration of a solution.

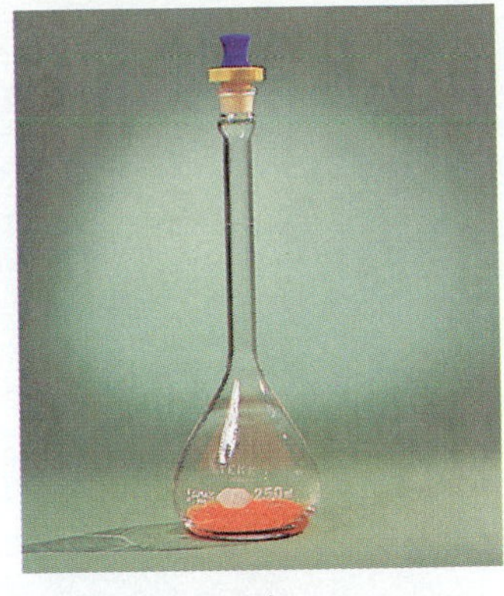

(a)

(b)

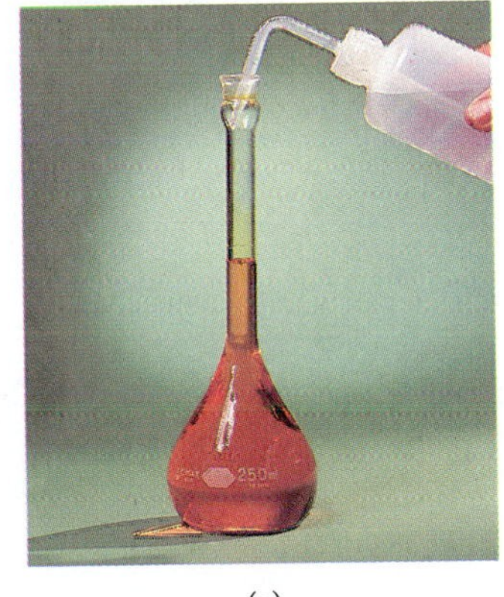

(c)

(a) A flask containing 7.35 g of potassium dichromate. (b) Distilled water is added to the flask to dissolve the orange salt. (c) The volume is carefully adjusted to the calibration line (250 mL) and the resulting solution is 0.100 M $K_2Cr_2O_7$.

EXAMPLE EXERCISE 14.10

What is the volume of 12.0 *M* hydrochloric acid that contains 7.30 g of HCl solute (36.5 g/mol)?

Solution: Let's express the unknown quantity as milliliters of acid. The relevant given value is 7.30 g HCl. The molarity and molar mass provide unit factors.

$$7.30 \text{ g HCl} \times \frac{\text{unit}}{\text{factors}} = \text{mL acid}$$

The molar mass is applied to convert from grams to moles. Since the solution concentration is 12.0 *M*, there are 12.0 mol of solute in each 1000 mL of acid solution. We can write the unit factors as

$$\frac{12.0 \text{ mol HCl}}{1000 \text{ mL acid}} \quad \text{and} \quad \frac{1000 \text{ mL acid}}{12.0 \text{ mol HCl}}$$

We will select the second unit factor to cancel units properly.

$$7.30 \cancel{\text{ g HCl}} \times \frac{1 \cancel{\text{ mol HCl}}}{36.5 \cancel{\text{ g HCl}}} \times \frac{1000 \text{ mL acid}}{12.0 \cancel{\text{ mol HCl}}} = 16.7 \text{ mL acid}$$

SELF-TEST EXERCISE

What is the volume of 6.00 *M* hydrochloric acid that contains 10.0 g of HCl solute (36.5 g/mol)?

Answer: 45.7 mL acid

EXAMPLE EXERCISE 14.11

What is the mass of H_2SO_4 (98.1 g/mol) in 50.0 mL of 6.00 *M* sulfuric acid?

Solution: The unknown quantity is grams of H_2SO_4. The relevant given value is 50.0 mL acid. The molarity and molar mass provide unit factors.

$$50.0 \text{ mL acid} \times \frac{\text{unit}}{\text{factors}} = \text{g } H_2SO_4$$

Since the solution concentration is 6.00 *M*, there are 6.00 mol of solute in each 1000 mL of acid solution. We can write the unit factors as

$$\frac{6.00 \text{ mol } H_2SO_4}{1000 \text{ mL acid}} \quad \text{and} \quad \frac{1000 \text{ mL acid}}{6.00 \text{ mol } H_2SO_4}$$

Choose the first factor to cancel units properly. The molar mass is applied to convert from moles to grams.

$$50.0 \cancel{\text{mL acid}} \times \frac{6.00 \cancel{\text{mol } H_2SO_4}}{1000 \cancel{\text{mL acid}}} \times \frac{98.1 \text{ g } H_2SO_4}{1 \cancel{\text{mol } H_2SO_4}} = 29.4 \text{ g } H_2SO_4$$

SELF-TEST EXERCISE

What is the mass of H_2SO_4 (98.1 g/mol) in 25.0 mL of 18.0 *M* sulfuric acid?

Answer: 44.1 g H_2SO_4

14.10 Molal Concentration and Colligative Properties

To write a pair of unit factors given the molal concentration of a solution.

To calculate the unknown quantity for a solution, given the molar mass of the solute and any two of the following: mass of solute, mass of solution, molal concentration.

To calculate the unknown quantity for a solution, given the molal freezing point constant and any three of the following: mass of solute, molar mass, mass of solution, freezing point lowering.

The freezing point of water is 0.00°C and the normal boiling point is 100.00°C. At 25°C, the vapor pressure of water is 23.8 mm Hg. When a nonvolatile solute is added to water, the freezing point, boiling point, and vapor pressure change. The dissolved solute particles (1) lower the freezing point, (2) raise the boiling point, and (3) lower the vapor pressure of the water.

The amount of change in freezing point, boiling point, and vapor pressure depends on the number of moles of solute particles in solution. It does not matter what the solute is, only the number of moles of solute particles. Lowering the freezing point, raising the boiling point, and lowering the vapor pressure, all of which depend on the number of solute particles, are referred to as **colligative properties**.

colligative property The properties of a solution that are affected by the number of solute particles in solution; for example, freezing point lowering, boiling point raising, and vapor pressure lowering.

Consider what happens if 1.00 mol of glucose, $C_6H_{12}O_6$, is dissolved in 1.00 kg of water. (1) The 0.00°C freezing point is lowered by 1.86°C to −1.86°C. (2) The 100.00°C boiling point is raised by 0.52°C to 100.52°C. The 23.8 mm Hg vapor pressure at 25°C is lowered by 0.4 mm Hg to 23.4 mm Hg. If the solution contained 2.00 mol of glucose, the changes would be twice as great. If 1.00 mol of solute ionizes in solution to give two moles of particles, the effect is the same as 2.00 mol of nonionized solute. For example, 1.00 mol of ionized NaCl has the same effect on the freezing point, boiling point, and vapor pressure as 2.00 mol of nonionized $C_6H_{12}O_6$.

When calculating the changes in colligative properties, the concentration of solute must be expressed as the molal concentration. The amount of change in freezing point, boiling point, or vapor pressure is proportional to the molal concentration of the solution. The molal concentration is the number of moles of solute dissolved in each kilogram of solvent; it is called **molality** (symbol ***m***). Molality can be expressed as follows:

molality (*m*) A solution concentration expression that relates the moles of solute dissolved in each kilogram of solvent.

$$\frac{\text{moles of solute}}{\text{kilogram of solvent}} = m$$

To see how to determine the molality of a solute, suppose we dissolve 18.0 g of $C_6H_{12}O_6$ (180.0 g/mol) in 1 kg of water. The solution has 0.100 mol $C_6H_{12}O_6$ per kg of water. That is,

$$\frac{18.0\ \cancel{\text{g } C_6H_{12}O_6}}{1.00\ \text{kg } H_2O} \times \frac{1\ \text{mol } C_6H_{12}O_6}{180.0\ \cancel{\text{g } C_6H_{12}O_6}} = 0.100\ m$$

Writing Unit Factors

We can solve molality calculations by unit analysis. As with molarity, the molal concentration is a unit factor. For example, a 0.100 *m* solution of $C_6H_{12}O_6$ contains 0.100 mol of solute in each kilogram of solvent. We can write a pair of unit factors:

$$\frac{0.100\ \text{mol } C_6H_{12}O_6}{1\ \text{kg } H_2O} \quad \text{and} \quad \frac{1\ \text{kg } H_2O}{0.100\ \text{mol } C_6H_{12}O_6}$$

Solving Molality Calculations

We can now apply unit factors derived from molal concentration to solution calculations. If a solution is prepared by dissolving 241 g sucrose, $C_{12}H_{22}O_{11}$, in 955 g of water, what is the molality of the solution?

In this problem, we begin by comparing the moles of sucrose solute to the mass of water. Second, we use the molar mass of $C_{12}H_{22}O_{11}$ (342.0 g/mol) to convert to moles of sucrose. The molal concentration is obtained as follows:

$$\frac{241\ \cancel{\text{g } C_{12}H_{22}O_{11}}}{955\ \cancel{\text{g } H_2O}} \times \frac{1\ \text{mol } C_{12}H_{22}O_{11}}{342.0\ \cancel{\text{g } C_{12}H_{22}O_{11}}} \times \frac{1000\ \cancel{\text{g } H_2O}}{1\ \text{kg } H_2O} = \frac{0.738\ \text{mol } C_{12}H_{22}O_{11}}{1\ \text{kg } H_2O}$$

The concentration of the solution is 0.738 *m*. The following example exercise further illustrates calculations based on the molal concentration of a solution.

EXAMPLE EXERCISE 14.12

What is the mass of ethanol, C_2H_5OH (46.0 g/mol), dissolved in 15.0 kg of water if the alcohol solution is 5.00 *m*?

Solution: The unknown quantity is grams of C_2H_5OH. The relevant given value is 15.0 kg H_2O. Setting up the unit analysis solution to the problem, we have

$$15.0\ \text{kg } H_2O \times \frac{\text{unit}}{\text{factors}} = \text{g } C_2H_5OH$$

Since the solution concentration is 5.00 *m*, there are 5.00 mol of solute in each 1 kg of H_2O. We can write the unit factors as

$$\frac{5.00 \text{ mol } C_2H_5OH}{1 \text{ kg } H_2O} \quad \text{and} \quad \frac{1 \text{ kg } H_2O}{5.00 \text{ mol } C_2H_5OH}$$

Select the first unit factor to cancel kg of H_2O. The molar mass of C_2H_5OH is used to convert from moles to grams. Thus,

$$15.0 \cancel{\text{kg } H_2O} \times \frac{5.00 \cancel{\text{mol } C_2H_5OH}}{1 \cancel{\text{kg } H_2O}} \times \frac{46.0 \text{ g } C_2H_5OH}{1 \cancel{\text{mol } C_2H_5OH}} = 3450 \text{ g } C_2H_5OH$$

SELF-TEST EXERCISE

What is the mass of water required to dissolve 10.0 g of glucose, $C_6H_{12}O_6$ (180.0 g/mol), in order to prepare a 0.100 *m* solution?

Answer: 0.556 kg H_2O

Freezing Point Lowering

One of the colligative properties is the ability of a dissolved solute to lower the freezing point of a pure solvent. For example, dissolving a solute in water lowers the freezing point below 0.0°C. A practical application of this principle is putting rock salt on wintery roads to melt the ice and snow. The rock salt solute, NaCl, lowers the freezing point several degrees. The ice melts at temperatures below zero, such as at −10°C. Another application is adding antifreeze to automobile radiators. The antifreeze lowers the freezing point of water in the cooling system. Since the coolant does not freeze, it does not expand and crack the engine block.

During the freezing process, a solvent loses energy and its molecules are attracted to one another. When a nonvolatile solute is added, its particles hinder the attraction between the solvent molecules. Thus, more energy must be lost from the solvent, and the freezing point is lowered. The amount of freezing point lowering is different for each solvent. That is, every solvent has its own response to a solute. The specific effect of a solute on a given solvent is the **molal freezing point constant** (symbol K_f). K_f can be applied as a unit factor and it has the units of degrees Celsius per molal, °C/*m*. Table 14.6 lists the molal freezing point constant for four common solvents.

molal freezing point constant (K_f) The number of degrees Celsius that a nonvolatile, nonionized solute lowers the freezing point of a one molal solution; the units are °C/m.

In addition to practical applications, quantitative experiments on freezing point lowering are carried out in the laboratory. The relationship of molality to the change in freezing point is given by the equation

$$m\, K_f = \Delta T_f$$

where ΔT_f is the freezing point lowering, *m* is the solution molality, and K_f is the molal freezing point constant for the solvent. The following examples illustrate this relationship.

EXAMPLE EXERCISE 14.13

Calculate the freezing point lowering of an antifreeze solution that contains 5.50 mol of ethylene glycol dissolved in each kilogram of water.

Solution: The unknown quantity is the freezing point lowering, ΔT_f. Since the antifreeze contains 5.50 mol of solute per kg solvent, the concentration is 5.50 *m*.

From Table 14.6, the value of K_f for water is 1.86°C/m. Using the above equation to solve the problem,

$$5.50\ \cancel{m} \times \frac{1.86°\text{C}}{\cancel{m}} = \Delta T_f$$

$$= 10.2°\text{C}$$

The freezing point of water in the antifreeze is lowered 10.2°C. Therefore, this antifreeze solution does not freeze until the temperature drops to −10.2°C.

SELF-TEST EXERCISE

Find the (a) freezing point lowering and the (b) freezing point of a solution that contains 0.750 mol of sugar dissolved in 1.25 kg of acetic acid.

Answers: **(a)** 2.34°C; **(b)** 14.3°C (16.6°C − 2.34°C)

Table 14.6 Molal Freezing Point Constants

Solvent	Molar Mass (g/mol)	Freezing Point (°C)	Molal Freezing Point Constant, K_f (°C/m)
water	18.0	0.0	1.86
acetic acid	60.0	16.6	3.90
ethyl alcohol	46.0	−117.3	1.99
benzene	78.0	5.5	5.12

EXAMPLE EXERCISE 14.14

A solution is prepared by dissolving 2.50 g of a unknown compound in 100.0 g of the solvent benzene. If the freezing point of benzene is lowered from 5.5°C to 4.0°C, what is the molar mass of the unknown?

Solution: To calculate the molar mass of the solute, we must first find the molality of the solution. The freezing point lowering of benzene is 5.5°C − 4.0°C = 1.5°C. In Table 14.6, we find that the value of K_f is 5.12°C/m. Using K_f as a unit factor, we can obtain the molality as follows:

$$1.5°\cancel{\text{C}} \times \frac{m}{5.12°\cancel{\text{C}}} = \frac{0.29 \text{ mol unknown}}{1 \text{ kg benzene}}$$

Since 2.50 g of the unknown compound are dissolved in 100.0 g of solvent and the molality is 0.29 mol solute per kg of benzene, we have

$$\frac{2.50 \text{ g unknown}}{100.0 \cancel{\text{ g benzene}}} \times \frac{1 \cancel{\text{ kg benzene}}}{0.29 \text{ mol unknown}} \times \frac{1000 \cancel{\text{ g benzene}}}{1 \cancel{\text{ kg benzene}}} = \frac{86 \text{ g unknown}}{\text{mol}}$$

The molar mass of the unknown organic compound is 86 g/mol.

SELF-TEST EXERCISE

A solution is prepared by dissolving 5.45 g of an unknown compound in 50.0 g of ethyl alcohol. If the freezing point of the alcohol is lowered from −117.3°C to −120.5°C, what is the molar mass of the compound?

Answer: 68 g/mol

14.11 Colloids

To distinguish between a solution and colloid.

When a solute is dissolved in a solvent, the solute particles are surrounded by the solvent molecules. The solute distributes itself uniformly throughout the solvent and shows no tendency to concentrate in any one place. The solute particles do not settle from the solution, even when the solution is centrifuged. The solute particles are small enough to pass through filter paper and through most membranes.

Colloids are similar to solutions in that their individual particles are dispersed throughout a gas, liquid, or solid. Under ordinary conditions, the dispersed particle in a colloid does not readily separate from the dispersing medium. Unlike solute particles, however, the large colloid particles demonstrate the **Tyndall effect** by scattering light (Figure 14.7).

Tyndall effect The phenomenon of scattering light by colloidal-size particles in solution.

true solution A homogeneous mixture in which the dispersed particles are less than 1 nm in diameter.

colloid A homogeneous mixture in which the diameter of the dispersed particles ranges from 1 to 100 nm.

The size of the dispersed particle distinguishes a solution from a colloid. A **true solution** contains solute particles that are smaller than one nanometer (1 nm) in diameter. A **colloid**, sometimes referred to as a colloidal dispersion, contains particles ranging in size from 1 to 100 nm. Colloids and solutions are both stable and their particles do not separate spontaneously.

In theory, ions and molecules remain in solution because of a strong solute—solvent interaction. Because of their larger size, colloidal particles are less strongly attracted to the dispersing medium. Colloidal particles will separate when spun in a centrifuge. Colloidal particles do not, however, settle spontaneously and are small enough to pass through ordinary filter paper. Colloidal particles are too large to pass through a natural cell membrane or a synthetic membrane like cellophane.

Kidney dialysis is a practical application of colloidal particles not passing through membranes. Blood is passed through an artificial kidney, which contains a semipermeable membrane and dialysis solution. Protein molecules, nucleic acids, and cells are retained by the blood because of their large size. Small toxin molecules as well as electrolytes pass through the membrane and out of the blood. The necessary electrolytes are restored to the blood from the dialyzing solution. After 2 to 3 hours of dialysis, a kidney patient will have blood that is "cleaned" of small toxic molecules.

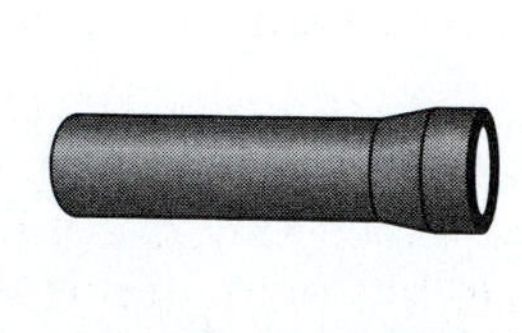
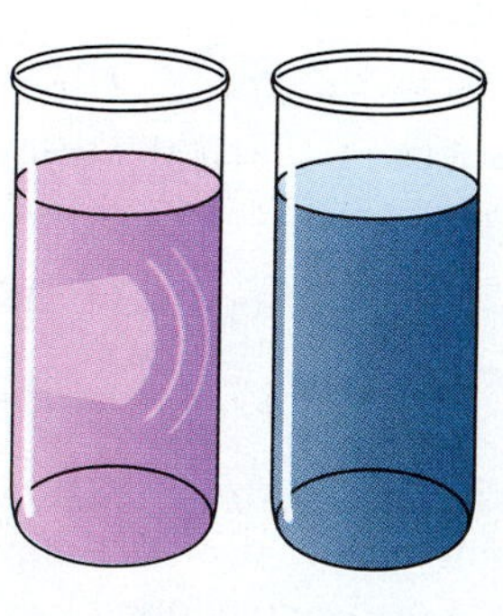
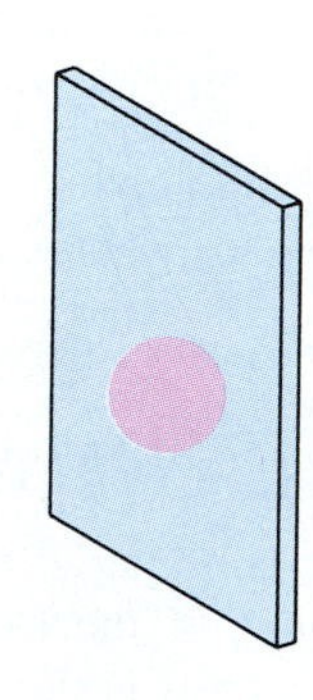

Figure 14.7 Tyndall Effect Why can we see the light beam in one glass and not in the other? Answer: In the left glass is a colloid with particles large enough to scatter light. On the right is a true solution, which does not scatter light.

Summary

Sections 14.1–14.2 In this chapter, we studied different types of solutions. A **solution** consists of a **solute** dissolved in a **solvent**. The most common solutions are formed from a solid, liquid, or gas dissolved in water. If the solute is a gas, its solubility is affected by temperature and pressure. Raising the temperature of the solution decreases the solubility of the dissolved gas. On the other hand, raising the partial pressure of the gas above the solution increases the amount of dissolved gas in the solution.

Section 14.2 We can usually predict the miscibility of two liquid solvents. The ***like dissolves like* rule** states that two liquids are **miscible** if they are both **polar** or both **nonpolar solvents**. It implies that a polar solvent and a nonpolar solvent are **immiscible**. Although it is an oversimplification, common polar solvents are small molecules containing one or more oxygen atoms, for example, H_2O or CH_3OH.

Section 14.3 We can often predict whether or not a solid compound is soluble in a given solvent. From the *like dissolves like* rule, a polar solvent will dissolve a polar compound and a nonpolar solvent will dissolve a nonpolar compound. Ionic compounds are extreme examples of polar compounds and are therefore not soluble in nonpolar solvents. As we learned earlier in Section 8.9, some ionic compounds are water soluble, but many are only slightly soluble.

Sections 14.4–14.5 In the process of dissolving a solid compound, solvent molecules attack particles that make up the compound and take them into solution. The solute particles are surrounded by solvent molecules that form a **solvent cage**. Molecular compounds have only nonmetal atoms. The fundamental particle is a molecule that is hydrated after dissolving in aqueous solution. Ionic compounds contain metal and nonmetal ions. Both the positive metal ion and the negative nonmetal ion are individually hydrated. The partially negative oxygen atom in a water molecule is attracted to the positive metal ion. The partially positive hydrogen end of the water molecule is attracted to the nonmetal ion. The rate of dissolving is increased by *heating the solution, stirring the solution,* or *grinding the solute*.

Sections 14.6–14.7 Generally, raising the temperature of a solution increases the solubility of a solid compound. By referring to a graph of solubility, we can determine the amount of dissolved solute in 100 g of water at a given temperature. If a solution contains the maximum amount of dissolved solute possible at a given temperature, the solution is said to be **saturated**. If the concentration of the solution is less than its maximum solubility, it is said to be **unsaturated**. Under special circumstances, it is possible for a solution concentration to exceed its maximum solubility. Such a solution is unstable and is said to be **supersaturated**.

Sections 14.8–14.9 Most often, solution concentration is quantitatively expressed as **mass/mass percent** (m/m%) or **molarity** (M). Mass percent concentration is the mass in grams of dissolved solute in 100 grams of solution. Molar concentration is the number of moles of dissolved solute in 1 liter of solution.

Section 14.10 Molality (m) is a concentration expression for the number of moles of dissolved solute in 1 kilogram of solvent. The molal concentration is associated with the **colligative properties** of a solution. The number of degrees a solute lowers the freezing point of a solvent is a characteristic property of the solvent called the **molal freezing point constant**.

Section 14.11 A **colloid** is similar to a **true solution** in that its particles are dispersed throughout another phase. In colloidal dispersions, the other phase contains particles having diameters between 1 and 100 nm. Like true solutions, colloids do not settle and cannot be filtered. Colloidal particles can, however, be separated in a centrifuge and are too large to pass through cell membranes.

Key Terms

Select the key term that corresponds to the following definitions.

____ **1.** the component of a solution that is the lesser quantity

____ **2.** the component of a solution that is the greater quantity

____ **3.** a general term for a solute dissolved in a solvent

____ **4.** the solubility of a gas in a liquid is proportional to the partial pressure of the gas above the liquid

____ **5.** the separation of partial positive and negative charge in a molecule resulting from a polar covalent bond

____ **6.** the overall effect of one or more dipoles operating in a molecule

____ **7.** a dissolving liquid composed of polar molecules

____ **8.** a dissolving liquid composed of nonpolar molecules

____ **9.** a general principle that solute–solvent interaction is greatest when the polarities of the solute and the solvent are similar

____ **10.** refers to two liquids that are soluble in one another

____ **11.** refers to two liquids that are not soluble in one another, and if mixed, separate into two layers

____ **12.** refers to a cluster of solvent molecules surrounding a solute molecule or ion in solution

____ **13.** refers to the maximum amount of solute that can be dissolved in a solvent at a specified temperature; usually expressed in grams of solute per 100 g of solvent

____ **14.** a solution that contains the maximum amount of dissolved solute that will dissolve at a given temperature

____ **15.** a solution that contains less than the maximum amount of dissolved solute that will dissolve at a given temperature

____ **16.** a solution that contains more dissolved solute than will ordinarily dissolve at a given temperature

____ **17.** a solution concentration expression that relates the mass of solute in grams dissolved in each 100 grams of solution

____ **18.** a solution concentration expression that relates the moles of solute dissolved in each liter of solution

____ **19.** a solution concentration expression that relates the moles of solute dissolved in each kilogram of solvent

____ **20.** the properties of a solution that are affected by the number (not the type) of solute particles in solution; for example, freezing point lowering, boiling point raising, and vapor pressure lowering

____ **21.** the number of degrees Celsius that a nonvolatile solute lowers the freezing point of a 1.00-molal solution; the units are °C/m

(a) colligative property *(Sec. 14.10)*
(b) colloid *(Sec. 14.11)*
(c) dipole *(Sec. 14.2)*
(d) Henry's law *(Sec. 14.1)*
(e) immiscible *(Sec. 14.2)*
(f) *like dissolves like* rule *(Sec. 14.2)*
(g) mass/mass percent (m/m%) *(Sec. 14.8)*
(h) miscible *(Sec. 14.2)*
(i) molal freezing point constant (K_f) *(Sec. 14.10)*
(j) molality (m) *(Sec. 14.10)*
(k) molarity (M) *(Sec. 14.9)*
(l) net dipole *(Sec. 14.2)*
(m) nonpolar solvent *(Sec. 14.2)*
(n) polar solvent *(Sec. 14.2)*
(o) saturated *(Sec. 14.7)*
(p) solubility *(Sec. 14.6)*
(q) solute *(Sec. 14.2)*
(r) solution *(Sec. 14.2)*
(s) solvent *(Sec. 14.2)*
(t) solvent cage *(Sec. 14.4)*
(u) supersaturated *(Sec. 14.7)*
(v) true solution *(Sec. 14.11)*
(w) Tyndall effect *(Sec. 14.11)*
(x) unsaturated *(Sec. 14.7)*

____ 22. a homogeneous mixture in which the dispersed particles are less than 1 nm in diameter

____ 23. a homogeneous mixture in which the diameter of the dispersed particles ranges from 1 to 100 nm

____ 24. the scattering of light by colloidal-size particles

Exercises

Gases in a Liquid Solution (Sec. 14.1)

1. Indicate whether the solubility of ammonia gas in water increases or decreases for each of the following.
   (a) temperature of the solution changes from 20° to 0°C
   (b) pressure of NH_3 changes from 754 to 775 mm Hg
2. Indicate whether the solubility of hydrogen chloride gas in water increases or decreases for each of the following.
   (a) temperature of the solution changes from 20° to 50°C
   (b) pressure of HCl changes from 29.0 to 27.5 in. Hg
3. If 1.45 g of carbon dioxide dissolve in a liter of champagne at 1.00 atm pressure, what is the solubility of carbon dioxide at 10.0 atm?
4. If the solubility of nitrogen is 1.90 cm^3 per 100 cm^3 of blood at 1.00 atm, what is the solubility of nitrogen in a scuba diver's blood at a depth of 185 ft where the pressure is 6.50 atm?
5. The solubility of chlorine gas is 0.63 g Cl_2/100 g of water at 25°C and 760 mm Hg. What is the solubility of chlorine gas in water at 25°C and 1200 mm Hg?
6. The solubility of nitrous oxide gas is 0.121 g N_2O/100 g water at 20°C and 1.00 atm. What is the partial pressure of nitrous oxide required to dissolve 1.18 g of the gas in 100 g of water at 20°C?

Liquids in a Liquid Solution (Sec. 14.2)

7. State whether the following combinations of solvents are miscible or immiscible.
   (a) polar solvent + polar solvent
   (b) polar solvent + nonpolar solvent
8. State whether the following combinations of solvents are miscible or immiscible.
   (a) nonpolar solvent + polar solvent
   (b) nonpolar solvent + nonpolar solvent
9. Predict whether the following solvents are polar or nonpolar.
   (a) water, H_2O (b) hexane, C_6H_{14}
   (c) acetone, C_3H_6O (d) chloroform, $CHCl_3$
10. Predict whether the following solvents are polar or nonpolar.
   (a) isopropyl alcohol, C_3H_7OH
   (b) pentane, C_5H_{12}
   (c) xylene, $C_6H_4(CH_3)_2$
   (d) trichloroethane, $C_2H_3Cl_3$
11. Predict whether the following solvents are miscible or immiscible with water.
   (a) heptane, C_7H_{16}
   (b) methyl alcohol, CH_3OH
   (c) methyl ethyl ketone, C_4H_8O
   (d) toluene, C_7H_8
12. Predict whether the following solvents are miscible or immiscible with hexane, C_6H_{14}.
   (a) ethyl alcohol, C_2H_5OH
   (b) chloroform, $CHCl_3$
   (c) trichloroethylene, C_2HCl_3
   (d) acetic acid, $HC_2H_3O_2$
13. In the laboratory, how could you quickly determine if an unknown liquid is polar or nonpolar?
14. An oil and vinegar salad dressing separates into two layers. Explain why the two liquids are immiscible using the *like dissolves like* rule.

Solids in a Liquid Solution (Sec. 14.3)

15. State whether the following combinations of solute and solvent are generally soluble or insoluble.
   (a) polar solute + polar solvent
   (b) nonpolar solute + polar solvent
   (c) ionic solute + polar solvent
16. State whether the following combinations of solute and solvent are generally soluble or insoluble.
   (a) polar solute + nonpolar solvent
   (b) nonpolar solute + nonpolar solvent
   (c) ionic solute + nonpolar solvent
17. Predict whether the following compounds are soluble or insoluble in water.
   (a) naphthalene, $C_{10}H_8$
   (b) potassium hydroxide, KOH
   (c) calcium acetate, $Ca(C_2H_3O_2)_2$
   (d) trichlorotoluene, $C_7H_5Cl_3$
   (e) glycine, $C_2H_5NO_2$
   (f) lactic acid, $HC_3H_5O_3$

18. Predict whether the following compounds are soluble or insoluble in hexane, C_6H_{14}.
(a) trichloroethylene, C_2HCl_3
(b) iron(III) nitrate, $Fe(NO_3)_3$
(c) sulfuric acid, H_2SO_4
(d) dodecane, $C_{12}H_{26}$
(e) mothballs, $C_6H_4Cl_2$
(f) tartaric acid, $H_2C_4H_4O_6$

19. Predict whether the following vitamins are water soluble or fat soluble.
(a) vitamin B_1, $C_{12}H_{18}Cl_2N_4OS$
(b) vitamin B_3, $C_6H_6N_2O$
(c) vitamin B_6, $C_8H_{11}NO_3$
(d) vitamin C, $C_6H_8O_6$
(e) vitamin D, $C_{27}H_{44}O$
(f) vitamin K, $C_{31}H_{46}O_2$

20. Predict whether the following biochemical compounds are water soluble or fat soluble.
(a) cholesterol, $C_{27}H_{46}O$
(b) citric acid, $C_6H_8O_7$
(c) fructose, $C_6H_{12}O_6$
(d) glycine, $CH_2(NH_2)COOH$
(e) urea, $CO(NH_2)_2$
(f) alanine, $CH_3CH(NH_2)COOH$

The Dissolving Process (Sec. 14.4)

21. Diagram a molecule of fructose, $C_6H_{12}O_6$, dissolved in water.

22. Diagram a crystal of sucrose, $C_{12}H_{22}O_{11}$, dissolving in aqueous solution.

23. Diagram a formula unit of the following substances dissolved in water.
(a) lithium bromide, LiBr
(b) calcium chloride, $CaCl_2$

24. Diagram a formula unit of the following substances dissolved in water.
(a) cobalt(II) sulfate, $CoSO_4$
(b) nickel(II) nitrate, $Ni(NO_3)_2$

Rate of Dissolving (Sec. 14.5)

25. What three factors increase the rate of dissolving of a solid substance in solution?

26. Indicate whether each of the following increases, decreases, or has no effect on the rate of dissolving 10.0 g of sugar in a liter of water.
(a) using chilled rather than tap water
(b) shaking rather than not disturbing the solution
(c) using powdered rather than crystalline sugar
(d) using tap water rather than distilled water

Solubility and Temperature (Sec. 14.6)

27. How many grams of the following solutes can dissolve in 100 g of water at 20°C? (Refer to Figure 14.5.)
(a) NaCl (b) KCl

28. How many grams of the following solutes can dissolve in 100 g of water at 30°C?
(a) LiCl (b) $C_{12}H_{22}O_{11}$

29. Determine the maximum solubility of the following solid compounds at 40°C. (Refer to Figure 14.5.)
(a) NaCl (b) KCl

30. Determine the maximum solubility of the following solid compounds at 50°C.
(a) LiCl (b) $C_{12}H_{22}O_{11}$

31. What is the minimum temperature required to dissolve each of the following? (Refer to Figure 14.5.)
(a) 34 g NaCl in 100 g water
(b) 50 g KCl in 100 g water

32. What is the minimum temperature required to dissolve each of the following?
(a) 90 g LiCl in 100 g water
(b) 120 g $C_{12}H_{22}O_{11}$ in 100 g water

33. At what temperature is each of the following solutions saturated? (Refer to Figure 14.5.)
(a) 40 g NaCl/100 g water (b) 45 g KCl/100 g water

34. At what temperature is each of the following solutions saturated?
(a) 105 g LiCl/100 g water
(b) 140 g $C_{12}H_{22}O_{11}$/100 g water

Unsaturated, Saturated, Supersaturated Solutions (Sec. 14.7)

35. State whether the following solutions are saturated, unsaturated, or supersaturated. (Refer to Figure 14.6.)
(a) 120 g $NaC_2H_3O_2$ in 100 g of water at 50°C
(b) 120 g $NaC_2H_3O_2$ in 100 g of water at 80°C
(c) 120 g $NaC_2H_3O_2$ in 100 g of water at 90°C

36. State whether the following solutions are saturated, unsaturated, or supersaturated. (Refer to Figure 14.5.)
(a) 105 g $C_{12}H_{22}O_{11}$/100 g H_2O at 25°C
(b) 120 g $C_{12}H_{22}O_{11}$/100 g H_2O at 50°C
(c) 130 g $C_{12}H_{22}O_{11}$/100 g H_2O at 45°C.

37. State whether the following solutions are saturated, unsaturated, or supersaturated. (Refer to Figure 14.5.)
(a) 40 g KCl/100 g H_2O at 20°C
(b) 40 g KCl/100 g H_2O at 35°C
(c) 40 g KCl/100 g H_2O at 50°C

38. State whether the following solutions are saturated, unsaturated, or supersaturated. (Refer to Figure 14.5.)
(a) 35 g NaCl/100 g H_2O at 0°C
(b) 35 g NaCl/100 g H_2O at 25°C
(c) 35 g NaCl/100 g H_2O at 100°C

39. The solubility of rock salt at 30°C is 40.0 g per 100 g of water. If a solution contains 10.0 g of rock salt in 25.0 g of water at 30°C, is the solution saturated, unsaturated, or supersaturated?

40. The solubility of sugar at 50°C is 100.0 g per 100 g of water. If a solution contains 95.0 g of sugar in 250 g of water at 50°C, is the solution saturated, unsaturated, or supersaturated?

41. Assume 100 g of LiCl is dissolved in 100 g of water at 100°C and the solution is allowed to cool to 20°C. (Refer to Figure 14.5.)
(a) How much solute remains in solution?
(b) How much solute crystallizes from solution?

42. Assume 120 g of $C_{12}H_{22}O_{11}$ is dissolved in 100 g of water at 100°C and the solution is allowed to cool to 20°C. (Refer to Figure 14.5.)
(a) How much solute remains in solution?
(b) How much solute crystallizes from solution?

Mass Percent Concentration (Sec. 14.8)

43. Calculate the mass/mass percent concentration for each of the following solutions.
(a) 1.25 g NaCl in 100.0 g solution
(b) 2.50 g $K_2Cr_2O_7$ in 95.0 g solution
(c) 10.0 g $CaCl_2$ in 250.0 g solution
(d) 65.0 g sugar in 125.0 g solution

44. Calculate the mass/mass percent concentration for each of the following solutions.
(a) 20.0 g KI in 100.0 g of water
(b) 2.50 g $AgC_2H_3O_2$ in 95.0 g of water
(c) 5.57 g $SrCl_2$ in 225.0 g of water
(d) 50.0 g sugar in 250.0 g of water

45. Write three pairs of unit factors for each of the following aqueous solutions given the mass/mass percent concentration.
(a) 1.50% KBr (b) 2.50% $AlCl_3$
(c) 3.75% $AgNO_3$ (d) 4.25% Li_2SO_4

46. Write three pairs of unit factors for each of the following aqueous solutions given the mass/mass percent concentration.
(a) 3.35% $MgCl_2$ (b) 5.25% $Cd(NO_3)_2$
(c) 6.50% Na_2CrO_4 (d) 7.25% $ZnSO_4$

47. What mass of solution contains the following amount of dissolved solute?
(a) 5.36 g of glucose in a 10.0% solution
(b) 25.0 g of sucrose in a 12.5% solution

48. What mass of solution contains the following amount of dissolved solute?
(a) 35.0 g of sulfuric acid in a 5.00% solution
(b) 10.5 g of acetic acid in a 4.50% solution

49. How many grams of solute are dissolved in the following solutions?
(a) 85.0 g of 2.00% $FeBr_2$ solution
(b) 105.0 g of 5.00% Na_2CO_3 solution

50. How many grams of solute are dissolved in the following solutions?
(a) 10.0 g of 6.00% KOH solution
(b) 50.0 g of 5.00% nitric acid, HNO_3

51. What mass of water is necessary to prepare each of the following solutions?
(a) 250.0 g of 0.90% saline solution
(b) 100.0 g of 5.00% sugar solution

52. What mass of water is necessary to prepare each of the following solutions?
(a) 250.0 g of 10.0% NaOH solution
(b) 100.0 g of 5.00% hydrochloric acid, HCl

Molar Concentration (Sec. 14.9)

53. Calculate the molar concentration for each of the following solutions.
(a) 1.50 g NaCl in 100.0 mL of solution
(b) 1.50 g $K_2Cr_2O_7$ in 100.0 mL of solution
(c) 5.55 g $CaCl_2$ in 125 mL of solution
(d) 5.55 g Na_2SO_4 in 125 mL of solution

54. Calculate the molar concentration for each of the following solutions.
(a) 1.00 g KCl in 75.0 mL of solution
(b) 1.00 g Na_2CrO_4 in 75.0 mL of solution
(c) 20.0 g $MgBr_2$ in 250.0 mL of solution
(d) 20.0 g Li_2CO_3 in 250.0 mL of solution

55. Write two pairs of unit factors for each of the following aqueous solutions.
(a) 0.100 *M* LiI (b) 0.100 *M* $NaNO_3$
(c) 0.500 *M* K_2CrO_4 (d) 0.500 *M* $ZnSO_4$

56. Write two pairs of unit factors for each of the following aqueous solutions.
(a) 0.150 *M* KBr (b) 0.150 *M* $Ca(NO_3)_2$
(c) 0.333 *M* $Sr(C_2H_3O_2)_2$ (d) 0.333 *M* NH_4Cl

57. What volume of each of the following solutions contains the indicated amount of dissolved solute?
(a) 10.0 g solute in 0.275 *M* NaF
(b) 10.0 g solute in 0.275 *M* $CdCl_2$
(c) 10.0 g solute in 0.408 *M* K_2CO_3
(d) 10.0 g solute in 0.408 *M* $Fe(ClO_3)_3$

58. What volume of each of the following solutions contains the indicated amount of dissolved solute?
(a) 2.50 g solute in 0.325 *M* KNO_3
(b) 2.50 g solute in 0.325 *M* $AlBr_3$
(c) 2.50 g solute in 1.00 *M* $Co(C_2H_3O_2)_2$
(d) 2.50 g solute in 1.00 *M* $(NH_4)_3PO_4$

59. What is the mass of solute dissolved in the indicated volume of each of the following solutions?
(a) 1.00 L of 0.100 *M* NaOH
(b) 1.00 L of 0.100 *M* $LiHCO_3$
(c) 25.0 mL of 0.500 *M* $CuCl_2$
(d) 25.0 mL of 0.500 *M* $KMnO_4$

60. What is the mass of solute dissolved in the indicated volume of each of the following solutions?
(a) 2.25 L of 0.200 *M* $FeCl_3$
(b) 2.25 L of 0.200 *M* KIO_4
(c) 50.0 mL of 0.295 *M* $ZnSO_4$
(d) 50.0 mL of 0.295 *M* $Ni(NO_3)_2$

61. What is the molar concentration of a saturated solution of calcium sulfate that contains 0.209 g of solute in 100 mL of solution?

62. What is the molar concentration of a saturated solution of calcium hydroxide that contains 0.185 g of solute in 100 mL of solution?

63. A normal hospital glucose solution is analyzed to check its concentration. A 10.0-mL sample, with a mass of 10.483 g, is evaporated to dryness. If the solid glucose residue has a mass of 0.524 g, what is the (a) mass percent concentration and (b) molar concentration of the glucose, $C_6H_{12}O_6$, solution?

64. A normal hospital saline solution is analyzed to confirm its concentration. A 50.0-mL sample, with a mass of 50.320 g, is evaporated to dryness. If the solid sodium chloride has a mass of 0.453 g, find the (a) mass/mass percent concentration and (b) molar concentration of the NaCl solution.

Molal Concentration and Colligative Properties (Sec. 14.10)

65. What is the molality of each of the following solutions?
   (a) 10.0 g of potassium fluoride, KF, is dissolved in 2.50 kg of water
   (b) 10.0 g of zinc sulfate, $ZnSO_4$, is dissolved in 375 g of water

66. What is the molality of each of the following solutions?
   (a) 55.0 g of ammonium chloride is dissolved in 2.50 kg of water
   (b) 55.0 g of calcium nitrate is dissolved in 375 g of water

67. What is the mass of sucrose, $C_{12}H_{22}O_{11}$, that must be dissolved in 6.5 kg of water in order to prepare a 2.00 *m* solution?

68. What is the mass of methanol, CH_3OH, that must be dissolved in 125 g of water in order to prepare a 0.500 *m* solution?

69. Calculate the freezing point lowering for a solution that contains 100.0 g of methanol, CH_3OH, dissolved in 500.0 g of water. What is the freezing point of the solution?

70. Calculate the freezing point lowering for a solution that contains 125 g of methanol, CH_3OH, dissolved in 1.15 kg of ethyl alcohol. What is the freezing point of the solution?

71. An aqueous solution contains 36.0 g of an unknown sugar dissolved in 100.0 g of water. If the freezing point is −3.72°C, what is the molar mass of the sugar?

72. A solution containing 7.50 g of an unknown compound dissolved in 75.0 g of acetic acid has a freezing point of 12.2°C. What is the molar mass of the unknown compound?

73. A solution is prepared by dissolving 4.50 g of an unknown compound in 100.0 g of ethyl alcohol. If the freezing point of the alcohol is lowered from −117.3 to −119.8°C, what is the molar mass of the compound?

74. A solution is prepared by dissolving 3.00 g of an unknown compound in 88.5 g of benzene. If the freezing point of benzene is lowered from 5.5 to 4.0°C, what is the molar mass of the unknown?

Colloids (Sec. 14.11)

75. In foggy weather, automobile headlights demonstrate the Tyndall effect. What is the approximate size of the water droplets in the fog?

76. At a concert in an auditorium, the stage lights demonstrate the Tyndall effect. What can you conclude about the air in the concert auditorium?

77. Indicate whether a solution or colloid produces the following observations.
   (a) dispersed particles separate in a centrifuge
   (b) dispersed particles demonstrate the Tyndall effect
   (c) dispersed particles pass through a semipermeable membrane

78. Indicate whether or not a colloid produces the following observations.
   (a) dispersed particles settle from solution
   (b) dispersed particles scatter light
   (c) dispersed particles separate with filter paper

General Exercises

79. Scuba divers can experience "the bends" if they rise to the surface too quickly. The nitrogen gas in the air they breathe is more soluble in the blood under the pressure at depth. The nitrogen is less soluble as the divers return to the surface. As the nitrogen leaves the blood, it can produce a painful sensation. The pressure at 125 ft is 4.68 atm. If the solubility is 0.0019 g N_2/100 g blood at normal pressure, what is the solubility at 125 ft?

80. Propose an explanation for why scuba divers who dive deeper than 125 ft use a special gas mixture of oxygen in helium. (Refer to exercise 79.)

81. Calculate the mass of sulfur dioxide gas in 1.00 L of saturated solution at 20°C. The solubility of SO_2 at 20°C is 22.8 g/100 mL.

82. Calculate the mass of chlorine gas in 500.0 mL of saturated solution at 20°C. The solubility of Cl_2 at 20°C is 0.63 g/100 mL.

83. Explain why bubbles form on the bottom and sides of a pan when water is heated.

84. Amyl alcohol, $C_5H_{11}OH$, is partially miscible with water. Explain why amyl alcohol is only partially soluble even though the molecule contains a polar −OH group.

85. Listed below are the solubilities of different alcohols in water. Propose an explanation for the decrease in solubility.

Alcohol	Solubility
ethanol, C_2H_5OH	miscible
propanol, C_3H_7OH	97.2 g/100 mL H_2O
butanol, C_4H_9OH	7.9 g/100 mL H_2O
hexanol, $C_6H_{13}OH$	0.59 g/100 mL H_2O
decanol, $C_{10}H_{21}OH$	immiscible

86. Predict whether water or carbon tetrachloride, CCl_4, is a better solvent for the following household substances.
(a) grease (b) maple syrup
(c) food coloring (d) gasoline

87. Identify the solutes and solvents in the following solutions.
(a) 80 proof vodka (40% ethyl alcohol in water)
(b) laboratory alcohol (95% ethyl alcohol in water)

88. Explain why grinding the solute increases the rate of dissolving of a solid substance in water.

89. If a household bleach solution contains 5.25% NaClO, what is the molarity of the sodium hypochlorite? (Assume that the solution density is 1.04 g/mL.)

90. If a vinegar solution contains 5.25% $HC_2H_3O_2$, what is the molarity of the acetic acid? (Assume that the solution density is 1.01 g/mL.)

91. After a 12-hour fast, normal blood glucose levels range from 70 to 90 milligrams per deciliter. If a patient's blood sample shows 0.0075 g of glucose in 10.0 mL of blood, is this a normal reading?

92. Explain why the aqueous 0.1 *m* solutions listed below have different freezing points.

Solution	Freezing Point (°C)
0.1 *m* sugar	−0.186
0.1 *m* NaCl	−0.372
0.1 *m* $BaCl_2$	−0.558

CHAPTER

15

Acids and Bases

Tropical fish are very sensitive to pH (acidity) changes in their environment. If the pH is too high or too low in the aquarium, this angel fish will not survive.

Acids and bases play an important role in our lives. The proper acidity of our blood and digestive fluids is vital to our well-being and carefully controlled by an elaborate buffering system. Many foods, including citric acid in citrus fruits and acetic acid in vinegar, are acidic. So are ascorbic acid in vitamin C and acetylsalicylic acid in aspirin. Baking soda and milk of magnesia are basic. We take antacid tablets that contain basic substances such as carbonates, bicarbonates, and hydroxides for an upset stomach. We adjust the acidity of our swimming pools with muriatic acid and fill our car batteries with sulfuric acid. We use a dilute solution of basic ammonia to clean floors and a concentrated solution of caustic sodium hydroxide to clean drains and ovens.

In this chapter we will specifically discuss many of these substances and generally describe the behavior of acids and bases.

15.1 Properties of Acids and Bases

To list properties of acids and bases.

To classify a solution with a given pH as strongly acidic, weakly acidic, neutral, weakly basic, or strongly basic.

An *acid* is any substance that produces hydrogen ions, H^+, in water. This fact lets us test a substance to see if it is acidic. In the laboratory, we can determine if a solution is acidic using blue litmus paper. Blue litmus paper turns red in the presence of hydrogen ions. Therefore, if we put a piece of blue litmus paper in a solution and the paper turns red, we know the solution is an acid. Litmus paper is made from a plant pigment impregnated in strips of paper. In fact, the colors of many plants are affected by the acidity of the soil in which they are grown. For example, the color of an orchid can vary from pale lavender to deep purple depending on the acidity of the soil. The color of a rose will fade from red to pink by placing it in an acid solution.

Acids have their own special properties. One property is that they have a sour taste. The tart taste of a lemon, an apple, or vinegar shows us that these are acidic foods. The taste buds that are sensitive to acids are located along the edge of our tongue. This is why we roll our tongue when we taste something that is sour. Another property of an acid is its pH value. The **pH** value expresses the acidity of a solution. A solution with a pH value of less than 7 is acidic. As the acidity increases, the pH

pH The concentration of hydrogen ion expressed on an exponential scale.

Table 15.1 Properties of Acids and Bases

Property	Acidic Solutions	Basic Solutions
taste	sour	bitter
feel	—	slippery, soapy
litmus paper	blue litmus turns red	red litmus turns blue
pH value	less than 7	greater than 7
neutralization reaction	react with bases to give a salt and water	react with acids to give a salt and water

value decreases. Thus, a solution having a pH of 3 is more acidic than a solution with a pH of 4.

A *base* is any substance that produces hydroxide ions, OH^-, in solution. Bases also have special properties. One is that they feel slippery or soapy to the touch. Another is that they have a bitter taste. The basic substance, unflavored milk of magnesia, has a bitter taste and may cause nauseousness. This sensation is due to the taste buds on the back of the tongue. When a basic substance contacts these taste buds, it often produces a response whereby we stick out our tongue and feel nauseous.

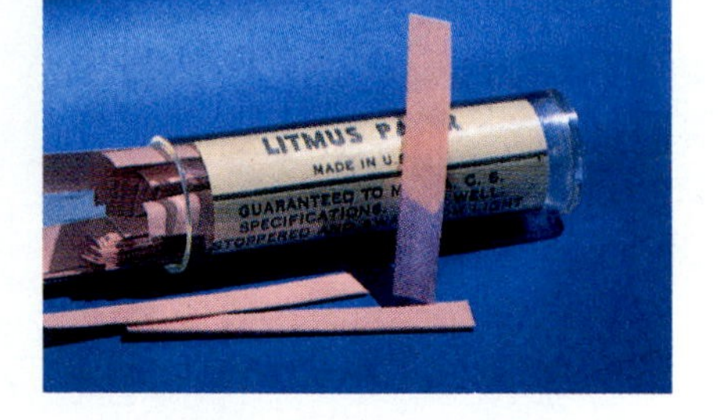

Strips of red litmus paper indicate a basic solution by turning blue.

In the laboratory, we can determine if a solution is basic using red litmus paper. Red litmus paper turns blue in the presence of hydroxide ion. The pH value of a basic solution is greater than 7. As the pH value increases above 7, the basicity increases. Thus, a solution with a pH of 11 is more basic than a solution having a pH of 10.

Another property of acids and bases is their neutralization reaction, which we studied in Section 8.11. An acid and a base react to produce a salt and water. For example, hydrochloric acid neutralizes a potassium hydroxide solution to give potassium chloride and water:

$$HCl(aq) + KOH(aq) \longrightarrow KCl(aq) + H_2O(l)$$

The properties of acids and bases are summarized in Table 15.1.

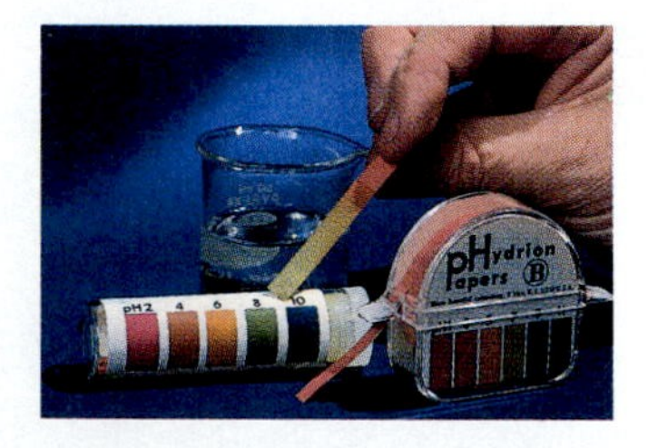

Broad-range indicator strips are used to measure the pH of a solution. The strip is dipped into a solution and the resulting color is compared to a color chart.

pH Scale

In Section 15.8, we will discuss the pH scale in detail. For now, we need to know that most solutions have a pH between 0 and 14. On the pH scale, acidic solutions have a pH less than 7, and basic solutions have a pH greater than 7. For example, a 1 *M* HCl solution has a pH of 0, and a 1 *M* NaOH solution has a pH of 14. A pH of 7 is considered neutral. Pure distilled water has a pH of 7.

On the basis of pH, a solution can be classified as strongly acidic, weakly acidic, neutral, weakly basic, or strongly basic. A strongly acidic solution has a pH between 0 and 2. A weakly acidic solution has a pH between 2 and 7. A weakly basic solution has a pH between 7 and 12. A strongly basic solution has a pH between 12 and 14. Table 15.2 lists common substances and their approximate pH.

The following example exercise illustrates the acid/base strength of a few common solutions.

EXAMPLE EXERCISE 15.1

Indicate whether the following solutions are considered to be strongly acidic, weakly acidic, neutral, weakly basic, or strongly basic.

(a) gastric juice, pH 1.5 **(b)** oven cleaner, pH 13.5

Table 15.2 The pH scale.

Acidic/Basic	pH	Example Solution
Strongly acidic	0	1 M HCl
	1	Stomach acid (1-3)
Weakly acidic	2	Lemon juice
	3	Vinegar, wine
	4	Grapes, orange juice
	5	Normal rain, coffee
	6	Milk, pH balanced shampoo
Neutral	7	Pure water
Weakly basic	8	Eggs, seawater
	9	Baking soda, antacids
	10	Milk of magnesia, soap
	11	Household ammonia
	12	Liquid bleach
Strongly basic	13	Drain cleaner
	14	1 M NaOH

(c) urine, pH 5.5
(d) saliva, pH 7.0
(e) blood, pH 7.32
(f) carbonated drink, pH 4.0

Solution: Refer to the guidelines in Table 15.2 to classify each solution.
(a) This gastric juice sample has a pH between 0 and 2. It is *strongly acidic*.
(b) Oven cleaner has a pH of 13.5. It is *strongly basic*.
(c) This urine specimen has a pH between 2 and 7. It is *weakly acidic*.
(d) This saliva specimen has a pH of 7 and is *neutral*.
(e) Blood has a pH of 7.32 and is *weakly basic*.
(f) A carbonated drink has a pH of 4.0 and is *weakly acidic*.

SELF-TEST EXERCISE

Indicate whether the following properties correspond to an acid or base.
(a) sour taste
(b) slippery feel
(c) turns blue litmus paper red
(d) pH greater than 7

Answers: **(a)** acid; **(b)** base; **(c)** acid; **(d)** base

15.2 Acid–Base Indicators

To state the color of the following indicators in a solution of given pH: phenolphthalein, methyl red, and bromthymol blue.

We mentioned in the previous section that litmus paper can be used to indicate whether a solution is acidic or basic. In addition, other paper test strips are commer-

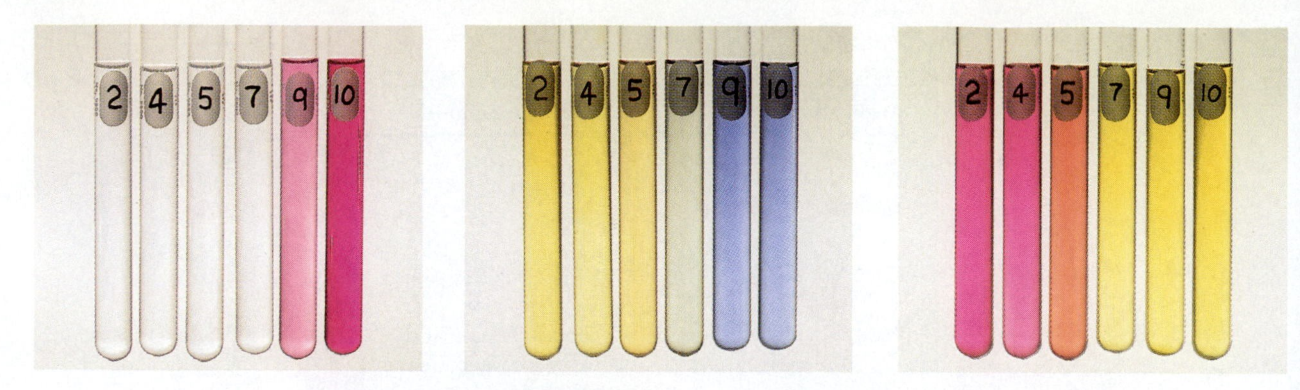

Figure 15.1 pH and Acid–Base Indicator Color The photograph shows the colors of three acid–base indicators at different pH values. Phenolphthalein (left) changes from colorless to pink at pH 9. Bromthymol blue (middle) turns blue at pH 7. Methyl red (right) changes from red to yellow at pH 5.

cially available that indicate the pH of a solution. These test strips are impregnated with substances that change color depending on the pH of the test solution. Examples of such substances are extracts from grape juice, red cabbage, and flower petals such as roses, violets, and orchids.

acid–base indicator A chemical substance that changes color according to the pH of the solution.

A solution that is pH sensitive and changes color is referred to as an **acid–base indicator**. Phenolphthalein, methyl red, and bromthymol blue are common indicators. In solutions having a pH of 9 or above, phenolphthalein is pink. In solutions having a pH below 9, the indicator is colorless. In solutions having a pH above 5, the indicator methyl red is yellow. In solutions with a pH below 5, the indicator is red. At a pH of 5, methyl red appears orange because only a portion of the indicator has been converted from yellow to red. That is, the indicator appears orange from an equal mixture of red and yellow. Dozens of acid–base indicators are available in the laboratory. A chemist selects an indicator on the basis of the pH at which it changes color (Figure 15.1). Table 15.3 lists three common indicators.

The following example exercise further illustrates the relationship of the color of an acid–base indicator and the pH of a given solution.

EXAMPLE EXERCISE 15.2

State the color of the acid–base indicator in each of the following solutions.

(a) A solution at pH 4 containing a drop of methyl red.
(b) A solution at pH 8 containing a drop of bromthymol blue.
(c) A solution at pH 10 containing a drop of phenolphthalein.

Table 15.3 Acid–Base Indicators

Indicator	Color Change	Color
methyl red	pH ~5	red, below pH 5 yellow, above pH 5
bromthymol blue	pH ~7	yellow, below pH 7 blue, above pH 7
phenolphthalein	pH ~9	colorless, below pH 9 pink, above pH 9

Solution: Refer to Table 15.3 to determine the color of each solution.
(a) A pH 4 solution containing methyl red indicator is *red.*
(b) A pH 8 solution containing bromthymol blue indicator is *blue.*
(c) A pH 10 solution containing phenolphthalein indicator is *pink.*

SELF-TEST EXERCISE

State the pH at which each of the following indicators changes color.
(a) methyl red **(b)** bromthymol blue
(c) phenolphthalein

Answers: **(a)** 5; **(b)** 7; **(c)** 9

15.3 Standard Solutions of Acids and Bases

To calculate the unknown quantity for a standard solution of acid or base, given any three of the following:
- **(a)** mass of the solid acid or solid base
- **(b)** molar mass of the solid acid or solid base
- **(c)** molarity of the acid or base
- **(d)** volume of the acid or base

A **standard solution** of an acid or base is a solution in which we know the concentration precisely, for example, three significant digits. Chemists use standard solutions to analyze substances, such as the neutralizing capacity of a commercial antacid tablet or the tartness of a wine. Standard solutions are also used in manufacturing processes to assure quality. Here, chemical samples are selected randomly and then analyzed using a standard solution. This important part of chemical manufacturing is called quality control (QC) or quality assurance (QA).

standard solution A solution whose concentration has been established precisely (usually to three or four significant digits).

To standardize a solution of acid, we can use a weighed quantity of a solid base. To standardize hydrochloric acid, for example, we often use solid sodium carbonate, Na_2CO_3. Let's find the molarity of hydrochloric acid if 25.50 mL of solution is required to neutralize 0.375 g of Na_2CO_3. The balanced equation for the reaction is

$$2\ HCl(aq) + Na_2CO_3(s) \longrightarrow 2\ NaCl(aq) + H_2O(l) + CO_2(g)$$

To calculate the molarity of the hydrochloric acid, we must first find the number of moles of HCl. From the balanced equation, we note that 2 moles of HCl react with 1 mole of Na_2CO_3 (106.0 g/mol). Thus,

$$0.375\ \cancel{g\ Na_2CO_3} \times \frac{1\ \cancel{mol\ Na_2CO_3}}{106.0\ \cancel{g\ Na_2CO_3}} \times \frac{2\ mol\ HCl}{1\ \cancel{mol\ Na_2CO_3}} = 0.00708\ mol\ HCl$$

To obtain the molarity of the acid, we divide the moles of HCl by the volume of HCl required to neutralize the sodium carbonate.

$$\frac{0.00708\ mol\ HCl}{25.50\ \cancel{mL\ solution}} \times \frac{1000\ \cancel{mL}}{1\ L} = \frac{0.277\ mol\ HCl}{1\ L\ solution}$$

The standard hydrochloric acid solution is 0.277 *M* HCl.

To standardize a solution of base, we can use a weighed quantity of a solid acid. To standardize aqueous sodium hydroxide, for example, we can use crystals of oxalic acid, $H_2C_2O_4$. We dissolve a weighed sample of $H_2C_2O_4$ in water and neutralize it with a measured volume of the basic solution. Example Exercise 15.3 illustrates the calculation for the concentration of a standard NaOH solution.

EXAMPLE EXERCISE 15.3

What is the molarity of a sodium hydroxide solution if 42.15 mL of NaOH are required to neutralize 0.424 g of solid oxalic acid, $H_2C_2O_4$ (90.0 g/mol)? The balanced equation for the reaction is

$$H_2C_2O_4(s) + 2\ NaOH(aq) \longrightarrow Na_2C_2O_4(aq) + 2\ H_2O(l)$$

Solution: To calculate the molarity of the NaOH, we must find the number of moles of NaOH. The mass of $H_2C_2O_4$ is 0.424 g and the molar mass is 90.0 g/mol. From the balanced equation, we see that 2 moles of NaOH react with 1 mole of $H_2C_2O_4$.

$$0.424\ \cancel{g\ H_2C_2O_4} \times \frac{1\ \cancel{mol\ H_2C_2O_4}}{90.0\ \cancel{g\ H_2C_2O_4}} \times \frac{2\ mol\ NaOH}{1\ \cancel{mol\ H_2C_2O_4}} = 0.00942\ mol\ NaOH$$

We obtain the molar concentration of the base from moles of NaOH divided by the volume of base required to neutralize the oxalic acid.

$$\frac{0.00942\ mol\ NaOH}{42.15\ \cancel{mL\ solution}} \times \frac{1000\ \cancel{mL}}{1\ L} = \frac{0.223\ mol\ NaOH}{1\ L\ solution}$$

The standard sodium hydroxide base solution is 0.223 *M* NaOH.

SELF-TEST EXERCISE

If an unknown sample of oxalic acid, $H_2C_2O_4$, is neutralized by 33.50 mL of 0.223 *M* NaOH, what is the mass of the sample? Refer to the balanced chemical equation of oxalic acid in the above example exercise.

Answer: 0.336 g $H_2C_2O_4$

Molar Mass of a Solid Acid or Base

One application of a standard solution is to determine the molar mass of a solid acid or base. For example, consider solid citric acid, which is added to some soft drinks to provide a tart taste. We begin by dissolving crystals of citric acid in water and neutralize the solution using a standard base. Given the mass of citric acid, the volume of standard base, and the balanced equation for the reaction, we can calculate the molar mass of citric acid. Example Exercise 15.4 illustrates the calculation for the molar mass of a solid acid.

EXAMPLE EXERCISE 15.4

Citric acid has three replaceable hydrogen ions and can be abbreviated H_3Cit. If 36.10 mL of 0.223 *M* NaOH neutralize a 0.515-g sample of citric acid, what is the molar mass of the acid? The balanced equation for the reaction is

$$H_3Cit(s) + 3\ \ NaOH(aq) \longrightarrow Na_3Cit(aq) + 3\ H_2O(l)$$

Solution: The first step is to find the moles of NaOH. The volume is 36.10 mL and the concentration is 0.223 *M*. The molarity of the NaOH solution can be used as a unit factor: 0.223 mol NaOH/1000 mL solution.

$$36.10 \text{ \sout{mL solution}} \times \frac{0.223 \text{ mol NaOH}}{1000 \text{ \sout{mL solution}}} = 0.00805 \text{ mol NaOH}$$

Next, we can find the moles of citric acid that reacted. From the balanced equation, 3 moles of NaOH react with 1 mole of H_3Cit.

$$0.00805 \text{ \sout{mol NaOH}} \times \frac{1 \text{ mol } H_3Cit}{3 \text{ \sout{mol NaOH}}} = 0.00268 \text{ mol } H_3Cit$$

The molar mass is defined as grams per mole; thus, we have

$$\frac{0.515 \text{ g } H_3Cit}{0.00268 \text{ mol } H_3Cit} = 192 \text{ g/mol}$$

The calculated molar mass of citric acid is 192 g/mol. The actual formula for citric acid is $H_3C_6H_5O_7$ (192.1 g/mol).

SELF-TEST EXERCISE

Benzoic acid has one replaceable hydrogen ion and can be abbreviated HBz. If 22.55 mL of 0.223 *M* NaOH neutralize a 0.613-g sample of benzoic acid, what is the molar mass of the acid? The balanced equation for the reaction is

$$HBz(s) + NaOH(aq) \longrightarrow NaBz(aq) + H_2O(l)$$

Answer: 122 g/mol HBz ($HC_7H_5O_2$)

15.4 Acid–Base Titrations

To calculate the unknown quantity for the titration of an acid or base solution, given any three of the following:

- **(a)** molarity of the acid
- **(b)** volume of the acid
- **(c)** molarity of the base
- **(d)** volume of the base

To express the molarity of an acid or base in terms of mass/mass percent concentration.

Vinegar tastes sour because it consists of acetic acid in water. To find the molar concentration of acetic acid in vinegar, we can analyze the vinegar sample using a standard base solution. In the laboratory, we can analyze the acetic acid by a titration method. A **titration** is a process whereby we deliver a measured volume of standard solution. To analyze a vinegar sample, we can titrate the acetic acid by delivering a measured volume of sodium hydroxide solution from a buret. We can use phenolphthalein as an indicator to signal when neutralization is achieved.

titration A laboratory procedure for delivering a measured volume of solution through a buret.

In our titration, a few extra drops of NaOH increase the pH dramatically, and the indicator changes from colorless to pink. When the indicator changes color, the

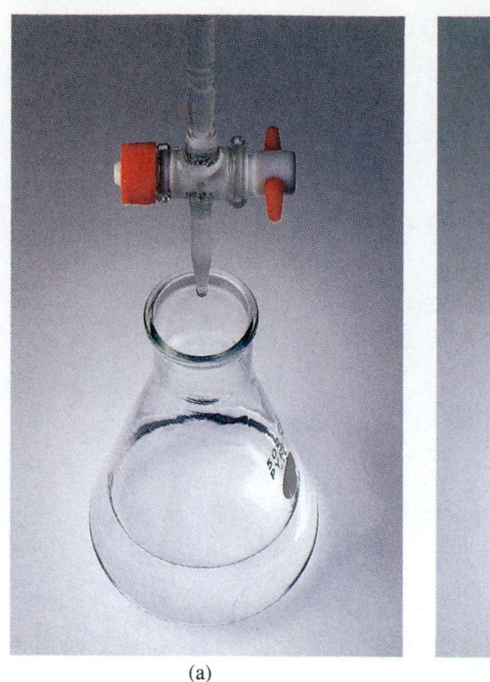

(a)

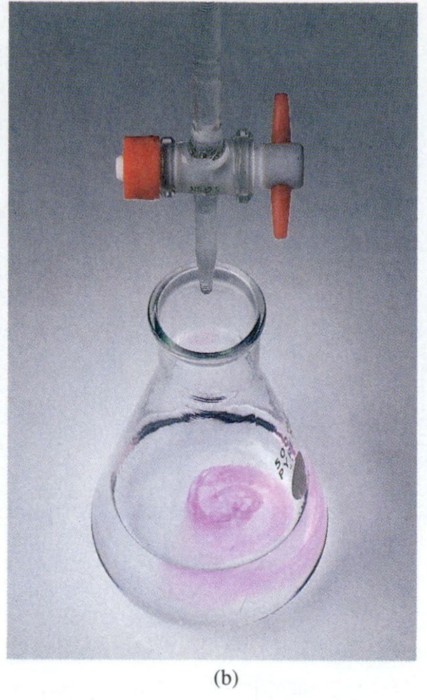

(b)

(c)

Figure 15.2 Titration of Acetic Acid with Sodium Hydroxide (a) The flask contains a sample of acetic acid and a drop of phenolphthalein indicator. (b) The buret delivers sodium hydroxide solution into the flask. (c) The titration is complete when the solution obtains a permanent pink color.

endpoint The stage in a titration when the indicator changes color.

titration is stopped. This point in the titration is the **endpoint**. Figure 15.2 illustrates the experimental procedure.

In an analysis, suppose a chemist uses a standard solution of sodium hydroxide to determine the concentration of commercial vinegar. The chemical technician obtains a sample of vinegar and titrates the acetic acid to a phenolphthalein endpoint with the NaOH. The technician then calculates the molar concentration of the vinegar sample. In practice, the concentration is usually expressed as a mass/mass percent (m/m%).

Consider the following analysis of a vinegar sample for acetic acid. A 10.0-mL sample of vinegar requires 37.55 mL of 0.223 *M* NaOH. The balanced equation for the reaction is

$$HC_2H_3O_2(aq) + NaOH(aq) \longrightarrow NaC_2H_3O_2(aq) + H_2O(l)$$

To find the molarity of the acetic acid, we first find the moles of NaOH. The volume of NaOH is 37.55 mL and the molar concentration is 0.223 *M*. The molarity of the NaOH can be written as the unit factor: 0.223 mol NaOH/1000 mL solution.

$$37.55 \text{ \sout{mL solution}} \times \frac{0.223 \text{ mol NaOH}}{1000 \text{ \sout{mL solution}}} = 0.00837 \text{ mol NaOH}$$

Next, we find the moles of acetic acid titrated. From the balanced equation we see that 1 mole of NaOH base neutralizes 1 mole of $HC_2H_3O_2$.

$$0.00837 \text{ \sout{mol NaOH}} \times \frac{1 \text{ mol } HC_2H_3O_2}{1 \text{ \sout{mol NaOH}}} = 0.00837 \text{ mol } HC_2H_3O_2$$

Finally, we can calculate the molarity of the acetic acid. The volume of the $HC_2H_3O_2$ solution is 10.0 mL. Thus,

$$\frac{0.00837 \text{ mol } HC_2H_3O_2}{10.0 \text{ mL solution}} \times \frac{1000 \text{ mL solution}}{1 \text{ L solution}} = \frac{0.837 \text{ mol } HC_2H_3O_2}{1 \text{ L solution}}$$

$$= 0.837\ M\ HC_2H_3O_2$$

Now let's convert the molar concentration of the acetic acid to mass/mass percent concentration. If the density of the vinegar is 1.01 g/mL, and the molar mass of acetic acid is 60.0 g/mol, we can proceed as follows:

$$\frac{0.837 \text{ mol } HC_2H_3O_2}{1000 \text{ mL solution}} \times \frac{60.0 \text{ g } HC_2H_3O_2}{1 \text{ mol } HC_2H_3O_2} \times \frac{1 \text{ mL solution}}{1.01 \text{ g solution}} \times 100$$

$$= \frac{4.97 \text{ g } HC_2H_3O_2}{100 \text{ g solution}}$$

$$= 4.97\%\ HC_2H_3O_2$$

Therefore, the concentration of acetic acid in the vinegar solution is 0.837 M and this is equivalent to 4.97% $HC_2H_3O_2$.

In a different analysis, a chemist analyzes a household cleaning solution for its ammonia content. After titrating a sample of ammonia solution with standard acid solution to a methyl red endpoint, the chemist calculates its concentration. Example Exercise 15.5 illustrates the analysis of the ammonia solution using a standardized HCl solution.

EXAMPLE EXERCISE 15.5

If 31.30 mL of 0.277 M HCl are used to titrate 5.00 mL of aqueous ammonia, what is the (a) molarity of NH_3, and (b) mass/mass percent concentration of NH_3? The density of the solution is 0.996 g/mL. The balanced equation is

$$HCl(aq) + NH_3(aq) \longrightarrow NH_4Cl(aq)$$

Solution: We will proceed by first calculating the molar concentration and then the mass/mass percent.

(a) To calculate the molarity of the NH_3, we must find the moles of NH_3. From the balanced equation, 1 mole of HCl neutralizes 1 mole of NH_3. Since the acid solution is 0.277 M HCl, we have the unit factor 0.277 mol HCl/1000 mL solution. Thus,

$$31.30 \text{ mL solution} \times \frac{0.277 \text{ mol HCl}}{1000 \text{ mL solution}} \times \frac{1 \text{ mol } NH_3}{1 \text{ mol HCl}} = 0.00867 \text{ mol } NH_3$$

We obtain the molar concentration by dividing moles of NH_3 by the volume titrated, that is, 5.00 mL.

$$\frac{0.00867 \text{ mol } NH_3}{5.00 \text{ mL solution}} \times \frac{1000 \text{ mL}}{1 \text{ L}} = \frac{1.73 \text{ mol } NH_3}{1 \text{ L solution}} = 1.73\ M\ NH_3$$

(b) To convert the 1.73 M NH_3 to percent concentration, we must use the unit factor 1.73 mol NH_3/1000 mL solution. The density of the solution is given, 0.996 g/mL, and the molar mass of ammonia is 17.0 g/mol. Let's convert the concentration as follows:

$$\frac{1.73\ \cancel{\text{mol NH}_3}}{1000\ \cancel{\text{mL solution}}} \times \frac{17.0\text{ g NH}_3}{1\ \cancel{\text{mol NH}_3}} \times \frac{1\ \cancel{\text{mL solution}}}{0.996\text{ g solution}} \times 100 = \frac{2.95\text{ g NH}_3}{100\text{ g solution}}$$

$$= 2.95\%\ \text{NH}_3$$

SELF-TEST EXERCISE

If 38.30 mL of 0.250 *M* NaOH are used to titrate 25.0 mL of phosphoric acid to a methyl red endpoint, what is (a) the molarity of the acid, and (b) the mass/mass percent concentration? The density of the phosphoric acid is 1.01 g/mL. The balanced equation for the reaction is

$$H_3PO_4(aq) + 3\ NaOH(aq) \longrightarrow Na_3PO_4(aq) + 3\ H_2O(l)$$

Answers: **(a)** 0.128 *M* H_3PO_4; **(b)** 1.24% H_3PO_4

Another application of a titration is to find the volume of base required to neutralize a given amount of acid. For example, we can find the volume of base needed to neutralize a sulfuric acid sample from battery acid. Example Exercise 15.6 illustrates the analysis of battery acid using previously standardized 0.223 *M* sodium hydroxide solution.

EXAMPLE EXERCISE 15.6

A 10.0-mL sample of battery acid is titrated with 0.223 *M* NaOH. If the acid is 0.555 *M*, what volume of sodium hydroxide is required for the titration? The balanced equation for the reaction is

$$H_2SO_4(aq) + 2\ NaOH(aq) \longrightarrow Na_2SO_4(aq) + 2\ H_2O(l)$$

Solution: The first step is to find the number of moles of sulfuric acid. The volume is 10.0 mL and the concentration is 0.555 *M*. We can use the molarity of the H_2SO_4 solution as a unit factor: 0.555 mol H_2SO_4/1000 mL solution.

$$10.0\ \cancel{\text{mL solution}} \times \frac{0.555\text{ mol H}_2\text{SO}_4}{1000\ \cancel{\text{mL solution}}} = 0.00555\text{ mol H}_2\text{SO}_4$$

Next, we find the moles of NaOH titrated. From the balanced equation we see that 2 moles of NaOH neutralize 1 mole of H_2SO_4.

$$0.00555\ \cancel{\text{mol H}_2\text{SO}_4} \times \frac{2\text{ mol NaOH}}{1\ \cancel{\text{mol H}_2\text{SO}_4}} = 0.0111\text{ mol NaOH}$$

Finally, we calculate the volume of NaOH. Since the concentration is 0.223 *M*, we have the unit factor 0.223 mol NaOH/1000 mL solution. To cancel units, we will apply the reciprocal unit factor: 1000 mL solution/0.223 mol NaOH.

$$0.0111\ \cancel{\text{mol NaOH}} \times \frac{1000\text{ mL solution}}{0.223\ \cancel{\text{mol NaOH}}} = 49.8\text{ mL solution}$$

SELF-TEST EXERCISE

A 25.0-mL sample of swimming pool acid is titrated with 0.125 *M* $Ba(OH)_2$. If 50.0 mL of barium hydroxide is required for the titration, what is the molar concentration of the acid? The balanced equation for the reaction is

$$2\ HCl(aq) + Ba(OH)_2(aq) \longrightarrow BaCl_2(aq) + 2\ H_2O(l)$$

Answer: 0.500 *M* HCl

15.5 Arrhenius Acid–Base Theory

To state whether Arrhenius acids and bases are strong or weak, given the degree of ionization in aqueous solution.

To identify the Arrhenius acid and base that produce a given salt from a neutralization reaction.

In 1884, the Swedish chemist Svante Arrhenius proposed the first definitions for an acid and a base. An **Arrhenius acid** is a substance that ionizes in water to produce hydrogen ions. An **Arrhenius base** is a substance that dissociates in water to release hydroxide ions. Acids and bases are of varying strengths. The strength of an Arrhenius acid is measured by the degree of ionization in solution. **Ionization** is the process whereby the molecules in a molecular compound break apart to form cations and anions. The strength of an Arrhenius base is measured by the degree of dissociation in solution. **Dissociation** is the process whereby the ions in an ionic compound simply separate. Thus, a molecule of HCl ionizes into H^+ and Cl^-, while NaOH dissociates into Na^+ and OH^-.

Arrhenius acid A substance that yields hydrogen ions when dissolved in water.

Arrhenius base A substance that yields hydroxide ions when dissolved in water.

ionization The process of a polar molecular compound dissolving in water and forming positive and negative ions; for example, HCl dissolves in water to give hydrogen ions and chloride ions.

dissociation The process of an ionic compound dissolving in water and separating into positive and negative ions; for example, NaOH dissolves in water to give sodium ions and hydroxide ions.

Arrhenius Acids

An acid is considered to be either strong or weak depending on how much it ionizes. According to the Arrhenius definition, a strong acid ionizes extensively to release hydrogen ions in solution. Hydrochloric acid is considered a strong acid because it ionizes nearly 100%. Acetic acid is a weak acid because it is only slightly ionized, about 1%. Table 15.4 lists some common Arrhenius acids.

All Arrhenius acids have a hydrogen atom in a molecule attached by a polar covalent bond. When the acid molecule ionizes, it breaks this bond. In an aqueous solution, polar water molecules help the acid molecule ionize by pulling the hydrogen ion away from the acid molecule. The ionization process for hydrochloric acid and acetic acid is as follows:

$$HCl(aq) + H_2O(l) \longrightarrow H_3O^+(aq) + Cl^-(aq) \quad (\sim 100\%)$$

$$HC_2H_3O_2(aq) + H_2O(l) \longrightarrow H_3O^+(aq) + C_2H_3O_2^-(aq) \quad (\sim 1\%)$$

Although chemists refer to acids as solutions with hydrogen ions, the solutions actually contain H_3O^+. The H_3O^+, or **hydronium ion**, is formed when the hydrogen ion in aqueous solution attaches to a water molecule.

hydronium ion (H_3O^+) The ion that results when a hydrogen ion attaches to a water molecule by a coordinate covalent bond.

Table 15.4 Common Arrhenius Acids

Aqueous Acids	Percent Ionization	Acid Strength
hydrochloric acid, HCl(aq)	~100	strong
hydrofluoric acid, HF(aq)	~5	weak
nitric acid, HNO_3(aq)	~100	strong
nitrous acid, HNO_2(aq)	~5	weak
sulfuric acid, H_2SO_4(aq)	~100	strong
acetic acid, $HC_2H_3O_2$(aq)	~1	weak
carbonic acid, H_2CO_3(aq)	~1	weak
phosphoric acid, H_3PO_4(aq)	~1	weak

Hydrofluoric acid is a weak acid, but it is the only acid that attacks glass. The design shown was created by coating a glass surface with wax and removing waxy areas. The design was produced by dipping the glass into aqueous HF which etched the exposed glass areas.

Note: Recent studies have shown that hydrogen ions exist in dilute solution not only as H_3O^+ but also as $H_5O_2^+$ and $H_9O_4^+$. The formula $H_5O_2^+$ represents a hydrogen ion hydrated by two water molecules; $H_9O_4^+$ represents a hydrogen ion hydrated by four water molecules. For simplicity, we will usually designate the hydrogen ion in aqueous solution as $H^+(aq)$.

Table 15.5 Common Arrhenius Bases

Aqueous Bases	Percent Dissociation	Base Strength
sodium hydroxide, $NaOH(aq)$	~100	strong
potassium hydroxide, $KOH(aq)$	~100	strong
lithium hydroxide, $LiOH(aq)$	~100	strong
calcium hydroxide, $Ca(OH)_2(aq)$	~100	strong
barium hydroxide, $Ba(OH)_2(aq)$	~100	strong
ammonium hydroxide, $NH_4OH(aq)$*	~1	weak

*Ammonium hydroxide is prepared by dissolving ammonia gas in water. It is frequently referred to as an aqueous ammonia solution, $NH_3(aq)$, and sometimes as ammonia water, $NH_3{\cdot}H_2O$. The name ammonium hydroxide is somewhat misleading since there is no evidence for ammonium hydroxide molecules in solution. For ease of balancing neutralization equations, however, we will refer to an ammonia solution as ammonium hydroxide, $NH_4OH(aq)$.

Arrhenius Bases

Recall that a base is considered to be either strong or weak, depending on how much it dissociates. According to the Arrhenius definition, a strong base dissociates extensively to release hydroxide ions in solution. Sodium hydroxide is a strong base and dissociates nearly 100% in aqueous solution. Ammonium hydroxide is a weak base because it provides relatively few ions in solution. Table 15.5 lists some common Arrhenius bases.

If we dissolve sodium hydroxide, NaOH, in water, it gives aqueous NaOH. In the dissolving process, NaOH dissociates into aqueous sodium ions and hydroxide

CHEMISTRY CONNECTION □ HISTORICAL

Svante Arrhenius

▶ **Can you guess the grade that Arrhenius was awarded for his brilliant Ph.D. dissertation on the theory of ionic solutions?**

Svante Arrhenius was a child genius who taught himself to read by age 3. After graduating from high school at the top of his class, he enrolled at the University of Uppsala in his native Sweden. He majored in chemistry and did research on the passage of electricity through aqueous solutions.

At 22, Arrhenius began his doctoral work at the University of Stockholm. For his graduate thesis, he continued to pursue his interest in the behavior of solutions. In particular, he was puzzled that sodium chloride solutions conducted electricity, whereas sugar solutions did not. After carefully considering his observations, Arrhenius boldly proposed that a solution of sodium chloride conducts an electric current because it separates into charged particles in solution. Sugar does not conduct electricity because it does not form charged particles in solution.

Arrhenius also noted that the freezing point of water was lowered twice as much for a salt solution as it was for a sugar solution. He explained that when sodium chloride dissolves in solution, NaCl separates into sodium ions and chloride ions. When sugar dissolves in solution, it remains as molecules.

Arrhenius was aware, however, that there were difficulties with the concept of charged ions in solution. At the time, the scientific community considered atoms to be indivisible and electrically neutral particles.

In 1884, Arrhenius took his oral examination for his Ph.D. dissertation and defended the ionic theory. After a lengthy 4-hour defense, he was awarded the lowest possible passing grade. The dissertation committee was willing to recognize Arrhenius' intellect, but was not ready to accept the idea of ions in solution.

Arrhenius' career stagnated for almost a decade and found very little support. The ionic theory was eventually championed by the noted chemist, Wilhelm Ostwald, who invited Arrhenius to work with him in Germany. Gradually, evidence accumulated that supported the concept of ions. Most notably, the discovery of the electron proved the existence of charged particles.

Svante August Arrhenius (1859–1927)

By 1897, the ionic theory had gained limited credibility and Arrhenius had taken a position at the University of Stockholm. Two years later he published his classic paper "*On the Dissociation of Substances in Aqueous Solutions.*" After struggling for 20 years, Arrhenius was awarded the 1903 Nobel prize in chemistry for his ionic theory of solutions.

▶ **Arrhenius received a barely passing grade; his committee simply did not believe in the ionic theory of solutions.**

ions. Aqueous NH_4OH is a weak base and provides few hydroxide ions in solution. We can show that this process for the two Arrhenius bases is as follows:

$$NaOH(aq) \longrightarrow Na^+(aq) + OH^-(aq) \quad (\sim 100\%)$$

$$NH_4OH(aq) \longrightarrow NH_4^+(aq) + OH^-(aq) \quad (\sim 1\%)$$

The following example exercise illustrates the classification of acid–base strength based on the ability to donate ions into solution.

EXAMPLE EXERCISE 15.7

Classify each of the following solutions as strong or weak Arrhenius acids, given the degree of ionization.

(a) perchloric acid, $HClO_4(aq)$ ~100%
(b) hypochlorous acid, $HClO(aq)$ ~1%

Solution: To classify an aqueous solution as a strong or weak acid, we must be given the amount of ionization.

(a) Perchloric acid is extensively ionized and considered a *strong acid*. An aqueous solution is primarily $H^+(aq)$ and $ClO_4^-(aq)$.
(b) Hypochlorous acid is a *weak acid* because of its lack of ionization. An aqueous solution contains primarily molecules of HClO.

SELF-TEST EXERCISE

Classify each of the following solutions as strong or weak Arrhenius bases, given the degree of dissociation.

(a) magnesium hydroxide, $Mg(OH)_2(s)$ ~1%
(b) rubidium hydroxide, $RbOH(aq)$ ~100%

Answers: **(a)** weak base; **(b)** strong base

Neutralization Reactions

salt The product of a neutralization reaction in addition to water.

We learned in Section 8.11 that an acid neutralizes a base to give a **salt** and water. For example, hydrochloric acid reacts with sodium hydroxide to give a salt, sodium chloride, and water. The equation for the reaction is

$$HCl(aq) + NaOH(aq) \longrightarrow NaCl(aq) + H_2O(l)$$

This reaction produces the aqueous salt NaCl. Different acids or bases give other salts. If, for example, we completely neutralize sulfuric acid using aqueous sodium hydroxide, we obtain sodium sulfate. The equation is

$$H_2SO_4(aq) + 2\ NaOH(aq) \longrightarrow Na_2SO_4(aq) + 2\ H_2O(l)$$

The hydrogen ions from an acid are neutralized by the hydroxide ions from a base. The hydrogen ions and hydroxide ions combine to form water. We notice from the above neutralization reaction that Na_2SO_4 is the salt produced.

We can identify the Arrhenius acid and base that produce any given salt. This is because each salt will be composed of the cation from the reacting base and the anion from the acid. As an example, we can predict the neutralization reaction that produces the salt potassium acetate, $KC_2H_3O_2$. Since the salt contains potassium, the neutralized base must be potassium hydroxide, KOH. The acetate ion came from acetic acid, $HC_2H_3O_2$. The equation for the reaction is

$$HC_2H_3O_2(aq) + KOH(aq) \longrightarrow KC_2H_3O_2(aq) + H_2O(l)$$

Consider another example. Let's predict the neutralization reaction that produces the salt potassium phosphate, K_3PO_4. As in the previous example, the neutralized base must be potassium hydroxide, KOH. The phosphate ion, PO_4^{3-}, came from phosphoric acid, H_3PO_4. The equation for the reaction is

$$H_3PO_4(aq) + 3\ KOH(aq) \longrightarrow K_3PO_4(aq) + 3\ H_2O(l)$$

The following example exercise further illustrates the neutralization of acids and bases to give salts and water.

EXAMPLE EXERCISE 15.8

Determine the acid and base that produce each of the following salts. Write a balanced equation for the neutralization reaction.

(a) lithium fluoride, LiF(aq) **(b)** calcium sulfate, $CaSO_4$(aq)

Solution: We can refer to the above discussion as follows.

(a) The salt LiF is produced from the neutralization of LiOH and HF acid. The equation for the reaction is

$$HF(aq) + LiOH(aq) \longrightarrow LiF(aq) + H_2O(l)$$

(b) Calcium sulfate is produced from the neutralization of $Ca(OH)_2$ and H_2SO_4. The equation for the reaction is

$$H_2SO_4(aq) + Ca(OH)_2(aq) \longrightarrow CaSO_4(aq) + 2\ H_2O(l)$$

SELF-TEST EXERCISE

Determine the acid and base that produce each of the following salts. Write a balanced equation for the neutralization reaction.

(a) potassium iodide, KI(aq) **(b)** barium nitrate, $Ba(NO_3)_2$(aq)

Answers:

(a) HI is the acid and KOH is the base.

$$HI(aq) + KOH(aq) \longrightarrow KI(aq) + H_2O(l)$$

(b) HNO_3 is the acid and $Ba(OH)_2$ is the base.

$$2\ HNO_3(aq) + Ba(OH)_2(aq) \longrightarrow Ba(NO_3)_2(aq) + 2\ H_2O(l)$$

15.6 Brønsted–Lowry Acid–Base Theory

OBJECTIVES

To classify the substance acting as a Brønsted–Lowry acid or base in a given neutralization reaction.

To designate the stronger Brønsted–Lowry acid and base participating in a reversible reaction.

In 1923, the Danish chemist Johannes Brønsted and the English chemist Thomas Lowry independently proposed broader definitions than those of Arrhenius for an acid and a base. Brønsted and Lowry each defined an acid as a substance that is a

Brønsted–Lowry acid A substance that donates a proton in an acid–base reaction.

proton donor A term used interchangeably with hydrogen ion donor.

Brønsted–Lowry base A substance that accepts a proton in an acid–base reaction.

hydrogen ion donor. Whereas Arrhenius defined an acid as a substance that donates hydrogen ion in water, Brønsted–Lowry defines an acid as a substance that donates a hydrogen ion to any other substance. Since most hydrogen ions are actually protons, a **Brønsted–Lowry acid** is also referred to as a **proton donor**.

Recall Arrhenius' definition of a base as a hydroxide ion donor in water. We know that hydroxide ions neutralize hydrogen ions to form water. There are, however, many substances that neutralize hydrogen ions besides the hydroxide ion. Brønsted and Lowry proposed that a base was any substance that accepted a hydrogen ion. Thus, a **Brønsted–Lowry base** is any *proton acceptor*.

Although the Brønsted–Lowry definitions of acids and bases are a bit different from the Arrhenius definitions, an acid and base still neutralize each other. The following reactions illustrate the neutralization of a Brønsted–Lowry base with hydrochloric acid.

$$HCl(aq) + NaOH(aq) \longrightarrow NaCl(aq) + H_2O(l)$$

$$HCl(aq) + NH_3(aq) \longrightarrow NH_4Cl(aq)$$

$$HCl(aq) + H_2O(l) \longrightarrow H_3O^+(aq) + Cl^-(aq)$$

In the first equation, aqueous NaOH is accepting a proton. According to the Brønsted–Lowry definition, it is therefore a base. In the second equation, aqueous NH_3 is accepting a proton. It is also a base. As the third equation shows, even water can act as a Brønsted–Lowry base, because it can accept a hydrogen ion.

Unlike Arrhenius acids and bases that depend on hydrogen ions and hydroxide ions dissolved in water, Brønsted–Lowry acids and bases depend on a particular reaction. For example, aqueous $NaHCO_3$ can act as a *base* by accepting a proton. In a different reaction, aqueous $NaHCO_3$ can act as an *acid* by donating a proton. The following examples will illustrate.

$$HCl(aq) + NaHCO_3(aq) \longrightarrow NaCl(aq) + H_2CO_3(aq) \qquad \textbf{(1)}$$

$$NaOH(aq) + NaHCO_3(aq) \longrightarrow Na_2CO_3(aq) + H_2O(l) \qquad \textbf{(2)}$$

amphiprotic A substance that is capable of both accepting a proton and donating a proton in an acid–base reaction.

In reaction (1), $NaHCO_3$ is accepting a proton from HCl and is therefore acting as a Brønsted–Lowry base. In reaction (2), $NaHCO_3$ is donating a proton to NaOH and is therefore acting as a Brønsted–Lowry acid. A substance that is capable of both accepting a proton and donating a proton is said to be **amphiprotic** (Greek *amphi* means both types).

Reversible Acid-Base Reactions

reversible reaction A reaction that proceeds simultaneously in both the forward direction toward products as well as in the opposite direction toward reactants.

In many reactions the products can react with each other to reform the original reactants. A reaction that can proceed in both the forward and reverse directions is called a **reversible reaction**.

Brønsted–Lowry acids and bases can participate in reversible reactions. In that case, the products as well as the reactants act as Brønsted–Lowry acids and bases. The strength of a Brønsted–Lowry acid is related to its ability to *donate a proton* to another substance. The strength of a base is measured by its ability to *pull a proton* away from another substance. Thus, the acid and base on the reactant side of the equation compete with the acid and base on the product side of the equation. This competition is illustrated by the following example.

$$\underset{acid_1}{HF(aq)} + \underset{base_1}{NaC_2H_3O_2(aq)} \rightleftharpoons \underset{acid_2}{HC_2H_3O_2(aq)} + \underset{base_2}{NaF(aq)}$$

The size of the arrows indicates that the forward reaction is dominant. Therefore, $acid_1$ is stronger than $acid_2$ and $base_1$ is stronger than $base_2$. That is, aqueous

HF is a stronger acid than aqueous $HC_2H_3O_2$. It also follows that $NaC_2H_3O_2$ is a stronger base than NaF.

In a different neutralization reaction, the reverse process may be favored. If the reverse reaction is dominant, it is indicated by the longer arrow; for example,

$$\underset{\text{acid}_1}{HF(aq)} + \underset{\text{base}_1}{NaH_2PO_4(aq)} \rightleftharpoons \underset{\text{acid}_2}{H_3PO_4(aq)} + \underset{\text{base}_2}{NaF(aq)}$$

In this example, the reverse reaction is dominant. Therefore, acid_2 and base_2 are stronger than acid_1 and base_1. Thus, aqueous H_3PO_4 is a stronger acid than HF and aqueous NaF is a stronger base than NaH_2PO_4. Notice that NaH_2PO_4 and NaF are each being a proton acceptor and therefore are both acting as Brønsted–Lowry bases. In this example, H_3PO_4 and NaF can be referred to as a *conjugate acid* and a *conjugate base*, respectively.

15.7 Ionization of Water

OBJECTIVES

- To state each of the following for the ionization of pure water: the equilibrium constant expression, the value for K_w at 25°C, the molar hydrogen ion and hydroxide ion concentrations.
- To calculate the molar hydroxide ion concentration, $[OH^-]$, given the molar hydrogen ion concentration $[H^+]$.

What would happen if we tested the electrical conductivity of a metal key? In Section 3.7, we learned that metals are good conductors of heat and electricity. The key should therefore conduct an electrical current. In Figure 15.3, we in fact see that the key completes the circuit in the conductivity apparatus and the bulb lights up. The explanation is as follows. Electricity is the flow of electrons. Metal atoms, such

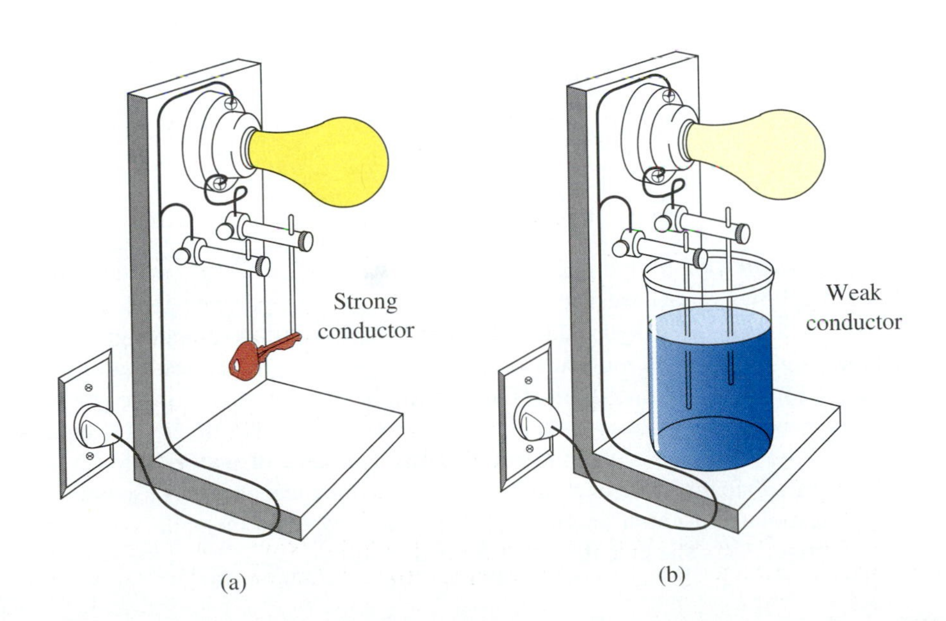

Figure 15.3 Conductivity Apparatus (a) Metals are good conductors of electricity. A metal key completes the electrical circuit and lights up the apparatus. (b) Pure water is a very poor conductor. It does, however, very weakly conduct an electric current.

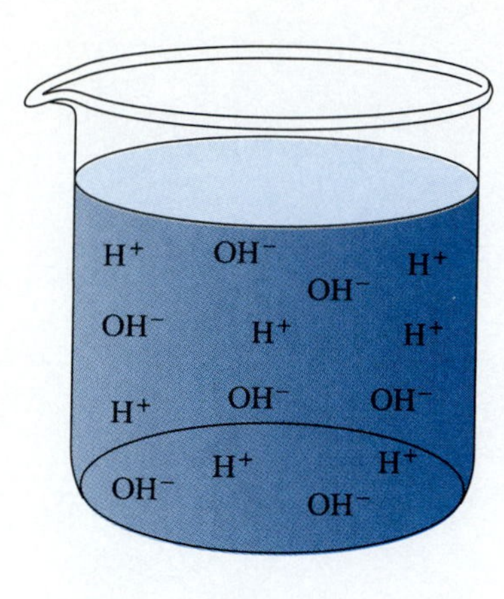

Figure 15.4 Ionization of Water Pure water ionizes to give a few hydrogen ions and hydroxide ions. Water is a very weak conductor of electricity because only one molecule in about 500,000,000 forms ions.

as those in the key, hold their electrons loosely. Therefore, an electrical apparatus can push electrons from one atom to another. Thus, electricity flows from one end of the metal key to the other as electrons move from atom to atom.

What happens when we test the electrical conductivity of pure water? In Figure 15.4, we see that pure water is a very poor conductor of electricity. The evidence suggests that it is difficult to push electrons from one water molecule to another. How then do we explain that some electricity does flow in pure water? The explanation comes from Arrhenius's observations in 1884 of solution conductivity.

Arrhenius found that salt solutions are good conductors of electricity. He correctly concluded that it was the *ions* in the salt solution that were responsible for electrical conductivity. Since water is a very poor conductor, we can conclude that there are only a few ions present in pure water. We can write a reversible reaction to illustrate the presence of ions in pure water:

$$H_2O(l) + H_2O(l) \rightleftharpoons \underset{\textit{hydronium ion}}{H_3O^+(aq)} + \underset{\textit{hydroxide ion}}{OH^-(aq)}$$

autoprotolysis A reversible reaction in pure water that produces hydronium ions and hydroxide ions.

chemical equilibrium A dynamic state for a reversible reaction in which the rates of the forward and reverse reactions are equal.

This reaction is called **autoprotolysis** (or autoionization) of water. From the kinetic theory, we know that water molecules in the liquid state move about striking one another. Although it happens only rarely, on occasion two water molecules collide with sufficient energy for their bonds to break apart. This bond breaking produces the hydronium ion, H_3O^+, and the hydroxide ion, OH^-. This is a dynamic process. While some molecules are breaking apart, other ions are combining to form water molecules. When some ions are forming at the same rate that other ions are combining, water is in a state of **chemical equilibrium**. At any given moment, only about two of every billion water molecules are present as ions. But apparently these few ions in water are sufficient to weakly conduct an electrical current (Figure 15.3).

We can write the reaction showing the collision of two water molecules more simply. That is, the ionization equilibrium in water can be shown as

$$H_2O(l) \rightleftharpoons \underset{\textit{hydrogen ion}}{H^+(aq)} + \underset{\textit{hydroxide ion}}{OH^-(aq)}$$

equilibrium constant of water (K_w) The product of the molar hydrogen ion concentration times the molar hydroxide ion concentration in water.

The concentration of hydrogen ions in pure water is 1.0×10^{-7} mole per liter at 25°C. It follows that if we know the concentration of hydrogen ions, we also know the concentration of hydroxide ions. Since the ionization of water gives one H^+ and one OH^- for each molecule of water that ionizes, the concentrations are equal. Therefore, the concentration of OH^- must also be 1.0×10^{-7} mole per liter at 25°C. Moreover, the product of the molar concentration of H^+ times OH^- equals a constant. This product is called the **equilibrium constant of water** (symbol K_w).

Let's calculate the value of K_w at 25°C. For convenience, chemists use brackets to symbolize molar concentration. Hence, $[H^+]$ is the symbol for the molar concentration of hydrogen ion. The calculation of K_w is as follows:

$$\text{If} \qquad [H^+] = 1.0 \times 10^{-7}$$

$$\text{and} \qquad [OH^-] = 1.0 \times 10^{-7}$$

$$\text{then,} \qquad [H^+][OH^-] = (1.0 \times 10^{-7})(1.0 \times 10^{-7})$$

$$\text{at 25°C,} \qquad K_w = [H^+][OH^-] = 1.0 \times 10^{-14}$$

In Chapter 16, we will discuss the concept of equilibrium in detail. For now, we will note that the ionization equilibrium in water shifts if there are changes in the hydrogen ion concentration. That is, if we increase the $[H^+]$, more OH^- is converted to H_2O molecules and the equilibrium shifts left. Suppose the $[H^+]$ increases from 10^{-7} to 10^{-5}. The increased hydrogen ion converts more of the hydroxide ions to water molecules, thus shifting the equilibrium to the left.

Conversely, suppose the $[H^+]$ decreases; then the $[OH^-]$ increases. When the $[H^+]$ decreases, the equilibrium shifts to the right. When the $[H^+]$ decreases from 10^{-7} to 10^{-8}, the equilibrium shifts to the right to provide more ions.

We can also explain this shift mathematically. When the $[H^+]$ decreases, the $[OH^-]$ must increase in order for the product of $[H^+][OH^-]$ to equal the equilibrium constant, K_w.

EXAMPLE EXERCISE 15.9

Given the following concentrations of $[H^+]$ in water, calculate the hydroxide ion concentration.

(a) $[H^+] = 1.4 \times 10^{-6}$ **(b)** $[H^+] = 5.2 \times 10^{-11}$

Solution: We are given that the product of the $[H^+][OH^-] = 1.0 \times 10^{-14}$.

(a) The $[H^+]$ is 1.4×10^{-6}, which is greater than 1.0×10^{-7}. Therefore, the $[OH^-]$ is calculated as follows.

$$K_w = [H^+][OH^-] = 1.0 \times 10^{-14}$$

$$1.4 \times 10^{-6}\,[OH^-] = 1.0 \times 10^{-14}$$

Dividing both sides of the equation by 1.4×10^{-6}, we have

$$[OH^-] = 7.1 \times 10^{-9}$$

(b) The $[H^+]$ is 5.2×10^{-11}, which is less than 1.0×10^{-7}. Therefore, the $[OH^-]$ is found as follows.

$$K_w = [H^+][OH^-] = 1.0 \times 10^{-14}$$

$$5.2 \times 10^{-11}\,[OH^-] = 1.0 \times 10^{-14}$$

Dividing both sides of the equation by 5.1×10^{-11}, we have

$$[OH^-] = 1.9 \times 10^{-4}$$

SELF-TEST EXERCISE

Given the following concentrations of $[OH^-]$ in water, calculate the hydrogen ion concentration.

(a) $[OH^-] = 7.5 \times 10^{-4}$ **(b)** $[OH^-] = 2.1 \times 10^{-10}$

Answers: **(a)** $[H^+] = 1.3 \times 10^{-11}$; **(b)** $[H^+] = 4.8 \times 10^{-5}$

15.8 The pH Concept

To perform calculations that relate integer pH values to the molar hydrogen ion concentration of a solution.

In Section 15.1, we introduced the term pH. Recall that pure water is neutral and has a pH of 7. We also know one property of an acid is that its pH is less than 7. As the pH value decreases, the solution becomes more acidic. For example, a solution of pH 2 is more acidic than a solution of pH 3. Basic solutions have pH values greater than 7. As the pH value increases, the solution becomes more basic. In fact, a solution of pH 12 is ten times more basic than a solution of pH 11.

Converting $[H^+]$ to pH

In water, the hydrogen ion concentration can vary from more than 1 *M* to less than 0.000 000 000 000 01 *M*. A pH scale is a convenient way to express this broad range of hydrogen ion concentration. The pH scale expresses the molar hydrogen ion concentration, $[H^+]$, as a power of 10. In other words, the pH is the negative logarithm of the molar hydrogen ion concentration. That is,

$$\text{pH} = -\log\,[H^+]$$

For example, if the molar hydrogen ion concentration, $[H^+]$, is 0.1 *M*, then

$$\text{pH} = -\log 0.1$$

We do not need to understand logarithms to solve pH problems. All we need to know is that if we express the hydrogen ion concentration as a power of 10 the logarithm is the exponent. We can write 0.1 *M* hydrogen ion concentration as a power of 10 (10^{-1}) and find the pH as follows:

$$\text{pH} = -\log 10^{-1}$$

$$= -(-1) = 1$$

Thus, the pH of the hydrogen ion concentration 0.1 *M* is 1. The following example exercise provides practice in converting between the molar hydrogen ion concentration of an aqueous solution and its pH.

EXAMPLE EXERCISE 15.10

Calculate the pH of the following solutions, given the molar hydrogen ion concentration.

(a) vinegar, $[H^+] = 0.001\ M$

(b) antacid, $[H^+] = 0.000\ 000\ 001\ M$

Solution:

(a) The pH of vinegar equals the negative log $[H^+]$; thus,

$$\text{pH} = -\log 0.001 = -\log 10^{-3}$$

$$= -(-3) = 3$$

The sour taste of vinegar indicates that an acid is present. The acid in vinegar is acetic acid, and the pH is approximately 3.

(b) The pH of the antacid equals the $-\log [H^+]$; hence,

$$pH = -\log 0.000\,000\,001 = -\log 10^{-9}$$
$$= -(-9) = 9$$

We calculated the pH of the antacid to express its acidity. Since we found the pH to be 9, we can classify it as a weakly basic solution.

SELF-TEST EXERCISE

Calculate the pH for apple juice that has a hydrogen ion concentration of 0.0001 *M*.

Answer: pH = 4

Converting pH to [H⁺]

We can calculate the hydrogen ion concentration given the pH of a solution. Let's rearrange the definition of pH to express the molar hydrogen ion concentration. The $[H^+]$ is equal to ten raised to a negative pH value. That is,

$$[H^+] = 10^{-pH}$$

For example, if a solution has a pH of 5, the $[H^+]$ is

$$[H^+] = 10^{-5} = 0.000\,01\ M$$

The following example exercise provides further practice in converting between the pH of an aqueous solution and its molar hydrogen ion concentration.

EXAMPLE EXERCISE 15.11

Calculate the molar hydrogen ion concentration given the pH of the following solutions.

(a) lime juice, pH = 2

(b) tomato juice, pH = 4

Solution:

(a) The hydrogen ion concentration of lime juice is equal to 10 raised to the negative value of the pH. Since the pH is 2,

$$[H^+] = 10^{-pH}$$
$$= 10^{-2} = 0.01\ M$$

A digital pH meter that measures hydrogen ion concentration. Pure, distilled water has a pH of 7.00.

CHEMISTRY CONNECTION □ ENVIRONMENTAL

Acid Rain

► Can you guess in which decade acid rain was first reported?

Compare the photographs of the Black Forest in Germany. Although only ten years elapsed between the times of the two photographs, the difference is striking. The decimation of the trees has been blamed on acid rain. In other instances, the disfiguring of marble statues and historic buildings has also been attributed to acid rain. Acid rain attacks a marble monument by dissolving away calcium carbonate in the stone.

The English chemist Robert Smith coined the term acid rain after studying the rainfall in London. He found that the air, heavily polluted from coal-burning, produced rain that was abnormally acidic. The term acid rain has persisted, and today it usually refers to rain having a pH of 5 or below.

The gases that most contribute to acid rain are oxides of sulfur and nitrogen. Sulfur dioxide and sulfur trioxide in the atmosphere are released mainly by industrial steel plants and electric power plants that burn low-grade coal which has a high sulfur content. Most of the oxides of nitrogen are emitted from automobiles. When oxides of sulfur and nitrogen are released into the air, they dissolve in atmospheric water vapor. After the moisture coalesces, rain falls as drops of sulfuric acid and nitric acid.

Normal rain and acid rain are both acidic, but their pH values are different. Normal rain has a pH about 5.5. It is acidic because carbon dioxide in the atmosphere dissolves in raindrops to form carbonic acid. Acid rain can be a hundred times more acidic than normal rain, and a pH of 2.8 has been recorded. In the northeastern United States, rainfall with a pH of 4 has been blamed on sulfur oxides released by heavy industry.

Acid rain is a global problem, and countries all over the world are attempting to reduce it. One way is to reduce the emission of environmental pollutants. Canada has begun a program to reduce its sulfur dioxide emissions by 50% within 10 years. The United States, Japan, and Germany are using high-grade coal in order to reduce their emissions.

Ultimately, the problem of acid rain may be minimized by alternative energy sources. For example, hydroelectric and solar power are potential sources of energy that do not pollute the atmosphere.

► Acid rain was first observed in the 1870s in London, England.

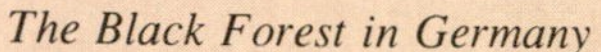

The Black Forest in Germany

(b) If the pH of tomato juice is 4, the $[H^+]$ is expressed as

$$[H^+] = 10^{-pH}$$
$$= 10^{-4} = 0.0001\ M$$

Notice that lime juice and tomato juice differ by only 2 pH units. The hydrogen ion concentration in lime juice is about 100 times greater than in tomato juice.

SELF-TEST EXERCISE

Calculate the hydrogen ion concentration for milk of magnesia that has a pH of 10.

Answer: $[H^+] = 1 \times 10^{-10}\ M$ (0.000 000 000 1 M)

The interconversion of pH and $[H^+]$ is readily accomplished using a calculator with a LOG key and a 10^x key. Appendix A explains how to use a calculator for pH problems.

15.9 Advanced pH and pOH Calculations

To perform calculations that relate fractional pH values to hydrogen ion concentrations that are not exact powers of 10.

To perform calculations that relate pOH values to the hydroxide ion concentration of a solution.

In the previous section, we defined pH as an exponential way of expressing the hydrogen ion concentration. We considered only whole-number pH values. Many chemical reactions must, however, be carefully controlled to a fraction of a pH unit. Biochemical processes, for example, are extremely sensitive to very small pH changes. In fact, the pH of our blood must be maintained within the narrow range of 7.3 to 7.5.

Converting $[H^+]$ to pH

Recall the mathematical definition of pH, that is, the negative logarithm of the molar hydrogen ion concentration:

$$pH = -\log [H^+]$$

Suppose we want to express the pH of a solution having a hydrogen ion concentration of 0.00015 M. The pH expression is

$$pH = -\log 0.00015$$
$$= -\log 1.5 \times 10^{-4}$$

A rule of logarithms is that the logarithm of a product is equal to the sum of the logarithm of each value. That is, $\log a \times b = \log a + \log b$. Let's apply this rule to the above hydrogen ion concentration:

$$pH = -(\log 1.5 + \log 10^{-4})$$
$$= -\log 1.5 - \log 10^{-4}$$
$$= -\log 1.5 - (-4)$$

We obtain the logarithms from a reference table or a calculator with a log key. Appendix G lists logarithms for numbers 1.0 to 9.9. It lists the log of 1.5 as 0.18. Appendix A shows the use of a log key.

$$\begin{aligned} pH &= -0.18 - (-4) \\ &= -0.18 + 4 = 3.82 \end{aligned}$$

For a solution having a hydrogen ion concentration of 0.00015 *M*, the pH is 3.82.

EXAMPLE EXERCISE 15.12

Calculate the pH for each of the following solutions, given the hydrogen ion concentration.

(a) stomach acid, $[H^+] = 0.020\ M$ **(b)** blood, $[H^+] = 0.000\ 000\ 047\ M$

Solution:

(a) We can calculate the pH of stomach acid as follows:

$$\begin{aligned} pH &= -\log 0.020 \\ &= -\log 2.0 \times 10^{-2} \\ &= -\log 2.0 - \log 10^{-2} \\ &= -0.30 - (-2) \\ &= \quad 1.70 \end{aligned}$$

The pH of stomach acid ranges from 1 to 3. Therefore, a pH of 1.70 is considered normal.

(b) We can calculate the pH of blood as follows:

$$\begin{aligned} pH &= -\log 0.000\,000\,047 \\ &= -\log (4.7 \times 10^{-8}) \\ &= -\log 4.7 - \log 10^{-8} \\ &= -0.67 - (-8) \\ &= \quad 7.33 \end{aligned}$$

The pH of blood must be maintained between 7.3 and 7.5. Hence, this sample is normal.

SELF-TEST EXERCISE

Calculate the pH for grape juice that has a hydrogen ion concentration of 0.000 089 *M*.

Answer: pH = 4.05

Converting pH to [H⁺]

Previously, we expressed $[H^+]$ mathematically by raising 10 to the negative pH value. That is,

$$[H^+] = 10^{-pH}$$

If, for example, a solution has a pH of 3.80, we can find the $[H^+]$ as follows:

$$[H^+] = 10^{-3.80}$$

A rule of exponents states that, to multiply two exponential numbers, add the powers together. That is, $10^a \times 10^b = 10^{a+b}$. Let's apply this rule to the above pH value of 3.80. For convenience, we can write the exponent as $0.20 + (-4)$. Thus,

$$[H^+] = 10^{0.20} \times 10^{-4}$$

To find the ordinary number that corresponds to a fractional exponent, we must obtain the antilogarithm, or antilog, from a reference table or calculator. In Appendix G, we see that the antilog of 0.20 is 1.6. Substituting gives

$$[H^+] = 1.6 \times 10^{-4}$$
$$= 0.000\,16\ M$$

For a solution having a pH of 3.80, the hydrogen ion concentration is 0.000 16 *M*.

EXAMPLE EXERCISE 15.13

Calculate the hydrogen ion concentration for each of the following solutions given the pH value.

(a) acid rain, pH = 3.68 **(b)** seawater, pH = 7.85

Solution:

(a) If acid rain has a pH of 3.68, we can find the $[H^+]$ as follows:

$$[H^+] = 10^{-3.68}$$

Applying the rules of exponents, we have,

$$[H^+] = 10^{0.32} \times 10^{-4}$$

Using a reference table or calculator with a log key, we find that the antilog of 0.32 is 2.1.

$$[H^+] = 2.1 \times 10^{-4}$$
$$= 0.000\,21\ M$$

Rain is considered normal if the pH is 5.5 to 7. Rainfall below pH 5.5 is considered "acid rain".

(b) The seawater sample has a pH of 7.85. We can find the $[H^+]$ as follows:

$$[H^+] = 10^{-7.85}$$
$$= 10^{0.15} \times 10^{-8}$$
$$= 1.4 \times 10^{-8}$$
$$= 0.000\,000\,014\ M$$

The pH of seawater samples varies, but typically ranges from 7.8 to 8.3.

SELF-TEST EXERCISE

Calculate the hydrogen ion concentration for a sample of wine having a pH of 3.25.

Answer: $[H^+] = 5.6 \times 10^{-4}\ M$ (0.000 56 *M*)

pOH

In the same way that pH expresses the acidity of a solution, **pOH** expresses the basicity. The mathematical definition of pOH is the negative logarithm of the molar hydroxide ion concentration. Thus,

$$pOH = -\log [OH^-]$$

Suppose we wish to express the pOH of a solution having a hydroxide ion concentration of 0.033 *M*. The pOH expression is

$$\begin{aligned} pOH &= -\log 0.033 \\ &= -\log 3.3 \times 10^{-2} \\ &= -\log 3.3 - \log 10^{-2} \\ &= -0.52 - (-2) \\ &= \ \ 1.48 \end{aligned}$$

The pOH of a solution having a hydroxide ion concentration of 0.033 *M* is 1.48. Note that if the pOH is low, for example 1.48, the hydroxide ion concentration is high. This means the solution is *strongly basic.*

Converting pOH to [OH⁻]

Just as we learned to convert pH values to $[H^+]$, we can convert pOH values to $[OH^-]$. To do so, we simply raise 10 to the negative pOH value. That is,

$$[OH^-] = 10^{-pOH}$$

As an example, let's find the $[OH^-]$ of a basic solution having a pOH of 3.43. We begin by raising 10 to the negative 3.43 power. We get

$$\begin{aligned} [OH^-] &= 10^{-3.43} \\ &= 10^{0.57} \times 10^{-4} \\ &= 3.7 \times 10^{-4} \\ &= 0.000\ 37\ M \end{aligned}$$

A solution with a pOH of 3.43 has a hydroxide ion concentration of 0.000 37 *M* and is considered to be *weakly basic.*

pH, pOH, and pK_w

In preceding sections, we learned how to find the pH and pOH of solutions. These values are related. Therefore, if we know one, we can find the other. In Section 15.8, we calculated the $[OH^-]$ of a solution given the $[H^+]$ and the value of the ionization constant, K_w. The overall relationship is

$$K_w = [H^+][OH^-] = 1.0 \times 10^{-14}$$

The p in pH and pOH is a mathematical symbol for the negative logarithm. Let's take the negative logarithm of the preceding relationship:

$$\begin{aligned} -\log K_w &= -\log([H^+][OH^-]) = -(\log 1.0 \times 10^{-14}) \\ -\log K_w &= -\log[H^+] - \log[OH^-] = -(0.0 - 14) \\ pK_w &= pH + pOH = 14.0 \end{aligned}$$

Table 15.6 Summary of pH and pOH Relationships

$$pH = -\log [H^+] \text{ and } [H^+] = 10^{-pH}$$
$$pOH = -\log [OH^-] \text{ and } [OH^-] = 10^{-pOH}$$
$$pH + pOH = pK_w = 14$$

Notice that the sum of the pH and pOH is equal to 14. Since that is always the case, if we know the pH of a solution, we can find the pOH by subtracting the pH from 14.

In this section, we performed calculations for pH and pOH. Ordinarily, the acidity of both acids and bases are expressed as a pH value. At times, when working with bases, the concentration of the hydroxide is reported as a pOH value. The overall relationship between pH and pOH is summarized in Table 15.6.

15.10 Strong and Weak Electrolytes

OBJECTIVES

To classify a substance in aqueous solution as a strong or weak acid, a strong or weak base, a soluble or slightly soluble ionic compound.

To write substances in aqueous solution as ionized or nonionized, given the electrolyte strength.

In Section 15.7, we explained that water is a weak conductor of electricity because it is slightly ionized. In the laboratory we can use a conductivity apparatus to determine if a substance in aqueous solution is a strong or weak conductor of electricity. If an aqueous solution is a good conductor, it is called a **strong electrolyte**. If an aqueous solution is a poor conductor, it is called a **weak electrolyte**. As a demonstration, we can test the electrical conductivity of hydrochloric acid and acetic acid using the apparatus in Figure 15.5. From the observations, we can conclude that hydrochloric acid is highly ionized because it is a strong electrolyte. Acetic acid is only slightly ionized because it is a weak electrolyte. Similar experiments using the conductivity apparatus demonstrate the electrolyte behavior of aqueous solutions.

strong electrolyte An aqueous solution that is a good conductor of electricity; for example, solutions of strong acids, strong bases, and soluble salts.

weak electrolyte An aqueous solution that is a poor conductor of electricity; for example, solutions of weak acids, weak bases, and slightly soluble salts.

Degree of Ionization in Solutions

By testing the conductivity of aqueous solutions, we can measure the degree of ionization. Strong electrolytes are highly ionized. Weak electrolytes are slightly ionized. Soluble ionic compounds dissolve by dissociating into ions. Recall Table 8.2, which listed the solubility rules for ionic compounds. Ionic compounds listed as insoluble are actually very slightly soluble. A sufficient amount of insoluble compound dissolves in solution to give a weak electrolyte. Table 15.7 lists examples of strong and weak electrolytes.

From conductivity testing experiments, we can determine the degree of ionization in solution. That is, we can distinguish between a highly ionized strong electrolyte and a slightly ionized weak electrolyte. Strong electrolytes yield ions in aqueous solution. For example, sodium chloride and calcium chloride highly dissociate into ions. To clarify the process, let's write equations for the formation of those ions.

$$NaCl(aq) \longrightarrow Na^{+}(aq) + Cl^{-}(aq)$$

$$CaCl_2(aq) \longrightarrow Ca^{2+}(aq) + 2\ Cl^{-}(aq)$$

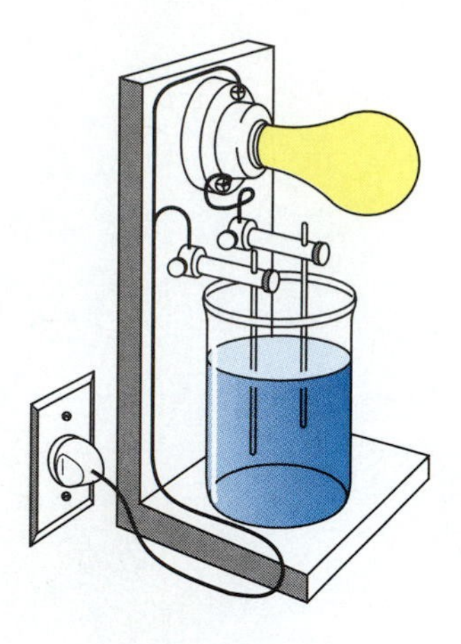

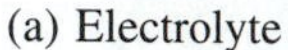

(a) Electrolyte

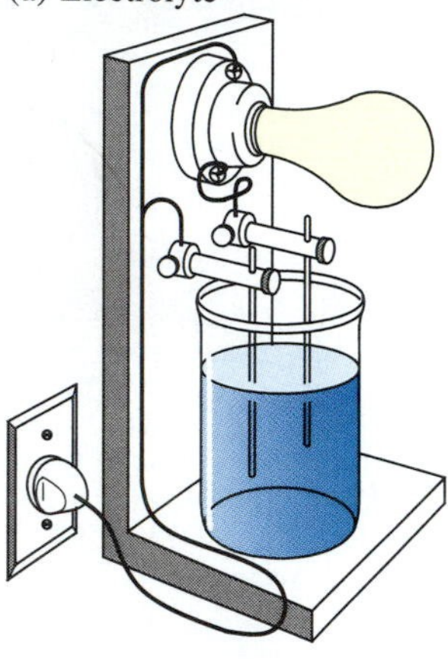

(b) Weak electrolyte

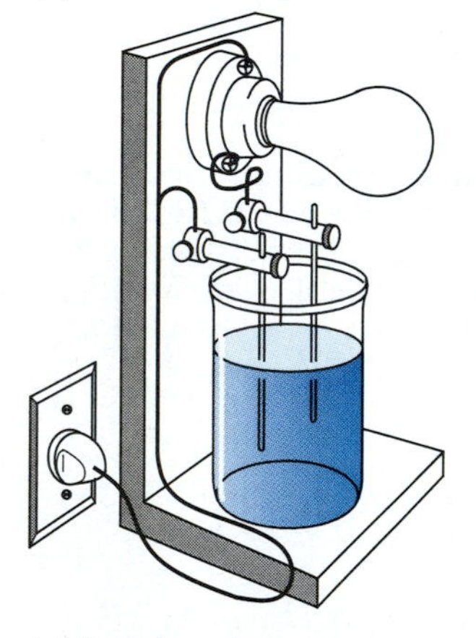

(c) Nonelectrolyte

Figure 15.5 Apparatus for Testing Conductivity (a) An aqueous solution of hydrochloric acid is a strong electrolyte and the bulb glows brightly. (b) An aqueous solution of acetic acid is a weak electrolyte and the bulb glows dimly. (c) A solution of alcohol is a nonelectrolyte and the bulb does not light.

Table 15.7 Strong and Weak Electrolytes

Strong Electrolyte	*Weak Electrolyte*
Strong Acids	***Weak Acids***
hydrochloric acid, $HCl(aq)$	hydrofluoric acid, $HF(aq)$
nitric acid, $HNO_3(aq)$	nitrous acid, $HNO_2(aq)$
sulfuric acid, $H_2SO_4(aq)$	sulfurous acid, $H_2SO_3(aq)$
perchloric acid, $HClO_4(aq)$	acetic acid, $HC_2H_3O_2(aq)$
	carbonic acid, $H_2CO_3(aq)$
	phosphoric acid, $H_3PO_4(aq)$
	most other acids
Strong Bases	***Weak Bases***
sodium hydroxide, $NaOH(aq)$	ammonium hydroxide, $NH_4OH(aq)$
potassium hydroxide, $KOH(aq)$	most other bases
lithium hydroxide, $LiOH(aq)$	
calcium hydroxide, $Ca(OH)_2(aq)$	
strontium hydroxide, $Sr(OH)_2(aq)$	
barium hydroxide, $Ba(OH)_2(aq)$	
Soluble Salts	***Very Slightly Soluble Salts***
sodium chloride, $NaCl(aq)$	silver chloride, $AgCl(s)$
potassium carbonate, $K_2CO_3(aq)$	calcium carbonate, $CaCO_3(s)$
copper(II) sulfate, $CuSO_4(aq)$	barium sulfate, $BaSO_4(s)$

Given the electrolyte strength of an aqueous solution, we can show the amount of ionization that has taken place. Strong electrolytes are highly ionized, so we write their chemical formulas in the ionized form. Weak electrolytes are only slightly ionized, so we write their formulas in the nonionized form. The nonionized formula is sometimes referred to as the molecular form. This is not quite accurate, since ionic compounds exist as ions, not molecules. Example Exercise 15.14 illustrates formula writing for electrolytes in aqueous solution.

EXAMPLE EXERCISE 15.14

Given the electrolyte strength for each of the following aqueous solutions, write the substance in the ionized or nonionized form.

(a) $HNO_3(aq)$, strong **(b)** $Ni(OH)_2(s)$, weak
(c) $Al(C_2H_3O_2)_3(aq)$, strong

Solution:

(a) Nitric acid is a strong electrolyte; therefore, it is highly ionized. It is written in the ionized form as $H^+(aq)$ and $NO_3^-(aq)$.

(b) Nickel hydroxide is a weak electrolyte and slightly dissociates into a nickel ion and two hydroxide ions. We write the substance in aqueous solution as $Ni(OH)_2(s)$.

(c) Aluminum acetate is a strong electrolyte and is highly dissociated in solution. It is written as $Al^{3+}(aq)$ and $3\ C_2H_3O_2^-(aq)$.

SELF-TEST EXERCISE

Given the electrolyte strength for each of the following aqueous solutions, write the substance in the ionized or nonionized form.

(a) $HCHO_2(aq)$, weak

(b) $Ba(OH)_2(aq)$, strong

(c) $PbBr_2(s)$, weak

Answers: **(a)** $HCHO_2(aq)$; **(b)** $Ba^{2+}(aq)$ and 2 $OH^-(aq)$; **(c)** $PbBr_2(s)$

In the examples, we considered aqueous solutions to be either strong or weak electrolytes. Some substances, however, are nonelectrolytes. That is, they do not conduct electricity at all. Examples include organic liquids such as alcohols and hydrocarbons. Since pure distilled water is a very weak electrolyte, it is also sometimes considered a nonelectrolyte.

15.11 Net Ionic Equations

To write a net ionic equation for a chemical reaction, given the names or formulas of the reactants and products.

In the previous section, we learned how to write substances in aqueous solution as ionized or nonionized. The concept of ionization lets us show ionic solutions more accurately. In Section 8.3 we learned to write balanced chemical equations. Now we can write more accurate chemical equations by showing strong electrolytes in the ionized form.

Let's consider the neutralization reaction of hydrochloric acid and sodium hydroxide. The equation for the reaction is

$$HCl(aq) + NaOH(aq) \longrightarrow NaCl(aq) + H_2O(l)$$

By writing the ionized form of the strong acid, strong base, and soluble salt, we can show the reaction more accurately. Each substance in the **total ionic equation** is written as it predominantly exists in solution. Table 15.7 shows highly ionized substances. Strong electrolytes include HCl, NaOH, and NaCl. Thus, the total ionic equation is

total ionic equation A chemical equation that writes highly ionized substances in the ionic form and slightly ionized substances in the nonionized form.

$$H^+(aq) + \cancel{Cl^-(aq)} + \cancel{Na^+(aq)} + OH^-(aq) \longrightarrow \cancel{Na^+(aq)} + \cancel{Cl^-(aq)} + H_2O(l)$$

Notice that $Na^+(aq)$ and $Cl^-(aq)$ appear on both sides of the equation. These ions are called **spectator ions**. They are in the solution but do not participate in the reaction. We can simplify the total ionic equation by eliminating spectator ions. The result is a **net ionic equation** that shows only the substances undergoing reaction. The net ionic equation for the above reaction is

spectator ions Ions that are in aqueous solution but do not participate in a reaction nor appear as reactants or products in the net ionic equation.

$$H^+(aq) + OH^-(aq) \longrightarrow H_2O(l)$$

net ionic equation A chemical equation that portrays an ionic reaction after spectator ions have been canceled from the total ionic equation.

Notice that we do not see the actual acid and base that were neutralized. This net ionic equation indicates that a strong acid and strong base reacted to give water. In fact, the net ionic equation is identical for all strong acid and strong base reactions that yield a soluble salt.

Let's state the procedure to use for writing net ionic equations. Keep in mind that the net ionic equation gives us a good picture of substances undergoing reaction. The nonionized equation shows the actual substances that produce the reaction.

1. Complete and balance the nonionized (molecular) equation. Refer to the guidelines in Section 8.3 for completing and balancing equations.
2. Convert the nonionized equation to the total ionic equation. Write strong electrolytes in the ionized form and leave weak electrolytes in the nonionized form.
3. Cancel spectator ions to obtain the net ionic equation.
   (a) If canceling spectator ions eliminates all species, there is no reaction.
   (b) If the coefficients can be simplified, do so in order to have the simplest whole-number relationship.
4. Check (√) each ion or atom on both sides of the equation. The total charge (positive or negative) on the reactants side of the equation must equal the total charge on the products side of the equation.

The following examples illustrate the above procedure for writing balanced net ionic equations.

EXAMPLE EXERCISE 15.15

Write a net ionic equation for the reaction of nitric acid and aqueous sodium hydrogen carbonate, given the nonionized equation:

$$HNO_3(aq) + NaHCO_3(aq) \longrightarrow NaNO_3(aq) + H_2O(l) + CO_2(g)$$

Solution: First, we must verify that the nonionized equation is balanced. In this case, the coefficients are all one and the equation is balanced. Second, we must determine which of the species are strong electrolytes and hence highly ionized. From Table 15.6, we find that HNO_3 is a strong acid; $NaHCO_3$ and $NaNO_3$ are soluble ionic compounds. Water and CO_2 gas are written as molecules. The total ionic equation is

$$H^+(aq) + \cancel{NO_3^-(aq)} + \cancel{Na^+(aq)} + HCO_3^-(aq) \longrightarrow \cancel{Na^+(aq)} + \cancel{NO_3^-(aq)} + H_2O(l) + CO_2(g)$$

Third, we obtain the net ionic equation by canceling spectator ions. In this example, Na^+ and NO_3^- are in the aqueous solution but do not participate in the reaction. The net ionic equation is

$$H^+(aq) + HCO_3^-(aq) \longrightarrow H_2O(l) + CO_2(g)$$

Finally, we quickly check to verify that the net ionic equation is balanced. The equation is balanced because (a) the number of atoms of each element is the same on each side, and (b) the total charge on each side is identical. In this case the net charge is zero.

SELF-TEST EXERCISE

Write a net ionic equation for the reaction of aqueous solutions of strontium iodide and sodium carbonate, given the nonionized equation:

$$SrI_2(aq) + Na_2CO_3(aq) \longrightarrow SrCO_3(s) + NaI(aq)$$

Answer: $Sr^{2+}(aq) + CO_3^{2-}(aq) \longrightarrow SrCO_3(s)$

EXAMPLE EXERCISE 15.16

Write a net ionic equation for the reaction of aqueous solutions of potassium bromide and copper(II) acetate, given the nonionized equation:

$$KBr(aq) + Cu(C_2H_3O_2)_2(aq) \longrightarrow CuBr_2(aq) + KC_2H_3O_2(aq)$$

Solution: First, we notice that the nonionized equation is not balanced. Therefore, we will use coefficients as follows:

$$2\ KBr(aq) + Cu(C_2H_3O_2)_2(aq) \longrightarrow CuBr_2(aq) + 2\ KC_2H_3O_2(aq)$$

Second, we determine the species that are strong electrolytes. From Table 8.2, we find that all these compounds are soluble. The total ionic equation is

$$\cancel{2\ K^+(aq)} + \cancel{2\ Br^-(aq)} + \cancel{Cu^{2+}(aq)} + \cancel{2\ C_2H_3O_2^-(aq)} \longrightarrow$$
$$\cancel{Cu^{2+}(aq)} + \cancel{2\ Br^-(aq)} + \cancel{2\ K^+(aq)} + \cancel{2\ C_2H_3O_2^-(aq)}$$

Third, we cancel the spectator ions. In this example, all the ions are spectator ions, and there is no net ionic equation. Thus,

$$KBr(aq) + Cu(C_2H_3O_2)_2(aq) \longrightarrow NR$$

Since there is no reaction, the two aqueous solutions simply mix together.

SELF-TEST EXERCISE

Write a net ionic equation for the reaction of iron metal and hydrofluoric acid, given the nonionized equation:

$$Fe(s) + HF(aq) \longrightarrow FeF_3(aq) + H_2(g)$$

Answer: $2\ Fe(s) + 6\ HF(aq) \rightarrow 2\ Fe^{3+}(aq) + 6\ F^-(aq) + 3\ H_2(g)$

Summary

Section 15.1 In this chapter we studied **acids** and **bases**. The properties of acids include tasting sour, turning blue litmus paper red, having a pH less than 7, and neutralizing bases to give a salt and water. Bases have a bitter taste, feel slippery, turn red litmus paper blue, and have a pH greater than 7. A solution having a **pH** of 0 to 2 is considered strongly acidic. One having a pH of 2 to 7 is weakly acidic. Pure water has a pH of 7, which is neutral. A solution having a pH of 7 to 12 is weakly basic. One having a pH of 12 to 14 is strongly basic.

Section 15.2 Most of us are familiar with litmus paper, which can indicate an acid or base. In addition to litmus paper, there are dozens of **acid-base indicators** available as solutions. Three of the most common are methyl red, bromthymol blue, and phenolphthalein. Methyl red changes color at pH 5, bromthymol blue at pH 7, and phenolphthalein at pH 9.

Section 15.3 The concentration of a **standard solution** is known accurately to at least three significant digits. Standard solutions are used to analyze the amount of acid or base in a given sample. For example, a chemist can determine the concentration of acetic acid in vinegar or the amount of baking soda in an antacid tablet.

Section 15.4 A **titration** is a laboratory procedure for analyzing the amount of acid or base in a solution. A measured sample of acid or base solution is delivered using a buret. An indicator is used to signal the neutralization point in the titration.

When the **endpoint** is reached, the indicator changes color and the titration is stopped. The titration data allow the chemist to calculate the amount of acid or base.

Section 15.5 In 1884, Arrhenius gave definitions for acids and bases. An **Arrhenius acid** is a substance that ionizes in water to give hydrogen ions. Hydrochloric acid is a strong acid because it ionizes extensively. Acetic acid is a weak acid since it provides only a few hydrogen ions. An **Arrhenius base** is a substance that gives hydroxide ions in water. Sodium hydroxide is dissociated in water and is a strong base. By definition, a **salt** is obtained from the neutralization of an acid and a base. Given the salt, we can predict the acid–base reaction. For example, potassium nitrate is obtained from the reaction of potassium hydroxide and nitric acid.

Section 15.6 In 1923, Brønsted and Lowry proposed broader definitions of acids and bases. A **Brønsted–Lowry acid** is a **proton donor**. A **Brønsted–Lowry base** is a proton acceptor. A substance that is capable of both donating and accepting a proton is **amphiprotic**. Although most reactions take place in water, according to Brønsted–Lowry the reactions can take place in any solvent. Furthermore, many acid–base reactions are **reversible reactions**. That is, an acid reacts with a base to give a second acid–base pair. The stronger acid–base pair is indicated by the longer arrow.

Section 15.7 On the basis of conductivity experiments, water is found to be a very weak electrolyte. Water autoionizes to give a hydrogen ion and hydroxide ion. The $[H^+]$ and $[OH^-]$ are each 1×10^{-7}. The **equilibrium constant of water, K_w**, is equal to 1×10^{-14} at 25°C. Since K_w is a constant, when the $[H^+]$ increases, the $[OH^-]$ decreases. Conversely, when the $[H^+]$ decreases, the $[OH^-]$ increases.

Section 15.8 The **pH** value expresses the $[H^+]$ on an exponential scale. That is, a solution of pH 1 is ten times more acidic than a solution of pH 2. To calculate the pH of a solution, we express the $[H^+]$ as a negative power of 10. For example, if the $[H^+]$ is 10^{-2} *M*, then the pH is 2. Conversely, to find the $[H^+]$, given the pH, we raise ten to the negative pH. For example, if the pH is 3, the $[H^+]$ is 10^{-3} *M*; that is, 0.001 *M*.

Section 15.9 To calculate the pH of solutions that are not exact powers of 10, we use logarithms. The pH is equal to the $-\log [H^+]$. Similarly, the $[OH^-]$ of basic solutions can be expressed by the **pOH**. Inasmuch as the $[H^+]$ and $[OH^-]$ are related to K_w, it follows that $pH + pOH = 14$.

Section 15.10 Experimentally, it is observed that a strong acid, a strong base, and a soluble ionic compound are each examples of a **strong electrolyte**. Therefore, these solutions are highly ionized. Hydrochloric acid is a strong acid and is better represented in solution as $H^+(aq)$ and $Cl^-(aq)$. The soluble ionic compound lithium nitrate is better represented as $Li^+(aq)$ and $NO_3^-(aq)$. A weak acid, a weak base, and a slightly soluble ionic compound are each examples of a **weak electrolyte**. Therefore, they give few ions in aqueous solution. Acetic acid is a weak acid and in aqueous solution should be written as the nonionized $HC_2H_3O_2(aq)$. The very slightly soluble ionic compound iron(III) hydroxide should be written as the nonionized $Fe(OH)_3(s)$.

Section 15.11 A **net ionic equation** shows a solution reaction more accurately. The first step is to balance the nonionized equation. Second, write a **total ionic equation** for the reaction. Strong electrolytes are written in the ionic form and weak electrolytes in the nonionized form. Third, cancel **spectator ions** that are identical on both sides of the equation. Finally, verify that the net ionic equation is balanced and simplify the coefficients when possible.

Key Terms

Select the key term that corresponds to the following definitions.

____ 1. a laboratory procedure for delivering a measured volume of solution through a buret
____ 2. a chemical substance that changes color according to the pH of the solution
____ 3. the stage in a titration when the indicator changes color
____ 4. a solution whose concentration has been established precisely (usually by titration to three or four significant digits)
____ 5. a substance that yields hydrogen ions when dissolved in water
____ 6. the ion that best represents the hydrogen ion in aqueous solution
____ 7. a substance that yields hydroxide ions when dissolved in water
____ 8. the process of a polar molecular compound dissolving in water and forming positive and negative ions; for example, HCl dissolves in water to give hydrogen ions and chloride ions
____ 9. the process of an ionic compound dissolving in water and forming positive and negative ions; for example, NaOH dissolves in water to give sodium ions and hydroxide ions
____ 10. a term used interchangeably with hydrogen ion donor
____ 11. a product from a neutralization reaction in addition to water
____ 12. a substance that donates a proton in an acid–base reaction
____ 13. a substance that accepts a proton in an acid–base reaction
____ 14. a substance that is capable of both accepting a proton and donating a proton in an acid–base reaction
____ 15. a reaction that proceeds in both the forward direction toward products as well as in the opposite direction toward reactants
____ 16. a reversible reaction in pure water that produces hydronium and hydroxide ions
____ 17. the state of a reversible reaction when the forward and reverse reactions are proceeding at the same rate
____ 18. the product of the molar hydrogen ion concentration times the molar hydroxide ion concentration
____ 19. the molarity of hydrogen ion expressed on an exponential scale; the negative logarithm of the molar hydrogen ion concentration
____ 20. the molarity of hydroxide ion expressed on an exponential scale; the negative logarithm of the molar hydroxide ion concentration
____ 21. an aqueous solution containing a substance that strongly conducts an electric current; for example, strong acids, strong bases, soluble salts
____ 22. an aqueous solution containing a substance that slightly conducts an electric current; for example, weak acids, weak bases, slightly soluble salts
____ 23. a chemical equation that portrays highly ionized substances written in the ionic form and weakly ionized substances in the molecular form
____ 24. those ions that are in aqueous solution but do not participate in a reaction or appear as reactants or products in the net ionic equation
____ 25. a chemical equation that portrays an ionic reaction after spectator ions have been canceled from the total ionic equation.

(a) acid–base indicator *(Sec. 15.2)*
(b) amphiprotic *(Sec. 15.6)*
(c) Arrhenius acid *(Sec. 15.5)*
(d) Arrhenius base *(Sec. 15.5)*
(e) autoprotolysis *(Sec. 15.7)*
(f) Brønsted–Lowry acid *(Sec. 15.6)*
(g) Brønsted–Lowry base *(Sec. 15.6)*
(h) chemical equilibrium *(Sec. 15.7)*
(i) dissociation *(Sec. 15.5)*
(j) endpoint *(Sec. 15.4)*
(k) equilibrium constant of water (K_w) *(Sec. 15.7)*
(l) hydronium ion (H_3O^+) *(Sec. 15.5)*
(m) ionization *(Sec. 15.5)*
(n) net ionic equation *(Sec. 15.11)*
(o) pH *(Sec. 15.1)*
(p) pOH *(Sec. 15.9)*
(q) proton donor *(Sec. 15.6)*
(r) reversible reaction *(Sec. 15.6)*
(s) salt *(Sec. 15.5)*
(t) spectator ions *(Sec. 15.11)*
(u) standard solution *(Sec. 15.3)*
(v) strong electrolyte *(Sec. 15.10)*
(w) titration *(Sec. 15.4)*
(x) total ionic equation *(Sec. 15.11)*
(y) weak electrolyte *(Sec. 15.10)*

Exercises

Properties of Acids and Bases (Sec. 15.1)

1. State at least three general properties of acids.

2. State at least three general properties of bases.

3. Classify the following foods as acidic, basic, or neutral.
(a) egg white, pH 7.9 (b) sour milk, pH 6.2
(c) maple syrup, pH 7.0 (d) lime juice, pH 1.8
(e) champagne, pH 3.8 (f) tomato juice, pH 4.1

4. Classify the following 0.1 *M* solutions as strongly acidic, weakly acidic, neutral, weakly basic, or strongly basic.
(a) 0.1 *M* NaOH, pH 13.0
(b) 0.1 *M* NaCl, pH 7.0
(c) 0.1 *M* Na_2CO_3, pH 11.7
(d) 0.1 *M* $NaHCO_3$, pH 8.3
(e) 0.1 *M* H_2CO_3, pH 3.7
(f) 0.1 *M* HNO_3, pH 1.0

Acid–Base Indicators (Sec. 15.2)

5. Given the pH of the following solutions containing a drop of methyl red indicator, state the color.
(a) pH 3 (b) pH 7

Many flowers are natural acid–base indicators. After the indicator is extracted (right), the dark-red rose becomes pale pink (left).

6. Given the pH of the following solutions containing a drop of bromthymol blue indicator, state the color.
(a) pH 5 (b) pH 9

7. Given the pH of the following solutions containing a drop of phenolphthalein indicator, state the color.
(a) pH 7 (b) pH 11

8. What is the color of methyl red indicator in a solution of pH 5?

9. What is the color of bromthymol blue indicator in a solution of pH 7?

10. What is the color of phenolphthalein indicator in pure water?

Standard Solutions of Acids and Bases (Sec. 15.3)

11. What is the molarity of nitric acid if 41.25 mL of HNO_3 is required to neutralize 0.689 g of sodium carbonate?

$$2\ HNO_3(aq) + Na_2CO_3(s) \longrightarrow 2\ NaNO_3(aq) + H_2O(l) + CO_2(g)$$

12. What is the molarity of sulfuric acid if 32.35 mL of H_2SO_4 is required for the neutralization of 0.750 g of sodium hydrogen carbonate?

$$H_2SO_4(aq) + 2\ NaHCO_3(s) \longrightarrow Na_2SO_4(aq) + 2\ H_2O(l) + 2\ CO_2(g)$$

13. Calculate the molarity of hydrochloric acid if 20.95 mL of solution is required to titrate 1.550 g of sodium oxalate, $Na_2C_2O_4$.

$$2\ HCl(aq) + Na_2C_2O_4(s) \longrightarrow H_2C_2O_4(aq) + 2\ NaCl(aq)$$

14. What is the molarity of a sodium hydroxide solution if 28.85 mL of NaOH reacts with 0.506 g of oxalic acid, $H_2C_2O_4$?

$$H_2C_2O_4(s) + NaOH(aq) \longrightarrow NaHC_2O_4(aq) + H_2O(l)$$

15. Calculate the molarity of a lithium hydroxide solution if 29.00 mL of LiOH is required to neutralize 0.627 g of oxalic acid, $H_2C_2O_4$.

$$H_2C_2O_4(s) + 2\ LiOH(aq) \longrightarrow Li_2C_2O_4(aq) + 2\ H_2O(l)$$

16. What is the molarity of a barium hydroxide solution if 24.65 mL is required to neutralize 1.655 g of potassium hydrogen phthalate, KHP (204.2 g/mol)?

$$2\ KHP(s) + Ba(OH)_2(aq) \longrightarrow K_2BaP_2(aq) + 2\ H_2O(l)$$

17. Vitamin C has the chemical name ascorbic acid and can be abbreviated HAsc. If 30.95 mL of 0.176 *M* NaOH neutralizes 0.959 g of ascorbic acid, what is the molar mass of vitamin C?

$$HAsc(s) + NaOH(aq) \longrightarrow NaAsc(aq) + H_2O(l)$$

18. Tartaric acid, H_2Tart, occurs naturally in grapes. If 28.15 mL of 0.295 *M* NaOH neutralizes 0.623 g of acid, what is the molar mass of tartaric acid?

$$H_2Tart(s) + 2\ NaOH(aq) \longrightarrow Na_2Tart(aq) + 2\ H_2O(l)$$

19. Alanine is an amino acid having one replaceable hydrogen ion. If 21.05 mL of 0.145 *M* NaOH neutralizes 0.272 g of alanine, calculate the molar mass of the amino acid.

20. THAM is an organic base used to standardize acid solutions. If 0.297 g of THAM requires 21.35 mL of 0.115 *M* HCl to achieve neutralization, what is the molar mass? (*Note:* One mole of THAM neutralizes one mole of acid.)

Acid-Base Titrations (Sec. 15.4)

21. The titration of a 25.0-mL sample of hydrochloric acid required 22.15 mL of 0.155 *M* sodium hydroxide. What is the molarity of the hydrochloric acid?

$$HCl(aq) + NaOH(aq) \longrightarrow NaCl(aq) + H_2O(l)$$

22. If 34.45 mL of 0.100 *M* perchloric acid is required to titrate 50.0 mL of calcium hydroxide, what is the molarity of the base?

$$2\ HClO_4(aq) + Ca(OH)_2(aq) \longrightarrow Ca(ClO_4)_2(aq) + 2\ H_2O(l)$$

23. If 34.45 mL of 0.210 *M* NaOH is required to titrate 28.55 mL of phosphoric acid, what is the molarity of the acid?

$$H_3PO_4(aq) + 3\ NaOH(aq) \longrightarrow Na_3PO_4(aq) + 3\ H_2O(l)$$

24. How many milliliters of 0.175 *M* nitric acid are required to neutralize 25.0 mL of 0.119 *M* ammonium hydroxide?

$$HNO_3(aq) + NH_4OH(aq) \longrightarrow NH_4NO_3(aq) + H_2O(l)$$

25. How many milliliters of 0.122 *M* sulfuric acid are required to completely neutralize 41.05 mL of 0.165 *M* KOH?

$$H_2SO_4(aq) + 2\ KOH(aq) \longrightarrow K_2SO_4(aq) + 2\ H_2O(l)$$

26. If a 10.0-mL sample of 0.225 *M* nitrous acid is titrated with 0.100 *M* barium hydroxide to complete neutralization, what volume of base is required?

$$2\ HNO_2(aq) + Ba(OH)_2(aq) \longrightarrow Ba(NO_2)_2(aq) + 2\ H_2O(l)$$

27. Given the molarity and density for each of the following acid solutions, calculate the mass/mass percent concentration.
   (a) 6.00 *M* HCl (d = 1.10 g/mL)
   (b) 1.00 *M* $HC_2H_3O_2$ (d = 1.01 g/mL)
   (c) 0.500 *M* HNO_3 (d = 1.01 g/mL)
   (d) 3.00 *M* H_2SO_4 (d = 1.18 g/mL)
   (e) 0.631 *M* H_3PO_4 (d = 1.03 g/mL)

28. Given the molarity and density for each of the following base solutions, calculate the mass/mass percent concentration.
   (a) 3.00 *M* NaOH (d = 1.12 g/mL)
   (b) 0.500 *M* KOH (d = 1.02 g/mL)
   (c) 6.00 *M* NH_3 (d = 0.954 g/mL)
   (d) 1.00 *M* Na_2CO_3 (d = 1.10 g/mL)
   (e) 0.250 *M* $K_2C_2O_4$ (d = 1.03 g/mL)

Arrhenius Acid-Base Theory (Sec. 15.5)

29. Classify each of the following Arrhenius acids as strong or weak, given the degree of ionization.
   (a) chloric acid, $HClO_3(aq)$ ~100%
   (b) hypoiodous acid, HIO(aq) ~1%
   (c) hydrobromic acid, HBr(aq) ~100%
   (d) benzoic acid, $HC_7H_5O_2(aq)$ ~1%

30. Classify each of the following Arrhenius bases as strong or weak, given the degree of ionization.
   (a) zinc hydroxide, $Zn(OH)_2(s)$ ~1%
   (b) lithium hydroxide, LiOH(aq) ~100%
   (c) manganese(II) hydroxide, $Mn(OH)_2(s)$ ~1%
   (d) strontium hydroxide, $Sr(OH)_2(aq)$ ~100%

31. Classify each of the following as an Arrhenius acid, Arrhenius base, or salt.
   (a) HClO(aq) (b) KOH(aq)
   (c) $K_2SO_4(aq)$ (d) $Ba(OH)_2(aq)$
   (e) $HNO_3(aq)$ (f) $Mg(NO_3)_2(aq)$
   (g) $Ca(OH)_2(aq)$ (h) $H_2SO_3(aq)$

32. Identify the Arrhenius acid, Arrhenius base, and salt in each of the following neutralization reactions.
   (a) $HI(aq) + NaOH(aq) \rightarrow NaI(aq) + H_2O(l)$
   (b) $HC_2H_3O_2(aq) + LiOH(aq) \rightarrow LiC_2H_3O_2(aq) + H_2O(l)$
   (c) $2\ HClO_3(aq) + Ba(OH)_2(aq) \rightarrow Ba(ClO_3)_2(aq) + 2\ H_2O(l)$
   (d) $H_2SO_4(aq) + 2\ KOH(aq) \rightarrow K_2SO_4(aq) + 2\ H_2O(l)$

33. Determine the acid and base that were neutralized to produce each of the following aqueous salts.
   (a) potassium bromide, KBr(aq)
   (b) barium chloride, $BaCl_2(aq)$
   (c) calcium nitrate, $Ca(NO_3)_2(aq)$
   (d) sodium phosphate, $Na_3PO_4(aq)$
   (e) cobalt(II) sulfate, $CoSO_4(aq)$
   (f) lithium carbonate, $Li_2CO_3(aq)$

34. Write a balanced equation for the neutralization reaction that produced the following aqueous salts.
   (a) sodium fluoride, NaF(aq)
   (b) magnesium iodide, $MgI_2(aq)$
   (c) potassium nitrite, $KNO_2(aq)$
   (d) zinc perchlorate, $Zn(ClO_4)_2(aq)$
   (e) bismuth sulfate, $Bi_2(SO_4)_3(aq)$
   (f) sodium formate, $NaCHO_2(aq)$

Brønsted-Lowry Acid-Base Theory (Sec. 15.6)

35. Identify the Brønsted–Lowry acid and base in each of the following neutralization reactions.
   (a) $HC_2H_3O_2(aq) + LiOH(aq) \rightarrow LiC_2H_3O_2(aq) + H_2O(l)$
   (b) $NaCN(aq) + HBr(aq) \rightarrow NaBr(aq) + HCN(aq)$
   (c) $2\ HClO_4(aq) + K_2CO_3(aq) \rightarrow 2\ KClO_4(aq) + H_2O(l) + CO_2(g)$
   (d) $2\ NH_3(aq) + H_2SO_4(aq) \rightarrow (NH_4)_2SO_4(aq)$

36. Identify the Brønsted–Lowry acid and base in each of the following neutralization reactions.
   (a) $HI(aq) + H_2O(l) \rightarrow H_3O^+(aq) + I^-(aq)$
   (b) $HC_2H_3O_2(aq) + HS^-(aq) \rightarrow H_2S(aq) + C_2H_3O_2^-(aq)$
   (c) $HCO_3^-(aq) + OH^-(aq) \rightarrow CO_3^{2-}(aq) + H_2O(l)$
   (d) $NO_2^-(aq) + HClO_4(aq) \rightarrow HNO_2(aq) + ClO_4^-(aq)$

37. Designate the stronger Brønsted–Lowry acid and base in each of the following reversible reactions.
(a) $H_2SO_3(aq) + NaHS(aq) \rightleftharpoons NaHSO_3(aq) + H_2S(aq)$
(b) $HF(aq) + Na_2SO_4(aq) \rightleftharpoons NaHSO_4(aq) + NaF(aq)$
(c) $H_2PO_4^-(aq) + NH_3(aq) \rightleftharpoons NH_4^+(aq) + HPO_4^{2-}(aq)$
(d) $H_3O^+(aq) + HSO_4^-(aq) \rightleftharpoons H_2SO_4(aq) + H_2O(aq)$

38. Designate the stronger Brønsted–Lowry acid and base in each of the following reversible reactions.
(a) $HNO_2(aq) + NaC_2H_3O_2(aq) \rightleftharpoons NaNO_2(aq) + HC_2H_3O_2(aq)$
(b) $H_2S(aq) + NaF(aq) \rightleftharpoons NaHS(aq) + HF(aq)$
(c) $H_2CO_3(aq) + HPO_4^{2-}(aq) \rightleftharpoons HCO_3^-(aq) + H_2PO_4^-(aq)$
(d) $HSO_4^-(aq) + Cl^-(aq) \rightleftharpoons SO_4^{2-}(aq) + HCl(aq)$

Ionization of Water (Sec. 15.7)

39. Indicate each of the following for the ionization of pure water.
(a) the equilibrium ionization reaction
(b) the equilibrium constant expression, K_w
(c) the value for K_w at 25°C
(d) the molar hydrogen ion concentration at 25°C
(e) the molar hydroxide ion concentration at 25°C

40. Given the following concentrations of $[H^+]$ in water, calculate the concentration of hydroxide ion.
(a) $[H^+] = 2.5 \times 10^{-2}$ (b) $[H^+] = 1.7 \times 10^{-5}$
(c) $[H^+] = 6.2 \times 10^{-7}$ (d) $[H^+] = 4.6 \times 10^{-12}$

41. State whether the hydrogen ion concentration in water increases or decreases with the following changes.

$$H_2O(l) \rightleftharpoons H^+(aq) + OH^-(aq)$$

(a) increase $[OH^-]$ (b) decrease $[OH^-]$

42. State whether the hydroxide ion concentration in water increases or decreases with the following changes.
(a) increase $[H^+]$ (b) decrease $[H^+]$

The pH Concept (Sec. 15.8)

43. Calculate the pH of the following solutions, given the molar hydrogen ion concentration.
(a) soft drink, $[H^+]$ = 0.001 *M*
(b) coffee, $[H^+]$ = 0.000 01 *M*
(c) egg white, $[H^+]$ = 0.000 000 01 *M*
(d) sour milk, $[H^+]$ = 0.000 001 *M*

44. Calculate the molar hydrogen ion concentration given the pH of the following solutions.
(a) shampoo, pH = 6
(b) pH balanced shampoo, pH = 6
(c) phosphate detergent, pH = 9
(d) nonphosphate detergent, pH = 11

Advanced pH and pOH Calculations (Sec. 15.9)

45. Calculate the pH for each of the following food items given the hydrogen ion concentration.
(a) carrots, $[H^+]$ = 0.000 007 9 *M*
(b) peas, $[H^+]$ = 0.000 000 39 *M*
(c) milk, $[H^+]$ = 0.000 000 30 *M*
(d) eggs, $[H^+]$ = 0.000 000 016 *M*

46. Calculate the hydrogen ion concentration for each of the following biological solutions.
(a) gastric juice, pH = 1.80 (b) urine, pH = 4.75
(c) saliva, pH = 6.55 (d) blood, pH = 7.50

47. Calculate the pOH for each of the following solutions, given the hydroxide ion concentration.
(a) $[OH^-]$ = 0.11 *M*
(b) $[OH^-]$ = 0.000 55 *M*
(c) $[OH^-]$ = 0.000 31 *M*
(d) $[OH^-]$ = 0.000 000 000 66 *M*

48. Calculate the hydroxide ion concentration for each of the following solutions, given the pOH value.
(a) pOH = 0.90 (b) pOH = 1.62
(c) pOH = 4.55 (d) pOH = 5.20

49. Given the pH of the following solutions, express the pOH.
(a) 0.50 *M* HCl, pH = 0.30
(b) 0.50 *M* $HC_2H_3O_2$, pH = 2.52

50. Given the pOH of the following solutions, express the pH.
(a) 0.25 *M* NaOH, pOH = 0.60
(b) 0.25 *M* NH_4OH, pOH = 2.67

Strong and Weak Electrolytes (Sec. 15.10)

51. Given the solution conductivity, classify the following as a strong or weak acid, strong or weak base, or soluble or very slightly soluble ionic compound.
(a) HBrO(aq), weak (b) $Sr(NO_3)_2(aq)$, strong
(c) HI(aq), strong (d) $HC_2H_3O_2(aq)$, weak
(e) KOH(aq), strong (f) $Sr(OH)_2(aq)$, strong
(g) $HClO_4(aq)$, strong (h) $NH_4OH(aq)$, weak
(i) $KC_2H_3O_2(aq)$, strong (j) $PbI_2(s)$, weak

52. Indicate whether aqueous solutions of the following substances contain relatively many ions or few ions.
(a) strong acids (b) weak acids
(c) strong bases (d) weak bases
(e) soluble ionic compounds (f) slightly soluble ionic compounds

53. Given the electrolyte strength for each of the following, write the substance in the ionized or nonionized form as it occurs in aqueous solution.
(a) $HClO_3(aq)$, strong (b) $Ca(NO_3)_2(aq)$, strong
(c) HBr(aq), strong (d) $HClO_2(aq)$, weak
(e) LiOH(aq), strong (f) $Al(OH)_3(s)$, weak
(g) $HNO_2(aq)$, weak (h) $HCHO_2(aq)$, weak
(i) $Cd(C_2H_3O_2)_2(aq)$, strong (j) $K_2SO_4(aq)$, strong

54. Refer to the common electrolytes in Table 15.7 and write each of the following in the ionized or nonionized form as it occurs in aqueous solution.
   (a) hydrochloric acid
   (b) hydrofluoric acid
   (c) hypochlorous acid
   (d) perchloric acid
   (e) strontium hydroxide
   (f) ammonium hydroxide
   (g) copper(II) sulfate
   (h) barium sulfate

Net Ionic Equations (Sec. 15.11)

55. List the four steps for writing balanced net ionic equations.

56. What is the term for those ions that appear in the total ionic equation, but are not present in the net ionic equation?

57. Write a balanced net ionic equation for each of the following reactions. Refer to Table 15.7 for electrolyte ionization information.
   (a) $HCl(aq) + KOH(aq) \rightarrow KCl(aq) + H_2O(l)$
   (b) $HC_2H_3O_2(aq) + Ca(OH)_2(aq) \rightarrow Ca(C_2H_3O_2)_2(aq) + H_2O(l)$
   (c) $HF(aq) + Li_2CO_3(aq) \rightarrow LiF(aq) + H_2O(l) + CO_2(g)$
   (d) $HNO_3(aq) + KHCO_3(aq) \rightarrow KNO_3(aq) + H_2O(l) + CO_2(g)$
   (e) $H_2SO_4(aq) + Ba(OH)_2(aq) \rightarrow BaSO_4(s) + H_2O(l)$

58. Write a balanced net ionic equation for each of the following reactions. Refer to Table 15.7 for electrolyte ionization information.
   (a) $AgNO_3(aq) + KI(aq) \rightarrow AgI(s) + KNO_3(aq)$
   (b) $BaCl_2(aq) + K_2CrO_4(aq) \rightarrow BaCrO_4(s) + KCl(aq)$
   (c) $Zn(NO_3)_2(aq) + NaOH(aq) \rightarrow Zn(OH)_2(s) + NaNO_3(aq)$
   (d) $MgSO_4(aq) + NH_4OH(aq) \rightarrow Mg(OH)_2(s) + (NH_4)_2SO_4(aq)$
   (e) $HC_2H_3O_2(aq) + Sn(s) \rightarrow Sn(C_2H_3O_2)_2(aq) + H_2(g)$

General Exercises

59. The acid–base indicator methyl orange has a color change range from pH 3.2 to 4.4. It appears red in strongly acidic solutions and yellow in basic solutions. Predict the color of the indicator in a solution having a pH of 3.8.

60. The acid–base indicator bromcresol green changes color from pH 3.8 to 5.4. It appears yellow in strongly acidic solutions and blue in basic solutions. Explain why the indicator appears green in a solution having a pH of 4.6.

61. A drop of methyl red indicator and a drop of phenolphthalein are both added to a beaker of distilled water. What is the resulting color of the water?

62. Cream of tartar is found in baking powder and has one replaceable hydrogen ion. If 42.10 mL of standard 0.100 *M* $Ba(OH)_2$ neutralizes 1.585 g of cream of tartar, what is its molar mass?

63. Given the following pair of equations, identify the amphiprotic substance.

$$H^+(aq) + H_2PO_4^-(aq) \longrightarrow H_3PO_4(aq)$$
$$OH^-(aq) + H_2PO_4^-(aq) \longrightarrow HPO_4^{2-}(aq) + H_2O(l)$$

64. Given the following pair of equations, identify the amphiprotic substance.

$$HCl(aq) + H_2O(l) \longrightarrow H_3O^+(aq) + Cl^-(aq)$$
$$NH_2^-(aq) + H_2O(l) \longrightarrow NH_3(aq) + OH^-(aq)$$

65. In 1 mL of water there are 3×10^{22} molecules of H_2O. At 25°C, how many water molecules exist as ions? (*Hint:* In 1 billion water molecules, two molecules are ionized.)

66. Write the autoprotolysis equilibrium reaction for pure water.

67. Consider the K_w equilibrium in an aqueous solution to answer the following:
   (a) If the pH of a solution increases, does the hydroxide ion concentration increase or decrease?
   (b) If the pOH of a solution increases, does the hydrogen ion concentration increase or decrease?

68. Which is more basic, a solution of pOH 3 or a solution of pOH 5?

69. Distinguish between a nonelectrolyte and a weak electrolyte.

70. Write the net ionic equation for the reaction of aqueous solutions of iron(II) nitrate and nickel(II) sulfate.

CHAPTER

16

Chemical Equilibrium

The depletion of ozone in the stratosphere above Antarctica is threatening the emperor penguin population. Ozone shields the earth from harmful solar radiation and is in equilibrium with oxygen in the upper atmosphere.

We learned about chemical changes in Section 3.8 and chemical equations in Section 8.2. In those chapters, we usually assumed a reaction continues until the reactants are consumed. In reality, however, most reactions occur as a dynamic reversible process. A reversible reaction is a dynamic process that occurs in opposing directions at the same time. For example, in a beaker of ice and water at 0°C, ice is melting and water is freezing at the same time. We can indicate a process is reversible and dynamic with two arrows pointing in opposite directions. That is,

$$\text{ice} \underset{\text{freezing}}{\overset{\text{melting}}{\rightleftharpoons}} \text{water}$$

Another example of a dynamic and reversible process is a reaction that gives a precipitate (see Figure 16.1). Initially, insoluble particles form a precipitate from the solution. The precipitate is slightly soluble, however, and some particles that precipitate may also dissolve and go back into solution. Since particles are continuing to crystallize as other particles are dissolving, this process is dynamic and reversible. We can indicate this relationship as

$$\text{dissolved substance in solution} \underset{\text{dissolving}}{\overset{\text{crystallizing}}{\rightleftharpoons}} \text{precipitate}$$

Figure 16.1 Precipitation, a Dynamic Reversible Process The reaction of lead(II) nitrate and potassium iodide produces PbI_2 particles. Even after the reaction appears to be complete, particles are dissolving from the yellow precipitate while other particles are crystallizing from the solution.

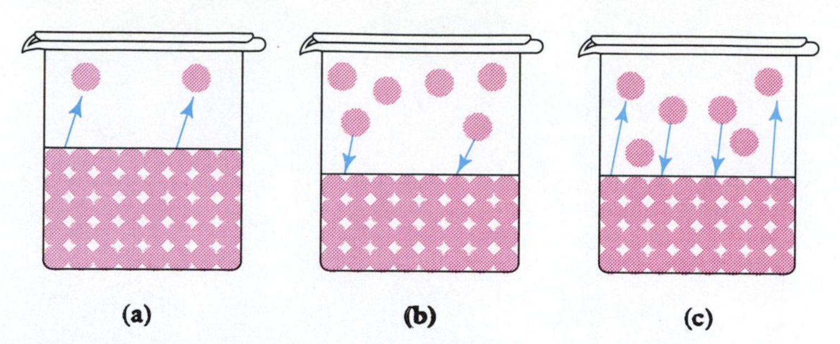

Figure 16.2 Evaporation, a Dynamic Reversible Process
(a) Water in a closed container begins to evaporate. (b) After a while, some water vapor condenses to a liquid. (c) When the rate of evaporation equals the rate of condensation, the system is in equilibrium. Note that evaporation and condensation go on continuously. This is an example of a dynamic reversible process.

Most chemical reactions are dynamic reversible processes. To see why, consider what happens in a solution. At first, reactants undergo a reaction to form products. Almost simultaneously, the products can react with each other to give the original reactants. Initially, the rate of the forward reaction is rapid. Then, as the amount of reactants decreases through reaction, the amount of products increases. As a consequence, the rate of the forward reaction slows down and the rate of the reverse reaction speeds up. When the rate of the forward reaction is equal to the rate of the reverse reaction, the system is said to be in a state of equilibrium. We indicate this process as follows:

$$\text{reactants} \underset{\text{reverse reaction}}{\overset{\text{forward reaction}}{\rightleftharpoons}} \text{products}$$

It is important to keep in mind that although the chemical reaction is in a state of equilibrium, it does not imply the amounts of reactants and products are equal. It also does not mean that there is no longer a reaction. Rather, chemical equilibrium is both a dynamic *and* reversible process. *That is, the forward and reverse reactions are proceeding at the same time. The process of evaporation and condensation are dynamic and reversible, as shown in Figure 16.2.*

16.1 Collision Theory of Reaction Rates

OBJECTIVES

To state the effect of collision frequency, collision energy, and geometry of molecules on the rate of a chemical reaction.

To state the effect of concentration, temperature, and catalyst on the rate of a chemical reaction.

We now know that a chemical system is at equilibrium when the rates of the forward and reverse reactions are equal. To better understand the process, let's visualize the change from reactants to products. Chemists have proposed a theoretical model, which states that molecules must collide in order to react. If a collision is successful, new molecules are formed. In other words, in a successful collision, chemical bonds

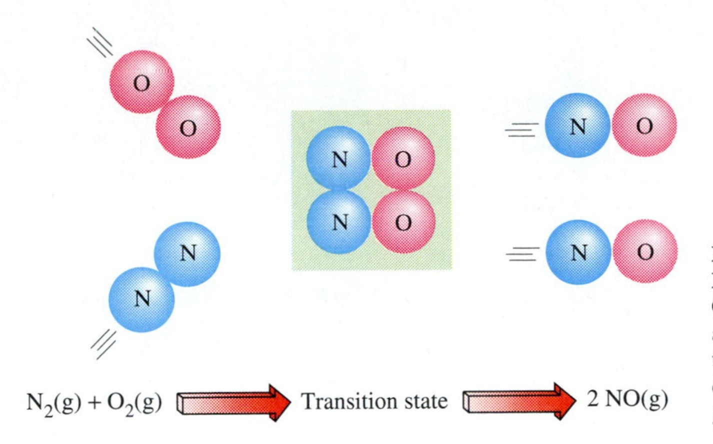

Figure 16.3 Reaction of Nitrogen and Oxygen Gaseous molecules of nitrogen and oxygen collide to form the transition state. In a successful collision, molecules of N_2 and O_2 give two molecules of NO.

are broken and new bonds are formed. This model is referred to as the **collision theory** of reaction rates. Moreover, when molecules collide, they go through a transition state in which bonds are rearranged. Figure 16.3 illustrates the collision theory by showing the reaction of nitrogen and oxygen to form nitric oxide.

collision theory The principle that the rate of a chemical reaction is regulated by the collision frequency, collision energy, and the orientation of molecules striking each other.

At the molecular level, certain factors affect the rate of reaction. That is, they affect the rate of effective collisions. According to collision theory, three factors influence the rate of effective collisions: (1) frequency, (2) energy, and (3) orientation of the molecules.

1. *Collision frequency*. If we can increase the collision frequency, that is, the frequency with which molecules collide, we can increase the rate of the reaction. To increase collision frequency, we can increase the concentration of molecules. As the concentration of molecules increases, there is a higher probability that the molecules will collide. Another way to increase collision frequency is to increase the temperature. If the temperature increases, the molecules have a higher kinetic energy and velocity. This higher velocity causes an increase in the frequency of molecules colliding.
2. *Collision energy*. Even though molecules collide, they may not react. For a reaction to occur, the molecules have to collide with sufficient energy so that bonds are broken and new bonds are formed. If we increase temperature, the kinetic energy of the molecules increases. As the molecular energy increases, the molecules move at a higher velocity and the resulting collisions are more energetic. The rate of a reaction proceeds faster as the energy of collision becomes greater.
3. *Collision geometry*. Even though energetic molecules collide, this does not guarantee that a reaction occurs. For a reaction to occur, the molecules must be oriented in a favorable position. That is, the orientation or geometry of the collision must allow the molecules to strike each other effectively in order to break bonds. Otherwise, the colliding molecules simply bounce off one another.

The effect of collision geometry on the reaction of nitrogen monoxide and ozone is shown in Figure 16.4.

Effects of Concentration, Temperature, and Catalysts

So far, we have described the factors that influence the rate of reaction on the molecular level. Here we will discuss how we can increase the rate of a chemical

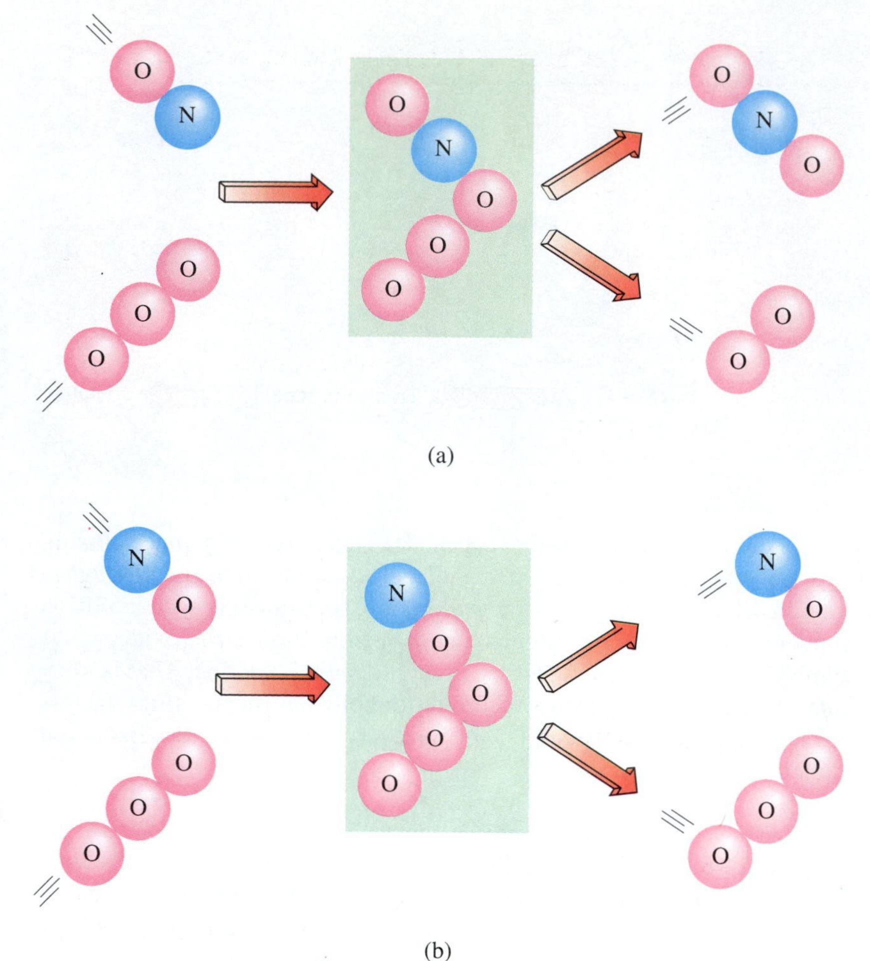

Figure 16.4 Effective versus Ineffective Molecular Collisions (a) Effective collision of gaseous NO and O_3 molecules gives NO_2 and O_2. (b) Ineffective collision of NO and O_3 molecules does not give products. If molecules do not have the correct orientation, they simply bounce off each other without forming new compounds.

reaction in the laboratory. Three experimental factors have been identified that affect the rate of reaction. They are concentration of reactants, temperature of the reactions, and the presence of a catalyst. Let's consider each of these factors.

1. *Reactant concentration.* When we increase the concentration of the reactant, the molecules are closer together and collide more frequently. It therefore follows that the rate of a chemical reaction proceeds faster as the concentration of the reactants increases.
2. *Reaction temperature.* If we increase the temperature, the molecular energy increases. This affects the reaction in two ways. First, the collision frequency increases and the reaction speeds up. Second, more molecules have enough energy to break bonds and form new ones. Thus, as temperature increases, the rate of reaction increases,because of collision frequency and collision energy.
3. *Catalyst.* If we introduce a catalyst, the rate of chemical reaction increases. One way a catalyst speeds up a reaction is by orienting reactant molecules to make collisions more effective. In reactions involving gases, a metal is often used as a catalyst. For example, Pt, Ni, and Zn metals are used to increase the rate at which hydrogen gas undergoes reaction. In living systems, enzymes act as catalysts that speed up complex biochemical reactions.

16.2 Energy Profiles of Chemical Reactions

To draw the general energy profile for each of the following:

(a) an endothermic reaction
(b) an exothermic reaction
(c) a catalyzed reaction

To label each of the following on an energy profile:

(a) vertical and horizontal axes
(b) reactants and products
(c) transition state
(d) energy of activation, E_{act}
(e) heat of reaction, ΔH

We are all aware that automobiles pollute our atmosphere with oxides of nitrogen. In air, however, nitrogen and oxygen gases are essentially unreactive. But at the temperature and pressure inside an automobile engine, the two gases produce nitrogen monoxide, a pollutant. To react, gaseous molecules must collide with sufficient energy to achieve a transition state before going on to form products. Since the transition state is at a higher energy than either reactants or products, it acts as an energy barrier. Figure 16.5 shows an analogy to help you understand the energy changes for a chemical reaction.

Endothermic and Exothermic Reaction Profiles

Some reactions proceed by releasing heat energy, while others proceed by consuming heat energy. As an example, consider the formation of the automobile pollutant nitrogen monoxide from nitrogen and oxygen gases. The reaction is

$$N_2(g) + O_2(g) + \text{heat} \longrightarrow 2\ NO(g)$$

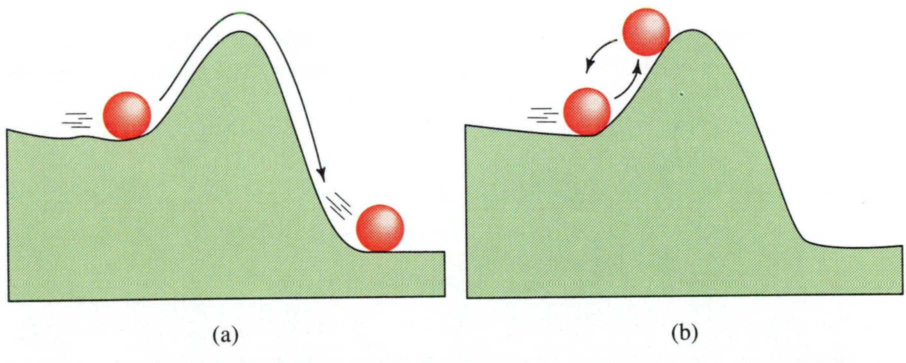

Figure 16.5 Energy Barrier Analogy for a Chemical Reaction The rolling ball is analogous to the energy change of molecules in a reaction. In (a) the ball has enough kinetic energy to go over the potential energy barrier. This symbolizes a reaction in which the energy of molecules is sufficient to give products. In (b) the rolling ball does not have enough energy to go over the hill. This symbolizes no reaction, in which the energy of molecules is not sufficient to give products.

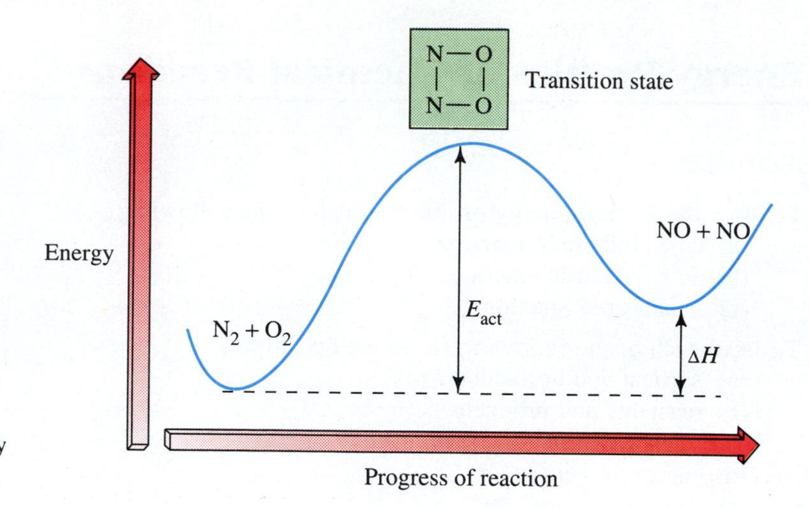

Figure 16.6 Reaction Profile of an Endothermic Reaction Notice that the product of the reaction, NO, is at a higher energy level than the reactants, $N_2 + O_2$.

reaction profile A graphical representation of the energy of a reaction as a function of the time of reaction; also called an energy diagram.

transition state The highest point on the reaction profile where reactant and product molecules have maximum energy.

activation energy (E_{act}) The energy necessary for reactants to achieve the transition state and form products.

heat of reaction (ΔH) The difference in heat energy between the reactants and the products for a given chemical reaction.

Such a reaction, which absorbs heat energy in the process of going to completion, is an endothermic reaction.

We can follow the progress of a chemical reaction by drawing a **reaction profile** (or energy diagram) showing the changes in energy. At the top of the reaction profile, the system has the highest potential energy. This stage of the process is called the **transition state**. A reaction profile for the reaction of nitrogen and oxygen is shown in Figure 16.6. The energy required for the reactants to achieve the transition state is called the **activation energy** (symbol E_{act}). The energy difference between reactants and products is termed the **heat of reaction** (symbol ΔH).

As an example of a reaction that proceeds by liberating heat energy, again consider that automobiles produce nitrogen monoxide, NO. Also consider that air contains ozone, O_3. The ozone, O_3, in air converts the NO to NO_2, a brown gaseous component of smog.

$$NO(g) + O_3(g) \longrightarrow NO_2(g) + O_2(g) + \text{heat}$$

As you see, the formation of nitrogen dioxide releases heat. Chemical processes that release heat energy are exothermic reactions. As before, we can construct a reaction profile to follow the changes in energy for this reaction. Figure 16.7 illustrates the energy difference between reactants and products for an exothermic reaction.

For reactants to form products, they must have sufficient energy to overcome the activation energy barrier. The higher the activation energy, the slower the reaction

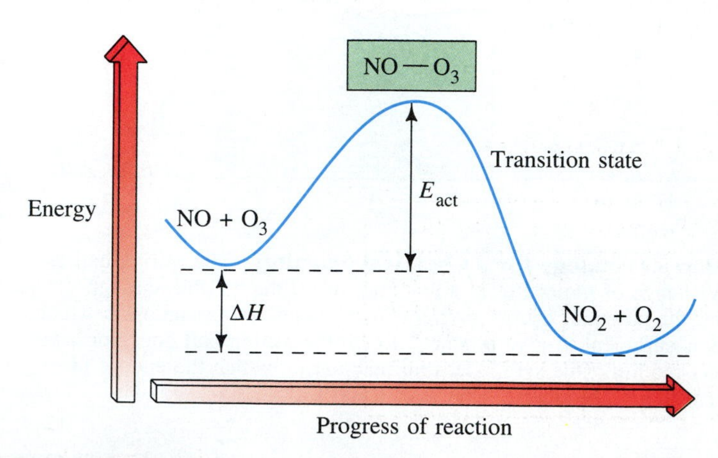

Figure 16.7 Reaction Profile of an Exothermic Reaction Notice that the products of the reaction, NO_2 and O_2, are at a lower energy level than the reactants, $NO + O_3$.

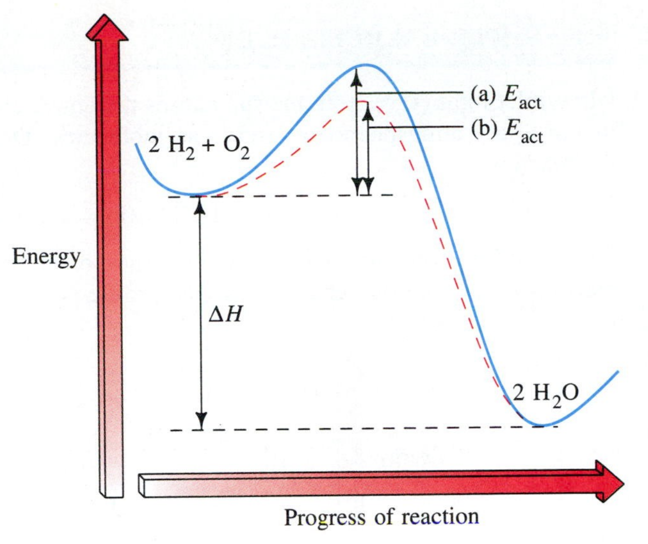

Figure 16.8 Effect of a Catalyst on the Activation Energy (a) Reaction profile for an uncatalyzed reaction. (b) Use of zinc metal as a catalyst lowers the energy of activation. Notice that a catalyst does not affect the energy released. The heat of reaction, ΔH, is constant.

proceeds. The reaction is slower because there are fewer molecules with enough energy to achieve the transition state.

To help a reaction proceed at a faster rate, we use catalysts. Recall that in Section 8.2 we defined a catalyst as a substance that speeds up the rate of a reaction without being consumed. Now we can define a **catalyst** as a substance that allows a reaction to proceed faster by lowering the energy of activation. As an example, consider that water is produced from the exothermic reaction of hydrogen and oxygen gases. The equation for the reaction is

catalyst A substance that allows a reaction to proceed faster by lowering the energy of activation.

$$2\ H_2(g) + O_2(g) \longrightarrow 2\ H_2O(g) + \text{heat}$$

Although the reaction produces heat energy, it is very slow at normal temperatures. When a mixture of hydrogen and oxygen gases is exposed to a flame or electrical spark, the reaction is instantaneous. In fact, the instantaneous release of a large amount of energy creates an explosion. At room temperature, however, a powdered metal, such as zinc dust, acts as a catalyst and causes the reaction to occur rapidly without explosion. Figure 16.8 illustrates the reaction profiles of catalyzed and uncatalyzed reactions of hydrogen and oxygen gases.

Note that the heat of reaction, ΔH, is independent of the catalyst. This means the rate of a reaction is not related to the heat of reaction. Table 16.1 shows the rate and heat of reaction for the formation of water from hydrogen and oxygen gases.

The following example exercise further illustrates the relationship of reactants and products and the reaction profile.

Table 16.1 Reaction of Hydrogen and Oxygen Gases

Catalyst	Rate of Reaction	Heat of Reaction*
none	very slow	57.8 kcal (242 kJ)
spark	explosive	57.8 kcal (242 kJ)
zinc dust	rapid	57.8 kcal (242 kJ)

* Energy released per mole of water produced at 25°C.

EXAMPLE EXERCISE 16.1

Ultraviolet light (UV) from the sun causes the conversion of oxygen gas molecules in the upper atmosphere to ozone gas molecules. Draw the energy diagram for the reaction:

$$3\ O_2(g) + 17.0\ \text{kcal} \xrightarrow{UV} 2\ O_3(g)$$

Solution: Since energy is necessary for the conversion of oxygen to ozone, the reaction is endothermic and the reaction profile is

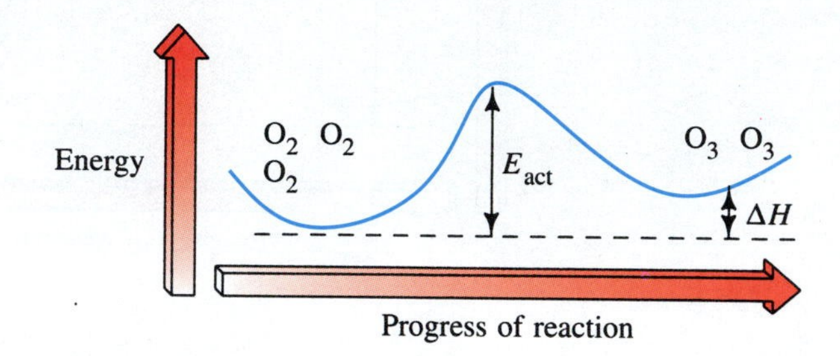

SELF-TEST EXERCISE

Nitrosyl bromide, NOBr, decomposes to give nitric oxide and bromine gas. Draw the energy diagram for the reaction:

$$2\ NOBr(g) \xrightarrow{UV} 2\ NO(g) + Br_2(g) + \text{heat}$$

Answer: Since the reaction is exothermic, the reaction profile is

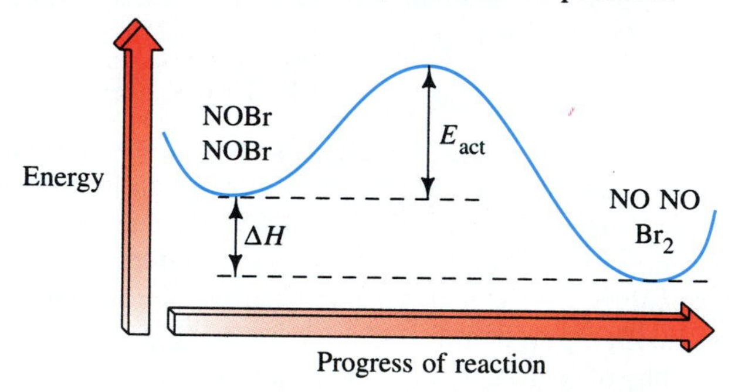

16.3 Rates of Reaction

To explain the meaning of the rate equation for a given chemical reaction.

To calculate the change in the rate of reaction, given the rate equation and the change in reactant concentration.

rate of reaction The change in concentration of a reactant or product with respect to time.

The **rate of reaction** refers to the change in reactant concentration with time. For most reactions, lowering the temperature slows down the reaction. For example, at cold temperatures, such as in refrigeration, the rate that microorganisms in apple cider convert sugar to ethyl alcohol, producing hard cider, is slow. At colder temperatures, the rate for the further conversion of hard cider to vinegar, or acetic acid, is also slower. The conversions are as follows:

$$\underset{(sugar)}{\text{apple cider}} \longrightarrow \underset{(alcohol)}{\text{hard cider}} \longrightarrow \underset{(acetic\ acid)}{\text{vinegar}}$$

Rate Equations

In the early 1800s, it was first proposed that reactions occur more readily as the concentrations of reactants increase. Gradually, it became clear that the rate of a chemical reaction increases proportionally with increases in the concentration of reactants. We can, in fact, express the reaction rate as being proportional ($\propto$) to the concentration of reactants. That is,

$$\text{rate} \propto [A]^x[B]^y$$

In the rate expression, **[A]** and **[B]** represent the molar concentrations of reactants A and B. The exponents x and y are the powers of the concentrations of the reactants. For simple one-step reactions, the exponents x and y are often the same as the coefficients of A and B in the balanced chemical equation. Depending on the mechanism for the reaction, x and y can be whole numbers, fractions, or zero. In all cases, x and y must be determined experimentally. The value of x, for example, can be determined by varying [A] and observing the change in reaction rate. The value of y is found by measuring the reaction rate for different molar concentrations of B.

[A] The symbol for the molar concentration of a chemical species A.

We can write the rate expression for a reaction as an equation. Thus, by introducing a proportionality constant (k), we have the **rate equation** for the reaction. It is

rate equation The general relationship between the reaction rate and reactant concentration; that is, rate = $k[A]^x[B]^y$.

$$\text{rate} = k[A]^x[B]^y$$

The symbol k is the specific rate constant for the reaction. Although k is a constant, its value can vary with changes in temperature.

Let's examine a specific reaction where the rate equation has been determined experimentally. Look at NO, which is converted to NO_2 as follows:

$$2\ NO(g) + O_2(g) \longrightarrow 2\ NO_2(g)$$

After experimenting with the reaction at varying concentrations of NO and O_2, we obtain this rate equation:

$$\text{rate} = k[NO]^2[O_2]$$

Since the exponent of $[O_2]$ is understood to be 1, if $[O_2]$ is doubled, the reaction rate doubles. Multiplying $[O_2]$ in the rate equation by 2 gives us

$$\text{rate} = k[NO]^2[O_2] \times 2$$

If $[O_2]$ is halved, the rate of reaction is half as fast. Dividing $[O_2]$ in the rate equation by 2 gives us

$$\text{rate} = \frac{k[NO]^2[O_2]}{2}$$

Since the exponent of [NO] in the rate equation is 2, if [NO] is doubled, the rate of reaction quadruples. Multiplying [NO] in the rate equation by 2 gives us

$$\text{rate} = k[NO]^2[O_2] \times 4$$

If the concentration of nitrogen monoxide is halved, the rate of reaction decreases fourfold. Dividing [NO] in the rate equation by 2 gives us

$$\text{rate} = \frac{k[NO]^2[O_2]}{4}$$

16.4 Law of Chemical Equilibrium

To describe the concept of dynamic equilibrium for a reversible reaction.

To write an equilibrium constant expression for any general reversible reaction.

We said that a **reversible reaction** has achieved a state of **chemical equilibrium** when the rate of the forward reaction equals the rate of the reverse reaction. If we call the rate of forward reaction rate_f and the rate of reverse reaction rate_r, we can write the equilibrium as an equation.

$$\text{rate}_f = \text{rate}_r$$

We can measure the rate of a chemical reaction by the changes in the concentration of the reactant or product. We can measure the rate of a forward reaction by the decrease in reactant concentration per unit time. We can also measure it by the increase in product concentration per unit time. Thus,

$$\text{rate}_f = \frac{\text{decrease [reactant]}}{\text{unit time}}$$

or

$$\text{rate}_f = \frac{\text{increase [product]}}{\text{unit time}}$$

As an example, in the upper atmosphere oxygen molecules (O_2) are converted to ozone (O_3). The ultraviolet light from the sun is responsible for the conversion of O_2 to O_3. This reaction is dynamic and reversible. In the reverse reaction, O_3 decomposes to give O_2. The equation for this reversible reaction is

$$\underset{\textit{oxygen}}{3\ O_2(g)} \rightleftharpoons \underset{\textit{ozone}}{2\ O_3(g)}$$

closed system A chemical reaction that is isolated and can be studied independently from the surrounding environment.

We can simulate the oxygen/ozone reaction in a closed system in the laboratory. A **closed system** is any isolated reaction that can be studied independently of its surrounding environment. We begin by pumping oxygen gas into a glass sphere and then focus ultraviolet light on the gas. Initially, the reaction is quite rapid. As the amount of O_2 decreases, however, the forward reaction slows down. At the same time, the reverse reaction gradually increases. In other words, as the reaction proceeds, the rate of ozone decomposition increases. When the system reaches equilibrium, we have

$$\text{rate}_f\ (O_2\ \text{reaction}) = \text{rate}_r\ (O_3\ \text{reaction})$$

Chemistry Connection □ Environmental

The Ozone Hole

► Can you guess what animal population has been threatened as a result of the ozone hole over Antarctica?

All living systems are protected from the sun's harmful radiation by a layer of ozone molecules in the upper atmosphere. However, this ozone layer that shields the earth's fragile life forms is being depleted. For some time, we have known about a large hole in the ozone over Antarctica. Recently, scientists have found other holes in the ozone over heavily populated areas. Ozone depletion has been detected over Russia, Europe, Canada, and the United States.

In the stratosphere, about 30 km above the earth, ozone molecules absorb high-energy ultraviolet radiation from the sun. A hole in the ozone layer is of great concern because we would no longer have protection from the sun's powerful rays. It is well known that ultraviolet radiation can cause skin cancer, cataracts, and weaken our immune system. The sun's harmful rays can also affect agriculture and cause lower crop yields.

In 1974, scientists at the University of California at Irvine first suggested that chlorofluorocarbons (CFCs) could cause depletion of the ozone layer. To date, over 10 million tons of CFCs have been manufactured. CFCs are used mostly as propellents in aerosol cans and as coolants in refrigeration and air conditioning units. These CFC compounds, such as Freon-12 (CF_2Cl_2), are quite inert. When released into the environment, they eventually diffuse up to the ozone layer.

It is generally agreed that CFCs destroy ozone molecules in a series of steps. First, solar ultraviolet radiation strips a chlorine atom away from a CFC molecule. The resulting chlorine atom has an unpaired electron (Cl·) and is referred to as a free radical.

$$\underline{\text{Step 1:}}\quad \text{CFC} \xrightarrow{\text{UV}} \text{Cl}\cdot$$

A chlorine free radical is very reactive and, through a complex mechanism, acts as a catalyst for the decomposition of ozone molecules, O_3, into oxygen.

$$\underline{\text{Step 2:}}\quad 2\,O_3 \xrightarrow{\text{Cl}\cdot} 3\,O_2$$

Moreover, this process can occur repeatedly and a single chlorine free radical can destroy scores of ozone molecules.

Ozone depletion is a worldwide problem, and several nations are working to solve it. Germany, Denmark, and the Netherlands have recently announced they will halt production and ban the use of CFCs. The United States has resolved to phase out CFCs but it is expensive to retrofit existing refrigeration units. However, the air conditioning units in many new homes and automobiles are being equipped with non-CFC gases.

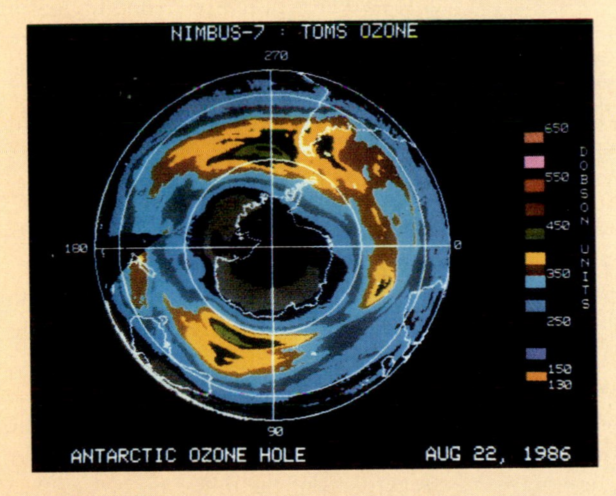

A map of the stratosphere over the southern hemisphere. Each color represents an ozone concentration and the black spot over the South Pole indicates a region of low concentration.

► The penguin population has decreased and may become extinct if ozone depletion persists over Antarctica.

Figure 16.9 Reaction Rates as a Function of Time (a) The rate of reaction for oxygen gas decreases with time. (b) As the reaction proceeds, the rate of ozone decomposition increases. (c) At equilibrium, the rates of the forward reaction and reverse reaction are the same.

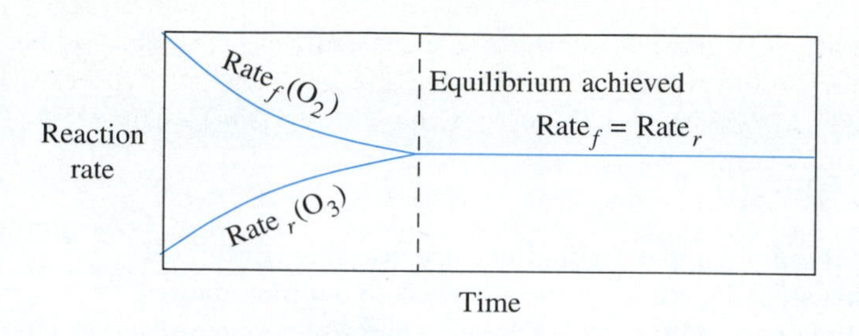

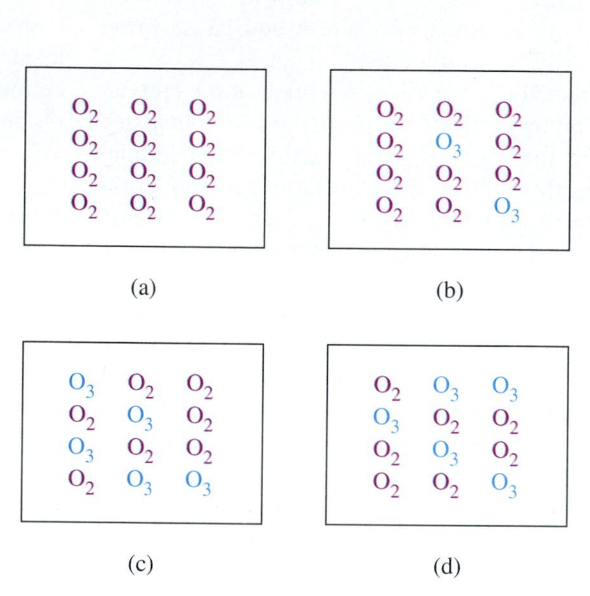

Figure 16.10 Model for Achieving Equilibrium (a) The vessel contains only O_2 gas molecules. (b) As the reaction proceeds, O_3 is formed. (c) After a time, the reaction reaches equilibrium. (d) Sometime later the ratio of O_2 to O_3 is the same as in (c). Equilibrium is a dynamic process, but the amounts of reactant and product do not change.

Figure 16.9 shows the changes in the rates of the forward and reverse reactions as the oxygen/ozone reaction progresses.

Figure 16.10 provides a model to help us understand the equilibrium process. Initially, a closed vessel contains only oxygen molecules (a). As the reaction moves forward, ozone is formed (b). As the concentration of ozone increases, the rate of decomposition of ozone increases. Equilibrium is approached as the forward reaction slows down and the reverse reaction speeds up (c). When the forward and reverse rates reach a steady state, the reaction is at equilibrium (d).

Law of Chemical Equilibrium

law of mass action The principle that a change in the amount of mass of a substance participating in a reversible reaction produces a shift in the equilibrium; also called the mass law.

In 1864, Cato Guldberg and Peter Waage observed that a change in the mass of a substance participating in a reversible reaction produced a shift in the equilibrium. This principle became known as the **law of mass action** or, simply, the mass law. Chemists then studied shifts in the following type of reversible reaction:

$$aA + bB \rightleftharpoons cC + dD$$

They experimentally observed that the ratio of the product concentration (raised to powers *c* and *d*) to the reactant concentration (raised to powers *a* and *b*) was always a constant value. In other words, regardless of the initial masses of A, B, C, and D, at equilibrium the ratio of molar concentrations equals a constant. The equilibrium constant is K_{eq} and for the general reversible reaction is

$$K_{eq} = \frac{[C]^c[D]^d}{[A]^a[B]^b}$$

where [A], [B], [C], and [D] represent the molar concentrations of reactants and products, and a, b, c, and d correspond to the coefficients in the balanced chemical equation. This relationship, which applies to every reversible reaction, is known as the **law of chemical equilibrium.**

law of chemical equilibrium The principle that the concentrations of the products of a reversible reaction divided by the concentrations of the reactants, each raised to a coefficient power from the balanced equation, are constants.

Although the law of chemical equilibrium was first proposed on the basis of experimental evidence, it can be derived theoretically from reaction rates. For the above general reversible reaction, the rate equations for the forward and reverse reactions are

$$\text{rate}_f = k_f[\text{A}]^x[\text{B}]^y$$

$$\text{rate}_r = k_r[\text{C}]^w[\text{D}]^z$$

By definition, the forward and reverse rates are equal when the reaction is at equilibrium. Setting the two rate expressions equal to each other, we have

$$\text{rate}_f = \text{rate}_r$$

Substituting gives us

$$k_f[\text{A}]^x[\text{B}]^y = k_r[\text{C}]^w[\text{D}]^z$$

Rearranging gives

$$\frac{k_f}{k_r} = \frac{[\text{C}]^w[\text{D}]^z}{[\text{A}]^x[\text{B}]^y}$$

The ratio of k_f/k_r is incorporated into the equilibrium constant, K_{eq}, for the reaction. Thus, the **general equilibrium expression** can be written as

general equilibrium expression The relationship of the reactants and the product concentrations having the form $K_{eq} = [\text{C}]^c[\text{D}]^d/[\text{A}]^a[\text{B}]^b$

$$K_{eq} = \frac{[\text{C}]^c[\text{D}]^d}{[\text{A}]^a[\text{B}]^b}$$

Let's write the equilibrium constant expression for the following reversible reaction:

$$2\ \text{A} \rightleftharpoons \text{B}$$

To write the equilibrium expression, K_{eq}, we place the concentration of the product in the numerator and the concentration of the reactant in the denominator. This gives us

$$K_{eq} = \frac{[\text{B}]}{[\text{A}]}$$

The coefficients in the balanced equation appear as exponents in the K_{eq} expression. The coefficient of A is 2. The coefficient of B is understood to be 1. These values, 2 and 1, are written as a power above the corresponding concentration. Thus,

$$K_{eq} = \frac{[\text{B}]}{[\text{A}]^2}$$

Example Exercise 16.2 provides additional practice in writing K_{eq} expressions for general cases of reversible reactions.

EXAMPLE EXERCISE 16.2

Write the equilibrium constant expression for each of the following.

(a) $\text{A} + 3\ \text{B} \rightleftharpoons 2\ \text{C}$ **(b)** $4\ \text{A} + 5\ \text{B} \rightleftharpoons 4\ \text{C} + 6\ \text{D}$

Solution: Let's proceed in two steps. First, substitute each species into the K_{eq} expression. Second, apply the coefficients of the reaction as the power of the species molar concentration.

(a) We substitute the product concentration into the numerator and reactant concentrations into the denominator. This gives us

$$K_{eq} = \frac{[C]}{[A][B]}$$

The coefficients of A, B, and C are 1, 3, and 2, respectively. Thus, the general equilibrium expression is

$$K_{eq} = \frac{[C]^2}{[A][B]^3}$$

(b) The ratio of product concentration to reactant concentration is

$$K_{eq} = \frac{[C][D]}{[A][B]}$$

The coefficients are 4, 5, 4, and 6, respectively. Thus,

$$K_{eq} = \frac{[C]^4[D]^6}{[A]^4[B]^5}$$

Although we are writing expressions for general reversible reactions, each of these examples corresponds to an actual equilibrium system.

SELF-TEST EXERCISE

Write the equilibrium expression for the following general reaction.

$$A + 2\,B \rightleftharpoons 3\,C + D$$

Answer: $K_{eq} = \dfrac{[C]^3[D]}{[A][B]^2}$

Figure 16.11 Analogy for Chemical Equilibrium
Automobiles can travel back and forth between two cities although their populations remain constant. In an analogous way, molecules can travel back and forth between reactants and products although their amounts remain constant.

An interesting analogy for understanding equilibrium is the flow of traffic in and out of a large city such as New York or San Francisco (Figure 16.11). Visualize commuters entering and leaving the city. This process is ongoing, but the population of the cities remains relatively constant. By analogy, reactions at equilibrium are dynamic but the amounts of reactants and products are constant.

16.5 Concentration Equilibrium Constant, K_c

OBJECTIVES

To write the equilibrium constant expression for a reversible reaction in the gaseous state.

To use experimental data to calculate the concentration equilibrium constant, K_c, for a reversible reaction.

In Section 16.4, we developed the concept of a general equilibrium constant expression, K_{eq}. Now, we can consider the specific case of reversible reactions in the gaseous state. This specific case of the general expression is

$$K_c = \frac{[C]^c[D]^d}{[A]^a[B]^b}$$

where K_c is the **concentration equilibrium constant** and A, B, C, and D are gases.

concentration equilibrium constant (K_c) The equilibrium constant that relates the concentrations of gases participating in a reversible reaction.

For example, consider the following equilibrium:

$$2\ SO_2(g) + O_2(g) \rightleftharpoons 2\ SO_3(g)$$

First, we substitute the concentration of the product into the numerator and the concentrations of the reactants into the denominator of the equilibrium expression. The respective coefficients are 2, 1, and 2. Thus, the equilibrium expression for the reaction is

$$K_{eq} = \frac{[SO_3]^2}{[SO_2]^2[O_2]}$$

As another example, consider hydrogen gas, which can be manufactured by passing steam over hot charcoal. The reversible reaction produces carbon monoxide and hydrogen as follows:

$$C(s) + H_2O(g) \rightleftharpoons CO(g) + H_2(g)$$

Notice that charcoal is a solid, but all the other species are gases. When the reaction is studied experimentally, it is found that the amount of charcoal has no effect on the equilibrium. The concentration of solid charcoal is considered unity, that is, [1]. Since carbon has no effect on the equilibrium, [C] is constant and incorporated into the equilibrium constant, K_c. This gives us

$$K_c = \frac{[CO][H_2]}{[H_2O]}$$

In most reversible reactions involving K_c, all the reactants and products are in the gaseous phase. A system in which all the substances are in the same physical

homogeneous equilibrium A type of equilibrium in which all the participating species are in the same state.

heterogeneous equilibrium A type of equilibrium in which one or more of the participating species is in a different state.

state is an example of **homogeneous equilibrium**. On occasion, as in the above example, a reactant or product may be in the solid or liquid state. When two physical states are present in the same system, we have a **heterogeneous equilibrium**. Note that, in all cases of gaseous equilibria, pure liquids and solids do not appear in the K_c expression.

EXAMPLE EXERCISE 16.3

Write the equilibrium constant expression for each of the following.

(a) $2\ NO(g) + 2\ H_2(g) \rightleftharpoons N_2(g) + 2\ H_2O(g)$

(b) $NH_4NO_3(s) \rightleftharpoons N_2O(g) + 2\ H_2O(g)$

Solution: Let's substitute into the K_c expression as follows.

(a) The product concentration appears in the numerator and the reactant concentration in the denominator. Each concentration is raised to a power corresponding to the coefficient in the balanced equation.

$$K_c = \frac{[N_2][H_2O]^2}{[NO]^2[H_2]^2}$$

(b) In this reaction, NH_4NO_3 is a solid. Since ammonium nitrate is not in the gaseous state, it does not appear in the equilibrium expression; therefore, we have

$$K_c = [N_2O][H_2O]^2$$

SELF-TEST EXERCISE

Write the equilibrium constant expression for the gaseous state conversion of ammonia to nitrogen monoxide. That is,

$$4\ NH_3(g) + 5\ O_2(g) \rightleftharpoons 4\ NO(g) + 6\ H_2O(g)$$

Answer: $K_c = \dfrac{[NO]^4[H_2O]^6}{[NH_3]^4[O_2]^5}$

Experimental Determination of K_c

Although we can calculate an equilibrium constant from theoretical considerations, we can also determine it from experimental data. One of the most thoroughly investigated reactions is the formation of hydrogen iodide from hydrogen gas and iodine vapor. It is

$$\underset{\text{colorless gas}}{H_2(g)} + \underset{\text{purple vapor}}{I_2(g)} \rightleftharpoons \underset{\text{colorless gas}}{2\ HI(g)}$$

Using an instrumental method, we can measure the molar concentration of each gas at equilibrium. At 425°C, the equilibrium constant for the reaction is determined to be 55.2. Thus,

$$K_c = \frac{[HI]^2}{[H_2][I_2]} = 55.2$$

Note that the K_c value for the reaction at 425°C is 55.2 regardless of the original concentrations of each gas. If, for example, we start with $[H_2] = 1.000$, $[I_2] = 1.000$, and $[HI] = 0$, we obtain 55.2 for K_c. If we start with $[H_2] = 0.500$,

Table 16.2 Experimental Determination of K_c

Experiment	Initial Concentration $[H_2]$	$[I_2]$	$[HI]$	Equilibrium Concentration $[H_2]$	$[I_2]$	$[HI]$	Calculated K_c* at 425°C $\frac{[HI]^2}{[H_2][I_2]}$
1	1.000	1.000	0	0.212	0.212	1.576	55.2
2	0	0	2.000	0.212	0.212	1.576	55.2
3	0.500	0.500	0	0.106	0.106	0.788	55.2
4	0	0	1.000	0.106	0.106	0.788	55.2

* The units for an equilibrium constant are usually omitted. In this case, the units cancel and K_c is dimensionless.

$[I_2] = 0.500$, and $[HI] = 0$, we also obtain 55.2 for K_c. In fact, even if we approach equilibrium from the opposite direction, $[H_2] = 0$, $[I_2] = 0$, and $[HI] = 1.000$, we obtain the same value for K_c. Table 16.2 presents the data for four different experimental studies of the HI equilibrium.

Note that in Table 16.2 the equilibrium constant has the same value, 55.2, regardless of the initial conditions. Another way to think of this point is to visualize a closed vessel containing the three gases H_2, I_2, and HI in equilibrium. If we change the amount of any species, the reaction shifts forward or backward in such a way that the equilibrium constant remains constant. The following example exercise illustrates the calculation of an equilibrium constant from experimental data.

EXAMPLE EXERCISE 16.4

Automobile pollution gives a mixture of nitrogen oxides in the atmosphere. One of the reversible reactions is

$$2\ NO(g) + O_2(g) \rightleftharpoons 2\ NO_2(g)$$

Given the equilibrium concentrations for each gas at 25°C, what is the K_c?

$$[NO] = 1.5 \times 10^{-11}$$

$$[O_2] = 8.9 \times 10^{-3}$$

$$[NO_2] = 2.2 \times 10^{-6}$$

Solution: We first write the equilibrium expression for the reaction.

$$K_c = \frac{[NO_2]^2}{[NO]^2[O_2]}$$

Second, we substitute the concentration values into the expression.

$$K_c = \frac{[2.2 \times 10^{-6}]^2}{[1.5 \times 10^{-11}]^2[8.9 \times 10^{-3}]}$$

$$= 2.4 \times 10^{12}$$

The very large K_c value for this reaction indicates that the equilibrium overwhelmingly favors the formation of NO_2. However, the K_c does not indicate the rate of reaction. Under these conditions, the rate of conversion of NO to NO_2 is slow.

SELF-TEST EXERCISE

Given the equilibrium concentrations for the gas mixture at 100°C, calculate the value of K_c for the following reaction.

$$\underset{0.00140\ M}{N_2O_4(g)} \rightleftharpoons \underset{0.0172\ M}{2\ NO_2(g)}$$

Answer: $K_c = 0.211$

In Section 12.5, we learned that gas pressure is proportional to the concentration of molecules. Therefore, we can write an equilibrium expression for a reversible gaseous reaction on the basis of partial pressures. The general expression for the equilibrium reaction is

$$K_p = \frac{pC^c pD^d}{pA^a pB^b}$$

where pA, pB, pC, and pD are the partial pressures of the gases participating in the equilibrium reaction. K_p is the partial pressure equilibrium constant.

16.6 Gaseous State Equilibria Shifts

To apply Le Chatelier's principle to reversible reactions in the gaseous state.

Henri Louis Le Chatelier (1850–1936) was academically trained in France as a mining engineer, like his father before him. When he graduated from college, he joined the faculty of a mining school as professor of chemistry. His first interest was in the study of flames with the intention of preventing mine explosions. Eventually, he studied heat transfer, and in 1888 stated the principle for which he is famous: *Every change of one of the factors of an equilibrium brings about a rearrangement of the system in such a direction as to minimize the original change.*

Le Chatelier's principle The statement that an equilibrium, when stressed by a change of concentration, temperature, or pressure, will shift to relieve the stress.

For chemical reactions, the factors are concentration, temperature, and pressure. Changes in any of these factors shift the equilibrium. **Le Chatelier's principle** is usually applied to chemical systems as follows: If a reversible reaction is stressed by a change of concentration, temperature, or pressure, the equilibrium will shift to relieve the stress.

Let's apply Le Chatelier's principle to reversible reactions in the gaseous state. Consider the NO_2–N_2O_4 equilibrium in smog. If we place NO_2 gas in a closed container, a reversible reaction occurs. It is

$$\underset{brown}{2\ NO_2(g)} \rightleftharpoons \underset{colorless}{N_2O_4(g)} + \text{heat}$$

Effect of Temperature

The brownish gas nitrogen dioxide reacts to give dinitrogen tetraoxide plus heat. We can view this equilibrium condition as a chemical "teeter-totter." If we increase

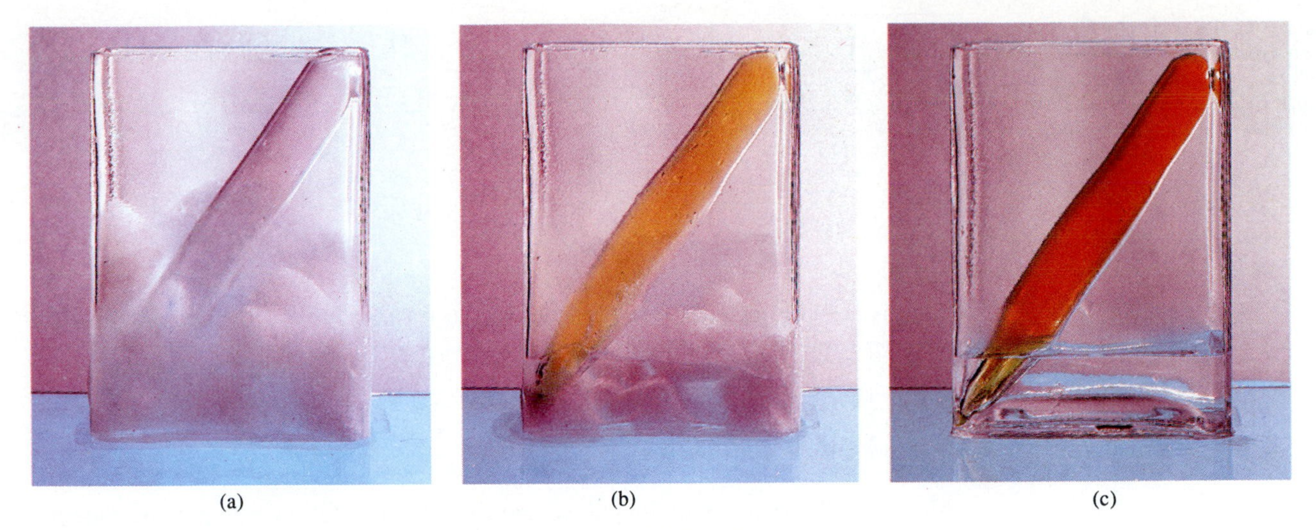

Figure 16.12 The $N_2O_4 \rightleftharpoons 2\ NO_2$ Equilibrium (a) If N_2O_4 is frozen, it is colorless. (b) If N_2O_4 is heated to a gas, some of the colorless $N_2O_4(g)$ produces brown $NO_2(g)$. (c) If heated further, more of the colorless $N_2O_4(g)$ changes to $NO_2(g)$ giving a dark-brown color.

the heat, the teeter-totter shifts to the left. That is, the equilibrium shifts away from that side of the reaction where the heat is. If we cool the reaction, the teeter-totter shifts to the right. The equilibrium shifts to the right and heat is released, which raises the temperature. Figure 16.12 illustrates the shifts in equilibrium.

Effect of Concentration

For any equilibrium system, we can experimentally shift the reaction forward or backward by changing the concentration of the reactant or product. For example, we can increase the NO_2 concentration for

$$2\ NO_2(g) \rightleftharpoons N_2O_4(g)$$

In this case, we shift the equilibrium to the right and decrease the excess NO_2. If we decrease the NO_2 concentration, the equilibrium shifts to the left, thereby increasing the NO_2 concentration and reducing the concentration of N_2O_4.

Effect of Pressure

For some gaseous equilibrium systems, we can experimentally shift the reaction forward or backward by changing the volume. If we stress a gaseous system by decreasing the volume, the equilibrium will shift, if possible, to produce fewer molecules. In the smog equilibria, there are two molecules of NO_2 for every one molecule of N_2O_4. If we decrease the volume of the container, the gas pressure increases. The equilibrium shifts to the right to relieve the stress. That is, the equilibrium shifts to decrease the pressure by producing fewer molecules. Figure 16.13 illustrates this process.

Now consider what happens to the NO_2–N_2O_4 equilibrium if we introduce inert argon gas into the container. If the number of gas molecules increase, the total gas pressure increases. The partial gas pressures of NO_2 and N_2O_4, however, do not change. Since the partial gas pressures of the NO_2–N_2O_4 equilibrium remain constant,

Figure 16.13 Effect of Pressure on Equilibrium (a) The initial NO_2–N_2O_4 equilibrium. (b) The NO_2–N_2O_4 equilibrium after a decrease in volume. Notice that two molecules of NO_2 have produced one molecule of N_2O_4. Thus, the number of molecules and gas pressure have both been decreased.

N_2O_4	NO_2
NO_2	NO_2
NO_2	NO_2
N_2O_4	

(a)

NO_2	NO_2
N_2O_4	N_2O_4
N_2O_4	NO_2

(b)

there is no shift in equilibrium. Example Exercise 16.5 further illustrates Le Chatelier's principle applied to equilibrium shifts.

EXAMPLE EXERCISE 16.5

Methane, CH_4, reacts with limited oxygen and releases heat according to the equation

$$2\ CH_4(g) + O_2(g) \rightleftharpoons 2\ CO(g) + 4\ H_2(g) + \text{heat}$$

Predict the direction of the equilibrium shift for each of the following stresses.

(a) [CH_4] increases
(b) [CO] increases
(c) [O_2] decreases
(d) [H_2] decreases
(e) temperature increases
(f) temperature decreases
(g) pressure decreases
(h) volume decreases
(i) inert radon gas is added
(j) a catalyst is added

Solution: Let's apply Le Chatelier's principle to each of the stresses.

(a) If [CH_4] increases, the equilibrium shifts to the *right* to relieve the stress.
(b) If [CO] increases, the equilibrium shifts to the *left* to decrease the additional CO concentration.
(c) If [O_2] decreases, the equilibrium shifts to the *left* to reduce the stress.
(d) If [H_2] decreases, the equilibrium shifts to the *right* to restore the H_2 concentration.
(e) Notice that the reaction is exothermic. If the temperature increases, the equilibrium shifts to the *left*.
(f) If the temperature decreases, the equilibrium shifts to the *right*.
(g) Note that there are three reactant molecules but six product molecules. If the pressure decreases, there are fewer molecules. To relieve the stress, we must restore the pressure. Thus, the equilibrium shifts to the *right*, giving more molecules.
(h) If we decrease the volume, the pressure increases. To reduce the change in pressure, we need fewer molecular collisions. Thus, the equilibrium shifts to the *left*, decreasing the number of molecules.
(i) Adding inert radon gas does not affect the partial pressure of the equilibrium gases. Therefore, there is *no shift* in equilibrium.
(j) Adding a catalyst increases the rate of reaction in both directions but has no effect on the amount of reactant or product. There is *no shift* in equilibrium.

SELF-TEST EXERCISE

Predict the direction of the equilibrium shift for each of the stresses on the following gaseous system.

$$N_2O_4(g) + heat \rightleftharpoons 2\ NO_2(g)$$

(a) $[N_2O_4]$ increases
(b) $[NO_2]$ increases
(c) $[N_2O_4]$ decreases
(d) $[NO_2]$ decreases
(e) temperature increases
(f) temperature decreases
(g) pressure increases
(h) volume increases
(i) a catalyst is added
(j) inert xenon gas is added

Answers: **(a)** shifts right; **(b)** shifts left; **(c)** shifts left; **(d)** shifts right; **(e)** shifts right; **(f)** shifts left; **(g)** shifts left; **(h)** shifts right; **(i)** no shift; **(j)** no shift

16.7 Aqueous Solution Equilibria

To state three forces that drive a partially reversible reaction in aqueous solution to completion.

So far we have only discussed equilibrium in the gaseous state. Reversible reactions also occur in aqueous solution. Three forces drive a reaction in aqueous solution to completion. They are (1) the formation of a gas, (2) the formation of an insoluble precipitate, and (3) the formation of a weak electrolyte.

1. *Formation of a gas*. When a gas is released from an aqueous solution, it helps drive a reversible reaction to completion. For example, consider the reaction of hydrochloric acid and aqueous sodium bicarbonate:

$$HCl(aq) + NaHCO_3(aq) \rightleftharpoons NaCl(aq) + H_2O(l) + CO_2(g)$$

Notice that all species are aqueous except $CO_2(g)$. Carbon dioxide is released as a gas and escapes. Thus, CO_2 does not participate effectively in the equilibrium with the aqueous species. The concentration of the gas is decreased, and there is a Le Chatelier shift toward the products.

2. *Formation of an insoluble precipitate*. When an insoluble substance precipitates from an aqueous solution, it drives a reversible reaction toward the products. For example, consider the reaction of hydrochloric acid and aqueous silver nitrate:

$$HCl(aq) + AgNO_3(aq) \rightleftharpoons AgCl(s) + HNO_3(aq)$$

Notice that all species are in the aqueous phase except AgCl(s). Silver chloride is insoluble and precipitates from solution. This is an example of a heterogeneous equilibrium. The formation of solid AgCl helps drive the slightly reversible reaction to completion.

3. *Formation of a weak electrolyte*. Weak electrolytes include weak acids, weak bases, and water. For example, the reaction of hydrochloric acid and aqueous sodium hydroxide yields a salt and water:

$$HCl(aq) + NaOH(aq) \rightleftharpoons NaCl(aq) + H_2O(l)$$

The reason the formation of water is a driving force for this reaction is more easily understood if we write an ionic equation. That is,

$$H^+(aq) + Cl^-(aq) + Na^+(aq) + OH^-(aq) \rightleftharpoons Na^+(aq) + Cl^-(aq) + H_2O(l)$$

In this aqueous solution reaction, all the species exist as ions with the exception of water. Water exists as molecules and has little tendency to ionize. Therefore, H_2O molecules strongly resist the reverse reaction. The equilibrium is thereby driven toward completion. Now let's consider the formation of a weak acid. The reaction of hydrochloric acid and aqueous sodium acetate gives the following:

$$HCl(aq) + NaC_2H_3O_2(aq) \rightleftharpoons NaCl(aq) + HC_2H_3O_2(aq)$$

The formation of the weak acid, $HC_2H_3O_2$, is a driving force for the reaction. If we write the ionic equation for the reaction, we clearly see the role of the weak acid. That is,

$$H^+(aq) + Cl^-(aq) + Na^+(aq) + C_2H_3O_2^-(aq) \rightleftharpoons Na^+(aq) + Cl^-(aq) + HC_2H_3O_2(aq)$$

In this aqueous solution reaction, all the species exist as ions with the exception of $HC_2H_3O_2$. Acetic acid is a weak acid. It exists as molecules in aqueous solution and has little tendency to ionize. The molecules of $HC_2H_3O_2$ resist the reverse reaction. The equilibrium is maintained in favor of the product side of the reaction.

The following example exercise further illustrates the three factors that drive a reversible reaction toward completion.

EXAMPLE EXERCISE 16.6

Which of the following, when formed as a reaction product, will drive an equilibrium in aqueous solution to completion?

(a) $H_2O(l)$
(b) $O_2(g)$
(c) $AgNO_3(aq)$
(d) $CaCO_3(s)$
(e) $HC_2H_3O_2(aq)$
(f) $HCl(aq)$
(g) $NaOH(aq)$
(h) $NH_4OH(aq)$

Solution: To act as a driving force for the equilibrium, the product must be either a gas, a precipitate, or a weak electrolyte.

(a) H_2O is a weak electrolyte, so it *drives* the reaction toward completion.
(b) O_2 is a gas, so it *drives* the reaction toward completion.
(c) $AgNO_3$ is a soluble salt, so it has *no effect* on the equilibrium.
(d) $CaCO_3$ is an insoluble salt, so it *drives* the reaction toward completion.
(e) $HC_2H_3O_2$ is a weak acid, so it *drives* the reaction toward completion.
(f) HCl is a strong acid, so it has *no effect* on the equilibrium.
(g) NaOH is a strong base, so it has *no effect* on the equilibrium.
(h) NH_4OH is a weak base, so it *drives* the reaction toward completion.

SELF-TEST EXERCISE

Indicate the substance responsible for driving the following aqueous solution reactions to completion.

(a) $H_2SO_4(aq) + Zn(s) \rightleftharpoons ZnSO_4(aq) + H_2(g)$
(b) $H_2CO_3(aq) + BaCl_2(aq) \rightleftharpoons BaCO_3(s) + 2\ HCl(aq)$
(c) $HNO_3(aq) + NaOH(aq) \rightleftharpoons NaNO_3(aq) + H_2O(l)$
(d) $NH_4NO_3(aq) + KOH(aq) \rightleftharpoons KNO_3(aq) + NH_4OH(aq)$

Answers: **(a)** H_2 (gas); **(b)** $BaCO_3$ (precipitate); **(c)** H_2O (weak electrolyte); **(d)** NH_4OH (weak electrolyte)

16.8 Ionization Equilibrium Constant, K_i

To write the equilibrium constant expression for a weak acid or weak base.

To use experimental data to calculate the ionization equilibrium constant, K_i, for a weak acid or weak base.

In Section 16.4, we developed the concept of a general equilibrium constant expression, K_{eq}. Here we will develop an equilibrium constant expression for a weak acid or base. This is characterized by a weak acid or base in equilibrium with its ions.

We will begin with an aqueous solution of a weak acid. For example, acetic acid in aqueous solution yields the following:

$$HC_2H_3O_2(aq) + H_2O(l) \rightleftharpoons H_3O^+(aq) + C_2H_3O_2^-(aq)$$

We can simplify the reversible reaction as follows:

$$HC_2H_3O_2(aq) \rightleftharpoons H^+(aq) + C_2H_3O_2^-(aq)$$

In general, the equilibrium for weak electrolytes lies overwhelmingly to the left. Typically, less than 1% of the parent molecules ionize. In the above reaction, very little ionization of acetic acid occurs. We indicate this by the length of the arrows in the symbol: $\rightleftharpoons$ Although about 1% of the molecules form ions at any one time, 99% do not. Thus, the reverse reaction is strongly favored.

When we substitute the acetic acid equilibrium into the general equilibrium constant expression, we obtain

$$K_i = \frac{[H^+][C_2H_3O_2^-]}{[HC_2H_3O_2]}$$

where K_i is the **ionization constant** for the weak acid. The ionization constant for acetic acid is small. That is, $K_i = 1.75 \times 10^{-5}$.

ionization constant (K_i) The equilibrium constant that relates concentrations of species in slightly ionized weak acids or bases in aqueous solution.

EXAMPLE EXERCISE 16.7

Write the equilibrium constant expression for the following weak acids.

(a) $HNO_2(aq) \rightleftharpoons H^+(aq) + NO_2^-(aq)$
(b) $H_2CO_3(aq) \rightleftharpoons H^+(aq) + HCO_3^-(aq)$
(c) $H_3PO_4(aq) \rightleftharpoons H^+(aq) + H_2PO_4^-(aq)$

Solution: Let's substitute into the K_i expression as follows.

(a) We substitute the ion concentrations into the numerator and the molecular acid concentration into the denominator.

$$K_i = \frac{[H^+][NO_2^-]}{[HNO_2]}$$

(b) Carbonic acid, H_2CO_3, ionizes to give H^+ and HCO_3^-. Substituting into the ionization equilibrium expression, we have

$$K_i = \frac{[H^+][HCO_3^-]}{[H_2CO_3]}$$

(c) Phosphoric acid, H_3PO_4, ionizes to give H^+ and $H_2PO_4^-$. Substituting into the K_i expression, we have

$$K_i = \frac{[H^+][H_2PO_4^-]}{[H_3PO_4]}$$

SELF-TEST EXERCISE

Write the equilibrium constant expression for the following weak acid.

$$HClO_2(aq) \rightleftharpoons H^+(aq) + ClO_2^-(aq)$$

Answer: $K_i = \dfrac{[H^+][ClO_2^-]}{[HClO_2]}$

An aqueous solution of a weak base behaves in a similar fashion as an aqueous solution of a weak acid. That is, a weak base is in equilibrium with its ions. For example, aqueous ammonium hydroxide gives the following equilibrium:

$$NH_4OH(aq) \rightleftharpoons NH_4^+(aq) + OH^-(aq)$$

Notice that the reactant side of the equilibrium is overwhelmingly favored. There is little ionization of the base. Substituting into the ionization equilibrium constant expression, we have

$$K_i = \frac{[NH_4^+][OH^-]}{[NH_4OH]}$$

Coincidentally, the ionization constant for ammonium hydroxide is comparable to that of acetic acid, that is, $K_i = 1.8 \times 10^{-5}$.

Experimental Determination of K_i

In the laboratory, we can calculate the equilibrium constant for a weak acid or weak base from pH measurements. For example, a pH meter can be used to measure the $[H^+]$ in a solution (see Section 15.8). The K_i value is then calculated from the hydrogen ion concentration.

Let's calculate the ionization constant, K_i, for acetic acid. In an experiment a 0.100 *M* solution of acetic acid is found to have a hydrogen ion concentration of 0.00134 *M*. We begin by analyzing the equilibrium concentrations. For each acetic acid molecule that ionizes, we obtain one H^+ and one $C_2H_3O_2^-$. Therefore, the $[H^+]$ must equal the $[C_2H_3O_2^-]$, that is, 0.00134 *M*. The reaction is

$$\begin{array}{ccccc} HC_2H_3O_2(aq) & \rightleftharpoons & H^+(aq) & + & C_2H_3O_2^-(aq) \\ 0.100\ M & & 0.00134\ M & & 0.00134\ M \end{array}$$

Next, we write the equilibrium constant expression by placing the ion concentrations in the numerator and the molar concentration of the weak acid in the denominator. This gives us

$$K_i = \frac{[H^+][C_2H_3O_2^-]}{[HC_2H_3O_2]}$$

Substituting for each concentration and neglecting the small amount of acetic acid ionization, we have

$$K_i = \frac{[0.00134][0.00134]}{[0.100]}$$

$$= 1.80 \times 10^{-5}$$

The ionization constant for acetic acid is determined to be 1.80×10^{-5}. The units of K_i are usually omitted. However, in this example they are mol/L.

16.9 Weak Acid–Base Equilibria Shifts

To apply Le Chatelier's principle to aqueous solutions of weak acids or weak bases.

In Section 16.6, we discussed Le Chatelier's principle in relation to the equilibria of reversible reactions in the gaseous state. The principle also applies to the equilibria of aqueous solutions. That is, a change in the concentration of any species in an ionic equilibrium will cause the equilibrium to shift in order to relieve the stress.

To see how to apply Le Chatelier's principle to aqueous solutions of weak acids and bases, reconsider the acetic acid equilibrium in aqueous solution. Recall that acetic acid is a weak acid. We indicate the equilibrium as follows:

$$HC_2H_3O_2(aq) \rightleftharpoons H^+(aq) + C_2H_3O_2^-(aq)$$

If we increase the $HC_2H_3O_2$ concentration, we stress the left side of the equilibrium. That is, there will be more acetic acid molecules. The result is that more $HC_2H_3O_2$ molecules ionize, which increases the H^+ and $C_2H_3O_2^-$ concentrations. The net effect of increasing the $[HC_2H_3O_2]$ is a Le Chatelier equilibrium shift to the right.

We can increase the H^+ concentration by adding HNO_3, a strong acid. This places a stress on the right side of the equation. The equilibrium shifts to the left to relieve the stress of more hydrogen ions. After the shift restores equilibrium, the concentrations of H^+ and $HC_2H_3O_2$ have increased, while the $C_2H_3O_2^-$ concentration has decreased.

We can also indicate a change of $[H^+]$ in terms of pH. In Section 14.8, we learned that $[H^+]$ and pH are inversely related. That is, if the pH increases, the $[H^+]$ decreases. Therefore, if the pH increases in acetic acid, the $[H^+]$ decreases. The equilibrium then shifts to the right.

Let's increase the $C_2H_3O_2^-$ concentration by adding $NaC_2H_3O_2$. Sodium acetate dissolves in the solution giving $C_2H_3O_2^-$. This places a stress on the right side of the equation, which shifts the equilibrium to the left. The net effect of adding $NaC_2H_3O_2$ is an increase in the $HC_2H_3O_2$ and $C_2H_3O_2^-$ concentrations and a decrease in the H^+ concentration.

We can also stress an equilibrium system indirectly. For example, we can add a strong base to neutralize an acid. Adding NaOH to the system neutralizes H^+. The effect is to lower the H^+ concentration, which shifts the equilibrium to the right. After the equilibrium shift, the concentrations of H^+ and $HC_2H_3O_2$ have decreased, while the $C_2H_3O_2^-$ concentration has increased.

As another example, consider what happens if we add $NaNO_3$ to aqueous acetic acid. Since sodium nitrate is a soluble salt, it dissolves, giving Na^+ and NO_3^- in solution. Neither of these ions participates in the acetic acid equilibrium. Thus, there is no shift in the system.

EXAMPLE EXERCISE 16.8

Predict the direction of equilibrium shift for each of the following stresses on an aqueous solution of hydrofluoric acid. The equilibrium is

$$HF(aq) \rightleftharpoons H^+(aq) + F^-(aq)$$

(a) [HF] increases
(b) $[H^+]$ increases
(c) $[F^-]$ decreases
(d) NaF is added
(e) NaCl is added
(f) NaOH is added
(g) pH increases

Solution: Let's apply Le Chatelier's principle to each of the stresses.

(a) If [HF] increases, the equilibrium shifts to the *right* to relieve the stress.
(b) If $[H^+]$ increases, the equilibrium shifts to the *left* to decrease the additional hydrogen ion concentration.
(c) If $[F^-]$ decreases, the equilibrium shifts to the *right* to restore the fluoride ion concentration.
(d) If NaF is added, the concentration of F^- increases. The equilibrium shifts to the *left* to reduce the stress.
(e) If NaCl is added, the salt dissolves, giving Na^+ and Cl^- in solution. Since neither of these ions participates in the equilibrium, there is *no shift* in the system.
(f) If NaOH is added, the solution is provided with basic OH^-. The hydroxide ion neutralizes the H^+ to give water. Hence, the $[H^+]$ decreases and the equilibrium shifts to the *right* to compensate for the stress.
(g) If the pH increases, by definition the $[H^+]$ decreases. The equilibrium shifts to the *right* to restore the hydrogen ion concentration.

SELF-TEST EXERCISE

Predict the direction of equilibrium shift for each of the following stresses on an aqueous solution of hydrocyanic acid. The equilibrium is

$$HCN(aq) \rightleftharpoons H^+(aq) + CN^-(aq)$$

(a) [HCN] increases
(b) $[H^+]$ increases
(c) $[CN^-]$ decreases
(d) KCN is added
(e) KCl is added
(f) KOH is added
(g) pH decreases

Answers: **(a)** shifts right; **(b)** shifts left; **(c)** shifts right; **(d)** shifts left; **(e)** no shift; **(f)** shifts right; **(g)** shifts left

16.10 Solubility Product Equilibrium Constant, K_{sp}

To write the equilibrium constant expression for a slightly soluble ionic compound.

To use experimental data to calculate the solubility product constant, K_{sp}, for a slightly soluble ionic compound.

In 1864, Guldberg and Waage first proposed the law of mass action, which led to the law of chemical equilibrium. Their proposal was based on the study of insoluble ionic compounds in aqueous solution. In particular, they studied insoluble barium carbonate in an aqueous solution of potassium sulfate. Interestingly, after a while, the system was found to contain insoluble barium sulfate as well as the insoluble barium carbonate. Guldberg and Waage proposed a reversible reaction for the precipitates in aqueous solution. It was

$$BaCO_3(s) + K_2SO_4(aq) \rightleftharpoons BaSO_4(s) + K_2CO_3(aq)$$

We can explain this equilibrium if we suggest that $BaCO_3$ precipitate is slightly soluble. That is,

$$BaCO_3(s) \rightleftharpoons Ba^{2+}(aq) + CO_3^{2-}(aq)$$

The length of the arrows indicates that the equilibrium overwhelmingly favors insoluble $BaCO_3$. When $BaCO_3$ does dissociate into ions, Ba^{2+} can recombine with SO_4^{2-} to give insoluble $BaSO_4$ precipitate. Therefore, we conclude that seemingly insoluble precipitates in aqueous solution are actually *very slightly soluble*. Moreover, these insoluble ionic compounds are in dynamic equilibrium with their constituent ions.

Now, let's look at lead(II) iodide, another seemingly insoluble ionic compound. In aqueous solution, PbI_2 precipitate shows the following solubility equilibrium:

$$PbI_2(s) \rightleftharpoons Pb^{2+}(aq) + 2\ I^-(aq)$$

Notice that PbI_2 dissociates into one Pb^{2+} and two I^-. Let's substitute this reversible reaction into the general K_{eq} expression. This gives

$$K_{eq} = \frac{[Pb^{2+}][I^-]^2}{[PbI_2]}$$

If the aqueous solution is saturated with insoluble PbI_2, the amount of precipitate has no effect on the dissociation equilibrium. That is, as long as there is solid precipitate, the concentrations of Pb^{2+} and I^- do not change.

We can rearrange the K_{eq} expression for PbI_2 to obtain the following:

$$K_{eq}[PbI_2] = [Pb^{2+}][I^-]^2$$

If the solution is saturated, the amount of precipitate is irrelevant. We can simplify the equation by designating that $K_{eq}[PbI_2]$ is equal to a new equilibrium constant, K_{sp} called the **solubility product constant**. Thus,

$$K_{sp} = [Pb^{2+}][I^-]^2$$

solubility product constant (K_{sp}) The equilibrium constant that relates the ion concentrations of slightly dissociated ionic compounds in aqueous solution.

When an insoluble compound is in aqueous solution, only the ions from the dissociation appear in the solubility product expression. The precipitate never appears in the K_{sp} expression. Since precipitates are only slightly soluble, we would expect solubility product constants to be small. The K_{sp} for PbI_2 is 8.7×10^{-9}.

Experimental Determination of K_{sp}

We can use several methods to find the K_{sp} value for a slightly soluble ionic compound. The most direct method uses an ion-selective electrode. Similar to the hydrogen ion electrode in a pH meter, ion electrodes can accurately measure very low concentrations of specific ions. Figure 16.14 shows an application of ion-selective electrodes.

If we find the ion concentrations in aqueous solution experimentally, we can calculate a value for the solubility product constant. For example, the hydroxide ion concentration in a saturated solution of milk of magnesia, $Mg(OH)_2$, is found to be 0.00032 *M*. To find the solubility product constant, let's begin by analyzing the equilibrium. For each formula unit of $Mg(OH)_2$ that dissociates, we obtain one Mg^{2+} and two OH^-. Therefore, the $[Mg^{2+}]$ equals half the $[OH^-]$, or 0.00016 *M*. Thus,

$$\begin{array}{ccccc} Mg(OH)_2(s) & \rightleftharpoons & Mg^{2+}(aq) & + & 2\ OH^-(aq) \\ & & 0.00016\ M & & 0.00032\ M \end{array}$$

The equilibrium constant expression for the reaction is

$$K_{sp} = [Mg^{2+}][OH^-]^2$$

After substituting the ion concentrations into the K_{sp} expression, we have

$$\begin{aligned} K_{sp} &= [1.6 \times 10^{-4}][3.2 \times 10^{-4}]^2 \\ &= 1.6 \times 10^{-11} \end{aligned}$$

The solubility product constant for magnesium hydroxide is 1.6×10^{-11}.

Figure 16.14 Ion Selective Electrodes Instruments are available that are similar to pH meters. These instruments have electrodes that measure the molar concentrations of ions other than the hydrogen ion. The technician shown is monitoring the concentration of ions in the Mississippi River.

16.11 Dissociation Equilibria Shifts

OBJECTIVES

To apply Le Chatelier's principle to a saturated solution containing a slightly soluble ionic compound.

In Section 16.9, we discussed Le Chatelier's principle as it applies to weak acid and weak base equilibria. Similarly, Le Chatelier's principle applies to solubility equilibria. That is, if we change the concentration of any ions participating in dissociation equilibrium, there will be a shift to relieve the stress.

Let's consider the dissociation of an antacid tablet in aqueous solution. The tablet contains aluminum hydroxide, $Al(OH)_3$, which is slightly soluble. The dissociation is as follows:

$$Al(OH)_3(s) \rightleftharpoons Al^{3+}(aq) + 3\ OH^-(aq)$$

Assume that the solution is already saturated. That is, Al^{3+} and OH^- are at maximum concentration. Therefore, if we add more aluminum hydroxide to the solution, it does not stress the system. Thus, there is no shift in equilibrium.

Suppose we add solid $AlCl_3$ to the solution to increase the Al^{3+} concentration. The solid dissolves to give Cl^- and more Al^{3+}. The aluminum ion stresses the right side of the reaction. The equilibrium shifts to the left to reduce the stress. Simultaneously, some of the hydroxide ions precipitate as $Al(OH)_3$, thus decreasing the concentration of OH^-.

If we add solid Na_3PO_4, the phosphate ion precipitates the Al^{3+} as insoluble $AlPO_4$. It thereby stresses the right side of the equation. There will be a shift to the right to restore the equilibrium. Simultaneously, some of the precipitate dissociates, thus increasing the concentration of OH^-.

Consider the effect of adding NaCl to the aqueous system. Sodium chloride dissolves to give sodium ions and chloride ions. Since neither Na^+ nor Cl^- participate in the equilibrium, NaCl does not stress the system. Thus, there is no effect on the reaction and no shift in equilibrium. The concentrations of Al^{3+} and OH^- remain constant.

Let's consider a less obvious stress on the equilibrium system. We increase the H^+ concentration by adding HCl, a strong acid. Hydrochloric acid ionizes to give H^+ and Cl^-. Although neither ion is shown in the equation, H^+ stresses the system. Remember that H^+ neutralizes OH^- to give water. Thus, the hydrogen ion ties up the hydroxide ion in the form of a water molecule. That is,

$$H^+(aq) + OH^-(aq) \longrightarrow H_2O(l)$$

In effect, the concentration of OH^- decreases. This stresses the right side of the equation. The equilibrium shifts to the right. To reestablish the equilibrium, more $Al(OH)_3$ dissolves and the concentration of Al^{3+} increases.

EXAMPLE EXERCISE 16.9

Predict the direction of equilibrium shift for each of the following stresses on a saturated solution of aqueous silver chloride.

$$AgCl(s) \rightleftharpoons Ag^+(aq) + Cl^-(aq)$$

(a) $[Ag^+]$ increases
(b) $[Cl^-]$ decreases
(c) $AgNO_3$ is added
(d) $NaNO_3$ is added
(e) NaI is added
(f) AgCl is added
(g) pH decreases

Solution: Apply Le Chatelier's principle to each of the stresses.

(a) If $[Ag^+]$ increases, the equilibrium shifts to the *left* to relieve the stress. The $[Cl^-]$ decreases.

(b) If $[Cl^-]$ decreases, the equilibrium shifts to the *right* to decrease the stress. The $[Ag^+]$ increases.

(c) If $AgNO_3$ is added, the concentration of Ag^+ increases. There is a shift to the *left* to restore the equilibrium.

(d) If $NaNO_3$ is added, it dissolves, giving Na^+ and NO_3^- in solution. Since neither of these ions participates in the equilibrium, there is *no shift* in the system.

(e) If NaI is added, it dissolves, giving Na^+ and I^- in solution. The iodide ion combines with Ag^+ to give insoluble AgI. Since the $[Ag^+]$ decreases, more AgCl dissolves to compensate for the stress. The equilibrium shifts to the *right*.

(f) Since the solution is already saturated, Ag^+ and Cl^- are at maximum concentration. Thus, additional AgCl in the solution causes *no shift* in equilibrium.

(g) As the pH decreases, the $[H^+]$ increases. The hydrogen ion is not a factor, directly or indirectly, so there is *no shift* in equilibrium.

SELF-TEST EXERCISE

Predict the direction of equilibrium shift for each of the following stresses on a saturated solution of aqueous lead(II) chromate.

$$PbCrO_4(s) \rightleftharpoons Pb^{2+}(aq) + CrO_4^{2-}(aq)$$

(a) $[Pb^{2+}]$ decreases
(b) $[CrO_4^{2-}]$ increases
(c) $Pb(NO_3)_2$ is added
(d) $PbCrO_4$ is added
(e) K_2CrO_4 is added
(f) $LiNO_3$ is added
(g) NaI is added
(h) pH decreases

Answers: **(a)** shifts right; **(b)** shifts left; **(c)** shifts left; **(d)** no shift; **(e)** shifts left; **(f)** no shift; **(g)** shifts right (PbI_2 is insoluble); **(h)** no shift

Summary

Section 16.1 In Chapter 16, we introduced the **collision theory** of reaction rates. On a molecular level, the rate of reaction is regulated by the collision frequency, collision energy, and orientation of the molecules. It follows that we can speed up a reaction by (1) increasing the concentration of reactants, (2) raising the temperature, or (3) adding a **catalyst**.

Section 16.2 The progress of a reaction can be diagrammed by constructing a **reaction profile**. The reaction profile graphs the change in energy as the reactants are converted to products. For molecules to collide effectively and form products, they must have enough energy to achieve the **transition state**. The energy necessary to reach the transition state is called the **activation energy**, E_{act}. The difference between the energy of the reactants and the products is the **heat of reaction** ΔH.

Section 16.3 The **rate of reaction** is the ratio of the decrease in reactant concentration with time. The **rate equation** for a reaction is written as rate $= k[A]^x[B]^y$, where [A] and [B] are the molar concentrations of reactants. The exponents x and y of the concentrations in the rate expression are always determined experimentally. The values of x and y can be whole numbers, fractions, or zero.

Sections 16.4 through 16.11 In time, reversible reactions eventually reach a state of **chemical equilibrium**. That is, the rates of the forward and reverse reactions are equal. The **law of chemical equilibrium** states that the product concentrations divided by the reactant concentrations, each raised to a coefficient power from the

Table 16.3 Selected Types of Equilibria

Type of Equilibrium	Equilibrium Expression
General Equilibrium	
$aA + bB \rightleftharpoons cC + dD$	$K_{eq} = \frac{[C]^c[D]^d}{[A]^a[B]^b}$
Gaseous Equilibrium	
$CH_4(g) + 2\,O_2(g) \rightleftharpoons CO_2(g) + 2\,H_2O(g)$	$K_c = \frac{[CO_2][H_2O]^2}{[CH_4][O_2]^2}$
Ionization Equilibrium	
$H_2S(aq) \rightleftharpoons H^+(aq) + HS^-(aq)$	$K_i = \frac{[H^+][HS^-]}{[H_2S]}$
Solubility Product Equilibrium	
$PbCl_2(s) \rightleftharpoons Pb^{2+}(aq) + 2\,Cl^-(aq)$	$K_{sp} = [Pb^{2+}][Cl^-]^2$

Table 16.4 Summary of Equilibrium Shifts

Change of Conditions	Effect on Equilibrium
Concentration Increases (all systems)	
for a reactant	*shifts to the right*
for a product	*shifts to the left*
Concentration Decreases (all systems)	
for a reactant	*shifts to the left*
for a product	*shifts to the right*
Temperature Increases (gaseous systems)	
for an endothermic reaction	*shifts to the right*
for an exothermic reaction	*shifts to the left*
Temperature Decreases (gaseous systems)	
for an endothermic reaction	*shifts to the left*
for an exothermic reaction	*shifts to the right*
Volume Increases or Pressure Decreases (gaseous systems)	
more reactant molecules	*shifts to the left*
more product molecules	*shifts to the right*
molecules of reactant = product	*no shift*
Volume Decreases or Pressure Increases (gaseous systems)	
more reactant molecules	*shifts to the right*
more product molecules	*shifts to the left*
molecules of reactant = product	*no shift*

balanced chemical equation, is equal to a constant, K_{eq}. The **general equilibrium expression** applies to various chemical systems, as illustrated in Table 16.3. Once a reversible reaction is at equilibrium, a change of concentration, temperature, or pressure can cause a stress to the system. According to **Le Chatelier's principle**, the equilibrium will shift in order to relieve the stress (see Table 16.4).

Key Terms

Select the key term that corresponds to the following definitions.

____ **1.** the principle that the rate of a chemical reaction is regulated by the frequency of molecules striking each other and reacting

____ **2.** a graphical representation of the energy of a reaction as a function of the time of reaction; also called an energy diagram

____ **3.** the highest point on the reaction profile

____ **4.** the energy necessary for reactants to achieve the transition state and overcome the energy barrier for forming products

____ **5.** the difference in heat energy between the reactants and the products for a given chemical reaction

____ **6.** a chemical reaction that takes place by liberating heat energy

____ **7.** a chemical reaction that takes place by absorbing heat energy

____ **8.** a substance that allows a reaction to proceed faster by lowering the energy of activation

____ **9.** the change in concentration of a reactant or product with respect to time

____ **10.** the symbol for the molar concentration of species A

____ **11.** the principle that a change in the mass of a substance participating in a reversible reaction produces a shift in the equilibrium; also termed mass law

____ **12.** the general relationship between the reaction rate and reactant concentration; that is, rate $= k[A][B]$

____ **13.** a type of reaction in which the products react with each other in order to form the original reactants

____ **14.** a dynamic state for a reversible reaction in which the rates of the forward and reverse reactions are equal

____ **15.** any chemical equilibrium that is isolated and can be studied independently of the surrounding environment

____ **16.** a type of equilibrium in which all the participating species are in the same state; for example, an equilibrium between gaseous reactants and products

____ **17.** a type of equilibrium in which one or more of the participating species is in a different state; for example, an equilibrium between an insoluble precipitate and its ions in aqueous solution

____ **18.** the statement that the concentrations of the products of a reversible reaction divided by the concentrations of the reactants, each raised to a coefficient power from the balanced equation, is a constant

____ **19.** the relationship of reactant and product concentrations having the form

$$K_{eq} = \frac{[C]^c[D]^d}{[A]^a[B]^b}$$

(a) [A] *(Sec. 16.3)*
(b) activation energy E_{act} *(Sec. 16.2)*
(c) catalyst *(Sec. 16.2)*
(d) chemical equilibrium *(Sec. 16.4)*
(e) closed system *(Sec. 16.4)*
(f) collision theory *(Sec. 16.1)*
(g) concentration equilibrium constant (K_c) *(Sec. 16.5)*
(h) endothermic reaction *(Sec. 16.2)*
(i) exothermic reaction *(Sec. 16.2)*
(j) general equilibrium expression *(Sec. 16.4)*
(k) heat of reaction (ΔH) *(Sec. 16.2)*
(l) heterogeneous equilibrium *(Sec. 16.5)*
(m) homogeneous equilibrium *(Sec. 16.5)*
(n) ionization constant (K_i) *(Sec. 16.8)*
(o) law of chemical equilibrium *(Sec. 16.4)*
(p) law of mass action *(Sec. 16.4)*
(q) Le Chatelier's Principle *(Sec. 16.6)*
(r) rate equation *(Sec. 16.3)*
(s) rate of reaction *(Sec. 16.3)*
(t) reaction profile *(Sec. 16.2)*
(u) reversible reaction *(Sec. 16.4)*
(v) solubility product constant (K_{sp}) *(Sec. 16.10)*
(w) transition state *(Sec. 16.2)*

_____ **20.** the statement that an equilibrium, when stressed by a change of concentration, temperature, or pressure, will shift to relieve the stress

_____ **21.** the equilibrium constant that relates concentrations of gases in a reversible reaction

_____ **22.** the equilibrium constant that relates concentrations of slightly ionized weak acids or bases in aqueous solution

_____ **23.** the equilibrium constant that relates the concentration of slightly dissociated ionic compounds in aqueous solution

Exercises

Collision Theory of Reaction Rates (Sec. 16.1)

1. State the three factors that influence the rate of effective molecular collisions.

2. When the temperature is increased, the rate of effective collisions increases for two reasons. State the two reasons.

3. Draw a diagram showing effective collision geometry between a hydrogen molecule and a chlorine molecule.

4. Draw a diagram showing an ineffective collision geometry between a hydrogen molecule and an iodine molecule.

5. State the effect on the rate of reaction for each of the following.
(a) increase the concentration of a reactant
(b) decrease the temperature of the reaction
(c) add a catalyst

6. The Haber process uses a metal oxide catalyst to produce ammonia gas. Does the catalyst increase the amount of ammonia? Explain.

7. Why will a spark ignite a coal dust explosion in a mine and not cause charcoal to react explosively in a barbecue?

8. Why do methane gas and chlorine gas react rapidly in sunlight and very slowly in the laboratory?

Energy Profiles of Chemical Reactions (Sec. 16.2)

9. Phosphorus pentachloride is used in the transistor industry to dope silicon chips. Draw the energy profile for the following reaction.

$$PCl_5(g) + \text{heat} \rightleftharpoons PCl_3(g) + Cl_2(g)$$

10. Ozone slowly decomposes in the atmosphere to give oxygen gas. Draw the energy profile for the reaction.

$$2\ O_3(g) \rightleftharpoons 3\ O_2(g) + \text{heat}$$

11. Draw the energy profile for the following endothermic reaction.

$$H_2(g) + I_2(g) \rightleftharpoons 2\ HI(g)$$

Label the axes and indicate the reactants, products, transition state, activation energy, and energy of reaction.

12. Draw the energy profile for the following exothermic reaction.

$$H_2(g) + Cl_2(g) \rightleftharpoons 2\ HCl(g)$$

Label the axes and indicate the reactants, products, transition state, activation energy, and energy of reaction.

13. State the effect of a catalyst on the energy of activation, E_{act}.

14. State the effect of a catalyst on the heat of reaction, ΔH.

15. Consider the energy profile for a reversible endothermic reaction. Is the E_{act} greater for the forward or reverse reaction?

16. Consider the energy profile for a reversible exothermic reaction. Is the E_{act} greater for the forward or reverse reaction?

Rates of Reaction (Sec. 16.3)

17. In many cases, the rate of reaction approximately doubles for each 10°C increase in temperature. If a certain gaseous state reaction requires 120 seconds at 20°C, how long will it take at 50°C?

18. In many cases, the reaction rate is approximately halved for each 10°C decrease in temperature. If a certain gaseous state reaction requires 120 seconds at 20°C, how long will it take at 0°C?

19. The iodine clock is a demonstration where two colorless solutions turn deep blue when mixed. The rate equation for the reaction is rate $= k[A][B]$. If A and B turn deep blue in 60.0 seconds after mixing, how long will the reaction take for the following changes in concentration?
(a) the [A] increases twofold
(b) the [A] decreases twofold
(c) the [B] increases fourfold
(d) the [B] decreases threefold

20. The nitrogen dioxide in smog converts carbon monoxide to carbon dioxide. Interestingly, the rate depends only on nitrogen dioxide: rate $= k[NO_2]^2$. Under controlled laboratory conditions, the reaction requires 10.0 minutes to reach equilibrium. How long will the reaction require for the following changes in concentration?

(a) the $[NO_2]$ is doubled
(b) the $[NO_2]$ is halved
(c) the $[NO_2]$ is tripled
(d) the $[NO_2]$ is quartered

21. The gaseous reaction of nitrogen monoxide and chlorine proceeds according to the following rate equation: rate = $k[NO]^2[Cl_2]$. Find the change in reaction rate for each of the following changes of concentration.
(a) [NO] increases from 1.00 *M* to 1.50 *M*
(b) [NO] decreases from 0.950 *M* to 0.750 *M*
(c) $[Cl_2]$ increases from 1.00 *M* to 3.00 *M*
(d) $[Cl_2]$ decreases from 0.650 *M* to 0.125 *M*

22. Chloroform, $CHCl_3$, is converted to carbon tetrachloride in the presence of chlorine gas. The rate equation for the reaction is rate = $k[CHCl_3][Cl_2]^{1/2}$. What happens to the reaction rate for the following changes of concentration?
(a) $[CHCl_3]$ increases from 0.100 *M* to 0.500 *M*
(b) $[CHCl_3]$ decreases from 0.250 *M* to 0.100 *M*
(c) $[Cl_2]$ increases from 0.360 *M* to 0.480 *M*
(d) $[Cl_2]$ decreases from 0.200 *M* to 0.115 *M*

Law of Chemical Equilibrium (Sec. 16.4)

23. Define the rate of a forward reaction in terms of the (a) reactant concentration and (b) product concentration.

24. Define chemical equilibrium using the symbols $rate_f$ and $rate_r$.

25. Which of the following statements is true regarding the general equilibrium expression?
(a) The K_{eq} expression was originally determined by experiment.
(b) The K_{eq} expression can be derived theoretically from rate equations.

26. Which of the following statements is true regarding the general equilibrium expression?
(a) The value of K_{eq} is independent of temperature.
(b) The K_{eq} expression contains only substances in the same physical state.

27. Write the general equilibrium constant expression for each of the following.
(a) $2\ A \rightleftharpoons C$
(b) $A + 2\ B \rightleftharpoons 3\ C$
(c) $2\ A + 3\ B \rightleftharpoons 4\ C + D$

28. Write the general equilibrium constant expression for each of the following.
(a) $3\ A \rightleftharpoons 2\ C$
(b) $A + B \rightleftharpoons 2\ C$
(c) $3\ A + 5\ B \rightleftharpoons C + 4\ D$

Concentration Equilibrium Constant, K_c (Sec. 16.5)

29. Does a substance in the solid state appear in the equilibrium expression for a gaseous state reaction?

30. Does a substance in the pure liquid state appear in the equilibrium expression for a gaseous state reaction?

31. Write the equilibrium constant expression for each of the following reversible reactions.
(a) $H_2(g) + F_2(g) \rightleftharpoons 2\ HF(g)$
(b) $4\ NH_3(g) + 7\ O_2(g) \rightleftharpoons 4\ NO_2(g) + 6\ H_2O(g)$
(c) $ZnCO_3(s) \rightleftharpoons ZnO(s) + CO_2(g)$

32. Write the equilibrium constant expression for each of the following reversible reactions.
(a) $H_2(g) + Br_2(g) \rightleftharpoons 2\ HBr(g)$
(b) $4\ HCl(g) + O_2(g) \rightleftharpoons 2\ Cl_2(g) + 2\ H_2O(g)$
(c) $CO(g) + 2\ H_2(g) \rightleftharpoons CH_3OH(l)$

33. Given the equilibrium concentrations for each gas at 850°C, calculate the value of K_c for the manufacture of sulfur trioxide.

$$2\ SO_2(g) + O_2(g) \rightleftharpoons 2\ SO_3(g)$$
1.75 *M* 1.50 *M* 2.25 *M*

34. Given the equilibrium concentrations for each gas at 500°C, calculate the value of K_c for the manufacture of ammonia.

$$N_2(g) + 3\ H_2(g) \rightleftharpoons 2\ NH_3(g)$$
0.400 *M* 1.20 *M* 0.195 *M*

Gaseous State Equilibria Shifts (Sec. 16.6)

35. Weather conditions affect the smog equilibrium in the atmosphere. What happens to the harmful nitrogen dioxide concentration on (a) hot, sunny days and (b) cool, overcast days?

$$N_2O_4(g) + heat \overset{UV}{\rightleftharpoons} 2\ NO_2(g)$$

36. The conditions for producing ammonia industrially are 500°C and 300 atm. What happens to the ammonia concentration if (a) temperature increases (b) pressure increases?

$$N_2(g) + 3\ H_2(g) \rightleftharpoons 2\ NH_3(g) + heat$$

37. The industrial process for producing carbon monoxide gas is to pass carbon dioxide over hot charcoal:

$$C(s) + CO_2(g) + heat \rightleftharpoons 2\ CO(g)$$

Predict the direction of equilibrium shift for each of the following stresses.
(a) $[CO_2]$ decreases
(b) [CO] decreases
(c) solid charcoal is added
(d) CO_2 gas is added
(e) nitrogen gas is added
(f) temperature increases
(g) pressure increases
(h) volume increases

38. The water–gas process for producing hydrogen reacts methane and steam at high temperature:

$$CH_4(g) + H_2O(g) + heat \rightleftharpoons CO(g) + 3\ H_2(g)$$

Predict the direction of equilibrium shift for each of the following stresses.
(a) $[CH_4]$ increases
(b) [CO] increases
(c) $[H_2O]$ decreases
(d) $[H_2]$ decreases
(e) xenon gas is added
(f) temperature decreases
(g) pressure decreases
(h) volume decreases

39. Coal-burning power plants release sulfur dioxide into the atmosphere. The SO_2 is converted to SO_3 by the pollutant nitrogen dioxide as follows:

$$SO_2(g) + NO_2(g) \rightleftharpoons SO_3(g) + NO(g) + \text{heat}$$

Sulfur trioxide dissolves in cloud moisture, eventually producing acid rain. Predict the direction of equilibrium shift for each of the following stresses.

(a) $[SO_2]$ decreases (b) $[SO_3]$ decreases
(c) $[NO_2]$ increases (d) $[NO]$ increases
(e) temperature decreases (f) pressure increases
(g) volume increases (h) catalyst added

40. Traces of formaldehyde in smog are responsible for an eye-burning sensation. Formaldehyde, CH_2O, results from the reaction of ozone and the hydrocarbon pollutant ethylene, C_2H_4, as follows:

$$2\ C_2H_4(g) + 2\ O_3(g) \rightleftharpoons 4\ CH_2O(g) + O_2(g) + \text{heat}$$

Predict the direction of equilibrium shift for each of the following stresses.

(a) $[C_2H_4]$ increases (b) $[CH_2O]$ increases
(c) $[O_3]$ decreases (d) $[O_2]$ decreases
(e) temperature increases (f) pressure decreases
(g) volume decreases (h) catalyst added

Aqueous Solution Equilibria (Sec. 16.7)

41. What are the three forces that drive a reversible reaction in aqueous solution to completion?

42. Formation of which of the following drives a reversible reaction in aqueous solution to completion?

(a) strong acid (b) weak acid
(c) strong base (d) weak base
(e) soluble salt (f) insoluble salt
(g) water (h) a gas

43. Which of the following reaction products will drive an equilibrium in aqueous solution to completion (if necessary, refer to Table 15.7)?

(a) $Li_2SO_4(aq)$ (b) $BaSO_4(s)$
(c) $H_2SO_3(aq)$ (d) $H_2SO_4(aq)$
(e) $KOH(aq)$ (f) $NH_3(aq)$

44. Which of the following reaction products will drive an equilibrium in aqueous solution to completion (if necessary, refer to Table 15.7)?

(a) $HNO_3(aq)$ (b) $HNO_2(aq)$
(c) $H_2O(l)$ (d) $H_2S(g)$
(e) $Na_2CO_3(aq)$ (f) $MgCO_3(s)$

45. Indicate the product that is responsible for driving the following partially reversible reactions in aqueous solution to completion.

(a) $2\ HCl(aq) + K_2CO_3(aq) \rightleftharpoons 2\ KCl(aq) + H_2O(l) + CO_2(g)$

(b) $HC_2H_3O_2(aq) + NH_4OH(aq) \rightleftharpoons NH_4C_2H_3O_2(aq) + H_2O(l)$

(c) $Pb(NO_3)_2(aq) + Na_2CrO_4(aq) \rightleftharpoons PbCrO_4(s) + 2\ NaNO_3(aq)$

46. Indicate the product that is responsible for driving the following partially reversible reactions in aqueous solution to completion.

(a) $H_2CO_3(aq) + BaCl_2(aq) \rightleftharpoons BaCO_3(s) + 2\ HCl(aq)$

(b) $NH_4NO_3(aq) + KOH(aq) \rightleftharpoons KNO_3(aq) + NH_4OH(aq)$

(c) $H_2SO_4(aq) + Ba(OH)_2(aq) \rightleftharpoons BaSO_4(s) + 2\ H_2O(l)$

Ionization Equilibrium Constant, K_i (Sec. 16.8)

47. Write the equilibrium constant expression for the following weak acids.

(a) $HCO_2H(aq) \rightleftharpoons H^+(aq) + CO_2H^-(aq)$
(b) $H_2C_2O_4(aq) \rightleftharpoons H^+(aq) + HC_2O_4^-(aq)$
(c) $H_3C_6H_5O_7(aq) \rightleftharpoons H^+(aq) + H_2C_6H_5O_7^-(aq)$

48. Write the equilibrium constant expression for the following weak bases.

(a) $NH_2OH(aq) + H_2O(l) \rightleftharpoons NH_3OH^+(aq) + OH^-(aq)$

(b) $C_6H_5NH_2(aq) + H_2O(l) \rightleftharpoons C_6H_5NH_3^+(aq) + OH^-(aq)$

(c) $(CH_3)_2NH(aq) + H_2O(l) \rightleftharpoons (CH_3)_2NH_2^+(aq) + OH^-(aq)$

49. Nitrous acid is used in the synthesis of selected organic nitrogen compounds. If the hydrogen ion concentration of a 0.125 *M* solution is 7.5×10^{-3}, what is the ionization constant for the acid?

50. Aqueous ammonium hydroxide is used as a household cleaning solution. If the hydroxide ion concentration of a 0.245 *M* solution is 2.1×10^{-3}, what is the ionization constant for the base?

51. Hydrofluoric acid is used to etch silicon oxide in the manufacture of transistors. If the pH of a 0.139 *M* HF solution is 2.00, what is the ionization constant of the acid?

52. Hydrazine is a weak base but is also used as a fuel for the space shuttle. If the pOH of a 0.769 *M* N_2H_4 solution is 3.00, what is the ionization constant of the base?

$$N_2H_4(aq) + H_2O(l) \rightleftharpoons N_2H_5^+(aq) + OH^-(aq)$$

Weak Acid–Base Equilibria Shifts (Sec. 16.9)

53. Given the chemical equation for the ionization of hydrofluoric acid:

$$HF(aq) \rightleftharpoons H^+(aq) + F^-(aq)$$

Predict the direction of equilibrium shift for each of the following stresses.

(a) increase [HF] (b) increase $[H^+]$
(c) decrease [HF] (d) decrease $[F^-]$
(e) add NaF solid (f) add HCl gas
(g) add NaOH solid (h) increase pH

54. Given the chemical equation for the ionization of nitrous acid:

$$HNO_2(aq) \rightleftharpoons H^+(aq) + NO_2^-(aq)$$

Predict the direction of equilibrium shift for each of the following stresses.

(a) decrease $[HNO_2]$ (b) decrease $[H^+]$
(c) increase $[HNO_2]$ (d) increase $[NO_2^-]$
(e) add KNO_2 solid (f) add KCl solid
(g) add KOH solid (h) increase pH

55. Given the chemical equation for the ionization of acetic acid:

$$HC_2H_3O_2(aq) \rightleftharpoons H^+(aq) + C_2H_3O_2^-(aq)$$

Predict the direction of equilibrium shift for each of the following stresses.

(a) increase $[HC_2H_3O_2]$ (b) increase $[H^+]$
(c) decrease $[HC_2H_3O_2]$ (d) decrease $[C_2H_3O_2^-]$
(e) add $NaC_2H_3O_2$ solid (f) add NaCl solid
(g) add NaOH solid (h) increase pH

56. Given the chemical equation for the ionization of ammonium hydroxide:

$$NH_4OH(aq) \rightleftharpoons NH_4^+(aq) + OH^-(aq)$$

Predict the direction of equilibrium shift for each of the following stresses.

(a) increase $[NH_4^+]$ (b) decrease $[OH^-]$
(c) increase $[NH_4OH]$ (d) decrease pH
(e) add NH_3 gas (f) add KCl solid
(g) add KOH solid (h) add NH_4Cl solid

Solubility Product Equilibrium Constant, K_{sp} (Sec. 16.10)

57. Write the solubility product expression for the following slightly soluble ionic compounds in aqueous solution.

(a) $MnCO_3(s) \rightleftharpoons Mn^{2+}(aq) + CO_3^{2-}(aq)$
(b) $Cd(CN)_2(s) \rightleftharpoons Cd^{2+}(aq) + 2\,CN^-(aq)$
(c) $Sb_2S_3(s) \rightleftharpoons 2\,Sb^{3+}(aq) + 3\,S^{2-}(aq)$

58. Write the solubility product expression for the following slightly soluble ionic compounds in aqueous solution.

(a) $AuI_3(s) \rightleftharpoons Au^{3+}(aq) + 3\,I^-(aq)$
(b) $Ag_2SO_4(s) \rightleftharpoons 2\,Ag^+(aq) + SO_4^{2-}(aq)$
(c) $Co_3(PO_4)_2(s) \rightleftharpoons 3\,Co^{2+}(aq) + 2\,PO_4^{3-}(aq)$

59. The cobalt ion concentration in a saturated solution of cobalt(II) sulfide, CoS, is 7.7×10^{-11} *M*. Calculate the K_{sp} for the slightly soluble ionic compound.

60. The fluoride ion concentration in a saturated solution of magnesium fluoride, MgF_2, is 2.3×10^{-3} *M*. Calculate the K_{sp} for the slightly soluble ionic compound.

61. The zinc ion concentration in a saturated solution of zinc phosphate, $Zn_3(PO_4)_2$, is 1.5×10^{-7} *M*. Calculate the K_{sp} for the slightly soluble compound.

62. The hydroxide ion concentration in a saturated solution of iron(III) hydroxide, $Fe(OH)_3$, is 2.2×10^{-10} *M*. Calculate the K_{sp} for the slightly soluble compound.

63. The K_{sp} values for $CaCO_3$ and CaC_2O_4 are 3.8×10^{-9} and 2.3×10^{-9}, respectively. Which calcium compound is more soluble?

64. The K_{sp} values for $MnCO_3$ and $Mn(OH)_2$ are 1.8×10^{-11} and 4.6×10^{-14}, respectively. Which manganese compound is more soluble?

Dissociation Equilibria Shifts (Sec. 16.11)

65. Teeth and bones are composed mainly of calcium phosphate, which dissociates as follows:

$$Ca_3(PO_4)_2(s) \rightleftharpoons 3\,Ca^{2+}(aq) + 2\,PO_4^{3-}(aq)$$

Predict the direction of equilibrium shift for each of the following stresses.

(a) increase $[Ca^{2+}]$ (b) increase $[PO_4^{3-}]$
(c) decrease $[Ca^{2+}]$ (d) decrease $[PO_4^{3-}]$
(e) add solid $Ca(NO_3)_2$ (f) add solid KNO_3

66. Given the chemical equation for the dissociation of cadmium sulfide:

$$CdS(s) \rightleftharpoons Cd^{2+}(aq) + S^{2-}(aq)$$

Predict the direction of equilibrium shift for each of the following stresses.

(a) increase $[Cd^{2+}]$ (b) increase $[S^{2-}]$
(c) decrease $[Cd^{2+}]$ (d) decrease $[S^{2-}]$
(e) add solid $Cd(NO_3)_2$ (f) add solid $NaNO_3$

67. Given the chemical equation for the dissociation of copper(II) hydroxide:

$$Cu(OH)_2(s) \rightleftharpoons Cu^{2+}(aq) + 2\,OH^-(aq)$$

Predict the direction of equilibrium shift for each of the following stresses.

(a) increase $[Cu^{2+}]$ (b) increase $[OH^-]$
(c) decrease $[Cu^{2+}]$ (d) decrease $[OH^-]$
(e) add solid $Cu(OH)_2$ (f) add solid NaOH
(g) add solid NaCl (h) add H^+

68. Given the chemical equation for the dissociation of strontium carbonate:

$$SrCO_3(s) \rightleftharpoons Sr^{2+}(aq) + CO_3^{2-}(aq)$$

Predict the direction of equilibrium shift for each of the following stresses.

(a) increase $[Sr^{2+}]$ (b) increase $[CO_3^{2-}]$
(c) decrease $[Sr^{2+}]$ (d) decrease $[CO_3^{2-}]$
(e) add solid $SrCO_3$ (f) add solid $Sr(NO_3)_2$
(g) add solid KNO_3 (h) add Ca^{2+}

General Exercises

69. Although the following are not examples of chemical equilibria, explain how each is a dynamic reversible process.

(a) a grandfather clock with a swinging pendulum
(b) a solar calculator charging and discharging

70. With regard to rate and concentration, what are the two characteristics of a system at equilibrium?

71. The rate of the following reaction depends only on the concentration of the nitrogen dioxide to the second power.

Write the rate equation and explain why [CO] is not found in the rate expression.

$$NO_2(g) + CO(g) \rightleftharpoons NO(g) + CO_2(g)$$

72. State the effect on the rate of reaction for each of the following.
(a) adding more liquid to a liquid–vapor equilibrium
(b) adding helium to a gaseous state equilibrium
(c) adding precipitate to a saturated solution

73. The N_2O_4–NO_2 reversible reaction in smog is experimentally found to have the following equilibrium concentrations at 100°C. Calculate K_c for the reaction.

$$N_2O_4(g) \rightleftharpoons 2\ NO_2(g)$$

$$4.5 \times 10^{-5}\ M \qquad 3.0 \times 10^{-3}\ M$$

74. For the previous equilibrium at 100°C, the partial pressure of N_2O_4 is 0.0014 atm and NO_2 is 0.092 atm. Calculate K_p for the reaction.

75. It has been suggested that Le Chatelier's principle is so general that it can be applied to human behavior. Consider the stress of final examinations and analyze any shifts in your behavior.

76. Given the equation for the ionization of water:

$$H_2O(l) \rightleftharpoons H^+(aq) + OH^-(aq)$$

Predict the direction of equilibrium shift for each of the following stresses.
(a) increase $[H^+]$
(b) increase $[OH^-]$
(c) decrease $[H^+]$
(d) decrease $[OH^-]$
(e) add liquid H_2SO_4
(f) add solid $Ba(OH)_2$
(g) add gaseous HCl
(h) add solid NaF
(i) add solid $FeCl_3$
(j) decrease pH

77. Give the units of the K_{sp} value for each of the following slightly soluble ionic compounds: (a) ZnS, and (b) $Zn(OH)_2$.

78. A saturated solution of tin(II) hydroxide, $Sn(OH)_2$, has a pOH of 8.47. Find the hydroxide ion concentration and calculate the K_{sp} value for the slightly soluble ionic compound.

79. A saturated solution of chromium(III) hydroxide, $Cr(OH)_3$, has a pH of 7.42. Find the hydroxide ion concentration and calculate the K_{sp} value for the slightly soluble ionic compound.

CHAPTER

17

Oxidation and Reduction

World-class athletes, like Michael Jordan, depend on biochemical reactions for their energy. These energy-producing processes involve the transfer of electrons and are called oxidation–reduction reactions.

In Section 8.4, we classified a chemical reaction as one of the following five types: combination, decomposition, single replacement, double replacement, or neutralization. Recall that a neutralization reaction involves the transfer of protons from an acid to a base. There is a similar type of reaction that involves the transfer of electrons from one substance to another. It is referred to as an oxidation–reduction reaction. *In a neutralization reaction, acids donate protons and bases accept protons. In an oxidation–reduction reaction, one substance donates electrons and is* oxidized, *while another substance accepts electrons and is* reduced.

Although we did not use the term oxidation–reduction, many of the reactions we have studied involved electron transfer. One example is when calcium metal is exposed to air—it oxidizes. That is, calcium reacts with oxygen from the air to give calcium oxide. The equation is

$$2\ Ca(s) + O_2(g) \longrightarrow 2\ CaO(s)$$

Since CaO *is an ionic compound, we can write the product as the ions* Ca^{2+} *and* O^{2-}. *Notice that, in the process of becoming a calcium ion, a calcium atom reacts by losing two electrons. By definition, we can say that the calcium is oxidized by oxygen. That is,*

$$Ca \longrightarrow Ca^{2+} + 2\ e^-$$

As the reaction proceeds, oxygen atoms are reduced and the calcium atoms are oxidized. Each oxygen atom reacts by gaining two electrons and becoming an oxide ion. That is,

$$O + 2\ e^- \longrightarrow O^{2-}$$

Let's consider another example. When sodium metal is heated and plunged into a cylinder of chlorine gas, there is an explosive reaction. The equation is

$$2\ Na(s) + Cl_2(g) \longrightarrow 2\ NaCl(s)$$

The product, NaCl, *is the ionic compound we know as table salt. Since sodium chloride is ionic, we can write* NaCl *as the ions* Na^+ *and* Cl^-. *The sodium metal atom reacts by losing an electron and becoming a sodium ion. By definition, sodium is oxidized by the chlorine gas. That is,*

$$Na \longrightarrow Na^+ + e^-$$

Simultaneously, chlorine atoms are reduced. Each chlorine atom reacts by gaining an electron and becoming a chloride ion. That is,

$$Cl + e^- \longrightarrow Cl^-$$

Previously, we considered the reaction of Na *and* Cl_2 *as a combination reaction. Since electrons are transferred in the process, we can now more correctly classify the reaction as an oxidation–reduction reaction.*

(a)

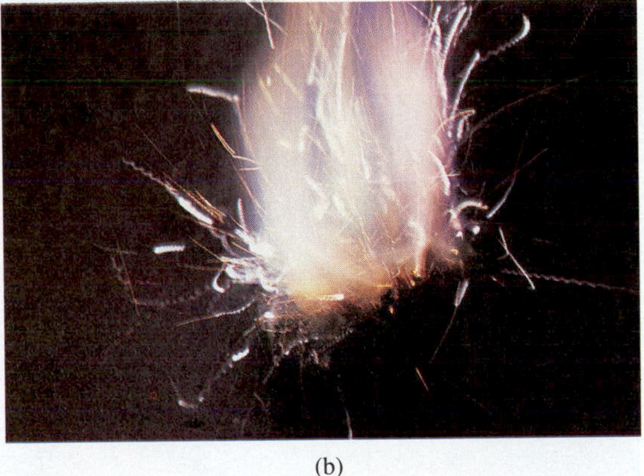

(b)

Figure 17.1 Oxidation-Reduction Reaction
(a) combustion of charcoal;
(b) burning magnesium metal;
(c) corrosion of steel pipe;
(d) tarnishing of silverware

(c)

(d)

In addition to combination reactions, decomposition and single-replacement reactions also involve an oxidation–reduction process. Figure 17.1 shows several different types of oxidation–reduction reactions.

17.1 Oxidation Numbers

OBJECTIVES

To assign an oxidation number for an element in each of the following:

- **(a)** metals and nonmetals
- **(b)** monoatomic and polyatomic ions
- **(c)** ionic and molecular compounds

To describe the number of electrons lost or gained by an atom, an **oxidation number** (also called an oxidation state) is assigned to each element. All elements in the free state are electrically neutral. Therefore, their oxidation numbers have a value of zero. For example, the elements Na, Mg, Al, Fe, and Si have an oxidation number of zero. Even nonmetals that exist as molecules, such as O_2 and Cl_2, have an oxidation number of zero.

oxidation number The value assigned to an atom in a molecule or ion in order to keep track of electrons gained or lost. An oxidation number can be positive or negative, but is zero for elements in the free state.

The oxidation number for any metal in the elemental state is zero.

The oxidation number of a monoatomic ion is equal to its ionic charge. For example, the oxidation number for Ca^{2+} is positive two (+2) and for Cl^- it is negative one (−1). Note the distinction between the way ionic charges and oxidation numbers are given. Ionic charges are indicated by the number followed by the sign, and oxidation numbers are indicated by the sign followed by the number. For example, the oxidation numbers for Fe, Fe^{2+}, and Fe^{3+} are 0, +2, and +3, respectively.

Next, we will see how to assign oxidation numbers to elements in a compound. In Chapter Six we learned that compounds are electrically neutral, so the sum of the oxidation numbers of the individual atoms is zero. For binary ionic compounds, the oxidation numbers for the metal and nonmetal correspond to their ionic charges. For example, in NaCl the oxidation number of Na^+ is +1 and of Cl^- it is −1. In $AlCl_3$, the oxidation number of aluminum, Al^{3+}, is +3 and for Cl^- it is −1.

In binary molecular compounds, the more electronegative element is assigned an oxidation number equal to the ionic charge on the free ion. Oxygen usually has an oxidation number of −2, and hydrogen usually has one of +1. After assigning

Table 17.1 Rules for Assigning Oxidation Numbers

1. Elements in an uncombined state have an oxidation number of zero.
2. Monoatomic ions have an oxidation number equal to their ionic charge.
3. Hydrogen is assigned an oxidation number of +1 in most compounds and ions. Metal hydrides such as LiH and CaH_2 are exceptions; the oxidation number is −1.
4. Oxygen is assigned an oxidation number of −2 in most compounds and ions. Peroxide compounds such as H_2O_2 are exceptions; the oxidation number of oxygen is −1.
5. In molecular compounds, the more electronegative element is assigned a negative oxidation state equal to its charge as an anion. For example, in carbon tetrachloride, CCl_4, the oxidation number of Cl is −1. Since there are four −1, the oxidation number of carbon is +4.
6. Ionic compounds are electrically neutral. Hence, the sum of the individual oxidation numbers is equal to zero.
7. Polyatomic ions are electrically charged. Therefore, the sum of the individual oxidation numbers is equal to the charge on the ion.

Table 17.2 Examples of Oxidation Numbers

Substance	Oxidation Number	
magnesium metal, Mg	Mg = 0	(rule 1)*
bromine liquid, Br_2	Br = 0	(rule 1)
potassium ion, K^+	K = +1	(rule 2)
sulfide ion, S^{2-}	S = −2	(rule 2)
water, H_2O	H = +1	(rule 3)
	O = −2	(rule 4)
sulfur hexafluoride, SF_6	F = −1	(rule 5)
	S = +6	(rule 5)
barium chloride, $BaCl_2$	Ba = +2	(rule 6)
	Cl = −1	(rule 6)
nitrate ion, NO_3^-	O = −2	(rule 4)
	N = +5	(rule 7)

* See Table 17.1.

oxidation numbers to oxygen and hydrogen, we can determine the value for the other nonmetal in a binary covalent compound. For example, in NO the oxidation number of nitrogen is +2. In NO_2, the oxidation number of nitrogen is +4. Rules for assigning oxidation numbers are listed in Table 17.1.

Illustrations of the rules are given in Table 17.2, which lists substances and their oxidation numbers.

The oxidation number for any uncombined nonmetal is zero (left to right: sulfur, white phosphorus—stored under water—bromine, and carbon).

Determining Oxidation Numbers

Now let's determine the oxidation number of an element in the compound oxalic acid, $H_2C_2O_4$. Oxalic acid, which is found naturally in spinach, creates a rough sensation on our teeth. To find the oxidation number of carbon in $H_2C_2O_4$, we proceed as follows. First, we assign the oxidation number of hydrogen to be +1. Second, we assign the oxidation number of oxygen to be -2. Since the compound is electrically neutral we can write the sum of the oxidation numbers (ox no) equal to zero. Thus,

$$2(\text{ox no H}) + 2(\text{ox no C}) + 4(\text{ox no O}) = 0$$

Substituting the known values for the oxidation numbers of H and O, we have

$$2(+1) + 2(\text{ox no C}) + 4(-2) = 0$$

Simplifying and solving for the oxidation number of carbon gives

$$+2 + 2(\text{ox no C}) + -8 = 0$$

$$2(\text{ox no C}) = +6$$

$$\text{ox no C} = +3$$

In this example, carbon has an oxidation number of $+3$. Example Exercise 17.1 provides illustrations of different oxidation numbers for carbon.

EXAMPLE EXERCISE 17.1

Calculate the oxidation number for carbon in each of the following.

(a) diamond, C
(b) Dry Ice, CO_2
(c) potash, K_2CO_3

Solution: Begin by recalling that elements in the free state and compounds are electrically neutral and have no charge.

(a) Carbon is found in the free state as diamond. Thus, the oxidation number of C is zero.

(b) Carbon is in Dry Ice, which is a molecular compound. Since we assign oxygen an oxidation number of -2, we can write

$$\text{ox no C} + 2(\text{ox no O}) = 0$$

$$\text{ox no C} + 2(-2) = 0$$

$$\text{ox no C} = +4$$

(c) Potash, K_2CO_3, is an important agricultural chemical. Since potassium is a Group IA/1 element, K^+ has an oxidation number of $+1$. For oxygen the oxidation number is -2. Therefore, we can write

$$2(\text{ox no K}) + \text{ox no C} + 3(\text{ox no O}) = 0$$

$$2(+1) + \text{ox no C} + 3(-2) = 0$$

$$\text{ox no C} = +4$$

SELF-TEST EXERCISE

Calculate the oxidation number for iodine in each of the following.

(a) iodine, I_2
(b) potassium iodide, KI
(c) hypoiodous acid, HIO(aq)
(d) sodium iodite, $NaIO_2$
(e) calcium iodate, $Ca(IO_3)_2$
(f) silver periodate, $AgIO_4$

Answers: **(a)** 0; **(b)** -1; **(c)** $+1$; **(d)** $+3$; **(e)** $+5$; **(f)** $+7$

Next, let's determine the oxidation number of an element in a polyatomic ion. Consider the dichromate ion, $Cr_2O_7^{2-}$, which is used in fireworks and safety matches. To find the oxidation number of chromium in the polyatomic ion, we proceed as

The oxidation number for chromium in K_2CrO_4 (*left*) and $K_2Cr_2O_7$ (*right*) is +6 in each compound.

follows. First, we assign oxygen an oxidation number of −2. Since the ion has an overall charge of 2−, we can write the sum of the oxidation numbers equal to −2. Thus,

$$2(\text{ox no Cr}) + 7(\text{ox no O}) = -2$$

Substituting the value for the oxidation number of O, we have

$$2(\text{ox no Cr}) + 7(-2) = -2$$

Simplifying and solving for the oxidation number of chromium gives

$$2(\text{ox no Cr}) + -14 = -2$$

$$2(\text{ox no Cr}) = +12$$

$$\text{ox no Cr} = +6$$

In the polyatomic ion $Cr_2O_7^{2-}$, chromium has an oxidation number of +6.

In Example Exercise 17.2, we will calculate various oxidation numbers of the transition metal manganese.

EXAMPLE EXERCISE 17.2

Calculate the oxidation number of manganese in each of the following ions.

(a) manganese ion, Mn^{2+}
(b) manganate ion, MnO_4^{2-}
(c) permanganate ion, MnO_4^-

Solution: Begin by recalling that the charge on the ion corresponds to the sum of the oxidation numbers.

(a) Manganese ion occurs as the monoatomic cation Mn^{2+}. The oxidation number of Mn is +2.

(b) Manganate ion is a polyatomic anion having a charge of 2−. After we assign oxygen an oxidation number of −2, we can write

$$\text{ox no Mn} + 4(\text{ox no O}) = -2$$

$$\text{ox no Mn} + 4(-2) = -2$$

$$\text{ox no Mn} = +6$$

(c) Permanganate ion is a polyatomic anion with a charge of 1−. After we assign oxygen an oxidation number of −2, we can write

$$\text{ox no Mn} + 4(\text{ox no O}) = -1$$

$$\text{ox no Mn} + 4(-2) = -1$$

$$\text{ox no Mn} = +7$$

SELF-TEST EXERCISE

Calculate the oxidation number of sulfur in each of the following ions.

(a) sulfide ion, S^{2-}
(b) sulfite ion, SO_3^{2-}
(c) sulfate ion, SO_4^{2-}
(d) thiosulfite ion, $S_2O_3^{2-}$

Answers: **(a)** −2; **(b)** +4; **(c)** +6; **(d)** +2

17.2 Oxidation–Reduction Reactions

OBJECTIVES

To identify the substance oxidized and the substance reduced in a given redox reaction.

To identify the oxidizing agent and reducing agent in a given redox reaction.

What does iron rusting, a magnesium flare burning, and silver tarnishing have in common? The answer is that all three reactions involve a transfer of electrons from a metal to oxygen. The iron atoms lose 3 electrons. The magnesium atoms lose 2 electrons. The silver atoms lose 1 electron. In addition, each oxygen atom gains 2 electrons. As we said earlier, a chemical change that involves the transfer of electrons is called an oxidation–reduction reaction, or simply a **redox reaction**.

redox reaction A chemical reaction that involves electron transfer between two reacting substances.

Oxidized and Reduced Substances

Heating copper metal with sulfur powder gives the following reaction:

$$2\ Cu(s) + S(s) \xrightarrow{\Delta} Cu_2S(s)$$

We can analyze the reaction as follows. The oxidation number of copper is changing from 0 to +1. Therefore, each copper atom is losing 1 electron. Simultaneously, the oxidation number of sulfur is changing from 0 to −2. Therefore, each sulfur atom gains 2 electrons. The process looks as follows:

$$Cu \xrightarrow{\text{loses 1 } e^-} Cu^+$$

$$Cu \xrightarrow{\text{loses 1 } e^-} Cu^+$$

$$S \xrightarrow{\text{gains 2 } e^-} S^{2-}$$

Since electrons are transferred from the metal to the nonmetal, this is an example of a redox reaction. Notice that both copper and sulfur undergo a change in oxidation number. That is,

$$\overset{0}{2\ Cu(s)} + \overset{0}{S(s)} \longrightarrow \overset{+1}{Cu_2}\overset{-2}{S}(s)$$

Oxidizing and Reducing Agents

The process of **oxidation** is characterized by the loss of electrons. The process of **reduction** is characterized by the gain of electrons. In the above example, copper is oxidized and sulfur is reduced. An **oxidizing agent** is a substance that causes the oxidation process. When sulfur reacts with copper, it causes Cu to be oxidized. Hence, S is the oxidizing agent. A **reducing agent** is a substance that brings about the reduction process. When copper reacts with sulfur, S is reduced from 0 to −2. Thus, Cu is the reducing agent.

oxidation A process in which a substance undergoes an increase in oxidation number. A process characterized by losing electrons.

reduction A process in which a substance undergoes a decrease in oxidation number. A process characterized by gaining electrons.

oxidizing agent The substance reduced in a redox reaction.

reducing agent The substance oxidized in a redox reaction.

We can illustrate that redox process by the following diagram:

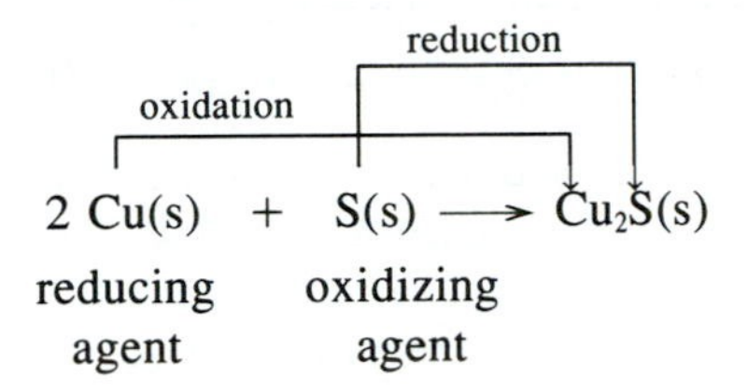

In a redox reaction, one of the substances increases its oxidation number while another substance decreases its oxidation number. In most instances, the maximum positive oxidation number is +7, and the maximum negative oxidation number is −4. Figure 17.2 illustrates the relationship of oxidation numbers for a redox reaction.

The following example exercise further illustrates substances acting as agents for oxidation or reduction.

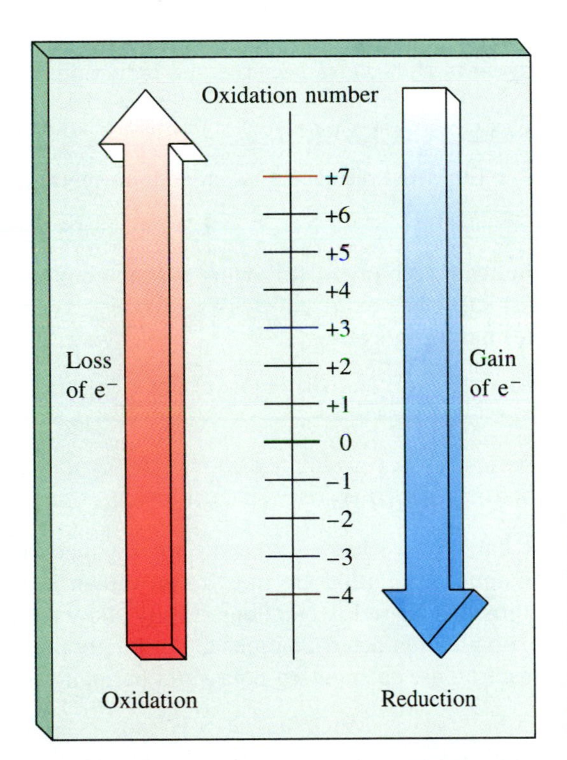

Figure 17.2 Oxidation and Reduction The process of oxidation involves the loss of electrons (oxidation number increases). The process of reduction involves the gain of electrons (oxidation number decreases). Oxidation numbers usually range from +7 to −4.

EXAMPLE EXERCISE 17.3

An oxidation–reduction reaction occurs when a stream of hydrogen gas is passed over hot copper(I) sulfide:

$$Cu_2S(s) + H_2(g) \longrightarrow 2\ Cu(s) + H_2S(g)$$

Indicate each of the following for the above redox reaction.

(a) substance oxidized **(b)** substance reduced
(c) oxidizing agent **(d)** reducing agent

Solution: By definition, the substance oxidized is identified by the loss of electrons. That is, its oxidation number increases. The substance reduced gains electrons and its oxidation number decreases. After assigning oxidation numbers to each atom, we have the following:

0 +1
+1 0

$$Cu_2S(s) + H_2(g) \longrightarrow 2\,Cu(s) + H_2S(g)$$

Notice that the oxidation number of sulfur remained constant (−2). The oxidation number of copper, however, decreased from +1 to 0. Thus, Cu_2S was *reduced*. The oxidation number of hydrogen increased from 0 to +1. Thus, H_2 was *oxidized*.

We can represent the redox process as follows:

oxidation
reduction

$$Cu_2S(s) + H_2(g) \longrightarrow 2\,Cu(s) + H_2S(g)$$

oxidizing agent   reducing agent

The *oxidizing agent* is Cu_2S because it causes hydrogen to be oxidized. The *reducing agent* is H_2 because it causes the reduction of Cu^+ in Cu_2S to the element copper.

SELF-TEST EXERCISE

A redox reaction occurs when sodium metal reacts with water:

$$2\,Na(s) + 2\,H_2O(l) \longrightarrow 2\,NaOH(aq) + H_2(g)$$

Indicate each of the following for the above redox reaction.

(a) substance oxidized **(b)** substance reduced
(c) oxidizing agent **(d)** reducing agent

Answers: **(a)** Na; **(b)** H_2O; **(c)** H_2O; **(d)** Na

A red rose (top) undergoes a redox reaction with colorless SO_2 gas. As the reaction proceeds, the color changes from red to pink (bottom).

Ionic Equations

Recall that we learned to write ionic equations in Section 15.11. Redox reactions in aqueous solution are most often shown in the ionic form. That's because ionic equations for redox reactions readily show us the change in oxidation number. For example, iron determinations are routinely performed by a redox titration. A typical reaction uses a standard potassium permanganate solution as follows:

$$5\,Fe^{2+}(aq) + MnO_4^-(aq) + 8\,H^+(aq) \longrightarrow 5\,Fe^{3+}(aq) + Mn^{2+}(aq) + 4\,H_2O(l)$$

Although the reaction appears complex, we can systematically analyze the equation for the substance oxidized and reduced. First, notice that iron is changing its oxidation number from +2 to +3. Since Fe^{2+} loses 1 electron, it is being oxidized. Because Fe^{2+} is oxidized, it is the reducing agent in this reaction.

Next, we identify the substance being reduced. Ordinarily, we can rule out a change in oxidation number for hydrogen and oxygen if they are not in the free state. When we calculate the oxidation number of manganese in MnO_4^-, we find that the value is +7. On the product side of the equation, the oxidation number of Mn^{2+} is +2. Since the permanganate ion gains 5 electrons, MnO_4^- is the substance reduced. Since MnO_4^- is reduced, it is the oxidizing agent for Fe^{2+}.

We can write a simplified unbalanced equation to show the overall redox process as follows:

+7 reduction +2

+2 oxidation +3

$$Fe^{2+}(aq) + MnO_4^-(aq) \longrightarrow Fe^{3+}(aq) + Mn^{2+}(aq)$$

reducing agent   oxidizing agent

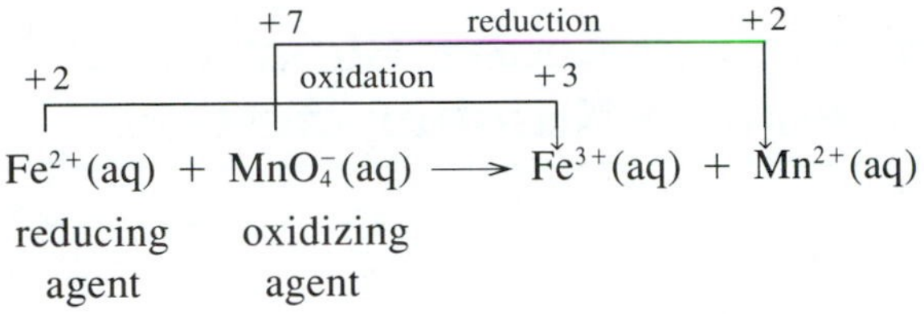

The following example exercise further illustrates substances acting as agents for oxidation or reduction.

EXAMPLE EXERCISE 17.4

The amount of iodine in a solution can be determined by a redox method using a standard sulfite solution as follows:

$$I_2(aq) + SO_3^{2-}(aq) + H_2O(l) \longrightarrow 2\ I^-(aq) + SO_4^{2-}(aq) + 2\ H^+(aq)$$

Indicate each of the following for the above reaction:

(a) substance oxidized
(b) substance reduced
(c) oxidizing agent
(d) reducing agent

Solution: We first notice that iodine is converted to iodide ion. Since iodine is changing from the free state to a negative ion, it is gaining electrons. Therefore, I_2 is being reduced and is functioning as the oxidizing agent.

The reducing agent is not as obvious. If we calculate the oxidation number for sulfur in SO_3^{2-} and SO_4^{2-}, we find a change from +4 to +6. Thus, sulfite loses 2 electrons, and SO_3^{2-} is the substance oxidized. Since SO_3^{2-} loses electrons, it is the reducing agent in the reaction. We can diagram the overall redox process as follows:

+4 oxidation +6

0 reduction −1

$$I_2(aq) + SO_3^{2-}(aq) \longrightarrow I^-(aq) + SO_4^{2-}(aq)$$

oxidizing agent   reducing agent

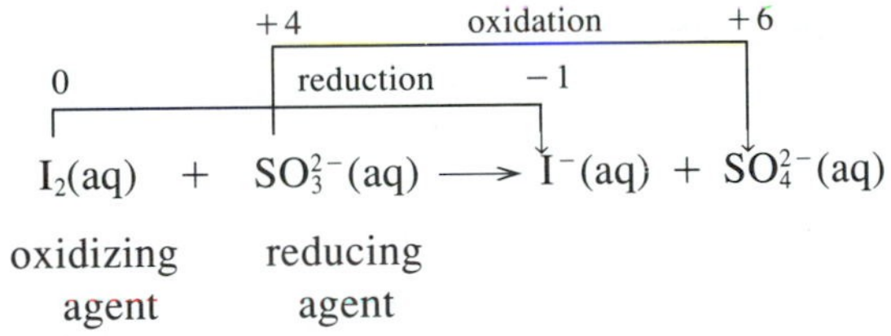

SELF-TEST EXERCISE

A redox reaction occurs between the tin(II) ion and the iodate ion as follows:

$$6\ H^+(aq) + 3\ Sn^{2+}(aq) + 2\ IO_3^-(aq) \longrightarrow 3\ Sn^{4+}(aq) + 2\ I^-(aq) + 3\ H_2O(l)$$

Indicate each of the following for the above redox reaction.

(a) substance oxidized
(b) substance reduced
(c) oxidizing agent
(d) reducing agent

Answers: **(a)** Sn^{2+}; **(b)** IO_3^-; **(c)** IO_3^-; **(d)** Sn^{2+}

It is helpful to know the oxidation number of an element in order to determine the substances oxidized and reduced. In some instances, the oxidation number value is obvious. In others, it must be calculated. With practice, you can readily obtain the oxidation number even for complex substances.

17.3 Balancing Redox Equations: Oxidation Number Method

To write a balanced chemical equation for a redox reaction using the oxidation number method.

In Section 8.3, we learned to balance chemical equations by inspection. We can also balance simple redox reactions by inspection. We cannot, however, balance complex redox reactions by this method.

Recall that in Section 8.3 we suggested some general guidelines to balance chemical equations systematically. In Table 17.3, we now offer guidelines for balancing redox equations using the *oxidation number method.*

Single-Replacement Redox Reactions

Let's apply the guidelines in Table 17.3 to a single-replacement redox reaction. Specifically, we will consider the reaction of a metal and aqueous salt solution. For example, copper metal reacts with aqueous silver nitrate according to the following unbalanced equation:

$$Cu(s) + AgNO_3(aq) \longrightarrow Cu(NO_3)_2(aq) + Ag(s)$$

Table 17.3 Guidelines for Balancing Redox Equations by the Oxidation Number Method

Guideline
1. Inspect the reactants and products to determine the substances undergoing a change in oxidation number. In some cases, it is not obvious and you must calculate the oxidation number. **(a)** Write the oxidation numbers above the element that is oxidized and the element that is reduced. **(b)** Diagram the number of electrons lost by the oxidized substance and gained by the reduced substance. **2.** Balance each element in the equation using a coefficient. Keep in mind that the total electron loss by oxidation must equal the total electron gain by the reduction process. **(a)** Place a coefficient in front of the oxidized substance that corresponds to the number of electrons gained by the reduced substance. **(b)** Place a coefficient in front of the reduced substance that corresponds to the number of electrons lost by the oxidized substance. **(c)** Balance the remaining elements. **3.** After balancing the equation, verify that the coefficients are correct. **(a)** Place a check (√) above the symbol of each element to indicate that the number of atoms is the same for reactants and products. **(b)** Make sure that for ionic equations the total charge on the reactant side of the equation equals the total charge on the product side.

In this reaction the oxidation number of copper metal is increasing from 0 to +2. Simultaneously, the oxidation number of silver is decreasing from +1 to 0. Thus, Cu loses 2 electrons and silver gains 1 electron. We can diagram the loss and gain of electrons as follows:

$$\overset{0}{Cu(s)} + \overset{+1}{Ag}NO_3(aq) \longrightarrow \overset{+2}{Cu}(NO_3)_2(aq) + \overset{0}{Ag(s)}$$

(Cu: 0 → +2, loses 2 e^-; Ag: +1 → 0, gains 1 e^-)

Since copper loses 2 e^-, and silver gains only 1 e^-, two silver atoms are required to balance the transfer of electrons. Therefore, we place the coefficient 2 in front of each silver substance. Since $AgNO_3$ gains only 1 e^-, the coefficient of each copper substance is 1. The balanced equation for the reaction is

$$Cu(s) + 2\ AgNO_3(aq) \longrightarrow Cu(NO_3)_2(aq) + 2\ Ag(s)$$

After placing the coefficients, we verify that the equation is balanced by checking off (√) each element in the reaction. Since the nitrate ion did not change, we can check off the polyatomic ion as a single unit. Thus,

$$\overset{\surd}{Cu}(s) + 2\ \overset{\surd}{Ag}\overset{\surd}{NO_3}(aq) \longrightarrow \overset{\surd}{Cu}(\overset{\surd}{NO_3})_2(aq) + 2\ \overset{\surd}{Ag}(s)$$

Since all the elements are balanced, we have verified that this is a balanced equation. Example Exercise 17.5 provides a more challenging illustration of balancing a redox reaction.

EXAMPLE EXERCISE 17.5

An industrial blast furnace reduces iron ore, Fe_2O_3, to molten iron. Balance the equation for the reaction using the oxidation number method.

$$Fe_2O_3(s) + CO(g) \longrightarrow Fe(l) + CO_2(g)$$

Solution: In this reaction, the oxidation number of iron is decreasing from +3 in Fe_2O_3 to 0 in Fe. At the same time, the oxidation number of carbon is increasing from +2 to +4. Thus, each Fe atom gains 3 electrons, while each C atom loses 2 electrons. We can diagram the redox process as follows:

$$\overset{+3}{Fe_2}O_3(s) + \overset{+2}{C}O(g) \longrightarrow \overset{0}{Fe}(l) + \overset{+4}{C}O_2(g)$$

(Fe: +3 → 0, gains 3 e^-; C: +2 → +4, loses 2 e^-)

Since the number of electrons gained and lost must be equal, we find the lowest common multiple. In this case it is six. Since each CO loses 2 e^-, we place the coefficient 3 in front of CO and CO_2. This gives us

$$Fe_2O_3(s) + 3\ CO(g) \longrightarrow Fe(l) + 3\ CO_2(g)$$

Since each Fe gains 3 e^-, the coefficient in front of each iron atom is 2. Since Fe_2O_3 has two iron atoms, it does not require the coefficient. We must only place a 2 in front of the Fe. Thus,

$$Fe_2O_3(s) + 3\ CO(g) \longrightarrow 2\ Fe(l) + 3\ CO_2(g)$$

Now, verify that this is the balanced equation for the reaction. Check (√) each element in the reaction:

$$\overset{\surd}{Fe_2}\overset{\surd}{O_3}(s) + 3\ \overset{\surd\surd}{CO}(g) \longrightarrow 2\ \overset{\surd}{Fe}(l) + 3\ \overset{\surd\surd}{CO_2}(g)$$

Since all the elements are balanced, we have verified that this is a balanced redox equation.

SELF-TEST EXERCISE

Balance the following redox equations by the oxidation number method.

$$Cl_2O_5(g) + CO(g) \longrightarrow Cl_2(g) + CO_2(g)$$

Answer: $Cl_2O_5(g) + 5\ CO(g) \longrightarrow Cl_2(g) + 5\ CO_2(g)$

Ionic Equations

Now let's consider a redox reaction written as an ionic equation. As an example, we'll consider a copper penny reacting with concentrated nitric acid according to the following unbalanced equation:

$$Cu(s) + H^+(aq) + NO_3^-(aq) \longrightarrow Cu^{2+}(aq) + NO_2(g) + H_2O(l)$$

In this ionic equation, we see the oxidation number of copper metal is increasing from 0 to +2. Although it is not as evident, the oxidation number of nitrogen is decreasing from +5 to +4. Thus, Cu loses 2 electrons while each N gains 1 electron. We can diagram the loss and gain of electrons as follows:

+5 gains 1 e^- +4

0 loses 2 e^- +2

$$Cu(s) + NO_3^-(aq) \longrightarrow Cu^{2+}(aq) + NO_2(g)$$

We can balance the loss and gain of electrons by placing the coefficient 2 in front of each nitrogen substance. It is understood that the coefficient in front of each copper is 1. Thus,

$$Cu(s) + H^+(aq) + 2\ NO_3^-(aq) \longrightarrow Cu^{2+}(aq) + 2\ NO_2(g) + H_2O(l)$$

Notice that in this example H^+ and H_2O also take part in the reaction. Hence, they must also be balanced. We finish balancing the equation by inspection. Since there is a total of six oxygen atoms as reactants, there must be six oxygen atoms in the products. Therefore, we need a coefficient 2 in front of H_2O. After balancing the remaining hydrogen atoms in the reaction, we have

$$Cu(s) + 4\ H^+(aq) + 2\ NO_3^-(aq) \longrightarrow Cu^{2+}(aq) + 2\ NO_2(g) + 2\ H_2O(l)$$

Now, we verify that the equation is balanced by checking off (√) each element in the reaction. That is,

$$\overset{\checkmark}{Cu}(s) + 4\ \overset{\checkmark}{H^+}(aq) + 2\ \overset{\checkmark\checkmark}{NO_3^-}(aq) \longrightarrow \overset{\checkmark}{Cu^{2+}}(aq) + 2\ \overset{\checkmark\checkmark}{NO_2}(g) + 2\ \overset{\checkmark\checkmark}{H_2O}(l)$$

As a final verification, we check to be sure the total charge on the reactants equals the total charge on the products. In the above reaction, we have $+4 - 2 = +2$, on the reactants side. The charge on the products side is also +2. Thus, the equation is properly balanced.

EXAMPLE EXERCISE 17.6

A sample of sodium iodide is analyzed by titrating with a standard potassium dichromate solution. Write a balanced equation given the following ionic reaction.

$$I^-(aq) + H^+(aq) + Cr_2O_7^{2-}(aq) \longrightarrow I_2(aq) + Cr^{3+}(aq) + H_2O(l)$$

Solution: In the given ionic equation, the oxidation number of iodine is increasing from -1 to 0. Since I_2 is diatomic, there are 2 e^- transferred per molecule. Although it is not obvious, the oxidation number of chromium is decreasing from $+6$ to $+3$. Since $Cr_2O_7^{2-}$ has two chromium atoms, there are 6 e^- transferred per polyatomic ion. We can show the loss and gain of electrons as follows:

+6 gains 6 e^- +3

−1 loses 2 e^- 0

$$2\ I^-(aq) + Cr_2O_7^{2-}(aq) \longrightarrow I_2(aq) + 2\ Cr^{3+}(aq)$$

In order that the number of electrons lost be equal to the number of electrons gained (6), we must place a coefficient 3 in front of each iodine substance. The coefficient of $Cr_2O_7^{2-}$ is 1. The equation for the reaction is now written as

$$6\ I^-(aq) + H^+(aq) + Cr_2O_7^{2-}(aq) \longrightarrow 3\ I_2(aq) + 2\ Cr^{3+}(aq) + H_2O(l)$$

Next, we balance the H^+ and H_2O that take part in the reaction. Since there are seven oxygen atoms as reactants, we need a coefficient of 7 in front of H_2O. We then need a coefficient of 14 in front of the H^+. Thus, we have

$$6\ I^-(aq) + 14\ H^+(aq) + Cr_2O_7^{2-}(aq) \longrightarrow 3\ I_2(aq) + 2\ Cr^{3+}(aq) + 7\ H_2O(l)$$

We can verify that the equation is balanced by checking (√) each element. That is,

$$\overset{\checkmark}{6\ I^-}(aq) + \overset{\checkmark}{14\ H^+}(aq) + \overset{\checkmark\ \checkmark}{Cr_2O_7^{2-}}(aq) \longrightarrow \overset{\checkmark}{3\ I_2}(aq) + \overset{\checkmark}{2\ Cr^{3+}}(aq) + \overset{\checkmark\ \checkmark}{7\ H_2O}(l)$$

Now, let's verify the charge balance. The total ionic charge on the reactants' side of the equation is: $-6 + 14 - 2 = +6$. The net charge on both the reactants and products is $+6$; thus, the charge balance is verified.

SELF-TEST EXERCISE

Balance the following redox equations by the oxidation number method.

$$H^+(aq) + MnO_4^-(aq) + NO_2^-(aq) \longrightarrow Mn^{2+}(aq) + NO_3^-(aq) + H_2O(l)$$

Answer:

$$6\ H^+(aq) + 2\ MnO_4^-(aq) + 5\ NO_2^-(aq) \longrightarrow 2\ Mn^{2+}(aq) + 5\ NO_3^-(aq) + 3\ H_2O(l)$$

note We can write and balance oxidation–reduction reactions in either the molecular or ionic form. Since redox reactions are often complex, you are not expected to complete the equation before balancing. In general, you will be given the reactants and products for each redox reaction.

17.4 Balancing Redox Equations: Ion–Electron Method

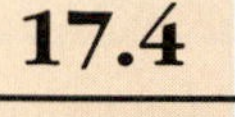

OBJECTIVES

To write a balanced ionic equation for a redox reaction using the ion–electron (half-reaction) method:

(a) in acidic solution
(b) in basic solution

In Section 15.11, we learned how to write net ionic equations. That is, we wrote ionized substances in solution as aqueous ions. For example, we wrote hydrochloric acid as HCl(aq). There is, however, a more accurate way to show ionized substances in an ionic equation. For example, HCl(aq) is shown as $H^+(aq)$ and $Cl^-(aq)$. In this section, we are going to balance ionic redox equations using a set procedure. It is referred to as the *ion–electron* or *half-reaction method.*

half-reaction A chemical equation with electrons that represents an oxidation process or a reduction process.

A **half-reaction** is a part of the redox reaction that shows the oxidation process or reduction process by itself. The ion–electron method systematically balances the oxidation half-reaction and the reduction half-reaction separately. As we did in Section 17.3, we can state general guidelines for balancing a redox reaction. They are given in Table 17.4.

Let's apply the guidelines in Table 17.4 to obtain a balanced chemical equation. For example, consider a copper penny that reacts with dilute nitric acid to give nitrogen monoxide gas. The ionic equation is as follows:

$$Cu(s) + NO_3^-(aq) \longrightarrow Cu^{2+}(aq) + NO(g)$$

Step 1: Write the oxidation and reduction half-reactions. Note the oxidation number of Cu increases, so Cu is oxidized and NO_3^- must be reduced. That is,

Table 17.4 Guidelines for Balancing Redox Equations by the Ion–Electron (Half-Reaction) Method

Guidelines
1. Write the half-reaction for both oxidation and reduction. **(a)** Identify the reactant that is oxidized and its product. **(b)** Identify the reactant that is reduced and its product.
2. Balance the atoms in each half-reaction using coefficients. **(a)** Balance all elements except oxygen and hydrogen. **(b)** Balance oxygen using H_2O. **(c)** Balance hydrogen using H^+. *Note:* For reactions in a basic solution, add OH^- to neutralize H^+. For example, 2 OH^- neutralizes 2 H^+ to give 2 H_2O. **(d)** Balance the ionic charges using e^-.
3. Multiply each half-reaction by a whole number so that the number of electrons lost by oxidation equals the number of electrons gained by reduction.
4. Add the two half-reactions together and cancel identical species, including electrons, on each side of the equation.
5. After balancing the equation, verify that the coefficients are correct. **(a)** Place a check mark (✓) above the symbol of each element to indicate that the number of atoms is the same for reactants and products. **(b)** Calculate the net ionic charge for the reactants and products to verify that they are equal.

Oxidation:	$Cu \longrightarrow Cu^{2+}$
Reduction:	$NO_3^- \longrightarrow NO$

Step 2: Balance each half-reaction for atoms and charge. For Cu, we need only add $2\ e^-$. For NO_3^-, we must balance the equation using H_2O, H^+, and e^-. The balanced half-reactions are

$$Cu \longrightarrow Cu^{2+} + 2\ e^-$$

$$3\ e^- + 4\ H^+ + NO_3^- \longrightarrow NO + 2\ H_2O$$

Step 3: Note that Cu loses $2\ e^-$ and NO_3^- gains $3\ e^-$. Thus, we multiply the Cu half-reaction by 3 and the NO_3^- by 2. This gives

$$3\ Cu \longrightarrow 3\ Cu^{2+} + 6\ e^-$$

$$6\ e^- + 8\ H^+ + 2\ NO_3^- \longrightarrow 2\ NO + 4\ H_2O$$

Step 4: Add the two half-reactions together and cancel the $6\ e^-$. We then have

$$3\ Cu \longrightarrow 3\ Cu^{2+} + \cancel{6\ e^-}$$

$$\cancel{6\ e^-} + 8\ H^+ + 2\ NO_3^- \longrightarrow 2\ NO + 4\ H_2O$$

$$3\ Cu + 8\ H^+ + 2\ NO_3^- \longrightarrow 3\ Cu^{2+} + 2\ NO + 4\ H_2O$$

Step 5: Check the atoms and ionic charges to verify that the equation is balanced. That is,

Atoms:	3 Cu, 8 H, 2 N, 6 O = 3 Cu, 2 N, 6 O, 8 H
Charges:	$+8 - 2 = +6$

Since the atoms and ionic charges are equal, the equation is correctly balanced.

Example Exercise 17.7 further illustrates balancing redox equations in acid using the half-reaction method.

EXAMPLE EXERCISE 17.7

Write a balanced ionic equation for the reaction of iron(II) sulfate and potassium permanganate in acidic solution. The ionic equation is

$$Fe^{2+}(aq) + MnO_4^-(aq) \longrightarrow Fe^{3+}(aq) + Mn^{2-}(aq)$$

Solution: Follow the procedure in Table 17.4 for balancing redox reactions by the ion–electron method.

Step 1: Write the oxidation and reduction half-reactions. First note that Fe^{2+} is being oxidized; therefore, MnO_4^- must be reduced. That is,

Oxidation:	$Fe^{2+} \longrightarrow Fe^{3+}$
Reduction:	$MnO_4^- \longrightarrow Mn^{2+}$

Step 2: Balance each half-reaction for mass and charge. We achieve a balance using H_2O, H^+, and e^- as necessary. Thus,

$$Fe^{2+} \longrightarrow Fe^{3+} + e^-$$

$$5\ e^- + 8\ H^+ + MnO_4^- \longrightarrow Mn^{2+} + 4\ H_2O$$

Step 3: Note that Fe^{2+} loses 1 e^- and MnO_4^- gains 5 e^-. The lowest common multiple is 5. Thus, we need only to multiply the Fe^{2+} half-reaction. This gives us

$$5\ Fe^{2+} \longrightarrow 5\ Fe^{3+} + 5\ e^-$$

$$5\ e^- + 8\ H^+ + MnO_4^- \longrightarrow Mn^{2+} + 4\ H_2O$$

Step 4: Add the two half-reactions together and cancel the 5 e^-. That is,

$$\cancel{5\ Fe^{2+}} \longrightarrow 5\ Fe^{3+} + \cancel{5\ e^-}$$

$$\cancel{5\ e^-} + 8\ H^+ + MnO_4^- \longrightarrow Mn^{2+} + 4\ H_2O$$

$$5\ Fe^{2+} + 8\ H^+ + MnO_4^- \longrightarrow 5\ Fe^{3+} + Mn^{2+} + 4\ H_2O$$

Step 5: Finally, check the atoms and ionic charges to verify that the equation is balanced. That is,

Atoms: 5 Fe, 8 H, 1 Mn, 4 O = 5 Fe, 1 Mn, 8 H, 4 O

Charges: $+10 + 8 - 1 = +15 + 2$

Since both the atoms and ionic charges are equal, the equation is balanced.

SELF-TEST EXERCISE

Write a balanced ionic equation for the reaction of sodium nitrite and potassium perchlorate in acidic solution. The ionic equation is

$$ClO_4^-(aq) + NO_2^-(aq) \longrightarrow Cl^-(aq) + NO_3^-(aq)$$

Answer: $ClO_4^- + 4\ NO_2^- \longrightarrow Cl^- + 4\ NO_3^-$

Example Exercise 17.8 illustrates balancing redox equations in basic solution using the half-reaction method.

EXAMPLE EXERCISE 17.8

Write a balanced ionic equation for the reaction of sodium iodide and potassium permanganate in basic solution. The ionic equation is

$$I^-(aq) + MnO_4^-(aq) \longrightarrow I_2(aq) + MnO_2(s)$$

Solution: Follow the procedure in Table 17.4 for balancing redox reactions by the ion–electron method.

Step 1: Write the oxidation and reduction half-reactions. First note that I^- loses electrons and is thus oxidized. Simultaneously, MnO_4^- is reduced to MnO_2. That is,

Oxidation: $I^- \longrightarrow I_2$

Reduction: $MnO_4^- \longrightarrow MnO_2$

Step 2: Balance each half-reaction for mass and charge. Again, using H_2O, H^+, and e^- as necessary, we have

$$2\ I^- \longrightarrow I_2 + 2\ e^-$$

$$3\ e^- + 4\ H^+ + MnO_4^- \longrightarrow MnO_2 + 2\ H_2O$$

Note that this redox reaction takes place in basic solution, not acid. Therefore, we follow guideline 2(d), which states that the H^+ should be neutralized with OH^-. In this case, we add 4 OH^- to each side.

$$3\,e^- + 4\,H^+ + 4\,OH^- + MnO_4^- \longrightarrow MnO_2 + 2\,H_2O + 4\,OH^-$$

$$3\,e^- + 4\,H_2O + MnO_4^- \longrightarrow MnO_2 + 2\,H_2O + 4\,OH^-$$

$$3\,e^- + 2\,H_2O + MnO_4^- \longrightarrow MnO_2 + 4\,OH^-$$

Step 3: Note that 2 I^- lose 2 e^- and MnO_4^- gains 3 e^-. To have an equal electron transfer, we multiply the I^- half-reaction by 3 and the MnO_4^- by 2. This gives

$$6\,I^- \longrightarrow 3\,I_2 + 6\,e^-$$

$$6\,e^- + 4\,H_2O + 2\,MnO_4^- \longrightarrow 2\,MnO_2 + 8\,OH^-$$

Step 4: Add the two half-reactions together and cancel the 6 e^-. That is,

$$\cancel{6\,I^-} \longrightarrow 3\,I_2 + \cancel{6\,e^-}$$

$$\cancel{6\,e^-} + 4\,H_2O + 2\,MnO_4^- \longrightarrow 2\,MnO_2 + 8\,OH^-$$

$$6\,I^- + 4\,H_2O + 2\,MnO_4^- \longrightarrow 3\,I_2 + 2\,MnO_2 + 8\,OH^-$$

Step 5: Check to verify that the equation is balanced:

Atoms: 6 I, 8 H, 2 Mn, 12 O = 6 I, 2 Mn, 12 O, 8 H

Charges: $-6 - 2 = -8$

Since both the atoms and ionic charges are equal, the equation is balanced.

SELF-TEST EXERCISE

Write a balanced ionic equation for the reaction of sodium bromite and potassium permanganate in basic solution. The ionic equation is

$$MnO_4^-(aq) + BrO_2^-(aq) \longrightarrow MnO_2(s) + BrO_4^-(aq)$$

Answer: $2\,H_2O + 4\,MnO_4^- + 3\,BrO_2^- \longrightarrow 4\,MnO_2 + 3\,BrO_4^- + 4\,OH^-$

17.5 Predicting Spontaneous Redox Reactions

To predict the stronger oxidizing agent and reducing agent, given examples from the activity series.

To predict whether a reaction is spontaneous or nonspontaneous, given the activity series.

In Section 15.6, we discussed the relative strength of Brønsted–Lowry acids and bases. Here we will see that the strength of an oxidizing agent is also relative to other substances.

First, consider a metal reacting in aqueous solution. For example, consider the reaction of zinc metal in aqueous copper sulfate. The net ionic equation for the reaction is

$$Zn\ (s) + Cu^{2+}(aq) \longrightarrow Zn^{2+}(aq) + Cu\ (s)$$

It is experimentally observed that the reaction of aqueous copper(II) sulfate and zinc metal proceeds spontaneously. We can therefore conclude that $Cu^{2+}(aq)$ has a greater tendency to undergo reduction than does $Zn^{2+}(aq)$. That is, if we compare the two half-reactions

$$Cu^{2+}(aq) + 2\ e^- \longrightarrow Cu(s)$$

$$Zn^{2+}(aq) + 2\ e^- \longrightarrow Zn(s)$$

we see the formation of Cu(s) has a greater tendency to occur than the formation of Zn(s).

By trying different combinations of metals and aqueous salt solutions, we can arrange a series of elements based on their ability to gain electrons. The strongest oxidizing agent is the substance most easily reduced. The strongest reducing agent is the substance most easily oxidized. Table 17.5 lists several half-reactions in order of their ability to undergo reduction. This list of half-reactions is called an **activity series**, or electromotive series. A substance higher in the table is more easily reduced. A substance lower in the table is less active and not as easily reduced. We can interpret the information given in Table 17.5 as follows.

activity series A list of half-reactions for metals and nonmetals arranged in order of their ability to undergo reduction; a relative scale of the strength of oxidizing agents.

Table 17.5 Relative Strength of Selected Oxidizing and Reducing Agents (1.00 *M* Aqueous Solutions at 25°C)

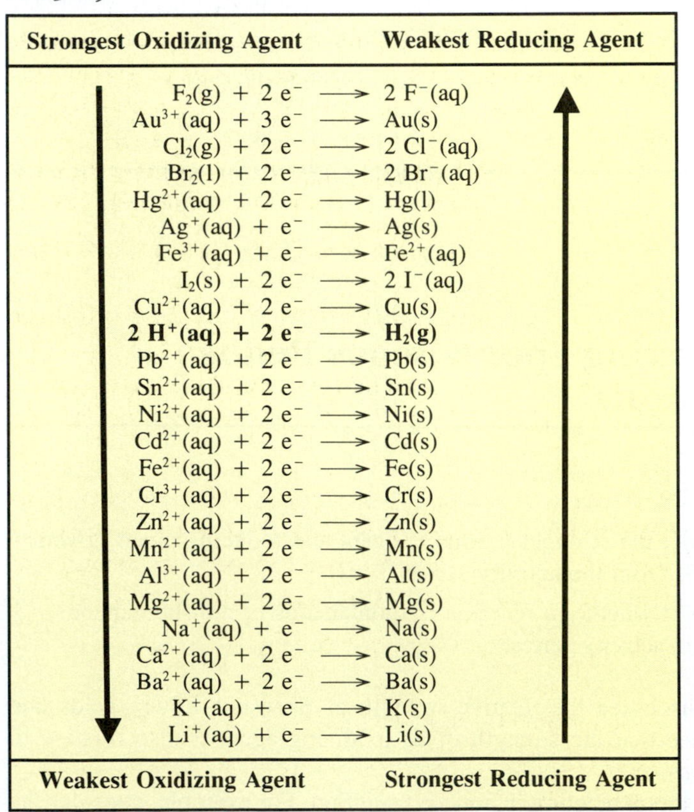

Strongest Oxidizing Agent		Weakest Reducing Agent
$F_2(g) + 2\ e^-$	$\longrightarrow$	$2\ F^-(aq)$
$Au^{3+}(aq) + 3\ e^-$	$\longrightarrow$	$Au(s)$
$Cl_2(g) + 2\ e^-$	$\longrightarrow$	$2\ Cl^-(aq)$
$Br_2(l) + 2\ e^-$	$\longrightarrow$	$2\ Br^-(aq)$
$Hg^{2+}(aq) + 2\ e^-$	$\longrightarrow$	$Hg(l)$
$Ag^+(aq) + e^-$	$\longrightarrow$	$Ag(s)$
$Fe^{3+}(aq) + e^-$	$\longrightarrow$	$Fe^{2+}(aq)$
$I_2(s) + 2\ e^-$	$\longrightarrow$	$2\ I^-(aq)$
$Cu^{2+}(aq) + 2\ e^-$	$\longrightarrow$	$Cu(s)$
$2\ H^+(aq) + 2\ e^-$	$\longrightarrow$	**$H_2(g)$**
$Pb^{2+}(aq) + 2\ e^-$	$\longrightarrow$	$Pb(s)$
$Sn^{2+}(aq) + 2\ e^-$	$\longrightarrow$	$Sn(s)$
$Ni^{2+}(aq) + 2\ e^-$	$\longrightarrow$	$Ni(s)$
$Cd^{2+}(aq) + 2\ e^-$	$\longrightarrow$	$Cd(s)$
$Fe^{2+}(aq) + 2\ e^-$	$\longrightarrow$	$Fe(s)$
$Cr^{3+}(aq) + 2\ e^-$	$\longrightarrow$	$Cr(s)$
$Zn^{2+}(aq) + 2\ e^-$	$\longrightarrow$	$Zn(s)$
$Mn^{2+}(aq) + 2\ e^-$	$\longrightarrow$	$Mn(s)$
$Al^{3+}(aq) + 2\ e^-$	$\longrightarrow$	$Al(s)$
$Mg^{2+}(aq) + 2\ e^-$	$\longrightarrow$	$Mg(s)$
$Na^+(aq) + e^-$	$\longrightarrow$	$Na(s)$
$Ca^{2+}(aq) + 2\ e^-$	$\longrightarrow$	$Ca(s)$
$Ba^{2+}(aq) + 2\ e^-$	$\longrightarrow$	$Ba(s)$
$K^+(aq) + e^-$	$\longrightarrow$	$K(s)$
$Li^+(aq) + e^-$	$\longrightarrow$	$Li(s)$
Weakest Oxidizing Agent		**Strongest Reducing Agent**

1. The ability of a substance to be reduced decreases from top to bottom in the table. The ability of a substance to be oxidized increases from bottom to top in the table.
2. A metal ion is reduced by any reducing agent lower than it in the table. For example, Au^{3+} can be reduced by Cl^-, Br^-, or Ag metal.
3. A nonmetal can oxidize any reducing agent below it in the table. For example, F_2 can oxidize any reducing agent, including Au, Cl^-, Br^-, Hg, Fe^{2+}, and Li metal.
4. An acid will react with any metal below H^+ in the table. It will not react with a metal above H^+. For example, an acid will react with Pb, Sn, and Al metal. An acid will not react with Cu, Ag, or Au.

From Table 17.5 we see that $Cu^{2+}(aq)$ is a stronger oxidizing agent than $Zn^{2+}(aq)$. We also see that Zn metal is a stronger reducing agent than Cu metal. Thus, we can explain the above reaction as follows:

$Cu^{2+}(aq)$	+	$Zn(s)$	$\longrightarrow$	$Cu(s)$	+	$Zn^{2+}(aq)$
stronger oxidizing agent		stronger reducing agent		*weaker* reducing agent		*weaker* oxidizing agent

Moreover, we see that the reverse process has very little tendency to occur and is nonspontaneous. We can therefore conclude that there is very little reaction between aqueous zinc ion and copper metal. Thus,

$Zn^{2+}(aq)$	+	$Cu(s)$	$\not\longrightarrow$	$Zn(s)$	+	$Cu^{2+}(aq)$
weaker oxidizing agent		*weaker* reducing agent		stronger reducing agent		stronger oxidizing agent

Now consider a nonmetal reaction in aqueous solution. For example, consider what happens if we bubble chlorine gas through an aqueous sodium bromide solution. The net ionic equation for the possible reaction is

$$Cl_2(g) + 2\ Br^-(aq) \longrightarrow 2\ Cl^-(aq) + Br_2(l)$$

From Table 17.5 we see that $Cl_2(g)$ is a stronger oxidizing agent than $Br_2(l)$. We also see that $Br^-(aq)$ is a stronger reducing agent than $Cl^-(aq)$. Thus, we can explain the above reaction as follows:

$Cl_2(g)$	+	$2\ Br^-(aq)$	$\longrightarrow$	$2\ Cl^-(aq)$	+	$Br_2(l)$
stronger oxidizing agent		stronger reducing agent		*weaker* reducing agent		*weaker* oxidizing agent

Since the reactants contain the stronger pair of oxidizing and reducing agents, the reaction takes place spontaneously. Is the reverse reaction spontaneous? Since the chloride ion and bromine liquid are the weaker pair of oxidizing and reducing agents, the reverse reaction is nonspontaneous. If a redox reaction is not spontaneous, we usually say there is no reaction. That is,

$2\ Cl^-(aq)$	+	$Br_2(l)$	$\not\longrightarrow$	$Cl_2(g)$	+	$2\ Br^-(aq)$
weaker reducing agent		*weaker* oxidizing agent		stronger oxidizing agent		stronger reducing agent

Example Exercise 17.9 provides additional practice in predicting whether or not a redox reaction occurs spontaneously.

EXAMPLE EXERCISE 17.9

Predict whether the following reaction is spontaneous or nonspontaneous.

$$Ni^{2+}(aq) + Sn(s) \longrightarrow Ni(s) + Sn^{2+}(aq)$$

Solution: Refer to the relative activity of each element in Table 17.5 to predict whether or not the reaction is spontaneous.

Table 17.5 lists $Ni^{2+}(aq)$ as a weaker oxidizing agent than $Sn^{2+}(aq)$. It lists Sn(s) as a weaker reducing agent than Ni(s). Hence, we can explain the above reaction as follows:

$Ni^{2+}(aq)$	+	Sn(s)	$\longrightarrow$	Ni(s)	+	$Sn^{2+}(aq)$
weaker oxidizing agent		*weaker* reducing agent		stronger reducing agent		stronger oxidizing agent

Since the reactants contain the weaker pair of oxidizing and reducing agents, the reaction is *nonspontaneous*. On the other hand, the reverse reaction between the stronger oxidizing and reducing agents does occur.

SELF-TEST EXERCISE

Refer to Table 17.5 and state whether the following reaction is spontaneous or nonspontaneous.

$$Cd(s) + Ni^{2+}(aq) \longrightarrow Ni(s) + Cd^{2+}(aq)$$

Answer: spontaneous

17.6 Voltaic Cells

To indicate the anode, cathode, oxidation half-reaction, and reduction half-reaction in spontaneous voltaic cells.

Batteries play an important role in our lives. They enable us to start an automobile simply by turning a switch to crank the engine. They extend the lives of heart patients who receive rhythmic electrical pulses from artificial pacemakers. They make our lives more comfortable by letting us use cordless electrical appliances. All these devices use electrical energy supplied by batteries. Although different devices use different types of batteries, all batteries use a spontaneous redox reaction as their source of electrical energy. Specifically, they convert chemical energy to electrical energy.

Redox Reactions in Voltaic Cells

electrochemistry The study of the interconversion of chemical energy and electrical energy by redox reactions.

The study of the conversion of chemical energy to electrical energy and (vice versa) from a redox reaction is called **electrochemistry**. As an example, consider the reaction of zinc metal in a copper(II) sulfate solution.

$$Zn(s) + CuSO_4(aq) \longrightarrow Cu(s) + ZnSO_4(aq)$$

Experimentally, we can physically separate the oxidation half-reaction from the reduction half-reaction. We do this by placing aqueous solutions of zinc sulfate

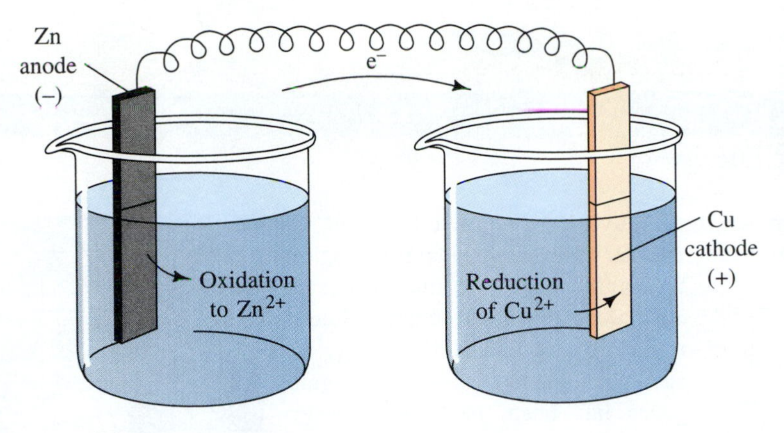

Figure 17.3 Separating a Redox Reaction into Half-Cells In the compartment on the left, zinc metal is being oxidized. In the compartment on the right, copper ion is being reduced. This redox process causes electrons to flow from the zinc electrode to the copper electrode. As the reaction proceeds, there is an increase in the number of positive zinc ions in the left half-cell and a decrease in positive copper(II) ions in the right half-cell.

and copper(II) sulfate in separate containers or compartments. Then, we place a Zn metal electrode into the first compartment and a Cu metal electrode into the second. The two electrodes are connected by a conducting wire, so electrons are free to travel between the two compartments of the **electrochemical cell**. Each compartment is called a **half-cell**. Figure 17.3 shows an electrochemical cell for the zinc and copper reaction.

electrochemical cell A general term for an apparatus containing two solutions with electrodes in separate compartments connected by a conducting wire.

Let's assume that the zinc half-cell contains 1.00 M $ZnSO_4$ and the copper compartment has 1.00 M $CuSO_4$. Initially, the ionic concentrations are as follows:

Zn half-cell: $[Zn^{2+}] = 1.00\ M, \quad [SO_4^{2-}] = 1.00\ M$

Cu half-cell: $[Cu^{2+}] = 1.00\ M, \quad [SO_4^{2-}] = 1.00\ M$

half-cell A portion of an electrochemical cell having a single electrode where either oxidation or reduction is occurring.

As the reaction proceeds, the concentration of positive Zn^{2+} ions in the left compartment becomes greater than 1.00 M. The concentration of Cu^{2+} ions in the right compartment becomes less than 1.00 M. The concentration of sulfate ion, SO_4^{2-}, remains constant at 1.00 M in each compartment. The result is that the left half-cell builds up a net positive charge as the right half-cell develops a net negative charge.

With the buildup of electrical charge in each compartment, the redox process comes to a halt. However, we can equalize the charge buildup by introducing a **salt bridge**. A salt bridge placed between each compartment allows ions to travel between the two half-cells. The excess negative sulfate ions in the copper solution can move to the zinc solution, which has excess Zn^{2+} ions. Since a salt bridge eliminates charge buildup, the reaction can continue spontaneously. For spontaneous reactions that produce electrical energy, the cell is termed a **voltaic cell** (or galvanic cell). Figure 17.4 shows the voltaic cell for the zinc and copper reaction.

salt bridge A porous device that allows ions to travel between two half-cells in order to maintain ionic charge balance in each compartment of an electrochemical cell.

voltaic cell An electrochemical cell in which a spontaneous redox reaction occurs and generates electrical energy.

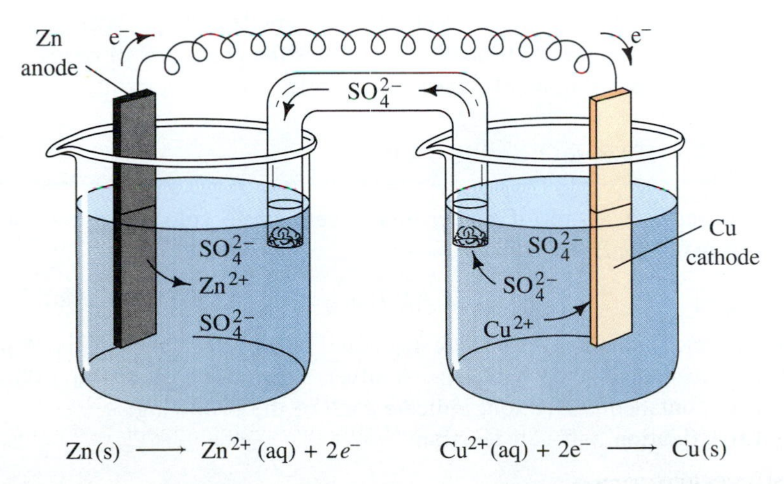

Figure 17.4 Voltaic Cell As the redox reaction proceeds, in the left compartment the number of positive zinc ions becomes greater than the number of negative sulfate ions. Conversely, in the right compartment, there are fewer positive copper ions than negative sulfate ions. The net result is that the left half-cell builds up a positive charge and the right half-cell a negative charge. The buildup of charge stops the redox process. However, if we connect the two half-cells by a salt bridge, the negative sulfate ions travel from the right half-cell to the left. A salt bridge reduces charge buildup, and the reaction continues spontaneously.

CHEMISTRY CONNECTION □ APPLIED

Hydrogen for Energy

▶ **Can you guess why helium gas, and not hydrogen gas, is used in lighter-than-air dirigibles?**

Among industrial nations there is an ever-increasing need to obtain more energy from nature without harming the environment. We are all aware that the world's petroleum reserves are dwindling, and although there is a substantial amount of coal available, burning coal contributes to acid rain. We have attempted to exploit nuclear energy, but there are technological as well as political problems. This combination of factors has led to a search for alternative energy sources. One of the most promising alternatives is the use of hydrogen as a fuel. Hydrogen gas is highly combustible and clean-burning. When hydrogen and oxygen gases react directly, water and heat energy are the only two products. The direct reaction of hydrogen and oxygen supplies the energy for launching the space shuttles.

A hydrogen-oxygen fuel cell is another example of using hydrogen to obtain energy. A fuel cell is a special type of electrochemical cell that produces electrical energy. A fuel cell using hydrogen and oxygen gases has been used to generate electricity on the Gemini and Apollo space missions. Unlike a lead storage battery in an automobile, a fuel cell does not require recharging. Rather, hydrogen and oxygen gases are supplied continuously.

In a hydrogen-oxygen fuel cell, hydrogen is oxidized in one compartment, while oxygen is reduced in the other. Each compartment is usually lined with graphite, which serves as the electrode. The electrochemical process is as follows:

Anode reaction (*oxidation*):

$$2\ H_2(g) + 4\ OH^-(aq) \longrightarrow 4\ H_2O(l) + 4\ e^-$$

Cathode reaction (*reduction*):

$$O_2(g) + 2\ H_2O(l) + 4\ e^- \longrightarrow 4\ OH^-(aq)$$

A typical space shuttle launch requires about 400,000 gallons of liquid hydrogen and 150,000 gallons of liquid oxygen.

In an electrochemical cell, the electrode at which oxidation occurs is called the **anode**. The electrode at which reduction occurs is called the **cathode**. For the cell in Figure 17.4, the Zn electrode is the anode and the Cu electrode is the cathode. Example Exercise 17.10 further illustrates the processes of oxidation and reduction that takes place in an electrochemical cell.

anode The electrode in an electrochemical cell at which oxidation occurs.

cathode The electrode in an electrochemical cell at which reduction occurs.

EXAMPLE EXERCISE 17.10

The redox reaction of nickel and silver nitrate solution occurs according to the following ionic equation:

$$Ni(s) + 2\ Ag^+(aq) \longrightarrow 2\ Ag(s) + Ni^{2+}(aq)$$

Assume the half-reactions are separated into two compartments. A nickel electrode is placed in 1.00 *M* $Ni(NO_3)_3$. A silver electrode is placed in 1.00 *M* $AgNO_3$. For the spontaneous reaction, indicate each of the following:

(a) oxidation half-cell reaction

After simplifying, we find that the overall equation for the fuel cell is:

$$2\ H_2(g) + O_2(g) \longrightarrow 2\ H_2O(l)$$

Hydrogen-oxygen fuel cells are very efficient and about 95% of the energy produced can be converted to electrical energy. In addition, fuel cells are nearly pollution free. As our need for energy escalates, along with our concern for the environment, hydrogen as a fuel becomes an attractive alternative energy source.

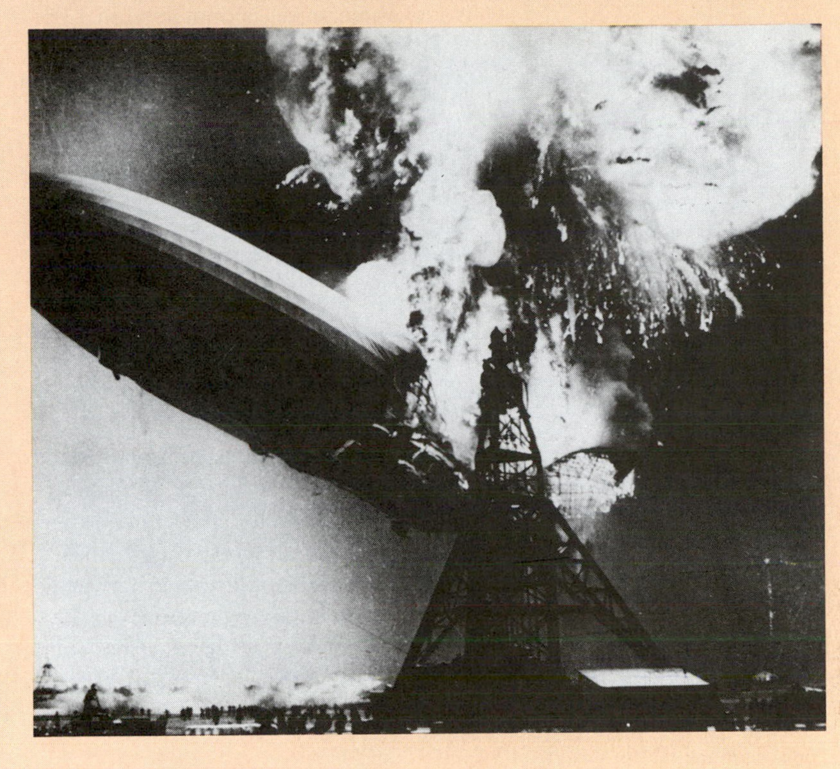

► **Helium is a nonflammable gas. Hydrogen is a flammable gas.**

(b) reduction half-cell reaction
(c) anode and cathode
(d) direction of electron flow
(e) direction of NO_3^- movement in the salt bridge

Solution: Referring to Table 17.5, we see that Ag^+ is higher in the activity series than Ni^{2+}. Therefore the process is spontaneous. Since Ni is being oxidized and Ag^+ is reduced, the half-cell processes are:

(a) Oxidation: $Ni \longrightarrow Ni^{2+} + 2\ e^-$
(b) Reduction: $Ag^+ + e^- \longrightarrow Ag$
(c) By definition, oxidation occurs at the anode. Thus, Ni is the *anode*. Reduction occurs at the cathode. Thus, Ag is the *cathode*.
(d) Since nickel is oxidized in the cell, the electrons flow from the Ni anode to the Ag cathode.
(e) As Ni is oxidized, the $[Ni^{2+}]$ becomes greater than 1.00 M. Thus, the Ni compartment gains a net positive charge. Simultaneously, the Ag compartment obtains

a net negative charge. For the cell to operate, a salt bridge must be used. The salt bridge allows NO_3^- to travel from the Ag compartment to the Ni compartment.

SELF-TEST EXERCISE

A nickel electrode is immersed in 1.00 *M* $NiSO_4$ while an iron electrode is in 1.00 *M* $FeSO_4$. A spontaneous redox reaction of iron and nickel sulfate solution occurs according to the following ionic equation:

$$Fe(s) + Ni^{2+}(aq) \longrightarrow Ni(s) + Fe^{2+}(aq)$$

Write the half-reactions occurring at the anode and cathode for the above voltaic cell which produces nickel metal.

Answer: Anode reaction: $Fe \longrightarrow Fe^{2+} + 2\,e^-$

Cathode reaction: $Ni^{2+} + 2\,e^- \longrightarrow Ni$

Batteries

battery A general term for any electrochemical cell that produces electrical energy spontaneously.

A **battery** is a general term for any electrochemical cell that spontaneously produces electrical energy. A battery is made up of one or more voltaic cells. For example, the ordinary lead storage battery in an automobile is a series of six cells, each of which produces about 2 volts. All six cells, operating together, generate 12 volts.

The batteries we use to power flashlights, electronic toys, portable shavers, and radios do not have electrodes immersed in an electrolyte. Rather, their electrodes are connected by a solid paste that conducts electricity. A battery that does not use an electrolyte solution is called a **dry cell**. The common *alkaline dry cell* battery was invented more than 100 years ago by the French chemist George Leclanché (1839–1882).

dry cell An electrochemical cell in which the anode and cathode reactions do not take place in aqueous solutions.

Figure 17.5 Common Batteries Batteries are voltaic cells that supply electrical energy. In the foreground are rechargeable nicad batteries, alkaline dry cells, and mercury dry cells. In the background is a lead storage battery.

A more compact version of the ordinary alkaline dry cell is the *mercury dry cell*. In this battery, a small zinc cup is filled with a paste of HgO and KOH. Oxidation takes place at the zinc anode and reduction at the steel cathode. The mercury dry cell is used to power small devices such as electronic wristwatches and hearing aids.

Although an ordinary dry cell is not rechargeable, a special type of dry cell called a nickel–cadmium battery, or *nicad battery*, can be recharged almost indefinitely. The nicad battery is used in rechargeable battery packs for calculators, portable stereos, televisions, and other electrical items.

A collection of various common batteries is shown in Figure 17.5.

17.7 Electrolytic Cells

To indicate the anode, cathode, oxidation half-reaction, and reduction half-reaction in a nonspontaneous electrochemical cell.

So far we have discussed electrochemical cells that operate spontaneously. Electrochemical cells that operate nonspontaneously are equally as important. If a reaction is nonspontaneous, an electrochemical cell can be supplied with direct electrical current to force the reaction to occur. If an electrochemical cell requires electric current for a redox reaction to occur, it is called an **electrolytic cell**. Electroplating is an electrolytic process. In that process, for example, chromium is electroplated onto steel automobile parts even though the reaction is not spontaneous.

electrolytic cell An electrochemical cell in which a nonspontaneous redox reaction occurs by the input of direct electric current.

Two important elements, sodium and chlorine, are produced by electrolysis. In the process, crystalline salt is first obtained from the evaporation of seawater. Next, the crystalline salt, NaCl, is placed in a crucible and heated until it melts. Then a pair of inert platinum metal electrodes are dipped into the molten NaCl and a direct current is applied. The electrical current drives the nonspontaneous reaction for the decomposition of NaCl. Thus,

Cathode reaction (reduction):

$$2\ Na^+ + \cancel{2\ e^-} \longrightarrow 2\ Na(s)$$

Anode reaction (oxidation):

$$2\ Cl^- \longrightarrow Cl_2(g) + \cancel{2\ e^-}$$

We obtain the overall equation for the electrolytic cell by adding the two half-reactions together and canceling electrons. Hence,

$$2\ Na^+ + 2\ Cl^- \longrightarrow 2\ Na(s) + Cl_2(g)$$

The following example exercise illustrates nonspontaneous electrochemical cell reactions.

EXAMPLE EXERCISE 17.11

Aluminum metal can be produced by passing direct electric current through bauxite, Al_2O_3, dissolved in the molten mineral cryolite, Na_3AlF_6. That is,

$$2\ Al_2O_3(l) \xrightarrow{\text{electricity}} 4\ Al(l) + 3\ O_2(g)$$

If carbon rods serve as the oxidation electrode and the reduction electrode, indicate each of the following for this nonspontaneous process:

(a) oxidation half-cell reaction
(b) reduction half-cell reaction
(c) anode and cathode
(d) direction of electron flow

Solution: Note that Al^{3+} is being reduced and Cl^- is oxidized. The half-cell processes are

(a) Oxidation: $2\ O^{2-} \longrightarrow O_2 + 4\ e^-$
(b) Reduction: $Al^{3+} + 3\ e^- \longrightarrow Al$
(c) By definition, oxidation occurs at the anode. We are not shown the electrodes, but the anode is where O_2 gas is released. Reduction occurs at the cathode where Al metal is produced.
(d) The electrons flow from the anode to the cathode.

SELF-TEST EXERCISE

Magnesium chloride is obtained from seawater. An electric current is passed through the molten salt to give Mg metal. Reduction occurs at a Pt electrode and oxidation at a C graphite electrode, according to the following reaction:

$$MgCl_2(l) \xrightarrow{\text{electricity}} Mg(l) + Cl_2(g)$$

Write the half-reactions occurring at the anode and cathode for the above electrolytic cell, which produces magnesium metal.

Answer: Oxidation (anode): $2\ Cl^- \longrightarrow Cl_2 + 2\ e^-$

Reduction (cathode): $Mg^{2+} + 2\ e^- \longrightarrow Mg$

Summary

Section 17.1 We began our discussion with familiar chemical reactions. An oxidation–reduction reaction, or redox reaction, is characterized by the transfer of electrons. When a substance is *oxidized*, the **oxidation number** increases. Conversely, when a substance is *reduced*, the oxidation number decreases.

The representative metals usually have two oxidation numbers, zero in the free state and another value as an ion. For example, sodium exists as Na and Na^+. Transition metals often have three or more oxidation numbers. In manganese compounds, for example, the oxidation numbers for Mn can be +7, +6, +4, +3, or +2. Nonmetals can also have several oxidation numbers. For Cl, the possible values include +7, +5, +3, +1, 0, and −1.

Section 17.2 In a **redox reaction**, oxidation and reduction are occurring at the same time. In the process of **oxidation**, a substance loses electrons. In the process of **reduction**, a substance gains electrons. The substance that loses electrons causes another substance to gain electrons and be reduced. Thus, the oxidized substance is referred to as the **reducing agent**. The substance that gains electrons is reduced and causes another substance to be oxidized. Hence, the reduced substance is called the **oxidizing agent**. The overall redox process is illustrated by the following diagram.

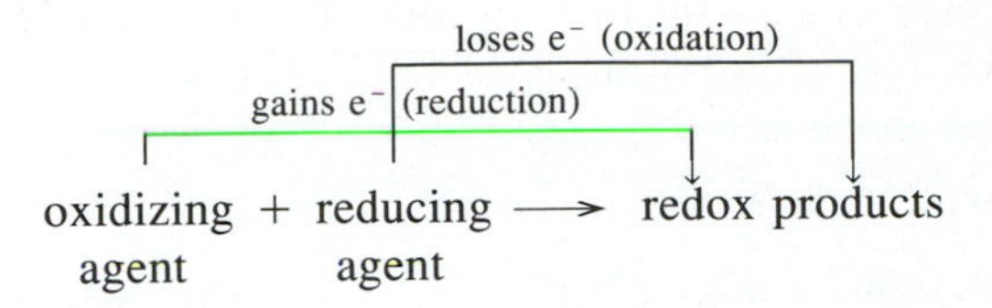

Section 17.3 Many reactions can be balanced by inspection. Redox reactions, however, are often more difficult to balance. The *oxidation number method* lets us balance redox reactions by determining the change in oxidation state for the oxidizing agent and the reducing agent. If the oxidizing agent gains 2 e^- and the reducing agent loses 1 e^-, the reducing agent must be multiplied by 2 to balance the electrons. The remaining elements in the equation can then be balanced by inspection.

Section 17.4 Redox reactions can also be balanced by the *half-reaction method*. First, a balanced **half-reaction** is written for oxidation and another half-reaction for reduction. Each half-reaction is then multiplied by the number of electrons lost or gained in the opposite half-reaction. This equalizes the number of electrons lost in oxidation and gained in reduction. Finally, the two balanced half-reactions are added together. A balanced redox equation is obtained by simplifying common substances on each side of the equation and canceling the electrons.

Section 17.5 Half-reactions can be arranged according to their ability to undergo reduction. A substance that is easily reduced is a strong oxidizing agent. A substance that is easily oxidized is a strong reducing agent. Redox reactions are somewhat reversible. It is the stronger oxidizing agent–reducing agent pair that drives the reaction. If the stronger pair is on the reactant side of the equation, the reaction proceeds spontaneously. If the stronger pair is the products, the reaction is nonspontaneous. To predict whether or not a reaction is spontaneous, we refer to the **activity series** table of half-reactions.

Section 17.6 Chemical and electrical energy are interconverted in an **electrochemical cell**. An electrochemical cell is composed of two **half-cell** compartments. Oxidation occurs in one compartment and reduction in the other. If the redox reaction occurs spontaneously, electricity is produced and the apparatus is called a **voltaic cell** or galvanic cell. If the reaction is nonspontaneous, electricity must be supplied and the apparatus is called an electrolytic cell. In an electrochemical cell, the **anode** is the electrode in the compartment where oxidation occurs. The **cathode** is the electrode in the reduction half-cell. To operate an electrochemical cell, the electrodes are connected by a conducting wire and the two half-cells can be joined by a **salt bridge**. A salt bridge helps maintain charge balance in each compartment by allowing ions to travel from one compartment to the other.

A **battery** is a device composed of one or more voltaic cells. The cell reactions are spontaneous, although the reverse nonspontaneous reaction can be forced to occur by the application of electricity. A **dry cell** is simply a special type of battery in which the conducting electrolyte is a solid paste.

Section 17.7 Whereas voltaic cells use chemical reactions to produce electrical energy, **electrolytic cells** require electrical energy to cause chemical reactions. The redox reactions occurring in an electrolytic cell are nonspontaneous. Electrolytic cells are important for producing active metals and some nonmetals. Metallic sodium, potassium, calcium, and others are obtained from electrolysis cells. Chlorine gas and bromine liquid are obtained commercially by passing a direct electric current through an electrolytic cell.

Key Terms

Select the key term that corresponds to the following definitions.

____ **1.** a chemical reaction in which an electron transfer takes place

____ **2.** a value assigned to an atom in a molecule or ion in order to keep track of electrons gained or lost; the value may be positive or negative and is zero in the free state

____ **3.** a chemical change in which a substance loses electrons or an increase in oxidation number occurs

____ **4.** a chemical change in which a substance gains electrons or a decrease in oxidation number occurs

____ **5.** the substance reduced in a redox reaction

____ **6.** the substance oxidized in a redox reaction

____ **7.** a chemical equation balanced with electrons that represents an oxidation process or a reduction process

____ **8.** a list of half-reactions for metals and nonmetals arranged in order of their ability to undergo reduction; a relative scale of the strength of oxidizing agents; also referred to as an electromotive series

____ **9.** the study of the interconversion of chemical energy and electrical energy from redox reactions

____ **10.** a general term for an apparatus containing two solutions with electrodes in separate compartments connected by a conducting wire

____ **11.** a portion of an electrochemical cell having a single electrode where either oxidation or reduction is occurring

____ **12.** an electrochemical cell in which a spontaneous redox reaction occurs and generates electrical energy

____ **13.** an electrochemical cell in which a nonspontaneous redox reaction occurs by the input of direct electric current

____ **14.** the electrode in an electrochemical cell at which oxidation occurs

____ **15.** the electrode in an electrochemical cell at which reduction occurs

____ **16.** a porous device that allows ions to travel between two half-cells in order to maintain ionic charge balance in each compartment

____ **17.** a general term for any electrochemical cell that produces electrical energy spontaneously

____ **18.** an electrochemical cell where the anode and cathode reactions do not take place in aqueous solutions

(a) activity series *(Sec. 17.5)*
(b) anode *(Sec. 17.6)*
(c) battery *(Sec. 17.6)*
(d) cathode *(Sec. 17.6)*
(e) dry cell *(Sec. 17.6)*
(f) electrochemical cell *(Sec. 17.6)*
(g) electrochemistry *(Sec. 17.6)*
(h) electrolytic cell *(Sec. 17.7)*
(i) half-cell *(Sec. 17.6)*
(j) half-reaction *(Sec. 17.4)*
(k) oxidation *(Sec. 17.2)*
(l) oxidation number *(Sec. 17.1)*
(m) oxidizing agent *(Sec. 17.2)*
(n) redox reaction *(Sec. 17.2)*
(o) reducing agent *(Sec. 17.2)*
(p) reduction *(Sec. 17.2)*
(q) salt bridge *(Sec. 17.6)*
(r) voltaic cell *(Sec. 17.6)*

Exercises

Oxidation Numbers (Sec. 17.1)

1. State the oxidation number for each of the following metals in their free state.
(a) Mg **(b)** Mn
(c) K **(d)** Zn

2. State the oxidation number for each of the following uncombined nonmetals.
(a) H_2 **(b)** He
(c) P_4 **(d)** I_2

3. State the oxidation number for the metal in each of the following monoatomic cations.
(a) Sr^{2+} **(b)** Sc^{3+}
(c) Ti^{4+} **(d)** Ag^{+}

4. State the oxidation number for the nonmetal in each of the following monoatomic anions.
(a) F^{-} **(b)** H^{-}
(c) P^{3-} **(d)** Te^{2-}

5. Calculate the oxidation number for silicon in the following compounds.
(a) SiO_2 (b) Si_2H_6
(c) Si_3N_4 (d) $CaSiO_3$

6. Calculate the oxidation number for nitrogen in the following compounds.
(a) NH_3 (b) N_2O_4
(c) Li_3N (d) KNO_3

7. Calculate the oxidation number for carbon in the following polyatomic ions.
(a) CO_3^{2-} (b) HCO_3^-
(c) CN^- (d) CNO^-

8. Calculate the oxidation number for sulfur in the following polyatomic ions.
(a) SO_3^{2-} (b) HSO_4^-
(c) $S_2O_3^{2-}$ (d) $S_2O_8^{2-}$

Oxidation–Reduction Reactions (Sec. 17.2)

9. Supply the term that corresponds to the following.
(a) a redox process characterized by electron loss
(b) a redox process characterized by electron gain

10. Supply the term that corresponds to the following.
(a) the substance that loses electrons in a redox reaction
(b) the substance that gains electrons in a redox reaction

11. Indicate the substance oxidized and the substance reduced in each of the following redox reactions.
(a) $Mn(s) + O_2(g) \rightarrow MnO_2(s)$
(b) $S(s) + O_2(g) \rightarrow SO_2(g)$
(c) $Cd(s) + F_2(g) \rightarrow CdF_2(s)$
(d) $Sr(s) + 2\ H_2O(l) \rightarrow Sr(OH)_2(aq) + H_2(g)$
(e) $Pb(s) + CuSO_4(aq) \rightarrow PbSO_4(aq) + Cu(s)$

12. Indicate the oxidizing agent and the reducing agent in each of the redox reactions in Exercise 11.

13. Indicate the substance oxidized and the substance reduced in each of the following redox reactions.
(a) $CuO(s) + H_2(g) \rightarrow Cu(s) + H_2O(l)$
(b) $Cl_2(g) + 2\ KBr(aq) \rightarrow Br_2(l) + 2\ KCl(aq)$
(c) $Ca(s) + 2\ H_2O(l) \rightarrow Ca(OH)_2(aq) + H_2(g)$
(d) $Mg(s) + 2\ HCl(aq) \rightarrow MgCl_2(aq) + H_2(g)$
(e) $PbO(s) + CO(g) \rightarrow Pb(s) + CO_2(g)$

14. Indicate the oxidizing agent and the reducing agent in each of the redox reactions in Exercise 13.

15. Indicate the substance oxidized and the substance reduced in each of the following redox reactions.
(a) $Al(s) + Cr^{3+}(aq) \rightarrow Al^{3+}(aq) + Cr(s)$
(b) $F_2(g) + 2\ Cl^-(aq) \rightarrow 2\ F^-(aq) + Cl_2(g)$
(c) $H_2O(l) + 2\ Fe^{3+}(aq) + SO_3^{2-}(aq) \rightarrow 2\ Fe^{2+}(aq) + SO_4^{2-}(aq) + 2\ H^+(aq)$
(d) $Sn^{2+}(aq) + 2\ Hg^{2+}(aq) \rightarrow Sn^{4+}(aq) + Hg_2^{2+}(aq)$
(e) $Cr^{2+}(aq) + AgI(s) \rightarrow Cr^{3+}(aq) + Ag(s) + I^-(aq)$

16. Indicate the oxidizing agent and the reducing agent in each of the redox reactions in Exercise 15.

17. Indicate the substance oxidized and the substance reduced in each of the following redox reactions.
(a) $2\ Br^-(aq) + Pt^{2+}(aq) \rightarrow Br_2(l) + Pt(s)$
(b) $TeO_2(s) + 4\ H^+(aq) + Zr(s) \rightarrow Te(s) + Zr^{4+}(aq) + 2\ H_2O(l)$
(c) $MnO_2(s) + SO_2(g) \rightarrow Mn^{2+}(aq) + SO_4^{2-}(aq)$
(d) $Pb^{4+}(aq) + H_2O_2(aq) \rightarrow Pb^{2+}(aq) + O_2(g) + 2\ H^+(aq)$
(e) $H_2O(l) + SO_3^{2-}(aq) + I_2(s) \rightarrow SO_4^{2-}(aq) + 2\ I^-(aq) + 2\ H^+(aq)$

18. Indicate the oxidizing agent and the reducing agent in each of the redox reactions in Exercise 17.

Balancing Redox Equations: Oxidation Number Method (Sec. 17.3)

19. Is it always true that the electron loss by oxidation equals the electron gain by reduction in a balanced redox equation?

20. Is it always true that the total ionic charge on the reactants equals the total ionic charge on the products in a balanced redox equation?

21. Write a balanced chemical equation for each of the following redox reactions using the oxidation number method.
(a) $Br_2(l) + NaI(aq) \rightarrow I_2(s) + NaBr(aq)$
(b) $PbS(s) + O_2(g) \rightarrow PbO(s) + SO_2(g)$
(c) $B_2O_3(s) + Cl_2(g) \rightarrow BCl_3(aq) + O_2(g)$

22. Write a balanced chemical equation for each of the following redox reactions using the oxidation number method.
(a) $Cl_2(g) + KI(aq) \rightarrow I_2(s) + KCl(aq)$
(b) $Sb(s) + HNO_3(aq) \rightarrow Sb_2O_5(s) + NO(g) + H_2O(l)$
(c) $Fe_2O_3(s) + CO(g) \rightarrow Fe(s) + CO_2(g)$

23. Write a balanced chemical equation for each of the following redox reactions using the oxidation number method.
(a) $MnO_4^-(aq) + I^-(aq) + H^+(aq) \rightarrow Mn^{2+}(aq) + I_2(s) + H_2O(l)$
(b) $Cu(s) + H^+(aq) + SO_4^{2-}(aq) \rightarrow Cu^{2+}(aq) + SO_2(g) + H_2O(l)$
(c) $Fe^{2+}(aq) + H_2O_2(aq) + H^+(aq) \rightarrow Fe^{3+}(aq) + H_2O(l)$

24. Write a balanced chemical equation for each of the following redox reactions using the oxidation number method.
(a) $Cr_2O_7^{2-}(aq) + Br^-(aq) + H^+(aq) \rightarrow Cr^{3+}(aq) + Br_2(l) + H_2O(l)$
(b) $MnO_4^-(aq) + SO_2(g) + H_2O(l) \rightarrow Mn^{2+}(aq) + SO_4^{2-}(aq) + H^+(aq)$
(c) $AsO_3^{3-}(aq) + MnO_4^-(aq) + H^+(aq) \rightarrow Mn^{2+}(aq) + AsO_3^-(aq) + H_2O(l)$

Balancing Redox Equations: Ion–Electron Method (Sec. 17.4)

25. Write a balanced half-reaction for each of the following, which take place in *acidic* solution.
(a) $F_2(g) \rightarrow 2\ F^-(aq)$
(b) $Hg(l) + Cl^-(aq) \rightarrow HgCl_2(s)$
(c) $BrO_3^-(aq) \rightarrow Br_2(l)$
(d) $H_2O_2(aq) \rightarrow H_2O(l)$
(e) $O_2(g) \rightarrow H_2O_2(aq)$

26. Write a balanced half-reaction for each of the following, which take place in *basic* solution.
(a) $ClO^-(aq) \rightarrow Cl^-(aq)$
(b) $MnO_4^-(aq) \rightarrow MnO_2(s)$
(c) $Ni(OH)_2(s) \rightarrow NiO_2(s)$
(d) $NO_2^-(aq) \rightarrow N_2O(g)$
(e) $NO_2^-(aq) \rightarrow NO_3^-(aq)$

27. Using the ion–electron method, write a balanced chemical equation for each of the following redox reactions occurring in *acidic* solution.
(a) $Zn(s) + NO_3^-(aq) \rightarrow Zn^{2+}(aq) + NO(g)$
(b) $Mn^{2+}(aq) + BiO_3^-(aq) \rightarrow MnO_4^-(aq) + Bi^{3+}(aq)$
(c) $MnO_4^-(aq) + SO_3^{2-}(aq) \rightarrow Mn^{2+}(aq) + SO_4^{2-}(aq)$
(d) $Sn^{2+}(aq) + IO_3^-(aq) \rightarrow Sn^{4+}(aq) + I_2(s)$
(e) $AsO_3^{3-}(aq) + Br_2(l) \rightarrow AsO_4^{3-}(aq) + Br^-(aq)$

28. Using the ion–electron method, write a balanced chemical equation for each of the following redox reactions occurring in *basic* solution.
(a) $MnO_4^-(aq) + S^{2-}(aq) \rightarrow MnO_2(s) + S(s)$
(b) $Cu(s) + ClO^-(aq) \rightarrow Cu^{2+}(aq) + Cl^-(aq)$
(c) $Cl_2(g) + BrO_2^-(aq) \rightarrow Cl^-(aq) + BrO_3^-(aq)$
(d) $Mn^{2+}(aq) + H_2O_2(aq) \rightarrow MnO_2(s) + H_2O(l)$
(e) $MnO_2(s) + O_2(g) \rightarrow MnO_4^-(aq) + H_2O(l)$

29. A disproportionation reaction is a special type of redox reaction in which a single element is both oxidized and reduced to give two different products. Balance the following disproportionation reactions in *acidic* solution.
(a) $Cl_2(g) \rightarrow Cl^-(aq) + HOCl(aq)$
(b) $HNO_2(aq) \rightarrow NO_3^-(aq) + NO(g)$
(c) $MnO_4^{2-}(aq) \rightarrow MnO_4^-(aq) + MnO_2(s)$

30. When a single element is simultaneously oxidized and reduced, the process is termed disproportionation. Balance the following disproportionation reactions, which occur in *basic* solution.
(a) $Cl_2(g) \rightarrow ClO_2^-(aq) + Cl^-(aq)$
(b) $Br_2(l) \rightarrow BrO_3^-(aq) + Br^-(aq)$
(c) $S(s) \rightarrow SO_3^{2-}(aq) + S^{2-}(aq)$

Predicting Spontaneous Redox Reactions (Sec. 17.5)

31. Refer to Table 17.5 and indicate which substance in the following pairs has the greatest tendency to be reduced.
(a) $Au^{3+}(aq)$ or $Hg^{2+}(aq)$
(b) $H^+(aq)$ or $H_2(g)$
(c) $Fe^{2+}(aq)$ or $I_2(s)$
(d) $Al^{3+}(aq)$ or $Br_2(l)$

32. Refer to Table 17.5 and indicate which substance in the following pairs has the greatest tendency to be oxidized.
(a) $Na(s)$ or $K(s)$
(b) $Ag(s)$ or $Hg(l)$
(c) $Fe^{2+}(aq)$ or $Fe(s)$
(d) $Br^-(aq)$ or $Cl^-(aq)$

33. Refer to Table 17.5 and indicate which substance in the following pairs is the stronger oxidizing agent.
(a) $F_2(g)$ or $Cl_2(g)$
(b) $Ag^+(aq)$ or $Br_2(l)$
(c) $Cu^{2+}(aq)$ or $H^+(aq)$
(d) $Mg^{2+}(aq)$ or $Mn^{2+}(aq)$

34. Refer to Table 17.5 and indicate which substance in the following pairs is the stronger reducing agent.
(a) $Cu(s)$ or $Hg(l)$
(b) $H_2(g)$ or $Ba(s)$
(c) $Al(s)$ or $I^-(aq)$
(d) $Cl^-(aq)$ or $H_2(g)$

35. Refer to Table 17.5 and state whether the following reactions are spontaneous or nonspontaneous.
(a) $Mg(s) + Sn^{2+}(aq) \rightarrow Mg^{2+}(aq) + Sn(s)$
(b) $H^+(aq) + I^-(aq) \rightarrow H_2(g) + I_2(s)$
(c) $H^+(aq) + Ni(s) \rightarrow H_2(g) + Ni^{2+}(aq)$
(d) $Hg^{2+}(aq) + Br^-(aq) \rightarrow Hg(l) + Br_2(l)$
(e) $Fe^{3+}(aq) + I^-(aq) \rightarrow Fe^{2+}(aq) + I_2(s)$

36. Refer to Table 17.5 and state whether the following reactions are spontaneous or nonspontaneous.
(a) $Br_2(l) + LiF(aq) \rightarrow F_2(g) + LiBr(aq)$
(b) $Al(NO_3)_3(aq) + Co(s) \rightarrow Co(NO_3)_2(aq) + Al(s)$
(c) $Cr(s) + H_2SO_4(aq) \rightarrow Cr_2(SO_4)_3(aq) + H_2(g)$
(d) $Au(s) + HCl(aq) \rightarrow AuCl_3(aq) + H_2(g)$
(e) $FeCl_3(aq) + NaI(aq) \rightarrow FeCl_2(aq) + NaCl(aq) + I_2(s)$

37. Refer to Exercise 35 and write a balanced chemical equation for each of the redox reactions. Assume that the reactions occur in acidic or neutral solution.

38. Refer to Exercise 36 and write a balanced chemical equation for each of the redox reactions. Assume that the reactions occur in acidic or neutral solution.

Voltaic Cells (Sec. 17.6)

39. Sketch the following voltaic cell showing the two compartments, the anode and cathode, the wire connecting the two electrodes, and a salt bridge.

$$Ni(s) + 2\,AgNO_3(aq) \longrightarrow 2\,Ag(s) + Ni(NO_3)_2(aq)$$

40. Diagram the direction of electron flow and the movement of nitrate ions in the salt bridge that connects the two half-cells in Exercise 39.

41. Sketch the following voltaic cell showing the two compartments, Pt anode and Pt cathode, the wire connecting the two electrodes, and a salt bridge.

$$Cl_2(g) + 2\,KBr(aq) \longrightarrow 2\,KCl(aq) + Br_2(l)$$

42. Diagram the direction of electron flow and the movement of potassium ions in the salt bridge that connects the two half-cells in Exercise 41.

43. The spontaneous redox reaction of tin and copper(II) sulfate solution occurs according to the following ionic equation:

$$Sn(s) + Cu^{2+}(aq) \longrightarrow Cu(s) + Sn^{2+}(aq)$$

Assume the half-reactions are separated into two compartments. A tin electrode is immersed in 1.00 *M* $SnSO_4$ and a copper electrode is placed in 1.00 *M* $CuSO_4$. For this voltaic cell, indicate each of the following:
(a) oxidation half-cell reaction
(b) reduction half-cell reaction
(c) anode and cathode
(d) direction of electron flow
(e) direction of SO_4^{2-} movement in the salt bridge

44. The spontaneous redox reaction of cobalt metal and iron(III) chloride solution occurs according to the following ionic equation:

$$Co(s) + 2\ Fe^{3+}(aq) \longrightarrow Co^{2+}(aq) + 2\ Fe^{2+}(aq)$$

Assume the half-reactions are separated into two compartments. A cobalt electrode is immersed in 1.00 *M* $CoCl_2$, while an iron electrode is placed in 1.00 *M* $FeCl_3$. For this voltaic cell, indicate each of the following:
(a) oxidation half-cell reaction
(b) reduction half-cell reaction
(c) anode and cathode
(d) direction of electron flow
(e) direction of Cl^- movement in the salt bridge

45. The spontaneous redox reaction of magnesium metal and manganese(II) nitrate solution occurs according to the following equation:

$$Mg(s) + Mn(NO_3)_2(aq) \longrightarrow Mn(s) + Mg(NO_3)_2(aq)$$

Assume the half-reactions are separated into two compartments. A magnesium electrode is immersed in 1.00 *M* $Mg(NO_3)_2$ and a manganese electrode is placed in 1.00 *M* $Mn(NO_3)_2$. For this voltaic cell, indicate each of the following:
(a) oxidation half-cell reaction
(b) reduction half-cell reaction
(c) anode and cathode
(d) direction of electron flow
(e) direction of NO_3^- movement in the salt bridge

46. The spontaneous redox reaction of hydrogen gas and iodine occurs according to the following equation:

$$H_2(g) + I_2(s) \longrightarrow 2\ H^+(aq) + 2\ I^-(aq)$$

Assume the half-reactions are separated into two compartments. A hydrogen gas Pt electrode is immersed in 1.00 *M* HI and a second Pt electrode is placed in 1.00 *M* KI/I_2 solution. For this voltaic cell, indicate each of the following:
(a) oxidation half-cell reaction
(b) reduction half-cell reaction
(c) anode and cathode
(d) direction of electron flow
(e) direction of I^- movement in the salt bridge

Electrolytic Cells (Sec. 17.7)

47. Sketch the following electrolytic cell showing the two compartments, the anode and cathode, the wire connecting the two electrodes, and a salt bridge.

$$Ni(s) + Cd(NO_3)_2(aq) \longrightarrow Cd(s) + Ni(NO_3)_2(aq)$$

48. Diagram the direction of electron flow and the movement of nitrate ions in the salt bridge that connects the two half-cells in Exercise 47.

49. Sketch the following electrolytic cell showing the two compartments, the anode and cathode, the wire connecting the two electrodes, and a salt bridge.

$$Cu(s) + 2\ HBr(aq) \longrightarrow CuBr_2(aq) + H_2(g)$$

50. Diagram the direction of electron flow and the movement of bromide ions in the salt bridge that connects the two half-cells in Exercise 49.

51. The nonspontaneous redox reaction of nickel and iron(II) sulfate solution can be forced to occur according to the following ionic equation:

$$Ni(s) + Fe^{2+}(aq) \longrightarrow Fe(s) + Ni^{2+}(aq)$$

Assume the half-reactions are separated into two compartments. A nickel electrode is immersed in 1.00 *M* $NiSO_4$ and an iron electrode is in 1.00 *M* $FeSO_4$. For this electrolytic cell, indicate each of the following:
(a) oxidation half-cell reaction
(b) reduction half-cell reaction
(c) anode and cathode
(d) direction of electron flow
(e) direction of SO_4^{2-} movement in the salt bridge

52. The nonspontaneous redox reaction of lead and zinc nitrate solution can be forced to occur according to the following ionic equation:

$$Pb(s) + Zn^{2+}(aq) \longrightarrow Zn(s) + Pb^{2+}(aq)$$

Assume the half-reactions are separated into two compartments. A zinc electrode is immersed in 1.00 *M* $Zn(NO_3)_2$, while a lead electrode is placed in 1.00 *M* $Pb(NO_3)_2$. For this electrolytic cell, indicate each of the following:
(a) oxidation half-cell reaction
(b) reduction half-cell reaction
(c) anode and cathode
(d) direction of electron flow
(e) direction of NO_3^- movement in the salt bridge

53. The nonspontaneous redox reaction of chromium and aluminum acetate solution can occur according to the following equation:

$$Cr(s) + Al(C_2H_3O_2)_3(aq) \longrightarrow Al(s) + Cr(C_2H_3O_2)_3(aq)$$

Assume the half-reactions are separated into two compartments. An aluminum electrode is immersed in 1.00 *M* $Al(C_2H_3O_2)_3$ and a chromium electrode is in 1.00 *M* $Cr(C_2H_3O_2)_3$. For this electrolytic cell, indicate each of the following:
(a) oxidation half-cell reaction
(b) reduction half-cell reaction
(c) anode and cathode
(d) direction of electron flow
(e) direction of $C_2H_3O_2^-$ movement in the salt bridge

54. The nonspontaneous redox reaction of chlorine gas and sodium fluoride can occur according to the following equation:

$$Cl_2(g) + 2\ NaF(aq) \longrightarrow 2\ NaCl(aq) + F_2(g)$$

Assume the half-reactions are separated into two compartments. A chlorine gas Pt electrode is immersed in 1.00 *M* NaCl and a second Pt electrode is placed in 1.00 *M* NaF. For this electrolytic cell, indicate each of the following:
(a) oxidation half-cell reaction

(b) reduction half-cell reaction
(c) anode and cathode
(d) direction of electron flow
(e) direction of Na^+ movement in the salt bridge

General Exercises

55. Calculate the oxidation number for sulfur in sodium thiosulfate, $Na_2S_2O_3$.
56. Calculate the oxidation number for sulfur in the tetrathionate ion $S_4O_6^{2-}$.
57. For an ionic redox reaction to be balanced, what two quantities must be equal on both the reactant and product side of the equation?
58. The reaction between cobalt metal and aqueous mercuric nitrite produces cobalt(II) nitrite and droplets of liquid mercury metal. Write a net ionic equation for the reaction and identify the oxidizing agent and reducing agent.
59. The reaction between zinc metal and sulfuric acid produces zinc sulfate and hydrogen gas. Write a net ionic equation for the reaction and identify the oxidizing agent and reducing agent.
60. The reaction between potassium metal and water produces potassium hydroxide and hydrogen gas. Write a net ionic equation for the reaction and identify the oxidizing agent and reducing agent.
61. A copper penny reacts and dissolves in concentrated nitric acid as follows:

$$Cu(s) + HNO_3(aq) \longrightarrow Cu(NO_3)_2(aq) + NO_2(g) + H_2O(l)$$

Write a balanced net ionic equation for the reaction and identify the oxidizing agent and reducing agent.
62. A redox titration is performed by law enforcement to routinely determine the percentage of blood alcohol. Ethyl alcohol, C_2H_5OH, is titrated by a standard dichromate solution as follows:

$$C_2H_5OH(aq) + Cr_2O_7^{2-}(aq) + H^+(aq) \longrightarrow HC_2H_3O_2(aq) + Cr^{3+}(aq) + H_2O(l)$$

Write a balanced net ionic equation for the reaction and identify the oxidizing agent and reducing agent.
63. State two advantages of obtaining energy from a hydrogen–oxygen fuel cell compared to burning fossil fuels.
64. What is the principal problem of obtaining energy from a hydrogen–oxygen fuel cell?

CHAPTER

18

Advanced Chemical Calculations

In the Tour de France, cyclists have spare frames and wheels since either can limit their progress in the race. By analogy, an insufficient amount of any reactant in a chemical reaction limits the progress of the reaction.

Throughout the preceding chapters, we have been building a mathematical foundation for solving chemistry problems. One important learned skill was how to perform calculations based on the unit analysis method of problem solving. Then, with each new topic, we applied that consistent and systematic approach to solving problems. In Chapter 7, we applied unit analysis to chemical formula calculations. In Chapter 9, we used the unit analysis method to tackle stoichiometry problems. Later, we modified the unit analysis approach to answer questions involving the gas laws and solution concentration.

In this chapter, we will integrate many of the preceding calculation topics. That is, topics previously covered individually will now be considered in combination. Specifically, we will incorporate two or more of these topics in the same problem: density, specific heat, mole concept, empirical formula, percentage composition, chemical reactions, stoichiometry, gas laws, and solution concentration.

Whenever appropriate, we will use relevant examples. For instance, we will see how chemists and chemical engineers routinely perform chemical calculations to determine the cost of manufacturing products. These calculations are critical to the production of chemicals for the agricultural, electronic, petrochemical, pharmaceutical, and textile industries. Each year the amount of chemicals used industrially are published, and over the last several years sulfuric acid has led the list of important chemicals. Figure 18.1 shows piles of sulfur for making sulfuric acid.

Figure 18.1 Blocks of Yellow Sulfur Sulfur is one of our most abundant resources and nearly 90% is converted into sulfuric acid. In 1990, the U.S. manufactured 45 million tons of H_2SO_4.

Chemical manufacturing is one of the most important industries in the United States. Each year billions of pounds of chemicals are produced for use at home and overseas. These chemicals are used for the manufacture of fertilizer, steel, plastics, glass, explosives, fibers, and other chemicals. Table 18.1 lists the top 10 chemicals ranked in order of their annual production.

Table 18.1 Top 10 U.S. Industrial Chemicals (1990)

Rank	Chemical	Annual Production (billions of pounds)
1	sulfuric acid, H_2SO_4	88.56
2	nitrogen, N_2	57.32
3	oxygen, O_2	38.99
4	ethylene, $CH_2{=}CH_2$	37.48
5	lime, CaO	34.80
6	ammonia, NH_3	33.92
7	phosphoric acid, H_3PO_4	24.35
8	sodium hydroxide, NaOH	23.38
9	propylene, $CH_3CH{=}CH_2$	22.12
10	chlorine gas, Cl_2	21.88

18.1 Advanced Problem Solving

OBJECTIVES

To state general techniques for solving chemical calculations, including: problem analysis, strategy maps, concept maps, algorithms, visualization, unit analysis, and algebraic analysis.

To this point, we have considered many of the individual aspects of problem solving. Now we will consider problems that involve two or more concepts. Moreover, you may have alternate methods for solving these problems. At this point, you have several options for solving a problem and need to seek out approaches that are reasonable to you. Now let's consider some of the alternative methods that have proven successful and are available to you.

Strategy Maps

To solve more complicated problems successfully, you will need to analyze the given information systematically. To do so, read the problem carefully and ask questions that bring it into focus. First, examine the problem to find the unknown quantity. Next, determine which of the given information is related to the unknown. Finally, plan a strategy that links the given value to the unknown. In summary, the three steps are as follows.

1. *Examine the problem to determine the unknown quantity.*
2. *Write down the given value related to the unknown quantity.*
3. *Plan a strategy that relates the given value to the unknown quantity.*

After reading the description of a problem, the first question you should always ask is, what is the unknown quantity that is to be calculated? The unknown quantity can be a mass, a volume, or a temperature. In most of the problems we have solved, the units of the unknown quantity have been specified. That is, the quantity you were asked to calculate is the mass in *grams,* the volume in *liters,* or a solution concentration in *moles per liter*. You will now encounter problems that may ask you to calculate an unknown quantity without specifying units. In these cases, choose any unit that is convenient. For example, if the quantity sought is a mass, you can express it in grams, kilograms, or even an English unit.

The second question you should ask is, what given value is related to the unknown quantity to be calculated? After the unknown quantity is stated, search the problem for a relevant given value.

A third question to be asked is, what steps are necessary to proceed from the given value to the unknown? The answer to this question is sometimes outlined in the form of a strategy map. A **strategy map** is simply a plan that involves two or more steps to convert the given value to the unknown quantity.

strategy map A problem-solving plan that relates a given value to an unknown quantity, usually shown by a series of arrows.

As an example, consider the following problem. What is the volume in liters of 32.0 g of oxygen gas at standard conditions? The strategy map for the problem is

$$\text{g } O_2 \longrightarrow \text{mol } O_2 \longrightarrow \text{L } O_2 \text{ (STP)}$$

To solve this problem, we must do two conversions. First, we convert the grams of O_2 to moles of O_2. Second, we convert the moles of O_2 to liters of O_2 at STP.

Concept Maps

It is sometimes helpful to draw a diagram that relates the various quantities in a given problem. In the above oxygen gas example, we can relate the mass of gaseous oxygen to its volume using the following diagram.

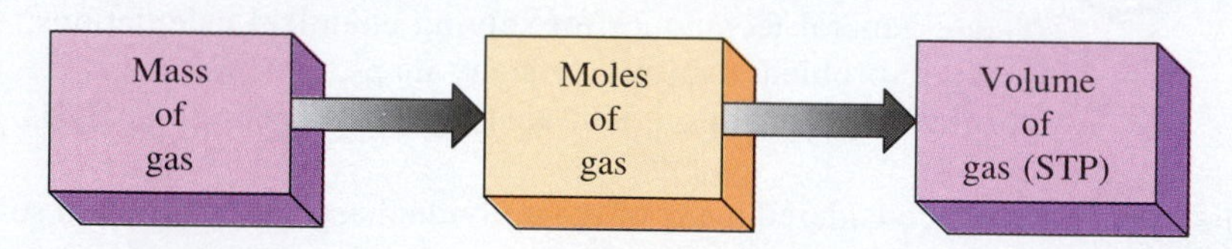

Recall that we first used this type of diagram to explain mole relationships in Chapter 7 and later to explain chemical equation relationships in Chapter 9. The diagrams helped us plan a solution to a problem. A diagram that relates quantities of substance is called a **concept map**. In Sections 18.2 and 18.3, we will introduce more powerful concept maps that apply to most stoichiometry calculations.

concept map A problem-solving plan that traces the relationship of various quantities, usually shown by interrelated boxes.

Algorithms

algorithm A problem-solving plan that relates a given quantity to an unknown quantity in a series of steps listed in a specific sequence.

A concept map may be too abstract to solve problems for some students. If this is the case, you may prefer using algorithms to help you solve problems. An **algorithm** is a step-by-step set of instructions for solving a problem. The three steps in the unit analysis method of problem solving, for example, are an algorithm. An algorithm accomplishes the same thing as a strategy map or a concept map. The difference is that it is a stepwise procedure, which may seem clearer and more effective for devising a successful strategy.

Let's write an algorithm for the following problem: What is the volume of 32.0 g of oxygen gas at standard conditions? An algorithm provides a pathway from the relevant given quantity to the unknown as a series of prescribed steps. In this problem the volume and mass of oxygen are related to the mole concept; this suggests the following algorithm:

Step 1: Compute the molar mass of O_2 from the periodic table.

Step 2: Convert the mass of oxygen to moles O_2.

Step 3: Recall that the molar volume of any gas at STP is 22.4 L.

Step 4: Convert the number of moles of oxygen to liters of O_2.

There is more than one way to write an algorithm for any given problem. Therefore, the number of steps can vary. For example, if you clearly understand this type of problem, you can write a shorter algorithm, such as the following.

Step 1: Convert the mass of oxygen to moles O_2.

Step 2: Convert the moles of oxygen to liters of O_2.

Example Exercise 18.1 offers additional practice in writing an algorithm for a chemical calculation.

EXAMPLE EXERCISE 18.1

Write an algorithm for the following problem: What is the number of chlorine molecules in 7.10 g of chlorine gas at STP?

Solution: We can write an algorithm similar to the one above. That is,

Step 1: Compute the molar mass of Cl_2 from the periodic table.

Step 2: Convert the mass of chlorine to moles Cl_2.

Step 3: Recall that the number of molecules in 1 mole is Avogadro's number, 6.02×10^{23}.

Step 4: Convert the number of moles of chlorine to molecules of Cl_2 using Avogadro's number.

Or, more simply, we can write the two-step algorithm:

Step 1: Convert the mass of chlorine to moles Cl_2.

Step 2: Convert the moles of chlorine to molecules of Cl_2.

SELF-TEST EXERCISE

Write an algorithm for the following problem: What is the mass of 5.00 L of oxygen gas at STP?

Answer: There is more than one possible algorithm for this problem; for example,

Step 1: Convert the volume of oxygen at STP to moles O_2.

Step 2: Find the molar mass of O_2 in the periodic table.

Step 3: Convert the number of moles of oxygen to mass of O_2 using the molar mass, 32.0 g/mol.

Or, more simply, we can write the two-step algorithm:

Step 1: Convert the volume of oxygen at STP to moles O_2.

Step 2: Convert the moles of oxygen to mass of O_2.

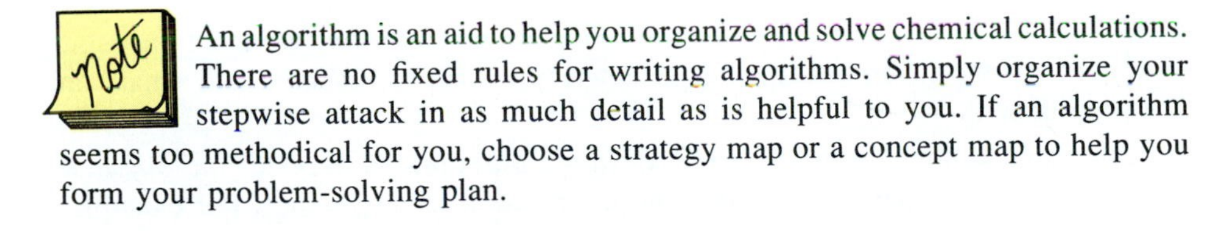

An algorithm is an aid to help you organize and solve chemical calculations. There are no fixed rules for writing algorithms. Simply organize your stepwise attack in as much detail as is helpful to you. If an algorithm seems too methodical for you, choose a strategy map or a concept map to help you form your problem-solving plan.

Visualization

Although you have never been on the moon, you probably would have no trouble describing its dusty surface or its many craters. We all have a mental picture of the lunar surface based on photographs and descriptions brought back by the Apollo astronauts. Many of the concepts in chemistry are similarly removed from our actual experience. These concepts become more clear through the process of **visualization**. One example is the visualization of atoms and molecules. We cannot actually see an atom, but we can picture an atom or group of atoms in a molecule. This process of forming mental pictures helps us understand concepts we cannot observe. Through our ability to form clear mental pictures, we make the idea of atoms more concrete and permanent.

visualization A problem-solving approach that involves the forming of mental pictures in order to make an abstract principle more concrete and permanent.

Figure 18.2 The Visualization Process Begin with a mental image of small molecules moving about a container. We can make the image stronger by picturing the container as a shiny metal container. Since a gas is mostly empty space, we see that the molecules are far apart.

Let's return to the problem of finding the volume of 32.0 g of oxygen gas. To use visualization to help solve the problem, begin by forming a mental picture of oxygen molecules moving about randomly in a stainless steel container. Next, visualize weighing the metal container empty, then with the gas to determine the mass of oxygen. Using the molar mass, we can calculate the number of moles of oxygen. As we picture the shiny stainless steel container, we can ask, how big is the container? Using the molar volume, we can find the moles of oxygen. Now, try to visualize the container while thinking about its volume in terms of liters.

The ability to visualize varies, but it is a skill that you can acquire and develop with practice. Refer to Figure 18.2, which illustrates the visualization process.

Unit Analysis

A technique with which you are very familiar by now is solving problems by **unit analysis**. The unit analysis method of problem solving is a powerful tool that can be applied to more complex problems as well. Recall that the format for unit analysis problem solving is

$$\underset{(2)}{\text{relevant given value}} \times \underset{(3)}{\frac{\text{unit}}{\text{factor(s)}}} = \underset{(1)}{\text{units in answer}}$$

To solve the problems in this chapter, we may need a more sophisticated approach. Consider these guidelines for using unit analysis in these more advanced problems.

1. *Many problems cannot be solved in a single unit analysis operation.* If the problem is complicated, we must break it down into two or more simpler problems. Before starting any calculations, plan an overall strategy to arrive at a final answer.
2. *If the unknown quantity is a single unit* (for example, cm, g, mL), *the given value should also be a single unit.* If the unknown quantity is a compound unit (for example, g/mL, g/mol, mol/L), the given value should also be a ratio. That is, the relevant given value is a ratio with units in both the numerator and the denominator.
3. *A problem may include given values that are not relevant.* We must then sort through the given information to determine what is essential and what can be ignored.
4. *A problem may omit some essential information that is necessary to obtain a solution.* You then have to find the information in a reference. For example, you can use the periodic table to find the molar mass of a substance. The appendices in most chemistry textbooks have reference tables. One of the best resources for data such as density, solubility, and specific heat is the *Handbook of Chemistry and Physics.* You can find this resource in most libraries or chemistry stockrooms.
5. *If the calculation is based on a chemical reaction, first write the equation for the reaction.* To solve any stoichiometry problem, you need a balanced chemical equation.
6. *Where possible, estimate the approximate answer before starting the calculations.* Consider the following unit analysis calculation:

$$1.95\ \text{g}\ \cancel{\text{AlCl}_3} \times \frac{1\ \cancel{\text{mol AlCl}_3}}{133.5\ \cancel{\text{g AlCl}_3}} \times \frac{3\ \cancel{\text{mol AgCl}}}{1\ \cancel{\text{mol AlCl}_3}} \times \frac{143.4\ \text{g AgCl}}{1\ \cancel{\text{mol AgCl}}} = \text{g AgCl}$$

To estimate the answer, first round off numbers. There are no absolute rules for rounding off data, but we can approximate the answer as follows. Round up 1.95 g to 2 g. Round off 143.4 g and 133.5 go to 100 g. That is,

$$2\ \text{g}\ \cancel{\text{AlCl}_3} \times \frac{1\ \cancel{\text{mol AlCl}_3}}{100\ \cancel{\text{g AlCl}_3}} \times \frac{3\ \cancel{\text{mol AgCl}}}{1\ \cancel{\text{mol AlCl}_3}} \times \frac{100\ \text{g AgCl}}{1\ \cancel{\text{mol AgCl}}} = 6\ \text{g AgCl}$$

The estimated answer is 6 g AgCl. After performing the calculations, check the final answer for agreement with your estimated value. If your answer does not seem reasonable, review your calculations. In this example, the calculated value is 6.28 g AgCl. There is good agreement between the estimated answer, 6 g, and the calculated value 6.28 g. Therefore, we have confidence that the answer is correct.

Algebraic Analysis

For most problems, the unit analysis method of problem solving is our first choice. There are some problems, however, for which unit analysis is not appropriate. Some relationships are more easily solved using **algebraic analysis**. In Section 2.8, we solved density problems using a unit analysis approach. We could have just as easily solved density problems using algebra.

algebraic analysis A systematic problem-solving approach that requires rearranging variables in an equation in order to calculate an unknown quantity.

The definition of density is mass per unit volume. Therefore, we can write the equation

$$d = \frac{m}{V}$$

Here d is the density, m is the mass, and V is the volume. If a problem states that the mass of a given solid is 27.0 g and the density is 2.70 g/cm^3, we can find the volume as follows:

$$2.70\ \text{g/cm}^3 = \frac{27.0\ \text{g}}{V}$$

Rearranging the equation and solving for the volume of the solid, we have

$$V = \frac{27.0\ \text{g}}{2.70\ \text{g/cm}^3} = 10.0\ \text{cm}^3$$

In Section 12.11, we introduced the ideal gas equation, $PV = nRT$. This is another good example where algebra may be more appropriate for solving a chemical calculation.

Note: Since individual chemical calculations are different, we cannot solve all problems the same way. Rather, we must choose a strategy from the techniques mentioned here. We can use strategy maps, concept maps, algorithms, visualization, unit analysis, and algebra to obtain an answer. As we progress in our understanding of chemistry, the problems become more diverse and we need different strategies to find the correct solutions to different problems.

18.2 Chemical Formula Calculations

To perform chemical formula calculations that relate moles of substance to the following quantities:

(a) number of entities (atoms, molecules, or formula units)
(b) mass of substance
(c) volume of gas
(d) volume of aqueous solution

In Chapter 7, we learned about mole calculations; in Chapter 14, we studied solution molarity calculations. In this chapter, we will relate all mole calculations. That is, we will relate molar mass, molar volume, and molar concentration. To gain an overall perspective of mole relationships, we can draw a concept map. The concept map in Figure 18.3 illustrates the relationship of various mole quantities.

Let's reinforce our understanding of mole calculations with the following example. We are given 1.00 g of ammonia gas, NH_3, which we dissolve in water to make 100.0 mL of solution. We want to calculate the number of ammonia molecules, the volume of NH_3 at STP, and the molar concentration of the dissolved gas. We see from the concept map in Figure 18.3 that we must first calculate the moles of ammonia. Using the periodic table, we find that the molar mass of NH_3 is 17.0 g/mol. Thus,

$$1.00\ \text{g}\ \cancel{NH_3} \times \frac{1\ \text{mol}\ NH_3}{17.0\ \text{g}\ \cancel{NH_3}} = \text{mol}\ NH_3$$
$$= 0.0588\ \text{mol}\ NH_3$$

Having found the number of moles of ammonia, we can calculate any of the

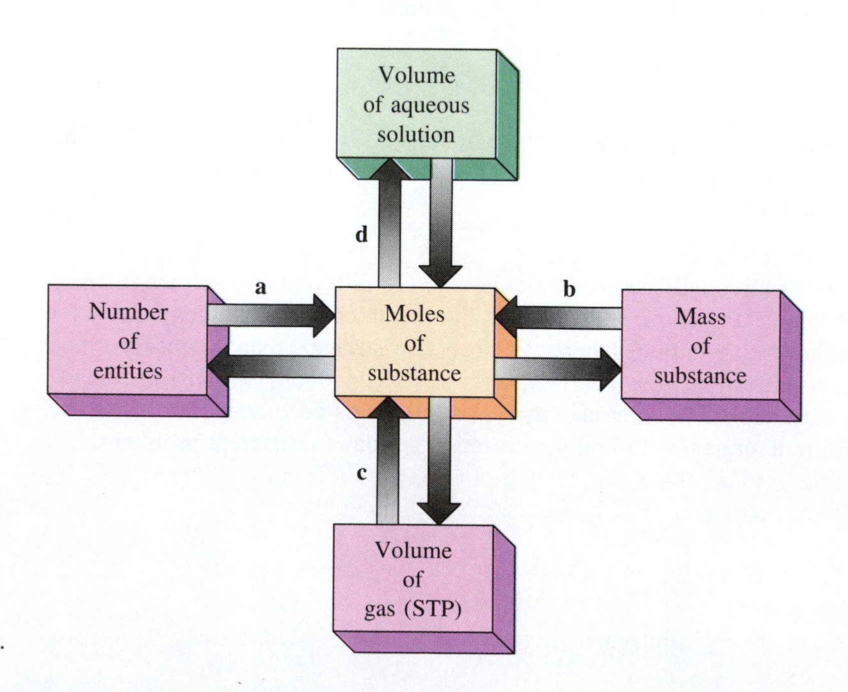

Figure 18.3 Mole Concept Map The concept map shows the various quantities related to the central mole concept. We can find the moles of substance from the (a) number of entities, (b) mass of substance, (c) volume of gas, or (d) volume and concentration of aqueous solution.

other quantities. For instance, in 1 mole of ammonia there is Avogadro's number of molecules. We can find the number of molecules in our given sample as follows:

$$0.0588\ \cancel{\text{mol NH}_3} \times \frac{6.02 \times 10^{23}\ \text{molecules NH}_3}{1\ \cancel{\text{mol NH}_3}} = \text{molecules NH}_3$$

$$= 3.54 \times 10^{22}\ \text{molecules NH}_3$$

We find the volume of ammonia gas at STP using the molar volume concept. One mole of ammonia, or any other gas at STP, occupies a volume of 22.4 liters. The volume of the given sample is

$$0.0588\ \cancel{\text{mol NH}_3} \times \frac{22.4\ \text{L NH}_3}{1\ \cancel{\text{mol NH}_3}} = \text{L NH}_3$$

$$= 1.32\ \text{L NH}_3$$

Finally, we will calculate the molar concentration of an ammonia solution that contains 1.00 g NH_3 dissolved in 100.0 mL of aqueous solution. Since the molarity of the solution has compound units, mol/L, the given value must also have compound units. The value in this example is 0.0588 mol NH_3 per 100.0 mL of solution. The molar concentration of the aqueous solution is

$$\frac{0.0588\ \text{mol NH}_3}{100.0\ \cancel{\text{mL solution}}} \times \frac{1000\ \cancel{\text{mL solution}}}{1\ \text{L solution}} = \frac{\text{mol NH}_3}{\text{L solution}}$$

$$= 0.588\ \text{mol NH}_3/\text{L solution}$$

$$= 0.588\ M\ \text{NH}_3$$

Example Exercise 18.2 provides more practice with mole calculations.

EXAMPLE EXERCISE 18.2

Given 50.0 mL of 0.500 *M* hydrochloric acid, calculate each of the following.

(a) grams of HCl gas dissolved in the acid solution
(b) liters of HCl gas (at STP) dissolved in the solution
(c) molecules of HCl gas dissolved in the solution

Solution: As usual, we must first determine the number of moles of HCl before we can calculate the other quantities. Recall that a concentration of 0.500 *M* corresponds to the unit factor 0.500 mol/1000 mL. Hence,

$$50.0\ \cancel{\text{mL solution}} \times \frac{0.500\ \text{mol HCl}}{1000\ \cancel{\text{mL solution}}} = 0.0250\ \text{mol HCl}$$

(a) From the periodic table, we find that the molar mass of HCl is 36.5 g/mol. The mass of HCl dissolved in the acid solution is therefore

$$0.0250\ \cancel{\text{mol HCl}} \times \frac{36.5\ \text{g HCl}}{1\ \cancel{\text{mol HCl}}} = 0.913\ \text{g HCl}$$

(b) The molar volume of any gas, including HCl, is 22.4 L/mol at STP. The volume of HCl gas at standard conditions is

$$0.0250\ \cancel{\text{mol HCl}} \times \frac{22.4\ \text{L HCl}}{1\ \cancel{\text{mol HCl}}} = 0.560\ \text{L HCl (STP)}$$

(c) The number of HCl molecules dissolved in the acid is found using Avogadro's number. To obtain the number of molecules, multiply the number of moles times 6.02×10^{23}.

$$0.0250 \, \cancel{\text{mol HCl}} \times \frac{6.02 \times 10^{23} \text{ molecules HCl}}{1 \, \cancel{\text{mol HCl}}} = 1.51 \times 10^{22} \text{ molecules HCl}$$

SELF-TEST EXERCISE

Given 100.0 mL of 0.105 *M* aqueous carbon dioxide, calculate the following.
(a) mass of CO_2 dissolved in the solution
(b) volume of CO_2 gas (at STP) dissolved in the solution
(c) molecules of CO_2 dissolved in the solution

Answers: **(a)** 0.462 g CO_2; **(b)** 0.235 L CO_2 (at STP); **(c)** 6.32×10^{21} molecules of CO_2

18.3 Chemical Equation Calculations

To perform chemical equation calculations for the following types of problems.
(a) mass-mass stoichiometry
(b) mass-volume stoichiometry
(c) volume-volume stoichiometry
(d) solution stoichiometry

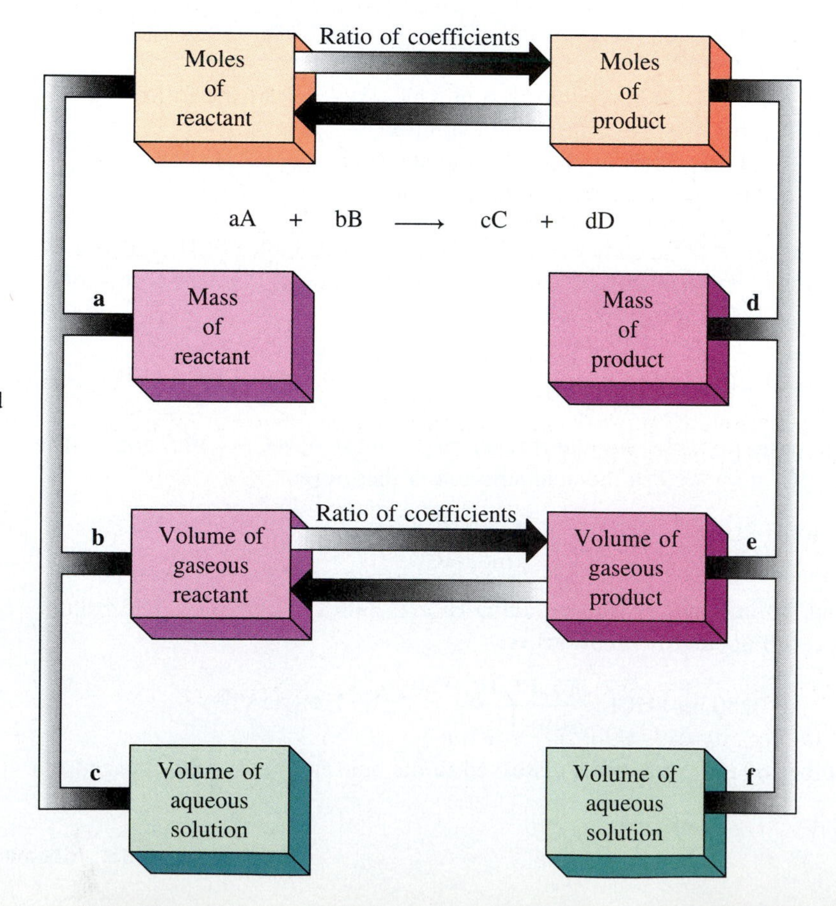

Figure 18.4 Stoichiometry Concept Map The map shows the quantities that are mole related by a balanced chemical equation. We can relate the (a) mass of the reactant, (b) volume of the gas, or (c) volume of the solution to the moles of the reactant. After converting to moles of a product, we can find the (d) mass of the product, (e) volume of the gas, or (f) volume of the solution. In a volume–volume problem, the reactants and products can be related in a single step using the coefficients of the balanced equation.

In Chapter 9, we first introduced chemical equation calculations. Later, in Section 15.4, we applied the rules of stoichiometry to acid-base reactions. In this section, we'll show the relationship of stoichiometry calculations to each other. That is, we'll learn how to solve mass-mass, mass-volume, volume-volume, and solution problems. To get an overall view of stoichiometric relationships, we can construct a concept map. Figure 18.4 is a stoichiometry concept map. It illustrates the relationship of the various quantities in a reaction.

Now let's reinforce our understanding of chemical equation calculations. Suppose we are given 4.24 g of sodium carbonate, Na_2CO_3, which we react with 0.150 *M* HCl. The equation for the reaction is

$$Na_2CO_3(s) + 2\ HCl(aq) \longrightarrow 2\ NaCl(aq) + H_2O(l) + CO_2(g)$$

Let's calculate the STP volume of CO_2 produced from the reaction. We see from the concept map that, first, we must convert the mass of Na_2CO_3 to moles. Second, we apply the coefficients of the equation to find the moles of CO_2. Third, we convert the moles of gas to a volume at STP. We can begin with a strategy map of the problem:

$$\text{g } Na_2CO_3 \longrightarrow \text{mol } Na_2CO_3 \longrightarrow \text{mol } CO_2 \longrightarrow \text{L } CO_2$$

Next, we show the unit analysis solution to the problem. By adding up the atomic masses from the periodic table, we find the molar mass of Na_2CO_3 is 106.0 g/mol. The value for molar volume is 22.4 L/mol. The calculation is as follows:

$$4.24\ \cancel{\text{g } Na_2CO_3} \times \frac{1\ \cancel{\text{mol } Na_2CO_3}}{106.0\ \cancel{\text{g } Na_2CO_3}} \times \frac{1\ \cancel{\text{mol } CO_2}}{1\ \cancel{\text{mol } Na_2CO_3}} \times \frac{22.4\ \text{L } CO_2}{1\ \cancel{\text{mol } CO_2}} = \text{L } CO_2$$

$$= 0.896\ \text{L } CO_2$$

Now we want to find the volume of hydrochloric acid that reacted to give the gas. Again, refer to the concept map. Then prepare a strategy map outline for the problem solution:

$$\text{g } Na_2CO_3 \longrightarrow \text{mol } Na_2CO_3 \longrightarrow \text{mol HCl} \longrightarrow \text{mL HCl}$$

As before, let's show the unit analysis solution to the problem. The molar mass of Na_2CO_3 is 106.0 g/mol. Since the hydrochloric acid is 0.150 *M* HCl, the unit factor is 0.150 mol HCl/1000 mL solution. The calculation is as follows:

$$4.24\ \cancel{\text{g } Na_2CO_3} \times \frac{1\ \cancel{\text{mol } Na_2CO_3}}{106.0\ \cancel{\text{g } Na_2CO_3}} \times \frac{2\ \cancel{\text{mol HCl}}}{1\ \cancel{\text{mol } Na_2CO_3}} \times \frac{1000\ \text{mL solution}}{0.150\ \cancel{\text{mol HCl}}}$$

$$= \text{mL solution}$$

$$= 533\ \text{mL solution}$$

Example Exercise 18.3 provides additional practice in solving chemical calculations involving the concentration of a solution.

EXAMPLE EXERCISE 18.3

Given that 37.5 mL of 0.100 *M* aluminum bromide react completely with 25.0 mL of silver nitrate solution:

$$AlBr_3(aq) + 3\ AgNO_3(aq) \longrightarrow 3\ AgBr(s) + Al(NO_3)_3(aq)$$

(a) What is the molarity of the $AgNO_3$ solution?
(b) What is the mass of AgBr precipitate?

Solution: We begin by verifying that the chemical equation is balanced. The equation is balanced because the coefficients are correct.
(a) We outline the plan for finding the molarity of aqueous $AgNO_3$:

$$\text{mL } AlBr_3 \longrightarrow \text{mol } AlBr_3 \longrightarrow \text{mol } AgNO_3 \longrightarrow M\ AgNO_3$$

From the balanced equation, we see that 1 mol of $AlBr_3$ reacts with 3 mol of $AgNO_3$. We calculate the moles of $AgNO_3$ as follows:

$$37.5\ \cancel{\text{mL } AlBr_3} \times \frac{0.100\ \cancel{\text{mol } AlBr_3}}{1000\ \cancel{\text{mL } AlBr_3}} \times \frac{3\text{ mol } AgNO_3}{1\ \cancel{\text{mol } AlBr_3}} = 0.0113\text{ mol } AgNO_3$$

The molarity of the $AgNO_3$ solution expresses the ratio of moles to volume. Since the answer requires compound units (mol/L), it is necessary to start with compound units (mol/mL). We are given that 25.0 mL of silver nitrate solution reacts. Therefore,

$$\frac{0.0113\text{ mol } AgNO_3}{25.0\ \cancel{\text{mL solution}}} \times \frac{1000\ \cancel{\text{mL}}}{1\text{ L}} = \frac{0.450\text{ mol } AgNO_3}{\text{L}}$$

$$= 0.450\ M\ AgNO_3$$

(b) To find the mass of precipitate, begin with a strategy map outline:

$$\text{mL } AlBr_3 \longrightarrow \text{mol } AlBr_3 \longrightarrow \text{mol } AgBr \longrightarrow \text{g } AgBr$$

From the balanced equation, we see 1 mol of $AlBr_3$ produces 3 mol of AgBr. We calculate the mass of AgBr as follows:

$$37.5\ \cancel{\text{mL } AlBr_3} \times \frac{0.100\ \cancel{\text{mol } AlBr_3}}{1000\ \cancel{\text{mL } AlBr_3}} \times \frac{3\ \cancel{\text{mol AgBr}}}{1\ \cancel{\text{mol } AlBr_3}} \times \frac{187.8\text{ g AgBr}}{1\ \cancel{\text{mol AgBr}}} = 2.11\text{ g AgBr}$$

SELF-TEST EXERCISE

Given that 27.5 mL of 0.210 *M* lithium iodide react completely with 0.133 *M* lead(II) nitrate solution:

$$Pb(NO_3)_2(aq) + 2\ LiI(aq) \longrightarrow PbI_2(s) + 2\ LiNO_3(aq)$$

(a) What volume of $Pb(NO_3)_2$ is required for complete precipitation?
(b) What is the mass of PbI_2 precipitate?

Answers: **(a)** 21.7 mL $Pb(NO_3)_2$; **(b)** 1.33 g PbI_2

Note: In Section 18.1, we considered several powerful techniques for analyzing and solving problems. This included strategy maps, concept maps, algorithms, visualization, unit analysis, and algebraic analysis. In Sections 18.2 and 18.3, we presented an overview of chemical formula calculations and stoichiometry. Now, we extend our understanding of problem solving to more complex problems, some of which will involve two or more chemical principles.

18.4 Limiting Reactant Concept

OBJECTIVES

To perform chemical equation calculations that involve a limiting reactant, given:

(a) the masses of two reactants

(b) the volumes of two gaseous reactants

In Chapter 9, we performed calculations using the information in a balanced chemical equation. For example, we considered a reaction between iron and sulfur according to the equation

$$Fe(s) + S(s) \longrightarrow FeS(s)$$

We solved typical stoichiometry problems such as the following: How many grams of FeS will be produced from the reaction of 10.0 g of iron? However, we assumed there was sufficient sulfur to react with the 10.0 g of iron. This assumption may not be true. What if there were not enough sulfur to react with the iron completely? In that case the sulfur would *limit* the amount of FeS produced. In a stoichiometry problem, the reactant that limits the amount of product is called the **limiting reactant**. Figure 18.5 is a visual model to help you understand the limiting reactant concept.

limiting reactant The substance in a chemical reaction that controls or limits the maximum amount of product formed.

Limiting Reactant Problems

Now that we have a visual model for the limiting reactant concept, let's examine data from an experiment and solve a problem. Suppose we heat 2.50 mol of iron with 3.00 mol of sulfur. How many moles of FeS are formed? We analyze the problem by considering the amount of each substance before and after the reaction.

Before the reaction begins, there is 2.50 mol of Fe, 3.00 mol of S, and 0.00 mol of FeS. According to the balanced equation, 1 mol of Fe reacts with 1 mol of S to give 1 mol of FeS. Therefore, 2.50 mol of Fe reacts with 2.50 mol of S to give 2.50 mol of FeS. At the outset, we had 3.00 mol of S. Therefore, sulfur is the excess

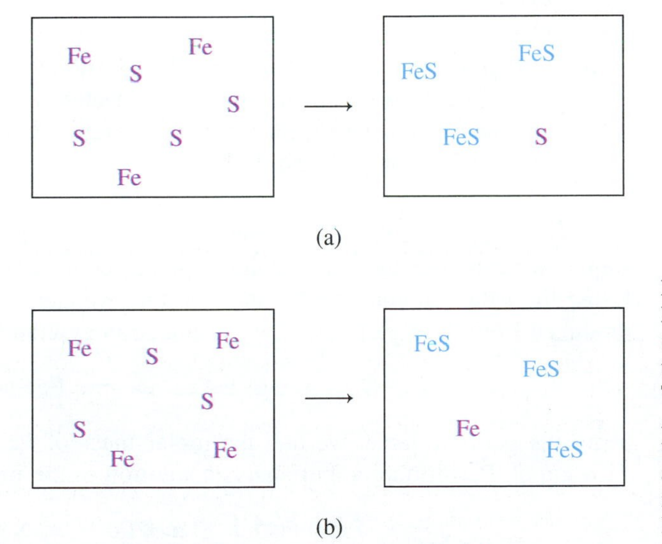

Figure 18.5 Model for Limiting Reactant Concept (a) As Fe and S react to give FeS, there is an excess amount of S. Thus, Fe is the limiting reactant. After the reaction, notice the unreacted S that remains. (b) In another reaction, the amount of Fe is in excess. Here, S is the limiting reactant.

Table 18.2 Summary for Experimental Synthesis of FeS

Conditions	mol Fe	mol S	mol FeS
before reaction	2.50	3.00	0.00
after reaction	0.00	0.50	2.50

reactant. After the reaction, the excess sulfur is 3.00 mol − 2.50 mol = 0.50 mol. Table 18.2 summarizes the amounts of each substance before and after the reaction.

Let's consider a more difficult example. Ferrous oxide is reduced to molten iron by heating it with aluminum metal. The reaction is

$$FeO(s) + Al(s) \longrightarrow Fe(l) + Al_2O_3(s)$$

Since all Stoichiometry problems require a chemical equation, we must balance the reactants and products. Thus,

$$3\ FeO(s) + 2\ Al(s) \longrightarrow 3\ Fe(l) + Al_2O_3(s)$$

Suppose we wish to find how much molten iron is produced from the reaction of 50.0 g of FeO with 25.0 g of Al. Since the masses of both reactants are given, it is not obvious whether FeO or Al limits the amount of product. There are several ways to solve limiting reactant problems. We will use the following approach.

1. *Calculate the maximum mass of the product that could be produced from the first reactant.*
   (a) calculate the moles of reactant
   (b) calculate the moles of product
   (c) calculate the mass of product
2. *Calculate the maximum mass of the product that could be produced from the second reactant.*
   (a) calculate the moles of reactant
   (b) calculate the moles of product
   (c) calculate the mass of product
3. *Determine the limiting reactant.* The limiting reactant is the reactant that produces the lesser amount of product. Determine the actual mass of product that can be obtained from the two reactants. It is the lesser of the masses obtained from steps 1 and 2.

Let's apply the three-step process to the ferrous oxide problem. According to step 1, we calculate the mass of the product obtained from the first reactant, FeO. Using the balanced equation for the reaction, we can calculate the amount of iron produced from 50.0 g of FeO. Let's use a strategy map to outline the solution:

$$\text{g FeO} \longrightarrow \text{mol FeO} \longrightarrow \text{mol Fe} \longrightarrow \text{g Fe}$$

From the periodic table, we find the molar mass of Fe is 55.8 g/mol and FeO is 71.8 g/mol. Performing a unit analysis solution to the problem, we have

$$50.0\ \cancel{\text{g FeO}} \times \frac{1\ \cancel{\text{mol FeO}}}{71.8\ \cancel{\text{g FeO}}} \times \frac{3\ \cancel{\text{mol Fe}}}{3\ \cancel{\text{mol FeO}}} \times \frac{55.8\ \text{g Fe}}{1\ \cancel{\text{mol Fe}}} = \text{g Fe}$$
$$= 38.9\ \text{g Fe}$$

CHEMISTRY CONNECTION □ APPLIED

Manufacturing Iron

▶ Can you state the difference between iron and steel?

The Iron Age followed the Bronze Age after it was discovered that iron could be made by heating iron ore with coal. Interestingly, iron is still obtained from a reaction that is similar.

Today, an industrial blast furnace makes iron by reducing iron ore. In the process, iron ore, coke (a form of coal), and limestone ($CaCO_3$) are added into the top of a heated furnace while a blast of hot air is blown in at the bottom of the furnace.

There are many reactions that take place in the process, but the main ones are as follows. First, oxygen in the hot air reacts with carbon in coke to produce carbon monoxide:

$$2\ C(s) + O_2(g) \longrightarrow 2\ CO(g)$$

The iron ore is an impure mixture of hematite, Fe_2O_3, and magnetite, Fe_3O_4. The iron ore is first reduced in the furnace to FeO. Next, the FeO is reduced to molten iron, which is collected at the bottom of the furnace. The reaction for the overall process is:

$$Fe_2O_3(s) + 3\ CO(g) \longrightarrow 2\ Fe(s) + 3\ CO_2(g)$$

Pouring molten iron into an industrial blast furnace

The iron ore also contains silicon compounds that must be removed during the process. One way that silicon compounds are removed is by reacting with CaO, which is produced from the reduction of limestone, $CaCO_3$. In the furnace, silicon compounds such as $CaSiO_3$ form and are referred to as *slag*. The slag is a liquid that floats on the molten iron and is easily removed.

The molten iron obtained from the blast furnace is not pure and is referred to as *pig iron*. Pig iron contains impurities such as carbon, silicon, and sulfur. In a second stage of the process, these impurities are removed by blowing oxygen gas through the molten pig iron. The impurities are converted to oxides and some are released as gases such as CO_2 and SO_2.

The iron obtained after the oxygen process contains a small amount of carbon. Manganese is then often added for strength and flexibility. The result is referred to as *steel*. By adding other metals, a large number of alloy steels can be produced. For example, stainless steel is an alloy that contains chromium and nickel.

▶ Iron is a pure element while steel is an alloy of iron with traces of carbon and manganese.

Next, using step 2, we calculate the mass of the product obtained from the second reactant. Using the balanced equation, we can calculate the amount of iron produced from 25.0 g of Al. The strategy map for solving the problem is

$$\text{g Al} \longrightarrow \text{mol Al} \longrightarrow \text{mol Fe} \longrightarrow \text{g Fe}$$

From the periodic table, we find that the molar mass of Al is 27.0 g/mol. The unit analysis solution to the problem is

$$25.0\ \cancel{\text{g Al}} \times \frac{1\ \cancel{\text{mol Al}}}{27.0\ \cancel{\text{g Al}}} \times \frac{3\ \cancel{\text{mol Fe}}}{2\ \cancel{\text{mol Al}}} \times \frac{55.8\ \text{g Fe}}{1\ \cancel{\text{mol Fe}}} = \text{g Fe}$$

$$= 77.5\ \text{g Fe}$$

Using step 3, we now compare the mass of product from each of the reactants. This gives us

$$50.0 \text{ g FeO} \longrightarrow 38.9 \text{ g Fe}$$

$$25.0 \text{ g Al} \longrightarrow 77.5 \text{ g Fe}$$

We see that FeO yields a smaller mass of product and therefore is the limiting reactant. Al is the excess reactant, and not all the aluminum metal is used by the reaction. The maximum yield of molten iron from 50.0 g of FeO and 25.0 g of Al is therefore 38.9 g of Fe.

Example Exercise 18.4 provides additional practice involving a limiting reactant calculation.

EXAMPLE EXERCISE 18.4

Given that 125.0 g of manganese dioxide reacts with 50.0 g of aluminum, find the limiting reactant and the mass of manganese metal produced from the reaction. The equation for the reaction is

$$3\ MnO_2(s) + 4\ Al(s) \longrightarrow 3\ Mn(l) + 2\ Al_2O_3(s)$$

Solution: We begin by verifying that the chemical equation is balanced. The coefficients are correct, so the equation is balanced. Step 1 is to calculate the maximum mass of Mn metal obtained from 125.0 g of MnO_2. We plan the solution as follows:

$$\text{g } MnO_2 \longrightarrow \text{mol } MnO_2 \longrightarrow \text{mol Mn} \longrightarrow \text{g Mn}$$

From the periodic table, we find that the molar mass of MnO_2 is 86.9 g/mol and Mn is 54.9 g/mol. The unit analysis solution to the problem is

$$125.0 \cancel{\text{g } MnO_2} \times \frac{1 \cancel{\text{mol } MnO_2}}{86.9 \cancel{\text{g } MnO_2}} \times \frac{3 \cancel{\text{mol Mn}}}{3 \cancel{\text{mol } MnO_2}} \times \frac{54.9 \text{ g Mn}}{1 \cancel{\text{mol Mn}}} = \text{g Mn}$$
$$= 79.0 \text{ g Mn}$$

In step 2 we calculate the maximum yield of product obtained from the second reactant. A strategy map for solving this problem is

$$\text{g Al} \longrightarrow \text{mol Al} \longrightarrow \text{mol Mn} \longrightarrow \text{g Mn}$$

From the periodic table, we find that the molar mass of Al is 27.0 g/mol. Starting with 50.0 g of Al, the unit analysis solution is

$$50.0 \cancel{\text{g Al}} \times \frac{1 \cancel{\text{mol Al}}}{27.0 \cancel{\text{g Al}}} \times \frac{3 \cancel{\text{mol Mn}}}{4 \cancel{\text{mol Al}}} \times \frac{54.9 \text{ g Mn}}{1 \cancel{\text{mol Mn}}} = \text{g Mn}$$
$$= 76.3 \text{ g Mn}$$

Step 3 is to compare mass of product from each reactant. This gives us

$$125.0 \text{ g } MnO_2 \longrightarrow 79.0 \text{ g Mn}$$

$$50.0 \text{ g Al} \longrightarrow 76.3 \text{ g Mn}$$

In this example, Al is the *limiting reactant* because it yields less mass of product. Not all of the manganese dioxide is used. Therefore, MnO_2 is the *excess reactant*. The maximum yield of Mn metal obtained from 125.0 g of MnO_2 and 50.0 g of Al is 76.3 g of Mn.

SELF-TEST EXERCISE

Given that 14.2 g of cobalt metal react with 50.0 g of sulfuric acid as follows:

$$2\ Co(s) + 3\ H_2SO_4(aq) \longrightarrow Co_2(SO_4)_3(s) + 3\ H_2(g)$$

(a) What is the limiting reactant?
(b) What is the mass of cobalt(III) sulfate precipitate?

Answers: **(a)** cobalt; **(b)** 48.9 g $Co_2(SO_4)_3$

Limiting Reactant Problems Involving Gases

Now let's try a limiting reactant problem that involves two reactants in the gaseous state. For example, consider ammonia, which is manufactured from the reaction of nitrogen and hydrogen gases using a metal catalyst. The balanced equation for the reaction is

$$N_2(g) + 3\ H_2(g) \xrightarrow{Fe/Al_2O_3} 2\ NH_3(g)$$

Suppose that 1.65 L of nitrogen gas reacts with 4.75 L of hydrogen gas at 500°C and 500 atm. Since the volumes of both reacting gases are given, it is not obvious whether N_2 or H_2 limits the amount of ammonia. If we assume that temperature and pressure remain constant, we can use the following general approach.

1. *Calculate the volume of product produced from the first reactant.* Recall that the coefficients of the equation are in the same ratio as the volumes of gases (Gay-Lussac's law of combining volumes).
2. *Calculate the volume of product produced from the second reactant.* (Again, use the coefficients in the balanced chemical equation to convert directly from reactants to products.)
3. *Determine the limiting reactant.* It is the reactant that produces the least volume of gas. Determine the maximum amount of gas that can be obtained from the two reactants. It is the lesser of the two volumes from steps 1 and 2.

Let's apply step 1 to calculate the volume of ammonia produced from 1.65 L of nitrogen gas. Using the balanced equation for the reaction, we can draw a strategy map for the solution. It is

$$L\ N_2 \longrightarrow L\ NH_3$$

From the balanced chemical equation, we see that 1 L of N_2 gives 2 L of NH_3. Therefore, the unit analysis solution to the problem is

$$1.65\ \cancel{L\ N_2} \times \frac{2\ L\ NH_3}{1\ \cancel{L\ N_2}} = L\ NH_3$$
$$= 3.30\ L\ NH_3$$

Now, we'll use step 2 and calculate the maximum volume of ammonia obtained from the second reactant. The strategy map is

$$L\ H_2 \longrightarrow L\ NH_3$$

From the balanced chemical equation, we note that 3 L of H_2 gives 2 L of NH_3.

Therefore, the unit analysis solution to the problem is

$$4.75\ \cancel{\text{L H}_2} \times \frac{2\ \text{L NH}_3}{3\ \cancel{\text{L H}_2}} = \text{L NH}_3$$

$$= 3.17\ \text{NH}_3$$

Using step 3, we compare the volume of product from each reactant. This gives

$$1.65\ \text{L N}_2 \longrightarrow 3.30\ \text{L NH}_3$$

$$4.75\ \text{L H}_2 \longrightarrow 3.17\ \text{L NH}_3$$

In this example, H_2 is the limiting reactant because it yields the lesser volume of product. Conversely, N_2 is the excess reactant, and not all the nitrogen gas is converted to ammonia. Thus, the maximum yield of NH_3 from 1.65 L of N_2 and 4.75 L of H_2 is 3.17 L NH_3.

Example Exercise 18.5 provides additional practice for a limiting reactant calculation involving gases.

EXAMPLE EXERCISE 18.5

Methane is the main component in natural gas and undergoes combustion with oxygen to give carbon dioxide and water. The chemical equation is

$$CH_4(g) + 2\ O_2(g) \xrightarrow{\text{spark}} CO_2(g) + 2\ H_2O(g)$$

If 25.0 mL of CH_4 react with 55.0 mL of O_2, what is the limiting reactant? Assuming constant conditions, what is the volume of CO_2 produced?

Solution: After checking the coefficients, we note that the equation is balanced. Step 1 is to calculate the maximum volume of CO_2 gas obtained from 25.0 mL of CH_4. We plan the solution as follows:

$$\text{mL CH}_4 \longrightarrow \text{mL CO}_2$$

From the equation, we observe that 1 volume of CH_4 gives 1 volume of CO_2. Therefore, the unit analysis solution to the problem is

$$25.0\ \cancel{\text{mL CH}_4} \times \frac{1\ \text{mL CO}_2}{1\ \cancel{\text{mL CH}_4}} = \text{mL CO}_2$$

$$= 25.0\ \text{mL CO}_2$$

Step 2 is to calculate the volume of product obtained from the second reactant. The outline for solving the problem is

$$\text{mL O}_2 \longrightarrow \text{mL CO}_2$$

From the chemical equation, we note that 2 volumes of O_2 give 1 volume of CO_2. The unit analysis solution to the problem is

$$55.0\ \cancel{\text{mL O}_2} \times \frac{1\ \text{mL CO}_2}{2\ \cancel{\text{mL O}_2}} = \text{mL CO}_2$$

$$= 27.5\ \text{mL CO}_2$$

Step 3 is to compare the volume of product from each reactant.

$$25.0\ \text{mL CH}_4 \longrightarrow 25.0\ \text{mL CO}_2$$

$$55.0\ \text{mL O}_2 \longrightarrow 27.5\ \text{mL CO}_2$$

The *limiting reactant* is CH_4 because it yields less volume of product. On the other hand, O_2 is the *excess reactant* and not all the oxygen gas reacts. The maximum volume of CO_2 produced from 25.0 mL of CH_4 and 55.0 mL of O_2 is therefore 25.0 mL of CO_2.

SELF-TEST EXERCISE

Ethane undergoes combustion with oxygen to give carbon dioxide and water. The equation for the reaction is

$$2\ C_2H_6(g) + 7\ O_2(g) \xrightarrow{\text{spark}} 4\ CO_2(g) + 6\ H_2O(g)$$

If 10.0 L of C_2H_6 react with 25.0 L of O_2, what is the limiting reactant? Assuming constant conditions, what is the volume of CO_2 produced?

Answer: The limiting reactant is O_2, which gives 14.3 L of CO_2.

In this section, we applied the limiting reactant concept to reactants in the solid state and to reactants in the gaseous state. There are also other types of limiting reactant problems. One type is the reaction of a metal and a gas. For example, if zinc metal reacts with chlorine gas, either the metal or gas can limit the amount of product. The chemical equation is

$$Zn(s) + Cl_2(g) \longrightarrow ZnCl_2(s)$$

Another variation on the limiting reactant concept is the reaction of a metal and a volume of aqueous acid. For example, if zinc reacts with 0.500 *M* sulfuric acid, either the zinc or the acid can be the limiting reactant. That is,

$$Zn(s) + H_2SO_4(aq) \longrightarrow ZnSO_4(aq) + H_2(g)$$

Although these two examples differ from the preceding illustrations, the basic problem-solving approach would be the same.

1. *Calculate the quantity of product from the first reactant.*
2. *Calculate the quantity of product from the second reactant.*
3. *Determine the limiting reactant.* The limiting reactant is the one that produces the least amount of product. The maximum amount of product that can be obtained is the lesser of these two quantities.

18.5 Thermochemical Stoichiometry

To write thermochemical equations for endothermic and exothermic reactions.

To perform chemical equation calculations that involve the heat of reaction, ΔH.

At this point, we know about obtaining information from a balanced chemical equation. We also know that *endothermic reactions* proceed by absorbing heat energy and *exothermic reactions* by releasing heat energy. We will now define the amount

heat of reaction (ΔH) The amount of heat energy absorbed or released by a given substance according to the balanced chemical equation.

thermochemical equation A balanced chemical equation that incorporates the amount of heat energy involved in the chemical reaction.

of heat energy gained or lost for a chemical change as the **heat of reaction** (symbol **ΔH**). Our next step is to relate the amount of heat energy absorbed or released by a given reaction to a balanced chemical equation.

Thermochemical Equations

A chemical equation that indicates the amount of heat energy involved in the change is a **thermochemical equation**. A thermochemical equation can refer to an endothermic or an exothermic reaction. For example, a method for producing fluorine gas is the reaction of hydrogen fluoride and chlorine gases. That is,

$$2\ HF(g) + Cl_2(g) \longrightarrow 2\ HCl(g) + F_2(g)$$

The reaction is endothermic and requires 85.5 kilocalories of heat energy per mole of fluorine produced. By convention, the heat of reaction for an endothermic reaction has a positive value. It is written as

$$\Delta H = +85.5\ \text{kcal/mol}$$

The corresponding thermochemical equation is

$$2\ HF(g) + Cl_2(g) + 85.5\ \text{kcal} \longrightarrow 2\ HCl(g) + F_2(g)$$

Now let's consider an exothermic reaction. Methanol, CH_3OH, is an autoracing fuel. It undergoes combustion as follows:

$$2\ CH_3OH(g) + 3\ O_2(g) \longrightarrow 2\ CO_2(g) + 4\ H_2O(g)$$

The reaction releases a large amount of heat energy, 323.4 kcal per mole of methanol. Since the balanced equation involves 2 mol of CH_3OH, the heat from the reaction is 646.8 kcal (2×323.4 kcal). By convention, the heat of reaction for an exothermic reaction has a negative value, It is written as

$$\Delta H = -646.8\ \text{kcal/mol}$$

The thermochemical equation for the combustion of methanol is

$$2\ CH_3OH(g) + 3\ O_2(g) \longrightarrow 2\ CO_2(g) + 4\ H_2O(g) + 646.8\ \text{kcal}$$

In review, the steps for writing thermochemical equations are as follows.

1. Write the balanced chemical equation for the reaction.
2. For endothermic reactions (ΔH is a *positive value*), add the heat of reaction to the reactant's side of the equation.
3. For exothermic reactions (ΔH is *a negative value*), add the heat of reaction to the product's side of the equation.

Example Exercise 18.6 provides additional practice in writing thermochemical equations.

EXAMPLE EXERCISE 18.6

Write a balanced thermochemical equation for each of the following.

(a) The reaction of hydrogen and iodine gases is endothermic and gives hydrogen iodide gas. The heat of reaction is 6.2 kcal/mol of HI.

(b) The reaction of hydrogen and oxygen gases is exothermic and gives water vapor. The heat of reaction is -57.8 kcal/mol of H_2O.

Solution: First, we must write a balanced chemical equation based on the description of the reaction.

(a) Hydrogen and iodine gases combine as follows:

$$H_2(g) + I_2(g) \longrightarrow 2\ HI(g)$$

Since the balanced equation involves 2 mol of HI, the heat of reaction is 2×6.2 kcal $= 12.4$ kcal. The thermochemical equation for the formation of HI is endothermic. Hence,

$$H_2(g) + I_2(g) + 12.4\ \text{kcal} \longrightarrow 2\ HI(g)$$

(b) Hydrogen and oxygen gases combine as follows:

$$2\ H_2(g) + O_2(g) \longrightarrow 2\ H_2O(g)$$

Since the balanced equation involves 2 mol of H_2O, the heat of reaction is 2×-57.8 kcal $= -115.6$ kcal. The thermochemical equation for the formation of H_2O is exothermic. Thus,

$$2\ H_2(g) + O_2(g) \longrightarrow 2\ H_2O(g) + 115.6\ \text{kcal}$$

SELF-TEST EXERCISE

Write a balanced thermochemical equation for the reaction of hydrogen and chlorine gases. (Given: $\Delta H = -22.1$ kcal/mol of hydrogen chloride, HCl.)

Answer: $H_2(g) + Cl_2(g) \longrightarrow 2\ HCl(g) + 44.2\ \text{kcal}$

Thermochemical Stoichiometry

Stoichiometry relates quantities of substance in a balanced chemical equation. **Thermochemical stoichiometry** relates quantities of substance to the heat of reaction. Using this information, we can solve problems involving chemical equations that produce or consume heat. For example, let's determine how much heat a recreational vehicle releases. When propane burns, it releases heat according to the equation

thermochemical stoichiometry The relationship of quantities of substance to the heat of reaction according to a balanced chemical equation.

$$C_3H_8(g) + 5\ O_2(g) \longrightarrow 3\ CO_2(g) + 4\ H_2O(g) + 530.6\ \text{kcal}$$

If a recreational vehicle burns 75.0 g of C_3H_8, how much heat is released in the process? Since 1 mol of C_3H_8 releases 530.6 kcal of heat, we can outline our strategy for solving the problem as follows:

$$\text{g } C_3H_8 \longrightarrow \text{mol } C_3H_8 \longrightarrow \text{kcal}$$

As usual, we can effectively use unit analysis to solve the problem. By adding the atomic masses from the periodic table, the molar mass of C_3H_8 is determined to be 44.0 g/mol. The solution is as follows:

$$75.0\ \cancel{\text{g } C_3H_8} \times \frac{1\ \cancel{\text{mol } C_3H_8}}{44.0\ \cancel{\text{g } C_3H_8}} \times \frac{530.6\ \text{kcal}}{1\ \cancel{\text{mol } C_3H_8}} = \text{kcal}$$

$$= 904\ \text{kcal}$$

Example Exercises 18.7 and 18.8 illustrate variations of the previous thermochemical calculation.

EXAMPLE EXERCISE 18.7

Oxygen and acetylene, C_2H_2, are used in oxyacetylene welding. The equation for the reaction is

$$2\ C_2H_2(g) + 5\ O_2(g) \longrightarrow 4\ CO_2(g) + 2\ H_2O(g) + 626.2\ \text{kcal}$$

Calculate the volume of acetylene (at STP) that must react in order to release 500.0 kcal of heat energy.

Oxyacetylene welding uses the heat of combustion from oxygen and acetylene gases.

Solution: We begin by verifying that the chemical equation is balanced. The equation is balanced because the coefficients are correct. Next, we outline the plan for finding the volume of C_2H_2. This gives us

$$\text{kcal} \longrightarrow \text{mol } C_2H_2 \longrightarrow \text{L } C_2H_2\ (\text{STP})$$

From the thermochemical equation, we see that 626.2 kcal of heat are released for every 2 mol of C_2H_2 that react. At STP, 1 mol of C_2H_2 occupies a volume of 22.4 liters. The unit analysis solution is

$$500.0\ \cancel{\text{kcal}} \times \frac{2\ \cancel{\text{mol } C_2H_2}}{626.2\ \cancel{\text{kcal}}} \times \frac{22.4\ \text{L } C_2H_2}{1\ \cancel{\text{mol } C_2H_2}} = \text{L } C_2H_2\ (\text{STP})$$
$$= 35.8\ \text{L } C_2H_2$$

SELF-TEST EXERCISE

How much heat is released from the combustion of 10.0 g of acetylene, C_2H_2? The heat of reaction $\Delta H = -313.1$ kcal/mol of acetylene.

$$2\ C_2H_2(g) + 5\ O_2(g) \longrightarrow 4\ CO_2(g) + 2\ H_2O(g)$$

Answer: 120 kcal

EXAMPLE EXERCISE 18.8

Limestone, $CaCO_3$, decomposes with heat according to the equation

$$CaCO_3(s) \xrightarrow{\Delta} CaO(s) + CO_2(g)$$

Calculate the mass of quicklime, CaO, produced by heating a large sample of $CaCO_3$ with 25.0 kcal of heat energy. (Given: $\Delta H = +42.1$ kcal/mol $CaCO_3$.)

Solution: Let's begin by writing the thermochemical equation for the reaction. Since the ΔH value is positive, the reaction is endothermic. That is,

$$CaCO_3(s) + 42.1 \text{ kcal} \longrightarrow CaO(s) + CO_2(g)$$

Next, we outline the plan for finding the mass of CaO. It is

$$\text{kcal} \longrightarrow \text{mol } CaCO_3 \longrightarrow \text{mol CaO} \longrightarrow \text{g CaO}$$

From the balanced equation, we see that 42.1 kcal are required for every 1 mol of $CaCO_3$ that decomposes. Moreover, 1 mol of $CaCO_3$ produces 1 mol of CaO. The molar mass of CaO is calculated from the periodic table to be 56.1 g/mol. The unit analysis solution to the problem is

$$25.0 \cancel{\text{kcal}} \times \frac{1 \cancel{\text{mol } CaCO_3}}{42.1 \cancel{\text{kcal}}} \times \frac{1 \cancel{\text{mol CaO}}}{1 \cancel{\text{mol } CaCO_3}} \times \frac{56.1 \text{ g CaO}}{1 \cancel{\text{mol CaO}}} = \text{g CaO}$$

$$= 33.3 \text{ g CaO}$$

SELF-TEST EXERCISE

How much heat is required to decompose $CaCO_3$ in order to give 10.0 L of CO_2 gas at STP conditions?

$$CaCO_3(s) + 42.1 \text{ kcal} \longrightarrow CaO(s) + CO_2(g)$$

Answer: 18.8 kcal

In this section, we chose the kilocalorie (kcal) as the unit of heat energy. A *kilocalorie* is the amount of heat energy required to raise the temperature of 1 kilogram of water 1°C. Thermochemical equations can also be written using kilojoule (kJ) as the unit of energy. Although the kcal unit is more appropriate for our discussion, the kJ unit is widely used to express heats of reaction. These two units of energy are easily converted using the relationship 1 kilocalorie (kcal) = 4.184 kilojoules (kJ).

18.6 Multiple-Reaction Stoichiometry

OBJECTIVES

To perform chemical equation calculations that involve two or more reactions.

In industy, it is important to perform chemical equation calculations to determine the cost of manufacturing a chemical. Moreover, many industrial processes involve a series of two or more chemical reactions. For example, the Ostwald process for making nitric acid, HNO_3, and the contact process for making sulfuric acid, H_2SO_4, each consists of three reactions. Fortunately, the stoichiometry techniques we have learned can be extended to multiple reactions.

One of the most important industrial chemicals is sodium carbonate, Na_2CO_3. Its common name is soda ash. Soda ash is mined directly from the rich deposits in Wyoming and the Mojave Desert in California. It is also chemically manufactured. In 1869, Ernest Solvay invented the industrial method for producing soda ash, which bears his name. The first reaction in the Solvay process is to convert carbon dioxide and aqueous ammonia to ammonium bicarbonate.

$$CO_2(g) + NH_3(g) + H_2O(l) \longrightarrow NH_4HCO_3(aq) \qquad \textbf{(1)}$$

In the second reaction, ammonium bicarbonate undergoes a double-replacement reaction to give sodium bicarbonate, $NaHCO_3$ (baking soda), that is reasonably insoluble in the solution:

$$NH_4HCO_3(aq) + NaCl(aq) \longrightarrow NaHCO_3(s) + NH_4Cl(aq) \qquad \textbf{(2)}$$

The third reaction decomposes baking soda thermally to give soda ash, Na_2CO_3:

$$2\ NaHCO_3(s) \longrightarrow Na_2CO_3(s) + CO_2(g) + H_2O(g) \qquad \textbf{(3)}$$

Let's suppose we wish to calculate the mass of soda ash produced from 75.0 g of carbon dioxide gas. We begin by analyzing the related quantities in the problem. We can outline the relationship of CO_2 to Na_2CO_3 as follows:

$$\text{mol } CO_2 \longrightarrow \text{mol } NH_4HCO_3 \longrightarrow \text{mol } NaHCO_3 \longrightarrow \text{mol } Na_2CO_3$$

Starting with CO_2, we carefully followed the element carbon as it was converted from one substance to another. In this problem we wish to calculate the mass of Na_2CO_3 derived from 75.0 g of CO_2.

In a multiple-reaction problem, we need to apply additional unit factors. That is, we need an extra unit factor to find the moles of substance in each separate reaction. An abbreviated plan for solving the problem is therefore

$$75.0 \text{ g } CO_2 \times \text{unit factors} = \text{g } Na_2CO_3$$

The general unit analysis approach that applies to single-reaction-problem is, however, still valid. The unit analysis solution is

$$75.0 \text{ g } \cancel{CO_2} \times \frac{1 \cancel{\text{mol } CO_2}}{44.0 \cancel{\text{g } CO_2}} \times \frac{1 \cancel{\text{mol } NH_4HCO_3}}{1 \cancel{\text{mol } CO_2}} \times \frac{1 \cancel{\text{mol } NaHCO_3}}{1 \cancel{\text{mol } NH_4HCO_3}}$$

$$\times \frac{1 \cancel{\text{mol } Na_2CO_3}}{2 \cancel{\text{mol } NaHCO_3}} \times \frac{106.0 \text{ g } Na_2CO_3}{1 \cancel{\text{mol } Na_2CO_3}} = \text{g } Na_2CO_3$$

$$= 90.3 \text{ g } Na_2CO_3$$

In practice, the CO_2 produced in (3) is recycled and additional soda ash is produced.

Notice that five unit factors were required to solve this problem. Ordinarily, mass-mass stoichiometry problems require three unit factors. In this case, there were three equations to consider. Thus, two extra unit factors were required for the solution. Example Exercise 18.9 illustrates multiple-reaction stoichiometry for the manufacture of nitric acid from ammonia.

EXAMPLE EXERCISE 18.9

Nitric acid is prepared industrially from the conversion of ammonia. There are three steps in the conversion, referred to as the Ostwald process. They are

$$4\ NH_3(g) + 5\ O_2(g) \xrightarrow{Pt/850°C} 4\ NO(g) + 6\ H_2O(g)$$

$$2\ NO(g) + O_2(g) \longrightarrow 2\ NO_2(g)$$

$$3\ NO_2(g) + H_2O(l) \longrightarrow 2\ HNO_3(aq) + NO(g)$$

What is the mass of HNO_3 (63.0 g/mol) produced, assuming 10.0 L of NH_3 at STP is completely converted to nitric acid?

Solution: We can plan our strategy for the relationship of NH_3 to HNO_3 as follows:

$$\text{L } NH_3 \longrightarrow \text{mol } NH_3 \longrightarrow \text{mol NO} \longrightarrow \text{mol } NO_2$$

$$\text{g } HNO_3 \longleftarrow \text{mol } HNO_3 \longleftarrow$$

Starting with NH_3, we followed the conversion of nitrogen to another substance containing the nitrogen atom. We are asked to calculate the mass of HNO_3 from 10.0 L of NH_3. Therefore, let's show the general format for the unit analysis calculation. It is

$$10.0\text{ L } NH_3 \times \text{unit factors} = \text{g } HNO_3$$

We see from the strategy map that five unit factors are required for the conversion. The complete unit analysis solution to the problem is as follows:

$$10.0\ \cancel{\text{L } NH_3} \times \frac{1\ \cancel{\text{mol } NH_3}}{22.4\ \cancel{\text{L } NH_3}} \times \frac{4\ \cancel{\text{mol NO}}}{4\ \cancel{\text{mol } NH_3}} \times \frac{2\ \cancel{\text{mol } NO_2}}{2\ \cancel{\text{mol NO}}} \times \frac{2\ \cancel{\text{mol } HNO_3}}{3\ \cancel{\text{mol } NO_2}}$$

$$\times \frac{63.0\text{ g } HNO_3}{1\ \cancel{\text{mol } HNO_3}} = \text{g } HNO_3$$

$$= 18.8\text{ g } HNO_3$$

The conversion of 10.0 L of NH_3 to 18.8 g of HNO_3 requires three separate reactions. We have assumed that in each reaction we have obtained a 100% yield. For most industrial processes, the actual yield is always less. The Ostwald process is approximately 90% efficient.

SELF-TEST EXERCISE

A large quantity of sulfur is obtained industrially from "sour" natural gas containing hydrogen sulfide. First, H_2S is burned in air to give sulfur dioxide. Second, the SO_2 is reacted with additional hydrogen sulfide as follows:

$$2\ H_2S(g) + 3\ O_2(g) \longrightarrow 2\ SO_2(g) + 2\ H_2O(g)$$

$$SO_2(g) + 2\ H_2S(g) \longrightarrow 3\ S(l) + 2\ H_2O(g)$$

Calculate the volume oxygen gas at STP that is required to produce 1.00 kg of elemental surfur.

Answer: 349 L O_2 at STP

Multiple-reaction stoichiometry problems involve two or more balanced chemical equations. These stoichiometry problems can therefore relate any of the quantities we have previously studied for single reactions, such as mass to mass, mass to volume, or volume to volume. Although multiple-reaction problems are more sophisticated, the calculations vary only in the additional unit factors in the conversion.

18.7 Advanced Problem-Solving Examples

To perform calculations that involve two or more chemical principles.

The chemical calculations we have solved up to now involved only one chemical principle. Therefore, we have been able to solve problems systematically by using a set of rules. Most of the time, we have applied the three basic steps in the unit analysis method of problem solving. Now it is time to tackle more challenging problems. The problems that follow involve two or more chemical principles. To solve them, we will have to analyze each problem carefully and then develop a strategy for its solution.

Although we have studied only introductory chemical principles, there is a very large number of combinations and variations of these principles. The problem examples that follow have been selected to illustrate the most important concepts we have studied. These examples by no means represent all the possible problems you may encounter.

Illustration 1: Metric System and Density

Let's consider a calculation that involves the metric system and density. In the following problem you will need to perform a metric-English conversion and apply the density concept.

EXAMPLE EXERCISE 18.10

Lithium is the lightest metallic element. Its density is 0.534 g/mL. What is the mass of lithium in a 1.00-inch cube of the metal?

Solution: We always begin a problem analysis by determining the quantity asked for. Since the quantity is mass, we will choose units of grams. Next, we notice that the density is in metric units and the volume is in English units. Let's plan the solution as two separate problems. First, we convert the volume from English to metric units.

$$\text{in.}^3 \text{ Li} \longrightarrow \text{cm}^3 \text{ Li} \longrightarrow \text{mL Li}$$

Second, we convert the volume of Li to a mass using the principle of density.

$$\text{mL Li} \longrightarrow \text{g Li}$$

The cube of lithium is 1.00 in. on each side. Therefore, the volume of the cube is

1.00 in. × 1.00 in. × 1.00 in. = 1.00 in.^{3}. Recall that 1.00 in. = 2.54 cm and 1.00 cm^3 = 1 mL. We can now perform the following unit conversion:

$$1.00 \text{ in.}^3 \times \frac{(2.54 \text{ cm})^3}{(1.00 \text{ in.})^3} \times \frac{1.00 \text{ mL}}{1.00 \text{ cm}^3} = \text{mL Li}$$

$$1.00 \text{ in.}^3 \times \frac{16.4 \text{ cm}^3}{1.00 \text{ in.}^3} \times \frac{1.00 \text{ mL}}{1.00 \text{ cm}^3} = 16.4 \text{ mL Li}$$

Since the density of lithium is 0.534 g/mL, 1 mL = 0.534 g. Converting from volume to mass, we have

$$16.4 \text{ mL} \times \frac{0.534 \text{ g}}{1.00 \text{ mL}} = 8.76 \text{ g Li}$$

For comparison, gold is one of the most dense metals in the periodic table. A 1.00-inch cube of gold weighs more than 300 g, more than 30 times the mass of the lithium!

SELF-TEST EXERCISE

Calculate the mass of water in an ice cube measuring 2.00 in. on a side. The density of ice is 0.917 g/cm^3.

Answer: The strategy for solving this problem is as follows:

$$\text{in.}^3\ H_2O \longrightarrow \text{cm}^3\ H_2O \longrightarrow \text{g}\ H_2O$$

The ice cube has a volume of 8.00 in.^{3}, or 131 cm^3; its mass is 120 g (1.20×10^2 g).

Illustration 2: Measurement and Mole Concept

Consider a problem that involves measurement and the mole concept. In the following example exercise, we will use the concentration of lead in gasoline and the mole concept to calculate the number of lead atoms released into the environment.

EXAMPLE EXERCISE 18.11

In 1991, the EPA set a maximum limit of 0.10 g lead in 1 gallon of gasoline. If an automobile travels 2186 miles from Chicago to San Francisco, averaging 25.0 miles per gallon, how many lead atoms enter the environment? (Assume all the lead in the gasoline used is released into the environment.)

Solution: We begin by asking what quantity is asked for. In this case, it is the number of lead atoms. Given the miles, mileage, and concentration, we can find the mass of lead as follows:

$$\text{mi} \longrightarrow \text{gal} \longrightarrow \text{g Pb}$$

We can use the mole concept to convert from mass to the number of Pb atoms; that is,

$$\text{g Pb} \longrightarrow \text{mol Pb} \longrightarrow \text{Pb atoms}$$

From the gas mileage, we have the relationship 1 gal = 25.0 mi. From the lead concentration, we have 0.10 g Pb = 1 gal. Applying unit analysis,

$$2186 \text{ mi} \times \frac{1 \text{ gal}}{25.0 \text{ mi}} \times \frac{0.10 \text{ g Pb}}{1 \text{ gal}} = 8.7 \text{ g Pb}$$

One mole of lead contains Avogadro's number of lead atoms. We can refer to the periodic table to find that 1 mol Pb = 207.2 g. Thus,

$$8.7 \cancel{\text{g Pb}} \times \frac{1 \cancel{\text{mol Pb}}}{207.2 \cancel{\text{g Pb}}} \times \frac{6.02 \times 10^{23} \text{ Pb atoms}}{1 \cancel{\text{mol Pb}}} = 2.5 \times 10^{22} \text{ Pb atoms}$$

In humans, a concentration of only 0.05 mg of lead in one liter of urine is an indication of lead poisoning.

SELF-TEST EXERCISE

A compact car travels 375 miles with an average gas mileage of 48.0 mi/gal. If 1.00 gallon of gasoline produces 8184 g of CO_2, how many liters of carbon dioxide at STP are released during the trip? (Given: 1 mi = 1.61 km)

Answer: The strategy for solving this problem is as follows:

$$\text{mi} \longrightarrow \text{gal} \longrightarrow \text{g CO}_2 \longrightarrow \text{mol CO}_2 \longrightarrow \text{L CO}_2$$

On the trip, 32,600 L (STP) of carbon dioxide are released.

Illustration 3: Empirical Formula and Ideal Gas Law

Let's consider a calculation that involves an empirical formula and the ideal gas law. We will first find the empirical formula of an organic vapor and then use a gas law calculation to find the actual formula of the compound.

EXAMPLE EXERCISE 18.12

Cyclohexane is used as a paint remover and in fungicidal preparations. An instrumental analysis of the compound gave 85.7% C and 14.3% H. If the cyclohexane vapor in a 0.500-L flask at 100°C and 0.995 atm weighed 1.37 g, what is the molecular formula of cyclohexane?

Solution: In this problem we are asked to find the molecular formula. Recall that the molecular formula corresponds to a number of repeating empirical formula units; that is $(C_xH_y)_n$. We must first determine the empirical formula from the percentage composition. If we assume a 100-g sample of cyclohexane, there are 85.7 g of C and 14.3 g of H. We can calculate the empirical formula as follows:

$$85.7 \cancel{\text{g C}} \times \frac{1 \text{ mol C}}{12.0 \cancel{\text{g C}}} = 7.14 \text{ mol C}$$

$$14.3 \cancel{\text{g H}} \times \frac{1 \text{ mol H}}{1.01 \cancel{\text{g H}}} = 14.2 \text{ mol H}$$

The empirical formula is in the ratio of $C_{7.14}H_{14.2}$. Dividing by the smaller mole value and simplifying, we have

$$C_{\frac{7.14}{7.14}}H_{\frac{14.2}{7.14}} = C_1H_{1.99} \approx C_1H_2$$

Next, let's calculate the molar mass of cyclohexane from the vapor density information. In this example, it is more convenient to use an algebraic approach. We can use the equation $PV = nRT$. Recall that R is a constant with a value of 0.0821 atm · L/mol · K. Substituting the given information into the equation, we have

$$0.995 \text{ atm} \times 0.500 \text{ L} = n \times 0.0821 \text{ atm} \cdot \text{L/mol} \cdot \text{K} \times 373 \text{ K}$$

Rearranging and solving, we have

$$n = 0.0162 \text{ mol}$$

We are asked to find the molar mass, that is the units of g/mol. The mass of vapor in the flask is given, 1.37 g. Therefore,

$$\frac{1.37 \text{ g}}{0.0162 \text{ mol}} = 84.6 \text{ g/mol}$$

Since the empirical formula is CH_2, the molecular formula is $(CH_2)_n$. The molar mass of the empirical formula unit, CH_2, is 14.0 g/mol. We found the molar mass of cyclohexane to be 84.6 g/mol. Thus, the number of repeating empirical formula units is

$$\text{cyclohexane:} \quad \frac{(CH_2)_n}{CH_2} = \frac{84.6 \text{ g/mol}}{14.0 \text{ g/mol}}$$

$$n = 6.04 \approx 6$$

Since the value of n is 6, the molecular formula for cyclohexane is $(CH_2)_6$ and is written C_6H_{12}.

SELF-TEST EXERCISE

The percentage composition of a gaseous fuel is 92.3% C and 7.7% H. If 1.33 g of the gas occupies a volume of 1.25 L at 20°C and 750 mm Hg, what is the (a) empirical formula and (b) molecular formula of the gas?

Answers: **(a)** CH; **(b)** C_2H_2

Illustration 4: Stoichiometry and Gas Laws

Consider a problem that involves chemical equation calculation and the combined gas law. This problem is more complex than mass–volume stoichiometry because the gas produced is not at standard conditions.

EXAMPLE EXERCISE 18.13

Phosphine, PH_3, is an extremely poisonous gas that can cause convulsions, coma, and even death. The reaction of calcium phosphide and water gives phosphine gas and aqueous calcium hydroxide.

$$Ca_3P_2(s) + H_2O(l) \longrightarrow PH_3(g) + Ca(OH)_2(aq)$$

If 3.50 L of phosphine gas is collected at 749 mm Hg and 25°C, what mass of calcium phosphide underwent reaction?

Solution: In every stoichiometry problem, we must begin with a balanced chemical equation. Balancing the above equation, we have

$$Ca_3P_2(s) + 6\ H_2O(l) \longrightarrow 2\ PH_3(g) + 3\ Ca(OH)_2(aq)$$

Stoichiometry problems involving gases must be at STP, or corrected to STP conditions. In this example, the PH_3 gas is not at standard conditions. Before starting the stoichiometry calculation, we must correct the volume of gas to 760 mm Hg and 0°C. In Section 12.9, we solved gas law problems using a modified unit analysis approach. Let's begin with a data table showing the gas at initial and final conditions.

	P	V	T
initial	749 mm Hg	3.50 L	25 + 273 = 298 K
final	760 mm Hg	V_{STP}	273 K

The unit analysis format for gas law calculations is

$$3.50 \text{ L} \times P_{\text{factor}} \times T_{\text{factor}} = \text{L at STP}$$

Since the gas pressure increases at STP conditions, the volume decreases. The temperature decreases, so the volume must therefore also decrease. Thus, the unit factor solution is

$$3.50 \text{ L} \times \frac{749 \cancel{\text{mm Hg}}}{760 \cancel{\text{mm Hg}}} \times \frac{273 \cancel{\text{K}}}{298 \cancel{\text{K}}} = 3.16 \text{ L}$$

In this problem, we are asked to find the mass of Ca_3P_2. We note that the volume of PH_3 is given. Thus, our plan is as follows:

$$\text{L } PH_3 \longrightarrow \text{mol } PH_3 \longrightarrow \text{mol } Ca_3P_2 \longrightarrow \text{g } Ca_3P_2$$

From the balanced equation, we see that 2 mol of PH_3 are obtained from every 1 mol of Ca_3P_2. From the periodic table, we compute the molar mass of Ca_3P_2 (182.3 g/mol). Let's calculate the mass of Ca_3P_2 as follows:

$$3.16 \cancel{\text{L } PH_3} \times \frac{1 \cancel{\text{mol } PH_3}}{22.4 \cancel{\text{L } PH_3}} \times \frac{1 \cancel{\text{mol } Ca_3P_2}}{2 \cancel{\text{mol } PH_3}} \times \frac{182.3 \text{ g } Ca_3P_2}{1 \cancel{\text{mol } Ca_3P_2}} = 12.9 \text{ g } Ca_3P_2$$

This is a complex problem involving several principles. However, by systematically applying the guidelines for solving simple problems, we successfully solved a difficult problem.

SELF-TEST EXERCISE

The reaction of calcium metal and water gives aqueous calcium hydroxide and hydrogen gas. If 1.05 g of calcium reacts completely and the hydrogen gas is collected over water at 759 mm Hg and 20°C, what is the actual volume of dry gas? (The vapor pressure of water at 20°C is 23.8 mm Hg.)

$$Ca(s) + H_2O(l) \longrightarrow Ca(OH)_2(aq) + H_2(g)$$

Answer: 651 mL (0.651 L) H_2

Illustration 5: Molarity and Ionic Solutions

Let's consider a calculation that involves the molarity of an ionic solution. We will analyze the nitrate ion concentration in two aqueous solutions before and after mixing the two solutions.

EXAMPLE EXERCISE 18.14

If 25.0 mL of 0.500 *M* potassium nitrate solution are mixed with 75.0 mL of 0.125 *M* aluminum nitrate, what is the resulting molar concentration of the nitrate ion? (You can assume that the solution volumes are additive and that the total volume is 100.0 mL.)

Solution: We are asked to find the molarity of nitrate ion after mixing aqueous solutions of KNO_3 and $Al(NO_3)_3$. Let's begin with equations for the dissociation of ions in each solution. They are

$$KNO_3(aq) \longrightarrow K^+(aq) + NO_3^-(aq)$$

$$Al(NO_3)_3(aq) \longrightarrow Al^{3+}(aq) + 3\ NO_3^-(aq)$$

We are asked to find molarity, which has the units of mol/L. We know that the volume after mixing is 100.0 mL. Therefore, we have to find the moles of NO_3^- from each solution. The calculations are as follows:

$$25.0\ \cancel{\text{mL solution}} \times \frac{0.500\ \cancel{\text{mol KNO}_3}}{1000\ \cancel{\text{mL solution}}} \times \frac{1\ \text{mol NO}_3^-}{1\ \cancel{\text{mol KNO}_3}} = \text{mol NO}_3^-$$

$$= 0.0125\ \text{mol NO}_3^-$$

$$75.0\ \cancel{\text{mL solution}} \times \frac{0.125\ \cancel{\text{mol Al(NO}_3)_3}}{1000\ \cancel{\text{mL solution}}} \times \frac{3\ \text{mol NO}_3^-}{1\ \cancel{\text{mol Al(NO}_3)_3}} = \text{mol NO}_3^-$$

$$= 0.0281\ \text{mol NO}_3^-$$

After combining the two solutions, the total moles of NO_3^- is

$$0.0125\ \text{mol NO}_3^- + 0.0281\ \text{mol NO}_3^- = 0.0406\ \text{mol NO}_3^-$$

The molar concentration of nitrate ion is found by dividing the total moles of NO_3^- by the total solution volume, 100.0 mL. This gives us

$$\frac{0.0406\ \text{mol NO}_3^-}{100.0\ \cancel{\text{mL solution}}} \times \frac{1000\ \cancel{\text{mL}}}{1\ \text{L}} = \frac{\text{mol NO}_3^-}{\text{L solution}}$$

$$= 0.406\ M\ \text{NO}_3^-$$

SELF-TEST EXERCISE

If 50.0 mL of 0.100 *M* sodium chloride solution are mixed with 100.0 mL of 0.150 *M* calcium chloride, what is the resulting molar concentration of the chloride ion? (Assume that the solution volumes are additive and the resulting total volume 150.0 mL.)

Answer: 0.233 M Cl^-

Illustration 6: Thermochemical Stoichiometry and Specific Heat

Let's consider a chemical equation calculation that involves heat of reaction and specific heat. This problem illustrates how the heat of reaction from a magnesium flare can be used to raise the temperature of a given quantity of water.

EXAMPLE EXERCISE 18.15

Magnesium flares produce a bright light and heat when the metal ignites with oxygen to give magnesium oxide. If the heat from a 25.0 g Mg flare is added to 2050 g of water at 21.5°C, what is the final temperature of the water? ($\Delta H = -143.8$ kcal/mol of Mg.)

Solution: In every stoichiometry problem, we must begin with a balanced chemical equation. From the reaction, we can write

$$2\ Mg(s) + O_2(g) \longrightarrow 2\ MgO(s)$$

Since $\Delta H = -143.8$ kcal/mol of Mg, the balanced thermochemical equation for 2 mol of magnesium is

$$2\ Mg(s) + O_2(g) \longrightarrow 2\ MgO(s) + 287.6\ kcal$$

We can outline our strategy for finding the heat released as follows:

$$g\ Mg \longrightarrow mol\ Mg \longrightarrow kcal$$

Using unit analysis to solve the problem, we have

$$25.0\ \cancel{g\ Mg} \times \frac{1\ \cancel{mol\ Mg}}{24.3\ \cancel{g\ Mg}} \times \frac{287.6\ kcal}{2\ \cancel{mol\ Mg}} = kcal$$

$$= 148\ kcal$$

This problem asks for the final temperature of the water. We must calculate the temperature change and recall that the specific heat of water is 1.00 cal/g · °C. If 148 kcal of heat released from the Mg reaction are used to raise the temperature of 2050 g of water, we have

$$\frac{148\ \cancel{kcal}}{2050\ \cancel{g\ H_2O}} \times \frac{1000\ \cancel{cal}}{1\ \cancel{kcal}} \times \frac{1\ \cancel{g\ H_2O} \times °C}{1.00\ \cancel{cal}} = °C$$

$$= 72.2°C$$

The temperature change is 72.2°C. Therefore, the final temperature is 21.5°C + 72.2°C = 93.7°C.

SELF-TEST EXERCISE

A fine iron powder instantly reacts with oxygen gas to give iron(III) oxide. If the heat from the reaction of 5.00 g of iron heats a tank of water from 25.0°C to 35.0°C, what is the mass of the water?

$$4\ Fe(s) + 3\ O_2(g) \longrightarrow 2\ Fe_2O_3(s) + 393.0\ kcal$$

Answer: 879 g of H_2O

Illustration 7: Multiple Reaction Stoichiometry

As a final example, let's consider a calculation involving two reactions. What makes this problem different, and somewhat more difficult, is that it does not ask for an unknown quantity. Instead, the problem asks for you to decide which of two substances is a more effective antacid tablet.

EXAMPLE EXERCISE 18.16

Sodium bicarbonate and aluminum hydroxide are found in antacid tablets. Suppose you have to decide whether a 1.00-g $NaHCO_3$ tablet or a 1.00-g $Al(OH)_3$ tablet would neutralize the most 0.100 *M* HCl stomach acid.

Solution: In every stoichiometry problem, we begin with a balanced chemical equation. There are two reactions, so we must write two balanced equations:

$$NaHCO_3(s) + HCl(aq) \longrightarrow NaCl(aq) + H_2O(l) + CO_2(g)$$

$$Al(OH)_3(s) + 3\ HCl(aq) \longrightarrow AlCl_3(aq) + 3\ H_2O(l)$$

Since we are asked to find whether 1.00 g of $NaHCO_3$ or 1.00 g of $Al(OH)_3$ neutralizes the most acid, we have two problems. First, let's calculate the volume of 0.100 *M* HCl consumed by the $NaHCO_3$. Second, let's find the volume of acid consumed by the $Al(OH)_3$. Thus,

$$\text{g NaHCO}_3 \longrightarrow \text{mol NaHCO}_3 \longrightarrow \text{mol HCl} \longrightarrow \text{mL HCl}$$

First, from the balanced equation we see that 1 mol of $NaHCO_3$ reacts with 1 mol of HCl. The molar mass of $NaHCO_3$ is computed from the values in the periodic table, 84.0 g/mol. We can calculate the volume of acid as follows:

$$1.00\ \text{g}\ \cancel{\text{NaHCO}_3} \times \frac{1\ \cancel{\text{mol NaHCO}_3}}{84.0\ \cancel{\text{g NaHCO}_3}} \times \frac{1\ \cancel{\text{mol HCl}}}{1\ \text{mol}\ \cancel{\text{NaHCO}_3}} \times \frac{1000\ \text{mL HCl}}{0.100\ \cancel{\text{mol HCl}}} = \text{mL HCl} = 119\ \text{mL HCl}$$

Second, from the balanced equation we see that 1 mol of $Al(OH)_3$ reacts with 3 mol of HCl. The molar mass of $Al(OH)_3$ is computed to be 78.0 g/mol. The milliliters of acid that react are

$$1.00\ \text{g}\ \cancel{\text{Al(OH)}_3} \times \frac{1\ \cancel{\text{mol Al(OH)}_3}}{78.0\ \cancel{\text{g Al(OH)}_3}} \times \frac{3\ \cancel{\text{mol HCl}}}{1\ \cancel{\text{mol Al(OH)}_3}} \times \frac{1000\ \text{mL HCl}}{0.100\ \cancel{\text{mol HCl}}} = \text{mL HCl} = 385\ \text{mL HCl}$$

The $Al(OH)_3$ neutralizes many more milliliters of stomach acid than the $NaHCO_3$. The aluminum hydroxide tablet is the more effective antacid.

SELF-TEST EXERCISE

Two antacid tablets are taken and one contains magnesium hydroxide and the other aluminum hydroxide. If each tablet contains 500 mg of antacid, what is the total volume of 0.100 *M* HCl neutralized by both tablets?

Answer: 172 mL + 192 mL = 364 mL of 0.100 *M* HCl

In this section on advanced problem solving, we examined just a few examples involving two or more chemical principles. Actually, there are thousands of possible chemical calculations based on the principles we have studied. The point to remember is that many problems cannot be solved by a standard procedure. Nevertheless, to solve the difficult problems, use the same tools and guidelines you used to solve simpler ones. Plan the strategy for solving an advanced problem using unit analysis, algebra, or some other tool. As you practice solving problems, you will develop skill in deciding the most appropriate method for finding a solution.

In the final analysis, three steps apply to both simple and advanced problems. To find the solution to a problem, always ask the following:

1. *What quantity is the problem asking for in the answer?*
2. *What given information is relevant to the answer?*
3. *How can I plan an overall strategy for obtaining an answer from the given information?*

Summary

Section 18.1 We began our discussion with an overview of problem solving. Each time we approach a problem, we should ask three questions. What is the quantity asked for in the answer? What given information is relevant? What strategy can be used to convert the given information to the answer? To design the solution plan, we can use a **strategy map**, a **concept map**, an **algorithm**, and **visualization**. Most often, we will use **unit analysis** to solve a problem. Some problems, however, are better suited to an **algebraic analysis**.

Section 18.2 The central concept in all chemical formula calculations is the mole. Recall that a mole is related to each of the following: (1) Avogadro's number of entities, (2) molar mass, (3) molar volume, and (4) molar concentration. The mole concept can be used to determine the number of particles, the mass of a substance, the volume of a gas, or the volume of an aqueous solution.

Section 18.3 Stoichiometry relates quantities of substance based on a balanced chemical equation. One type of stoichiometry relates a given amount of reactant to a product. Another type of stoichiometry asks for the amount of reactant that yields a certain amount of product. Given a reactant or a product, we can relate (1) two masses of substance, (2) two volumes of gaseous substances, (3) two volumes of aqueous solution, or various combinations thereof.

Section 18.4 In more difficult stoichiometry problems, we cannot assume that the reactants are available in sufficient quantities. That is, the amount of one reactant is often in excess. The substance that limits the product yield is referred to as the **limiting reactant**.

Section 18.5 Every chemical reaction is accompanied by a change in heat energy. If a reaction absorbs heat, it is said to be endothermic. If a reaction releases heat, it is said to be exothermic. The **heat of reaction** for an endothermic reaction is positive ($\Delta H = +$value). The heat of reaction for an exothermic reaction is negative ($\Delta H = -$value). A **thermochemical equation** includes the heat of reaction. For example, oxygen causes metallic iron to rust: $\Delta H = -196.5$ kcal/mol Fe_2O_3. Since ΔH is negative, the reaction is exothermic and the thermochemical equation is written as

$$4\ Fe(g) + 3\ O_2(g) \longrightarrow 2\ Fe_2O_3(g) + 393.0\ kcal$$

Notice that two moles of Fe_2O_3 are produced. Therefore, we must double the ΔH value to balance the the thermochemical equation.

Section 18.6 Many industrial processes involve a series of two or more reactions. To perform stoichiometric calculations, we must relate a common quantity in all the reactions. For example, sulfuric acid is manufactured from elemental sulfur in a series of three reactions. We can relate the element to the acid by keeping track of each of the intermediate products. In this case, sulfur is converted to sulfuric acid as follows:

$$S \longrightarrow SO_2 \longrightarrow SO_3 \longrightarrow H_2SO_4$$

Quantities involving multiple reactions are related by extending the basic rules of stoichiometry. In the second step in a stoichiometry problem, moles of reactant are related to moles of product. For multiple reactions, this step is repeated for each additional reaction.

Section 18.7 This chapter presented some difficult introductory chemical calculations. For example, you were asked to find answers to problems solving two or more principles. Previously, problem solving was limited to a single principle and a specific method. Here you were asked to design a plan for finding a solution to a complex problem.

Key Terms

Select the key term that corresponds to the following definitions.

____ **1.** a problem-solving plan that involves one or more steps to relate a given value to an unknown quantity, usually shown by a series of arrows; for example, g $H_2 \longrightarrow$ mol $H_2 \longrightarrow$ molecules H_2

____ **2.** a problem-solving plan that traces the relationship of various quantities, usually shown in interrelated boxes; for example,

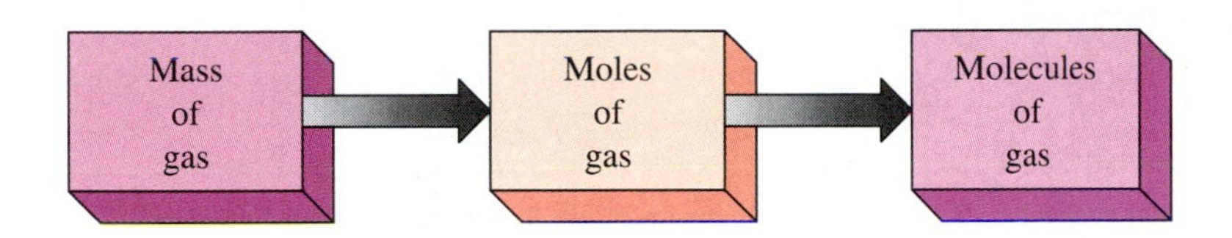

____ **3.** a problem-solving plan that relates a given quantity to an unknown quantity in a series of steps listed in a specific sequence; for example,

Step 1: Convert the mass of hydrogen to moles H_2 using the molar mass of hydrogen gas.
Step 2: Convert the number of moles of H_2 to the number of molecules of H_2 using Avogadro's number.

____ **4.** a problem-solving approach that involves the forming of mental pictures in order to make an abstract principle more concrete and permanent; for example, the process of forming a mental picture of gas molecules in a container

____ **5.** a systematic problem-solving approach that employs unit factors in order to convert the units of the given value to the units of the unknown

____ **6.** a systematic problem-solving approach that requires rearranging variables in an equation in order to calculate an unknown quantity

____ **7.** the substance in a chemical reaction that controls the maximum amount of product formed

____ **8.** the amount of heat energy absorbed or released for a given reaction according to the balanced chemical equation

____ **9.** a balanced chemical equation that incorporates the amount of heat energy involved in the chemical reaction

____ **10.** the relationship of quantities of substance to the heat of reaction according to a balanced chemical equation

(a) algebraic analysis *(Sec. 18.1)*
(b) algorithm *(Sec. 18.1)*
(c) concept map *(Sec. 18.1)*
(d) heat of reaction (ΔH) *(Sec. 18.5)*
(e) limiting reactant *(Sec. 18.4)*
(f) strategy map *(Sec. 18.1)*
(g) thermochemical equation *(Sec. 18.5)*
(h) thermochemical stoichiometry *(Sec. 18.5)*
(i) unit analysis *(Sec. 18.1)*
(j) visualization *(Sec. 18.1)*

Exercises

Advanced Problem Solving (Sec. 18.1)

1. Answer each of the following questions regarding problem analysis.
   - **(a)** What is the first step in solving any problem?
   - **(b)** What is the second step in solving any problem?
   - **(c)** What two quantities are connected by a strategy map?
   - **(d)** What two quantities are connected by a concept map?
   - **(e)** What is the process of writing systematic steps for solving a problem?
   - **(f)** What is the process of forming mental pictures to make a problem more concrete?

2. Answer each of the following questions regarding problem-solving techniques.
   - **(a)** What is unique about the relevant given quantity if the answer has compound units, for example, g/mol?
   - **(b)** What must be written before solving any stoichiometry problem?
   - **(c)** What happens to given information that is not relevant to the solution of a problem?
   - **(d)** What are three common references that can provide information that may not be included in the problem description?
   - **(e)** What types of problems are more easily solved by algebra than the unit analysis method?
   - **(f)** What should be done after setting up a problem but before starting the calculations? What should be done after calculating a numerical answer?

3. State the physical quantity measured by each of the following units.
   - **(a)** gram **(b)** second
   - **(c)** cubic centimeter **(d)** inch
   - **(e)** pound **(f)** kelvin
   - **(g)** kilogram **(h)** milliliter
   - **(i)** moles per liter **(j)** degree Celsius
   - **(k)** decimeter **(l)** grams per liter
   - **(m)** quart **(n)** microsecond

4. State a common metric unit(s) for each of the following physical quantities.
   - **(a)** length **(b)** mass
   - **(c)** volume **(d)** time
   - **(e)** temperature **(f)** heat energy
   - **(g)** density **(h)** solution concentration

5. A popular computer application can transfer information between a word-processing document, a data base, and a spreadsheet. Draw a concept map for the application program.

6. The board of directors at a large college decided on the following administrative structure. The president of the college is to report to the board of directors. In turn, the college president supervises three vice-presidents. The chief financial officer is to report directly to the board. Draw a concept map for this college administration.

7. Draw a concept map that relates each of the following mole quantities.
   - **(a)** molecules of sulfur dioxide gas and volume of SO_2 at STP
   - **(b)** volume of nitrogen gas at STP and mass of N_2

8. Draw a concept map that relates each of the following mole quantities.
   - **(a)** mass of chloroform and molecules of $CHCl_3$
   - **(b)** volume of aqueous carbon dioxide solution and mass of CO_2

9. Write an algorithm for converting each of the following mole quantities.
   - **(a)** atoms of helium gas to volume of He at STP
   - **(b)** volume of chlorine gas at STP to mass of Cl_2

10. Write an algorithm for converting each of the following mole quantities.
    - **(a)** mass of copper to atoms of Cu
    - **(b)** volume and molarity of aqueous sodium chloride solution to mass of NaCl solute

11. Draw a concept map that relates each of the following stoichiometric quantities.
    - **(a)** mass of reactant and mass of product
    - **(b)** mass of reactant and volume of gaseous product

12. Draw a concept map that relates each of the following stoichiometric quantities.
    - **(a)** volume of gaseous reactant and volume of gaseous product
    - **(b)** volume of aqueous solution reactant and volume of gaseous product

13. Write an algorithm for each of the following stoichiometric conversions.
    - **(a)** mass of reactant to mass of product
    - **(b)** mass of reactant to volume of gaseous product at STP

14. Write an algorithm for each of the following stoichiometric conversions.
    - **(a)** volume of gaseous reactant to volume of gaseous product
    - **(b)** volume of aqueous solution reactant to volume of gaseous product

15. Estimate an approximate numerical answer for each of the following
    - **(a)** $$1.550\ \text{g}\ \cancel{Cr_2O_3} \times \frac{1\ \cancel{\text{mol}\ Cr_2O_3}}{152.0\ \cancel{\text{g}\ Cr_2O_3}} \times \frac{2\ \text{mol Cr}}{1\ \cancel{\text{mol}\ Cr_2O_3}} \times \frac{52.0\ \text{g Cr}}{1\ \cancel{\text{mol Cr}}} = \text{g Cr}$$
    - **(b)** $$42.0\ \cancel{\text{mL}\ HNO_3} \times \frac{0.195\ \cancel{\text{mol}\ HNO_3}}{1000\ \cancel{\text{mL}\ HNO_3}} \times \frac{1\ \cancel{\text{mol Ba(OH)}_2}}{2\ \cancel{\text{mol}\ HNO_3}} \times \frac{1000\ \text{mL Ba(OH)}_2}{0.105\ \cancel{\text{mol Ba(OH)}_2}} = \text{mL Ba(OH)}_2$$

(c) $21.5 \text{ \sout{mL BiCl_3}} \times \frac{0.115 \text{ \sout{mol BiCl_3}}}{1000 \text{ \sout{mL BiCl_3}}}$

$\times \frac{3 \text{ \sout{mol Hg_2Cl_2}}}{2 \text{ \sout{mol BiCl_3}}} \times \frac{471.2 \text{ g } Hg_2Cl_2}{1 \text{ \sout{mol Hg_2Cl_2}}}$

$= \text{g } Hg_2Cl_2$

16. Use your calculator to find the answers to the problems in Exercise 15. Compare your approximate estimate to the calculated answer.

Chemical Formula Calculations (Sec. 18.2)

17. Given that 0.0142 mol of hydrogen chloride dissolves in 47.5 mL of solution, calculate each of the following.
(a) grams of HCl gas dissolved in the solution
(b) liters of HCl gas (STP) dissolved in the solution
(c) molecules of HCl gas dissolved in the solution
(d) molar concentration of the hydrochloric acid solution

18. Given that 0.0755 mol of ammonia is dissolved in 0.155 L of solution, calculate each of the following.
(a) grams of NH_3 gas dissolved in the solution
(b) liters of NH_3 gas (at STP) dissolved in the solution
(c) molecules of NH_3 gas dissolved in the solution
(d) molar concentration of the ammonia solution

19. Given that 1.00 g of hydrogen fluoride is dissolved in 100.0 mL of solution, calculate each of the following.
(a) liters of HF gas at (STP) dissolved in the solution
(b) molecules of HF gas dissolved in the solution
(c) molar concentration of the hydrofluoric acid solution

20. Given that 2.00 g of liquid hydrazine, N_2H_4, is dissolved in 0.250 L of solution, calculate each of the following.
(a) milliliters of N_2H_4 (d = 1.02 g/mL) dissolved in the solution
(b) molecules of N_2H_4 dissolved in the solution
(c) molar concentration of the hydrazine solution

21. Given that 1.00 L of carbon dioxide at STP is dissolved in 500.0 mL of solution, calculate each of the following.
(a) grams of CO_2 gas (at STP) dissolved in the solution
(b) molecules of CO_2 gas dissolved in the solution
(c) molar concentration of carbonic acid, H_2CO_3

22. Given that 555 mL of sulfur dioxide at STP is dissolved in 0.250 L of solution, calculate each of the following.
(a) grams of SO_2 gas (at STP) dissolved in the solution
(b) molecules of SO_2 gas dissolved in the solution
(c) molar concentration of sulfurous acid, H_2SO_3

23. Given that 2.22×10^{22} molecules of hydrogen sulfide are dissolved in 450.0 mL of solution, calculate each of the following.
(a) liters of H_2S gas (STP) dissolved in the solution
(b) grams of H_2S gas dissolved in the solution
(c) molar concentration of the hydrosulfuric acid, H_2S, solution

24. Given tht 1.12×10^{23} molecules of sulfuric trioxide are dissolved in 0.325 L of solution, calculate each of the following.
(a) liters of SO_3 gas (STP) dissolved in the solution
(b) grams of SO_3 gas dissolved in the solution
(c) molar concentration of the sulfuric acid, H_2SO_4, solution

Chemical Equation Calculations (Sec. 18.3)

25. Methane gas, CH_4, reacts with oxygen to give 5.00 g of water according to the unbalanced equation

$$CH_4(g) + O_2(g) \longrightarrow CO_2(g) + H_2O(l)$$

(a) What is the mass of methane that reacted?
(b) What is the STP volume of oxygen that underwent reaction?

26. Ammonia gas, NH_3, reacts with oxygen using a platinum catalyst to give 25.0 L of nitrogen monoxide gas at STP according to the unbalanced equation

$$NH_3(g) + O_2(g) \xrightarrow{\text{Pt/825°C}} NO(g) + H_2O(l)$$

(a) What is the STP volume of ammonia that underwent reaction?
(b) What is the mass of water that was produced?

27. Hydrochloric acid reacts with 0.466 g of aluminum metal to give hydrogen gas according to the unbalanced equation

$$Al(s) + HCl(aq) \longrightarrow AlCl_3(aq) + H_2(g)$$

(a) What is the volume of hydrogen gas at STP?
(b) What volume of 0.100 M HCl is required for complete reaction?

28. A sample of potassium hydrogen carbonate decomposes to give 255 mL of carbon dioxide at STP according to the unbalanced equation

$$KHCO_3(s) \longrightarrow K_2CO_3(s) + H_2O(g) + CO_2(g)$$

(a) What is the mass of $KHCO_3$ that decomposes?
(b) What is the mass of K_2CO_3 produced?

29. Given that 50.0 mL of 0.100 M magnesium bromide reacts completely with 13.9 mL of silver nitrate solution according to the unbalanced equation

$$MgBr_2(aq) + AgNO_3(aq) \longrightarrow AgBr(s) + Mg(NO_3)_2(aq)$$

(a) What is the molarity of the $AgNO_3$ solution?
(b) What is the mass of AgBr precipitate?

30. Given that 24.0 mL of 0.170 M sodium iodide reacts completely with 0.209 M mercury(II) nitrate solution according to the unbalanced equation

$$Hg(NO_3)_2(aq) + NaI(aq) \longrightarrow HgI_2(s) + NaNO_3(aq)$$

(a) What volume of $Hg(NO_3)_2$ is required for complete precipitation?
(b) What is the mass of HgI_2 precipitate?

Limiting Reactant Concept (Sec. 18.4)

31. If an egg ranch collects 551 eggs on a given day and there are 45 egg cartons available, how many cartons of eggs can be shipped? (Assume 12 eggs per carton.)

32. An American sports car manufacturer produces 9555 automobile chassis in a single year. If the assembly plant has 38,000 tires available, how many complete automobiles can be shipped? (Assume four tires per car.)

33. Given that 25.0 L of sulfur dioxide reacts with 25.0 L of oxygen gas according to the unbalanced equation

$$SO_2(g) + O_2(g) \xrightarrow{V_2O_5/650°C} SO_3(g)$$

(a) Assuming constant conditions, what is the volume of SO_3?
(b) What is the limiting reactant?

34. Given that 45.0 mL of nitrogen gas react with 65.0 mL of oxygen gas according to the unbalanced equation

$$N_2(g) + O_2(g) \longrightarrow NO_2(g)$$

(a) Assuming constant conditions, what is the volume of NO_2?
(b) What is the limiting reactant?

35. Given that 1.00 g of aluminum hydroxide reacts with 25.0 mL of 0.500 *M* sulfuric acid according to the unbalanced equation

$$Al(OH)_3(s) + H_2SO_4(aq) \longrightarrow Al_2(SO_4)_3(aq) + H_2O(l)$$

(a) What is the mass of H_2O produced?
(b) What is the limiting reactant?

36. Given that 36.5 mL of 0.266 *M* calcium acetate reacts with 25.0 mL of 0.385 *M* sodium carbonate according to the unbalanced equation

$$Ca(C_2H_3O_2)_2(aq) + Na_2CO_3(aq) \longrightarrow CaCO_3(s) + NaC_2H_3O_2(aq)$$

(a) What is the mass of $CaCO_3$ precipitate formed?
(b) What is the limiting reactant?

Thermochemical Stoichiometry (Sec. 18.5)

37. For endothermic reactions, is the ΔH value positive or negative?

38. For exothermic reactions, is the ΔH value positive or negative?

39. Write a balanced thermochemical equation for each of the following.
(a) Nitrous oxide (laughing gas), N_2O, is prepared from the careful decomposition of solid ammonium nitrate. The exothermic reaction (47.8 kcal/mol nitrous oxide) also produces water vapor.
(b) The photosynthetic conversion of carbon dioxide and water to glucose, $C_6H_{12}O_6$, and oxygen gas is endothermic. The heat of reaction is 674.0 kcal/mol glucose.

40. Write a balanced thermochemical equation for each of the following.
(a) The decomposition of water vapor gives hydrogen and oxygen gases ($\Delta H = +57.8$ kcal/mol water).
(b) Tetraphosphorus decaoxide is prepared from the direct reaction of red phosphorus, P_4, and oxygen gas ($\Delta H = -696.4$ kcal/mol P_4O_{10}).

41. Ethanol, C_2H_5OH, is blended with gasoline to make gasohol. The thermochemical equation for the combustion of ethanol is

$$C_2H_5OH(g) + 3\ O_2(g) \longrightarrow 2\ CO_2(g) + 3\ H_2O(g) + 66.4 \text{ kcal}$$

(a) What mass of C_2H_5OH must burn in order to release 1000.0 kcal?
(b) What is the volume of carbon dioxide produced at STP in order to release 1000.0 kcal?

42. Calcium sulfate is decomposed according to the following equation:

$$CaSO_4(s) \longrightarrow CaO(s) + SO_3(g)$$

(a) What mass of CaO is produced by heating excess $CaSO_4$ with 27.5 kcal? (Given: $\Delta H = +96.4$ kcal/mol $CaSO_4$.)
(b) What is the volume of sulfur trioxide evolved at STP if 27.5 kcal of heat are supplied?

Multiple-Reaction Stoichiometry (Sec. 18.6)

43. Carbon monoxide gas is used in the blast furnace process to convert iron ore to impure pig iron. The CO_2 produced is used as a fire extinguisher and to make Dry Ice. Starting with coal, carbon undergoes the following reactions.

$$2\ C(s) + O_2(g) \longrightarrow 2\ CO(g)$$

$$2\ CO(g) + O_2(g) \longrightarrow 2\ CO_2(g)$$

Starting with 25.0 g of carbon and excess oxygen gas, calculate:
(a) the mass of carbon dioxide produced
(b) the volume of carbon dioxide produced at STP
(c) the mass of oxygen consumed in the process

44. In some regions of the country, water contains hydrogen sulfide, which gives a rotten egg odor. Chlorine gas is used to purify drinking water and remove the sulfide. The sulfur obtained can then react with fluorine gas to give sulfur hexafluoride, a gas used on special electrical circuits. The reactions are as follows.

$$8\ H_2S(aq) + 8\ Cl_2(g) \longrightarrow 16\ HCl(aq) + S_8(s)$$

$$S_8(s) + 24\ F_2(g) \longrightarrow 8\ SF_6(g)$$

Starting with 0.950 L of Cl_2 (STP) and excess fluorine gas, calculate:
(a) the mass of sulfur hexafluoride produced
(b) the volume of sulfur hexafluoride produced at STP
(c) the volume of aqueous 0.0265 *M* H_2S that reacted

45. Sulfuric acid is the single most important industrial chemical. About 40 million tons are manufactured each year. Elemental sulfur is converted to sulfuric acid by the contact process as follows:

$$S(s) + O_2(g) \longrightarrow SO_2(g)$$

$$2\ SO_2(g) + O_2(g) \xrightarrow{V_2O_5} 2\ SO_3(g)$$

$$SO_3(g) + H_2O(l) \longrightarrow H_2SO_4(l)$$

Starting with 1.00 kg of sulfur and excess oxygen gas, calculate:

(a) the mass of sulfur trioxide produced
(b) the STP volume of sulfur trioxide produced
(c) the mass of sulfuric acid produced assuming a 55.0% process yield

46. Iron ore is converted to pig iron in an industrial blast furnace. The pig iron, in turn, is converted to carbon steel by high-temperature oxidation. The blast furnace process takes place in a series of three reactions.

$$2\ Fe_2O_3(s) + CO(g) \xrightarrow{200°C} 2\ Fe_3O_4(s) + CO_2(g)$$

$$Fe_3O_4(s) + CO(g) \xrightarrow{700°C} 3\ FeO(s) + CO_2(g)$$

$$FeO(g) + CO(g) \xrightarrow{1200°C} Fe(l) + CO_2(g)$$

Starting with 1.00 kg of iron(III) oxide and excess carbon monoxide gas, calculate:

(a) the mass of iron(II) oxide produced
(b) the mass of molten iron produced
(c) the mass of molten iron produced assuming a 70.0% process yield

47. The Raschig process converts ammonia to the important chemical hydrazine, N_2H_4. Steering rockets on the Space Shuttle are powered by the reaction of hydrazine and dinitrogen tetraoxide. The reactions for making hydrazine and for the steering rockets propulsion are as follows.

$$2\ NH_3(aq) + NaOCl(aq) \longrightarrow N_2H_4(aq) + NaCl(aq) + H_2O(l)$$

$$2\ N_2H_4(l) + N_2O_4(l) \longrightarrow 3\ N_2(g) + 4\ H_2O(g)$$

Starting with 50.0 mL of 6.00 *M* ammonia and excess other reagents, calculate:

(a) the mass of nitrogen gas produced
(b) the volume of nitrogen gas produced at STP
(c) the mass of water vapor produced
(d) the volume of water vapor produced at STP

48. Hydrogen peroxide, H_2O_2, is prepared industrially from the reaction of oxygen and isopropyl alcohol, C_3H_7OH. Hydrogen peroxide reacts explosively with hydrazine, N_2H_4, and is used as a rocket fuel. The reactions for manufacturing hydrogen peroxide and for rocket propulsion are as follows.

$$C_3H_7OH(l) + O_2(g) \longrightarrow H_2O_2(l) + C_3H_6O(l)$$

$$2\ H_2O_2(l) + N_2H_4(l) \longrightarrow N_2(g) + 4\ H_2O(g)$$

Starting with 50.0 mL of isopropyl alcohol (d = 0.786 g/mL) and excess other reagents, calculate:

(a) the mass of nitrogen gas produced
(b) the volume of nitrogen gas produced at STP
(c) the mass of water vapor produced
(d) the volume of water vapor produced at STP

Advanced Problem-Solving Examples (Sec. 18.7)

49. Calculate the mass of nitrogen dioxide gas occupying a volume of 2.50 L at 35°C and 0.974 atm pressure.

50. Calculate the molar mass of an unknown gas if a vapor density experiment yielded a value of 1.45 g/L at 100°C and 752 mm Hg.

51. A 34.5-g sample of calcium chloride is dissolved in 500.0 mL of solution. Find the molar concentration of the calcium ions and chloride ions in solution. (Assume that the salt is 100% ionized.)

52. What is the molar chloride ion concentration that results from mixing 100.0 mL of 0.156 *M* lithium chloride and 150.0 mL of 0.225 *M* barium chloride?

53. How many milliliters of 0.100 *M* hydrochloric acid react with excess zinc metal in order to collect 50.0 mL of dry hydrogen gas over water at STP?

54. What volume of 0.150 *M* hydrochloric acid reacts with excess lead(II) nitrate solution in order to yield 1.88 g of lead(II) chloride precipitate?

55. Given 0.150 L of nitrous oxide at 25°C and 749 mm Hg pressure, calculate the number of N_2O molecules. What is the mass of the gas?

56. Given 6.55 g of carbon dioxide gas, calculate the volume of gas at 75°C and 0.750 atm pressure. What is the number of gas molecules?

General Exercises

57. State the term that corresponds to each of the following descriptions of important chemical principles.

(a) the amount of mass of substance in a given volume
(b) the energy necessary to raise the temperature of 1 g of any substance 1°C
(c) a unit of energy necessary to heat 1 kg of water 1°C
(d) a unit of energy corresponding to a 1 kg mass moving at a velocity of 1 m/s
(e) the number of atoms in exactly 12 g of carbon-12
(f) the amount of substance that contains the same number of entities as there are atoms in exactly 12 g of carbon-12
(g) the mass of 1 mole of substance expressed in grams
(h) the volume of 1 mole of any gas at standard conditions
(i) a chemical formula that expresses the simplest ratio of elements in an ionic or molecular compound
(j) a chemical formula that expresses the actual ratio of elements in a molecular compound
(k) the statement that the total mass of reactants is equal to the total mass of products for any given reaction
(l) the calculations that relate quantities of substance involved in a chemical reaction according to the balanced equation
(m) the statement that the pressure exerted by a gas is inversely proportional to the volume and directly proportional to the Kelvin temperature

(n) the equation that relates the pressure and volume of a gas to the product of the moles of gas and temperature

(o) the proportionality constant R in the relationship $PV = nRT$; the value of R is 0.0821 atm · L/mol · K

(p) the concentration expression for the grams of solute dissolved in 100 grams of solution

(q) the concentration expression for the moles of solute dissolved in 1 liter of solution

(r) the concentration expression for the moles of solute dissolved in 1 kilogram of solvent

58. Seawater contains approximately 1.22×10^{10} atoms of gold per milliliter. How many kilograms of seawater must be evaporated to obtain 12.0 g of gold? (Assume that the density of seawater is 1.05 g/mL.)

59. The volume of the oceans is 1.4×10^{21} liters. Calculate the mass of chlorine in all seawater given that the mass percent concentration of chloride ion is 1.90%. (Assume that the density of seawater is 1.05 g/mL.)

60. Liquid bromine can be prepared by passing chlorine gas through an aqueous solution of sodium bromide. How many milliliters of bromine are produced from the reaction of 10.0 g of Cl_2 and 10.0 g of NaBr? (Given: density of liquid Br_2 is 3.12 g/mL.)

61. Mannitol is an artificial sweetener found in sugarless gum. The percentage composition is 39.6% carbon, 7.7% hydrogen, and 52.8% oxygen, and the density is 1.47 g/mL. If mannitol contains 4.93×10^{21} molecules per milliliter, what is the molecular formula of mannitol?

62. Heating solid potassium chlorate with manganese dioxide catalyst produces solid potassium chloride and oxygen gas. If 455 mL of *wet* oxygen are collected over water at 23.0°C and 766 mm Hg, what mass of potassium chlorate was decomposed?

63. What is the molar sodium ion concentration resulting from mixing of 50.0 mL of 0.100 *M* sodium chloride and 50.0 mL of 0.200 *M* sodium sulfate?

64. Express the following heats of reaction in kilojoules, given the ΔH value in kilocalories.

(a) $\Delta H = +90.3$ kcal/mol NO

(b) $\Delta H = +1209$ kcal/mol SF_6

(c) $\Delta H = -90.8$ kcal/mol HgO

(d) $\Delta H = -124.4$ kcal/mol $AgNO_3$

65. A sample of sodium carbonate is treated with 50.0 mL of 0.345 *M* HCl. The excess hydrochloric acid is titrated with 15.9 mL of 0.155 *M* NaOH. Calculate the mass of the sodium carbonate sample.

$$Na_2CO_3(s) + 2\ HCl(aq) \longrightarrow 2\ NaCl(aq) + H_2O(l) + CO_2(g)$$

$$HCl(aq) + NaOH(aq) \longrightarrow NaCl(aq) + H_2O(l)$$

66. The chloride in a sample of $BaCl_2$ is precipitated with 50.0 mL of 0.100 *M* $AgNO_3$. The excess silver nitrate is titrated with 17.0 mL of 0.125 *M* K_2CrO_4. Calculate the mass of the barium chloride sample.

$$BaCl_2(s) + 2\ AgNO_3(aq) \longrightarrow 2\ AgCl(s) + Ba(NO_3)_2(aq)$$

$$2\ AgNO_3(aq) + K_2CrO_4(aq) \longrightarrow Ag_2CrO_4(s) + 2\ KNO_3(aq)$$

Appendices

The Scientific Calculator

A hand calculator is essential for many of the calculations in this text. Given a choice of calculators, choose a scientific calculator rather than a business calculator. Most scientific calculators have the following keys that you will find helpful while performing typical chemical calculations.

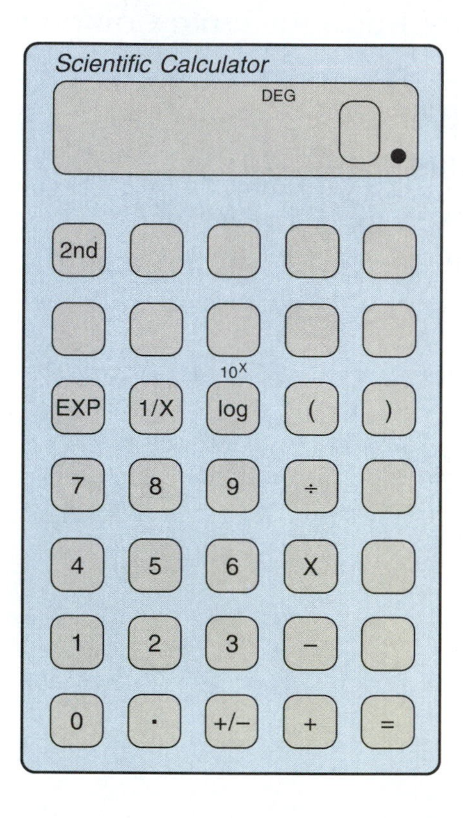

Arithmetic Operations

Basic Function Keys: + − × ÷ =

Example	Key In	Display
(a) 85.8 + 6.43	**85.8** + **6.43** =	*92.23*
(b) 297 − 11.04	**297** − **11.04** =	*285.96*
(c) 0.882 × 6.02	**.882** × **6.02** =	*5.30964*
(d) 768 ÷ 0.16	**768** ÷ **.16** =	*4800.*

Second Function Key: 2nd *or* SHIFT *or* INV

Scientific calculators have many preprogrammed functions. Usually there are more functions than keys. Some of the keys serve two functions. The first function of a key is printed directly on the keypad. The second function of a key is usually printed above the keypad. To access the second function, first press the **2nd** key and then press the function key. The **2nd** function key may also be called a **SHIFT** key or an **INV** key.

Reciprocal Key: 1/X *or* X^{-1}

The reciprocal of a number is 1 divided by that number. For example, the reciprocal of 4 is 1/4, which appears in the display as 0.25. The reciprocal of 100 is 1/100 or 0.01. To obtain a reciprocal of a number, simply enter the number and press the **1/X** key.

Change Sign Key: +/−

This key is used to change the sign of a number to the opposite sign. If a number is positive, the change of sign key will make it negative and a negative sign will appear. If the number is negative, this key will change it to a positive number and the negative sign will disappear.

Exponential Operations

Exponent Key: EXP *or* EE

Using exponents is a shorthand way of expressing very large and very small values. A positive exponent indicates a value that is greater than 1, while a negative exponent indicates a value that is less than 1 (Sections 1.6 and 1.7).

To enter an exponential number, first enter the numerical portion and then press the **EXP** key, followed by the exponent. If the exponent is negative, touch the change sign key after pressing the **EXP** key. The following examples illustrate how to enter exponential numbers into a calculator.

Example	Key In	Display
(a) 5.87×10^{-7}	**5.87** [EXP] **7** [+/–]	*5.87^{-07}*
(b) 1.29×10^{2}	**1.29** [EXP] **2**	*1.29^{02}*
(c) 7×10^{-4}	**7** [EXP] **4** [+/–]	*7.$^{-04}$*
(d) 6.02×10^{23}	**6.02** [EXP] **23**	*6.02^{23}*

Note: Never enter the **× 10** portion of an exponential number. The [EXP] key serves this function.

Note: When the exponent is small, some calculators will automatically display the value in numerical form. For example, a calculator may display 1.29×10^2 as *129*, rather than *1.29^{02}*. If your calculator has a **SCI** or **MODE** key, it will be possible to program your calculator to have a display that is always in scientific notation. In fact, you may be able to fix the number of significant digits shown in the display. If your calculator has an **FSE** key, you can toggle among fixed decimal (**F**), scientific notation (**S**), and engineering notation (**E**). For details about your particular calculator, refer to the instruction booklet or ask your instructor.

Chain Calculations

A calculation that requires more than one operation is referred to as a chain calculation. The following example exercise provides practice in using your calculator to solve problems involving chain calculations and exponents. For further practice, do the end-of-chapter exercises for Section 1.8.

EXAMPLE EXERCISE

Use your calculator to solve the following problems and express the answer in calculator notation (refer to Section 1.8 as necessary).

(a) $(5.76 \times 10^{-4}) \times (3.15 \times 10^{2})$

(b) $(1.02) \times 10^{6}) \div (3.13 \times 10^{-21})$

(c) $(9.53 \times 10^{-7}) \times \dfrac{(6.02 \times 10^{23})}{4.95 \times 10^{17}}$

(d) $1.98 \times 10^{9} \times \dfrac{2.34 \times 10^{5}}{8.67 \times 10^{-8}}$

Solution:

Key In	Display
(a) **5.76** [EXP] **4** [+/–] [×] **3.15** [EXP] **2** [=]	*1.81^{-01}*
(b) **1.02** [EXP] **6** [÷] **3.13** [EXP] **21** [+/–] [=]	*3.26^{26}*
(c) **9.53** [EXP] **7** [+/–] [×] **6.02** [EXP] **23** [÷] **4.95** [EXP] **17** [=]	*1.16^{00}*
(d) **1.98** [EXP] **9** [×] **2.34** [EXP] **5** [÷] **8.67** [EXP] **8** [+/–] [=]	*5.34^{21}*

(a) 1.81 E–01; **(b)** 3.26 E26; **(c)** 1.16; **(d)** 5.34 E21

SELF-TEST

Use your calculator to solve the following problems and express the answer in calculator notation.

(a) $(9.41 \times 10^{-18}) \times (6.98 \times 10^{5})$ **(b)** $(1.67 \times 10^{21}) \div (2.32 \times 10^{-6})$

(c) $8.59 \times 10^{-7} \times \dfrac{7.36 \times 10^{27}}{6.32 \times 10^{21}}$ **(d)** $\dfrac{1}{4.45 \times 10^{-17}} \times \dfrac{5.92 \times 10^{5}}{1.45 \times 10^{20}}$

Answers: **(a)** 6.57 E−12; **(b)** 7.20 E26; **(c)** 1.00; **(d)** 9.17 E01

Logarithmic Operations

Base 10 Logarithm Key: LOG

The base 10 logarithm of a number is the power to which 10 must be raised in order to equal that number. In this text, we will use logarithms to calculate the pH of solutions (see Section 15.9). The following examples show the conversion of $[H^+]$ to pH; that is, $-\log [H^+]$.

Example	Key In	Display
(a) $[H^+] = 3.83 \times 10^{-5}$	**3.83** EXP **5** +/− LOG +/−	*4.42*
(b) $[H^+] = 1.59 \times 10^{-3}$	**1.59** EXP **3** +/− LOG +/−	*2.80*
(a) pH = 4.42 (b) pH = 2.80		

Base 10 Antilogarithm Key: 10^x

The inverse logarithm, or antilogarithm, is the reverse operation of finding the logarithm of a number. You will encounter this operation when converting pH into the molar hydrogen ion concentration of an acid solution. The following examples show the conversion of pH to $[H^+]$

Example	Key In	Display
(a) pH = 8.15	**8.15** +/− 2nd 10^x	*7.08^{-09}*
(b) pH = 1.55	**1.55** +/− 2nd 10^x	*0.0282*
(a) $[H^+] = 7.08 \times 10^{-9}$; (b) $[H^+] = 0.0282$		

Units of Measurement

English

Mass

12 troy ounces (t oz)	=	1 troy pound (t lb)
16 ounces (oz)	=	1 pound (lb)
2000 pounds (lb)	=	1 ton (ton)

Length

12 inches (in.)	=	1 foot (ft)
3 feet (ft)	=	1 yard (yd)
5280 feet (ft)	=	1 mile (mi)

Volume

32 fluid ounces (fl oz)	=	1 quart (qt)
2 pints (pt)	=	1 quart (qt)
4 quarts (qt)	=	1 gallon (gal)

Pressure

Unit	Standard Pressure
atmosphere	1.00 atm
inches of mercury	29.9 in. Hg
centimeters of mercury	76.0 cm Hg
millimeters of mercury	760 mm Hg
torr	760 torr
pounds per square inch	14.7 psi
kilopascal	101 kPa

English-Metric Equivalents

Mass: 1 pound (lb)	=	454 grams (g)
Volume: 1 quart (qt)	=	946 milliliters (mL)
Length: 1 inch (in.)	=	2.54 centimeters (cm)
Time: 1 second (sec)	=	1 second (s)

Temperature	Absolute Zero
$°F = \frac{9}{5}°C + 32$	$-459.67°F$
$°C = \frac{5}{9}(°F - 32)$	$-273.15°C$
$K = °C + 273$	0 K

Energy

1 calorie (cal)	=	4.184 joules (J)
1 kilocalorie (kcal)	=	4.184 kilojoules (kJ)
	=	1000 calories (cal)
	=	1 Calorie (Cal)

Physical Constants

Avogadro's number	=	6.02×10^{23}
Molar volume of a gas at STP	=	22.4 L/mol
Ideal gas constant, *R*	=	0.0821 atm · L/mol · K
Mass of proton	=	1.0073 amu
Mass of neutron	=	1.0087 amu
Mass of electron	=	0.00055 amu
Velocity of light	=	3.00×10^8 m/s

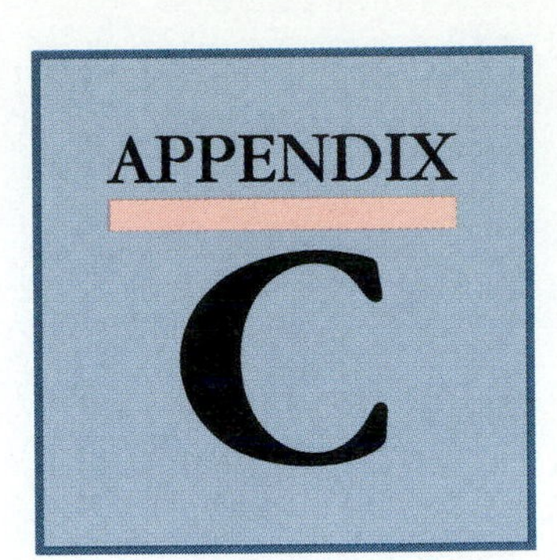

Activity Series

Li
K
Ba
Sr
Ca
Na
Mg
Al
Mn
Zn
Fe
Cd
Co
Ni
Sn
Pb
(**H**)
Cu
Ag
Hg
Au

Solubility Rules for Ionic Compounds

Compounds containing the following ions are generally *soluble* in water:

1. Alkali metal ions and ammonium ions, Li^+, Na^+, K^+, NH_4^+
2. Acetate ion, $C_2H_3O_2^-$
3. Nitrate ion, NO_3^-
4. Halide ions (X), Cl^-, Br^-, I^- (AgX, Hg_2X_2, and PbX_2 are insoluble exceptions)
5. Sulfate ion, SO_4^{2-} ($SrSO_4$, $BaSO_4$, and $PbSO_4$ are insoluble exceptions)

Compounds containing the following ions are generally *insoluble*[†] in water:

6. Carbonate ion, CO_3^{2-} (see rule 1 exceptions, which are soluble)
7. Chromate ion, CrO_4^{2-} (see rule 1 exceptions, which are soluble)
8. Phosphate ion, PO_4^{3-} (see rule 1 exceptions, which are soluble)
9. Sulfide ion, S^{2-} (CaS, SrS, BaS, and rule 1 exceptions are soluble)
10. Hydroxide ion, OH^- [$Ca(OH)_2$, $Sr(OH)_2$, $Ba(OH)_2$, and rule 1 exceptions are soluble]

[†] Actually, these compounds are slightly soluble or very slightly soluble in water.

Properties of Water

Density of H_2O:	0.99987 g/mL at 0°C 1.00000 g/mL at 4°C 0.99707 g/mL at 25°C
Heat of fusion at 0°C:	80.0 cal/g (335 J/g)
Heat of vaporization at 100°C:	540 cal/g (2260 J/g)
Specific heat of ice:	0.50 cal/g·°C (2.1 J/g·°C)
of water:	1.00 cal/g·°C (4.18 J/g·°C)
of steam:	0.48 cal/g·°C (2.0 J/g·°C)
Ionization constant, K_w:	1.00×10^{-14} at 25°C

Vapor Pressure of Water

Temperature (°C)	Vapor Pressure (mm Hg)	Temperature (°C)	Vapor Pressure (mm Hg)	Temperature (°C)	Vapor Pressure (mm Hg)
0	4.6	21	18.7	35	42.2
5	6.5	22	19.8	40	55.3
10	9.2	23	21.1	45	71.9
12	10.5	24	22.4	50	92.5
14	12.0	25	23.8	55	118.0
16	13.6	26	25.2	60	149.4
17	14.5	27	26.7	70	233.7
18	15.5	28	28.4	80	355.1
19	16.5	29	30.0	90	525.8
20	17.5	30	31.8	100	760.0

Logarithm Table

N	0.0	0.1	0.2	0.3	0.4	0.5	0.6	0.7	0.8	0.9
1	0.00	0.04	0.08	0.11	0.15	0.18	0.20	0.23	0.26	0.28
2	0.30	0.32	0.34	0.36	0.38	0.40	0.42	0.43	0.45	0.46
3	0.48	0.49	0.51	0.52	0.53	0.54	0.56	0.57	0.58	0.59
4	0.60	0.61	0.62	0.63	0.64	0.65	0.66	0.67	0.68	0.69
5	0.70	0.71	0.72	0.72	0.73	0.74	0.75	0.76	0.76	0.77
6	0.78	0.79	0.79	0.80	0.81	0.81	0.82	0.83	0.83	0.84
7	0.85	0.85	0.86	0.86	0.87	0.88	0.88	0.89	0.89	0.90
8	0.90	0.91	0.91	0.92	0.92	0.93	0.93	0.94	0.94	0.95
9	0.95	0.96	0.96	0.97	0.97	0.98	0.98	0.99	0.99	1.00

To obtain the logarithm of a number (*N*) between 1 and 10, use the vertical column for the first digit and the horizontal row for the second digit. For example, the log of 4.5 is 0.65.

For a brief explanation of logarithms and the way to compute them on your scientific calculator, see Appendix A.

Glossary of Key Terms

A

[A] The symbol for the molar concentration of a chemical species A. *(Sec. 16.3)*

absolute zero The theoretical temperature at which the kinetic energy of a gas is zero. *(Sec. 12.10)*

accuracy Refers to the error in the results obtained from an experiment. *(Sec. 9.8)*

acid A hydrogen-containing compound that releases hydrogen ions (H^+) when dissolved in water. *(Secs. 6.10 and 8.11)*

acid–base indicator A chemical substance that changes color according to the pH of the solution. *(Sec. 15.2)*

acid salt An ionic compound that results from the partial neutralization of an acid; the compound contains one or more hydrogen atoms bonded to the anion, for example, $NaHCO_3$ and NaH_2PO_4. *(Sec. 6.10)*

actinide series The elements with atomic numbers 90 to 103. *(Sec. 5.3)*

activation energy (E_{act}) The energy necessary for reactants to achieve the transition state in order to form products. *(Sec. 16.2)*

active metal A metal that is sufficiently active to react with water at 25°C. *(Sec. 8.8)*

activity Refers to the number of nuclei in a radioactive sample which disintegrate per unit time; for example, 500 disintegrations per minute (500 dpm). *(Sec. 19.4)*

activity series A relative order of elements arranged by their ability to undergo reaction; also called an electromotive series. *(Sec. 8.7)* A list of half-reactions for metals and nonmetals arranged in order of their ability to undergo reduction; a relative scale of the strength of oxidizing agents. *(Sec. 17.5)*

actual yield The amount of substance experimentally measured in a laboratory procedure. *(Sec. 9.7)*

addition reaction A chemical reaction in which an unsaturated compound adds a molecule, such as H—H or Br—Br, across a double or triple bond. *(Sec. 20.3)*

algebraic analysis A systematic problem-solving approach that requires rearranging variables in an equation in order to calculate an unknown quantity. *(Sec. 18.1)*

algorithm A problem-solving plan that relates a given quantity to an unknown quantity in a series of steps listed in a specific sequence. *(Sec. 18.1)*

alkali metals The Group IA/1 elements, excluding hydrogen. *(Sec. 5.3)*

alkaline earth metals The Group IIA/2 elements. *(Sec. 5.3)*

alkyl group (R—) An alkane with a hydrogen atom removed. *(Sec. 20.2)*

alpha particle (α) A nuclear radiation identical to a helium-4 nucleus. *(Sec. 19.1)*

amphiprotic A substance that is capable of either accepting a proton or donating a proton in acid-base reactions. *(Sec. 15.6)*

anhydrous Refers to a compound that does not contain water. *(Sec. 13.10)*

anion A negatively charged ion. *(Sec. 6.1)*

anode The electrode in an electrochemical cell at which oxidation occurs. *(Sec. 17.6)*

aqueous solution A solution of a substance dissolved in water. *(Secs. 6.1 and 8.2)*

aromatic hydrocarbon A hydrocarbon containing a benzene ring. *(Sec. 20.4)*

Arrhenius acid A substance that yields hydrogen ions when dissolved in water. *(Sec. 15.5)*

Arrhenius base A substance that yields hydroxide ions when dissolved in water. *(Sec. 15.5)*

aryl group (Ar—) An aromatic group with a hydrogen atom removed. *(Sec. 20.2)*

atmospheric pressure The pressure exerted by the gas molecules in air; at sea level this pressure supports a 760-mm column of mercury. *(Sec. 12.2)*

atom The smallest particle that represents an element. *(Sec. 3.4)*

atomic mass The weighted average mass of all the naturally occurring isotopes of an element. *(Sec. 4.6)*

atomic mass unit (amu) A unit of mass equal to exactly 1/12 the mass of a C-12 atom. *(Sec. 4.5)*

atomic notation A symbolic method for expressing the composition of an atomic nucleus; the mass number and atomic number are indicated to the left of the chemical symbol for the element. *(Sec. 4.4)*

atomic nucleus A region of very high density in the center of the atom. *(Sec. 4.3)*

atomic number (Z) A number characteristic of an element that indicates the number of protons found in the nucleus of one of its atoms. *(Secs. 4.4 and 19.2)*

autoprotolysis A reversible reaction in pure water that produces hydronium ions and hydroxide ions. *(Sec. 15.7)*

Avogadro's law The principle that equal volumes of gases, at the same temperature and pressure, contain equal numbers of molecules. The principle applies to equal volumes of the same gas or different gases. *(Secs. 7.8 and 9.1)*

Avogadro's number (*N*) The number of atoms, molecules, or formula units that constitute 1 mole of a given substance; 6.02×10^{23} individual particles. *(Sec. 7.1)*

B

Balmer formula A mathematical equation for calculating the emitted wavelength of light from an excited hydrogen atom when an electron drops to the second energy level. *(Sec. 4.7)*

barometer An instrument for measuring atmospheric pressure. *(Sec. 12.2)*

base A hydroxide containing compound that releases hydroxide ions (OH^-) when dissolved in water. *(Secs. 6.10 and 8.11)*

battery A general term for any electrochemical cell that produces electrical energy spontaneously. *(Sec. 17.6)*

beta particle (β) A nuclear radiation identical to an electron. *(Sec. 19.1)*

binary acid A compound containing hydrogen and a nonmetal dissolved in water. *(Sec. 6.1)*

binary ionic Refers to a compound with one metal and one nonmetal. *(Sec. 6.1)*

binary molecular Refers to a compound containing two nonmetals. *(Sec. 6.1)*

binding energy The energy corresponding to the mass defect that holds nucleons together in a nucleus. *(Sec. 19.8)*

Bohr atom A model of the atom that pictures the electron circling the nucleus in an orbit of specific energy. *(Secs. 4.7 and 10.1)*

boiling point (bp) The temperature at which the vapor pressure of a liquid is equal to the atmospheric pressure. *(Sec. 12.3)*

bond angle The angle formed by two atoms bonded to the central atom in a molecule. *(Secs. 11.4 and 13.7)*

bond energy The amount of energy required to break a given bond in 1 mole of substance in the gaseous state. *(Sec. 11.2)*

bond length The distance between the nuclei of two atoms that are covalently bonded. *(Sec. 11.3)*

bonding electrons The valence electrons in a molecule that are shared between two atoms. *(Sec. 11.4)*

Boyle's law At constant temperature, the pressure and volume of a gas are inversely proportional. *(Sec. 12.6)*

breeder reactor A nuclear reactor that converts a nonfissionable isotope into a fissionable isotope, for example, converting U-238 into Pu-239. *(Sec. 19.10)*

Brønsted–Lowry acid A substance that donates a proton in an acid–base reaction. *(Sec. 15.6)*

Brønsted–Lowry base A substance that accepts a proton in an acid–base reaction. *(Sec. 15.6)*

C

calculator notation A method of displaying exponential numbers without using superscripts; for example, 7.75 E–0.8. *(Sec. 1.8)*

Calorie (Cal) A nutritional unit of heat energy equal to 1 kilocalorie. *(Sec. 2.10)*

calorie (cal) The amount of heat required to raise the temperature of 1 g of water 1°C. *(Sec. 2.10)*

carbonyl group The structural unit composed of a carbon–oxygen double bond found in aldehydes, ketones, and other organic compounds. *(Sec. 20.5)*

carboxyl group The functional group in a carboxylic acid, —COOH. *(Sec. 20.10)*

catalyst A substance that increases the rate of reaction but can be recovered without being permanently changed. *(Sec. 8.2)* A substance that allows a reaction to proceed faster by lowering the energy of activation. *(Sec. 16.2)*

cathode The electrode in an electrochemical cell at which reduction occurs. *(Sec. 17.6)*

cathode ray A stream of negative particles produced in a cathode-ray tube. *(Sec. 4.2)*

cation A positively charged ion. *(Sec. 6.1)*

Celsius degree (°C) The basic unit of temperature in the metric system. *(Sec. 2.9)*

centimeter (cm) A common unit of length in the metric system of measurement that is equal to one-hundredth of a meter. *(Sec. 1.1)*

chain reaction A fission reaction whereby the neutrons produced initiate a second reaction, which in turn initiates a third reaction, and so on. *(Sec. 19.7)*

Charles' law At constant pressure, the volume and Kelvin temperature of a gas are directly proportional. *(Sec. 12.7)*

chemical bond The attraction between two atoms or two ions. *(Sec. 11.1)*

chemical change The process of undergoing a change in chemical formula or chemical composition. *(Sec. 3.8)*

chemical equation A shorthand representation using formulas and symbols to describe a chemical change. *(Sec. 8.2)*

chemical equilibrium A dynamic state for a reversible reaction in which the rates of the forward and reverse reactions are equal. *(Secs. 15.7 and 16.4)*

chemical formula An abbreviation for the name of a chemical compound that indicates the number of atoms of each element; for example, H_2O is the formula for water. *(Sec. 3.6)*

chemical property A property of a substance that cannot be observed without changing the chemical formula of the substance. *(Sec. 3.7)*

chemical reaction The process of undergoing a chemical change. *(Sec. 8.1)*

chemical symbol An abbreviation for the name of a chemical element; for example, Cu is the symbol for copper. *(Sec. 3.4)*

class of compounds A family of compounds in which all the members have a similar structural feature (that is, an atom or group of atoms) and similar chemical properties. *(Sec. 20.5)*

closed system A chemical reaction or substance that is isolated and can be studied independently from the surrounding environment. *(Sec. 16.4)*

coefficient A digit placed in front of a chemical formula in order to balance a chemical equation. *(Sec. 8.3)*

colligative property A property of a solution that is affected by the number (not the type) of solute particles in solution; for example, freezing point lowering, boiling point raising, and vapor pressure lowering. *(Sec. 14.10)*

collision theory The principle that the rate of a chemical reaction is regulated by the collision frequency, collision energy, and the orientation of molecules striking each other. *(Sec. 16.1)*

colloid A homogeneous mixture in which the diameter of the dispersed particles ranges from 1 to 100 nm. *(Sec. 14.11)*

combination reaction A type of reaction in which two substances react to produce a single compound. *(Sec. 8.4)*

combined gas law The pressure exerted by a gas is inversely proportional to its volume and directly proportional to its Kelvin temperature. *(Sec. 12.9)*

combustion reaction A chemical reaction in which a substance reacts rapidly with oxygen to produce heat; for example, a hydrocarbon can undergo combustion with oxygen to give carbon dioxide and water. *(Sec. 20.2)*

compound A pure substance that can be broken down into two or more simpler substances by chemical reaction. *(Sec. 3.3)*

concentration equilibrium constant (K_c) The equilibrium constant that relates the concentrations of gases participating in a reversible reaction. *(Sec. 16.5)*

concept map A problem-solving plan that traces the relationship of various quantities, usually shown by interrelated boxes. *(Sec. 18.1)*

continuous spectrum A broad, uninterrupted band of radiant energy. *(Sec. 4.7)*

coordinate covalent bond A bond in which an electron pair is shared but both electrons have been donated by a single atom. *(Sec. 11.7)*

core The portion of the atom that includes the nucleus and inner electrons that are not available for bonding; also termed the kernel of the atom. *(Sec. 5.10)*

core notation A method of writing electron configuration where all the inner electrons are represented by a noble gas symbol in brackets followed by valence electrons; for example, $[Ne]3s^2$. *(Sec. 5.11)*

covalent bond A chemical bond characterized by the sharing of one or more pairs of valence electrons. *(Sec. 11.1)*

critical mass The minimum amount of fissionable substance necessary to sustain a chain reaction. *(Sec. 19.7)*

crystalline solid A substance in the solid state that contains particles that repeat in a regular geometric pattern. *(Sec. 13.4)*

cubic centimeter (cm^3) A unit of volume occupied by a cube 1 centimeter on a side; a volume exactly equal to 1 milliliter. *(Sec. 2.6)*

D

Dalton's law of partial pressures The pressure exerted by a mixture of gases is equal to the sum of the pressures exerted by each gas in the mixture. *(Sec. 12.4)*

decay series See radioactive decay series.

decomposition reaction A type of reaction in which a single compound decomposes into two or more substances. *(Sec. 8.4)*

deionized water Water purified by removing ions using an ion exchange method; also termed demineralized water. *(Sec. 13.11)*

delta (δ) notation A method of indicating the partial positive charge (δ^+) and partial negative charge (δ^-) in a polar covalent bond. *(Sec. 11.5)*

density (*d*) The amount of mass in a unit volume of matter. *(Sec. 2.8)*

deuterium The isotope of hydrogen with one neutron in the nucleus. *(Sec. 19.8)*

diatomic molecule A particle composed of two nonmetal atoms that are covalently bonded. *(Sec. 11.6)*

dipole A directional shift of an electron pair in a polar covalent bond to give a molecule with regions of partial positive and negative charge. *(Secs. 13.3 and 14.2)*

directly proportional Refers to two related variables; if one variable doubles, the other variable doubles; if one variable triples, the other triples; and so on. *(Sec. 12.7)*

dispersion forces Intermolecular attraction based on temporary dipoles in molecules; also called London forces. *(Sec. 13.3)*

dissociation The process of an ionic compound dissolving in water and separating into positive and negative ions; for example, NaOH dissolves in water to give Na^+ and OH^-. *(Sec. 15.5)*

distilled water Water purified by boiling water and collecting the condensed vapor. *(Sec. 13.11)*

double bond A bond between two atoms composed of two electron pairs. A double bond is represented as two dashes between the symbols of two atoms. *(Sec. 11.4)*

double replacement reaction A type of reaction in which two cations in different compounds exchange anions. *(Sec. 8.4)*

dry cell An electrochemical cell in which, in general, the anode and cathode reactions do not take place in aqueous solutions. *(Sec. 17.6)*

ductile The property of a metal that allows it to be drawn into a wire. *(Sec. 3.7)*

E

elastic collision Refers to gas molecules that collide without losing energy. *(Sec. 12.10)*

electrochemical cell A general term for an apparatus containing two solutions with electrodes in separate compartments that are connected by a conducting wire and salt bridge. *(Sec. 17.6)*

electrochemistry The study of the interconversion of chemical and electrical energy by redox reactions. *(Sec. 17.6)*

electrolysis A nonspontaneous chemical reaction produced from the passage of electricity through an aqueous solution, or molten salt. *(Sec. 13.9)*

electrolytic cell An electrochemical cell in which a nonspontaneous redox reaction occurs by the input of direct electric current. *(Sec. 17.7)*

electron (e^-) A subatomic particle having a negligible mass and a charge of one minus. *(Sec. 4.2)*

electron capture (EC) A nuclear decay process whereby a heavy isotope attracts an inner electron into its nucleus. *(Sec. 19.2)*

electron configuration A shorthand description of the arrangement of electrons by sublevels according to increasing energy. *(Sec. 4.9)*

electron dot formula A representation of an atom and its valence electrons that shows the chemical symbol surrounded by a dot for each valence electron. *(Sec. 5.8)* A representation of a molecule or polyatomic ion that shows the chemical symbol of each atom surrounded by a dot for each bonding or nonbonding electron; also called a Lewis diagram. *(Sec. 11.4)*

electronegativity The ability of an atom to attract a pair of electrons in a chemical bond. *(Sec. 11.5)*

electron sea model A theory for metallic crystals that explains the electrical conductivity of metals. *(Sec. 13.5)*

electron shell A collection of orbitals that have the same value of *n*; for example, the 3*s*, 3*p*, and 3*d* orbitals comprise the third electron shell. *(Sec. 10.4)*

electron spin A property of an electron that is similar to a charged particle spinning on an axis through its center. The electron spin creates a small magnetic field that either aligns with or against an external magnetic field. *(Sec. 10.5)*

electron subshell A collection of orbitals that have the same value of *n* and *l*; for example, the $2p_x$, $2p_y$, and $2p_z$ orbitals comprise the 2*p* subshell. *(Sec. 10.4)*

element A pure substance that cannot be broken down any further by ordinary chemical reaction. *(Sec. 3.6)*

emission line spectrum A collection of narrow slits of light that results from excited atoms of a given element releasing energy. *(Sec. 4.7)*

empirical formula The chemical formula of a compound that expresses the simplest ratio of the atoms in a molecule, or ions in a formula unit. *(Sec. 7.6)*

endothermic reaction A reaction that absorbs heat energy. *(Secs. 8.1 and 16.2)*

endpoint The stage in a titration when the indicator changes color. *(Sec. 15.4)*

energy level An orbit of specific energy that electrons occupy at a fixed distance from the nucleus; designated 1, 2, 3, 4 . . . . *(Sec. 4.7)*

energy sublevel An electron energy level resulting from the splitting of a principal energy level; designated *s, p, d, f* . . . . *(Sec. 4.8)*

English system A nondecimal system of measurement which does not have a basic unit for length, mass, or volume. *(Sec. 2.1)*

equilibrium constant of water (K_w) The product of the molar hydrogen ion concentration times the molar hydroxide ion concentration in water. *(Sec. 15.7)*

error Expresses the difference between the experimental result and the theoretical value for an analysis. *(Sec. 9.8)*

exothermic reaction A reaction that releases heat energy. *(Secs. 8.1 and 16.2)*

exponent A number written as a superscript that indicates a value is multiplied times itself; for example, $10^4 = 10 \times 10 \times 10 \times 10$, or $cm^3 = cm \times cm \times cm$. *(Sec. 1.6)*

F

Fahrenheit degree (°F) A basic unit of temperature in the English system. *(Sec. 2.9)*

formula unit The smallest representative particle in an ionic compound. *(Sec. 6.4)* The simplest representative entity in a compound held together by ionic bonds. *(Sec. 11.1)*

fourth quantum number (*s*) The quantum number that describes the spin of an electron in an orbital. The allowed values of *s* are $+\frac{1}{2}$ or $-\frac{1}{2}$. *(Sec. 10.5)*

frequency The number of times a light wave travels a complete cycle in 1 second. *(Sec. 4.7)*

functional group An atom or group of atoms that gives a class of compounds its characteristic chemical properties. *(Sec. 20.5)*

G

gamma ray (γ) A nuclear radiation identical to the radiation in the high-energy region of the electromagnetic spectrum. *(Sec. 19.1)*

gas density The ratio of mass per unit volume for a gas; usually expressed in grams per liter. *(Sec. 7.8)*

gas pressure A physical quantity that measures the frequency and energy of molecular collisions on a surface. The force per unit area exerted by a gas. *(Sec. 12.2)*

Gay-Lussac's law At constant volume, the pressure exerted by a gas is directly proportional to its Kelvin temperature. *(Sec. 12.8)*

general equilibrium expression The relationship of the reactants and the products concentrations having the form $K_{eq} = [C]^c[D]^d/[A]^a[B]^b$. *(Sec. 16.4)*

gram (g) The basic unit of mass in the metric system. *(Secs. 1.1 and 2.1)*

group A vertical column in the periodic table; a family of elements with similar properties. *(Sec. 5.1)*

H

half-cell A portion of an electrochemical cell having a single electrode where either oxidation or reduction occurs. *(Sec. 17.6)*

half-life ($t_{1/2}$) The amount of time required for 50% of a given sample of a specific radioactive isotope to decay. *(Sec. 19.4)*

half-reaction A balanced chemical equation that represents an oxidation process or a reduction process separately. *(Sec. 17.4)*

halogens The Group VIIA/17 elements. *(Sec. 5.3)*

hard water Water containing a variety of cations and anions, such as Ca^{2+}, Mg^{2+}, Fe^{3+}, CO_3^{2-}, SO_4^{2-}, and PO_4^{3-}. *(Sec. 13.11)*

heat The flow of energy from a system at a higher temperature to another system at a lower temperature. Heat is a measure of the total energy in a system. *(Sec. 2.10)*

heat of fusion (H_{fusion}) The heat required to convert a solid to a liquid at its melting point. *(Sec. 13.6)*

heat of reaction (ΔH) The difference in heat energy between the reactants and the products for a given chemical reaction. *(Secs. 16.2 and 18.5)*

heat of vaporization (H_{vapor}) The heat required to convert a liquid to a gas at its boiling point. *(Sec. 13.6)*

heavy water (D_2O) A molecule of water in which the hydrogen-1 atoms are replaced by hydrogen-2 atoms. *(Sec. 13.8)*

Henry's law The principle that the solubility of a gas in a liquid is proportional to the partial pressure of the gas above the liquid. *(Sec. 14.1)*

heterogeneous equilibrium A type of equilibrium in which one or more of the participating species is in a different state; for example, an equilibrium between a gas and a solid, or an insoluble precipitate and its ions in aqueous solution. *(Sec. 16.5)*

heterogeneous mixture Matter having variable composition and indefinite properties; matter composed of two or more substances that can be separated using physical methods. *(Sec. 3.3)*

homogeneous equilibrium A type of equilibrium in which all the participating species are in the same state; for example, an equilibrium between reactants and products in the gaseous state. *(Sec. 16.5)*

homogeneous mixture Matter having variable composition, but definite and consistent properties; examples include gas mixtures, solutions, and alloys. *(Sec. 3.3)*

homogeneous substance Matter having constant composition and definite and consistent properties. *(Sec. 3.3)*

Hund's rule The principle which states that electrons in a given subshell occupy different orbitals of the same energy. That is, electrons in a given subshell remain unpaired as long as orbitals of the same energy are available. *(Sec. 10.6)*

hydrate A compound that contains a specific number of water molecules per formula unit in a crystalline compound. *(Sec. 13.9)*

hydrocarbon An organic compound containing only hydrogen and carbon. *(Sec. 20.1)*

hydrocarbon derivative An organic compound containing carbon, hydrogen, and another element such as oxygen, nitrogen, or a halogen. *(Sec. 20.1)*

hydrogen bond An intermolecular bond such as that between hydrogen and oxygen in two molecules of water. *(Sec. 13.3)*

hydronium ion (H_3O^+) The ion that results when a hydrogen ion attaches to a water molecule by a coordinate covalent bond. The hydronium ion is the predominant form of the hydrogen ion in aqueous acid solution. *(Sec. 15.5)*

hydroxyl group The functional group in an alcohol, —OH. *(Sec. 20.7)*

I

IUPAC nomenclature The system of rules set forth by the International Union of Pure and Applied Chemistry for naming chemical compounds. *(Sec. 6.1)*

ideal gas A hypothetical gas that perfectly obeys the kinetic theory. *(Sec. 12.10)*

ideal gas constant The proportionality constant R in the equation $PV = nRT$. *(Sec. 12.11)*

ideal gas equation The relationship that $PV = nRT$ for an ideal gas. *(Sec. 12.11)*

immiscible Refers to two liquids that are not soluble in one another, and if mixed, separate into two layers. *(Sec. 14.2)*

inner transition elements The elements in the lanthanide and actinide series. *(Sec. 5.3)*

inorganic compound A compound not containing the element carbon. *(Sec. 6.1)*

instrument A device for recording a measurement, such as length, mass, volume, time, or temperature. *(Sec. 1.1)*

International System (SI) A sophisticated scientific system of measurement having seven base units. *(Sec. 2.11)*

inversely proportional Refers to two related variables; if one variable doubles, the other variable is reduced to one half; if one quadruples, the other is reduced to one-fourth; and so on. *(Sec. 12.6)*

ion An atom (or group of atoms) that bears a charge as the result of gaining or losing valence electrons. *(Sec. 5.9)*

ionic bond A chemical bond characterized by the attraction between a cation and anion. *(Sec. 11.1)*

ionic charge Refers to the positive charge on an atom that has lost electrons or the negative charge on an atom that has gained electrons. *(Sec. 5.10)*

ionic solid A crystalline solid composed of ions that repeat in a regular pattern. *(Sec. 13.5)*

ionization The process of a polar molecular compound dissolving in water and forming positive and negative ions; for example, HCl dissolves in water to give hydrogen ions and chloride ions. *(Sec. 15.5)*

ionization constant (K_i) The equilibrium constant that relates concentrations of species in slightly ionized weak acids or bases in aqueous solution. *(Sec. 16.8)*

ionization energy The amount of energy necessary to remove an electron from a neutral atom in the gaseous state. *(Sec. 5.9)*

isoelectronic Refers to two or more ions (or ions and an atom) having the same electron configuration; for example, O^{2-} and Mg^{2+} each have 10 electrons and their electron configurations are identical to the noble gas neon. *(Sec. 5.10)*

isotopes Atoms having the same atomic number but a different mass number. Atoms of the same element that have a different number of neutrons in the nucleus. *(Sec. 4.4)*

K

Kelvin unit (K) The basic unit of temperature in the SI system. *(Sec. 2.9)*

kinetic energy The energy associated with the mass and velocity of a particle. The kinetic energy of a particle is equal to one-half its mass times its velocity squared. *(Sec. 3.2)*

kinetic theory of gases A set of assumptions that describes gas molecules demonstrating perfectly consistent behavior. *(Sec. 12.10)*

L

lanthanide series The elements with atomic numbers 58 to 71. *(Sec. 5.3)*

Latin system A naming system that designates the variable charge on a metal cation using an -ic or -ous suffix attached to the stem of the Latin name. *(Sec. 6.2)*

law of chemical equilibrium The principle that the concentrations of the products of a reversible reaction divided by the concentrations of the reactants, each raised to a coefficient power from the balanced equation, is a constant. *(Sec. 16.4)*

law of combining volumes The principle that volumes of gases that combine in a chemical reaction, at the same temperature and pressure, are in the ratio of small whole numbers; also called Gay-Lussac's law of combining volumes. *(Sec. 9.1)*

law of conservation of energy The principle which states that energy can neither be created nor destroyed. *(Sec. 3.10)*

law of conservation of mass The principle which states that mass can neither be created nor destroyed. *(Sec. 3.9)* The statement that the total mass of products from a chemical reaction must equal the sum of the masses of reactants. *(Sec. 9.1)*

law of conservation of mass and energy The principle from which it follows that the total mass and energy of substances before a chemical change is equal to the total mass and energy after a chemical change. *(Sec. 3.11)*

law of constant composition The principle which states that a compound always contains the same elements in the same proportion by mass. *(Sec. 3.6)*

law of mass action The principle that a change in the amount of a substance participating in a reversible reaction produces a shift in the equilibrium; also called the mass law. *(Sec. 16.4)*

Le Chatelier's principle The statement that an equilibrium when stressed by a change of concentration, temperature, or pressure, will shift to relieve the stress. *(Sec. 16.6)*

***like dissolves like* rule** The general principle that solute–solvent interaction is greatest when the polarity of the solute is similar to that of the solvent. *(Sec. 14.2)*

light The portion of the radiant energy spectrum that is visible. A general term for any form of radiant energy, gamma rays through microwaves. *(Sec. 4.7)*

limiting reactant The substance in a chemical reaction which controls or limits the maximum amount of product formed. *(Sec. 18.4)*

liter (L) The basic unit of volume in the metric system; a volume equal to a cube 10 cm on a side. *(Sec. 2.1)*

M

malleable The property of a metal that allows it to be hammered or machined into a foil. *(Sec. 3.7)*

mass The quantity of matter in an object. The measurement is obtained using a balance. *(Sec. 1.1)*

mass defect The mass difference between the mass of a nucleus and the sum of the masses of the individual nucleons. *(Sec. 19.8)*

mass/mass percent (m/m%) A solution concentration expression that relates the mass of solute in grams dissolved in each 100 grams of solution. *(Sec. 14.8)*

mass–mass problem A type of stoichiometry calculation that relates the masses of substances in a balanced chemical equation. *(Sec. 9.3)*

mass number (*A*) A number that represents the total of the numbers of protons and neutrons in the nucleus of a given isotope. *(Secs. 4.4 and 19.2)*

mass spectrometer An instrument used to determine the atomic mass of an isotope relative to carbon-12. *(Sec. 4.5)*

mass–volume problem A type of stoichiometry calculation that relates the mass of a substance to the volume of a gaseous substance in a balanced chemical equation. *(Sec. 9.3)*

measurement A numerical value with attached units that expresses a physical quantity such as length, mass, or volume. *(Sec. 1.1)*

metal An element that is usually shiny in appearance, has a high density and a high melting point, and is a good conductor of heat and electricity. *(Sec. 3.5)*

metal oxide A compound that forms an alkaline solution when it dissolves in water; also called a basic oxide. *(Sec. 13.9)*

metallic solid A crystalline solid composed of metal atoms that repeat in a regular pattern. *(Sec. 13.5)*

meter (m) The basic unit of length in the metric system of measurement. *(Sec. 2.1)*

metric system A decimal system of measurement using prefixes and a basic unit to express physical quantities such as length, mass, and volume. *(Sec. 2.1)*

milliliter (mL) A common unit of volume in the metric

system of measurement that is equal to one-thousandth of a liter. *(Sec. 1.1)*

miscible Refers to two or more liquids that are soluble in one another *(Sec. 14.2)*

molal freezing point constant (K_f) The number of degrees Celsius that a nonvolatile solute lowers the freezing point of a 1.00 molal solution; the units are °C/*m*. *(Sec. 14.10)*

molality (*m*) A solution concentration expression that relates the moles of solute dissolved in each kilogram of solvent. *(Sec. 14.10)*

molar mass (MM) The mass of 1 mole of pure substance expressed in grams; the mass of Avogadro's number of atoms, molecules, or formula units. *(Secs. 7.3 and 9.1)*

molar volume The volume occupied by 1 mole of any gas at standard conditions of temperature and pressure; at 0°C and 1.00 atm the volume of 1 mole of any gas is 22.4 L. *(Secs. 7.8 and 9.1)*

molarity (*M*) A solution concentration expression that relates the moles of solute dissolved in each liter of solution. *(Sec. 14.9)*

mole (mol) The amount of substance containing the same number of entities as there are atoms in exactly 12 g of carbon-12. *(Sec. 7.2)*

molecular formula The chemical formula of a compound that expresses the actual number of atoms present in one molecule. *(Sec. 7.7)*

molecular solid A solid, often crystalline, composed of molecules that repeat, in a regular pattern. *(Sec. 13.5)*

molecule The smallest particle that represents a compound. *(Sec. 3.2)* The smallest representative entity of an element or compound held together by covalent bonds. *(Sec. 11.1)*

mole–mole problem A type of calculation that relates the moles of two substances that appear in a balanced chemical equation. *(Sec. 9.2)*

momentum The quantity that expresses the mass of a particle times velocity; for example, the momentum of an electron is equal to its mass times its velocity. *(Sec. 10.3)*

monoatomic ion A single atom bearing a positive or negative charge as the result of gaining or losing valence electrons. *(Secs. 6.1 and 11.8)*

N

net dipole The overall effect of one or more dipoles in a molecule. *(Secs. 13.7 and 14.2)*

net ionic equation A chemical equation that portrays an ionic reaction after spectator ions have been canceled from the total ionic equation. *(Sec. 15.11)*

neutralization reaction A type of reaction in which an acid and a base react to produce a salt and water. *(Secs. 6.10 and 8.4)*

neutron (n^0) A subatomic particle having an approximate mass of 1 amu and a charge of zero. *(Sec. 4.3)*

noble gases The relatively unreactive Group VIIIA/18 elements. *(Sec. 5.1)*

nonbonding electrons The valence electrons in a molecule that are not shared. *(Sec. 11.4)*

nonmetal An element that is generally dull in appearance, has a low density and low melting point, and is not a good conductor of heat and electricity. *(Sec. 3.5)*

nonmetal oxide A compound that forms an acidic solution when it dissolves in water; also called an acidic oxide. *(Sec. 13.9)*

nonpolar covalent bond A bond in which one or more pairs of electrons are shared equally. *(Sec. 11.6)*

nonpolar solvent A liquid composed of nonpolar molecules. *(Sec. 14.2)*

nonsignificant digits The digits in a measurement that exceed the uncertainty of the instrument. *(Sec. 1.3)*

nuclear equation A shorthand representation using atomic notation to describe a nuclear change. *(Sec. 19.2)*

nuclear fission A nuclear reaction in which a large nucleus splits into two or more smaller nuclei. *(Sec. 19.7)*

nuclear fusion A nuclear reaction in which two smaller nuclei combine into a single larger nucleus. *(Sec. 19.9)*

nuclear reaction A high-energy change involving atomic nuclei. *(Sec. 19.2)*

nucleon A general term for a nuclear particle, that is, either a proton or a neutron. *(Sec. 19.2)*

O

octet rule The statement that each atom in a compound must attain eight valence electrons in order to be energetically stable. Hydrogen is an exception to the rule and requires only two electrons. *(Sec. 11.1)*

orbital A region in space surrounding the nucleus of an atom in which there is a high probability (~95%) of finding an electron with a given energy. An orbital describes the energy state of the electron. *(Sec. 10.3)*

orbital energy diagram An energy profile of the orbitals in a given subshell relative to the orbitals in other subshells. *(Sec. 10.6)*

organic chemistry The study of carbon-containing compounds. *(Sec. 20.1)*

organic compound A compound containing the element carbon. *(Sec. 6.1)*

oxidation A process in which a substance undergoes

an increase in oxidation number. A process characterized by losing electrons. *(Sec. 17.2)*

oxidation number The value assigned to an atom in a molecule or ion in order to keep track of electrons gained or lost. An oxidation number can be positive or negative but is zero for elements in the free state. *(Sec. 17.1)*

oxidizing agent A substance that causes the oxidation of another substance in a redox reaction. The substance reduced in a redox reaction. *(Sec. 17.2)*

P

parent-daughter isotopes Refers to the relationship of a decaying isotope and the resulting isotope that is produced. *(Sec. 19.3)*

partial pressure The pressure exerted by a single gas in a mixture of two or more gases. *(Sec. 12.4)*

Pauli exclusion principle The statement that no two electrons in an atom can have the same set of four quantum numbers. *(Sec. 10.7)*

percent (%) The ratio of a single quantity compared to all quantities in a group (times 100); parts per hundred parts. *(Sec. 1.11)*

percent yield The ratio of the actual yield compared to the theoretical yield, all times 100. *(Sec. 9.7)*

percentage composition An expression for the ratio of the mass of a single element compared to the mass of a given compound containing the element, all times 100. *(Sec. 7.5)*

period A horizontal row in the periodic table; a series of elements with properties that vary from metallic to nonmetallic. *(Sec. 5.1)*

periodic law The principle which states that properties of the elements recur in a repeating pattern when the elements are arranged according to increasing atomic number. *(Sec. 5.2)*

pH The molarity of hydrogen ion expressed on an exponential scale; the negative logarithm of the molar hydrogen ion concentration. *(Sec. 15.1)*

photoelectric effect The phenomenon of a metal ejecting an electron when struck by a photon of sufficient energy. *(Sec. 10.2)*

photon A unit of radiant energy that corresponds to the particle nature of light. *(Sec. 10.2)*

physical change The process of undergoing a change without altering the chemical formula of a substance; for example, a change in physical state. *(Sec. 3.8)*

physical property A property that can be observed without changing the chemical formula of a substance. *(Sec. 3.7)*

physical state Refers to the condition of a substance existing as a solid, liquid, or gas. *(Sec. 3.1)*

pOH The molarity of hydroxide ion expressed on an exponential scale; the negative logarithm of the molar hydroxide ion concentration. *(Sec. 15.9)*

polar covalent bond A bond in which one or more pairs of electrons are shared unequally. *(Sec. 11.5)*

polar solvent A liquid composed of polar molecules. *(Sec. 14.2)*

polyatomic ion A unit containing two or more atoms bearing a positive or negative charge. *(Sec. 6.1)* A unit containing two or more atoms covalently bonded together and bearing a positive or negative charge as the result of losing or gaining valence electrons. *(Sec. 11.8)*

positron A nuclear radiation identical in mass, but opposite in charge, to that of an electron. The term positron is derived from "positive electron." *(Sec. 19.2)*

potential energy The stored energy that matter possesses owing to its position or chemical composition. *(Sec. 3.10)*

power of 10 A positive or negative exponent of 10. *(Sec. 1.6)*

precipitate An insoluble substance produced by a reaction in aqueous solution. *(Sec. 8.1)*

precision Refers to the range in the results obtained from an experiment. *(Sec. 9.8)*

principal energy level A main energy level composed of sublevels. *(Sec. 4.8)*

principal quantum number (n) The quantum number that describes the size and energy of a particular orbital. The allowed values of n are 1, 2, 3, ... *(Sec. 10.3)*

product A substance resulting from a chemical reaction. *(Sec. 8.2)*

proton (p^+) A subatomic particle having an approximate mass of 1 amu and a relative charge of one plus. *(Sec. 4.2)*

proton donor A term used interchangeably with hydrogen ion donor. *(Sec. 15.6)*

Q

quantum (*pl.* quanta) A unit of radiant energy that is emitted when an electron in an excited atom drops to a lower energy level. *(Secs. 4.7 and 10.2)*

quantum energy level A general term that refers to an allowed energy state for an electron in an atom; also called a quantum level. *(Sec. 10.3)*

quantum mechanical atom A sophisticated model of the atom that describes an electron of given energy in terms of its probability of being found in a particular location about the nucleus. *(Sec. 10.3)*

quantum notation A method for describing the energy and location of an electron in an atom using a set of four quantum numbers, that is, (n, l, m, s). *(Sec. 10.7)*

quantum theory A theory that specifically explains the gain or loss of energy by electrons in atoms. A theory that generally explains energy changes in objects by small specific increments. *(Sec. 10.2)*

R

radiant energy spectrum Light energy extending from short-wavelength gamma rays through long-wavelength microwaves; also called the electromagnetic spectrum. *(Sec. 4.7)*

radioactive decay series The stepwise disintegration of a radioactive isotope until a stable nucleus is reached. *(Sec. 19.3)*

radioactivity The emission of particles or energy from an unstable atomic nucleus. *(Sec. 19.1)*

radioisotope An isotope that is unstable and eventually disintegrates by emitting radiation. An atom of a specific radioactive isotope is sometimes referred to as a radionuclide. *(Sec. 19.5)*

range Expresses the spread of results for an experiment having two or more trials. *(Sec. 9.8)*

rare earth elements The elements with atomic numbers 21, 39, 57, and 58 through 71. *(Sec. 5.3)*

rate equation The general relationship between the reaction rate and reactant concentration, that is, rate $= k\,[A]^x\,[B]^y$. *(Sec. 16.3)*

rate of reaction The change in concentration of a reactant or product with respect to time. *(Sec. 16.3)*

reactant A substance undergoing a chemical reaction. *(Sec. 8.2)*

reaction profile A graphical representation of the energy of a reaction as a function of the time of reaction; also called an energy diagram. *(Sec. 16.2)*

real gas A gas such as hydrogen or oxygen that does not exhibit ideal behavior; all gases that actually exist. *(Sec. 12.10)*

reciprocal The relationship of a fraction and its inverse; for example, 2/3 and 3/2, or 1 m/100 cm and 100 cm/1 m. *(Sec. 1.9)*

redox reaction A chemical reaction that involves electron transfer between two reacting substances. *(Sec. 17.2)*

reducing agent A substance that causes the reduction of another substance in a redox reaction. The substance oxidized in a redox reaction. *(Sec. 17.2)*

reduction A process in which a substance undergoes a decrease in oxidation number. A process characterized by gaining electrons. *(Sec. 17.2)*

representative elements The Group A (1, 2, 13–18) elements in the periodic table; also termed main group elements. *(Sec. 5.3)*

reversible reaction A reaction that proceeds simultaneously in both the forward direction toward products as well as in the opposite direction toward reactants. *(Secs. 15.6 and 16.4)*

rounding off The process of eliminating digits that are not significant. *(Sec. 1.3)*

S

salt An ionic compound produced from the reaction of an acid and a base that does not contain the hydrogen ion or the hydroxide ion. *(Sec. 6.10)* The product of a neutralization reaction in addition to water. *(Secs. 8.11 and 15.5)*

salt bridge A porous device that allows ions to travel between two half-cells in order to maintain ionic charge balance in each compartment of an electrochemical cell. *(Sec. 17.6)*

saturated hydrocarbon A hydrocarbon containing only single bonds. *(Sec. 20.1)*

saturated solution A solution that contains the maximum amount of dissolved solute that will dissolve at a given temperature. *(Sec. 14.7)*

scientific notation A method for expressing very large or small numbers by placing the decimal after the first significant digit and setting the magnitude using a power of 10. *(Sec. 1.7)*

second quantum number (l) The quantum number that describes the shape of an orbital. The allowed values of l are $0, 1, 2, 3, \ldots, n - 1$. *(Sec. 10.3)*

semimetal An element that is generally metallike in appearance and has properties that are between that of a metal and a nonmetal; also called a metalloid. *(Sec. 3.5)*

significant digits The digits in a measurement that are known with certainty plus one digit that is estimated; also called significant figures. *(Sec. 1.2)*

single bond A bond between two atoms composed of one electron pair. A single bond is represented as a single dash between the symbols of two atoms. *(Sec. 11.4)*

single replacement reaction A type of reaction in which a more active element displaces a less active element from a solution or a compound. *(Sec. 8.4)*

soft water Water containing sodium ions and a variety of anions. *(Sec. 13.11)*

solubility Refers to the maximum amount of solute that dissolves in a solvent at a specified temperature; usually expressed in grams of solute per 100 g of solvent. *(Sec. 14.6)*

solubility product constant (K_{sp}) The equilibrium constant that relates the molar ion concentrations of slightly dissociated ionic compounds in aqueous solution. *(Sec. 16.10)*

solute The component of a solution that is the lesser quantity. *(Sec. 14.2)*

solution A general term for a solute dissolved in a solvent. A solution is an example of a homogeneous mixture. *(Sec. 14.2)*

solvent The component of a solution that is the greater quantity. *(Sec. 14.2)*

solvent cage Refers to a cluster of solvent molecules surrounding a solute molecule or ion in solution. *(Sec. 14.4)*

specific gravity (sp gr) The ratio of the density of a substance compared to the density of water at 4°C; a unitless expression. *(Sec. 2.8)*

specific heat The amount of heat required to raise the temperature of 1 g of any substance 1°C. *(Sec. 2.10)*

spectator ions Those ions that are in aqueous solution but do not participate in a reaction or appear as reactants or products in the net ionic equation. *(Sec. 15.11)*

spectral lines The individual narrow slits of light in an emission line spectrum. *(Sec. 4.7)*

standard conditions Refers to a temperature of 0°C and a pressure of 1.00 atmosphere for a gas; also referred to as STP conditions. *(Secs. 9.1 and 12.9)*

standard solution A solution whose concentration has been established accurately (usually by titration to three or four significant digits). *(Sec. 15.3)*

standard temperature and pressure (STP) Refers to a gas at 0°C and 1.00 atmosphere pressure which have been arbitrarily chosen as standard conditions. *(Sec. 7.8)*

Stock system A naming system that designates the variable charge on a metal cation using Roman numerals in parentheses. *(Sec. 6.2)*

stoichiometry The relationship of quantities (mass of substance or volume of gas) in a chemical change according to the balanced chemical equation. *(Sec. 9.1)*

strategy map A problem solving plan that relates a given value to an unknown quantity, usually shown by a series of arrows. *(Sec. 18.1)*

strong electrolyte An aqueous solution that is a good conductor of electricity; for example, strong acids, strong bases, and soluble salts. *(Sec. 15.10)*

structural formula A representation of a molecule or polyatomic ion that shows each atom connected by a dash for each pair of bonding electrons. *(Sec. 11.4)* A formula that shows the bonding of atoms in a molecule. *(Sec. 20.2)*

structural isomers Compounds with the same molecular formula but different structural formulas. *(Sec. 20.2)*

sublimation A direct change of state from a solid to a gas without forming a liquid. *(Sec. 3.1)*

subscript A digit in a chemical formula that represents the number of atoms or ions occurring in a substance. *(Sec. 8.3)*

substitution reaction A chemical reaction in which one atom replaces another atom in a molecule. *(Sec. 20.2)*

supersaturated solution A solution that contains more dissolved solute than will ordinarily dissolve at a given temperature. *(Sec. 14.7)*

surface tension The intermolecular attraction that causes a liquid droplet to have a minimum surface area. *(Sec. 13.2)*

T

temperature A measure of the average energy in a system. *(Sec. 2.10)*

ternary ionic Refers to a compound containing three elements including at least one metal. *(Sec. 6.1)*

ternary oxyacid A compound containing hydrogen, a nonmetal, and oxygen dissolved in water. *(Sec. 6.1)*

theoretical yield The amount of product that is calculated from given amounts of reactants. *(Sec. 9.7)*

thermochemical equation A balanced chemical equation that incorporates the amount of heat energy involved in the chemical reaction. *(Sec. 18.5)*

thermochemical stoichiometry The relationship of quantities of substances to the heat of reaction according to a balanced chemical equation. *(Sec. 18.5)*

third quantum number (m) The quantum number that describes the orientation of orbitals having the same value of n and l; also referred to as the magnetic quantum number. The allowed values of m are $+l, \ldots 0, \ldots -l$. *(Sec. 10.5)*

titration A laboratory procedure for delivering a measured volume of solution through a buret. *(Sec. 15.4)*

torr A unit of pressure equal to 1 mm Hg. *(Sec. 12.2)*

total ionic equation A chemical equation that writes highly ionized substances in the ionic form and slightly ionized substances in the nonionized form. *(Sec. 15.11)*

transition elements The Group B (3 to 12) elements in the periodic table. *(Sec. 5.3)*

transition state The highest point on the reaction profile where molecules of reactants and products have maximum energy. *(Sec. 16.2)*

transmutation The conversion of one element into another by bombarding a target nucleus with a particle or energy. *(Sec. 19.6)*

transuranium elements The elements beyond atomic number 92. All the elements following uranium are synthetic and do not occur naturally except for traces from nuclear reactions. *(Sec. 5.3)*

triple bond A bond between two atoms composed of three electron pairs. A triple bond is represented as three dashes between the symbols of two atoms. *(Sec. 11.4)*

tritium The isotope of hydrogen with two neutrons in the nucleus. *(Sec. 19.9)*

true solution A homogeneous mixture in which the dispersed particles are less than 1 nm in diameter. *(Sec. 14.11)*

Tyndall effect The phenomenon of scattering light by colloidal-size particles in solution. *(Sec. 14.11)*

U

uncertainty The degree of precision or exactness in an instrumental measurement. *(Sec. 1.1)*

uncertainty principle The statement that it is impossible to precisely measure both the location and momentum of a small particle at the same time. *(Sec. 10.3)*

unit analysis A systematic method of problem solving that converts a given value to an unknown value by applying one or more unit factors. *(Secs. 2.3 and 18.5)*

unit analysis method A systematic procedure for solving problems that proceeds to a desired value from a related given value by the conversion of units. *(Sec. 1.10)*

unit equation A simple statement of two equivalent values written as an equation; for example, 1 m = 100 cm. *(Secs. 1.9 and 2.2)*

unit factor A ratio of two quantities that are equivalent and used to convert from one unit to another; for example, 1 m/100 cm. *(Secs. 1.9 and 2.2)*

unsaturated hydrocarbon A hydrocarbon containing a double bond or triple bond. *(Sec. 20.1)*

unsaturated solution A solution that contains less than the maximum amount of dissolved solute that will dissolve at a given temperature. *(Sec. 14.7)*

V

vacuum A volume that does not contain gas molecules or any other matter. *(Sec. 12.2)*

valence electrons The electrons that occupy the outermost *s* and *p* sublevels of an atom. *(Sec. 5.7)* The electrons in the highest *s* and *p* subshells in an atom that undergo reactions and form chemical bonds. *(Sec. 11.1)*

vapor pressure The pressure exerted by molecules in the gaseous state in dynamic equilibrium with the same molecules in the liquid state; for example, water molecules above liquid water. *(Secs. 12.3 and 13.2)*

viscosity The resistance of a liquid to flow as a result of intermolecular attraction. *(Sec. 13.2)*

visible spectrum Light energy that is observed as violet, blue, green, yellow, orange, or red; the radiant energy spectrum from approximately 400 to 700 nm. *(Sec. 4.7)*

visualization A problem-solving approach that involves the forming of mental pictures to make an abstract principle more concrete and permanent. *(Sec. 18.1)*

voltaic cell An electrochemical cell in which a spontaneous redox reaction occurs and generates electrical energy; also referred to as a galvanic cell. *(Sec. 17.6)*

volume by displacement A method for determining the volume of an object by measuring the increase in liquid level when the object is immersed in water. *(Sec. 2.7)* A technique for determining the volume of a gas by measuring the amount of water it displaces. *(Sec. 12.4)*

volume-volume problem A type of stoichiometry calculation that relates the volumes of two gases (at the same temperature and pressure) according to a balanced chemical equation. *(Sec. 9.3)*

W

water of hydration Water molecules bound in a hydrate; also called the water of crystallization. *(Sec. 13.10)*

wavelength The distance a light wave travels to complete one cycle. *(Sec. 4.7)*

weak electrolyte An aqueous solution that is a poor conductor of electricity; for example, weak acids, weak bases, and slightly soluble salts. *(Sec. 15.10)*

weight The gravitational force of attraction between an object and the planet on which it is measured. The measurement is obtained using a scale.*(Sec. 1.1)*

Answers to Key Term Exercises

Chapter 1

1. g, **2.** b, **3.** d, **4.** h, **5.** e, **6.** p, **7.** f, **8.** t, **9.** o, **10.** m, **11.** i, **12.** c, **13.** k, **14.** a, **15.** n, **16.** r, **17.** s, **18.** l, **19.** q, **20.** j

Chapter 2

1. f, **2.** n, **3.** j, **4.** m, **5.** h, **6.** l, **7.** s, **8.** t, **9.** r, **10.** d, **11.** u, **12.** e, **13.** o, **14.** i, **15.** q, **16.** g, **17.** c, **18.** k, **19.** p, **20.** b, **21.** a

Chapter 3

1. w, **2.** z, **3.** l, **4.** s, **5.** k, **6.** i, **7.** j, **8.** f, **9.** h, **10.** r, **11.** t, **12.** y, **13.** q, **14.** g, **15.** a, **16.** p, **17.** v, **18.** d, **19.** b, **20.** u, **21.** e, **22.** c, **23.** x, **24.** n, **25.** m, **26.** o

Chapter 4

1. h, **2.** j, **3.** u, **4.** s, **5.** d, **6.** e, **7.** q, **8.** c, **9.** p, **10.** b, **11.** r, **12.** a, **13.** v, **14.** i, **15.** z, **16.** o, **17.** w, **18.** y, **19.** m, **20.** g, **21.** l, **22.** x, **23.** f, **24.** n, **25.** t, **26.** k

Chapter 5

1. q, **2.** g, **3.** p, **4.** b, **5.** c, **6.** h, **7.** o, **8.** n, **9.** a, **10.** s, **11.** t, **12.** i, **13.** r, **14.** u, **15.** d, **16.** v, **17.** f, **18.** l, **19.** j, **20.** m, **21.** k, **22.** e

Chapter 6

1. k, **2.** p, **3.** l, **4.** g, **5.** t, **6.** h, **7.** d, **8.** f, **9.** u, **10.** i, **11.** c, **12.** n, **13.** q, **14.** s, **15.** m, **16.** j, **17.** a, **18.** e, **19.** o, **20.** r, **21.** b

Chapter 7

1. b, **2.** g, **3.** e, **4.** i, **5.** c, **6.** h, **7.** f, **8.** j, **9.** a, **10.** d

Chapter 8

1. h, **2.** g, **3.** r, **4.** q, **5.** f, **6.** u, **7.** i, **8.** n, **9.** m, **10.** d, **11.** c, **12.** b, **13.** p, **14.** j, **15.** k, **16.** t, **17.** l, **18.** o, **19.** a, **20.** e, **21.** s

Chapter 9

1. e, **2.** c, **3.** p, **4.** f, **5.** i, **6.** k, **7.** g, **8.** h, **9.** r, **10.** j, **11.** o, **12.** b, **13.** q, **14.** l, **15.** d, **16.** n, **17.** a, **18.** m

Chapter 10

1. a, **2.** n, **3.** r, **4.** k, **5.** l, **6.** u, **7.** g, **8.** p, **9.** h, **10.** o, **11.** m, **12.** s, **13.** d, **14.** b, **15.** t, **16.** e, **17.** c, **18.** f, **19.** i, **20.** j, **21.** q

Chapter 11

1. e, **2.** y, **3.** s, **4.** n, **5.** g, **6.** m, **7.** o, **8.** i, **9.** c, **10.** b, **11.** d, **12.** q, **13.** k, **14.** w, **15.** a, **16.** v, **17.** j, **18.** x, **19.** l, **20.** r, **21.** t, **22.** h, **23.** f, **24.** p, **25.** u

Chapter 12

1. k, **2.** b, **3.** c, **4.** t, **5.** u, **6.** w, **7.** d, **8.** h, **9.** r, **10.** x, **11.** v, **12.** i, **13.** p, **14.** e, **15.** f, **16.** l, **17.** g, **18.** q, **19.** m, **20.** s, **21.** j, **22.** a, **23.** o, **24.** n

Chapter 13

1. x, **2.** y, **3.** w, **4.** e, **5.** f, **6.** o, **7.** c, **8.** p, **9.** s, **10.** r, **11.** i, **12.** k, **13.** l, **14.** b, **15.** t, **16.** m, **17.** h, **18.** q, **19.** u, **20.** n, **21.** z, **22.** a, **23.** j, **24.** v, **25.** d, **26.** g

Chapter 14

1. q, **2.** s, **3.** r, **4.** d, **5.** c, **6.** l, **7.** n, **8.** m, **9.** f,

10. h, **11.** e, **12.** t, **13.** p, **14.** o, **15.** x, **16.** u, **17.** g, **18.** k, **19.** j, **20.** a, **21.** i, **22.** v, **23.** b, **24.** w

Chapter 15

1. w, **2.** a, **3.** j, **4.** u, **5.** c, **6.** l, **7.** d, **8.** m, **9.** i, **10.** q, **11.** s, **12.** f, **13.** g, **14.** b, **15.** r, **16.** e, **17.** h, **18.** k, **19.** o, **20.** p, **21.** v, **22.** y, **23.** x, **24.** t, **25.** n

Chapter 16

1. f, **2.** t, **3.** w, **4.** b, **5.** k, **6.** i, **7.** h, **8.** c, **9.** s, **10.** a, **11.** p, **12.** r, **13.** u, **14.** d, **15.** e, **16.** m, **17.** l, **18.** o, **19.** j, **20.** q, **21.** g, **22.** n, **23.** v

Chapter 17

1. n, **2.** l, **3.** k, **4.** p, **5.** m, **6.** o, **7.** j, **8.** a, **9.** g, **10.** f, **11.** i, **12.** r, **13.** h, **14.** b, **15.** d, **16.** q, **17.** c, **18.** e

Chapter 18

1. f, **2.** c, **3.** b, **4.** j, **5.** i, **6.** a, **7.** e, **8.** d, **9.** g, **10.** h

Answers to Selected Exercises

Chapter 1

1. No instrument gives an ''exact measurement'' because all instruments have uncertainty.

3. **(a)** length; **(b)** volume; **(c)** mass; **(d)** volume; **(e)** time; **(f)** volume

5. **(a)** 6.4 to 6.6 cm; **(b)** 0.50 to 0.52 g; **(c)** 9.9 to 10.1 mL; **(d)** 35.4 to 35.6 s

7. **(a)** 1; **(b)** 3; **(c)** 3; **(d)** 4; **(e)** 4; **(f)** 1; **(g)** 3; **(h)** 4

9. **(a)** 31.5; **(b)** 214,000; **(c)** 5.16; **(d)** 77.5; **(e)** 0.0184; **(f)** 0.000 000 485; **(g)** 2.57×10^5; **(h)** 5.70×10^{-2}

11. **(a)** 31.7 cm; **(b)** 55.5 cm; **(c)** 1.3 g; **(d)** 23.1 g; **(e)** 54.0 mL; **(f)** 21 mL

13. **(a)** 7.67 cm^2; **(b)** 10.1 cm^2; **(c)** 28 cm^2; **(d)** 400 cm^2; **(e)** 3 cm^3; **(f)** 14 cm^3; **(g)** 12.0 cm^3; **(h)** 17 cm^3

15. **(a)** 3^2; **(b)** 2^6; **(c)** 2^{-3}; **(d)** 3^{-5}; **(e)** 10^3; **(f)** 10^6; **(g)** 10^{-3}; **(h)** 10^{-4}

17. **(a)** 10^0; **(b)** 10^{-1}; **(c)** 10^6; **(d)** 10^{-4}; **(e)** 10^{10}; **(f)** 10^{-6}; **(g)** 10^{17}; **(h)** 10^{-15}

19. **(a)** 8.0916×10^7; **(b)** 1.5×10^{-8}; **(c)** 3.356×10^{14}; **(d)** 9.27×10^{-13}

21. 2.69×10^{22} atoms

23. **(a)** 4.4 E08, 4.4×10^8; **(b)** 2.26 E07, 2.26×10^7; **(c)** 2.61 E09, 2.61×10^9; **(d)** 5.41 E18, 5.41×10^{18}; **(e)** 4.86 E−07, 4.86×10^{-7}; **(f)** 6.98 E02, 6.98×10^2; **(g)** 3.64 E00, 3.64×10^0; **(h)** 4.42 E00, 4.42×10^0; **(i)** 5.32 E−02, 5.32×10^{-2}; **(j)** 3.84 E28, 3.84×10^{28}

25. **(a)** 1 dollar = 10 dimes, 10 dimes/1 dollar and 1 dollar/10 dimes; **(b)** 1 nickel = 5 pennies, 1 nickel/5 pennies and 5 pennies/1 nickel; **(c)** 1 dollar = 4 quarters, 1 dollar/4 quarters and 4 quarters/1 dollar; **(d)** 10 pennies = 1 dime, 10 pennies/1 dime and 1 dime/10 pennies; **(e)** 2 nickels = 1 dime, 2 nickels/1 dime and 1 dime/2 nickels; **(f)** 1 half-dollar = 10 nickels, 10 nickels/1 half-dollar and 1 half-dollar/10 nickels; **(g)** 50 pennies = 1 half-dollar, 50 pennies/1 half-dollar and 1 half-dollar/50 pennies; **(h)** 25 pennies = 1 quarter, 25 pennies/1 quarter and 1 quarter/25 pennies; **(i)** 1 quarter = 5 nickels, 5 nickels/1 quarter and 1 quarter/5 nickels; **(j)** 5 dimes = 1 half-dollar, 5 dimes/1 half-dollar and 1 half-dollar/5 dimes

27. **(a)** 40; **(b)** \$215.00; **(c)** 108 pencils; **(d)** 25 reams; **(e)** \$487.50; **(f)** \$42,550.00

29. **(a)** 1.73%; **(b)** 39.0% O^+, 34.0% A^+, 8.5% B^+, 3.5% AB^+, 7.0% O^-, 6.0% A^-, 1.5% B^-, 0.5% AB^-; **(c)** 1.96 g; **(d)** 1080 mL; **(e)** 40.0%; **(f)** 12.0%

31. 94.99% copper, 5.01% zinc

33. 1.5 ± 0.1 cm, 1.45 ± 0.05 cm

35. 10.0 mL **37.** 23.0 atomic mass units **39.** 258.1 g

41. **(a)** 3.52×10^6; **(b)** 4.16×10^5; **(c)** 1.70×10^1; **(d)** 1.25×10^{-1}

43. 320,000,000,000,000,000,000 **45.** 1.67356×10^{-24} g

47. 200 pounds **49.** 8.06×10^{-2} cm^3

51. 3.5×10^{13} hours **53.** 373 grams

55. 1 troy pound = 373 g, 1 avoirdupois pound = 454 g Thus, 1 pound (avoirdupois) of feathers weighs more than 1 pound (troy) of gold.

Chapter 2

1. **(a)** true; **(b)** true; **(c)** true; **(d)** true

3. **(a)** meter, m; **(b)** second, s; **(c)** kilogram, kg; **(d)** degree Celsius, °C; **(e)** liter, L; **(f)** calorie, cal

5. **(a)** μg; **(b)** μm; **(c)** km; **(d)** ms; **(e)** dL; **(f)** kcal; **(g)** cL; **(h)** cm^3; **(i)** ng; **(j)** m^2

7. **(a)** 1000 mm = 1 m; **(b)** 10 dL = 1 L; **(c)** 1 g = 10^6 μg; **(d)** 10 dm = 1 m; **(e)** 1 kL = 1000 L; **(f)** 10^9 ns = 1 s

9. **(a)** 1 m = 10^9 nm; **(b)** 10^6 μm = 1 m; **(c)** 100 cg = 1 g; **(d)** 1 mL = 1 cm^3; **(e)** 1 L = 10 dL; **(f)** 1000 cal = 1 kcal

11. Step 1: Write down the units of the unknown. Step 2: Write down the relevant given value. Step 3: Apply one or more unit factors.

13. **(a)** 1550 m; **(b)** 1.06 g; **(c)** 48.6 cg; **(d)** 0.125 L; **(e)** 18.85 dL; **(f)** 1×10^{-7} s; **(g)** 388 mm; **(h)** 94,600 cal; **(i)** 0.125 ms; **(j)** 3.15×10^2 g

15. **(a)** 2.54 cm = 1 in.; **(b)** 454 g = 1 lb; **(c)** 946 mL = 1 qt; **(d)** 946 cm^3 = 1 qt

17. **(a)** 170 cm; **(b)** 34 in.; **(c)** 459 g; **(d)** 0.079 lb; **(e)** 473 mL; **(f)** 0.79 qt

19. 22 km/L **21.** 370 m/s **23.** 15,800 cm^3 **25.** 1.26 cm

27. **(a)** 1000 mL = 1 L; **(b)** 1000 cm^3 = 1 L; **(c)** 1 mL = 1 cm^3; **(d)** 16.4 cm^3 = 1 $in.^3$

29. 6.80 L **31.** 10.5 mL

33. **(a)** float; **(b)** float; **(c)** float; **(d)** float; **(e)** sink; **(f)** sink

35. **(a)** 170 g; **(b)** 28.9 g; **(c)** 1.6 g; **(d)** 780 g; **(e)** 2.82×10^5 g

37. **(a)** 0.788 g/mL; **(b)** 0.714 g/mL; **(c)** 2.8 g/mL; **(d)** 8.47 g/mL; **(e)** 19.3 g/mL

39. **(a)** 32°F; **(b)** 273 K; **(c)** 0°C

41. **(a)** 38°C; **(b)** −137°C; **(c)** 1185°C; **(d)** −40°C

43. **(a)** 315 K; **(b)** 253 K; **(c)** 768 K; **(d)** 88 K

45. Temperature is a measure of the average kinetic energy of a system, and heat is a measure of the total energy of a system.

47. 19,000 cal **49.** 67.5 cal **51.** 0.0308 cal/g·°C

53. 56.5 g **55.** 59.2°C

57. **(a)** 2.15×10^{-8} Eg; **(b)** 20 L; **(c)** 98.3 am; **(d)** 0.104 s

59. 1.36×10^4 kg/m^3

61. approximately 450 ft by 75 ft by 45 ft

63. **(a)** infinite (exact, by definition); **(b)** 3; **(c)** infinite (exact, by definition); **(d)** 3

65. 436 m/s **67.** 2.17×10^4 cm/cycle **69.** 50 disks

71. 62.4 lb/ft^3 **73.** 19.8 g

75. **(a)** −130°N; **(b)** −3230°N

77. **(a)** 2.5×10^4 cal; **(b)** 1.0×10^5 J

79. 11.9°C **81.** 3.08×10^{16} m

Chapter 3

1. **(a)** definite and fixed; **(b)** indefinite and fixed; **(c)** indefinite and variable

3. **(a)** negligible; **(b)** restricted; **(c)** unrestricted

5. **(a)** vaporizing; **(b)** sublimation; **(c)** freezing; **(d)** melting; **(e)** condensing; **(f)** deposition

7. Temperature is a measure of the average kinetic energy of a substance. **9.** **(a)** increase; **(b)** increase **11.** equal to

13. equal to

15. **(a)** see Glossary; **(b)** see Glossary; **(c)** see Glossary

17. **(a)** compound; **(b)** compound; **(c)** element; **(d)** mixture; **(e)** mixture; **(f)** element; **(g)** mixture; **(h)** mixture **19.** oxygen, silicon, aluminum

21. **(a)** Br; **(b)** O; **(c)** Sb; **(d)** Li; **(e)** Ar; **(f)** Mg; **(g)** F; **(h)** Na; **(i)** Bi; **(j)** Ni; **(k)** He; **(l)** Te; **(m)** H; **(n)** Sn; **(o)** Ca; **(p)** Pt; **(q)** C; **(r)** K; **(s)** I; **(t)** Ti; **(u)** Fe; **(v)** Xe; **(w)** Kr; **(x)** Zn

23. **(a)** 83; **(b)** 47; **(c)** 48; **(d)** 11; **(e)** 80; **(f)** 29; **(g)** 19; **(h)** 50

25. B, Si, Ge, As, Sb, Te

27. **(a)** solid; **(b)** gas; **(c)** solid; **(d)** solid; **(e)** solid; **(f)** solid; **(g)** solid; **(h)** solid; **(i)** liquid; **(j)** liquid

29. **(a)** metal; **(b)** semimetal; **(c)** nonmetal; **(d)** metal; **(e)** metal; **(f)** nonmetal; **(g)** metal; **(h)** nonmetal; **(i)** nonmetal; **(j)** metal

31. 5:1:4

33. **(a)** 9 carbon atoms, 8 hydrogen atoms, 4 oxygen atoms; **(b)** 12 carbon atoms, 22 hydrogen atoms, 11 oxygen atoms; **(c)** 3 carbon atoms, 8 hydrogen atoms, 3 oxygen atoms; **(d)** 6 carbon atoms, 3 hydrogen atoms, 10 oxygen atoms, 3 nitrogen atoms; **(e)** 4 carbon atoms, 10 hydrogen atoms, 1 oxygen atom

35. **(a)** 72; **(b)** 26; **(c)** 48; **(d)** 10; **(e)** 20

37. **(a)** physical; **(b)** chemical; **(c)** physical; **(d)** chemical; **(e)** physical

39. **(a)** chemical; **(b)** physical; **(c)** chemical; **(d)** physical; **(e)** chemical; **(f)** physical; **(g)** physical; **(h)** physical; **(i)** physical; **(j)** chemical

41. **(a)** physical; **(b)** physical; **(c)** physical; **(d)** physical; **(e)** chemical; **(f)** physical; **(g)** physical

43. **(a)** chemical; **(b)** chemical; **(c)** physical; **(d)** chemical; **(e)** physical; **(f)** physical; **(g)** physical; **(h)** chemical; **(i)** chemical; **(j)** physical

45. 3.94 g **47.** 0.055 g

49. On a roller-coaster ride the potential energy is greatest when the car is at its highest point. As the car descends, it loses potential energy and gains kinetic energy, having its greatest kinetic energy at its lowest point. As the car ascends, it loses kinetic energy and gains potential energy.

51. **(a)** potential energy; **(b)** kinetic energy; **(c)** potential energy; **(d)** kinetic energy; **(e)** potential energy; **(f)** kinetic energy; **(g)** kinetic energy; **(h)** potential energy; **(i)** potential energy; **(j)** kinetic energy

53. 807 cal

55. heat, light, chemical, electrical, mechanical, nuclear

57. E = energy, m = mass, c = speed of light

59. see Glossary

61. At high temperatures, plasma is formed when electrons are stripped from atoms. The sun at 6000°C sends out solar flares, that is, plasma.

63. $KE = \frac{1}{2}mv^2$, KE = kinetic energy of particle, m = mass of particle, v = velocity of particle

65. **(a)** homogeneous; **(b)** heterogeneous; **(c)** homogeneous; **(d)** homogeneous; **(e)** heterogeneous

67. 3.5 kcal

69. **(a)** mercury, Hg; **(b)** sodium, Na; **(c)** copper, Cu; **(d)** potassium, K; **(e)** antimony, Sb; **(f)** lead, Pb; **(g)** iron, Fe; **(h)** gold, Au; **(i)** tin, Sn; **(j)** silver, Ag

Chapter 4

1. see Section 4.1

3. **(a)** atoms are divisible; **(b)** elements may have isotopes

5. the electron **7.** 1−, 1/1836 the mass of a proton

9. electrons

11. Rutherford concluded that atoms could not be "plum-pudding" spheres as Thomson had suggested. Rutherford proposed that the mass of an atom exists only in a tiny fraction of its total volume. Furthermore, the mass is located at the center of the atom (nucleus) and has a positive charge.

13. Protons and neutrons are in the center of the atom with electrons revolving in orbits about the nucleus, much like the planets around the sun.

15. 1−, 1+, 0 **17.** **(a)** 2; **(b)** 16; **(c)** 5; **(d)** 24; **(e)** 8; **(f)** 28; **(g)** 14; **(h)** 30

19. **(a)** (1n° 1p⁺)) 1 e⁻ **(b)** (1n° 2p⁺)) 2 e⁻

(c) (4n° 3p⁺)) 3 e⁻ **(d)** (7n° 6p⁺)) 6 e⁻

(e) (8n° 8p⁺)) 8 e⁻ **(f)** (10n° 10p⁺)) 10 e⁻

21. No, the atomic number determines the element.

23.

Atomic Notation	Atomic Number	Mass Number	Number of Protons	Number of Neutrons	Number of Electrons
$^{4}_{2}He$	2	4	2	2	2
$^{13}_{6}C$	6	13	6	7	6
$^{21}_{10}Ne$	10	21	10	11	10
$^{28}_{14}Si$	14	28	14	14	14
$^{40}_{18}Ar$	18	40	18	22	18
$^{50}_{22}Ti$	22	50	22	28	22

25. carbon-12, 12 amu exactly

27. **(a)** decreased; **(b)** increased; **(c)** decreased; **(d)** increased

29. The atomic mass of lithium, 6.941 amu, is the *weighted* average of the naturally occurring isotopic masses (^{6}Li = 6.015 amu, ^{7}Li = 7.016 amu).

31. simple average = 3.5 g, weighted average = 3 g

33. 107.87 amu **35.** 47.88 amu **37.** 24.31 amu **39.** Cl-35

41. Radioactive; only the mass of its most stable isotope is given.

43. violet **45.** 450 nm **47.** see Figure 4-14

49. heat and light **51.** **(a)** 3 to 2; **(b)** 4 to 2; **(c)** 5 to 2

53. 5 to 2 **55.** 490 nm **57.** 430 nm

59. The lines in the emission spectrum of hydrogen suggested the existence of energy levels.

61. **(a)** 1*s*; **(b)** 2*s* 2*p*; **(c)** 3*s* 3*p* 3*d*; **(d)** 4*s* 4*p* 4*d* 4*f*

63. **(a)** 2; **(b)** 2; **(c)** 6; **(d)** 6; **(e)** 10; **(f)** 10; **(g)** 14; **(h)** 14

65. **(a)** 2; **(b)** 8; **(c)** 18; **(d)** 32

67. 1*s* 2*s* 2*p* 3*s* 3*p* 4*s* 3*d* 4*p* 5*s* 4*d* 5*p*

69. **(a)** $1s^2$; **(b)** $1s^2\,2s^2$; **(c)** $1s^2\,2s^2\,2p^2$; **(d)** $1s^2\,2s^2\,2p^6\,3s^2$; **(e)** $1s^2\,2s^2\,2p^6\,3s^2\,3p^4$; **(f)** $1s^2\,2s^2\,2p^6\,3s^2\,3p^6\,4s^1$; **(g)** $1s^2\,2s^2\,2p^6\,3s^2\,3p^6\,4s^2\,3d^7$; **(h)** $1s^2\,2s^2\,2p^6\,3s^2\,3p^6\,4s^2\,3d^{10}\,4p^6\,5s^2\,4d^{10}$

71. **(a)** Li; **(b)** Si; **(c)** Ti; **(d)** Sr **73.** atomic nuclei

75. 9.09×10^{-28} g **77.** the earth **79.** 70.9 amu

81. No, neon always emits a reddish-orange light. Other gases, for example helium, argon, and krypton, emit different colors.

83. **(a)** ultraviolet; **(b)** visible; **(c)** infrared; **(d)** infrared

85. 120 nm **87.** 1900 nm

89. The 3*d* electrons are filling the 3*d* sublevel in the fourth row of the periodic table.

Chapter 5

1. bromine, strontium **3.** Cu, Ag

5. increasing atomic weight **7.** increasing atomic number

9. chemical groups or chemical families

11. representative elements **13.** inner transition elements

15. nonmetals

17. **(a)** IA/1; **(b)** IIA/2; **(c)** VIIA/17; **(d)** VIIIA/18

19. lanthanides **21.** rare earth elements

23. **(a)** 1; **(b)** 11; **(c)** 13; **(d)** 3; **(e)** 15; **(f)** 5; **(g)** 17; **(h)** 7

25. **(a)** Ge; **(b)** Na; **(c)** I; **(d)** Pm; **(e)** Lu; **(f)** Ba; **(g)** Zr; **(h)** He

27. increases 29. increases 31. (a) Na; (b) P; (c) Ca; (d) Kr; (e) Rb; (f) As; (g) Pb; (h) I

33. predicted atomic radius of K ~ 0.230 nm; predicted density of Rb ~ 1.38 g/mL; predicted melting point of Cs ~ 14.5°C

35. predicted atomic radius of Cr ~ 0.125 nm; predicted density of Mo ~ 13.20 g/mL; predicted melting point of W ~ 3377°C

37. (a) Li_2O; (b) CaO; (c) Ga_2O_3; (d) SnO_2

39. (a) CdO; (b) ZnS; (c) HgS; (d) CdSe

41. (a) SO_3; (b) TeO_3; (c) SeS_3; (d) TeS_3

43. the *s* sublevel 45. the *d* sublevel 47. the 4*f* sublevel

49. (a) 1*s*; (b) 3*s*; (c) 4*f*; (d) 4*p*; (e) 5*s*; (f) 2*p*; (g) 5*p*; (h) 6*s*; (i) 6*p*; (j) 6*d*

51. (a) $1s^2\,2s^1$; (b) $1s^2\,2s^2\,2p^5$; (c) $1s^2\,2s^2\,2p^6\,3s^2$; (d) $1s^2\,2s^2\,2p^6\,3s^2\,3p^3$; (e) $1s^2\,2s^2\,2p^6\,3s^2\,3p^6\,4s^2$; (f) $1s^2\,2s^2\,2p^6\,3s^2\,3p^6\,4s^2\,3d^5$; (g) $1s^2\,2s^2\,2p^6\,3s^2\,3p^6\,4s^2\,3d^{10}\,4p^1$; (h) $1s^2\,2s^2\,2p^6\,3s^2\,3p^6\,4s^2\,3d^{10}\,4p^6\,5s^1$; (i) $1s^2\,2s^2\,2p^6\,3s^2\,3p^6\,4s^2\,3d^{10}\,4p^6\,5s^2\,4d^5$; (j) $1s^2\,2s^2\,2p^6\,3s^2\,3p^6\,4s^2\,3d^{10}\,4p^6\,5s^2\,4d^{10}\,5p^6$

53. (a) 1; (b) 3; (c) 5; (d) 7

55. (a) 1; (b) 3; (c) 5; (d) 7; (e) 2; (f) 4; (g) 6; (h) 8

57. (a) H·; (b) ·B·; (c) ·N:; (d) :F:; (e) Ca·; (f) ·Si·; (g) ·O:; (h) :Ar: 59. decreases 61. VIIIA/18

63. (a) Mg; (b) S; (c) Sn; (d) N; (e) Sb; (f) Br; (g) Ge; (h) P; (i) Cl; (j) As

65. (a) 1+; (b) 2+; (c) 3+; (d) 4+

67. (a) 1+; (b) 2+; (c) 3+; (d) 4+; (e) 4−; (f) 2−; (g) 2−; (h) 1−

69. (a) yes; (b) yes; (c) no; (d) yes; (e) yes; (f) no

71. (a) $1s^2\,2s^2\,2p^6$; (b) $1s^2\,2s^2\,2p^6\,3s^2\,3p^6$; (c) $1s^2\,2s^2\,2p^6\,3s^2\,3p^6\,3d^6$; (d) $1s^2\,2s^2\,2p^6\,3s^2\,3p^6\,4s^2\,3d^{10}\,4p^6\,5s^2\,4d^{10}\,5p^6$

73. (a) $1s^2\,2s^2\,2p^6$; (b) $1s^2\,2s^2\,2p^6\,3s^2\,3p^6$; (c) $1s^2\,2s^2\,2p^6$; (d) $1s^2\,2s^2\,2p^6\,3s^2\,3p^6\,4s^2\,3d^{10}\,4p^6\,5s^2\,4d^{10}\,5p^6$

75. scandium, gallium, germanium

77. (a) IA; (b) IB; (c) IIIB; (d) IIIA; (e) VB; (f) VA

79. 0.284 nm, 2.21 g/mL, 17.9°C

81. When alkali metal atoms lose one electron they assume a stable noble gas electron configuration. Alkaline earth elements must lose two electrons to assume a noble gas configuration.

83. Hydrogen atoms have a single electron that must be stripped away from the proton nucleus. The alkali metals have core electrons that shield the valence electrons from the nucleus.

Chapter 6

1. (a) ternary ionic; (b) binary ionic; (c) binary molecular; (d) binary acid; (e) ternary oxyacid

3. (a) monoatomic anion; (b) polyatomic cation; (c) polyatomic anion; (d) monoatomic cation

5. (a) potassium ion; (b) barium ion; (c) silver ion; (d) cadmium ion

7. (a) mercury(II) ion; (b) copper(II) ion; (c) iron(II) ion; (d) cobalt(III) ion

9. (a) cuprous ion; (b) ferric ion; (c) stannous ion; (d) plumbic ion

11. (a) fluoride ion; (b) iodide ion; (c) oxide ion; (d) phosphide ion

13. (a) hypochlorite ion; (b) sulfite ion; (c) acetate ion; (d) carbonate ion

15. (a) OH^-; (b) NO_2^-; (c) $Cr_2O_7^{2-}$; (d) HCO_3^-

17. (a) NaCl; (b) $AlBr_3$; (c) Ag_2O; (d) Bi_2O_3; (e) SnI_4

19. (a) KNO_3; (b) $(NH_4)_2Cr_2O_7$; (c) $Al_2(SO_3)_3$; (d) $Bi(ClO)_3$

21. (a) $Sr(NO_2)_2$; (b) $Zn(MnO_4)_2$; (c) $CaCrO_4$; (d) $Cr(ClO_4)_3$

23. (a) magnesium oxide; (b) silver bromide; (c) cadmium chloride; (d) aluminum sulfide

25. (a) copper(II) oxide; (b) iron(II) oxide; (c) mercury(II) oxide; (d) tin(II) oxide

27. (a) cuprous oxide; (b) ferric oxide; (c) mercurous oxide; (d) stannic oxide

29. (a) potassium permanganate; (b) strontium perchlorate; (c) barium chromate; (d) cadmium cyanide

31. (a) copper(II) sulfate; (b) iron(II) chromate; (c) mercury(II) nitrite; (d) lead(II) acetate

33. (a) cuprous sulfate; (b) ferric chromate; (c) mercurous nitrite; (d) plumbic acetate

35. (a) disulfur dichloride; (b) diphosphorus trioxide; (c) dinitrogen oxide; (d) tricarbon dioxide

37. (a) NO_2; (b) CCl_4; (c) IBr; (d) H_2S

39. (a) hydrofluoric acid; (b) hydrobromic acid; (c) hydroselenic acid

41. (a) chlorous acid; (b) nitric acid; (c) phosphoric acid; (d) sulfurous acid

43. (a) $HC_2H_3O_2(aq)$; (b) HClO(aq); (c) $H_3PO_3(aq)$; (d) $HClO_3(aq)$

45. (a) calcium hydrogen phosphate; (b) lithium dihydrogen phosphate; (c) zinc hydrogen carbonate; (d) nickel(II) hydrogen sulfite

47. (a) RbCl; (b) BeO

49. (a) Fr_2SO_4; (b) $Ba(BrO_3)_2$

51. **(a)** HBrO(aq); **(b)** HIO_4(aq)

53. **(a)** 0; **(b)** 2+; **(c)** 3+; **(d)** 0 **55.** (b), (c)

57.

Ions	Ag^+	Hg_2^{2+}	Al^{3+}	Sn^{4+}
Br^-	$AgBr$ silver bromide	Hg_2Br_2 mercury(I) bromide	$AlBr_3$ aluminum bromide	$SnBr_4$ tin(IV) bromide
F^-	AgF silver fluoride	Hg_2F_2 mercury(I) fluoride	AlF_3 aluminum fluoride	SnF_4 tin(IV) fluoride
O^{2-}	Ag_2O silver oxide	Hg_2O mercury(I) oxide	Al_2O_3 aluminum oxide	SnO_2 tin(IV) oxide
N^{3-}	Ag_3N silver nitride	$(Hg_2)_3N_2$ mercury(I) nitride	AlN aluminum nitride	Sn_3N_4 tin(IV) nitride

59.

Ions	K^+	Cd^{2+}	Cr^{3+}	Bi^{3+}
CrO_4^{2-}	K_2CrO_4 potassium chromate	$CdCrO_4$ cadmium chromate	$Cr_2(CrO_4)_3$ chromium(III) chromate	$Bi_2(CrO_4)_3$ bismuth chromate
MnO_4^-	$KMnO_4$ potassium permanganate	$Cd(MnO_4)_2$ cadmium permanganate	$Cr(MnO_4)_3$ chromium (III) permanganate	$Bi(MnO_4)_3$ bismuth permanganate
ClO_4^-	$KClO_4$ potassium perchlorate	$Cd(ClO_4)_2$ cadmium perchlorate	$Cr(ClO_4)_3$ chromium(III) perchlorate	$Bi(ClO_4)_3$ bismuth perchlorate
SO_3^{2-}	K_2SO_3 potassium sulfite	$CdSO_3$ cadmium sulfite	$Cr_2(SO_3)_3$ chromium(III) sulfite	$Bi_2(SO_3)_3$ bismuth sulfite

61. **(a)** -ide; **(b)** -ic acid; **(c)** -ite; **(d)** -ous acid; **(e)** -ate; **(f)** -ic acid

63. **(a)** H_2O_2; **(b)** NaClO; **(c)** NaOH; **(d)** $NaHCO_3$

65. **(a)** boron trifluoride; **(b)** silicon tetrachloride; **(c)** diarsenic pentaoxide; **(d)** diantimony trioxide

67. $Ca(C_2H_3O_2)_2$ **69.** $NaHC_2O_4$

Chapter 7

1. **(a)** 1.0 amu; **(b)** 6.9 amu; **(c)** 12.0 amu; **(d)** 31.0 amu; **(e)** 40.1 amu; **(f)** 65.4 amu; **(g)** 74.9 amu; **(h)** 238.0 amu

3. (a), (b), (c), and (d) are each 6.02×10^{23}

5. **(a)** 2.02×10^{23}; **(b)** 6.74×10^{22}; **(c)** 1.17×10^{24}; **(d)** 2.02×10^{23}

7. **(a)** 0.0689 mol; **(b)** 0.00550 mol; **(c)** 6.96×10^{-4} mol; **(d)** 1.35 mol

9. **(a)** 200.6 g/mol; **(b)** 28.1 g/mol; **(c)** 63.5 g/mol; **(d)** 79.0 g/mol

11. **(a)** 175.3 g/mol; **(b)** 110.3 g/mol; **(c)** 232.8 g/mol; **(d)** 452.8 g/mol

13. **(a)** 98.3 g; **(b)** 0.540 g; **(c)** 1.75 g

15. **(a)** 2.31×10^{22}; **(b)** 8.84×10^{21}; **(c)** 5.66×10^{22}

17. **(a)** 1.50×10^{-23} g; **(b)** 3.82×10^{-23} g; **(c)** 9.79×10^{-23} g; **(d)** 1.24×10^{-22} g

19. 60.9% C, 4.4% H, 34.7% O

21. 67.3% C, 6.9% H, 21.1% O, 4.6% N

23. 30.2% C, 5.0% H, 20.2% S, 44.6% Cl

25. 13.6% Na, 35.5% C, 4.7% H, 8.3% N, 37.9% O

27. SnO_2 **29.** HgO **31.** Co_2S_3

33. **(a)** MnF_2; **(b)** CuCl; **(c)** $SnBr_2$; **(d)** TlI

35. C_2HCl_3 **37.** $C_9H_8O_4$ **39.** $C_6H_{10}O_4$ **41.** $C_2H_6O_2$

43. CHCl, $C_6H_6Cl_6$ **45.** $C_{10}H_{14}N_2$

47. $T = 0°C$, $P = 1.00$ atm

49.

Gas	Molecules	Mass	Vol. at STP
fluorine, F_2	6.02×10^{23}	38.0 g	22.4 L
hydrogen fluoride, HF	6.02×10^{23}	20.0 g	22.4 L
silicon tetrafluoride, SiF_4	6.02×10^{23}	104 g	22.4 L
oxygen difluoride, OF_2	6.02×10^{23}	54.0 g	22.4 L

51. **(a)** 0.902 g/L; **(b)** 3.17 g/L; **(c)** 2.05 g/L; **(d)** 5.71 g/L

53. **(a)** 30.0 g/mol; **(b)** 27.6 g/mol; **(c)** 121 g/mol; **(d)** 45.9 g/mol;

55. **(a)** 1.60 g; **(b)** 18.1 g; **(c)** 0.291 g

57. **(a)** 2.69×10^{20}; **(b)** 1.89×10^{21}; **(c)** 2.69×10^{22}

59.

Gas	Molecules	Atoms	Mass	Vol. at STP
N_2	1.35×10^{23}	2.71×10^{23}	6.31 g	5.04 L
NO_2	1.35×10^{23}	4.06×10^{23}	10.4 g	5.04 L
NO	1.35×10^{23}	2.71×10^{23}	6.75 g	5.04 L
N_2O_4	1.35×10^{23}	8.13×10^{23}	20.7 g	5.04 L

61. both are 6.02×10^{23}

63. The number of Ni atoms is 51,000,000 times greater!

65. A mole of moles weighs 6×10^{22} kg. Therefore, the earth weighs about 100 times more than the mole of moles!

67. Ga_2O_3 **69.** 2.99×10^{-23} cm^3

71. 2.11×10^{22} atoms C **73.** 11 atoms C

75. 6.03×10^{23} atoms/mol

Chapter 8

1. All observations listed are evidence for a chemical reaction.

3. **(a)** $Sn(s) + O_2(g) \rightarrow SnO_2(s)$;
(b) $ZnCO_3(s) \rightarrow ZnO(s) + CO_2(g)$;
(c) $Mg(s) + Cd(NO_3)_2(aq) \rightarrow Mg(NO_3)_2(aq) + Cd(s)$;
(d) $LiBr(aq) + AgNO_3(aq) \rightarrow AgBr(s) + LiNO_3(aq)$;
(e) $HC_2H_3O_2(aq) + KOH(aq) \rightarrow KC_2H_3O_2(aq) + H_2O(l)$;

5. (a), (e)

7. **(a)** $4\ Co(s) + 3\ O_2(g) \rightarrow 2\ Co_2O_3(s)$;
(b) $2\ CsClO_3(s) \rightarrow 2\ CsCl(s) + 3\ O_2(g)$;
(c) $Cu(s) + 2\ AgC_2H_3O_2(aq) \rightarrow Cu(C_2H_3O_2)_2(aq) + 2\ Ag(s)$;
(d) $Pb(NO_3)_2(aq) + LiCl(aq) \rightarrow PbCl_2(s) + 2\ LiNO_3(aq)$;
(e) $3\ H_2SO_4(aq) + 2\ Al(OH)_3(aq) \rightarrow Al_2(SO_4)_3(aq) + 6\ H_2O(l)$

9. **(a)** $2\ Pb(s) + O_2(g) \rightarrow 2\ PbO(s)$;
(b) $2\ LiNO_3(s) \rightarrow 2\ LiNO_2(s) + O_2(g)$;
(c) $Mg(s) + 2\ HC_2H_3O_2(aq) \rightarrow Mg(C_2H_3O_2)_2(aq) + H_2(g)$;
(d) $Hg_2(NO_3)_2(aq) + 2\ NaBr(aq) \rightarrow Hg_2Br_2(s) + 2\ NaNO_3(aq)$;
(e) $H_2CO_3(aq) + 2\ NH_4OH(aq) \rightarrow (NH_4)_2CO_3(aq) + 2\ HOH(l)$

11. **(a)** combination; **(b)** decomposition;
(c) single replacement; **(d)** double replacement;
(e) neutralization

13. **(a)** combination; **(b)** decomposition;
(c) single replacement; **(d)** double replacement;
(e) neutralization

15. **(a)** $4\ Cu(s) + O_2(g) \rightarrow 2\ Cu_2O(s)$;
(b) $2\ Sn(s) + O_2(g) \rightarrow 2\ SnO(s)$;
(c) $4\ Fe(s) + 3\ O_2(g) \rightarrow 2\ Fe_2O_3(s)$;
(d) $Ti(s) + O_2(g) \rightarrow TiO_2(s)$

17. **(a)** $Hg(l) + Cl_2(g) \rightarrow HgCl_2(s)$;
(b) $3\ Pb(s) + 2\ P(s) \rightarrow Pb_3P_2(s)$;
(c) $2\ Fe(s) + 3\ F_2(g) \rightarrow 2\ FeF_3(s)$; **(d)** $Co(s) + S(s) \rightarrow CoS(s)$

19. **(a)** $2\ Cd + O_2 \rightarrow 2\ CdO$; **(b)** $2\ Al + 3\ Cl_2 \rightarrow 2\ AlCl_3$;
(c) $Sr + S \rightarrow SrS$; **(d)** $3\ Ba + N_2 \rightarrow Ba_3N_2$

21. **(a)** $2\ Sr + O_2 \rightarrow 2\ SrO$; **(b)** $2\ Bi + 3\ S \rightarrow Bi_2S_3$;
(c) $2\ Li + Br_2 \rightarrow 2\ LiBr$; **(d)** $Mg + Cl_2 \rightarrow MgCl_2$

23. **(a)** $2\ AgHCO_3(s) \rightarrow Ag_2CO_3(s) + H_2O(g) + CO_2(g)$;
(b) $Cu(HCO_3)_2(s) \rightarrow CuCO_3(s) + H_2O(g) + CO_2(g)$

25. **(a)** $CuCO_3(s) \rightarrow CuO(s) + CO_2(g)$;
(b) $MnCO_3(s) \rightarrow MnO(s) + CO_2(g)$

27. **(a)** $2\ PbO_2(s) \rightarrow 2\ PbO(s) + O_2(g)$;
(b) $2\ AgNO_3(s) \rightarrow 2\ AgNO_2(s) + O_2(g)$

29. **(a)** $2\ KHCO_3 \rightarrow K_2CO_3 + H_2O + CO_2$;
(b) $Zn(HCO_3)_2 \rightarrow ZnCO_3 + H_2O + CO_2$;
(c) $Li_2CO_3 \rightarrow Li_2O + CO_2$;
(d) $CdCO_3 \rightarrow CdO + CO_2$

31. **(a)** no reaction; **(b)** reaction; **(c)** no reaction;
(d) reaction

33. **(a)** reaction; **(b)** reaction; **(c)** reaction; **(d)** reaction

35. **(a)** reaction; **(b)** no reaction; **(c)** reaction;
(d) no reaction

37. **(a)** no reaction; **(b)** reaction;
(c) reaction; **(d)** no reaction

39. **(a)** $Cu(s) + Al(NO_3)_3(aq) \rightarrow$ *NR*;
(b) $2\ Al(s) + 3\ Cu(NO_3)_2(aq) \rightarrow 2\ Al(NO_3)_3(aq) + 3\ Cu(s)$;
(c) $Cd(s) + FeSO_4(aq) \rightarrow$ *NR*;
(d) $Fe(s) + CdSO_4(aq) \rightarrow FeSO_4(aq) + Cd(s)$

41. **(a)** $Cd(s) + 2\ HCl(aq) \rightarrow CdCl_2(aq) + H_2(g)$;
(b) $Mn(s) + 2\ HC_2H_3O_2(aq) \rightarrow Mn(C_2H_3O_2)_2(aq) + H_2(g)$;
(c) $Zn(s) + H_2SO_4(aq) \rightarrow ZnSO_4(aq) + H_2(g)$;
(d) $Mg(s) + H_2CO_3(aq) \rightarrow MgCO_3(s) + H_2(g)$

43. **(a)** $2\ Li(s) + 2\ H_2O(l) \rightarrow 2\ LiOH(aq) + H_2(g)$;
(b) $Pb(s) + H_2O(l) \rightarrow$ *NR*; **(c)** $Co(s) + H_2O(l) \rightarrow$ *NR*;
(d) $Ca(s) + 2\ H_2O(l) \rightarrow Ca(OH)_2(aq) + H_2(g)$

45. **(a)** $Zn(s) + Pb(NO_3)_2(aq) \rightarrow Zn(NO_3)_2(aq) + Pb(s)$;
(b) $Co(s) + NiSO_4(aq) \rightarrow CoSO_4(aq) + Ni(s)$;
(c) $Ni(s) + SnSO_4(aq) \rightarrow NiSO_4(aq) + Sn(s)$;
(d) $Sn(s) + LiC_2H_3O_2(aq) \rightarrow$ *NR*;
(e) $Zn(s) + 2\ HCl(aq) \rightarrow ZnCl_2(aq) + H_2(g)$

47. **(a)** soluble; **(b)** soluble; **(c)** soluble; **(d)** soluble;
(e) soluble; **(f)** insoluble; **(g)** soluble; **(h)** soluble

49. **(a)** $MgSO_4(aq) + BaCl_2(aq) \rightarrow BaSO_4(s) + MgCl_2(aq)$;
(b) $2\ AlBr_3(aq) + 3\ Na_2CO_3(aq) \rightarrow Al_2(CO_3)_3(s) + 6\ NaBr(aq)$;
(c) $3\ NiSO_4(aq) + 2\ Li_3PO_4(aq) \rightarrow Ni_3(PO_4)_2(s) + 3\ Li_2SO_4(aq)$

51. **(a)** $MnSO_4(aq) + 2\ NH_4OH(aq) \rightarrow Mn(OH)_2(s) + (NH_4)_2SO_4(aq)$;
(b) $ZnCl_2(aq) + Hg_2(NO_3)_2(aq) \rightarrow Hg_2Cl_2(s) + Zn(NO_3)_2(aq)$

53. **(a)** $2\,HCl(aq) + Ca(OH)_2(aq) \rightarrow CaCl_2(aq) + 2\,HOH(l)$; **(b)** $H_2SO_4(aq) + 2\,LiOH(aq) \rightarrow Li_2SO_4(aq) + 2\,HOH(l)$; **(c)** $2\,HNO_2(aq) + Sr(OH)_2(aq) \rightarrow Sr(NO_2)_2(aq) + 2\,HOH(l)$

55. **(a)** $NaOH(aq) + HNO_3(aq) \rightarrow NaNO_3(aq) + HOH(l)$; **(b)** $3\,Ba(OH)_2(aq) + 2\,H_3PO_4(aq) \rightarrow Ba_3(PO_4)_2(s) + 6\,HOH(l)$

57. **(a)** $2\,SO_2(g) + O_2(g) \rightarrow 2\,SO_3(g)$; **(b)** $F_2(g) + 2\,NaBr(aq) \rightarrow Br_2(l) + 2\,NaF(aq)$; **(c)** $Cl_2(g) + H_2O(l) \rightarrow HCl(aq) + HClO(aq)$; **(d)** $PCl_5(s) + 4\,H_2O(l) \rightarrow H_3PO_4(aq) + 5\,HCl(aq)$; **(e)** $Sb_2S_3(s) + 6\,HCl(aq) \rightarrow 2\,SbCl_3(aq) + 3\,H_2S(aq)$

59. **(a)** $CH_4(g) + 2\,O_2(g) \rightarrow CO_2(g) + 2\,H_2O(g)$; **(b)** $2\,CH_4O(l) + 3\,O_2(g) \rightarrow 2\,CO_2(g) + 4\,H_2O(g)$; **(c)** $C_3H_8(g) + 5\,O_2(g) \rightarrow 3\,CO_2(g) + 4\,H_2O(g)$; **(d)** $2\,C_3H_8O(l) + 9\,O_2(g) \rightarrow 6\,CO_2(g) + 8\,H_2O(g)$

61. $4\,HCl(g) + O_2(g) \rightarrow 2\,Cl_2(g) + 2\,H_2O(g)$

63. $4\,NH_3(g) + 5\,O_2(g) \rightarrow 4\,NO(g) + 6\,H_2O(g)$; $2\,NO(g) + O_2(g) \rightarrow 2\,NO_2(g)$; $3\,NO_2(g) + H_2O(l) \rightarrow 2\,HNO_3(aq) + NO(g)$

Chapter 9

1. numbers of molecules, numbers of moles, volumes of gases

3. **(a)** 10; **(b)** 4; **(c)** 6; **(d)** 4

5. **(a)** 2(101.1) g = 2(85.1) g + 1(32.0) g; **(b)** 2(27.0) g + 3(32.1) g = 150.3 g

7. 0.250 mol O_2, 0.500 mol H_2O

9. 0.500 mol Cl_2, 0.333 mol $FeCl_3$

11. 1.50 mol NH_3, 2.25 mol H_2O, 1.88 mol O_2

13. **(a)** mass–mass; **(b)** volume–volume; **(c)** mass–volume

15. 2.94 g ZnO **17.** 5.21 g $BiCl_3$ **19.** 2.09 g Ag

21. 6.39 g Hg **23.** 1.68 g $Ca_3(PO_4)_2$ **25.** 366 mL CO_2

27. 262 mL CO_2 **29.** 0.244 g Mg **31.** 0.167 g H_2O_2

33. 1.00 L O_2 **35.** **(a)** 19.3 mL N_2 **(b)** 57.8 mL H_2

37. 50.0 L SO_3 **39.** 93.1% **41.** 106%

43. **(a)** −2.0%; **(b)** 1.3% **45.** **(a)** 0.8%; **(b)** 3.5%

47. **(a)** 0.929 g Cr_2O_3; **(b)** 137 mL N_2 **49.** 0.593 L H_2S

51. **(a)** 16.4 g H_2O; **(b)** 15,300 mL CO_2

Chapter 10

1. **(a)** true; **(b)** false **3.** violet

5. **(a)** 1 quantum; **(b)** 1 quantum; **(c)** 100 quanta; **(d)** 100 quanta; **(e)** 500 quanta; **(f)** 500 quanta

7. The Bohr atom successfully explained the spectral lines in the emission spectrum of hydrogen.

9. **(a)** continuous; **(b)** quantized

11. **(a)** continuous; **(b)** quantized

13. The photons of infrared light are not sufficiently energetic, whereas the photons of ultraviolet light have sufficient energy to dislodge electrons.

15. An *orbit* is the path traveled by an electron about the nucleus of an atom, according to the Bohr model. An *orbital* is a region about the nucleus in which there is a high probability of finding an electron with a given energy, according to the quantum mechanical model.

17. see Glossary

19. **(a)** $n = 2$; **(b)** $n = 3$; **(c)** $n = 3$; **(d)** $n = 5$

21. (b) and (c) each have identical pairs

23. **(a)**

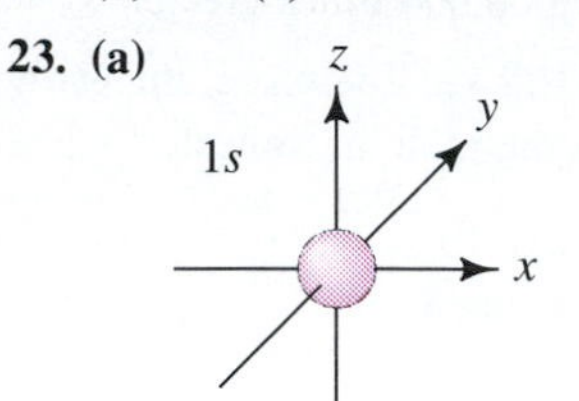

(b)

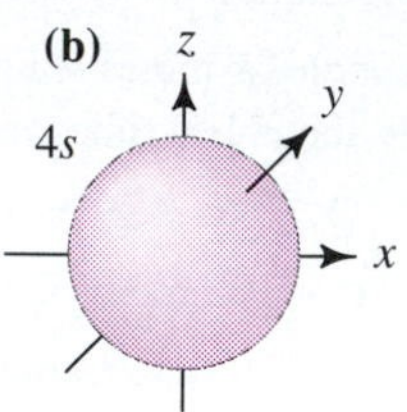

(c)

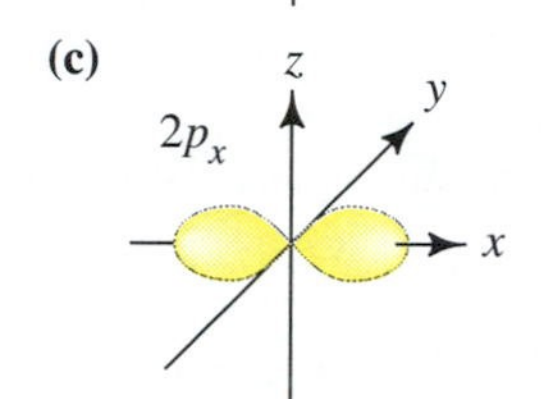

(d)

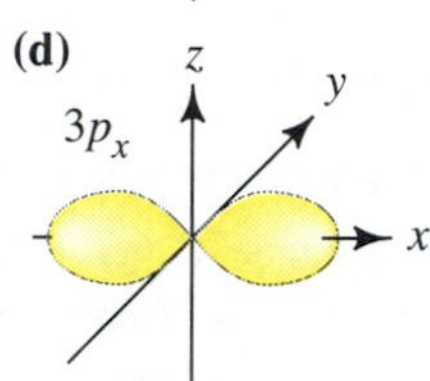

(e)

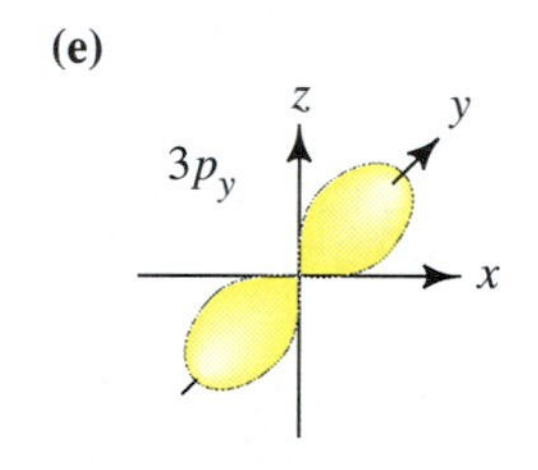

(f)

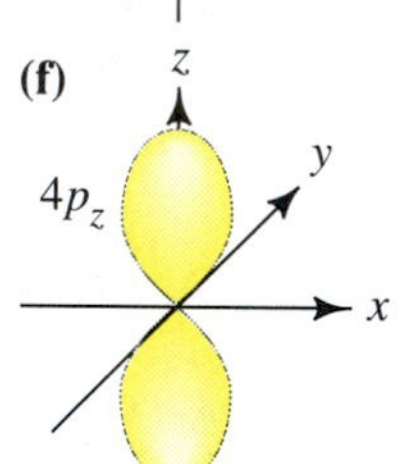

25. **(a)** $3s$; **(b)** $3p_x$; **(c)** same; **(d)** same **27.** **(a)** $5s$; **(b)** $4p$

29. **(a)** 2; **(b)** 2; **(c)** 2; **(d)** 2 **31.** **(a)** 2; **(b)** 2; **(c)** 2; **(d)** 2

33. **(a)** 1; **(b)** 3; **(c)** 5; **(d)** 7 **35.** **(a)** 1; **(b)** 2; **(c)** 3; **(d)** 4

37. **(a)** 2; **(b)** 6; **(c)** 10; **(d)** 14

39. **(a)** 2; **(b)** 8; **(c)** 18; **(d)** 32 **41.** $n = 1, 2, 3, \ldots$

43. **(a)** $l = 0$; **(b)** $l = 0, 1$; **(c)** $l = 0, 1, 2$; **(d)** $l = 0, 1, 2, 3$

45. **(a)** $m = 0$; **(b)** $m = +1, 0, -1$; **(c)** $m = +2, +1, 0, -1, -2$; **(d)** $m = +3, +2, +1, 0, -1, -2, -3$

47. **(a)** $m = 0$; **(b)** $m = +1, 0, -1$; **(c)** $m = +2, +1, 0, -1, -2$; **(d)** $m = +3, +2, +1, 0, -1, -2, -3$

49. **(a)** $s = +\frac{1}{2}, -\frac{1}{2}$; **(b)** $s = +\frac{1}{2}, -\frac{1}{2}$; **(c)** $s = +\frac{1}{2}, -\frac{1}{2}$; **(d)** $s = +\frac{1}{2}, -\frac{1}{2}$

51. orbital energy and size

53. orbital orientation

55. **(a)** $1s^2\,2s^2\,2p^2$; **(b)** $1s^2\,2s^2\,2p^6\,3s^2\,3p^3$; **(c)** $1s^2\,2s^2\,2p^6\,3s^2\,3p^6\,4s^1$; **(d)** $1s^2\,2s^2\,2p^6\,3s^2\,3p^6\,4s^2\,3d^7$

57. 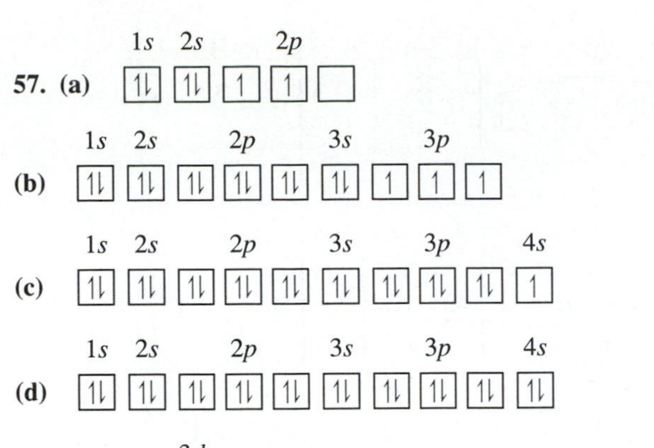

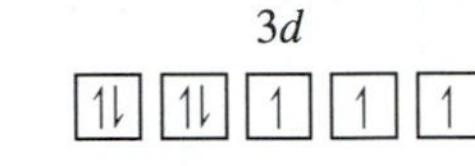

59. 2 unpaired e–

61. 1*s* [⇅]

63. 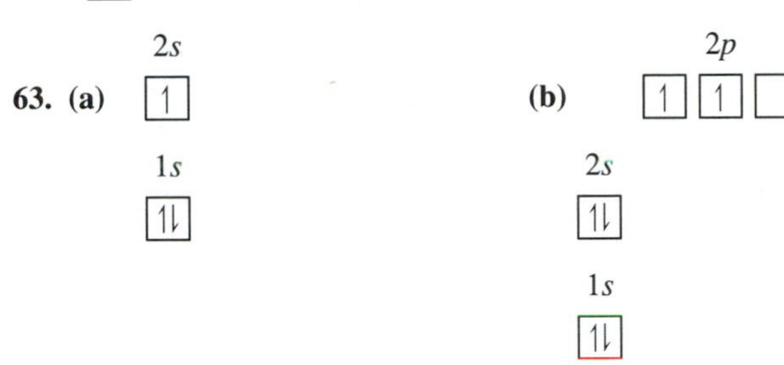

65. **(a)** $(2, 1, +1, +\frac{1}{2})$; **(b)** $(2, 1, -1, -\frac{1}{2})$; **(c)** $(3, 0, 0, -\frac{1}{2})$; **(d)** $(2, 1, +1, -\frac{1}{2})$

67. **(a)** $(3, 0, 0, +\frac{1}{2})$; **(b)** $(2, 1, 0, -\frac{1}{2})$; **(c)** $(3, 2, +2, +\frac{1}{2})$; **(d)** $(4, 1, -1, -\frac{1}{2})$

69. **(a)** Li; **(b)** S; **(c)** Nb; **(d)** I

71. **(a)** atom; **(b)** molecule or formula unit; **(c)** photon; **(d)** electron

73. automatic door openers; television, stereo, and garage door remote controls

75. Since these orbitals overlap and share some of the same volume, the electrons in the 2*s* and 2*p* orbitals can be found in the same region about the nucleus.

77. If chromium promotes one of its two 4*s* electrons into a vacant 3*d* orbital, the resulting 4*s* and 3*d* orbital sets are all half-filled. This is unusually stable.

79. **(a)** paramagnetic; **(b)** no; **(c)** paramagnetic; **(d)** no

Chapter 11

1. An ionic bond results from the attraction between a positively-charged cation and a negatively-charged anion.

3. magnesium: 2, 0; sulfur: 6, 8

5. **(a)** covalent; **(b)** ionic; **(c)** covalent; **(d)** ionic

7. **(a)** molecule; **(b)** formula unit; **(c)** molecule; **(d)** molecule; **(e)** molecule; **(f)** formula unit

9. **(a)** atom; **(b)** molecule; **(c)** molecule; **(d)** atom **(e)** molecule; **(f)** formula unit

11. **(a)** 1+; **(b)** 1+; **(c)** 4+; **(d)** 4+

13. **(a)** 1−; **(b)** 1−; **(c)** 2−; **(d)** 3−

15. **(a)** $1s^2$; **(b)** $1s^2\ 2s^2\ 2p^6\ 3s^2\ 3p^6$; **(c)** $1s^2\ 2s^2\ 2p^6\ 3s^2\ 3p^6$; **(d)** $1s^2\ 2s^2\ 2p^6\ 3s^2\ 3p^6\ 4s^2\ 3d^{10}\ 4p^6\ 5s^2\ 4d^{10}\ 5p^6\ 6s^2\ 4f^{14}\ 5d^{10}\ 6p^6$

17. **(a)** $1s^2\ 2s^2\ 2p^6\ 3s^2\ 3p^6$; **(b)** $1s^2\ 2s^2\ 2p^6\ 3s^2\ 3p^6\ 4s^2\ 3d^{10}\ 4p^6\ 5s^2\ 4d^{10}\ 5p^6$; **(c)** $1s^2\ 2s^2\ 2p^6\ 3s^2\ 3p^6$; **(d)** $1s^2\ 2s^2\ 2p^6\ 3s^2\ 3p^6$

19. **(a)** He; **(b)** Ar; **(c)** Ar; **(d)** Rn

21. **(a)** Ar; **(b)** Kr; **(c)** Ar; **(d)** Kr

23. **(a)** Li atom; **(b)** Mg atom; **(c)** F ion; **(d)** O ion

25. **(a)** false; **(b)** false; **(c)** false; **(d)** false; **(e)** false

27. **(a)** the sum of the atomic radii; **(b)** the sum of the atomic radii

29. **(a)** false; **(b)** false; **(c)** false; **(d)** false; **(e)** false

31.

		Electron dot	Structural
(a)	2 e⁻	H:H	H—H
(b)	14 e⁻	:F:F:	F—F
(c)	8 e⁻	H:Br:	H—Br
(d)	8 e⁻	H:N:H (with H above N)	H—N—H (with H above N)
(e)	32 e⁻	:Cl:C:Cl: (with :Cl: above and below C)	Cl—C—Cl (with Cl above and below C)
(f)	20 e⁻	:F:O:F:	F—O—F

33. (a) 24 e⁻ H:Ö:N::Ö: (with :Ö: above N) H—O—N=O (with O above N)

(b) 18 e⁻ :Ö:S::Ö: O—S=O

(c) 12 e⁻ H:C::C:H (H H above) H—C=C—H (H H above)

(d) 10 e⁻ H:C:::C:H H—C≡C—H

(e) 14 e⁻ H:Ö:Cl: H—O—Cl

(f) 18 e⁻ H:Ö:N::Ö: H—O—N=O

35. decreases 37. nonmetals

39. (a) Cl; (b) O; (c) B; (d) Se; (e) F; (f) P

41. (a) 0.7; (b) 1.9; (c) 0.5; (d) 1.2; (e) 1.0

43. (a) δ⁺ H—S δ⁻; (b) δ⁻ O—S δ⁺;
(c) δ⁻ Cl—B δ⁺; (d) δ⁺ S—Cl δ⁻;
(e) δ⁺ N—F δ⁻;

45. (a), (b), (d), and (e) are nonpolar

47. (a), (b), (c), (f), (g), (i), (j)

49. H:Ö:Cl: H:Ö:Cl—Ö:

51. H:N:H (H below) [H:N:H (H above and below)]⁺

53. [:Ö:N::Ö:]⁻ [:Ö:N::Ö: (with :Ö: above N)]⁻

55. [:Ö:S:Ö: (with :Ö: above and below S)]²⁻ [:Ö:S:Ö—H (with :Ö: above and below S)]⁻

57. (a) 8 e⁻ [:Ö:H]⁻ [O—H]⁻

(b) 14 e⁻ [:I:Ö:]⁻ [I—O]⁻

(c) 20 e⁻ [:Ö:Cl:Ö:]⁻ [O—Cl—O]⁻

(d) 8 e⁻ [H:P:H (H above and below)]⁺ [H—P—H (H above and below)]⁺

(e) 32 e⁻ [:Ö:I:Ö: (with :Ö: above and below I)]⁻ [O—I—O (O above and below)]⁻

59. (a) 14 e⁻ [:Cl:Ö:]⁻ [Cl—O]⁻

(b) 20 e⁻ [:Ö:Br:Ö:]⁻ [O—Br—O]⁻

(c) 26 e⁻ [:Ö:I:Ö: (with :Ö: above I)]⁻ [O—I—O (O above)]⁻

(d) 8 e⁻ [H:Ö:H (H above)]⁺ [H—O—H (H above)]⁺

(e) 32 e⁻ [:Ö:Cl:Ö: (with :Ö: above and below Cl)]⁻ [O—Cl—O (O above and below)]⁻

61. (a) solid; (b) crystalline; (c) high; (d) high;
(e) conductor; (f) fast

63. (a) ionic; (b) molecular; (c) ionic; (d) molecular;
(e) ionic; (f) molecular

65. Sodium has lost its only electron in the 3*s* sublevel, which is larger than the 2*p* sublevel. In addition, the ion has one more proton than electron; thus, the remaining electrons are drawn closer to the nucleus.

67. The properties of the elements Fe and O_2 are not related to the properties of the compound Fe_2O_3.

69. (a) CaI_2; (b) RaO; (c) GaF_3; (d) Ba_3P_2

71. (a) $Al_2(CO_3)_3$; (b) $Sr(OH)_2$; (c) Ag_3PO_4;
(d) $Cd(NO_3)_2$

73. The properties of the elements N_2 and O_2 are not related to the properties of the compound NO_2.

75. (a) 8 e⁻ H:Sb:H (H below) H—Sb—H (H below)

(b) 32 e⁻ :Cl:Ge:Cl: (with :Cl: above and below Ge) Cl—Ge—Cl (Cl above and below)

77. After the formation of chemical bonds, the valence electrons from individual atoms are delocalized throughout the entire molecule.

79. :Ö:S::Ö:, :Ö::S:Ö:

81. (a) 24 e⁻

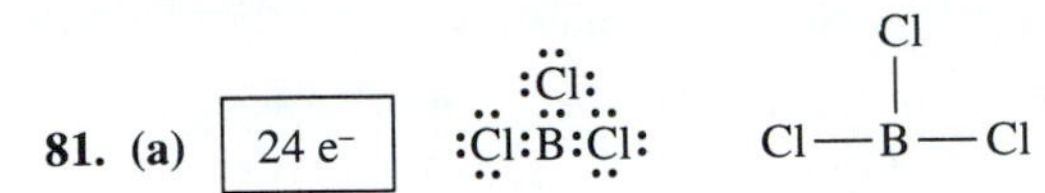

(b) 40 e⁻

:Br:
:Br:P:Br:
:Br: :Br:

Br
Br—P—Br
Br Br

(c) 48 e⁻

:F: :F:
:F:S:F:
:F: :F:

F F
F—S—F
F F

83. H^+, $H:^-$

Chapter 12

1. see Section 12.1

3. (a) 1.00 atm; (b) 760 mm Hg; (c) 760 torr; (d) 76.0 cm Hg

5. (a) 3990 mm Hg; (b) 3990 torr; (c) 399 cm Hg; (d) 157 in. Hg

7. (a) 0.963 atm; (b) 732 mm Hg; (c) 73.2 cm Hg; (d) 732 torr

9. see Section 12.3

11. As the temperature of a liquid increases, so does its vapor pressure.

13. 56°C **15.** see Figure 12.7 **17.** 742 torr **19.** 752 torr

21. volume, temperature, and number of molecules

23. (a) decreases; (b) increases; (c) increases

25. (a) decreases; (b) increases; (c) decreases

27.

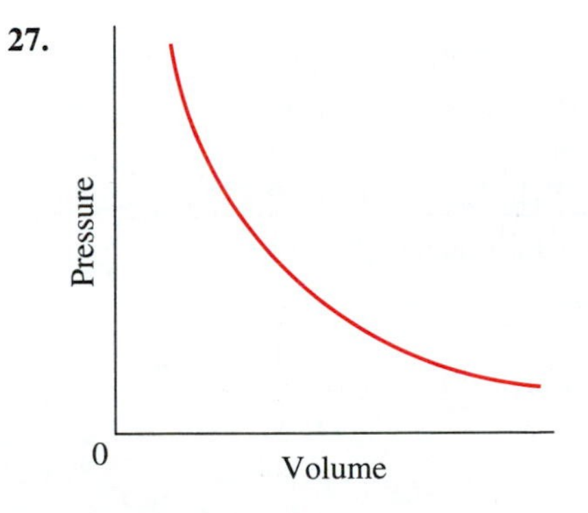

29. 0.286 atm **31.** 17.0 psi

33.

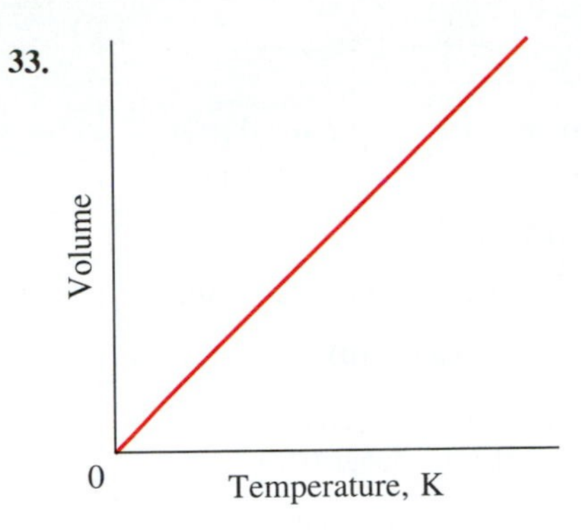

35. 363 mL **37.** 109 cm^3

39.

Pressure

0 Temperature, K

41. 1230 torr **43.** 104 cm Hg **45.** 94.3 mL **47.** 5.31 L

49. 7820 torr **51.** 1290 mm Hg **53.** 607°C

55. see Section 12.10 **57.** high temperature, low pressure

59. (a) all the same; (b) all the same; (c) He; (d) Ar

61. 0 torr **63.** 245 atm **65.** 2.78 mol **67.** 48.0 g/mol

69. 4.27 g **71.** 62.4 torr · L/mol · K **73.** 37,000 lb

75. A liquid boils when its vapor pressure equals the external pressure. When the vacuum pump lowers the external pressure to that of the vapor pressure of the water, the water will begin to boil.

77. ~ 60°C **79.** 38.5 mL **81.** 12.3 atm **83.** 5.11 L

85. number of molecules

87. Helium is less dense than the nitrogen it replaces because its molecules are lighter. Lighter molecules travel faster (at the same temperature), resulting in the higher-pitched voice.

Chapter 13

1. see Section 13.1

3. (a) solid; (b) gas; (c) solid; (d) gas; (e) liquid; (f) gas

5. see Figure 13.2

7. The viscosity of a liquid is a measure of its resistance to flow. Since the intermolecular attraction is high in water, the viscosity of water is higher than other molecules of the same size.

9. **(b), (c), (d)**

11. **(a)** dispersion; **(b)** hydrogen bond; **(c)** dipole; **(d)** dipole

13. **(a)** C_2H_5Cl **(b)** CH_3OCH_3

15. **(a)** CH_3COOH; **(b)** C_2H_5OH **17.** see Section 13.4

19. **(a)** solid; **(b)** liquid; **(c)** solid; **(d)** solid; **(e)** liquid; **(f)** liquid

21. table salt (NaCl), ice (H_2O), copper (Cu)

23. **(a)** ions; **(b)** molecules; **(c)** metal atoms

25. **(a)** metallic; **(b)** ionic; **(c)** molecular; **(d)** molecular

27.

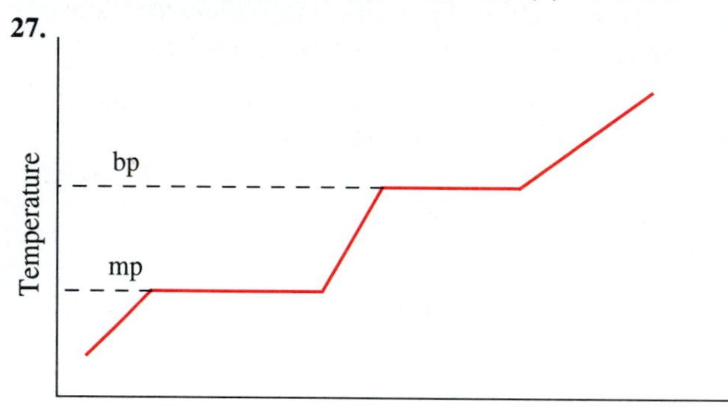

29. 1.00×10^4 cal or 10.0 kcal **31.** 15,400 cal

33. 82,800 cal **35.** 28,100 cal **37.** 75,200 cal

39. $H\!:\!\ddot{\underset{..}{O}}\!:\!H$ and H—O—H

41. 104.5° **43.** $O^{\delta-}$ with δ^+H and $H\,\delta^+$ **45.** H—F····H—F (H-bond)

47. float **49.** **(a)** H_2O; **(b)** H_2Se **51.** **(a)** H_2O; **(b)** H_2Se

53. **(a)** increase; **(b)** increase; **(c)** increase; **(d)** increase

55. Heavy water is colorless and odorless, but toxic to animals and humans.

57. $2\ H_2O(l) \xrightarrow{elec} 2\ H_2(g) + O_2(g)$

59. **(a)** $2\ Li(s) + 2\ H_2O(l) \rightarrow 2\ LiOH(aq) + H_2(g)$;
(b) $2\ Rb(s) + 2\ H_2O(l) \rightarrow 2\ RbOH(aq) + H_2(g)$;
(c) $Na_2O(s) + H_2O(l) \rightarrow 2\ NaOH(aq)$;
(d) $Cs_2O(s) + H_2O(l) \rightarrow 2\ CsOH(aq)$;
(e) $CO_2(g) + H_2O(l) \rightarrow H_2CO_3(aq)$;
(f) $P_2O_5(s) + 3\ H_2O(l) \rightarrow 2\ H_3PO_4(aq)$

61. **(a)** $2\ C_3H_6(g) + 9\ O_2(g) \rightarrow 6\ CO_2(g) + 6\ H_2O(g)$;
(b) $C_3H_6O(g) + 4\ O_2(g) \rightarrow 3\ CO_2(g) + 3\ H_2O(g)$;
(c) $2\ HF(aq) + Ca(OH)_2(aq) \rightarrow CaF_2(aq) + 2\ HOH(l)$;
(d) $H_2CO_3(aq) + 2\ KOH(aq) \rightarrow K_2CO_3(aq) + 2\ HOH(l)$;
(e) $Na_2Cr_2O_7 \cdot 2\ H_2O(s) \rightarrow Na_2Cr_2O_7(s) + 2\ H_2O(g)$;
(f) $Ca(NO_3)_2 \cdot 4\ H_2O(s) \rightarrow Ca(NO_3)_2(s) + 4\ H_2O(g)$

63. **(a)** magnesium sulfate heptahydrate;
(b) cobalt(III) cyanide trihydrate;
(c) manganese(II) sulfate monohydrate;
(d) sodium dichromate dihydrate;
(e) strontium nitrate hexahydrate;
(f) cobalt(II) acetate tetrahydrate;
(g) copper(II) sulfate pentahydrate;
(h) chromium(III) nitrate nonahydrate

65. **(a)** 40.5%; **(b)** 10.9%; **(c)** 51.1%; **(d)** 28.3%; **(e)** 10.7%; **(f)** 30.8%

67. **(a)** $NiCl_2 \cdot 2\ H_2O$; **(b)** $Sr(NO_3)_2 \cdot 6\ H_2O$; **(c)** $CrI_3 \cdot 9\ H_2O$; **(d)** $Ca(NO_3)_2 \cdot 4\ H_2O$

69. **(a)** Ca^{2+}, Mg^{2+}, Fe^{3+}, Na^+; **(b)** Na^+; **(c)** negligible; **(d)** negligible

71. Cations found in tap water (Ca^{2+}, Mg^{2+}, Fe^{3+}) react with soap anions to form a precipitate.

73. **(a)** ~ 97%; **(b)** ~ 1%

75. A liquid that contains one or more dissolved substances, for example, ethyl alcohol dissolved in water.

77. Sulfuric acid, H_2SO_4, is very polar, so we can predict that it behaves somewhat like water. That is, owing to surface tension sulfuric acid "raindrops" on Venus will form spheres as raindrops do on earth.

79. 79,700 cal

81. The estimated values for water in the absence of hydrogen bonding are mp ~ −110°C, bp ~ −80°C, H_{fusion} ~ 540 cal/mol, H_{vapor} ~ 4300 cal/mol

83. $Mg^{2+}(aq) + Na\ soap(aq) \rightarrow Mg(soap)_2(s) + 2\ Na^+(aq)$

Chapter 14

1. **(a)** increases; **(b)** increases **3.** 14.5 g/L

5. 0.99 g/100 g water **7.** **(a)** miscible; **(b)** immiscible

9. **(a)** polar; **(b)** nonpolar; **(c)** polar; **(d)** nonpolar

11. **(a)** immiscible; **(b)** miscible; **(c)** miscible; **(d)** immiscible

13. Add several drops of unknown liquid into a test tube containing water, which is a polar solvent. If the unknown liquid is also polar, it will be miscible. If the liquid is nonpolar, it will be immiscible and form two layers in the test tube.

15. **(a)** soluble; **(b)** insoluble; **(c)** soluble

17. **(a)** insoluble; **(b)** soluble; **(c)** soluble; **(d)** insoluble; **(e)** soluble; **(f)** soluble

19. **(a)** water soluble; **(b)** water soluble; **(c)** water soluble; **(d)** water soluble; **(e)** fat soluble; **(f)** fat soluble

21. H_2O---$C_6H_{12}O_6$---H_2O (with H_2O above and below $C_6H_{12}O_6$)

23. (a) 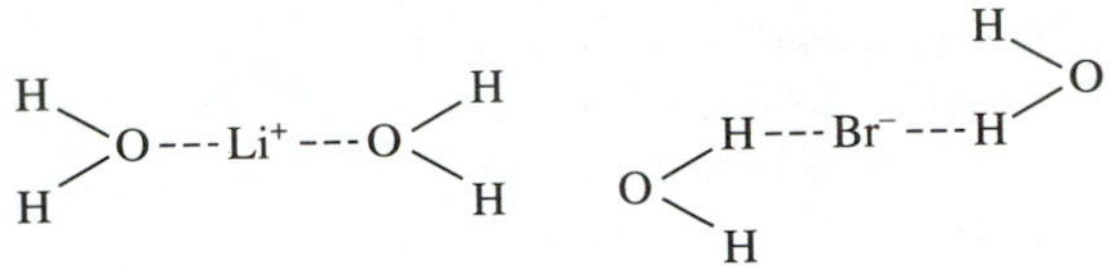 $H_2O\text{---}Li^{+}\text{---}OH_2$ $H_2O\text{---}H\text{---}Br^{-}\text{---}H\text{---}OH$

(b) 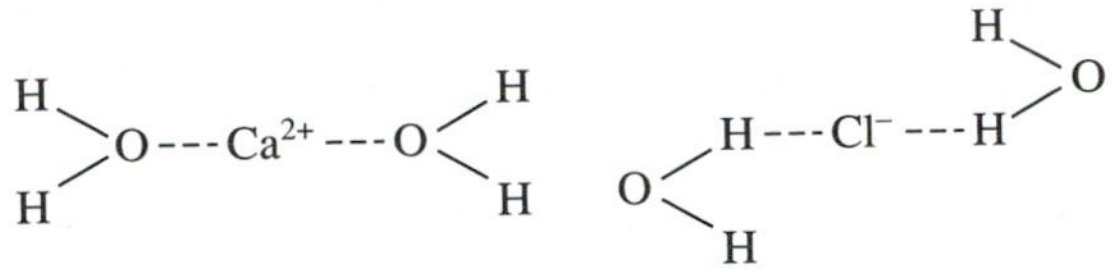 $H_2O\text{---}Ca^{2+}\text{---}OH_2$ $H_2O\text{---}H\text{---}Cl^{-}\text{---}H\text{---}OH$

25. Heating the solution, stirring the solution, grinding the solute

27. (a) ~ 35 g NaCl/100 g H_2O; (b) ~ 35 g KCl/100 g H_2O

29. (a) ~ 36 g NaCl/100 g H_2O; (b) ~ 41 g KCl/100 g H_2O

31. (a) ~ 0°C; (b) ~ 70°C

33. (a) ~ 100°C; (b) ~ 50°C

35. (a) supersaturated; (b) saturated; (c) unsaturated

37. (a) supersaturated; (b) saturated; (c) unsaturated

39. saturated

41. (a) ~ 80 g; (b) ~ 20 g

43. (a) 1.25%; (b) 2.63%; (c) 4.00%; (d) 52.0%

45. (a) $\frac{1.50\text{ g KBr}}{100.00\text{ g solution}}, \frac{98.50\text{ g }H_2O}{100.00\text{ g solution}}, \frac{1.50\text{ g KBr}}{98.50\text{ g }H_2O}$

(b) $\frac{2.50\text{ g }AlCl_3}{100.00\text{ g solution}}, \frac{97.50\text{ g }H_2O}{100.00\text{ g solution}}, \frac{2.50\text{ g }AlCl_3}{97.50\text{ g }H_2O}$

(c) $\frac{3.75\text{ g }AgNO_3}{100.00\text{ g solution}}, \frac{96.25\text{ g }H_2O}{100.00\text{ g solution}}, \frac{3.75\text{ g }AgNO_3}{96.25\text{ g }H_2O}$

(d) $\frac{4.25\text{ g }Li_2SO_4}{100.00\text{ g solution}}, \frac{95.75\text{ g }H_2O}{100.00\text{ g solution}}, \frac{4.25\text{ g }Li_2SO_4}{95.75\text{ g }H_2O}$

47. (a) 53.6 g; (b) 200 g (2.00×10^2 g)

49. (a) 1.70 g $FeBr_2$; (b) 5.25 g Na_2CO_3

51. (a) 247.8 g H_2O; (b) 95.00 g H_2O

53. (a) 0.256 *M*; (b) 0.0510 *M*; (c) 0.400 *M*; (d) 0.312 *M*

55. (a) $\frac{0.100\text{ mol LiI}}{1\text{ L solution}}, \frac{1\text{ L solution}}{0.100\text{ mol LiI}}$

$\frac{0.100\text{ mol LiI}}{1000\text{ mL solution}}, \frac{1000\text{ mL solution}}{0.100\text{ mol LiI}}$

(b) $\frac{0.100\text{ mol }NaNO_3}{1\text{ L solution}}, \frac{1\text{ L solution}}{0.100\text{ mol }NaNO_3}$

$\frac{0.100\text{ mol }NaNO_3}{1000\text{ mL solution}}, \frac{1000\text{ mL solution}}{0.100\text{ mol }NaNO_3}$

(c) $\frac{0.500\text{ mol }K_2CrO_4}{1\text{ L solution}}, \frac{1\text{ L solution}}{0.500\text{ mol }K_2CrO_4}$

$\frac{0.500\text{ mol }K_2CrO_4}{1000\text{ mL solution}}, \frac{1000\text{ mL solution}}{0.500\text{ mol }K_2CrO_4}$

(d) $\frac{0.500\text{ mol }ZnSO_4}{1\text{ L solution}}, \frac{1\text{ L solution}}{0.500\text{ mol }ZnSO_4}$

$\frac{0.500\text{ mol }ZnSO_4}{1000\text{ mL solution}}, \frac{1000\text{ mL solution}}{0.500\text{ mol }ZnSO_4}$

57. (a) 0.866 L; (b) 0.198 L; (c) 0.177 L; (d) 0.0800 L

59. (a) 4.00 g; (b) 6.79 g; (c) 1.68 g; (d) 1.98 g

61. 0.0153 mol/L **63.** (a) 5.00%; (b) 0.291 *M*

65. (a) 0.0688 *m* (b) 0.165 *m* **67.** 4400 g

69. 11.6°C, −11.6°C **71.** 180 g/mol (1.80×10^2 g/mol)

73. 35 g/mol **75.** 1 to 100 nm

77. (a) colloid; (b) colloid; (c) solution

79. 0.0089 g N_2/100 g blood **81.** 228 g SO_2

83. Air is less soluble in hot water than in cold tap water. Therefore, air forms bubbles and leaves the solution.

85. The polar —OH on an alcohol can hydrogen bond with water, causing it to be soluble. As the nonpolar C_xH_y— portion of the molecule increases in size, the overall molecule becomes less polar and eventually immiscible with water.

87. (a) solute: alcohol, solvent: water; (b) solute: water, solvent: alcohol

89. 0.733 *M*

91. Yes, the observed reading corresponds to 75 mg per deciliter.

Chapter 15

1. see Table 15.1

3. (a) basic; (b) acidic; (c) neutral; (d) acidic; (e) acidic; (f) acidic

5. (a) red; (b) yellow

7. (a) colorless; (b) pink **9.** green (yellow + blue)

11. 0.315 *M* **13.** 1.104 *M* **15.** 0.479 *M* **17.** 176 g/mol

19. 89.2 *M* **21.** 0.137 *M* **23.** 0.0844 *M*

25. 27.8 mL

27. (a) 19.9%; (b) 5.94%; (c) 3.12%; (d) 24.9%; (e) 6.00%

29. (a) strong; (b) weak; (c) strong; (d) weak

31. (a) acid; (b) base; (c) salt; (d) base; (e) acid; (f) salt; (g) base; (h) acid

33. (a) HBr and KOH; (b) HCl and $Ba(OH)_2$; (c) HNO_3 and $Ca(OH)_2$; (d) H_3PO_4 and NaOH; (e) H_2SO_4 and $Co(OH)_2$; (f) H_2CO_3 and LiOH

35. (a) acid: $HC_2H_3O_2$, base LiOH; (b) acid: HBr, base: NaCN; (c) acid: $HClO_4$, base K_2CO_3; (d) acid: H_2SO_4, base: NH_3

37. **(a)** stronger acid: H_2SO_3, stronger base: NaHS;
(b) stronger acid: $NaHSO_4$, stronger base: NaF;
(c) stronger acid: $H_2PO_4^-$, stronger base: NH_3;
(d) stronger acid: H_2SO_4, stronger base: H_2O

39. **(a)** $H_2O(l) \rightleftharpoons H^+(aq) + OH^-(aq)$;
(b) $K_w = [H^+][OH^-]$; **(c)** $K_w = 1.0 \times 10^{-14}$;
(d) 1.0×10^{-7} M H^+; **(e)** 1.0×10^{-7} M OH^-

41. **(a)** decrease $[H^+]$; **(b)** increase $[H^+]$

43. **(a)** 3; **(b)** 5; **(c)** 8; **(d)** 6

45. **(a)** 5.10; **(b)** 6.41; **(c)** 6.52; **(d)** 7.80

47. **(a)** 0.96; **(b)** 3.26; **(c)** 3.51; **(d)** 9.18

49. **(a)** 13.70; **(b)** 11.48

51. **(a)** weak acid; **(b)** soluble ionic compound;
(c) strong acid; **(d)** weak acid; **(e)** strong base;
(f) strong base; **(g)** strong acid; **(h)** weak base;
(i) soluble ionic compound;
(j) slightly soluble ionic compound

53. **(a)** $H^+(aq)$ and $ClO_3^-(aq)$; **(b)** $Ca^{2+}(aq)$ and 2 $NO_3^-(aq)$;
(c) $H^+(aq)$ and $Br^-(aq)$; **(d)** $HClO_2(aq)$;
(e) $Li^+(aq)$ and $OH^-(aq)$; **(f)** $Al(OH)_3(aq)$; **(g)** $HNO_2(aq)$;
(h) $HCHO_2(aq)$; **(i)** $Cd^{2+}(aq)$ and 2 $C_2H_3O_2^-(aq)$;
(j) 2 $K^+(aq)$ and $SO_4^{2-}(aq)$

55. see Section 15.11

57. **(a)** $H^+(aq) + OH^-(aq) \rightarrow H_2O(l)$;
(b) $HC_2H_3O_2(aq) + OH^-(aq) \rightarrow C_2H_3O_2^-(aq) + H_2O(l)$;
(c) $2\ HF(aq) + CO_3^{2-}(aq) \rightarrow 2\ F^-(aq) + H_2O(l) + CO_2(g)$;
(d) $H^+(aq) + HCO_3^-(aq) \rightarrow H_2O(l) + CO_2(g)$;
(e) $H^+(aq) + SO_4^{2-}(aq) + Ba^{2+}(aq) + OH^-(aq) \rightarrow$
$BaSO_4(s) + H_2O(l)$

59. orange **61.** yellow **63.** $H_2PO_4^-(aq)$ **65.** 6×10^{13}

67. **(a)** increase; **(b)** increase

69. A weak electrolyte solution conducts electricity slightly, while a nonelectrolyte solution does not conduct at all.

Chapter 16

1. **(a)** frequency of collisions; **(b)** energy of collisions;
(c) orientation of colliding molecules

3.

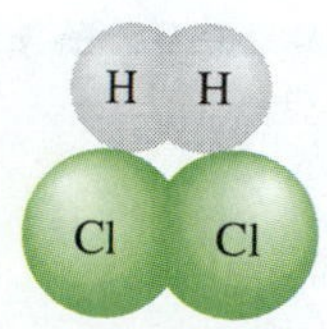

5. **(a)** rate increases; **(b)** rate decreases; **(c)** rate increases

7. In a coal mine, fine particles of coal dust are mixed in the air, providing much greater contact area between the reactants. The charcoal in the barbecue is in lumps or briquettes that have less surface area to contact the oxygen in the air.

9.

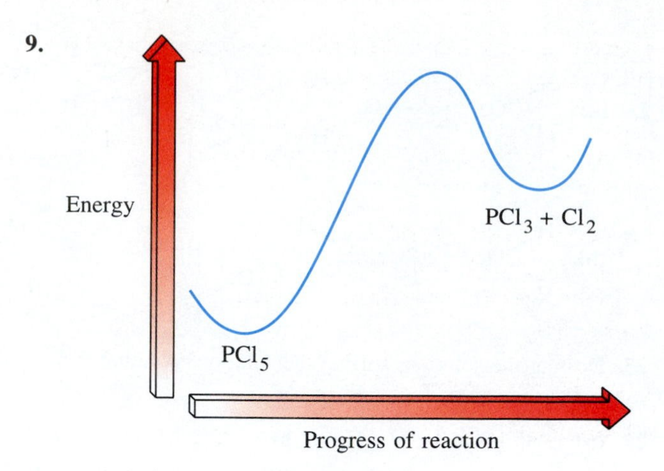

11.

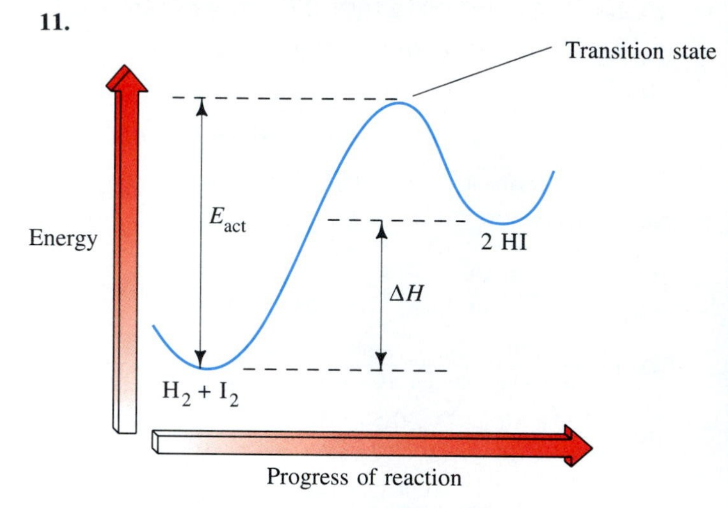

13. A catalyst lowers the energy of activation, E_{act}.

15. forward **17.** 15 s

19. **(a)** 30.0 s; **(b)** 120.0 s; **(c)** 15.0 s; **(d)** 180.0 s

21. **(a)** rate increases 2.25 times;
(b) rate decreases 0.623 times; **(c)** rate triples;
(d) rate decreases 0.192 times

23. **(a)** The rate of the forward reaction is measured by the rate of change in the decrease in concentration of reactant(s) in a given unit of time.
(b) The rate of the forward reaction is measured by the rate of change in the increase in concentration of product(s) in a given unit of time.

25. **(a)** true; **(b)** true

27. **(a)** $K_{eq} = [C]/[A]^2$; **(b)** $K_{eq} = [C]^3/[A][B]^2$;
(c) $K_{eq} = [C]^4[D]/[A]^2[B]^3$

29. no

31. **(a)** $K_c = [HF]^2/[H_2][F_2]$;
(b) $K_c = [NO_2]^4[H_2O]^6/[NH_3]^4[O_2]^7$; **(c)** $K_c = [CO_2]$

33. $K_c = 1.10$

35. **(a)** On hot days, the heat shifts the equilibrium to the right, creating more NO_2, less N_2O_4; **(b)** on cool, overcast days, the lack of heat shifts the equilibrium to the left, creating more N_2O_4, less NO_2.

37. **(a)** shifts left; **(b)** shifts right; **(c)** no shift; **(d)** shifts right; **(e)** no shift; **(f)** shifts right; **(g)** shifts left; **(h)** shifts right

39. **(a)** shifts left; **(b)** shifts right; **(c)** shifts right; **(d)** shifts left; **(e)** shifts right; **(f)** no shift; **(g)** no shift; **(h)** no shift

41. formation of a precipitate, a gas, a weaker electrolyte

43. **(b), (c), (f)**

45. **(a)** $CO_2(g)$ and $H_2O(l)$; **(b)** $H_2O(l)$; **(c)** $PbCrO_4(s)$

47. **(a)** $K_i = [H^+][CO_2H^-]/[HCO_2H]$; **(b)** $K_i = [H^+][HC_2O_4^-]/[H_2C_2O_4]$; **(c)** $K_i = [H^+][H_2C_6H_5O_7^-]/[H_3C_6H_5O_7]$

49. $K_i = 4.5 \times 10^{-4}$ **51.** $K_i = 7.2 \times 10^{-4}$

53. **(a)** shifts right; **(b)** shifts left; **(c)** shifts left; **(d)** shifts right; **(e)** shifts left; **(f)** shifts left; **(g)** shifts right; **(h)** shifts right

55. **(a)** shifts right; **(b)** shifts left; **(c)** shifts left; **(d)** shifts right; **(e)** shifts left; **(f)** no shift; **(g)** shifts right; **(h)** shifts right

57. **(a)** $K_{sp} = [Mn^{2+}][CO_3^{2-}]$; **(b)** $K_{sp} = [Cd^{2+}][CN^-]^2$; **(c)** $K_{sp} = [Sb^{3+}]^2[S^{2-}]^3$

59. $K_{sp} = 5.9 \times 10^{-21}$

61. $K_{sp} = 3.4 \times 10^{-35}$ **63.** $CaCO_3$

65. **(a)** shifts left; **(b)** shifts left; **(c)** shifts right; **(d)** shifts right; **(e)** shifts left; **(f)** no shift

67. **(a)** shifts left; **(b)** shifts left; **(c)** shifts right; **(d)** shifts right; **(e)** no shift; **(f)** shifts left; **(g)** no shift; **(h)** shifts right

69. **(a)** The pendulum is in constant motion and reverses its swing in opposing directions. **(b)** A solar calculator requires energy to operate (discharging). The reverse process (charging) is ongoing while exposed to light.

71. rate $= k[NO_2]^2$. The rate of reaction is only dependent on $[NO_2]$ so $[CO]$ does not appear in the rate expression.

73. $K_c = 0.20$

75. A possible answer is that the stress of final exams causes a shift toward concentration on course work and away from extracurricular activities.

77. **(a)** mol^2/L^2; **(b)** mol^3/L^3

79. $[OH^-] = 2.6 \times 10^{-7}$, $K_{sp} = 1.5 \times 10^{-27}$

Chapter 17

1. **(a)** 0; **(b)** 0; **(c)** 0; **(d)** 0

3. **(a)** +2; **(b)** +3; **(c)** +4; **(d)** +1

5. **(a)** +4; **(b)** +3; **(c)** +4; **(d)** +4

7. **(a)** +4; **(b)** +4; **(c)** +2; **(d)** +4

9. **(a)** oxidation; **(b)** reduction

11. **(a)** oxidized: Mn, reduced: O_2; **(b)** oxidized: S, reduced: O_2; **(c)** oxidized: Cd, reduced: F_2; **(d)** oxidized: Sr, reduced: H_2O **(e)** oxidized: Pb, reduced: $CuSO_4$

13. **(a)** oxidized: H_2, reduced: CuO; **(b)** oxidized: KBr, reduced: Cl_2; **(c)** oxidized: Ca, reduced: H_2O; **(d)** oxidized: Mg, reduced: HCl; **(e)** oxidized: CO, reduced: PbO

15. **(a)** oxidized: Al, reduced: Cr^{3+}; **(b)** oxidized: Cl^-, reduced: F_2; **(c)** oxidized: SO_3^{2-}, reduced: Fe^{3+}; **(d)** oxidized: Sn^{2+}, reduced: Hg^{2+}; **(e)** oxidized: Cr^{2+}, reduced: AgI

17. **(a)** oxidized: Br^-, reduced: Pt^{2+}; **(b)** oxidized: Zr, reduced: TeO_2; **(c)** oxidized: SO_2, reduced: MnO_2; **(d)** oxidized: H_2O_2, reduced: Pb^{4+}; **(e)** oxidized: SO_3^{2-}, reduced: I_2

19. yes **21.** **(a)** $Br_2(l) + 2\ NaI(aq) \rightarrow I_2(s) + 2\ NaBr(aq)$; **(b)** $2\ PbS(s) + 3\ O_2(g) \rightarrow 2\ PbO(s) + 2\ SO_2(g)$; **(c)** $2\ B_2O_3(s) + 6\ Cl_2(g) \rightarrow 4\ BCl_3(aq) + 3\ O_2(g)$

23. **(a)** $2\ MnO_4^-(aq) + 10\ I^-(aq) + 16\ H^+(aq) \rightarrow 2\ Mn^{2+}(aq) + 5\ I_2(s) + 8\ H_2O(l)$; **(b)** $Cu(s) + 4\ H^+(aq) + SO_4^{2-}(aq) \rightarrow Cu^{2+}(aq) + SO_2(g) + H_2O(l)$; **(c)** $2\ Fe^{2+}(aq) + H_2O_2(aq) + 2\ H^+(aq) \rightarrow 2\ Fe^{3+}(aq) + 2\ H_2O(l)$

25. **(a)** $F_2(g) + 2e^- \rightarrow 2\ F^-(aq)$; **(b)** $Hg(l) + 2\ Cl^-(aq) \rightarrow HgCl_2(s) + 2\ e^-$; **(c)** $2\ BrO_3^-(aq) + 12\ H^+(aq) + 10\ e^- \rightarrow Br_2(l) + 6\ H_2O(l)$; **(d)** $H_2O_2(aq) + 2\ H^+(aq) + 2\ e^- \rightarrow 2\ H_2O(l)$; **(e)** $O_2(g) + 2\ H^+(aq) + 2\ e^- \rightarrow H_2O_2(aq)$

27. **(a)** $3\ Zn(s) + 2\ NO_3^-(aq) + 8\ H^+(aq) \rightarrow 3\ Zn^{2+}(aq) + 2\ NO(g) + 4\ H_2O(l)$; **(b)** $2\ Mn^{2+}(aq) + 5\ BiO_3^-(aq) + 14\ H^+(aq) \rightarrow 2\ MnO_4^-(aq) + 5\ Bi^{3+}(aq) + 7\ H_2O(l)$; **(c)** $2\ MnO_4^-(aq) + 5\ SO_3^{2-}(aq) + 6\ H^+(aq) \rightarrow 2\ Mn^{2+}(aq) + 5\ SO_4^{2-}(aq) + 2\ H_2O(l)$; **(d)** $5\ Sn^{2+}(aq) + 2\ IO_3^-(aq) + 12\ H^+(aq) \rightarrow 5\ Sn^{4+}(aq) + I_2(s) + 6\ H_2O(l)$;

(e) $AsO_3^{3-}(aq) + Br_2(l) + H_2O(l) \rightarrow$
$AsO_4^{3-}(aq) + 2\ Br^-(aq) + 2\ H^+(aq)$

29. (a) $Cl_2(g) + H_2O(l) \rightarrow Cl^-(aq) + HOCl(aq) + H^+(aq)$;
(b) $3\ HNO_2(aq) \rightarrow NO_3^-(aq) + 2\ NO(g) + H_2O(l) + H^+(aq)$;
(c) $3\ MnO_4^{2-}(aq) + 4\ H^+(aq) \rightarrow$
$2\ MnO_4^-(aq) + MnO_2(s) + 2\ H_2O(l)$

31. (a) $Au^{3+}(aq)$; **(b)** $H^+(aq)$; **(c)** $I_2(s)$; **(d)** $Br_2(l)$

33. (a) $F_2(g)$; **(b)** $Br_2(l)$; **(c)** $Cu^{2+}(aq)$; **(d)** $Mn^{2+}(aq)$

35. (a) spontaneous; **(b)** nonspontaneous; **(c)** spontaneous; **(d)** nonspontaneous; **(e)** spontaneous

37. (a) $Mg(s) + Sn^{2+}(aq) \rightarrow Mg^{2+}(aq) + Sn(s)$;
(b) $2\ H^+(aq) + 2\ I^-(aq) \rightarrow H_2(g) + I_2(s)$;
(c) $2\ H^+(aq) + Ni(s) \rightarrow H_2(g) + Ni^{2+}(aq)$;
(d) $Hg^{2+}(aq) + 2\ Br^-(aq) \rightarrow Hg(l) + Br_2(l)$;
(e) $2\ Fe^{3+}(aq) + 2\ I^-(aq) \rightarrow 2\ Fe^{2+}(aq) + I_2(s)$

39.

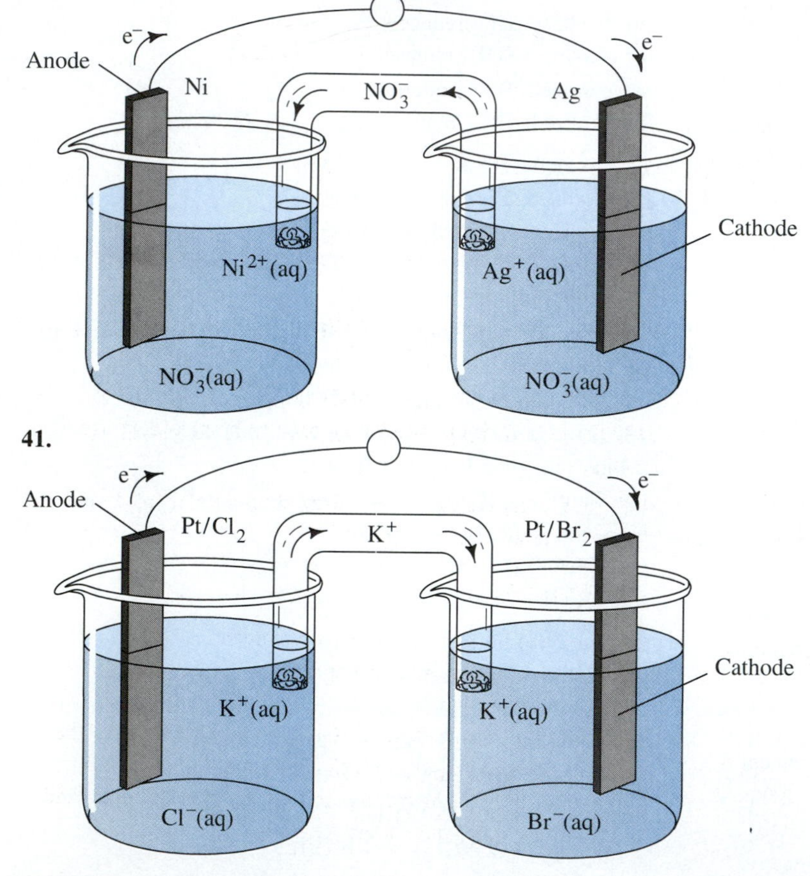

43. (a) $Sn(s) \rightarrow Sn^{2+}(aq) + 2\ e^-$;
(b) $Cu^{2+}(aq) + 2\ e^- \rightarrow Cu(s)$;
(c) anode: Sn electrode, cathode: Cu electrode;
(d) from anode to cathode;
(e) from reduction half-cell to oxidation half-cell (opposite of e^- flow)

45. (a) $Mg(s) \rightarrow Mg^{2+}(aq) + 2\ e^-$;
(b) $Mn^{2+}(aq) + 2\ e^- \rightarrow Mn(s)$;
(c) anode: Mg electrode, cathode: Mn electrode;
(d) from anode to cathode;
(e) from reduction half-cell to oxidation half-cell (opposite of e^- flow)

47.

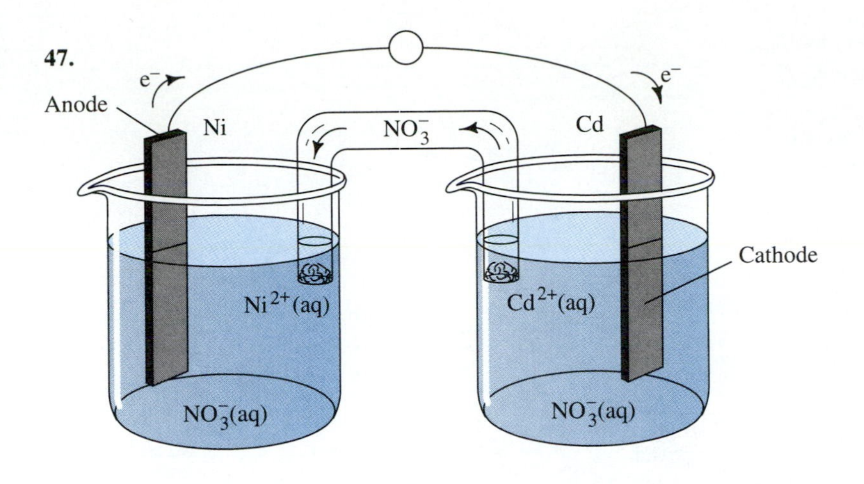

49.

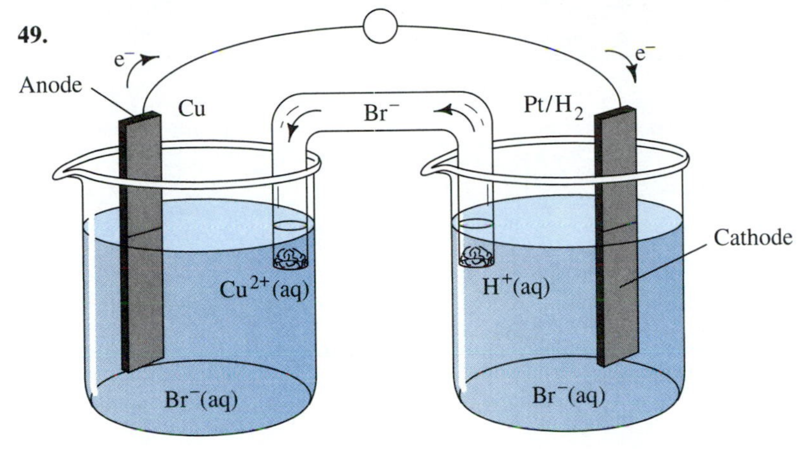

51. **(a)** $Ni(s) \rightarrow Ni^{2+}(aq) + 2\ e^-$;
(b) $Fe^{2+}(aq) + 2\ e^- \rightarrow Fe(s)$;
(c) anode: Ni electrode, cathode: Fe electrode;
(d) from anode to cathode;
(e) from reduction half-cell to oxidation half-cell (opposite of e^- flow)

53. **(a)** $Cr(s) \rightarrow Cr^{3+}(aq) + 3\ e^-$;
(b) $Al^{3+}(aq) + 3\ e^- \rightarrow Al(s)$;
(c) anode: Cr electrode, cathode: Al electrode;
(d) from anode to cathode;
(e) from reduction half-cell to oxidation half-cell (opposite of e^- flow)

55. +2

57. For a redox equation to be balanced, the total number of atoms of each element and the total ionic charge for reactants and products must be equal.

59. $Zn(s) + 2\ H^+(aq) \rightarrow Zn^{2+}(aq) + H_2(g)$; reducing agent: Zn, oxidizing agent: H^+

61. $Cu(s) + 4\ H^+(aq) + 2\ NO_3^-(aq) \longrightarrow Cu^{2+}(aq) + 2\ NO_2(g) + 2\ H_2O(l)$; reducing agent: Cu; oxidizing agent: HNO_3

63. Fuel cells are (a) more energy efficient and (b) do not produce atmospheric pollution.

Chapter 18

1. **(a)** Determine the units of the unknown.
(b) Write down the given value that is related to the unknown.
(c) A strategy map relates a given value to the units of the unknown; for example, L N_2 (STP) $\rightarrow$ mol SO_2 $\rightarrow$ g SO_2
(d) A concept map is more general than a strategy map and simply relates two quantities; for example,

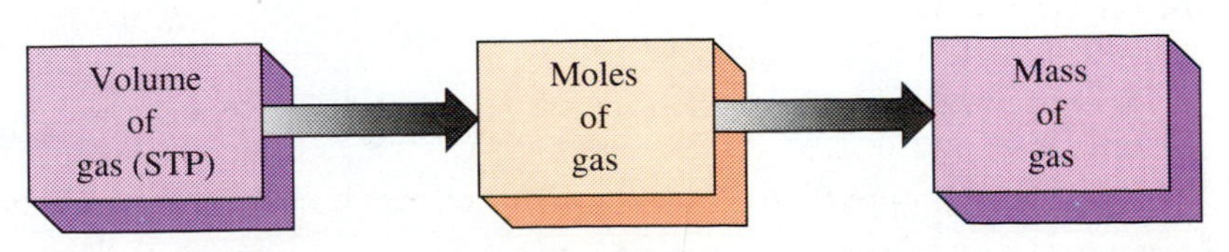

(e) An algorithm is a process for solving a problem by writing out a series of sequential steps in a calculation operation.

(f) Visualization is a process by which a concept becomes more concrete by attempting to form a mental picture.

3. **(a)** mass; **(b)** time; **(c)** volume; **(d)** length; **(e)** weight; **(f)** temperature; **(g)** mass; **(h)** volume; **(i)** concentration; **(j)** temperature; **(k)** length; **(l)** density; **(m)** volume; **(n)** time

5. A possible concept map for the integrated software program is

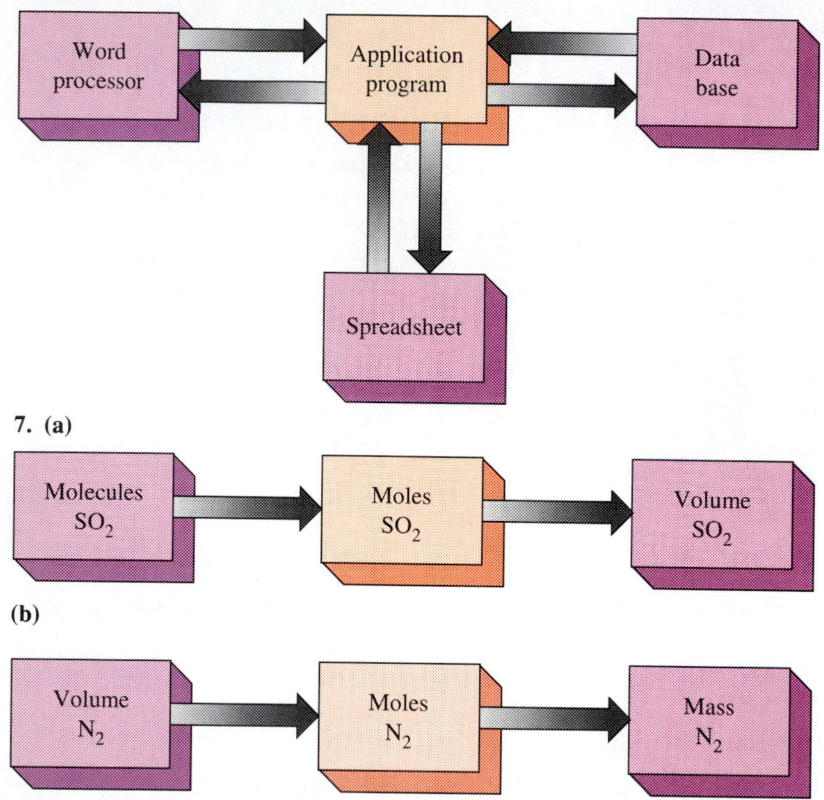

9. **(a)** Step 1: Convert atoms of He to mol He using Avogadro's number. Step 2: Convert mol He to volume of He (at STP) using molar volume.

(b) Step 1: Convert the volume of Cl_2 (at STP) to moles of Cl_2 using the molar volume. Step 2: Convert the moles of Cl_2 to mass of Cl_2 using the molar mass.

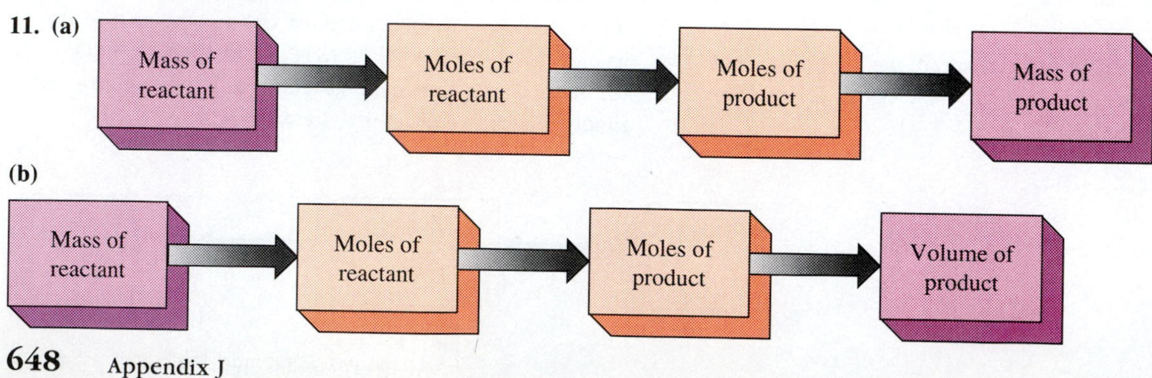

13. Consider the equation: aA + bB $\rightarrow$ cC + dD **(a)** Step 1: Convert the mass of reactant A to the moles of A using the molar mass of A. Step 2: Convert the moles of reactant A to the moles of product C using the coefficients from the balanced chemical equation. Step 3: Convert the moles of product C to the mass of C using the molar mass of C. **(b)** Step 1: Convert the mass of reactant A to the moles of A using the molar mass of A. Step 2: Convert the moles of reactant A to the moles of gaseous product D using the coefficients from the balanced equation. Step 3: Convert the moles of product D to the volume of D using the molar volume concept.

15. **(a)** ~1 g Cr; **(b)** ~40 mL $Ba(OH)_2$; **(c)** ~1.5 g Hg_2Cl_2

17. **(a)** 0.518 g HCl; **(b)** 0.318 L HCl; **(c)** 8.55×10^{21} molecules HCl; **(d)** 0.299 *M* HCl

19. **(a)** 1.12 L HF; **(b)** 3.01×10^{22} molecules HF; **(c)** 0.500 *M* HF

21. **(a)** 1.96 g CO_2; **(b)** 2.69×10^{22} molecules CO_2; **(c)** 0.0893 *M* H_2CO_3

23. **(a)** 0.826 L H_2S; **(b)** 1.26 g H_2S; **(c)** 0.0819 *M* H_2S

25. **(a)** 2.22 g CH_4; **(b)** 6.22 L O_2

27. **(a)** 0.580 L H_2; **(b)** 518 mL HCl solution

29. **(a)** 0.719 *M* $AgNO_3$; **(b)** 1.88 g AgBr **31.** 45 cartons

33. **(a)** 25.0 L SO_3; **(b)** SO_2

35. **(a)** 0.450 g H_2O; **(b)** H_2SO_4

37. positive

39. **(a)** $NH_4NO_3(s) \rightarrow N_2O(g) + 2\ H_2O(g) + 47.8$ kcal; **(b)** $6\ CO_2(g) + 6\ H_2O(l) + 674.0$ kcal $\rightarrow C_6H_{12}O_6(s) + 6\ O_2(g)$

41. **(a)** 693 g C_2H_5OH; **(b)** 675 L CO_2 at STP

43. **(a)** 91.7 g CO_2; **(b)** 46.7 L CO_2 at STP; **(c)** 66.6 g O_2

45. **(a)** 2.50 kg SO_3; **(b)** 698 L SO_3 at STP; **(c)** 1.68 kg H_2SO_4

47. **(a)** 6.30 g N_2; **(b)** 5.04 L N_2; **(c)** 5.40 g H_2O; **(d)** 6.72 L H_2O

49. 4.43 g NO_2

51. 0.622 *M* Ca^{2+}, 1.24 *M* Cl^- **53.** 44.6 mL HCl solution

55. 3.64×10^{21} molecules N_2O, 0.266 g N_2O

57. **(a)** density; **(b)** specific heat; **(c)** kilocalorie; **(d)** joule; **(e)** Avogadro's number; **(f)** mole; **(g)** molar mass; **(h)** molar volume; **(i)** empirical formula; **(j)** molecular formula; **(k)** conservation of mass law; **(l)** stoichiometry; **(m)** combined gas law; **(n)** ideal gas law; **(o)** gas constant; **(p)** mass/mass % concentration; **(q)** molarity; **(r)** molality

59. 2.8×10^{22} g Cl_2 **61.** $C_6H_{14}O_6$ **63.** 0.250 *M* Na^+

65. 0.787 g Na_2CO_3

Photo Credits

Key: COP = Chapter Opening Photo; UP = Unnumbered Photo; CCP = Chemistry Connections Photo; UPP = Update Photo

Page	Credit and Reference number
1	Synaptek Scientific Products, Inc., SPL/Photo Researchers (COP 1)
4	Ohaus Scale Corporation (Fig. 1.2A)
4	Ohaus Scale Corporation (Fig. 1.2B)
4	Sartorius Balances and Scales, Brinkmann Instruments, Inc. (Fig. 1.2C)
8	Texas Instruments (UP 1.1)
10	Apple Computer, Inc. (UP 1.2)
15	Kitt Peak National Observatory (UP 1.3, left)
15	CNRA/Science Source/Photo Researchers (UP 1.4, right)
18	Rich Treplow/Photo Researchers (UP 1.5)
20	Ted Clutter/Photo Researchers (UP 1.6)
21	Tom McHugh/Photo Researchers (UP 1.7)
23	Tom Bochsler/Photography Limited (CCP 1.1)
24	NASA (UP 1.8)
27	Texas Instruments (UP 1.9, left)
27	Dan McCoy/Rainbow (UP 1.10, right)
30	Focus on Sports (CO 2)
32	Cliché Bureau International des Poids et Mesures (Fig. 2.1A)
32	Cliché Bureau International des Poids et Mesures (Fig. 2.1B)
40	Focus on Sports (CCP 2.1)
53	Tom Bochsler/Photography Limited (UP 2.2)
62	Tom Bochsler/Photography Limited (UP 2.3)
63	Tom Bochsler/Photography Limited (UP 2.4)
64	Gianni Tortoli/Photo Researchers (COP 3)
65	U.S. Dept. of Navy, Naval Photographic Center (Fig. 3.1)
67	Tom Bochsler/Photography Limited (UP 3.1, top)
67	Tom Bochsler/Photography Limited (UP 3.2, middle)
67	NASA (UP 3.3, bottom)
70	Tom Bochsler/Photography Limited (UP 3.4)
71	Tom Bochsler/Photography Limited (UP 3.5)
72	Courtesy of ALCOA (CCP 3.1)
78	Lawrence Berkeley Laboratory (UPP 3.1)
84	Donald Clegg and Roxy Wilson (Fig. 3.12 A–C)
85	Richard Megna (UP 3.6)
87	H. E. LeMay, Jr. (Fig. 3.1A)
87	Paul Silverman/Fundamental Photographs (Fig. 3.1B)
87	Donald Clegg and Roxy Wilson (Fig. 3.1 C–D)
96	Tom Stack and Associates (UP 3.7)
97	Paul Silverman/Fundamental Photographs (UP 3.8)
98	Tom Bochsler/Photography Limited (UP 3.9)
100	The Metropolitan Museum of Art, anonymous gift in memory of Mr. & Mrs. A. B. Frank, 1981, (1981.159) (COP 4)
102	Library of Congress (UP 4.1)
105	The Granger Collection (Fig. 4.5)
129	Rob Atkins/The Image Bank (CCP 4.1, top left)
129	Richard Megna/Fundamental Photographs (CCP 4.2A & 4.2B)
137	David Pollack/The Stock Market (COP 5)
141	Historical Picture Service (CCP 5.1)
145	H. E. LeMay, Jr. (UP 5.1)
146	Dr. E. R. Degginger (UP 5.2)
171	S. Wess/Photo Researchers (COP 6)
174	Courtesy of C. Corwin (UPP 6.1)
175	Tom Bochsler/Photography Limited (UP 6.1)
182	D. A. Humphreys (UP 6.2)
185	Tom Bochsler/Photography Limited (UP 6.3)
187	D. A. Humphreys (UP 6.5, top)
187	Tom Bochsler/Photography Limited UP (6.4, bottom)
189	The Image Works (Fig. 6.4)
193	Donald Clegg and Roxy Wilson (CCP 6.1)
200	Tom Bochsler/Photography Limited (UP 6.6)
203	Charlotte Raymond/Science Source/Photo Researchers (COP 7)
208	Richard Megna/Fundamental Photographs (UP 7.2)
216	E. F. Smith Memorial Collection, Center for History of Chemistry, University of Pennsylvania (CCP 7.1)
227	Richard Megna/Fundamental Photographs (UP 7.1)
228	Tom Bochsler/Photography Limited (UP 7.3)
232	David Weintraub, 1985/Photo Researchers (COP 8)
233	Dr. E. R. Degginger (Fig. 8.1)
235	Richard Megna/Fundamental Photographs (Fig. 8.1A)
236	Lawrence Migdale/Photo Researchers (Fig. 8.1B)
236	Richard Megna/Fundamental Photographs (Fig. 8.1C)
236	Kristen Brochman/Fundamental Photographs (Fig. 8.1D)
244	Tom Bochsler/Photography Limited (UP 8.1)
245	Tom Bochsler/Photography Limited (UP 8.2)
246	Tom Bochsler/Photography Limited (UP 8.3)
249	Tom Bochsler/Photography Limited (UP 8.4)
250	Tom Bochsler/Photography Limited (UP 8.5)
251	Dr. E. R. Degginger (UP 8.7)
252	Tom Bochsler/Photography Limited (UP 8.8)
253	Tom Bochsler/Photography Limited (UP 8.6)
262	Tom Bochsler/Photography Limited (UP 8.9)
264	Tom Bochsler/Photography Limited (UP 8.10)
266	Focus on Sports (COP 9)
269	Courtesy of C. Marvin Lang; photo by Gary Shulfer (UP 9.1)
277	Richard Megna/Fundamental Photographs (UP 9.2)
283	Farmland Industries (CCP 9.1)
293	IBM Research (COP 10)
295	Courtesy Prof. Colin Humphreys, Univ. of Cambridge (Fig. 10.2)
296	Peter Sorula/Photo Researchers (CCP 10.1, left)
296	H. E. LeMay, Jr. (CCP 10.1, right)
298	Library of Congress (UP 10.1)
299	Four by Five/Superstock (Fig. 10.5, left)
299	Christian Steiner (Fig. 10.5, right)
300	Paul Ehrenfest/AIP Niels Bohr Library (UP 10.2)
305	Tom Bochsler/Photography Limited (UPP 10.3)
324	Science Source/Photo Researchers (COP 11)
328	Lane Penland, (c) 1981, Smithsonian Institution, Reprinted courtesy of Department of Mineral Sciences, Museum of Natural History (UP 11.1)
331	Tom Bochsler/Photography Limited (UP 11.2)
332	Courtesy of Morton Salt Division, Morton Thiokol, Inc. (UP 11.3)
340	Courtesy of the Bancroft Library, University of California, Berkeley (CCP 11.1)
350	Richard Megna/Fundamental Photographs (UP 11.40)
356	Reprinted courtesy of Department of Mineral Sciences, Museum of Natural History, Smithsonian Institution (UP 11.5)

Page	Credit and Reference number
357	Courtesy of Gary Schrobilgen, McMaster University (UP 11.6)
358	Bob Daemmrich/Stock, Boston (COP 12)
359	Tom Bochsler/Photography Limited (UP 12.1)
372	The Granger Collection (CCP 12.1)
393	Farrell Grehan/Photo Researchers (COP 13)
394	NASA (Fig. 13.1)
397	Kristen Brochmann/Fundamental Photographs (UP 13.1, margin)
397	(c) Michael P. Godomski/National Audobon Society/Photo Researchers (UP 13.2, bottom)
401	Fred J. Maroon/Photo Researchers (Fig. 13.6A)
401	Richard Megna/Fundamental Photographs (Fig. 13.6B)
402	Reynolds Metal Company (CCP 13.1)
403	Dr. E. R. Degginger (Fig. 13.8)
404	Tom Bochsler/Photography Limited (Fig. 13.9)
405	Reprinted courtesy of Department of Mineral Sciences, Museum of Natural History, Smithsonian Institution (UP 13.3 A–B)
415	Dr. E. R. Degginger (Fig. 13.17)
424	T. Kitchin/Tom Stack and Associates (UP 13.4)
427	Robert Torrez/Tony Stone Worldwide
432	Tom Bochsler/Photography Limited (UP 14.1)
435	Richard Megna/Fundamental Photographs (Fig. 14.4)
439	Richard Megna/Fundamental Photographs (UP 14.2 A–C)
440	Robert Mathena/Fundamental Photographs (CCP 14.1)
445	Tom Bochsler/Photography Limited (UP 14.3)
458	J. M. Labat/Jacana/Photo Researchers (COP 15)
460	Tom Bochsler/Photography Limited (UP 15.1, top)
460	Tom Bochsler/Photography Limited (UP 15.2, bottom)
462	Tom Bochsler/Photography Limited (Fig. 15.1)
466	Richard Megna/Fundamental Photographs (Fig. 15.2)
470	Richard Megna/Fundamental Photographs (UP 15.4)
471	E. F. Smith Memorial Collection, Center for the History of Chemistry, University of Pennsylvania (CCP 15.2)
479	Richard Megna/Fundamental Photographs (UP 15.5)
480	Régis/Bossu, Sygma (CCP 15.1, A & B)
492	Tom Bochsler/Photography Limited (UP 15.6)
496	Kjell B. Sandved/Photo Researchers (COP 16)
497	Lawrence Migdale/Photo Researchers (Fig. 16.1)
507	NASA/Goddard Space Center (CCP 16.1)
510	Royce Bair/Monkmeyer Press (Fig. 16.11)
515	Richard Megna/Fundamental Photographs (Fig. 16.12 A–C)
524	Bob Rashid/Monkmeyer Press (Fig. 16.14)
535	Focus on Sports (COP 17)
537	Will & Deni McIntyre/Photo Researchers (Fig. 17.1A)
537	Dr. E. R. Degginger (Fig. 17.1B)
537	Jack Dermid/Photo Researchers (Fig. 17.1C)
537	Robert Mathena/Fundamental Photographs (Fig. 17.1D)
538	Dr. E. R. Degginger (UP 17.1)
539	Dr. E. R. Degginger (UP 17.2)
541	Richard Megna/Fundamental Photographs (UP 17.3)
544	Tom Bochsler/Photography Limited (UP 17.4, 2)
558	NASA, Johnson Space Center (CCP 17.1)
559	UPI/Bettman Newsphotos (CCP 17.2)
560	Richard Megna/Fundamental Photographs (Fig. 17.5)
569	Nutan/Vandystadt/Photo Researchers (COP 18)
570	Tom Bochsler/Photography Limited (Fig. 18.1)
583	Samsung America, Inc. (CCP 18.1)
590	Lowell Georgia/Photo Researchers (UP 18.1)
609	Paul Conklin/Monkmeyer Press (COP 19)
610	Hugh Rogers/Monkmeyer Press (Fig. 19.1)
611	Richard Megna/Fundamental Photographs (UP 19.1, 2)
612	The Bettmann Archive (UP 19.2)
618	Dave Davidson/Tom Stack & Associates (Fig. 19.4)
621	The Granger Collection (UP 19.3)
622	Martin Dohan/Science Photo Library/Photo Researchers (CCP 19.1)
623	Vindicator, Inc. (Fig. 19.6)
625	Standford Linear Accelerator/U.S. Department of Energy (Fig. 19.7)
628	Argonne National Laboratory (Fig. 19.10)
641	Sinclair Stammers/Science Library/Photo Researchers (COP 20)
644	Courtesy of C. Corwin (Fig. 20.2 A–D)
646	Courtesy of C. Corwin (Fig. 20.3)
660	Courtesy of C. Corwin (Fig. 20.6 A–C)
661	Richard Megna/Fundamental Photographs (UP 20.1)
662	Courtesy of Corwin (Fig. 20.7 A–C)
663	Donald Clegg and Roxy Wilson (UP 20.2)
665	Courtesy of C. Corwin (Fig. 20.9 A–C)
666	Courtesy of C. Corwin (Fig. 20.10 A–C)
667	Courtesy of C. Corwin (Fig. 20.11 A–C)
668	Courtesy of C. Corwin (Fig. 20.12 A–C)
670	Courtesy of C. Corwin (Fig. 20.13 A–C)
670	Richard Megna/Fundamental Photographs (UP 20.3)
672	Courtesy of C. Corwin (Fig. 20.14 A–C)
672	Richard Megna/Fundamental Photographs (UP 20.4)
673	Courtesy of C. Corwin (Fig. 20.15 A–C)

Index

Note: Boldface pages locate definitions.

B

C

D

F

G

H

I

J

K

L

M

N

O

P

Q

R

S

T

COMMON MONOATOMIC CATIONS

— Stock System

Cation	Name	Cation	Name
Al^{3+}	aluminum ion	Pb^{4+}	lead(IV) ion
Ba^{2+}	barium ion	Li^{+}	lithium ion
Bi^{3+}	bismuth ion	Mg^{2+}	magnesium ion
Cd^{2+}	cadmium ion	Mn^{2+}	manganese(II) ion
Ca^{2+}	calcium ion	Hg_2^{2+}	mercury(I) ion
Co^{2+}	cobalt(II) ion	Hg^{2+}	mercury(II) ion
Co^{3+}	cobalt(III) ion	Ni^{2+}	nickel(II) ion
Cu^{+}	copper(I) ion	K^{+}	potassium ion
Cu^{2+}	copper(II) ion	Ag^{+}	silver ion
Cr^{3+}	chromium(III) ion	Na^{+}	sodium ion
H^{+}	hydrogen ion	Sr^{2+}	strontium ion
Fe^{2+}	iron(II) ion	Sn^{2+}	tin(II) ion
Fe^{3+}	iron(III) ion	Sn^{4+}	tin(IV) ion
Pb^{2+}	lead(II) ion	Zn^{2+}	zinc ion

— Latin System

Cation	Name	Cation	Name
Co^{2+}	cobaltous ion	Pb^{2+}	plumbous ion
Co^{3+}	cobaltic ion	Pb^{4+}	plumbic ion
Cu^{+}	cuprous ion	Hg_2^{2+}	mercurous ion
Cu^{2+}	cupric ion	Hg^{2+}	mercuric ion
Fe^{2+}	ferrous ion	Sn^{2+}	stannous ion
Fe^{3+}	ferric ion	Sn^{4+}	stannic ion

COMMON MONOATOMIC ANIONS

Anion	Name	Anion	Name
Br^{-}	bromide ion	N^{3-}	nitride ion
Cl^{-}	chloride ion	O^{2-}	oxide ion
F^{-}	fluoride ion	P^{3-}	phosphide ion
I^{-}	iodide ion	S^{2-}	sulfide ion

COMMON POLYATOMIC IONS

Cation	Name	Cation	Name
NH_4^{+}	ammonium ion	H_3O^{+}	hydronium ion

Anion	Name	Anion	Name
$C_2H_3O_2^{-}$	acetate ion	OH^{-}	hydroxide ion
CO_3^{2-}	carbonate ion	ClO^{-}	hypochlorite ion
ClO_3^{-}	chlorate ion	NO_3^{-}	nitrate ion
ClO_2^{-}	chlorite ion	NO_2^{-}	nitrite ion
CrO_4^{2-}	chromate ion	ClO_4^{-}	perchlorate ion
CN^{-}	cyanide ion	MnO_4^{-}	permanganate ion
$Cr_2O_7^{2-}$	dichromate ion	PO_4^{3-}	phosphate ion
HCO_3^{-}	hydrogen carbonate ion	SO_4^{2-}	sulfate ion
HSO_4^{-}	hydrogen sulfate ion	SO_3^{2-}	sulfite ion
HSO_3^{-}	hydrogen sulfite ion	$S_2O_3^{2-}$	thiosulfate ion